Lecture Notes in Computer Science 5993

Commenced Publication in 1973
Founding and Former Series Editors:
Gerhard Goos, Juris Hartmanis, and Jan van Leeuwen

Lecture Notes in Computer Science

Cathal Gurrin   Yulan He   Gabriella Kazai
Udo Kruschwitz   Suzanne Little
Thomas Roelleke   Stefan Rüger
Keith van Rijsbergen (Eds.)

Advances in Information Retrieval

32nd European Conference on IR Research, ECIR 2010
Milton Keynes, UK, March 28-31, 2010
Proceedings

 Springer

Volume Editors

Cathal Gurrin
Dublin City University, Ireland
E-mail: cgurrin@computing.dcu.ie

Yulan He
The Open University, Milton Keynes, UK
E-mail: y.he@open.ac.uk

Gabriella Kazai
Microsoft Research Ltd, Cambridge, UK
E-mail: v-gabkaz@microsoft.com

Udo Kruschwitz
University of Essex, Colchester, UK
E-mail: udo@essex.ac.uk

Suzanne Little
The Open University, Milton Keynes, UK
E-mail: s.little@open.ac.uk

Thomas Roelleke
Queen Mary University of London, UK
E-mail: thor@dcs.qmul.ac.uk

Stefan Rüger
The Open University, Milton Keynes, UK
E-mail: s.rueger@open.ac.uk

Keith van Rijsbergen
University of Glasgow, UK
E-mail: keith@dcs.gla.ac.uk

Library of Congress Control Number: Applied for

CR Subject Classification (1998): H.3, H.2, I.2, H.4, H.2.8, I.7, H.5

LNCS Sublibrary: SL 3 – Information Systems and Application, incl. Internet/Web and HCI

ISSN   0302-9743
ISBN-10   3-642-12274-4 Springer Berlin Heidelberg New York
ISBN-13   978-3-642-12274-3 Springer Berlin Heidelberg New York

springer.com

© Springer-Verlag Berlin Heidelberg 2010
Printed in Germany

Typesetting: Camera-ready by author, data conversion by Scientific Publishing Services, Chennai, India
Printed on acid-free paper   06/3180

Preface

These proceedings contain the papers presented at ECIR 2010, the 32nd European Conference on Information Retrieval. The conference was organized by the Knowledge Media Institute (KMi), the Open University, in co-operation with Dublin City University and the University of Essex, and was supported by the Information Retrieval Specialist Group of the British Computer Society (BCS-IRSG) and the Special Interest Group on Information Retrieval (ACM SIGIR). It was held during March 28-31, 2010 in Milton Keynes, UK.

ECIR 2010 received a total of 202 full-paper submissions from Continental Europe (40%), UK (14%), North and South America (15%), Asia and Australia (28%), Middle East and Africa (3%). All submitted papers were reviewed by at least three members of the international Program Committee. Out of the 202 papers 44 were selected as full research papers. ECIR has always been a conference with a strong student focus. To allow as much interaction between delegates as possible and to keep in the spirit of the conference we decided to run ECIR 2010 as a single-track event. As a result we decided to have two presentation formats for full papers. Some of them were presented orally, the others in poster format. The presentation format does not represent any difference in quality. Instead, the presentation format was decided after the full papers had been accepted at the Program Committee meeting held at the University of Essex. The views of the reviewers were then taken into consideration to select the most appropriate presentation format for each paper.

In addition to the full research papers, ECIR 2010 also received a total of 73 poster/demo submissions, each of which was reviewed by three members of the Program Committee. Out of the 73 poster/demo submissions, 23 posters and 5 demos were accepted.

The total of 72 accepted papers consists of 35 from Europe (49%), 14 from UK (19%), 16 from North and South America (22%), 7 from Asia and Australia (10%). The accepted contributions represent the state of the art in information retrieval (IR) and cover a diverse range of topics including NLP and text mining, Web IR, evaluation, multimedia IR, distributed IR and performance issues, IR theory and formal models, personalization and recommendation, domain-specific IR, cross-language IR, and user issues.

As in recent years, ECIR was preceded by a day of workshops and tutorials. This format has proven to be very successful and for ECIR 2010 we accepted four tutorials and five workshops. Following the success of previous BCS-IRSG Industry Days, we organized an Industry Day as part of ECIR 2010 that provided a topical selection of talks, panels and case studies from the world of search and information access, with the emphasis on content drawn from the IR practitioner community. Industry Day was held at the same venue as the main conference.

The BCS-IRSG in conjunction with the BCS has created an annual award to commemorate the achievements of Karen Spärck Jones. This annual award is to encourage and promote talented researchers who have endeavored to advance our understanding of IR and natural language processing with significant experimental contributions. To celebrate the commemorative event, the recipient of the award would be invited to present a keynote lecture at the BCS-IRSG's annual conference – the ECIR. The 2009 KSJ Award was given to Mirella Lapata. She shared her results of work in image and natural language processing for multimedia information retrieval at the conference.

We would like to thank another two invited speakers, Hinrich Schütze and Barry Smyth, for their stimulating contributions to the conference. We would also like to thank all authors who spent their time and effort to submit their work, and all the participants and student volunteers for their contributions and help. Our gratitude goes to the ECIR 2010 Program Committee members and other invited reviewers for their objective and thorough reviews of the submitted papers.

We are very grateful to the sponsors for generously providing financial support for the conference including Apriorie Ltd., Active Web Solutions, Exalead, Flax, Funnelback, GCHQ, Google, Information Retrieval Facility, Microsoft Research, and Yahoo! Research. In particular, Yahoo! Research sponsored the best-paper award and GCHQ sponsored the banquet in association with Bletchley Park.

We would also like to thank the Knowledge Media Institute (KMi) at the Open University for hosting the event, dealing with all local arrangements, and providing registration and financial management. Our special thanks go to Damian Dadswell (Web), Jane Wilde, Rachel Barnett, Aneta Tumilowicz and the Open University's Finance Division (budget and financial management).

Last but not least, we wish to convey our sincere thanks to Springer for providing excellent professional support in preparing this volume.

March 2010

Cathal Gurrin
Yulan He
Gabriella Kazai
Udo Kruschwitz
Suzanne Little
Thomas Roelleke
Stefan Rüger
Keith van Rijsbergen

Organization

ECIR 2010 was organized by the Knowledge Media Institute (KMi) of the Open University, the School of Computing of Dublin City University, and the Department of Computing and Electronic Systems of the University of Essex.

Organizing Committee

General Chair   Stefan Rüger, The Open University, UK
Program Chairs   Cathal Gurrin, Dublin City University, Ireland
     Udo Kruschwitz, The University of Essex, UK
Honorary Chair   Keith van Rijsbergen, University of Glasgow, UK
Publication Chair   Yulan He, The Open University, UK
Poster/Demo Chair   Gabriella Kazai, Microsoft Research, UK
Workshops/Tutorials Chair   Thomas Roelleke, Queen Mary, University of London, UK
Local Chair   Suzanne Little, The Open University, UK

Sponsoring Institutions

The Open University
Dublin City University
The University of Essex
British Computer Society - Information Retrieval Specialist Group(BCS-IRSG)
ACM Special Interest Group on Information Retrieval (ACM SIGIR)

Program Committee

Agosti, Maristella	University of Padua, Italy
Amati, Giambattista	Fondazione Ugo Bordoni, Italy
Amini, Massih-Reza	Universite Pierre et Marie Curie/CNRC, France
Azzopardi, Leif	University of Glasgow, UK
Baeza-Yates, Ricardo	Yahoo! Research, Spain
Bailey, Alex	Google, Switzerland
Barreiro, Alvaro	University of A Coruña, Spain
Bashir, Shariq	Vienna University of Technology, Austria
Basile, Pierpaolo	University of Bari, Italy
Basili, Roberto	University of Rome, Tor Vergata, Italy
Batista, David	University of Lisbon, Portugal
Beigbeder, Michel	Ecole des Mines de Saint-Etienne, France
Belkin, Nicholas	Rutgers University, USA
Bennett, Paul	Microsoft Research, USA
Berendt, Bettina	Katholieke Universiteit Leuven, Belgium
Bermingham, Adam	Dublin City University, Ireland
Bordogna, Gloria	CNR, Italy
Boscarino, Corrado	CWI, The Netherlands
Bothma, Theo	University of Pretoria, South Africa
Boughanem, Mohand	Universite de Toulouse-IRIT, France
Bran, Janez	J. Stefan Institute, Slovenia
Brigo, Riccardo	Microsoft, Bing STC Europe
Bruza, Peter	Queensland University of Technology, Australia
Buntine, Wray	NICTA, Australia
Cacheda, Fidel	University of A Coruña, Spain
Callan, Jamie	Carnegie Mellon University, USA
Caprani , Niamh	Dublin City University, Ireland
Caputo, Annalina	University of Bari, Italy
Cardoso, Nuno	University of Lisbon, Portugal
Carpineto, Claudio	Fondazione Ugo Bordoni, Italy
Carvalho, Paula	University of Lisbon, Portugal
Casoto, Paolo	University of Udine, Italy
Castillo Ocaranza, Carlos	Yahoo!, Spain
Chen, Yi	Dublin City University, Ireland
Chiaramella, Yves	CLIPS-IMAG, France
Chirita, Paul-Alexandru	Adobe Systems, Romania
Clarke, Charles	University of Waterloo, Canada
Clough, Paul	University of Sheffield, UK
Crestani, Fabio	University of Lugano, Switzerland
Croft, Bruce	University of Massachusetts, USA

Table of Contents

Recent Developments in Information Retrieval 1
Cathal Gurrin, Yulan He, Gabriella Kazai, Udo Kruschwitz,
Suzanne Little, Thomas Roelleke, Stefan Rüger, and
Keith van Rijsbergen

Invited Talks

Web Search Futures: Personal, Collaborative, Social 10
Barry Smyth

IR, NLP, and Visualization 11
Hinrich Schütze

Image and Natural Language Processing for Multimedia Information
Retrieval 12
Mirella Lapata

Regular Papers

NLP and Text Mining

A Language Modeling Approach for Temporal Information
Needs 13
Klaus Berberich, Srikanta Bedathur, Omar Alonso, and
Gerhard Weikum

Analyzing Information Retrieval Methods to Recover Broken Web
Links 26
Juan Martinez-Romo and Lourdes Araujo

Between Bags and Trees – Constructional Patterns in Text Used for
Attitude Identification 38
Jussi Karlgren, Gunnar Eriksson, Magnus Sahlgren, and
Oscar Täckström

Improving Medical Information Retrieval with PICO Element
Detection 50
Florian Boudin, Lixin Shi, and Jian-Yun Nie

The Role of Query Sessions in Extracting Instance Attributes from
Web Search Queries 62
Marius Paşca, Enrique Alfonseca, Enrique Robledo-Arnuncio,
Ricardo Martin-Brualla, and Keith Hall

Transliteration Equivalence Using Canonical Correlation Analysis   75
 Raghavendra Udupa and Mitesh M. Khapra

Web IR

Explicit Search Result Diversification through Sub-queries   87
 Rodrygo L.T. Santos, Jie Peng, Craig Macdonald, and Iadh Ounis

Interpreting User Inactivity on Search Results   100
 Sofia Stamou and Efthimis N. Efthimiadis

Learning to Select a Ranking Function............................   114
 Jie Peng, Craig Macdonald, and Iadh Ounis

Mining Anchor Text Trends for Retrieval   127
 Na Dai and Brian D. Davison

Predicting Query Performance via Classification   140
 Kevyn Collins-Thompson and Paul N. Bennett

Evaluation

A Case for Automatic System Evaluation   153
 *Claudia Hauff, Djoerd Hiemstra, Leif Azzopardi, and
 Franciska de Jong*

Aggregation of Multiple Judgments for Evaluating Ordered Lists.......   166
 Hyun Duk Kim, ChengXiang Zhai, and Jiawei Han

Evaluation and User Preference Study on Spatial Diversity   179
 Jiayu Tang and Mark Sanderson

News Comments: Exploring, Modeling, and Online Prediction   191
 Manos Tsagkias, Wouter Weerkamp, and Maarten de Rijke

Query Performance Prediction: Evaluation Contrasted with
Effectiveness   204
 *Claudia Hauff, Leif Azzopardi, Djoerd Hiemstra, and
 Franciska de Jong*

Multimedia IR

A Framework for Evaluating Automatic Image Annotation
Algorithms...   217
 *Konstantinos Athanasakos, Vassilios Stathopoulos, and
 Joemon M. Jose*

BASIL: Effective Near-Duplicate Image Detection Using Gene Sequence
Alignment 229
    Hung-sik Kim, Hau-Wen Chang, Jeongkyu Lee, and Dongwon Lee

Beyond Shot Retrieval: Searching for Broadcast News Items Using
Language Models of Concepts 241
    Robin Aly, Aiden Doherty, Djoerd Hiemstra, and Alan Smeaton

Ranking Fusion Methods Applied to On-Line Handwriting Information
Retrieval.. 253
    Sebastián Peña Saldarriaga, Emmanuel Morin, and
    Christian Viard-Gaudin

Distributed IR and Performance Issues

Improving Query Correctness Using Centralized Probably
Approximately Correct (PAC) Search.............................. 265
    Ingemar Cox, Jianhan Zhu, Ruoxun Fu, and Lars Kai Hansen

Learning to Distribute Queries into Web Search Nodes............... 281
    Marcelo Mendoza, Mauricio Marín, Flavio Ferrarotti, and
    Barbara Poblete

Text Clustering for Peer-to-Peer Networks with Probabilistic
Guarantees... 293
    Odysseas Papapetrou, Wolf Siberski, and Norbert Fuhr

XML Retrieval Using Pruned Element-Index Files.................... 306
    Ismail Sengor Altingovde, Duygu Atilgan, and Özgür Ulusoy

IR Theory and Formal Models

Category-Based Query Modeling for Entity Search 319
    Krisztian Balog, Marc Bron, and Maarten de Rijke

Maximum Margin Ranking Algorithms for Information Retrieval 332
    Shivani Agarwal and Michael Collins

Query Aspect Based Term Weighting Regularization in Information
Retrieval.. 344
    Wei Zheng and Hui Fang

Using the Quantum Probability Ranking Principle to Rank
Interdependent Documents 357
    Guido Zuccon and Leif Azzopardi

Wikipedia-Based Semantic Smoothing for the Language Modeling
Approach to Information Retrieval 370
    Xinhui Tu, Tingting He, Long Chen, Jing Luo, and Maoyuan Zhang

Personalization and Recommendation

A Performance Prediction Approach to Enhance Collaborative Filtering
Performance   382
   Alejandro Bellogín and Pablo Castells

Collaborative Filtering: The Aim of Recommender Systems and the
Significance of User Ratings   394
   Jennifer Redpath, David H. Glass, Sally McClean, and Luke Chen

Goal-Driven Collaborative Filtering – A Directional Error Based
Approach   407
   Tamas Jambor and Jun Wang

Personalizing Web Search with Folksonomy-Based User and Document
Profiles   420
   David Vallet, Iván Cantador, and Joemon M. Jose

Tripartite Hidden Topic Models for Personalised Tag Suggestion   432
   Morgan Harvey, Mark Baillie, Ian Ruthven, and Mark J. Carman

Domain-Specific IR and CLIR

Extracting Multilingual Topics from Unaligned Comparable Corpora . . .   444
   Jagadeesh Jagarlamudi and Hal Daumé III

Improving Retrievability of Patents in Prior-Art Search   457
   Shariq Bashir and Andreas Rauber

Mining OOV Translations from Mixed-Language Web Pages for Cross
Language Information Retrieval   471
   Lei Shi

On Foreign Name Search   483
   Jason Soo and Ophir Frieder

Promoting Ranking Diversity for Biomedical Information Retrieval
Using Wikipedia   495
   Xiaoshi Yin, Xiangji Huang, and Zhoujun Li

Temporal Shingling for Version Identification in Web Archives   508
   Ralf Schenkel

User Issues

Biometric Response as a Source of Query Independent Scoring in
Lifelog Retrieval   520
   Liadh Kelly and Gareth J.F. Jones

Enabling Interactive Query Expansion through Eliciting the Potential
Effect of Expansion Terms 532
　　Nuzhah Gooda Sahib, Anastasios Tombros, and Ian Ruthven

Evaluation of an Adaptive Search Suggestion System 544
　　Sascha Kriewel and Norbert Fuhr

How Different Are Language Models and Word Clouds? 556
　　Rianne Kaptein, Djoerd Hiemstra, and Jaap Kamps

Posters

Colouring the Dimensions of Relevance 569
　　Ulises Cerviño Beresi, Yunhyong Kim, Mark Baillie,
　　Ian Ruthven, and Dawei Song

On Improving Pseudo-Relevance Feedback Using Pseudo-Irrelevant
Documents 573
　　Karthik Raman, Raghavendra Udupa, Pushpak Bhattacharya, and
　　Abhijit Bhole

Laplacian Co-hashing of Terms and Documents 577
　　Dell Zhang, Jun Wang, Deng Cai, and Jinsong Lu

Query Difficulty Prediction for Contextual Image Retrieval 581
　　Xing Xing, Yi Zhang, and Mei Han

Estimating Translation Probabilities from the Web for Structured
Queries on CLIR 586
　　Xabier Saralegi and Maddalen Lopez de Lacalle

Using Weighted Tagging to Facilitate Enterprise Search 590
　　Shengwen Yang, Jianming Jin, and Yuhong Xiong

An Empirical Study of Query Specificity 594
　　Avi Arampatzis and Jaap Kamps

Semantically Enhanced Term Frequency 598
　　Christof Müller and Iryna Gurevych

Crowdsourcing Assessments for XML Ranked Retrieval 602
　　Omar Alonso, Ralf Schenkel, and Martin Theobald

Evaluating Server Selection for Federated Search 607
　　Paul Thomas and Milad Shokouhi

A Comparison of Language Identification Approaches on Short,
Query-Style Texts 611
　　Thomas Gottron and Nedim Lipka

Filtering Documents with Subspaces.................................. 615
  Benjamin Piwowarski, Ingo Frommholz, Yashar Moshfeghi,
  Mounia Lalmas, and Keith van Rijsbergen

User's Latent Interest-Based Collaborative Filtering 619
  Biyun Hu, Zhoujun Li, and Jun Wang

Evaluating the Potential of Explicit Phrases for Retrieval Quality 623
  Andreas Broschart, Klaus Berberich, and Ralf Schenkel

Developing a Test Collection for the Evaluation of Integrated Search ... 627
  Marianne Lykke, Birger Larsen, Haakon Lund, and Peter Ingwersen

Retrieving Customary Web Language to Assist Writers 631
  Benno Stein, Martin Potthast, and Martin Trenkmann

Enriching Peer-to-Peer File Descriptors Using Association Rules on
Query Logs 636
  Nazli Goharian, Ophir Frieder, Wai Gen Yee, and Jay Mundrawala

Cross-Language High Similarity Search: Why No Sub-linear Time
Bound Can Be Expected 640
  Maik Anderka, Benno Stein, and Martin Potthast

Exploiting Result Consistency to Select Query Expansions for Spoken
Content Retrieval 645
  Stevan Rudinac, Martha Larson, and Alan Hanjalic

Statistics of Online User-Generated Short Documents.................. 649
  Giacomo Inches, Mark J. Carman, and Fabio Crestani

Mining Neighbors' Topicality to Better Control Authority Flow 653
  Na Dai, Brian D. Davison, and Yaoshuang Wang

Finding Wormholes with Flickr Geotags 658
  Maarten Clements, Pavel Serdyukov, Arjen P. de Vries, and
  Marcel J.T. Reinders

Enhancing N-Gram-Based Summary Evaluation Using Information
Content and a Taxonomy 662
  Mijail Kabadjov, Josef Steinberger, Ralf Steinberger,
  Massimo Poesio, and Bruno Pouliquen

Demos

NEAT: News Exploration Along Time 667
  Omar Alonso, Klaus Berberich, Srikanta Bedathur, and
  Gerhard Weikum

Opinion Summarization of Web Comments   668
    Martin Potthast and Steffen Becker

EUROGENE: Multilingual Retrieval and Machine Translation Applied
to Human Genetics   670
    Petr Knoth, Trevor Collins, Elsa Sklavounou, and Zdenek Zdrahal

NETSPEAK—Assisting Writers in Choosing Words   672
    Martin Potthast, Martin Trenkmann, and Benno Stein

A Data Analysis and Modelling Framework for the Evaluation of
Interactive Information Retrieval   673
    Ralf Bierig, Michael Cole, Jacek Gwizdka, and Nicholas J. Belkin

Author Index   675

Recent Developments in Information Retrieval

Cathal Gurrin[1], Yulan He[2], Gabriella Kazai[3], Udo Kruschwitz[4],
Suzanne Little[2], Thomas Roelleke[5], Stefan Rüger[2], and Keith van Rijsbergen[6]

[1] School of Computing, Dublin City University, Ireland
cgurrin@computing.dcu.ie
[2] Knowledge Media Institute, The Open University, UK
{y.he,s.little,s.rueger}@open.ac.uk
[3] Microsoft Research Cambridge, UK
v-gabkaz@microsoft.com
[4] School of Computer Science and Electronic Engineering, University of Essex, UK
udo@essex.ac.uk
[5] Department of Computer Science, Queen Mary, University of London, UK
thor@dcs.qmul.ac.uk
[6] Department of Computing Science, University of Glasgow, UK
keith@dcs.gla.ac.uk

Abstract. This paper summarizes the scientific work presented at the
32nd European Conference on Information Retrieval. It demonstrates
that information retrieval (IR) as a research area continues to thrive
with progress being made in three complementary sub-fields, namely IR
theory and formal methods together with indexing and query representation issues, furthermore Web IR as a primary application area and
finally research into evaluation methods and metrics. It is the combination of these areas that gives IR its solid scientific foundations. The
paper also illustrates that significant progress has been made in other
areas of IR. The keynote speakers addressed three such subject fields,
social search engines using personalization and recommendation technologies, the renewed interest in applying natural language processing to
IR, and multimedia IR as another fast-growing area.

Introduction. The field of information retrieval (IR) has come a long way in
the last fifty years, and has enabled easier and faster access to large corpora of
unstructured data. Besides the continuing efforts in developing the core of IR
technologies, such as IR theory and formal models, ranking techniques, system
evaluations, etc, progress has also been made in revisiting established methods and creating new methodologies for multimedia retrieval, cross-lingual IR,
XML retrieval, and domain-specific IR such as biomedical IR, patent retrieval,
spatial IR etc. A noticeable trend is the renewed interest in applying natural
language processing (NLP) to IR problems. In addition, there have been tremendous growth and progress in addressing performance issues through distributed
IR and issues associated with users of IR systems such as personalization and
recommendation, social media, and interactive IR. The last year also saw an
increasing interest in adapting techniques from other disciplines and applying

C. Gurrin et al. (Eds.): ECIR 2010, LNCS 5993, pp. 1–9, 2010.

them innovatively in IR such as gene sequence alignment algorithm in biology and quantum theory in physics etc. The recent advancements in IR are enabling us to continue to push the boundaries of the field.

In the remainder we will briefly introduce the full papers presented at ECIR 2010. They are topically grouped. It needs of course be pointed out that deciding on any such grouping can be an art in its own right.

NLP and Text Mining. Berberich *et al.* [8] tackle the inherent uncertainty problem in temporal expressions in users' queries by integrating temporal expressions into a language modeling and thereby making them first-class citizens of the retrieval model and considering their inherent uncertainty. Martinez-Romo and Araujo [25] compare different techniques to automatically find candidate web pages to substitute broken links. They extract information from the anchor text, the content of the page containing the link, and the cache page in some digital library. Term frequencies and a language modeling approach are used to select terms to construct the queries submitted to a search engine, while co-occurrence measures and a language model approach are applied for ranking the final results. Karlgren *et al.* [20] propose to use non-terminological information, structural features of text, in addition to the presence of content terms, to find attitudinal expressions in written English text. They find that constructional features transfer well across different text collections. Boudin *et al.* [9] tackle the problem of identifying appropriate resources and searching for the best available evidence for medical treatment in evidence-based medicine (EBM) by exploring the incorporation of the PICO elements, Population/Problem (P), Intervention (I), Comparison (C) and Outcome (O), into the IR process. They found that the I and P elements can be used to enhance the retrieval process and thereby give significantly better retrieval effectiveness than the state-of-the-art methods. Pasca *et al.* [28] propose a weakly-supervised extraction method which exploits anonymized Web-search query sessions, as an alternative to isolated, individual queries, to acquire per-instance attributes (e.g., *top speed* for *chevrolet corvette*, or *population density* for *Brazil*). Inherent challenges associated with using sessions for attribute extraction, such as a large majority of within-session queries not being related to attributes, are overcome by using attributes globally extracted from isolated queries as an unsupervised filtering mechanism. Udupa and Khapra [40] address the problem of Transliteration Equivalence, i.e. determining whether a pair of words in two different languages are name transliterations or not, by considering name transliterations in two languages as two views of the same semantic object and compute a low-dimensional common semantic feature space using Canonical Correlation Analysis (CCA). Similarity of the words in the common semantic feature space forms the basis for classifying a pair of names as transliterations.

Web IR. Santos *et al.* [32] introduce xQuAD, a novel model for search result diversification that attempts to build such a diversified ranking by explicitly accounting for the relationship between documents retrieved for the original query and the possible aspects underlying this query, in the form of sub-queries. Stamou and Efthimiadis [36] study users inactivity on search results in relation to their pursued search goals and investigate the impact of displayed results on user clicking decisions. Their study examines two types of post-query user inactivity, pre-determined and post-determined, depending on whether the user started searching with a preset intention to look for answers only within the result snippets and did not intend to click through the results, or the user inactivity was decided after the user had reviewed the list of retrieved documents. Peng *et al.* [29] propose a novel Learning To Select framework that selectively applies an appropriate ranking function on a per-query basis. The use of divergence, which measures the extent that a document ranking function alters the scores of an initial ranking of documents for a given query, is employed as a query feature to identify similar training queries for an unseen query. The ranking function which performs the best on this identified training query set is then chosen for the unseen query. Dai and Davison [12] argue that historical trends of anchor text importance is important to web search. They propose a novel temporal anchor text weighting method to incorporate the trends of anchor text creation over time, which combines historical weights on anchor text by propagating the anchor text weights among snapshots over the time axis. Collins-Thompson and Bennett [10] investigate using topic prediction data, as a summary of document content, to compute measures of search result quality. Their findings suggest that class-based statistics can be computed efficiently online and using class predictions can offer comparable performance to full language models while reducing computation overhead.

Evaluation. Hauff *et al.* [16] perform a wider analysis of system ranking estimation methods on 16 TREC data sets which cover more tasks and corpora than previously tested. Their analysis reveals that the performance of system ranking estimation approaches varies across topics and can be improved by selecting the "right" subset of topics from a topic set. They also observe that the commonly experienced problem of underestimating the performance of the best systems is data set dependent and not inherent to system ranking estimation methods. Kim *et al.* [23] propose three new methods for aggregating multiple order judgments to evaluate ordered lists; weighted aggregation, combined ranking, frequent sequential pattern based aggregation. The first method assigns different weights on experts judgments based on the consensus and compute evaluation score based on weights. The second method finds combined ranking from multiple judgments and uses it for evaluation. The third method collects frequent sequential patterns and their frequencies from all the human judgments of ordering and scores an ordered list based on how well the list matches these frequent sequential patterns. Tang and Sanderson [37] conduct evaluation and user preference study on spatial diversity in the context of spatial information retrieval. They show the potentials of spatial diversity by not only the traditional evaluation metrics (precision and cluster recall), but also through a user preference study using Amazon Mechanical Turk. Tsagkias *et al.* [38] explore the news comments space, and compare the log-normal and the negative binomial distributions for modeling comments from various news agents. They also examine the feasibility of predicting the number of comments, based on the number of comments observed shortly after publication. Hauff *et al.* [15] control the quality of a query performance predictor in order to examine how the strength of the correlation achieved affects the effectiveness of an adaptive retrieval system. They show that many existing predictors fail to achieve a correlation strong enough to reliably improve the retrieval effectiveness in the Selective Query Expansion as well as the Meta-Search setting.

Multimedia IR. Athanasakos *et al.* [4] introduce a framework for the evaluation of image annotation models, which they use to evaluate two state-of-the-art Automatic Image Annotation (AIA) algorithms. They reveal that a simple SVM approach using Global MPEG-7 Features outperforms state-of-the-art AIA models across several collection settings. Kim *et al.* [22] propose a new algorithm, termed as "BlASted Image Linkage" (BASIL), which uses the popular gene sequence alignment algorithm BLAST in Biology in detecting near-duplicate images. They study how various image features and gene sequence generation methods (using gene alphabets such as A, C, G, and T in DNA sequences) affect the accuracy and performance of detecting near-duplicate images. Aly *et al.* [3] propose a retrieval framework for news story items consisting of multiple shots. The framework consists of a concept based language model which ranks news items with known occurrences of semantic concepts by the probability that an important concept is produced from the concept distribution of the news item and a probabilistic model of the uncertain presence, or risk, of these concepts. Peña Saldarriaga *et al.* [31] present an empirical study on the application of ranking fusion methods in the context of handwriting information retrieval. Retrieval approaches on texts obtained through handwriting recognition and recognition-free methods using word-spotting algorithms are combined to give a significant effectiveness improvement.

Distributed IR and Performance Issues. Cox *et al.* [11] propose a modification to a non-deterministic architecture for IR, the PAC architecture, by introducing a centralized query coordination node. To respond to a query, random sampling of computers is replaced with pseudo-random sampling using the query as a seed. Then, for queries that occur frequently, this pseudo-random sample is iteratively refined so that performance improves with each iteration. Experiments on the TREC-8 dataset demonstrate that for queries that occur 10 or more times, the performance of a non-deterministic PAC architecture can closely match that of a deterministic system. Mendoza *et al.* [26] investigate machine learning algorithms to distribute queries onto web search nodes and propose a logistic regression model to quickly predict the most pertinent search nodes for a given query. Papapetrou *et al.* [27] present a text clustering algorithm for peer-to-peer networks with high scalability by using a probabilistic approach for assigning documents to clusters. It enables a peer to compare each of its documents only with very few selected clusters, without significant loss of clustering quality. Altingovde *et al.* [2] propose using static index pruning techniques for obtaining more compact index files that can still result in comparable retrieval performance to that of a full index in XML retrieval.

IR Theory and Formal Models. Balog *et al.* [5] present a general probabilistic framework for category-based query modeling for entity search. They
focus on the use of category information and show the advantage of a category-
based representation over a term-based representation, and also demonstrate the
effectiveness of category-based expansion using example entities. Agarwal and
Collins [1] propose a family of ranking algorithms by optimizing variations of
the hinge loss used in support vector machines. The algorithms preserve the
simplicity of standard pair-wise ranking methods in machine learning, yet show
performance comparable to state-of-the-art IR ranking algorithms. Zheng and
Fang [43] study the incorporation of query term relations into existing retrieval
models by first developing a general strategy that can systematically integrate
a term weighting regularization function into existing retrieval functions, and
then proposing two specific regularization functions based on the guidance provided by constraint analysis. Zuccon and Azzopardi [44] explore whether the
new Quantum Probability Ranking Principle (QPRP), which implicitly captures dependencies between documents through "quantum interference", leads
to improved performance for subtopic retrieval, where novelty and diversity is
required. They show that QPRP consistently outperforms the previous ranking strategies such as Probability Ranking Principle, Maximal Marginal Relevance and Portfolio Theory. Tu *et al.* [39] presents a novel Wikipedia-based
semantic smoothing method that decomposes a document into a set of weighted
Wikipedia concepts and then maps those unambiguous Wikipedia concepts into
query terms. The mapping probabilities from each Wikipedia concept to individual terms are estimated through the EM algorithm. Document models based
on Wikipedia concept mapping are then derived.

Personalization and Recommendation. Bellogín and Castells [7] investigate
the adaptation of clarity-based query performance predictors to define predictors of neighbor performance in Collaborative Filtering (CF). A predictor is
proposed and introduced in a memory-based CF algorithm to produce a dynamic variant where neighbor ratings are weighted based on their predicted
performance. Redpath *et al.* [30] investigate the significance of user ratings in
recommender systems by considering their inclusion/exclusion in both the generation and evaluation of recommendations. They argue that it is important
to identify the aim of the system before evaluating the accuracy of a recommender algorithm since the use of ratings would generate different results when
different evaluation metrics are used. Jambor and Wang [18] argue that defining
an error function that is uniform across rating scales in collaborative filtering
is limited since different applications may have different recommendation goals
and thus error functions. They propose a flexible optimization framework that
can adapt to individual recommendation goals by introducing a weight function
to capture the cost (risk) of each individual predictions which can be learned
from the specified performance measures. Vallet *et al.* [41] propose to represent
a user profile in terms of social tags, manually provided by users in folksonomy
systems to describe, categorize and organize items of interest, and investigate a
number of novel techniques that exploit the users social tags to re-rank results

obtained with a Web search engine. Harvey *et al.* [14] extend the latent Dirichlet allocation topic model to include user data and use the estimated probability distributions to provide personalized tag suggestions to users.

Domain-Specific IR and CLIR. Jagarlamudi and Daumé [17] present a generative model called JointLDA which uses a bilingual dictionary to mine multilingual topics from an unaligned corpus. They speculate that the JointLDA model has better predictive power compared to the bag-of-word based translation model leaving the possibility for JointLDA to be preferred over bag-of-word model for cross-lingual IR applications. Bashir and Rauber [6] expand prior-art queries generated from query patents using query expansion with pseudo relevance feedback in patent retrieval. They propose a novel approach to automatically select better terms from query patents based on their proximity distribution with prior-art queries that are used as features for computing similarity measures. Shi [34] presents a method that automatically acquires a large quantity of OOV translations from the web by adaptively learning translation extraction patterns based on the observation that translation pairs on the same page tend to appear following similar layout patterns. Soo and Frieder [35] propose a language-independent foreign names search approach, called Segments, and compare it against traditional *n*-gram and Soundex based solutions. Yin *et al.* [42] propose a cost-based re-ranking method to promote ranking diversity for biomedical information retrieval which aims at finding passages that cover many different aspects of a query topic. Schenkel [33] presents temporal shingling, an extension of the well-established shingling technique for measuring how similar two snapshots of a page are, for version identification in web archives. The method considers the lifespan of shingles to differentiate between important updates that should be archived and transient changes that may be ignored.

User Issues. Kelly and Jones [21] present a novel query independent static biometric scoring approach for re-ranking result lists retrieved from a lifelog, which contains digital records captured from an individuals daily life, e.g. emails, web pages downloaded and SMSs sent or received, using a BM25 model for content and content + context data. They explored the utility of galvanic skin response (GSR) and skin temperature (ST) associated with past access to items as a measure of potential future significance of items. Gooda Sahib *et al.* [13] investigate a method of increasing the uptake of Interactive Query Expansion by displaying summary previews that allow searchers to view the impact of their expansion decisions in real time, engage more with suggested terms, and support them in making good expansion decision. Kriewel and Fuhr [24] describe an adaptive search suggestion system based on case-based reasoning techniques, and detail an evaluation of its usefulness in helping users employ better search strategies. Kaptein *et al.* [19] investigate what tag clouds and language modeling approaches can learn from each other, and specifically whether language modeling techniques can be used to generate tag or word clouds automatically from documents.

References

1. Agarwal, S., Collins, M.: Maximum margin ranking algorithms for information retrieval. In: Gurrin, C., et al. (eds.) ECIR 2010. LNCS, vol. 5993, pp. 332–343. Springer, Heidelberg (2010)
2. Altingovde, I.S., Atilgan, D., Ulusoy, O.: Xml retrieval using pruned element-index files. In: Gurrin, C., et al. (eds.) ECIR 2010. LNCS, vol. 5993, pp. 306–318. Springer, Heidelberg (2010)
3. Aly, R., Doherty, A., Hiemstra, D., Smeaton, A.: Beyond shot retrieval: Searching for broadcast news items using language models of concepts. In: Gurrin, C., et al. (eds.) ECIR 2010, vol. 5993, pp. 241–252. Springer, Heidelberg (2010)
4. Athanasakos, K., Stathopoulos, V., Jose, J.M.: A framework for evaluating automatic image annotation algorithms. In: Gurrin, C., et al. (eds.) ECIR 2010. LNCS, vol. 5993, pp. 217–228. Springer, Heidelberg (2010)
5. Balog, K., Bron, M., de Rijke, M.: Category-based query modeling for entity search. In: Gurrin, C., et al. (eds.) ECIR 2010. LNCS, vol. 5993, pp. 319–331. Springer, Heidelberg (2010)
6. Bashir, S., Rauber, A.: Improving retrievability of patents in prior-art search. In: Gurrin, C., et al. (eds.) ECIR 2010. LNCS, vol. 5993, pp. 457–470. Springer, Heidelberg (2010)
7. Bellogín, A., Castells, P.: A performance prediction approach to enhance collaborative filtering performance. In: Gurrin, C., et al. (eds.) ECIR 2010. LNCS, vol. 5993, pp. 382–393. Springer, Heidelberg (2010)
8. Berberich, K., Bedathur, S., Alonso, O., Weikum, G.: A language modeling approach for temporal information needs. In: Gurrin, C., et al. (eds.) ECIR 2010. LNCS, vol. 5993, pp. 13–25. Springer, Heidelberg (2010)
9. Boudin, F., Shi, L., Nie, J.Y.: Improving medical information retrieval with pico element detection. In: Gurrin, C., et al. (eds.) ECIR 2010. LNCS, vol. 5993, pp. 50–61. Springer, Heidelberg (2010)
10. Collins-Thompson, K., Bennett, P.N.: Predicting query performance via classification. In: Gurrin, C., et al. (eds.) ECIR 2010. LNCS, vol. 5993, pp. 140–152. Springer, Heidelberg (2010)
11. Cox, I., Zhu, J., Fu, R., Hansen, L.K.: Improving query correctness using centralized probably approximately correct (pac) search. In: Gurrin, C., et al. (eds.) ECIR 2010. LNCS, vol. 5993, pp. 265–280. Springer, Heidelberg (2010)
12. Dai, N., Davison, B.D.: Mining anchor text trends for retrieval. In: Gurrin, C., et al. (eds.) ECIR 2010. LNCS, vol. 5993, pp. 127–139. Springer, Heidelberg (2010)
13. Gooda Sahib, N., Tombros, A., Ruthven, I.: Enabling interactive query expansion through eliciting the potential effect of expansion terms. In: Gurrin, C., et al. (eds.) ECIR 2010. LNCS, vol. 5993, pp. 532–543. Springer, Heidelberg (2010)
14. Harvey, M., Baillie, M., Ruthven, I., Carman, M.: Tripartite hidden topic models for personalised tag suggestion. In: Gurrin, C., et al. (eds.) ECIR 2010. LNCS, vol. 5993, pp. 432–443. Springer, Heidelberg (2010)
15. Hauff, C., Azzopardi, L., Hiemstra, D., de Jong, F.: Query performance prediction: Evaluation contrasted with effectiveness. In: Gurrin, C., et al. (eds.) ECIR 2010. LNCS, vol. 5993, pp. 204–216. Springer, Heidelberg (2010)
16. Hauff, C., Hiemstra, D., Azzopardi, L., de Jong, F.: A case for automatic system evaluation. In: Gurrin, C., et al. (eds.) ECIR 2010. LNCS, vol. 5993, pp. 153–165. Springer, Heidelberg (2010)

17. Jagarlamudi, J., Daumé III, H.: Extracting multilingual topics from unaligned comparable corpora. In: Gurrin, C., et al. (eds.) ECIR 2010. LNCS, vol. 5993, pp. 444–456. Springer, Heidelberg (2010)

18. Jambor, T., Wang, J.: Goal-driven collaborative filtering - a directional error based approach. In: Gurrin, C., et al. (eds.) ECIR 2010. LNCS, vol. 5993, pp. 407–419. Springer, Heidelberg (2010)

19. Kaptein, R., Hiemstra, D., Kamps, J.: How different are language models and word clouds? In: Gurrin, C., et al. (eds.) ECIR 2010. LNCS, vol. 5993, pp. 556–568. Springer, Heidelberg (2010)

20. Karlgren, J., Eriksson, G., Sahlgren, M., Täckström, O.: Between bags and trees - constructional patterns in text used for attitude identification. In: Gurrin, C., et al. (eds.) ECIR 2010. LNCS, vol. 5993, pp. 38–49. Springer, Heidelberg (2010)

21. Kelly, L., Jones, G.J.F.: Biometric response as a source of query independent scoring in lifelog retrieval. In: Gurrin, C., et al. (eds.) ECIR 2010. LNCS, vol. 5993, pp. 520–531. Springer, Heidelberg (2010)

22. Kim, H.-s., Chang, H.-W., Lee, J., Lee, D.: Basil: Effective near-duplicate image detection using gene sequence alignment. In: Gurrin, C., et al. (eds.) ECIR 2010. LNCS, vol. 5993, pp. 229–240. Springer, Heidelberg (2010)

23. Kim, H.D., Zhai, C., Han, J.: Aggregation of multiple judgments for evaluating ordered lists. In: Gurrin, C., et al. (eds.) ECIR 2010. LNCS, vol. 5993, pp. 166–178. Springer, Heidelberg (2010)

24. Kriewel, S., Fuhr, N.: Evaluation of an adaptive search suggestion system. In: Gurrin, C., et al. (eds.) ECIR 2010. LNCS, vol. 5993, pp. 544–555. Springer, Heidelberg (2010)

25. Martinez-Romo, J., Araujo, L.: Analyzing information retrieval methods to recover broken web links. In: Gurrin, C., et al. (eds.) ECIR 2010. LNCS, vol. 5993, pp. 26–37. Springer, Heidelberg (2010)

26. Mendoza, M., Marín, M., Ferrarotti, F., Poblete, B.: Learning to distribute queries onto web search nodes. In: Gurrin, C., et al. (eds.) ECIR 2010. LNCS, vol. 5993, pp. 281–292. Springer, Heidelberg (2010)

27. Papapetrou, O., Siberski, W., Fuhr, N.: Text clustering for peer-to-peer networks with probabilistic guarantees. In: Gurrin, C., et al. (eds.) ECIR 2010. LNCS, vol. 5993, pp. 293–305. Springer, Heidelberg (2010)

28. Pasca, M., Alfonseca, E., Robledo-Arnuncio, E., Martin-Brualla, R., Hall, K.: The role of query sessions in extracting instance attributes from web search queries. In: Gurrin, C., et al. (eds.) ECIR 2010. LNCS, vol. 5993, pp. 62–74. Springer, Heidelberg (2010)

29. Peng, J., Macdonald, C., Ounis, I.: Learning to select a ranking function. In: Gurrin, C., et al. (eds.) ECIR 2010. LNCS, vol. 5993, pp. 114–126. Springer, Heidelberg (2010)

30. Redpath, J., Glass, D.H., McClean, S., Chen, L.: Collaborative filtering: The aim of recommender systems and the significance of user ratings. In: Gurrin, C., et al. (eds.) ECIR 2010. LNCS, vol. 5993, pp. 394–406. Springer, Heidelberg (2010)

31. Peña Saldarriaga, S., Morin, E., Viard-Gaudin, C.: Ranking fusion methods applied to on-line handwriting information retrieval. In: Gurrin, C., et al. (eds.) ECIR 2010. LNCS, vol. 5993, pp. 253–264. Springer, Heidelberg (2010)

32. Santos, R.L.T., Peng, J., Macdonald, C., Ounis, I.: Explicit search result diversification through sub-queries. In: Gurrin, C., et al. (eds.) ECIR 2010. LNCS, vol. 5993, pp. 87–99. Springer, Heidelberg (2010)

33. Schenkel, R.: Temporal shingling for version identification in web archives. In: Gurrin, C., et al. (eds.) ECIR 2010. LNCS, vol. 5993, pp. 508–519. Springer, Heidelberg (2010)

34. Shi, L.: Mining oov translations from mixed-language web pages for cross language information retrieval. In: Gurrin, C., et al. (eds.) ECIR 2010. LNCS, vol. 5993, pp. 471–482. Springer, Heidelberg (2010)

35. Soo, J., Frieder, O.: On foreign name search. In: Gurrin, C., et al. (eds.) ECIR 2010. LNCS, vol. 5993, pp. 483–494. Springer, Heidelberg (2010)

36. Stamou, S., Efthimiadis, E.N.: Interpreting user inactivity on search results. In: Gurrin, C., et al. (eds.) ECIR 2010. LNCS, vol. 5993, pp. 100–113. Springer, Heidelberg (2010)

37. Tang, J., Sanderson, M.: Evaluation and user preference study on spatial diversity. In: Gurrin, C., et al. (eds.) ECIR 2010. LNCS, vol. 5993, pp. 179–190. Springer, Heidelberg (2010)

38. Tsagkias, M., Weerkamp, W., de Rijke, M.: News comments: Exploring, modeling, and online prediction. In: Gurrin, C., et al. (eds.) ECIR 2010. LNCS, vol. 5993, pp. 191–203. Springer, Heidelberg (2010)

39. Tu, X., He, T., Chen, L., Luo, J., Zhang, M.: Wikipedia-based semantic smoothing for the language modeling approach to information retrieval. In: Gurrin, C., et al. (eds.) ECIR 2010. LNCS, vol. 5993, pp. 370–381. Springer, Heidelberg (2010)

40. Udupa, R., Khapra, M.M.: Transliteration equivalence using canonical correlation analysis. In: Gurrin, C., et al. (eds.) ECIR 2010. LNCS, vol. 5993, pp. 75–86. Springer, Heidelberg (2010)

41. Vallet, D., Cantador, I., Jose, J.M.: Personalizing web search with folksonomy-based user and document profiles. In: Gurrin, C., et al. (eds.) ECIR 2010. LNCS, vol. 5993, pp. 420–431. Springer, Heidelberg (2010)

42. Yin, X., Huang, X., Li, Z.: Promoting ranking diversity for biomedical information retrieval using wikipedia. In: Gurrin, C., et al. (eds.) ECIR 2010. LNCS, vol. 5993, pp. 495–507. Springer, Heidelberg (2010)

43. Zheng, W., Fang, H.: Query aspect based term weighting regularization in information retrieval. In: Gurrin, C., et al. (eds.) ECIR 2010. LNCS, vol. 5993, pp. 344–356. Springer, Heidelberg (2010)

44. Zuccon, G., Azzopardi, L.: Using the quantum probability ranking principle to rank interdependent documents. In: Gurrin, C., et al. (eds.) ECIR 2010. LNCS, vol. 5993, pp. 357–369. Springer, Heidelberg (2010)

Web Search Futures: Personal, Collaborative, Social

Barry Smyth

School of Computer Science & Informatics
University College Dublin, Ireland
barry.smyth@ucd.ie

Abstract. In this talk we will discuss where Web search may be heading, focusing on a number of large-scale research projects that are trying to develop the "next big thing" in Web search. We will consider some important recent initiatives on how to improve the quality of the Web search experience by helping search engines to respond to our individual needs and preferences. In turn, we will focus on some innovative work on how to take advantage of the inherently collaborative nature of Web search as we discuss recent attempts to develop so-called "social search engines".

Biography: Barry Smyth is a Professor of Computer Science in University College Dublin. Barry Graduated with a PhD in Computer Science from Trinity College Dublin in 1996 and his research interests include artificial intelligence, machine learning, case-based reasoning, and information retrieval with a core focus on recommender systems and personalization technologies. In 1999 Barry co-founded ChangingWorlds as a UCD campus company to commercialize personalization technologies for the mobile sector, helping to grow the company to more than 150 people before it was acquired by Amdocs Inc. in late 2008. Today Barry is the Director of CLARITY, the SFI-funded Centre for Sensor Web Technologies, and a joint initiative between University College Dublin, Dublin City University, and the Tyndall National Institute.

C. Gurrin et al. (Eds.): ECIR 2010, LNCS 5993, p. 10, 2010.
© Springer-Verlag Berlin Heidelberg 2010

IR, NLP, and Visualization

Hinrich Schütze

Institute for Natural Language Processing
University of Stuttgart, Germany
hs999@ifnlp.org

Abstract. In the last ten years natural language processing (NLP) has become an essential part of many information retrieval systems, mainly in the guise of question answering, summarization, machine translation and preprocessing such as decompounding. However, most of these methods are shallow. More complex natural language processing is not yet sufficiently reliable to be used in IR. I will discuss how new visualization technology and rich interactive environments offer new opportunities for complex NLP in IR.

Biography: Hinrich Schütze is best known for co-authoring the standard reference book on statistical natural language processing (http://nlp.stanford.edu/fsnlp/) (Google Scholar lists more than 4,700 citations of this book). His new book "Introduction to Information Retrieval" (http://nlp.stanford.edu/IR-book/information-retrieval-book.html) (co-authored with Chris Manning and Prabhakar Raghavan) was published in 2008 and has already been adopted by many IR courses throughout the world. Dr. Schütze obtained his PhD from Stanford University and has worked for a number of Silican Valley companies, including two large search engines and several text mining startups. He is currently Chair of Theoretical Computational Linguistics at the University of Stuttgart (http://www.ims.uni-stuttgart.de/~schuetze/).

C. Gurrin et al. (Eds.): ECIR 2010, LNCS 5993, p. 11, 2010.
© Springer-Verlag Berlin Heidelberg 2010

Image and Natural Language Processing for Multimedia Information Retrieval

Mirella Lapata

School of Informatics
University of Edinburgh
mlap@inf.ed.ac.uk

Abstract. Image annotation, the task of automatically generating description words for a picture, is a key component in various image search and retrieval applications. Creating image databases for model development is, however, costly and time consuming, since the keywords must be hand-coded and the process repeated for new collections. In this work we exploit the vast resource of images and documents available on the web for developing image annotation models without any human involvement. We describe a probabilistic framework based on the assumption that images and their co-occurring textual data are generated by mixtures of latent topics. Applications of this framework to image annotation and retrieval show performance gains over previously proposed approaches, despite the noisy nature of our dataset. We also discuss how the proposed model can be used for story picturing, i.e., to find images that appropriately illustrate a text and demonstrate its utility when interfaced with an image caption generator.

Biography: Mirella Lapata is a Reader at the School of Informatics at the University of Edinburgh. She received an MA from Carnegie Mellon University and a PhD in Natural Language Processing from the University of Edinburgh. She has held appointments at the Department of Computational Linguistics at Saarland University (Saarbruecken) and at the Department of Computer Science in Sheffield. She has worked on various problems in natural language processing, mostly with an emphasis on statistical methods and generation applications. Her current research interests include the use of integer linear programming for summarization, topic models for image retrieval, automated story generation, and probabilistic models of semantic representation. She has published broadly in leading NLP and AI conferences and journals and was two times recipient of the best paper award at EMNLP. She has received an Advanced Research Fellowship and several project grants from EPSRC (the UK Engineering and Physical Sciences Research Council) and has served on the programme committee of major NLP, AI and machine learning conferences.

C. Gurrin et al. (Eds.): ECIR 2010, LNCS 5993, p. 12, 2010.
© Springer-Verlag Berlin Heidelberg 2010

A Language Modeling Approach for Temporal Information Needs

Klaus Berberich, Srikanta Bedathur, Omar Alonso*, and Gerhard Weikum

Max-Planck Institute for Informatics, Saarbrücken, Germany
{kberberi,bedathur,oalonso,weikum}@mpi-inf.mpg.de

Abstract. This work addresses information needs that have a temporal dimension conveyed by a temporal expression in the user's query. Temporal expressions such as "in the 1990s" are frequent, easily extractable, but not leveraged by existing retrieval models. One challenge when dealing with them is their inherent uncertainty. It is often unclear which exact time interval a temporal expression refers to.

We integrate temporal expressions into a language modeling approach, thus making them first-class citizens of the retrieval model and considering their inherent uncertainty. Experiments on the New York Times Annotated Corpus using Amazon Mechanical Turk to collect queries and obtain relevance assessments demonstrate that our approach yields substantial improvements in retrieval effectiveness.

1 Introduction

Many information needs have a temporal dimension as expressed by a temporal phrase contained in the user's query. Existing retrieval models, however, often do not provide satisfying results for such *temporal information needs*, as the following example demonstrates.

Consider a sports journalist, interested in FIFA World Cup tournaments during the 1990s, who issues the query fifa world cup 1990s. A document stating that *France won the FIFA World Cup in 1998* would often not be found by existing retrieval models, despite its obvious relevance to the journalist's information need. The same holds for a document published in 1998 that mentions the *FIFA World Cup final in July*. This is because existing retrieval models miss the semantic connections between the temporal expressions "in 1998" and "in July" contained in the documents and the user's query temporal expression "1990s".

Improving retrieval effectiveness for such temporal information needs is an important objective for several reasons. First, a significant percentage of queries has temporal information needs behind them – about 1.5% of web queries were found to contain an explicit temporal expression (as reported in [1]) and about 7% of web queries have an implicit temporal intent (as reported in [2]). Notice that these numbers are based on general web queries – for *specific domains* (e.g., news or sports) or *expert users* (e.g., journalists or historians) we expect a larger fraction of queries to have a temporal information need behind them. Second,

---

* Current affiliation: Microsoft Corp.

C. Gurrin et al. (Eds.): ECIR 2010, LNCS 5993, pp. 13–25, 2010.
© Springer-Verlag Berlin Heidelberg 2010

thanks to improved digitization techniques and preservation efforts, many document collections, including the Web, nowadays contain documents that (i) were *published a long time ago* and (ii) *refer to different times*. Consider, as one such document collection, the archive of the New York Times that covers the years 1851–2009. Articles in this archive provide a contemporary but also retrospective account on events during that time period. When searching these document archives, the temporal dimension plays an important role.

Temporal expressions are frequent across many kinds of documents and can be extracted and resolved with relative ease. However, it is not immediately clear how they should be integrated into a retrieval model. The key problem here is that the actual meaning of many temporal expressions is uncertain, or more specifically, it is not clear which exact time interval they actually refer to. As an illustration, consider the temporal expression "in 1998". Depending on context, it may refer to a particular day in that year, as in the above example about the FIFA World Cup final, or, to the year as a whole as in the sentence *in 1998 Bill Clinton was President of the United States.*

Our approach, in contrast to earlier work [3, 4, 5], considers this uncertainty. It integrates temporal expressions, in a principled manner, into a language modeling approach, thus making them first-class citizens of the retrieval model.

Contributions made in this work are: (i) a novel approach that integrates temporal expressions into a language model retrieval framework and (ii) a comprehensive experimental evaluation on the New York Times Annotated Corpus [6], as a real-world dataset, for which we leverage the crowd-sourcing platform Amazon Mechanical Turk [7] to collect queries and obtain relevance assessments.

Organization. The rest of this paper is organized as follows. In Section 2, we introduce our model and notation. Section 3 describes how temporal expressions can be integrated into a language modeling approach. Conducted experiments and their results are described in Section 4. Section 5 puts our work in context with existing related research. Finally, we conclude in Section 6.

2   Model

In this work, we apply a discrete notion of time and assume the integers $\mathbb{Z}$ as our *time domain* with timestamps $t \in \mathbb{Z}$ denoting the number of time units (e.g., milliseconds or days) passed (to pass) since (until) a reference time-point (e.g., the UNIX epoch). These time units will be referred to as *chronons* in the remainder. We model a temporal expression T as a quadruple

$$T = (\, tb_l, \, tb_u, \, te_l, \, te_u \,) \, .$$

In our representation tb_l and tb_u are respectively a lower bound and upper bound for the begin boundary of the time interval – marking the time interval's earliest and latest possible begin time. Analogously, te_l and te_u are respectively a lower bound and upper bound for the end boundary of the time interval – marking the time interval's earliest and latest possible end time. Since the time interval is not necessarily known exactly, we hence capture lower and upper bounds for its boundaries. To give a concrete example, the temporal expression "in 1998" from the introduction is represented as

$$(\ 1998/01/01, \ 1998/12/31, \ 1998/01/01, \ 1998/12/31 \ ) \ .$$

This representation thus captures the uncertainty inherent to many temporal expressions – a temporal expression T can refer to any time interval $[b, \ e]$ having a begin point $b \in [tb_l, \ tb_u]$ and an end point $e \in [te_l, \ te_u]$ along with the constraint $b \leq e$. We consider these time intervals thus as our *elementary units of meaning* in this work. In the remainder, when we refer to the temporal expression T, we implicitly denote the set of time intervals that T can refer to. Note that for notational convenience we use the format YYYY/MM/DD to represent chronons – their actual values are integers as described above.

Let D denote our document collection. A document $d \in D$ is composed of its *textual part* d_{text} and its *temporal part* d_{time}. The textual part d_{text} is a bag of textual terms drawn from a vocabulary V. The temporal part d_{time} is a bag of temporal expressions.

Analogously, a query q also consists of a textual part q_{text} and a temporal part q_{time}. We distinguish two modes how we derive such a query from the user's input, which differ in how they treat temporal expressions extracted from the input. In the *inclusive* mode, the parts of the user's input that constitute a temporal expression are still included in the textual part of the query. In the *exclusive* mode, these are no longer included in the textual part. Thus, for the user input boston july 4 2002, as a concrete example, in the inclusive mode we obtain $q_{text} = \{\text{boston}, \text{july}, 4, 2002\}$, whereas we obtain $q_{text} = \{\text{boston}\}$ in the exclusive mode.

3   Language Models Integrating Temporal Expressions

With our formal model and notation established, we now turn our attention to how temporal expressions can be integrated into a language modeling approach and how we can leverage them to improve retrieval effectiveness for temporal information needs.

We use a query-likelihood approach and thus rank documents according to their estimated probability of generating the query. We assume that the textual and temporal part of the query q are generated independently from the corresponding parts of the document d, yielding

$$P(\ q \ | \ d \ ) = P(\ q_{text} \ | \ d_{text} \ ) \times P(\ q_{time} \ | \ d_{time} \ ) \ . \tag{1}$$

The first factor $P(\ q_{text} \ | \ d_{text} \ )$ can be implemented using an existing text-based query-likelihood approach, e.g., the original Ponte and Croft model [8]. In our concrete implementation, as detailed in Section 4, we employ a unigram language model with Jelinek-Mercer smoothing as described in Manning et al. [9].

For the second factor in (1), we assume that query temporal expressions in q_{time} are generated independently from each other, i.e.,

$$P(\ q_{time} \ | \ d_{time} \ ) = \prod_{Q \in q_{time}} P(\ Q \ | \ d_{time} \ ) \ . \tag{2}$$

For the generation of temporal expressions from a document d we use a two-step generative model. In the first step, a temporal expression T is drawn at uniform random from the temporal expressions contained in the document. In the second step, a temporal expression is generated from the temporal expression T just drawn. Under this model, the probability of generating the query temporal expression Q from document d is

$$P(Q \mid d_{time}) = \frac{1}{|d_{time}|} \sum_{T \in d_{time}} P(Q \mid T) . \qquad (3)$$

In the remainder of this section we describe two ways how $P(Q \mid T)$ can be defined. Like other language modeling approaches, our model is prone to the zero-probability problem – if one of the query temporal expressions has zero probability of being generated from the document, the probability of generating the query from this document is zero. To mitigate this problem, we employ Jelinek-Mercer smoothing, and estimate the probability of generating the query temporal expression Q from document d as

$$P(Q|d_{time}) = (1-\lambda)\cdot\frac{1}{|D_{time}|} \sum_{T \in D_{time}} P(Q|T) + \lambda\cdot\frac{1}{|d_{time}|} \sum_{T \in d_{time}} P(Q|T) \quad (4)$$

where $\lambda \in [0, 1]$ is a tunable mixing parameter, and D_{time} refers to the temporal part of the document collection treated as a single document.

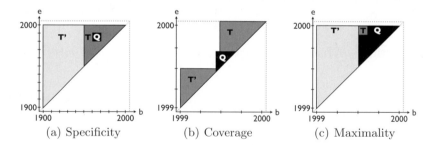

(a) Specificity   (b) Coverage   (c) Maximality

Fig. 1. Three requirements for a generative model

Before giving two possible definitions of $P(Q \mid T)$, we identify the following requirements that any definition of $P(Q \mid T)$ must satisfy (these are illustrated in Figure 1, where each temporal expression is represented as a two-dimensional region that encompasses compatible combinations of begin point b and end point e):

– **Specificity:** Given two temporal expressions T and T', we have

$$|T \cap Q| = |T' \cap Q| \wedge |T| \leq |T'| \Rightarrow P(Q \mid T) \geq P(Q \mid T') .$$

In other words, a query temporal expression is more likely to be generated from a temporal expression that closely matches it. Referring to Figure 1(a), the probability of generating Q (corresponding, e.g., to "from the 1960s until the 1980s") from T (corresponding, e.g., to "in the second half of the 20th century") is more than generating it from T' (corresponding, e.g., to "in the 20th century").

Coverage: Given two temporal expressions T and T', we have

$$|T| = |T'| \wedge |T \cap Q| \leq |T' \cap Q| \Rightarrow P(Q \,|\, T) \leq P(Q \,|\, T') \,.$$

In this requirement, we capture the notion that a larger overlap with the query temporal expression is preferred. In Figure 1(b), the overlap of Q (corresponding, e.g., to "in the summer of 1999") with T (corresponding, e.g., to "in the first half of 1999") is more than the overlap with T' (corresponding, e.g., to "in the second half of 1999"). Therefore, the latter temporal expression is preferable and should have a higher probability of generating Q.

– **Maximality:** $P(Q \,|\, T)$ should be maximal for $T = Q$, i.e.,

$$T \neq Q \Rightarrow P(Q \,|\, T) \leq P(Q \,|\, Q) \,.$$

This requirement captures the intuition that the probability of generating a query temporal expression from a temporal expression matching it exactly must be the highest. As shown in Figure 1(c), the probability of generating Q (corresponding, e.g., to "in the second half of 1999") from itself should be higher than the probability of generating it from T (corresponding, e.g., to "from July 1999 until December 1999") or T' (corresponding, e.g., to "in 1999"). Note that the maximality requirement can be derived by combining the requirements of specificity and coverage.

3.1   Uncertainty-Ignorant Language Model

Our first approach, referred to as LMT in the remainder, ignores the uncertainty inherent to temporal expressions. A temporal expression T can only generate itself, i.e.,

$$P(Q \,|\, T) = \mathbb{1}(T = Q) \,, \tag{5}$$

where $\mathbb{1}(T = Q)$ is an indicator function whose value assumes 1 iff $T = Q$ (i.e., $tb_l = qb_l \wedge tb_u = qb_u \wedge te_l = qe_l \wedge te_u = qe_u$). The approach thus ignores uncertainty, since it misses the fact that a temporal expression T and a query temporal expression Q may refer to the same time interval, although $T \neq Q$. It can easily be verified that this approach meets the above requirements[2].

Despite its simplicity the approach still profits from the extraction of temporal expressions. To illustrate this, consider the two temporal expressions "in the 1980s" and "in the '80s". Both share the same formal representation in our model, so that LMT can generate a query containing one of them from a document containing the other. In contrast, a text-based approach (i.e., one not paying special attention to temporal expressions), would not be aware of the semantic connection between the textual terms '80s and 1980s.

---

[2] Proofs and additional details are provided in our accompanying technical report [10].

3.2   Uncertainty-Aware Language Model

As explained in the introduction, for many temporal expressions the exact time interval that they refer to is uncertain. Our second approach LMTU explicitly considers this uncertainty. In detail, we define the probability of generating Q from the document d as

$$P(Q \mid T) = \frac{1}{|Q|} \sum_{[q_b, q_e] \in Q} P([q_b, q_e] \mid T), \tag{6}$$

where the sum ranges over all time intervals included in Q. The approach thus assumes equal likelihood for each time interval $[q_b, q_e]$ that Q can refer to. Intuitively, each time interval that the user may have had in mind when uttering Q is assumed equally likely. Recall that $|Q|$ denotes the huge but finite total number of such time intervals.

The probability of generating the time interval $[q_b, q_e]$ from a temporal expression T is defined as

$$P([q_b, q_e] \mid T) = \frac{1}{|T|} \mathbb{1}([q_b, q_e] \in T) \tag{7}$$

where $\mathbb{1}([q_b, q_e] \in T)$ is an indicator function whose value is 1 iff $[q_b, q_e] \in T$. For T we thus also assume all time intervals that it can refer to as equally likely. Putting (6) and (7) together we obtain

$$P(Q \mid T) = \frac{1}{|Q|} \sum_{[q_b, q_e] \in Q} \frac{1}{|T|} \mathbb{1}([q_b, q_e] \in T), \tag{8}$$

which can be simplified as

$$P(Q \mid T) = \frac{|T \cap Q|}{|T| \cdot |Q|}. \tag{9}$$

Both Q and T are inherently uncertain, i.e., it is not clear which time interval the user issuing the query and author writing the document had in mind when uttering Q and T, respectively. Having no further information, our model assumes equal likelihood for all possible time intervals that Q and T respectively can refer to. This definition meets our three requirements defined above[2].

Efficient Computation. For the practical applicability of this model, one important issue that needs addressing is the efficient evaluation of (9). Naïvely enumerating all time intervals that T and Q can refer to, before computing $|T \cap Q|$ is clearly not a practical solution. Consider again the temporal expression (1998/01/01, 1998/12/31, 1998/01/01, 1998/12/31). For a temporal resolution with chronons corresponding to days (hours) the total number of time intervals that this temporal expression can refer to is 66,795 (38, 373, 180). Fortunately, though, there is a more efficient way to evaluate (9).

Given $T = (tb_l, tb_u, te_l, te_u)$, we can compute $|T|$ by distinguishing two cases: (i) if $tb_u \leq te_l$ then $|T|$ can simply be computed as

$$(tb_u - tb_l + 1) \cdot (te_u - te_l + 1)$$

since any begin point b is compatible with any end point e, (ii) otherwise, if $tb_u > te_l$, we can compute $|T|$ as

$$|T| = \sum_{b=tb_l}^{tb_u} (te_u - max(b, te_l) + 1) \tag{10}$$
$$= (te_l - tb_l + 1) \cdot (te_u - te_l + 1) + (tb_u - te_l) \cdot (te_u - te_l + 1)$$
$$- 0.5 \cdot (tb_u - te_l) \cdot (tb_u - te_l + 1) .$$

This captures that only end points $e \geq max(b, te_l)$ are compatible with a fixed begin point b. The derivation of (10) is given in the appendix.

Let $Q = (qb_l, qb_u, qe_l, qe_u)$ be a query temporal expression. Using (10) we can determine $|Q|$. For computing $|Q \cap T|$ notice that each time interval $[b, e] \in Q \cap T$ fulfills $b \in [tb_l, tb_u] \cap [qb_l, qb_u]$ and $e \in [te_l, te_u] \cap [qe_l, qe_u]$. Therefore, $|T \cap Q|$ can be computed by considering the temporal expression

$$(max(tb_l, qb_l), min(tb_u, qb_u), max(te_l, qe_l), min(te_u, qe_u)) .$$

Thus, we have shown that the generative model underlying LMTU allows for efficient computation. When processing a query with a query temporal expression Q, we need to examine all temporal expressions T with $T \cap Q \neq \emptyset$ and the documents that contain them, as can be seen from (9). This can be implemented efficiently by keeping a small inverted index in main memory that keeps track of the documents that contain a specific temporal expression. Its lexicon, which consists of temporal expressions, can be organized, for instance, using interval trees to support the efficient identification of qualifying temporal expressions via interval intersection.

4   Experimental Evaluation

This section describes the experimental evaluation of our approach.

4.1   Setup and Datasets

Methods under Comparison in our experimental evaluation are:

- LM(γ) – Unigram language model with Jelinek-Mercer smoothing
- LMT-IN(γ, λ) – Uncertainty-ignorant method using inclusive mode
- LMT-EX(γ, λ) – Uncertainty-ignorant method using exclusive mode
- LMTU-IN(γ, λ) – Uncertainty-aware method using inclusive mode
- LMTU-EX(γ, λ) – Uncertainty-aware method using exclusive mode

Apart from our baseline LM, we thus consider all four combinations of (a) inclusive vs. exclusive mode (i.e., whether query terms constituting a temporal expression are part of q_{text}) and (b) uncertainty-ignorant vs. uncertainty-aware definition of $P(Q|T)$. The mixing parameters γ and λ control the Jelinek-Mercer smoothing used when generating the textual part and the temporal part of the query, respectively. We consider values in $\{0.25, 0.5, 0.75\}$ for each of them, giving us a total of 39 method configurations under comparison. Further, notice

		Sports	Culture
Day		boston red sox [october 27, 2004] ac milan [may 23, 2007]	kurt cobain [april 5, 1994] keith harring [february 16, 1990]
Month		stefan edberg [july 1990] italian national soccer team [july 2006]	woodstock [august 1994] pink floyd [march 1973]
Year		babe ruth [1921] chicago bulls [1991]	rocky horror picture show [1975] michael jackson [1982]
Decade		michael jordan [1990s] new york yankees [1910s]	sound of music [1960s] mickey mouse [1930s]
Century		la lakers [21st century] soccer [21st century]	academy award [21st century] jazz music [21st century]

		Technology	World Affairs
Day		mac os x [march 24, 2001] voyager [september 5, 1977]	berlin [october 27, 1961] george bush [january 18, 2001]
Month		thomas edison [december 1891] microsoft halo [june 2000]	poland [december 1970] pearl harbor [december 1941]
Year		roentgen [1895] wright brothers [1905]	nixon [1970s] iraq [2001]
Decade		internet [1990s] sewing machine [1850s]	vietnam [1960s] monica lewinsky [1990s]
Century		musket [16th century] siemens [19th century]	queen victoria [19th century] muhammed [7th century]

Fig. 2. Queries categorized according to their topic and temporal granularity

that our baseline LM, not aware of temporal expressions, always only considers q_{text} as determined using the inclusive mode.

Document Collection. As a dataset for our experimental evaluation we use the publicly-available *New York Times Annotated Corpus* [6] that contains about 1.8 million articles published in New York Times (NYT) between 1987 and 2007.

Queries. Since we target a specific class of information needs, query workloads used in benchmarks like TREC [11] are not useful in our setting. To assemble a query workload that captures users' interests and preferences, we ran two user studies on Amazon Mechanical Turk (AMT). In our first study, workers were provided with an entity related to one of the topics *Sports, Culture, Technology,* or *World Affairs* and asked to specify a temporal expression that fits the given entity. In our second study, users were shown a temporal expression corresponding to a *Day, Month, Year, Decade,* or *Century* and asked to add an entity related to one of the aforementioned topics. Among the queries obtained from our user studies, we selected the 40 queries shown in Figure 2. Queries are categorized according to their topic and temporal granularity, giving us a total of 20 query categories, each of which contains two queries.

Relevance Assessments were also collected using AMT. We computed top-10 query results for each query and each method configuration under comparison, pooled them, yielding a total of 1,251 query-document pairs. Each of these query-document pairs was assessed by five workers on AMT. Workers could state whether they considered the document *relevant* or *not relevant* to the query. To prevent spurious assessments, a third option (coined *I don't know*) was provided, which workers should select if they had insufficient information or knowledge to assess the document's relevance. Further, we asked workers to explain in their own words, why the document was relevant or not relevant. We found the feedback provided through the explanations extremely insightful. Examples of provided explanations are:

- roentgen [1895]: *Wilhelm Roentgen was alive in 1895 when the building in New York at 150 Nassau Street in downtown Manhattan, NYC was built, they do not ever intersect other than sharing the same timeline of existence for a short while.*
- nixon [1970s]: *This article is relevant. It is a letter to the editor in response to a column about 1970s-era Nixon drug policy.*
- keith harring [february 16, 1990]: *The article does not have any information on Keith Harring, only Laura Harring. Though it contains the keywords Harring and 1990, the article is obviously not what the searcher is looking for.*

Apart from that, when having to explain their assessment, workers seemed more thorough in their assessments. Per completely assessed query-document pair we paid $0.02 per assignment to workers. Workers chose relevant for 33%, not relevant for 63%, and the third option (i.e., *I don't know*) for 4% of the total 6,255 relevance assessments. Relevance assessments with the last option are ignored when computing retrieval-effectiveness measures below.

Implementation Details. We implemented all methods in Java. All data was kept in an Oracle 11g database. Temporal expressions were extracted using TARSQI [12]. TARSQI detects and resolves temporal expressions using a combination of hand-crafted rules and machine learning. It annotates a given input document using the TimeML [13] markup language. Building on TARSQI's output, we extracted range temporal expressions such as "from 1999 until 2002", which TARSQI does not yet support. Further, we added each article's publication date as an additional temporal expression. We map temporal expressions to our quadruple representation using milliseconds as chronons and the UNIX epoch as our reference time-point.

4.2 Experimental Results

We measure the retrieval effectiveness of the methods under comparison using Precision at k (P@k) and nDCG at k (N@k) as two standard measures. When computing P@k, we employ majority voting. Thus, a document is considered relevant to a query, if the majority of workers assessed it as relevant. When computing N@k, the average relevance grade assigned by workers is determined interpreting *relevant* as grade 1 and *not relevant* as grade 0, respectively.

Table 1. Retrieval effectiveness overall

	P@5	N@5	P@10	N@10
LM ($\gamma = 0.25$)	0.33	0.34	0.30	0.32
LM ($\gamma = 0.75$)	0.38	0.39	0.37	0.38
LMT-IN ($\gamma = 0.25,\ \lambda = 0.75$)	0.26	0.27	0.23	0.25
LMT-IN ($\gamma = 0.75,\ \lambda = 0.75$)	0.29	0.31	0.25	0.28
LMT-EX ($\gamma = 0.25,\ \lambda = 0.75$)	0.36	0.36	0.32	0.33
LMT-EX ($\gamma = 0.5,\ \lambda = 0.75$)	0.37	0.37	0.32	0.33
LMTU-IN ($\gamma = 0.25,\ \lambda = 0.75$)	0.41	0.42	0.37	0.37
LMTU-IN ($\gamma = 0.75,\ \lambda = 0.25$)	0.44	0.44	0.39	0.40
LMTU-EX ($\gamma = 0.25,\ \lambda = 0.75$)	0.53	0.51	0.49	0.49
LMTU-EX ($\gamma = 0.5,\ \lambda = 0.75$)	**0.54**	**0.52**	**0.51**	**0.49**

Overall. Table 1 gives retrieval-effectiveness figures computed using all queries and cut-off levels $k = 5$ and $k = 10$. For each of the five methods under comparison, the table shows the best-performing and worst-performing configuration with their respective parameter values γ and λ. The figures shown support the following three observations: (i) the exclusive mode outperforms the inclusive mode for both LMT and LMTU, (ii) LMT does not yield improvements over the baseline LM, but (iii) LMTU is at par with the baseline LM when the inclusive mode is used and outperforms it significantly when used with the exclusive mode. For LMTU-EX the worst configuration beats the best configuration of the baseline. Further, the worst and best configuration of LMTU-EX are close to each other demonstrating the method's robustness.

Table 2. Retrieval effectiveness by topic

	Sports		Culture		Technology		World Affairs	
	P@10	**N@10**	**P@10**	**N@10**	**P@10**	**N@10**	**P@10**	**N@10**
LM	0.33	0.33	0.39	0.38	0.27	0.32	0.50	0.49
LMT-IN	0.36	0.36	0.25	0.30	0.10	0.15	0.30	0.30
LMT-EX	0.46	0.44	0.33	0.34	0.12	0.17	0.38	0.38
LMTU-IN	0.46	0.44	0.41	0.42	0.21	0.27	0.48	0.48
LMTU-EX	**0.67**	**0.58**	**0.47**	**0.49**	**0.29**	**0.34**	**0.60**	**0.57**

By Topic. For the best-performing configuration of each method (as given in Table 1), we compute retrieval-effectiveness measures at cut-off level $k = 10$ and group them by topic. Table 2 shows the resulting figures. These support our above observations. In addition, we observe that all methods perform worst on queries from *Technology*. The best performance varies per method and measure.

Table 3. Retrieval effectiveness by temporal granularity

	Day		Month		Year		Decade		Century	
	P@10	**N@10**	**P@10**	**N@10**	**P@10**	**N@10**	**P@10**	**N@10**	**P@10**	**N@10**
LM	0.35	0.38	0.42	0.40	0.65	0.59	0.20	0.28	0.25	0.26
LMT-IN	0.18	0.22	0.20	0.21	0.55	0.50	0.23	0.30	0.20	0.24
LMT-EX	0.26	0.28	0.24	0.25	0.58	0.55	0.28	0.33	0.31	0.32
LMTU-IN	0.33	0.36	0.47	0.46	0.59	0.56	0.34	0.35	0.24	0.27
LMTU-EX	**0.43**	**0.44**	**0.50**	**0.50**	**0.69**	**0.64**	**0.56**	**0.54**	**0.36**	**0.35**

By Temporal Granularity. In analogy, we can group retrieval-effectiveness measurements at cut-off level $k = 10$ by temporal granularity – again considering only the best-performing configuration of each method. Table 3 gives the resulting figures. Again, LMTU-EX consistently achieves the best retrieval performance. We further observe significant variations in retrieval effectiveness across temporal granularities. For queries including a year, all methods achieve their best performance. The worst performance varies per method and measure.

Summary. Putting things together, there is a clear winner in our experimental evaluation. LMTU-EX consistently achieves the best retrieval performance. This demonstrates that (i) considering the uncertainty inherent to temporal expressions is essential and (ii) excluding terms that constitute a temporal expression from the textual part of the query is beneficial. These findings are confirmed by a second experiment on a snapshot of the English Wikipedia [14][2].

5   Related Work

We now put our work in context with existing prior research. Alonso et al. [15] highlight the importance of temporal information in Information Retrieval, and suggest the problem addressed in this work as one not yet satisfactorily supported by existing approaches.

Li and Croft [16] and Dakka et al. [17] both propose language models that take into account publication times of documents, in order to favor, for instance, more recent documents. Kanahuba and Nørvåg [18] and de Jong et al. [19] employ language models to date documents, i.e., determine their publication time. Del Corso et al. [20] address the problem of ranking news articles, taking into account publication times but also their interlinkage. Jones and Diaz [21] focus on constructing query-specific temporal profiles based on the publication times of relevant documents. Thus, all of the approaches mentioned are based on the publication times of documents. None of the approaches, though, considers temporal expressions contained in the documents' contents.

Baeza-Yates [4] is the earliest approach that considers temporal expressions contained in documents for retrieval purposes. It aims at searching information that refers to the future. The proposed retrieval model is focused on confidences associated with statements about the future, thus favoring relevant documents that are confident about their predictions regarding a future time of interest. Kalczynski et al. [5] study the human perception of temporal expressions and propose a retrieval model for business news archives that takes into account temporal expressions. Arikan et al. [3] integrate temporal expressions into a language modeling approach but ignore the aspect of uncertainty. Metzler et al. [2], most recently, identify so-called implicitly temporal queries and propose a method to bias ranking functions in favor of documents matching the user's implicit temporal intent.

The extraction of temporal expressions is a well-studied problem. For an overview of the current state of the art and a description of the TARSQI toolkit, we refer to Verhagen et al. [12, 22]. Our formal representation of temporal expressions as quadruples is adopted from Zhang et al. [23]. Koen and Bender [24] describe the Time Frames system that extracts temporal expressions and uses them to augment the user experience when reading news articles, for instance, by displaying a temporal context of concurrent events.

Several prototypes are available that make use of temporal expressions when searching the Web, most notably, Google's Timeline View [25] and Time-Search [26]. Details about their internals, though, have not been published.

Crowd-sourcing platforms such as AMT are becoming a common tool for conducting experiments in Information Retrieval. For a discussion of their benefits and guidelines on how to use them, we refer to Alonso et al. [27].

6 Conclusion

In this work, we have developed a novel approach that integrates temporal expressions into a language model retrieval framework, taking into account the uncertainty inherent to temporal expressions. Comprehensive experiments on a large corpus of New York Times articles with relevance assessments obtained using Amazon Mechanical Turk showed that our approach substantially improves retrieval effectiveness for temporal information needs.

Ongoing and Future Work. Our focus in this work has been on temporal information needs disclosed by an *explicit* temporal expression in the user's query. Often, as somewhat explored in [2], queries may not contain such an explicit temporal expression, but still have an associated *implicit* temporal intent. Consider a query such as bill clinton arkansas that is likely to allude to Bill Clinton's time as Governor of Arkansas between 1971 and 1981. Detecting and dealing with such queries is part of our ongoing research.

Acknowledgment

This work was partially supported by the EU within the 7th Framework Programme under contract 216267 "Living Web Archives (LiWA)".

References

[1] Nunes, S., et al.: Use of Temporal Expressions in Web Search. In: ECIR (2008)
[2] Metzler, D., et al.: Improving Search Relevance for Implicitly Temporal Queries. In: SIGIR (2009)
[3] Arikan, I., et al.: Time Will Tell: Leveraging Temporal Expressions in IR. In: WSDM (2009)
[4] Baeza-Yates, R.A.: Searching the future. In: SIGIR Workshop MF/IR (2005)
[5] Kalczynski, P.J., Chou, A.: Temporal document retrieval model for business news archives. Inf. Process. Manage. (2005)
[6] New York Times Annotated Corpus, http://corpus.nytimes.com
[7] Amazon Mechanical Turk, http://www.mturk.com
[8] Ponte, J.M., Croft, W.B.: A language modeling approach to information retrieval. In: SIGIR (1998)
[9] Manning, C.D., et al.: Introduction to Information Retrieval. Cambridge University Press, Cambridge (2008)
[10] Berberich, K., et al.: A Language Modeling Approach for Temporal Information Needs. Research Report MPI-I-2010-5-001
[11] Text REtrieval Conference, http://trec.nist.gov
[12] Verhagen, M., et al.: Automating Temporal Annotation with TARSQI. In: ACL (2005)
[13] TimeML Specification Language, http://www.timeml.org
[14] Wikipedia, http://www.wikipedia.org
[15] Alonso, O., et al.: On the value of temporal information in information retrieval. SIGIR Forum (2007)
[16] Li, X., Croft, W.B.: Time-based language models. In: CIKM (2003)
[17] Dakka, W., et al.: Answering general time sensitive queries. In: CIKM (2008)

[18] Kanhabua, N., Nørvåg, K.: Improving temporal language models for determining time of non-timestamped documents. In: ECDL (2008)
[19] de Jong, F. et al.: Temporal language models for the disclosure of historical text. In: AHC (2005)
[20] Corso, G.M.D., et al.: Ranking a stream of news. In: WWW (2005)
[21] Jones, R., Diaz, F.: Temporal profiles of queries. ACM Trans. Inf. Syst. (2007)
[22] Verhagen, M., Moszkowicz, J.L.: Temporal Annotation and Representation. In: Language and Linguistics Compass (2009)
[23] Zhang, Q., et al.: TOB: Timely Ontologies for Business Relations. In: WebDB (2008)
[24] Koen, D.B., Bender, W.: Time frames: Temporal augmentation of the news. IBM Systems Journal (2000)
[25] Google's Timeline View, http://www.google.com/experimental/
[26] TimeSearch History, http://www.timesearch.info
[27] Alonso, O., et al.: Crowdsourcing for relevance evaluation. SIGIR Forum (2008)

Appendix: Derivation of Equation 10

Recall that we assume $tb_u > te_l$.

$$|T| = \sum_{tb=tb_l}^{tb_u} (te_u - max(tb, te_l) + 1)$$

$$= \sum_{tb=tb_l}^{te_l} (te_u - max(tb, te_l) + 1) + \sum_{tb=te_l+1}^{tb_u} (te_u - max(tb, te_l) + 1)$$

$$= (te_l - tb_l + 1) \cdot (te_u - te_l + 1) + \sum_{tb=te_l+1}^{tb_u} (te_u - tb + 1)$$

$$= (te_l - tb_l + 1) \cdot (te_u - te_l + 1) + \sum_{c=1}^{tb_u-te_l} (te_u - c - te_l + 1)$$

$$= (te_l - tb_l + 1) \cdot (te_u - te_l + 1) + (tb_u - te_l) \cdot (te_u - te_l + 1) - \sum_{c=1}^{tb_u-te_l} c$$

$$= (te_l - tb_l + 1) \cdot (te_u - te_l + 1) + (tb_u - te_l) \cdot (te_u - te_l + 1) - 0.5 \cdot (tb_u - te_l) \cdot (tb_u - te_l + 1)$$

Analyzing Information Retrieval Methods to Recover Broken Web Links*

Juan Martinez-Romo and Lourdes Araujo

NLP & IR Group, UNED, Madrid 28040, Spain
{juaner,lurdes}@lsi.uned.es

Abstract. In this work we compare different techniques to automatically find candidate web pages to substitute broken links. We extract information from the anchor text, the content of the page containing the link, and the cache page in some digital library. The selected information is processed and submitted to a search engine. We have compared different information retrieval methods for both, the selection of terms used to construct the queries submitted to the search engine, and the ranking of the candidate pages that it provides, in order to help the user to find the best replacement. In particular, we have used term frequencies, and a language model approach for the selection of terms; and cooccurrence measures and a language model approach for ranking the final results. To test the different methods, we have also defined a methodology which does not require the user judgments, what increases the objectivity of the results.

Keywords: Information retrieval, link integrity, recommender system.

1 Introduction

Missing pages are very frequent on the web: many websites disappear while others are not properly maintained. Thus, broken links represent an important problem that affects the information access and the ranking of the search engines. There exist several validators to check the broken links of our pages. However, once we have detected a broken link, it is not always easy nor fast to find again the disappeared page. In this work, we try to analyze the recovery of broken links as an information retrieval problem for building an user recommendation system. In the same way that an user query to a search engine to find a Web page with the information he needs, our system captures all data related to the broken link context in order to find a new page that containing information similar to the missing one.

Our system checks the links of the page given as input. For those which are broken, the system proposes to the user a set of candidate pages to substitute the broken link. Figure 1 presents a scheme of the proposed system. The first step of our work has been the analysis of a large number of web pages and their

---

* This work has been partially supported by the Spanish Ministry of Science and Innovation within the project QEAVis-Catiex (TIN2007-67581-C02-01) and the Regional Government of Madrid under the Research Network MAVIR (S-0505/TIC-0267).

C. Gurrin et al. (Eds.): ECIR 2010, LNCS 5993, pp. 26–37, 2010.

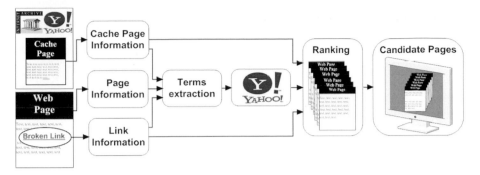

Fig. 1. Scheme of the system for automatic recovering of broken links

links in order to determine which ones are the most useful sources of information and which of them are the most appropriate in each case. Sometimes we can recover a broken link by entering the anchor text as an user query in a search engine. There are many works which have analyzed the importance of the anchor text and title like a source of information[1]. However, there are many cases in which the anchor text does not contain enough information to do that. In these cases, we can compose queries adding terms extracted from other sources of information to the anchor text of the broken link. For that, our system performs a form of query expansion[2], a well-known method to improve the performance of information retrieval systems. In our case, the original query is composed of the terms extracted from the anchor text, and the sources of expansion terms are the elements of the parent web page containing the broken link (text, title, url, etc), and also, if they exist, the elements of the cache page corresponding to the disappeared page that can be stored in a search engine (*Yahoo*) or web archive (*Wayback Machine*). In this work we have investigated the performance of different approaches to extract the expansion terms from the mentioned sources of information in the context of a link.

After the term extraction step, different expanded queries are submitted to the considered search engine, and the set of top ranked documents are retrieved. In order to tune the results, the pages recovered in this way are ranked according to relevance measures obtained by applying information retrieval (IR) techniques, and finally they are presented to the user.

In order to evaluate the different IR techniques considered, we have developed a methodology which mainly relies on the random selection of pages and the use of links that are not really broken to check how many are properly recovered.

The rest of this paper is organized as follows. We start off reviewing related work in the next section 2; section 3 describes the methodology we have followed; section 4 analyses how useful is the anchor text of the links to recover the page; section 5 studies the suitability of different sources of information to provide terms for the query expansion process; section 6 is devoted to describe the process to rank the candidate documents; section 7 presents the scheme resulting of the previous analysis; section 8 analyses the parameter setting of the proposed algorithm. Finally, section 9 draws the main conclusions of this work.

2   Background and Related Work

Despite the problem of broken links was considered the second most serious problem on the Web[3] many years ago, missing pages are still frequent when users are surfing the Internet. Previous works quantified this problem: Kahle[4] reported the expected life-time of a web page is 44 days. Koehler et al.[5] performed a longitudinal study of web page availability and found the random test collection of Urls eventually reached a "steady state" after approximately 67% of the Urls were lost over a 4-year period. Markwell et al.[3] monitored the resources of three authentic courses during 14 months, and 16.5% of the links had disappeared or were nonviable.

Most of previous attempts to recover broken links are based on information annotated in advance with the link[6,7]. Closest to our research[8], Nakamizo et al.[9,10] have developed a tool that finds new Urls of web pages after pages are moved. This tool outputs a list of web pages sorted by their plausibility of being link authorities. Klein et al.[11] save a small set of terms (lexical signature) derived from a document that capture the "aboutness" of that document, and they can be used to discover a similar Web page if it disappears in the future. Harrison et al.[12] presented Opal, a framework for interactively locating missing web pages, using the previous lexical signatures which are then used to search for similar versions of a Web page. Our work differs from previous proposals since it does not rely on any information about the links annotated in advance, and it can be applied to any web page. Previous works do not present a evaluation methodology, thus we have also defined a methodology which does not require the user judgments, what increases the objectivity of the results.

Many works appeared using different techniques in the TREC-10 Web Track (homepage finding task) to perform the proposed task, although our work differs in the following respects: (i) the size of the collection is much smaller than the whole Internet, (ii) the number of potential candidates is much more reduced, and (iii) most of papers used the Url depth as a main factor for the page selection.

3   Methodology

If we analyze the usefulness of the different sources of information directly employed on broken links, it is very difficult to evaluate the quality of the sources of information and the techniques used to extract the terms. Therefore, at this phase of analysis, we employ random web links, which are not really broken, and we called *pseudobroken* links. Thus we have the page at which they point and we are able to evaluate our recommendation. We take links from pages selected randomly (250 words are required) by means of successive requests to *www.randomwebsite.com*, a site that provides random web pages. After the best sources of information and information retrieval techniques are selected, we employ really broken links in order to test the whole system.

We consider that a link has been recovered if the system finds a web page which content is practically the same that the *pseudobroken* link. For that, we apply the vector space model [13], i.e we represent each page by a term vector and calculate the cosine distance between them (similarity). If this value is higher than 0.9, we consider that the page has been recovered. Lowering the similarity threshold (e.g. 0.8) very few additional links are recovered, and the number of wrong results increase.

4   Analyzing the Anchor Text in a Hyperlink

In many cases, terms which compose the anchor text of a hyperlink are the main source of information to identify the pointed page. To verify this theory we performed a study searching in Yahoo for the anchor text. Using the previously defined similarity, 41% of the links were recovered in the top ten results. In addition, 66% of the recovered links appear in the first position. These results prove that anchor text is a efficient source of information to recover a broken link.

Sometimes anchor terms provide little or no descriptive value. Let us imagine a link whose anchor text is "click here". In this case, finding the broken link might be impossible. For this reason it is very important to analyze these terms so as to be able to decide which tasks should be performed depending on their quantity and quality. Thus, the system carry out a recognition of named entities (persons, organizations or places) on the anchor text in order to extract certain terms whose importance is higher than the remaining ones. Several experiments have proved that the presence of any named entity in the anchor favors the recovery of the link. The most prominent result is the very small number of cases in which the correct document is recovered when the anchor consists of just a term and it is not a named entity.

5   The Page Text

The most frequent terms of a web page (after removing stop-words) are a way to characterize the main topic of the cited page. This technique requires the page text to be long enough. A clear example of utility of this information are the links to personal pages. The anchor of a link to a personal page is frequently formed by the name of that person. However, in many cases forename and surname do not identify a person in a unique way, specially if they are very common. If we perform a search using only the forename and the surname, the personal page of this person probably will not appear among the first pages retrieved. However, if we expand the query using some terms related to that person, that can be present at his web page, then his personal web page will go up to the first positions.

We have applied classical information retrieval techniques, described below, to extract the most representative terms from a page. After eliminating the stop words, we generate a ranked term list. The first ten terms of this list are used to compose ten expanded queries, one for each expansion term. Every query is formed by the anchor text and it is expanded with each of those terms. Finally, the first ten retrieved documents are taken in each case.

5.1   Frequency-Based Approaches to Select Terms

Frequency-Based are the most simple approaches to select expansion terms. We have considered two different criteria based on frequencies for term selection. The first one is the raw term frequency (TF) in the parent or cache page. There are some terms with very little or no discriminating power as descriptors of the page, despite they are frequent on it. The reason is that those terms are also frequent in many other documents of the considered collection. To take into account these cases, we apply the well-known *Tf-Idf* weighting scheme for a term, where *Idf*(t) is the inverse document frequency of that term. A dump of English Wikipedia articles[1] has been used as reference collection.

5.2   Language Model Approach

One of the most successful methods based on term distribution analysis uses the concept of Kullback-Liebler Divergence[14] to compute the divergence between the probability distributions of terms in the whole collection and the particular considered documents. The most suitable terms to expand the query are those with a high probability in the document, which is the source of terms, and a low probability in the whole collection. For the term t this divergence is:

$$KLD_{(PP,PC)}(t) = P_P(t)\log\frac{P_P(t)}{P_C(t)} \tag{1}$$

where $P_P(t)$ is the probability of the term t in the considered page, and $P_C(t)$ is the probability of the term t in the whole collection.

  Computing this measure requires a reference collection of documents. The relation between this reference collection and the analyzed document, is an important factor in the results obtained with this approach. Obviously, we can not use the whole web as reference collection. To study the impact of this factor on the results we have used two different collections of web pages indexed with Lucene: (i) English Wikipedia articles (3.6 million) dump (Enwiki) and (ii) Homepage at sites (4.5 million) in DMOZ Open Directory Project (ODP).

5.3   Comparing Different Approaches for the Extraction of Terms

Figure 2 show the results obtained using frequency, *Tf-Idf* and *KLD* for the extraction of expansion terms from the parent page and the cache page respectively. In the case of the parent page (Figure 2(i)) the frequency method performs slightly better than *Tf-Idf*, whereas for the cache page (Figure 2(ii)) it is the opposite. We can observe that when we use *KLD*, the results obtained with the Wikipedia as reference collection are better (the total number of correct recovered pages). The reason is probably that this collection provides a wider range of topics. Another important observation is that for extracting terms from the parent page, the results obtained with the methods based on frequencies are higher than *KLD*. On

---

[1] http://download.wikimedia.org/enwiki/

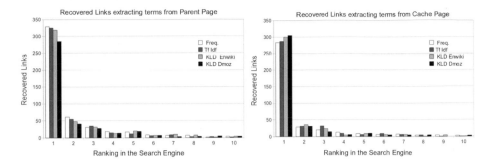

Fig. 2. Recovered links by expanding the query with (i) terms from the parent page and (ii) terms from the cache page. The data of the figure indicate the number of cases in which Yahoo has provided the correct page in the corresponding position when the system expands with the terms extracted applying different approaches.

Table 1. Analysis of the number of retrieved documents in the first position and in top 10 positions, according to if query expansion is used or not

Analysis	1st position	1-10 positions
No Expansion (anchor text)	253	380
Expansion (terms from page)	213	418

the contrary, *KLD* is the most suitable approach to extract terms from the cache. The reason is perhaps that in some cases the parent page content is not closely related to the page to recover, and thus, by refining the methods to select the most representative terms of the parent page does not improve the results. Accordingly, we have used the frequency method for extracting terms from the parent page, and we have used *KLD* with the English Wikipedia as reference collection for extracting terms from the cache page in the remaining experiments.

5.4 The Effect of Expansion on the Relationship between Precision-Recall

We performed a new experiment in order to study the effect of expansion in the relationship between precision-recall. In Table 1 can be observed that expansion considerably increases the number of recovered links ranked in the top ten positions (recall). In spite of this, the number of recovered links ranked in the first position is reduced (precision). Accordingly, we think that the most suitable mechanism is to apply both recovery approaches, and later ranking the whole set of results to present the user the most relevant web pages ranked in the top positions.

6   Ranking Methods

Once the system has retrieved a set of candidate pages to replace the broken link by combining both *No Expansion* and *Expansion* approaches, the system needs to present the results to the user in decreasing order of relevance. In order to establish the best ranking function for the candidate pages, we performed an analysis in order to compare different similarity approaches and elements from parent, cache and candidate pages.

6.1   Vector Space Model

The vector space model[13] is one of the approaches that we have applied to represent the documents. Methods based on term cooccurrence[15] have been used very frequently to identify semantic relationships among documents. In our experiments, we have used the well-known Tanimoto, Dice and Cosine cooccurrence coefficients to measure the similarity between the vectors representing the reference document D_1 and the candidate document D_2:

$$\text{Tanimoto}(\boldsymbol{D_1}, \boldsymbol{D_2}) = \frac{\boldsymbol{D_1 D_2}}{|\boldsymbol{D_1}|^2 + |\boldsymbol{D_2}|^2 - \boldsymbol{D_1 D_2}} \tag{2}$$

$$\text{Dice}(\boldsymbol{D_1}, \boldsymbol{D_2}) = \frac{2\boldsymbol{D_1 D_2}}{|\boldsymbol{D_1}|^2 + |\boldsymbol{D_2}|^2} \tag{3}$$

$$\text{Cosine}(\boldsymbol{D_1}, \boldsymbol{D_2}) = \frac{\boldsymbol{D_1 D_2}}{|\boldsymbol{D_1}||\boldsymbol{D_2}|} \tag{4}$$

6.2   Language Model Approach

We have also applied a language model approach to rank the set of candidate documents. In this case we look at the differences in the term distribution between two documents by computing the Kullback-Leibler divergence:

$$KLD(D_1||D_2) = \sum_{t \in D_1} P_{D_1}(t) log \frac{P_{D_1}(t)}{P_{D_2}(t)} \tag{5}$$

where $P_{D_1}(t)$ is the probability of the term t in the reference document, and $P_{D_2}(t)$ is the probability of the term t in the candidate document.

6.3   Performance of Ranking Methods

We have applied the approaches described above to different elements from the parent, cache and candidate pages. Specifically, we have studied the similarity among the following pairs of elements: (i) parent anchor text & candidate title, (ii) parent anchor text & candidate content, (iii) parent content & candidate content, and (iv) cache content & candidate content.

In addition to these comparisons, we also used the anchor text and the snippet of the candidate document but the results were not improved. We can observe in Figure 3(i) and Figure 3(ii) that the results obtained with KLD are worse than those obtained by using the cooccurrence measures, especially in Figure 3(i). In these Figures we are studying the similarity between a very short text, the anchor

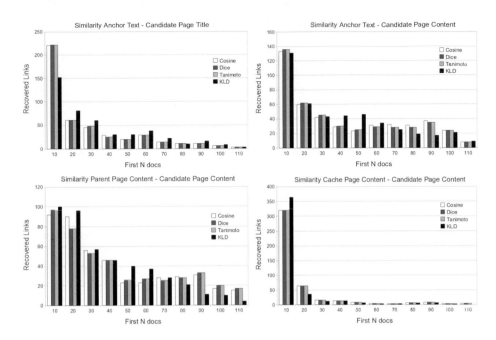

Fig. 3. Results of different approaches (Cosine, Dice, Tanimoto and Kullback-Liebler Divergence) applied to measure the similarity between: (i) the Anchor Text of the broken link and the Title of the candidate page, (ii) the Anchor Text of the broken link and the Content of the candidate page, (iii) the Content of the page where is the broken link and the Content of the candidate page, and (iv) the Content of the Cache page of the broken link and the Content of the candidate page. Results show the position of the best page recovered by the system after the ranking.

text, and other short text which is the title of the candidate page (Figure 3(i)) or the parent page content (Figure 3(ii)). On the contrary, *KLD* performs better than the cooccurrence measures in Figure 3(iii) and Figure 3(iv), where we are measuring the similarity between the content of two pages; the parent or the cache page, and the candidate page. Thus, we can conclude that *KLD* performs better that the cooccurrence methods only if it is applied to texts long enough, such as the content of a page. In Figure 3(iv) we can observe that, as expected, the best results are obtained when they are ranked by similarity with the cache page. However, in many cases this page is not available (near a 60%). Regarding the remaining results, it can be observed that the best results are obtained when the similarity between the anchor text of the broken link and the title of the candidate page is used, and by applying cooccurrence methods. According to these results, *KLD* will be used to rank the candidates pages with respect to the cache page if the system is able to retrieve the cache page. Otherwise, the system will use the similarity between the anchor text and the candidate page title, measured with a cooccurrence method, such a *Dice*, which performs slightly better in some cases.

7   Algorithm for Automatic Recovery of Links

The results of the analysis described in the previous sections suggest several criteria to decide for which cases there is enough information to try the retrieval
of the link and which sources of information to use. According to them, we propose the recovery process which appears in Figure 4. First of all, it is checked
whether the anchor number of terms is just one (`length(anchor) = 1`) and
whether it does not contain named entities (`NoNE(anchor)`). If both features
are found, the retrieval is only attempted provided the link of the missing page
appears in the cache of a search engine or web archive (`InCache(page)`), and
therefore we have reliable information to verify that the proposal presented to
the user can be useful. Otherwise, the user is informed that the recommendation
is not possible (`No_recovered`). If the page is in the cache, then the recovery is
performed, expanding the query (anchor terms) with terms extracted from the
cache using *KLD*. Then the results are ranked (by similarity between the candidate page and the cache page computed with *KLD*) and only if any of them
is sufficiently similar to the cache content (`similarity(docs, cache(page) >
0.9)`), the user is recommended this list of candidate documents. In the remain-
ing cases, that is, when the anchor has more than one term or when it contains
some named entity, the recovery is performed using the anchor terms and the
terms from the cache (applying *KLD*) or parent page (applying frequency selection). After that, all documents are grouped and ranked according to the cache
page (`rank(docs, cache_content_KLD)`) if it is available in a search engine or
web archive, or according to the similarity between the anchor text and the title
of the candidate page applying the Dice cooccurrence coefficient (`rank(docs,
anchor_title_Dice)`) otherwise.

7.1   Results Applying the System to Broken Links

The previous algorithm has been applied to a set of pages with broken links,
but they have only been used those that were present in a digital library (search
engine cache or web archive). The reason is that only in this case we can ob-
jectively evaluate the results. Thanks to the algorithm, the system recovered
553 from 748 broken links (74% of the total links). Table 2 shows a ranking of
recovered links. We have verified that in some cases the original page is found
(it has been moved to other web site) and in some other cases, the system retrieved pages with very similar content. We can observe the system is able to
provide useful replacements documents among the top 10 positions in 46% of
the recovered broken links, and among the 20 first ones in 70% of the cases.

8   Tuning the Algorithm Parameters

An important issue to investigate is the trade-off between the amount of information collected to recover the broken links, and the time required to do it.
We can expect that, in general, the more information available, the better the

```
if length(anchor) = 1 and NoNE(anchor) then
   if InCache(page) then
      docs = web_search(anchor + cache_terms_KLD)
      rank(docs, cache_content_KLD)
      if similarity(docs, cache(page) > 0.9) then
         user_recommendation(docs)
      else
         No_recovered
   else
      No_recovered
else
   docs = web_search(anchor)
   if InCache(page) then
      docs = docs + web_search(anchor + cache_terms_KLD)
      rank(docs, cache_content_KLD)
   else
      docs = docs + web_search(anchor + page_terms_Freq)
      rank(docs, anchor_title_Dice)
   user_recommendation(docs)
```

Fig. 4. Links Automatic Recovery Algorithm for broken links

Table 2. Number of recovered broken links (best candidate which content is very similar to the missing page) according to his cache similarity, among first N documents using the proposed algorithm

First N documents	Recovered Broken Links
1-10	254
10-20	134
20-50	122
50-110	43

recommendation for the broken link. However it is important to know the cost incurred for each increment of the collected data. The amount of information collected mainly depends on two parameters: the number of terms used to expand the query extracted from the anchor text, and the number of hits taken from the results of the search engine for each query. We performed several experiments to evaluate how these parameters affect the results. Figure 5(a) shows the number of recovered links according to the number of retrieved hits from the search engine. We can observe that the improvement is much smaller from 25-30 hits, especially when the expansion is not performed. Figure 5(b) shows the number of recovered links for different number of terms used in the query expansion process. In these experiments the number of hits retrieved from the search engine was set to 10. It is interesting to notice that the expansion approach beats the approach without expansion when expanding with 6 or more terms, and this improvement is quite small from 10 or 15 terms.

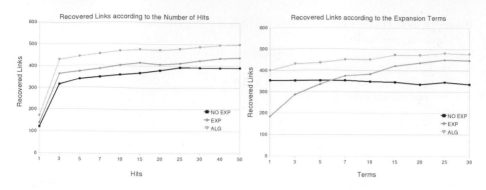

Fig. 5. Recovered Links according to the number of hits/terms used to carry out each query/expanded query. This figure shows the results using a method without expansion (NO EXP), an expansion method (EXP) and the combination of both methods (ALG).

Finally we have carried out an execution time analysis in order to determine the influence on time of the number of hits retrieved from the search engine and the number of terms used for expansion. According to obtained results and the previous results showed in Figure 5, the number of terms and hits has been set to 10, as a trade-off between the improvement in the performance and the execution time (around 17 seconds for every link).

9   Conclusions

In this paper we have analyzed different information retrieval methods for both, the selection of terms used to construct the queries submitted to the search engine, and the ranking of the candidate pages that it provides, in order to help the user to find the best replacement for a broken link. To test the sources, we have also defined a evaluation methodology which does not require the user judgments, what increases the objectivity of the results. We have also studied the effect of using terms from the page that contains the link, and a cache page stored in some search engine or web archive to expand the query formed by the anchor text. This study shows that the results are better when the query is expanded, than when the anchor text is used alone. Thus, query expansion reduce the ambiguity that would entail the limited quantity of anchor terms. We have compared different methods for the extraction of terms to expand the anchor text. Experiments have shown that the best results are obtained using a frequency approach if the cache page is not available, and a language model approach otherwise. We have decided to combine both methods and later reordering the obtained results by applying a relevance ranking in order to present to the user the best candidate pages at the beginning. We have also compared different approaches to rank the candidate documents: cooccurrence approaches and a language model divergence approach (KLD). The best results are obtained applying KLD when the cache page is available, otherwise a cooccurrence method

such as Dice applied between the anchor text of the broken link and the title of
the candidate page. This analysis has allowed us to design a strategy that has
been able to recover a page very similar to the missing one in 74% of the cases.
Moreover, the system was able to provide 46% from those recovered links in the
top ten of the results, and among the top 20 in 70%.

References

1. Craswell, N., Hawking, D., Robertson, S.: Effective site finding using link anchor
   information. In: SIGIR 2001: Proceedings of the 24th annual international ACM
   SIGIR conference on Research and development in information retrieval, pp. 250–
   257. ACM Press, New York (2001)
2. Efthimiadis, E.N.: Query expansion. Annual Review of Information Systems and
   Technology 31, 121–187 (1996)
3. Markwell, J., Brooks, D.W.: Broken links: The ephemeral nature of educational
   www hyperlinks. Journal of Science Education and Technology 11(2), 105–108
   (2002)
4. Kahle, B.: Preserving the internet. Scientific American 276(3), 82–83 (1997)
5. Koehler, W.: Web page change and persistence—a four-year longitudinal study. J.
   Am. Soc. Inf. Sci. Technol. 53(2), 162–171 (2002)
6. Ingham, D., Caughey, S., Little, M.: Fixing the broken-link problem: the w3objects
   approach. Comput. Netw. ISDN Syst. 28(7-11), 1255–1268 (1996)
7. Shimada, T., Futakata, A.: Automatic link generation and repair mechanism for
   document management. In: HICSS 1998: Proceedings of the Thirty-First Annual
   Hawaii International Conference on System Sciences, vol. 2, p. 226. IEEE Computer
   Society Press, Los Alamitos (1998)
8. Martinez-Romo, J., Araujo, L.: Recommendation system for automatic recovery
   of broken web links. In: Geffner, H., Prada, R., Machado Alexandre, I., David, N.
   (eds.) IBERAMIA 2008. LNCS (LNAI), vol. 5290, pp. 302–311. Springer, Heidelberg (2008)
9. Nakamizo, A., Iida, T., Morishima, A., Sugimoto, S., Kitagawa, H.: A tool to compute reliable web links and its applications. In: SWOD 2005: Proc. International
   Special Workshop on Databases for Next Generation Researchers, pp. 146–149.
   IEEE Computer Society, Los Alamitos (2005)
10. Morishima, A., Nakamizo, A., Iida, T., Sugimoto, S., Kitagawa, H.: Pagechaser:
    A tool for the automatic correction of broken web links. In: ICDE, pp. 1486–1488
    (2008)
11. Klein, M., Nelson, M.L.: Revisiting lexical signatures to (re-)discover web pages.
    In: Christensen-Dalsgaard, B., Castelli, D., Ammitzbøll Jurik, B., Lippincott, J.
    (eds.) ECDL 2008. LNCS, vol. 5173, pp. 371–382. Springer, Heidelberg (2008)
12. Harrison, T.L., Nelson, M.L.: Just-in-time recovery of missing web pages. In: HY-
    PERTEXT 2006: Proceedings of the seventeenth conference on Hypertext and
    hypermedia, pp. 145–156. ACM Press, New York (2006)
13. Manning, C.D., Raghavan, P., Schütze, H.: Introduction to Information Retrieval.
    Cambridge University Press, Cambridge (2008)
14. Cover, T.M., Thomas, J.A.: Elements of information theory. Wiley Interscience,
    New York (1991)
15. Rijsbergen, C.J.V.: A theoretical basis for the use of cooccurrence data in infor-
    mation retrieval. Journal of Documentation 33, 106–119 (1977)

Between Bags and Trees – Constructional Patterns in Text Used for Attitude Identification

Jussi Karlgren, Gunnar Eriksson, Magnus Sahlgren, and Oscar Täckström

Swedish Institute of Computer Science, Box 1263
SE-164 29 Kista, Sweden
{jussi,guer,mange,oscar}@sics.se
http://www.sics.se

Abstract. This paper describes experiments to use non-terminological information to find attitudinal expressions in written English text. The experiments are based on an analysis of text with respect to not only the vocabulary of content terms present in it (which most other approaches use as a basis for analysis) but also with respect to presence of structural features of the text represented by constructional features (typically disregarded by most other analyses). In our analysis, following a construction grammar framework, structural features are treated as occurrences, similarly to the treatment of vocabulary features. The constructional features in play are chosen to potentially signify opinion but are not specific to negative or positive expressions.

The framework is used to classify clauses, headlines, and sentences from three different shared collections of attitudinal data. We find that constructional features transfer well across different text collections and that the information couched in them integrates easily with a vocabulary based approach, yielding improvements in classification without complicating the application end of the processing framework.

1 Attitude Analysis Is Mostly Based on Lexical Statistics

Attitude analysis, opinion mining, or sentiment analysis, a subtask of information refinement from texts, has gained interest in recent years, both for its application potential and for the promise of shedding new light on hitherto unformalised aspects of human language usage: the expression of attitude, opinion, or sentiment is a quintessentially human activity. It is not explicitly conventionalised to the degree that many other aspects of language usage are.

Most attempts to identify attitudinal expression in text have been based on lexical factors. Resources such as SentiWordNet, the Opinion Finder subjectivity lexicon, or the General Inquirer lexicon are utilised or developed by most research groups engaged in attitude analysis tasks [4,18,14]. But attitude is not a solely lexical matter. Expressions with identical or near-identical terms can be more or less attitudinal by virtue of their form ("He blew me off" vs. "He blew off"); combinations of fairly attitudinally loaded terms may lack attitudinal power ("He has the best result, we cannot fail him" vs. "This is the best coffee, we

C. Gurrin et al. (Eds.): ECIR 2010, LNCS 5993, pp. 38–49, 2010.

cannot fail with it"); certain terms considered neutral in typical language use can have strong attitudinal loading in certain discourses or certain times ("Fifth Avenue", "9/11").

Our approach takes as its starting point the observation that lexical resources always are noisy, out of date, and most often suffer simultaneously from being both too specific and too general. Not only are lexical resources inherently some-what unreliable or costly to maintain, but they do not cover all the possibilites of expression afforded by human linguistic behaviour.

We believe that attitudinal expression in text is an excellent test case for general purpose approaches for processing of linguistic data. We have previously tested resource-thrifty approaches for annotation of textual materials, arguing that general purpose linguistic analysis together with appropriate background materials for training a general language model provide a more general, more portable, and more robust methodology for extracting information from text [10].

This paper reports a series of experiments to investigate the general effective-ness of structural features as carriers of information in text, applied to the task of attitude analysis.

2   Constructions as Characteristic Features of Utterances

Our hypothesis is that investigating utterances for presence of content-bearing words may be useful for identifying attitudinal expressions, but that finding structural features carries over easier from one topical area to another, from one discourse to another.

It has previously been suggested that attitude in text is carried by depen-dencies among words, rather than by keywords, cue phrases, or high-frequency words [1]. We agree, but in contrast with previous work, we explicitly incorporate *constructions* in our knowledge representation, not as relations between terms but as features in their own right, following a construction grammar framework [9,3].

Our claim is that the *pattern* of an utterance is a feature with the same onto-logical status as the terms that occur in the utterance: constructional features and lexical features both have conceptual meaning. Patterns are part of the sig-nal, not incidental to it. This claim, operationalised for experimental purposes, gives us a convenient processing model. Where the step from bag-of-words anal-yses to complete parse trees is both computationally daunting and brittle in face of fluid and changing data, we can within a constructional framework find mid-dle ground: we use observations of pattern occurrences as features similarly to how we use observations of word occurrences. An utterance will then not only be characterised as being a container for a number of words, but also a container for some observed patterns. Some previous approaches for using syntactic analys in large-scale text analysis have used segments of parse trees rather than the entire tree; however, the distinction between lexical features indicating content and syntactic features indicating relations between lexical items are central in those analyses, while our approach does not separate those categories of features.

TENSE SHIFT *It* **is** *this, I* **think***, that commentators* **mean** *mean when they* **say** *glibly That the "world* **changed***" after Sept 11.*

TIME ADVERBIAL *In Bishkek, they agreed to an informal meeting* **later this year***, most likely to be held in Russia.*

OBJECT CLAUSE *China could use the test as a political signal to show the US* **that it is a rising nuclear power** *at this tense moment.*

VERB CHAIN *"Money* **could be earned** *by selling recycled first-run fuel and separated products which retain over 50 per cent of unused uranium," Interfax news agency reported him as saying.*

Fig. 1. K examples

3   Combining Constructional and Lexical Features

The texts used in the present experiments are viewed as sequences of sentences: the sentence is taken as the basis of analysis, as a proxy for the utterance we view as the basis for attitudinal expression. All texts are preprocessed by a linguistic analysis toolkit, the Connexor Functional Dependency (FDG) parser, [17], to yield a lexical categorisation and morphological analysis of each word and a full dependency parse for each sentence. Our experimental features consist of three types: *content words* (I), *function words* (F) and *construction markers* (K) and are extracted from that analysis, making use of a mixture of low-level and abstract analysis results. Our features are general and not crucially bound to any specific analysis component, and no attitudinal lexical resources are employed to establish the constructions used in the further processes.

When observed in a sentence, the words from the *content word* class – nouns, adjectives, verbs (including verbal uses of participles), adverbs, abbreviations, numerals, interjections, and negation – form the feature class I for that sentence. Analogously, the *function words* – prepositions, determiners, conjunctions, pronouns – constitute the F feature class of the sentence.

Besides word occurrence based feature classes we introduce a further feature class intended to capture aspects of the constructions in employ in the sentence, *constructional features* (feature class K). Some markers, such as adverbial types and information about predicate and relative clauses are given directly by the FDG dependency or morphology analysis, others involved further processing steps to aggregate information from different levels of the analysis. The full list of K markers is given in Table 1.

The primary aim of the K features is to capture aspects of sentence composition, and therefore many features are concerned with clause types, and the way different types of clauses pattern in a sentence. Examples of such features, involving a certain amount of further processing, are occurrence of transitive and intransitive clauses, and temporality pattern of clauses. The full list of K markers for sentence composition is: TRIN, TRTR, PREDCLS, TRMIX, TNSSHIFT, OBJCLS, RELCLS.

Table 1. K, with occurrence statistics from a corpus of approx. 90,000 words in 4,306 sentences of newspaper text.

Form	K tag	Freq	Form	K tag	Freq
Non-transitive			Adverbial of quantity	AdvlQuant	85
clause	TRin	2919	Undetermined		
Transitive clause	TRtr	2350	prepositional phrase	PPUndet	22
Predicative clause	PredCls	1439	Negation	Neg	17
Transitivity mix	TRmix	1283	Verb chain	VChain	532
Tense shift	TnsShift	733	Verb particle	PartV	7
Object clause	ObjCls	351	Quantifier	Quant	69
Subordinating			Adjective modifier	AdjMod	82
conjunction	SubCnj	200	Base form adjective	AbsAdj	3417
Relative clause	RelCls	601	Present tense	TnsPres	2373
Adverbial of location	AdvlSpat	1367	Past tense	TnsPast	2145
Adverbial of time	AdvlTim	1110	Comparative adjective	KmpAdj	463
Sentence adverbial	AdvlSnt	973	Superlative adjective	SupAdj	241
Adverbial of manner	AdvlMan	608	Prepositional		
Adverbial of condition	AdvlCond	547	post-modifier	PpPomod	572
Clause-initial adverbial	AdvlClsIn	387	No K-traits	NoK	66

Another set of K features concerns the quality and type of adverbial, if present, in the sentence: AdvlSpat, AdvlTim, AdvlSnt, AdvlMan, AdvlCond, AdvlClsIn, AdvlQuant. Yet another type covers some morphological traits of sentences: the tense of verbs in clauses (obviously related to the tense pattern feature of the first subtype): TnsPres and TnsPast and the grade of occurring adjectives (AbsAdj, KmpAdj, and SupAdj), which we expected to correlate well with our intended task to identify opinionated and attitudinal sentences. The remainder of the K feature set consists of markers of various word dependency relations, related to internal phrase structure, coordination, or clause polarity. These tags are: SubCnj, PPUndet, Neg, PpPomod, VChain, AdjMod, Quant, PartV.

In the experiments below, all constructional features are treated as sentence features, exactly as the lexical features are treated, i.e., no coupling between the features and the words carrying them is made.

4   Test Data and Experimental Materials

For our experiments we use data from the NTCIR information retrieval evaluation challenge organised by NII, Tokyo, in its English section of the opinion identification task [12], the multi-perspective question answering (MPQA) test sentence set with assessed attitudinal sentences [15], and the 2007 Semantic Evaluation Affective Task (SEMEVAL) test set of news headlines [16], all of which have assessments by human judges. We use a lenient scoring scheme, scoring a sentence as attitudinal if two out of three NTCIR judges have marked it attitudinal; for the SEMEVAL data if the intensity score is over 50 or under -50. All

"It is this, I think, that commentators mean when they say glibly that the 'world changed' after Sept 11."
I be think commentator mean when **say** glibly world change sept 11
F it this i that they that the after
K AdvlTim, AdvlMan, ObjCls, PredCls, TRin, TRtr, TRmix, TnsPres, TnsPast, TnsShift

"Mr Cohen, beginning an eight-day European tour including a Nato defence ministers' meeting in Brussels today and tomorrow, said he expected further international action soon, though not necessarily military intervention."
I mr cohen begin eight-day european tour include nato defence minister meeting brussels today tomorrow **say** expect international action soon though not necessarily military intervention
F an a in and he further
K AdvlTim, AdvlSpat, ObjCls, TRtr, TnsPast

Fig. 2. Example attitude analyses of sentences. These sentence are taken from the NTCIR opinion analysis task data set. The first sentence is assessed by task judges to be a opinion carrier, the last non-opinion. The content word feature "say" is a strong marker for opinion but would yield the wrong categorisation in this case; our linear classifier correctly identified the first sentence as attitudinal and the last as non-attitudinal.

attitudinal sentences or headlines, irrespective of source, are assigned the class ATT and all other sentences assigned the class NOATT. Statistics for the collection are given in Table 2. Some sentences from the MPQA and NTCIR test sets, about ten in total, yielded no analyses in our system and are removed from the test set. These test sets are very different in character: the SEMEVAL set, which consists of news headlines rather than sentences from running text are different in structure; for each set, the assessors appear to have had different instructions. Our assumption is that this will approximate the variance that real-life tasks of this vein will encounter.

The data have been used by several research groups in various experiments. Our classifiers perform well enough to yield a tie with the reported best result from the shared opinion identification task of NTCIR as measured by F-score. (In the experiments described below we report precision and recall separately.)

Table 2. Test sentence statistics

	NTCIR 6	NTCIR 7	SEMEVAL	MPQA
Attitudinal	1392	1075	76	6021
Non-attitudinal	4416	3201	174	4982
Total	5808	4276	250	11003

5   Feature Strength Analysis

In order to gain some insight into which features show most utility for attitude identification we first performed some exploratory analyses on the NTCIR 6 and 7 test sets using NTCIR 6 as labeled training materials, testing on NTCIR 7 data, without using any other background material.

After training a SVM classifier for the ATT-NOATT distinction on all three (I, F, and K) feature types for the NTCIR data 2 245 features were utilised by the classifier. Most features were lexical, but we found that 17 K features were also used to discriminate between the classes. Table 3 list the 17 K markers in use with classifier-scored rank of importance, sub-grouping in the K set, and frequency of occurrence in the NTCIR-7 corpus. 85 of the remaining features were function words (F), and 2142 features were from the content word I set. The *rank* column in the table, given by the classifier, gives an indication of the relative importance of the features; the rank of feature set K in this given set is significantly higher than that of the I and F sets. (Mann Whitney U, p > 0.95).

We found that the relative scoring of the strongest features in the discrimination model scored certain of our manually chosen K features very highly compared to I and F features. Tense and transitivity measures, e.g., scored highly: "Tense shift", the strongest single K feature is found in sentences where the verbs of the main clause and the subordinate clause have different tense forms. This occurs often in sentences of utterance or cognition: "Noam Chomsky said$_{past}$ that what makes human language unique is$_{present}$ recursive centre embedding"; "M.A.K. Halliday believed$_{past}$ that grammar, viewed functionally, is$_{present}$ natural". The tense shift feature obviates the need of acquiring and maintaining lists of utterance, pronuncement, and cognition - categories which have obvious relation to attitudinal expression.

Another way to investigate the impact of different K traits is to study their occurrence in the sentences in the NTCIR-7 corpus. A matrix of K features and attitudinal status of sentences was constructed, and reduced to two dimensions by *correspondence analysis*, cf. [6]. This is a method similar to *principal components analysis*, but with the additional feature of placing the column and row variables on the same plane, and thus makes it possible to study the K features occurrence in sentences of different attitudinal type.

Figure 3 shows a plot of the result from the correspondence analysis, with non-opionated outliers and less informative K labels ignored. The proximity of two labels is a measure of their co-occurrence, and we can notice that some K markers predominately show up in non-attitudinal sentences, e.g. verb chains and negation. On the opinionated side of the plane we find K traits as clause objects, tense shift patterns, adjectives in superlative grade, and predicative clauses. We can also see that the opinionated and non-opinionated sentences are spread along the x-axis, the most important of the resulting two correspondence analysis dimensions, while the y-axis appears to involve the polarity of the sentence.

We have thus established that the K features carry signal value for predicting attitude in text. We now turn to investigating how much they might help in categorisation experiments. Rather than utilising the features directly from the

Table 3. Ranking of K features according to decision function weights, among the 2245 predictive features in the NTCIR-7 corpus

Rank	K tag	Type	Freq
75	TNSSHIFT	sentence composition	733
269	TNSPAST	morphology	2145
281	PARTV	phrase structure	7
290	TRMIX	sentence composition	1283
385	ADVLQUANT	adverbial	85
502	PREDCLS	sentence composition	1439
505	QUANT	phrase structure	69
680	TRIN	sentence composition	2919
686	NEG	polarity	17
746	ADVLTIM	adverbial	1110
780	TRTR	sentence composition	2350
813	PPPOMOD	phrase structure	572
969	PPUNDET	phrase structure	22
1055	KMPADJ	morphology	463
1105	VCHAIN	phrase structure	532
1673	TNSPRES	morphology	2373
2222	ADVLCOND	adverbial	547

test set, to avoid overtraining on the experimental sets we use a larger text collection to establish how similar the various features are in use.

6   Classification Experiment

Our main experiment is based on a background language representation built by analysis of a reasonable-sized general text collection. We use that model to establish similarities and differences between the sentences under analysis. Our general analysis procedure is to investigate how the utterance or sentence under consideration is related to language usage in the norm, either by deviation from the norm in some salient way, or by conforming with an identified model of usage.

In this experiment we use one year of newsprint from two Asian English-language news sources, the Korean Times and the Mainichi Daily, distributed as part of the NTCIR information retrieval evaluation challenge, including the opinion and attitude analysis task [7,12]. We also use one year of the Glasgow Herald, distributed as part of the CLEF information retrieval evaluation challenge [2]. Collection sizes are as shown in Table 4.

We segment these corpora into sentences and process each sentence to extract the features given above – I, F, K. We use this to build a cooccurrence-based first-order word space [11] using the Random Indexing framework [8] (1000 dimensions, two non-zero elements per index vector).

Each *sentence* is represented by a random position in the word space model. Each observed *feature*, I, F, and K alike, is given an initially empty *context*

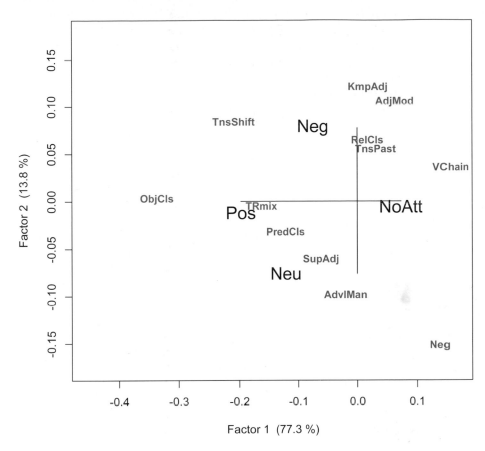

Fig. 3. K × Attitude. Black labels denote opinionated sentences with positive (Pos), negative (Neg), and neutral (Neu) polarity; NoAtt denotes non-opinionated sentences.

vector in the word space model. Each feature observed in a sentence has its context vector incremented by vector addition with the representation vector for that sentence. Features which occur only once in the data were removed. Each feature will then in its context vector carry information about every sentence it has occurred in, and its context vector will eventually grow to become similar with other features it has cooccurred with. This is a standard word space model, but here augmented with the K features. We use this word-and-feature space to be able to generalise from observed features in a sentence to other features and to establish similarities between sentences based on their feature values, even when there is little or no feature overlap.

Once the general language word-and-feature space model is built, each test sentence can be represented by the centroid of its feature set in the respective background space. Using these centroids, we used a support vector machine with a linear kernel to build a classifier for the ATT-NOATT distinction. The data were scaled to a range of approximately $[-1 \ldots 1]$ and standard settings for the LIBLINEAR library [5] were used; since the class sizes were unbalanced, the penalty cost parameter for the classes was set in inverse proportion to their size.

We then ran five-fold crossvalidation of on each set to establish classification performance for the feature sets. This test was performed for each of the I, F, and K sets, and combinations thereof, yielding seven feature combinations for each set of test sentences and each background collection. The results are shown in Table 5 (precision), Table 6 (recall), and Table 7 (F_1).

We find that the combination of all three feature sets gives consistently high results as regards precision. We find that the including the K feature set gives the best results for recall for many of the combinations, and frequently using it alone gives the best results. We also find that the results are stable across the three background collections.

The test sets behave quite differently, however. News headlines (the SEMEVAL set) quite obviously give less purchase for classification. The low performance reported for the SEMEVAL set for the F feature set is unsurprising, since the language given in news headlines typically is quite terse, with structural cue words omitted for brevity, and rather differnet from the language in the background materials.

In comparison with other reported results these results are just slightly better than the best reported result for NTCIR-6 (best reported F_1 score from proceedings: 46.5) [13] and NTCIR-7 (best reported F_1 score from proceedings: 49.7) [12]; for the SEMEVAL task (best reported F_1 score from proceedings:

Table 4. Background text materials

	Korean Times	Mainichi Daily	Glasgow Herald
Sentences	326 486	123 744	2 158 196
Characters	61M	25M	452M

Table 5. Precision (%) for the NTCIR data set (five-fold crossvalidation)

	NTCIR 6			NTCIR 7			MPQA			SEMEVAL		
	KT	MD	GH	KT	MD	GH	KT	MD	GH	KT	MD	GH
I	37.7	36.0	37.5	37.4	35.4	37.4	71.9	68.9	70.7	25.4	24.6	33.9
F	35.0	35.4	39.0	35.4	35.2	39.9	70.3	69.8	72.3	20.0	25.0	31.8
K	33.6	33.7	37.1	35.3	35.2	37.2	66.4	66.2	71.1	25.4	26.4	26.7
IF	39.9	38.4	40.2	40.7	39.3	40.1	**74.1**	**72.9**	73.8	31.3	28.8	33.6
IK	38.5	37.2	40.2	39.7	38.5	40.9	71.8	69.7	73.7	25.0	23.9	30.5
FK	37.1	37.2	40.2	38.0	37.5	40.4	70.1	69.8	74.4	30.7	28.4	29.5
IFK	**40.9**	**39.7**	**41.7**	**41.7**	**40.2**	**41.3**	73.7	72.5	**75.5**	**32.5**	**29.3**	**34.2**

Table 6. Recall (%) for the NTCIR data set (five-fold crossvalidation)

	NTCIR 6			NTCIR 7			MPQA			SEMEVAL		
	KT	MD	GH	KT	MD	GH	KT	MD	GH	KT	MD	GH
I	50.5	44.8	50.3	55.2	52.1	57.1	60.0	54.8	57.4	46.1	46.1	**56.6**
F	52.5	52.1	52.9	59.0	60.7	58.8	56.5	56.2	59.8	12.5	12.5	55.3
K	**60.3**	**58.5**	49.1	**64.8**	**65.4**	52.7	60.1	59.1	57.6	36.7	49.0	46.1
IF	54.3	52.1	53.5	59.2	58.0	58.0	61.5	58.1	61.9	**55.3**	52.6	54.0
IK	57.1	53.7	53.3	60.0	59.1	**59.8**	64.2	61.1	61.5	47.4	44.7	52.6
FK	55.8	54.6	54.0	59.9	59.3	58.9	58.9	58.3	62.6	51.3	52.6	51.3
IFK	59.4	56.6	**55.3**	62.3	61.6	59.1	**64.2**	**61.4**	**63.9**	51.3	**53.9**	54.0

Table 7. F_1 (%) for the NTCIR data set (five-fold crossvalidation)

	NTCIR 6			NTCIR 7			MPQA			SEMEVAL		
	KT	MD	GH	KT	MD	GH	KT	MD	GH	KT	MD	GH
I	43.2	39.9	46.1	44.6	42.2	45.2	65.4	61.0	63.4	32.7	32.1	**42.4**
F	42.0	42.2	44.9	44.2	44.6	47.5	62.6	62.3	65.4	15.4	16.7	40.4
K	43.2	42.8	42.3	45.7	45.7	43.6	63.1	62.4	63.7	30.0	34.3	33.8
IF	46.0	44.2	45.9	48.2	46.9	47.4	67.2	64.6	67.3	**40.0**	37.2	41.4
IK	46.0	44.0	45.9	47.8	46.6	**48.6**	67.8	65.1	67.0	32.7	31.2	38.6
FK	44.5	44.3	46.1	46.5	45.9	47.9	64.0	63.5	68.0	38.4	36.9	37.5
IFK	**48.4**	**46.7**	**47.5**	**49.9**	**48.7**	**48.6**	**68.6**	**66.5**	**69.2**	39.8	**38.0**	41.8

42.43) the best results from these experiments are comparable [16]. The MPQA corpus has been used in a large number of experiments and reported results vary, best F_1 scores being somewhat higher than the ones given here.

7   Conclusions

With increasingly sophisticated semantic relations being mined from data, processing must take a more sophisticated view of the linguistic signal than simple containers of topical words. Many approaches begin by assuming the structure of the linguistic data primarily to be relations between topical elements. In constructional approaches, the constructions, the combinational patterns, themselves are accorded presence in the signal – these experiments show that the patterns capture information, without the need for full analysis of dependencies between the word tokens. *Our level of analysis is between bags of words and parse trees in this respect.*

We find that representing constructions, even hand chosen constructions such as the ones given in this experiment, especially given unrelated general language background data, can provide a reliability which well matches or even surpasses that of word occurrence, the arguably primary carrier of information in the linguistic signal. The results are comparable to finely tuned experimental systems given by other research groups, trained for these specific experiments. *Constructions carry signal.*

Using constructions in parallel with word occurrence features not only has theoretical motivation based on the construction grammar theory, but also provides a convenient and familiar processing model and a straightforward extension for term based models. From a philological standpoint, a bottom-up approach to data analysis, examining the power of constructions as ontological items, would appear to be better motivated than basing information processing on descriptive language models, originally intended for description of human behaviour, for comparative studies of world languages, or for the scholarly instruction of foreign languages. *Constructions are easy to insert into a bag-of-words-based processing framework. Constructions do not promise more than they can deliver.*

Several of the constructions with greatest predictive effect for this example task were those that had the greatest scope: tense and transitivity patterns, predicative clauses. *Construction capture non-local information in sentences.*

The suggested K features might be one way of mimicking the contribution of constructions to the meaning of a sentence or utterance by combining these atomic markers of structure with the lexical items, without the need to represent the inter-relations between the two different sets. But the current implementation of the K traits idea has a number of drawbacks and limitations. The current set of K attributes was selected heuristically with task of attitude identification in mind. *Constructions should be learnt through data-driven methods, rather than selected heuristically.*

References

1. Bai, X., Padman, R., Airoldi, E.: On learning parsimonious models for extracting consumer opinions. In: Proceedings of HICSS 2005, the 38th Annual Hawaii International Conference on System Sciences, Washington, DC, USA. IEEE Computer Society, Los Alamitos (2005)
2. Braschler, M., Peters, C.: Cross-language evaluation forum: Objectives, results, achievements. In: Information Retrieval, pp. 7–31 (2004)
3. Croft, W.: Radical and typological arguments for radical construction grammar. In: Östman, Fried (eds.) [9]
4. Esuli, A., Sebastiani, F.: SentiWordNet: A publicly available lexical resource for opinion mining. In: Proceedings of the Fifth International Conference on Language Resources and Evaluation (LREC 2006) (2006)
5. Fan, R.-E., Chang, K.-W., Hsieh, C.-J., Wang, X.-R., Lin, C.-J.: LIBLINEAR: A library for large linear classification. Journal of Machine Learning Research 9, 1871–1874 (2008)
6. Greenacre, M.J.: Theory and applications of correspondence analysis. Academic Press, London (1984)
7. Kando, N.: Overview of the Seventh NTCIR Workshop. In: Proceedings of the 7th NTCIR Workshop Meeting on Evaluation of Information Access Technologies. NII, Tokyo (2008)

8. Kanerva, P., Kristofersson, J., Holst, A.: Random indexing of text samples for latent semantic analysis. In: Proceedings of the 22nd Annual Conference of the Cognitive Science Society, p. 1036. Erlbaum, Mahwah (2000)

9. Östman, J.-O., Fried, M. (eds.): Construction Grammars: Cognitive grounding and theoretical extensions. John Benjamins, Amsterdam (2005)

10. Sahlgren, M., Karlgren, J., Eriksson, G.: SICS: Valence annotation based on seeds in word space. In: Proceedings of Fourth International Workshop on Semantic Evaluations (SemEval 2007), Prague, Czech Republic (2007)

11. Schütze, H.: Word space. In: Hanson, S., Cowan, J., Giles, C. (eds.) Advances in Neural Information Processing Systems 5. Morgan Kaufmann Publishers, San Francisco (1993)

12. Seki, Y., Evans, D.K., Ku, L.-W., Sun, L., Chen, H.-H., Kando, N.: Overview of Multilingual Opinion Analysis Task at NTCIR-7. In: Proc. of the 7th NTCIR Workshop Meeting on Evaluation of Information Access Technologies. NII, Tokyo (2008)

13. Seki, Y., Evans, D.K., Ku, L.-W., Sun, L., Chen, H.-H., Kando, N., Lin, C.-Y.: Overview of opinion analysis pilot task at ntcir-6. In: Kando, N., Evans, D.K. (eds.) Proceedings of the Sixth NTCIR Workshop Meeting on Evaluation of Information Access Technologies, National Institute of Informatics, 2-1-2 Hitotsubashi, Chiyoda-ku, Tokyo 101-8430, Japan (May 2007)

14. Stone, P.: Thematic text analysis: new agendas for analyzing text content. In: Roberts, C. (ed.) Text Analysis for the Social Sciences, vol. 2. Lawrence Erlbaum Associates, Mahwah (1997)

15. Stoyanov, V., Cardie, C., Litman, D., Wiebe, J. (eds.): Evaluating an Opinion Annotation Scheme Using a New Multi-Perspective Question and Answer Corpus, Stanford University, California, March 2004. AAAI. AAAI Technical Report Series SS-04-07. AAAI Press, Menlo Park (2004) ISSN 978-1-57735-219-x

16. Strapparava, C., Mihalcea, R.: SemEval-2007 Task 14: Affective Text. In: Proceedings of the Fourth International Workshop on Semantic Evaluations (SemEval 2007), Prague, Czech Republic, June 2007, pp. 70–74. Association for Computational Linguistics (2007)

17. Tapanainen, P., Järvinen, T.: A non-projective dependency parser. In: Proc. of the 5th Conference on Applied Natural Language Processing, pp. 64–71 (1997)

18. Wilson, T., Wiebe, J., Hoffmann, P.: Recognizing contextual polarity in phrase-level sentiment analysis. In: Proceedings of HLT-EMNLP (2005)

Improving Medical Information Retrieval with PICO Element Detection

Florian Boudin, Lixin Shi, and Jian-Yun Nie

DIRO, Université de Montréal,
CP. 6128, succursale Centre-ville
Montréal, H3C 3J7 Quebec, Canada
{boudinfl,shilixin,nie}@iro.umontreal.ca

Abstract. Without a well formulated and structured question, it can be very difficult and time consuming for physicians to identify appropriate resources and search for the best available evidence for medical treatment in evidence-based medicine (EBM). In EBM, clinical studies and questions involve four aspects: Population/Problem (P), Intervention (I), Comparison (C) and Outcome (O), which are known as PICO elements. It is intuitively more advantageous to use these elements in Information Retrieval (IR). In this paper, we first propose an approach to automatically identify the PICO elements in documents and queries. We test several possible approaches to use the identified elements in IR. Experiments show that it is a challenging task to determine accurately PICO elements. However, even with noisy tagging results, we can still take advantage of some PICO elements, namely I and P elements, to enhance the retrieval process, and this allows us to obtain significantly better retrieval effectiveness than the state-of-the-art methods.

1 Introduction

Physicians are educated to formulate their clinical questions according to several well defined aspects in evidence-based medicine (EBM): Population/Problem (P), Intervention (I), Comparison (C) and Outcome (O), which are called PICO elements. The PICO structure is commonly used in clinical studies [7]. In many documents in medical literature, one can find the PICO structure, which is, however, often implicit and not explicitly annotated. To identify documents corresponding to a patient's state, physicians also formulate their clinical questions in PICO structure. For example, in the question *"In children with an acute febrile illness, what is the efficacy of single-medication therapy with acetaminophen or ibuprofen in reducing fever?"* one can identify the following elements: P ⇒ *"children with acute febrile illness"*, I ⇒ *"single-medication therapy with acetaminophen"*, C ⇒ *"ibuprofen"* and O ⇒ *"efficacy in reducing fever"*.

Using a well-formulated question according to the PICO structure facilitates searching for a precise answer within a large medical citation database [10]. However, using PICO structure in Information Retrieval (IR) is not as straightforward as it seems. It requires first the identification of the PICO elements in

C. Gurrin et al. (Eds.): ECIR 2010, LNCS 5993, pp. 50–61, 2010.

the documents, as well as in the question if these elements are not explicitly separated in it. Several studies have been carried out on identifying PICO elements in medical documents, and to use them in IR [5,4]. However, these studies are limited in several aspects. First, many studies on identification of PICO elements are limited to some segments of the medical documents (e.g. Method) [4], and in most cases, the test collection is very small (a few hundreds abstracts). It is difficult to see whether one can easily identify PICO elements in all parts of medical documents in a large collection. Secondly, there have been very few tests on IR using PICO elements [5]. This is due to the lack of a standard test collection with questions in PICO structure. IR tests have been carried out on small test collections, and in many cases, not compared to the traditional IR methods. It is not clear whether IR based on PICO structure is more effective than traditional IR approaches.

In this paper, we propose an approach to perform IR using PICO elements. The identification of these elements is cast as a classification task. A mixture of knowledge-based and statistical techniques is employed to extract discriminant features that once combined in a classifier will allow us to identify clinically relevant elements in MEDLINE abstracts. Using these detected elements, we show that the information retrieval process can be improved. In particular, it turns out that the I and P elements should be enhanced in retrieval. The remainder of this paper is organized as follows. In the next section, we give an overview of the related work. Then, we present our classification approach to identify PICO elements in documents. Next, IR experiments using these elements are reported. Finally, we draw some conclusions.

2   Previous Work

The first aspect of this study concerns the identification of PICO elements in medical documents. Several previous approaches have already proposed to categorize sentence types in medical abstracts using classification tools. [8] showed that Machine Learning can be applied to label structural information of sentences (i.e. Introduction, Method, Results and Conclusion). Thereafter, [5] presented a method that uses either manually crafted pattern-matching rules or a combination of basic classifiers to detect PICO elements in medical abstracts. Prior to that, biomedical concepts are labelled by Metamap [2] while relations between these concepts are extracted with SemRep [9], both tools being based on the Unified Medical Language System (UMLS). Using these methods, they obtained an accuracy of 80% for Population and Intervention, 86% for Problem and between 68% and 95% for Outcome. However, it is difficult to generalize this result, as the test was done on a very small dataset: 143 abstracts for outcome and 100 abstracts for other elements.

Recently, supervised classification was proposed by [6] to extract the number of trial participants. Results reported in this study show that the Support Vector Machine (SVM) algorithm achieves the best results with an f-measure of 86%. Again, it has to be noted that the testing data, which contains only

75 highly topic-related abstracts, is not representative of a real world task. In a later study, [4] extended this work to I and O elements using Conditional Random Fields (CRF). To overcome data sparseness, PICO structured abstracts were automatically gathered from MEDLINE to construct an annotated testing set (318 abstracts). This method showed promising results: f-measure of 83% for I and 84% for O. However, this study has been carried out in a limited context: elements are only detected within the Method section, while several other sections such as Aim, Conclusion, etc. are discarded. It is not clear whether the identification of PICO elements in the whole document can lead to the same level of performance. In this study, we do not restrict ourselves to some of the sections in documents, but try to identify elements in the whole documents.

On the retrieval aspect, there have been only a few studies trying to use PICO elements in IR and compare it to traditional methods. [5] is one of the few such studies. The method they describe consists in re-ranking an initial list of retrieved citations. To this end, the relevance of a document is scored by the use of detected PICO elements, in accordance with the principles of evidence-based medicine (i.e. quality of publication or task specificity are taken into consideration). Several other studies aimed to build a Question-Answering system for clinical questions [1]. But again, the focus has been set on the post-retrieval step, while the document retrieval step only uses a standard IR approach. In this paper, we argue that IR has much to gain by using PICO elements.

Although the retrieval effectiveness is reported in some studies using PICO elements, it is yet to be proved that a PICO-based retrieval approach will always produce better effectiveness than the traditional IR methods. In this study, we will examine the effect of using PICO elements in the retrieval process in several ways and compare them to the traditional IR models. In the next section, let us start with the first step: identifying PICO elements in medical documents.

3   Identification of PICO Elements in Documents

PICO elements are often implicitly described in medical documents. It is important to identify them automatically. One can use linguistic patterns for this. However, a pattern-based approach may require a large amount of manual work, and the robustness has yet to be proved on large dataset. In this study, we will rather use a more robust statistical classification approach, which requires a minimal amount of manual preparation. There may be two levels of classification: one can identify each PICO element in the document, whether it is described by a word, a phrase or a complete sentence; one can also make a coarser-grain annotation – to annotate a sentence as describing only one of the PICO elements. The second method is much simplified. Nevertheless, while the first classification is very difficult, the second one is easier to implement. Moreover, for the purpose of IR, a coarse-grain classification may be sufficient.

3.1   Construction of Training and Testing Data

Even for a coarse-grain classification task, we are still lack of a standard test collection with PICO annotated elements. This increases the difficulty of developing and testing an automatic tool that tags these elements. This is also the reason why previous studies have focused on a small set of abstracts for testing. We notice that many recent documents in PubMed[1] do contain explicit headings such as "PATIENTS", "SAMPLE" or "OUTCOMES", etc. The sentences under the "PATIENT" and "SAMPLE" headings describe the P elements, and those under the "OUTCOMES" heading describe the O elements. Below is a segment of a document extracted from PubMed (pmid 19318702):

*... **PARTICIPANTS:** 2426 nulliparous, non-diabetic women at term, with a singleton cephalic presenting fetus and in labour with a cervical dilatation of less than 6 cm. **INTERVENTION:** Consumption of a light diet or water during labour. **MAIN OUTCOME MEASURES:** The primary outcome measure was spontaneous vaginal delivery rate. Other outcomes measured included duration of labour ...*

We collect a set of roughly 260K abstracts from PubMed by stating the limitations: *published in the last 10 years, Humans, Clinical Trial, Randomized Controlled Trial, English*. Then, structured abstracts containing distinctive sentence headings are selected and these sentences marked with the corresponding PICO elements. We notice that both Intervention and Comparison elements belong to the same semantic group and are often described under the same heading. We then choose to group the corresponding segments into the same set. From the entire collection, three sets of segments have been extracted: Population/Problem (14 279 segments), Intervention/Comparison (9 095) and Outcome (2 394). Note that abstracts can also contain sentences under other headings, which we do not include in our extraction process. Therefore, it is possible that no Outcome is extracted from a document by our process. This conservative extraction approach allows us to obtain a dataset with as little noise as possible.

3.2   Features Used in Classification

Prior to classification, each sentence undergoes pre-processing treatments that replace words into their canonical forms. Alpha-numeric numbers are converted to numeric numbers while each word appearance in a series of manually crafted cue-words/verbs lists is investigated. The cue-words and cue-verbs are determined manually. Some examples are shown below:

Cue-verbs: *recruit* (P), *prescribe* (I), *assess* (O)
Cue-words: *group* (P), *placebo* (I), *mortality* (O)

On top of that, three semantic type lists, generated from the MeSH[2] ontology, are used to label terms in sentences. These lists are composed with entry terms corresponding to a selection of subgroups belonging to semantic types "Living Beings", "Disorders" and "Chemicals & Drugs". The final set of features we

---

[1] http://www.ncbi.nlm.nih.gov/pubmed/
[2] http://www.nlm.nih.gov/mesh/

where $p(t \mid M_Q)$ and $p(t \mid M_D)$ are (unigram) language models of the query and document respectively. Usually, the query model is simply estimated by Maximum Likelihood Estimation over the query words, while the document model is smoothed (e.g. using *Dirichlet* smoothing) to avoid zero probabilities problem.

4.1   Model Definitions

We propose several approaches that extend the basic LM approach to take into consideration the PICO element annotation. According to the PICO tagging, the content of queries and documents is divided into the following four fields: Population and Problem (P), Intervention/Comparison (I), Outcome (O), and Others (X). Let us use the following notation: $Q_{All} = Q_P + Q_I + Q_O + Q_X$ for the query Q and $D_{All} = D_P + D_I + D_O + D_X$ for the document D. In case of missing tagging information, the basic bag-of-words model is used.

4.1.1 Using PICO Tags in Queries

We try to assign an importance (weight) to each of the PICO elements. Intuitionally, the more important is a field, the higher should be its weight. We propose the following two models by adjusting the M_Q weighting:

Model-1T: adjusting weights on PICO element (term) level. The query model is re-defined as follows:

$$p_1(t \mid M_Q) = \gamma \cdot \frac{count(t, Q)}{\mid Q \mid} \cdot \left(1 + \sum_{E \in P,I,O} w_{Q,E} \cdot \delta(Q_E, t) \right) \qquad (2)$$

where $w_{Q,E}$ is the weight of query field E; $\delta(Q_E, t) = 1$ if $t \in Q_E$, 0 otherwise; γ is a normalization factor. The score function of this model, namely $score_{1T}$, is obtained by replacing the $p(t \mid M_Q)$ by $p_1(t \mid M_Q)$ in Equation (1).

Model-1F: adjusting weights on PICO field level. Four basic models for D_{ALL}, D_P, D_I and D_O are created. The final score is their weighted linear interpolation with $w_{Q,E}$:

$$score_{1F}(Q, D) = score(Q_{All}, D) + \sum_{E \in P,I,O} w_{Q,E} \cdot score(Q_E, D) \qquad (3)$$

4.1.2 Using PICO Tags in Documents

We assume each field in the tagged document has a different importance weight $w_{D,E}$. The document model is redefined as follows:

$$p_2(t \mid M_D) = \gamma \cdot \left(p(t \mid M_{D_{All}}) + \sum_{E \in P,I,O} w_{D,E} \cdot p(t \mid M_{D_E}) \right) \qquad (4)$$

where γ is a normalization factor, and $p(t \mid D_E)$ uses the Dirichlet smoothing function. We denote this model by **Model-2**, and the $score_2$ is obtained by replacing $p(t \mid M_D)$ by $p_2(t \mid M_D)$ in Equation (1).

4.1.3 Using PICO Tags in Both Queries and Documents

Model-3T: enhancement at the term level. The query model is redefined as in case 1 and document model is redefined as in case 2.

$$score_{3T}(Q, D) = \sum_{t \in Q} p_1(t \mid M_Q) \cdot \log p_2(t \mid M_D) \tag{5}$$

Model-3F: enhancement at the field level. This is the combination of **Model-2** and **Model-1F**.

$$score_{3F}(Q, D) = score_2(Q_{All}, D) + \sum_{E \in P, I, O} w_{Q,E} \cdot score_2(Q_E, D) \tag{6}$$

In all our models, there are a total of 6 weighting parameters, 3 for queries $(w_{Q,P}, w_{Q,I}, w_{Q,O})$ and 3 for documents $(w_{D,P}, w_{D,I}, w_{D,O})$.

4.2 Identifying Elements in Queries

PICO elements may be manually marked in queries by the user. This is, however, not a realistic situation. More likely, queries will be formulated as a free sentence or phrases. Identifying PICO elements in a query is different from what we did on documents because we need to classify smaller units. In this paper, we adopt a language model classification method [3], which is an extension to Naïve Bayes. The principle is straightforward: Let P, I and O be the classes. The score of a class c_i for a given term t is estimated by $p(t \mid c_i) \cdot p(c_i)$. The probability $p(c_i)$ can be estimated by the percentage of training examples belonging to class c_i and $p(t \mid c_i)$ by maximum likelihood with *Jelinek-Mercer* smoothing:

$$p_{JM}(t \mid c_i) = (1 - \lambda) \cdot p(t \mid c_i) + \lambda \cdot p(t \mid C) \tag{7}$$

where C is the whole collection and λ is smoothing parameter.

The above approach requires a set of classified data in order to construct the LM of each class. To this end, we use the sentences classified by the previously described approach (see Section 3). Usually, users prefer to select important terms as their queries. As a consequence, queries should contain more PICO elements than documents. Therefore, we assume that each query term belongs to one of the P, I, or O classes. Performance of the classification method is computed over a set of 52 queries (corpus described in Section 5) by comparison to a manual tagging and experimented on different values of the parameter λ. Best results are obtained for λ set to 0.5 with an f-measure of 77.8% for P, 68,3% for I and 50% for O.

5 IR Experiments

We gathered a collection of 151,646 abstracts from PubMed by searching for the keyword *"diabetes"* and stating the following limitations: *Humans* and *English language*. The average length of the documents is 276 words. The tagging time spent by our fusing strategy (see Section 3) was approximately one hour on a standard desktop computer. For queries, we use the Cochrane systematic reviews[4] on 10 clinical questions about "*diabetes*". All the references in the "Included" studies are judged to be relevant for the question. These included studies are selected by the reviewer(s) (the author(s) of the review article) and judged to be related to the clinical question. As these studies are published prior to the review article, we only try to retrieve documents published before the review's publication date. From the selected 10 questions, medical professionals (professors in family medicine) have formulated a set of 52 queries. Each query has been manually annotated according to the following elements, which extend the PICO structure: Population (P), Problem (Pr), Intervention (I), Comparison (C), Outcome (O), and Duration (D). However, in our experiments, we will use a simplified tagging: P and Pr are grouped together (as P), C and D are discarded. Below are some of the alternative formulations of queries for the question "*Pioglitazone for type 2 diabetes mellitus*":

In patients$^{(P)}$ | with type 2 diabetes$^{(Pr)}$ | does pioglitazone$^{(I)}$ | compared to placebo$^{(C)}$ | reduce stroke and myocardial infarction$^{(O)}$ | 2 year period$^{(D)}$

In patients$^{(P)}$ | with type 2 diabetes who have a high risk of macrovascular events$^{(Pr)}$ | does pioglitazone$^{(I)}$ | compared to placebo$^{(C)}$ | reduce mortality$^{(O)}$

The resulting testing corpus is composed of 52 queries (average length of 14.7 words) and 378 relevant documents. In our experiments, we will try to answer several questions: does the identification of PICO elements in documents and/or in queries helps in IR? and in the case of a positive answer, how should these elements be used in the retrieval process?

5.1   Baseline Methods

We first tested a naïve approach that matches the tagged elements in the query with the corresponding elements in the documents, i.e. each PICO tag defines a field, and terms are allowed to match within the same field. However, this approach quickly turns out to be too restrictive. This restriction is amplified by the low accuracy of PICO tagging. Therefore, we will not consider this method as baseline but the two following instead:

Boolean model: This is the search mode widely used in medical domain. Usually, a user will construct a Boolean query iteratively by adding and modifying terms in the query. We simulate this process by creating a conjunction of all the words. Queries created in this way may be longer than what a physician would construct. Boolean retrieval resulted in a MAP of 0.0887 and a P@10 of 0.1885.

Language model: This is one of the state-of-the-art approaches in current IR research. In this method, both a document and a query are considered as bag-of-words, and no PICO structure is considered. The LM approach resulted in a MAP of 0.1163 and a P@10 of 0.25. This is the baseline we will compare to.

---

[4] http://www.cochrane.org/reviews/

5.2   Using Document Tagging

In this first series of experiments, we consider the detected PICO elements in documents while the queries are considered as bag-of-words. During the retrieval process, each element E, $E \in \{P, I, O\}$, is boosted by a corresponding weight $w_{D,E}$. We begin by setting weights to 0.1 to see the impact of boosting each element alone. Table 2 shows that when these elements are enhanced, no noticeable improvement is obtained. We then try different combinations of weighting parameters from 0 to 0.9 by steps of 0.1. The best improvement remains very small ($w_{D,P} = 0.5/w_{D,I} = 0.2/w_{D,O} = 0$) and in most cases, we get worse results.

Table 2. MAP scores for Model-2 (without query tagging, $\star$: $w_{D,P} = 0.5$, $w_{D,I} = 0.2$)

Baseline	$w_{D,P} = 0.1$	$w_{D,I} = 0.1$	$w_{D,O} = 0.1$	Best*
0.1163	0.1168 (0.0%)	0.1161 (−0.2%)	0.1162 (−0.1%)	0.1169 (+0.5%)

The above results show that it is not useful to consider PICO elements only in documents, while using a query as bag-of-words. There may be several reasons for this. First, the accuracy of the automatic document tagging may be insufficient. Second, even if elements are correctly identified in documents, if queries are treated as bags-of-words then any PICO element can match with any identical word in the query, whether it describe the same element or not. In this sense, identifying elements only in documents is not very useful.

5.3   Using Both Query and Document Tagging

Now, we consider PICO tagging in both queries and documents. For simplicity, the same weight is used for queries and documents. In this series of tests, we use manual tagging for the queries and automatic tagging for documents. Results in Table 3 show the best figure we can obtain using this method. We can see that by properly setting the parameters, the retrieval effectiveness can be significantly improved, in particular when I elements are set to a relatively high weight, P elements to a medium one, and no enhancement to O. This seems to indicate that the I element is the most important in medical search (at least for the queries we considered). This is consistent with some previous studies on IR using PICO elements. In fact, [11] suggested firstly using I and P elements to construct Boolean queries; and only if too many results are obtained that other elements should be considered.

Table 3. Performance measures for Model-1T, Model-3T ($w_{\cdot,P} = 0.3/w_{\cdot,I} = 0.9/w_{\cdot,O} = 0$), Model-1F and Model-3F ($w_{\cdot,P} = 0.1/w_{\cdot,I} = 0.3/w_{\cdot,O} = 0$) ($\ddagger$: t.test < 0.01). Increase percentage over baseline is given in parentheses.

Measure	Model-1T	Model-3T	Model-1F	Model-3F
MAP	0.1442 (+24.0%‡)	0.1452 (+24.8%‡)	0.1514 (+30.2%‡)	0.1522 (+30.9%‡)
P@10	0.3173 (+26.9%‡)	0.3404 (+36.1%‡)	0.3538 (+42.7%‡)	0.3577 (+23.0%‡)

5.4   Determining Parameters

The question now is: can we determine reasonable weights automatically? We use cross-validation in this series of exepriments to test this. We have divided the 52 tagged queries into two groups: Q26A and Q26B. A grid search (from 0 to 1 by step of 0.1) is used to find the best parameters for Q26A, and test on Q26B, and vice versa. Results are shown in Table 4. The best parameters found for Q26A in **Model-1T** are $w_{Q,P} = 0.6/w_{Q,I} = 0.9/w_{Q,O} = 0$ (MAP = 0.1688, P@10 = 0.2269), and for Q26B are $w_{Q,P} = 0/w_{Q,I} = 0.9/w_{Q,O} = 0$ (MAP = 0.1301, P@10 = 0.4192). Similar for **Model-1F**, the best parameters for Q26A are $w_{Q,P} = 0.2/w_{Q,I} = 0.3/w_{Q,O} = 0$ (MAP = 0.1784, P@10 = 0.2308), and for Q26B are $w_{Q,P} = 0/w_{Q,I} = 0.3/w_{Q,O} = 0$ (MAP = 0.1350, P@10 = 0.4808). The experiments in Table 4 show that by cross-validation, we can determine parameters that lead to a retrieval accuracy very close to the optimal settings.

Table 4. Performance measures in cross-validation (train→test) for Model-1T and Model-1F, queries are manually annotated

Cross-validation	Measure	Baseline	Model-1T	Model-1F
Q26A→Q26B	MAP	0.1221	0.1566 (+28.2%‡)	0.1596 (+30.6%‡)
	P@10	0.1846	0.2154 (+16.7%‡)	0.2308 (+25.0%‡)
Q26B→Q26A	MAP	0.1104	0.1251 (+13.4%‡)	0.1341 (+21.5%‡)
	P@10	0.3154	0.4192 (+32.9%‡)	0.4769 (+51.2%‡)

5.5   Impact of Automatic Query Tagging

Previous results show that query tagging leads to better IR accuracy. The question is whether this task, if performed automatically, still leads to improvements. Compared to manual annotation, automatic query tagging also works well even with low tagging accuracy (Table 5). One explanation may be that the manual tagging is not always optimal. For example, the query *"In patients with type 2 diabetes$^{(P)}$; pioglitazone$^{(I)}$; reduce cardiovascular events adverse events mortality improve health related quality life$^{(O)}$"* is automatically tagged as *"patients type 2 diabetes cardiovascular health$^{(P)}$; pioglitazone reduce$^{(I)}$; events adverse events mortality improve related quality life$^{(O)}$"*. The average precision for this query is improved from 0.245 to 0.298. Intuitively, tagging *cardiovascular* as P seems to be better than O even if it is not necessarily more correct. However, one also has to consider the utilization of it. By marking *cardiovascular* as P, this concept will be more enhanced, which in this case turns out to be more beneficial.

Table 5. Performance measures for Model-1F ($w_{Q,P} = 0.1/w_{Q,I} = 0.3$)

Measure	Baseline	Manual	Automatic
MAP	0.1163	0.1514 (+30.2%)	0.1415 (+21.7%)
P@10	0.2500	0.3538 (+41.5%)	0.3038 (+21.5%)

6   Conclusion

PICO is a well defined structure widely used in many medical documents which can also be used to formulate clinical questions. However, few systems have been developed to allow physicians to use PICO structure effectively in their search. In this paper, we have investigated the utilization of PICO elements in medical IR. We first tried to identify these elements in documents and queries, then a series of models have been tested to compare different utilizations of them.

Our experiments on the identification of PICO elements showed that the task is very challenging. Our classification accuracy is relatively low. This may lead one to think that the identification result is not useable. However, our experiments on IR showed that significant improvements using PICO elements can be achieved, despite the relatively low accuracy. This shows that we do not need a perfect identification of PICO elements before using them. IR can tolerate a noisy identification result. The key problem is the correct utilization of the tagging results. In our experiments, we have found that enhancing some PICO elements in queries (and in documents) leads to better retrieval results. This is especially true for the I and P elements.

References

1. Andrenucci, A.: Automated Question-Answering Techniques and the Medical Domain. In: HEALTHINF, pp. 207–212 (2008)
2. Aronson, A.R.: Effective Mapping of Biomedical Text to the UMLS Metathesaurus: The MetaMap Program. In: AMIA Symposium (2001)
3. Bai, J., Nie, J.Y., Paradis, F.: Using language models for text classification. In: Asia Information Retrieval Symposium (AIRS), Beijing, China (2004)
4. Chung, G.: Sentence retrieval for abstracts of randomized controlled trials. BMC Medical Informatics and Decision Making 9(1), 10 (2009)
5. Demner-Fushman, D., Lin, J.: Answering clinical questions with knowledge-based and statistical techniques. Computational Linguistics 33(1), 63–103 (2007)
6. Hansen, M.J., Rasmussen, N.O., Chung, G.: A method of extracting the number of trial participants from abstracts describing randomized controlled trials. Journal of Telemedicine and Telecare 14(7), 354–358 (2008)
7. Hersh, W.R.: Information retrieval: a health and biomedical perspective. Springer, Heidelberg (2008)
8. McKnight, L., Srinivasan, P.: Categorization of Sentence Types in Medical Abstracts. In: AMIA Symposium (2003)
9. Rindflesch, T.C., Fiszman, M.: The interaction of domain knowledge and linguistic structure in natural language processing: interpreting hypernymic propositions in biomedical text. Journal of Biomedical Informatics 36(6), 462–477 (2003)
10. Schardt, C., Adams, M., Owens, T., Keitz, S., Fontelo, P.: Utilization of the PICO framework to improve searching PubMed for clinical questions. BMC Medical Informatics and Decision Making 7(1), 16 (2007)
11. Weinfeld, J.M., Finkelstein, K.: How to answer your clinical questions more efficiently. Family practice management 12(7), 37 (2005)

The Role of Query Sessions in Extracting Instance Attributes from Web Search Queries

Marius Paşca, Enrique Alfonseca, Enrique Robledo-Arnuncio,
Ricardo Martin-Brualla*, and Keith Hall

Google Inc.
{mars,ealfonseca,era,rmbrualla,kbhall}@google.com

Abstract. Per-instance attributes are acquired using a weakly supervised extraction method which exploits anonymized Web-search query sessions, as an alternative to isolated, individual queries. Examples of these attributes are *top speed* for *chevrolet corvette*, or *population density* for *brazil*). Inherent challenges associated with using sessions for attribute extraction, such as a large majority of within-session queries not being related to attributes, are overcome by using attributes globally extracted from isolated queries as an unsupervised filtering mechanism. In a head-to-head qualitative comparison, the ranked lists of attributes generated by merging attributes extracted from query sessions, on one hand, and from isolated queries, on another hand, are about 12% more accurate on average, than the attributes extracted from isolated queries by a previous method.

1   Introduction

Motivation: Early work on information extraction studies how to train supervised systems on small to medium-sized document collections, requiring relatively expensive, manual annotations of data [1]. More recently, some authors investigate the possibility of obtaining annotated corpora more easily, through the creation of semi-automatic annotations [2]. But as larger amounts of textual data sources have become available at lower computational costs, either directly as document collections or indirectly through the search interfaces of the larger Web search engines, information extraction has seen a shift towards large-scale acquisition of open-domain information [3]. In this framework, information at mainly three levels of granularity is extracted from text, with weak or no supervision: class instances (e.g., *vicodin*, *oxycontin*); associated class labels (e.g., *painkillers*), and relations. These last may hold between instances (e.g., *france-capital-paris*) or classes (e.g., *countries-capital-cities*) [4,5].

One type of relation that can be learned for classes and instances are their attributes (e.g., *side effects* and *maximum dose*), which capture quantifiable properties of their respective classes (e.g., *painkillers*) or instances (e.g., *oxycontin*), and thus serve as building blocks in the knowledge bases constructed around open-domain classes or instances. Consequently, a variety of attribute extraction methods mine textual data sources ranging from unstructured [6] or structured [7,8] text within Web documents, to human-compiled

---

* Contributions made during an internship at Google.

C. Gurrin et al. (Eds.): ECIR 2010, LNCS 5993, pp. 62–74, 2010.
© Springer-Verlag Berlin Heidelberg 2010

encyclopedia [9], in an attempt to extract, for a given class, a ranked list of attributes that is as comprehensive and accurate as possible.

Using Query Session: Although Web search query logs have already been used for automatically extracting instance attributes [10, 11], as far as we know query session information (indicating which queries are issued by the same user within a limited amount of time) has not been explored for finding instance attributes. Session data is richer than simple sets of individual queries, because sessions contain queries issued in sequence, and may thus be related to one another. This paper explores the use of search queries for automatically extracting instance attributes, and shows that simple algorithms can produce results that are competitive with the current state of the art. Furthermore, by combining the results of this new approach with previous work [10, 11] we are able to produce ranked lists of attributes with much higher precision. This is an interesting result, indicating that the kind of information contained in session logs is complementary to the one that can be obtained from single-query logs and web documents.

Applications: The special role played by attributes, among other types of relations, is documented in earlier work on language and knowledge representation [12, 13]. It inspired the subsequent development of text mining methods aiming at constructing knowledge bases automatically [14]. In Web search, the availability of instance attributes is useful for applications such as search result ranking and suggestion of related queries [15], and has also been identified to be a useful resource in generating product recommendations [16].

2 Previous Work

Query Sessions: A query session is a series of queries submitted by a single user within a small range of time [17, 18, 19]. Information stored in the session logs may include the text of the queries, together with some metadata: the time, the type of query (e.g. using the normal or the advance form), and some user settings, such as the Web browser used [17].

One of the primary uses of query sessions is the identification of related queries [20, 21]. In turn, the related queries can be used, for example, as query suggestions to help users refine their queries, or query substitutions for increasing recall in sponsored search.

Typical intra-session association metrics are the chi-square test and the correlation coefficient [17], the Mutual Information and Pointwise Mutual Information metrics [22], or the log-likelihood ratio (LLR) [23, 24]. High LLR values indicate that two queries are substitutable, i.e., they are close in meaning and for most practical purposes one could be replaced with the other, as with *baby trolley* and *baby cart*. It was shown [24] that, if one removes all substitutable query pairs from sessions, the remaining pairs that still have high LLR are associated queries, which refer to closely related, but different concepts, e.g., *ski* and *snowboard*. Other metrics take into account user clicks to relate queries that lead to clicks on the same results [25, 26]. There is also increasing interest on classifying the relationships between consecutive queries in sessions, in order to identify the user intent [27, 28, 29] when issuing two consecutive queries.

Query logs have been used in the past for obtaining semantic information [30, 31, 32, 33, 34]. The most similar work that we have found is [35], which learn *query aspects*. The main differences are that (a) [35] does not make the distinction between class labels and attributes, as both can be considered aspects of queries; and (b) it is focused on clustering the attributes in very few (one to three) maximally informative aspects, whereas this paper focuses on maximising precision for large sets of attributes.

Learning Instance Attributes: Previous work on attribute extraction uses a variety of types of textual data as sources for mining attributes. Taking advantage of structured and semi-structured text available within Web documents, the method introduced in [7] assembles and submits list-seeking queries to general-purpose Web search engines, and analyzes the retrieved documents to identify common structural (HTML) patterns around class labels given as input, and potential attributes. Similarly, layout (e.g., font color and size) and other HTML tags serve as clues to acquire attributes from either domain-specific documents such as those from product and auction Web sites [36], or from arbitrary documents [37]. As an alternative to Web documents, articles within online encyclopedia can also be exploited as sources of structured text for attribute extraction, as illustrated by previous work using infoboxes and category labels [38, 39, 40] associated with articles within Wikipedia.

Working with unstructured text within Web documents, the method described in [6] applies manually-created lexico-syntactic patterns to document sentences in order to extract candidate attributes, given various class labels as input. The candidate attributes are ranked using several frequency statistics. If the documents are domain-specific, such as documents containing product reviews, additional heuristically-motivated filters and scoring metrics can be used to extract and rank the attributes [41]. In [15], the extraction is guided by a small set of seed instances and attributes rather than manually-created patterns, with the purpose of generating training data and extract new pairs of instances and attributes from text.

Web search queries have also been considered as a textual data source for attribute extraction, using lexico-syntactic patterns [10] or seed attributes [11] to guide the extraction, and leading to attributes of higher accuracy than those extracted with equivalent techniques from Web documents [42].

3   Extraction Method

Extraction from Query Sessions: Intuitively, some of the search engine users interested in information about an instance $\mathcal{I}$ may attempt to search for different characteristics of $\mathcal{I}$ during the same search session, in order to collect more complete information. For example, someone looking for information about the president of the United States may start with a query containing just his name, *barack obama*, and proceed with other queries to get more results containing some of his most relevant attributes, such as his biography, early life, quotes, opinions, poll results, etc. Although clearly not every user is expected to behave this way, as long as at least some users do display this demeanour, it is possible to learn relevant attributes about various instances.

Web search queries are typically very short, containing 2.8 words on average [43], although there is recent evidence that the average length of the queries has grown over

For i from 1 to $n - 1$
    1. For j from $i + 1$ to n
        (a) If q_i is a prefix of q_j, strip the prefix from q_j and add the remainder as a candidate attribute for q_i.
        (b) Otherwise, stop the inner loop.

Fig. 1. Algorithm for collecting candidate attributes from a query session (i.e., a sequence of consecutive queries from the same user) of the form $\mathcal{S} = [q_1, q_2, ..., q_n]$

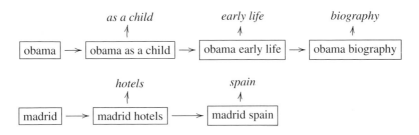

Fig. 2. Example fragments from query sessions, and candidate attributes extracted from them

time [44]. Queries may exhibit some linguistic structure, but this is typically simple: 71% of all query terms are proper or common nouns, with only 3% being prepositions, and almost 70% of the full queries are noun phrases [45]. Given their simple structure, queries seeking for information about an instance and an attribute in particular are most likely to occur in the simplest possible form, that is, the concatenation of the instance and the attribute, as in *barack obama biography*. Other forms, such as *biography of barack obama*, are equally valid but less likely to be submitted as queries.

The algorithm to collect candidate attributes from query sessions is shown in Figure 1. Figure 2 shows two examples of query sessions from which candidate attributes would be extracted: [as a child, early life, biography] for *obama*, and [hotels, spain] for *madrid*. After this step has been performed for all queries, candidate attributes within each instance are ranked among one another by their frequency.

Note that, if the session contains the two consecutive queries *[hotel]*, *[hotel madrid]*, then *madrid* will be extracted as a candidate attribute for the instance *hotel*. These cases will be taken care of in the filtering step.

Attribute Filtering: As can be seen in the previous example, the simple co-occurrence in consecutive queries far from guarantees that the extracted phrase is an attribute of the original query. In the example, *as a child* is a temporal restriction from a user interested in biographies, whereas *spain* is a different instance related to the original query through the country-capital relation.

In order to identify good attributes among inherently noisy phrases extracted from sessions, we have created automatically a whitelist that contains every phrase that appears as an attribute of at least one instance in the dataset associated with the attribute extraction method described in [10]. Any attribute from the ranked candidate lists that does not appear in the whitelist is removed. Table 1 shows the effect of this filtering for

Table 2. Set of 100 target instances, used in the evaluation of instance attribute extraction

Instances
17th century, accounting, alton towers, ancient greece, artificial intelligence, attention deficit disorder, beans, biodiesel, body language, brampton, brazil, cadmium, capri, cardboard, chhattisgarh, civil engineering, clay, cobol, communication skills, constantine, contemporary art, corporate governance, cortex, cricket, crisps, data warehousing, death penalty, decimals, delhi, dentist, digoxin, dns, electronic commerce, ferns, finland, forensics, fredericton, glycine, guinea pig, guitars, gurgaon, halogens, high blood pressure, hilary duff, instructional design, irrigation, jessica simpson, johnny depp, kidney stones, library science, lil romeo, majorca, manisha koirala, maya angelou, medicaid, medical records, methanol, mexico city, moon phases, nematodes, oil, pancho villa, pensacola, phosphorus, photography, physician assistant, podiatry, police brutality, prednisone, prose, qualitative research, railroads, reese witherspoon, refrigerator, reggaeton, resistors, richard branson, ritalin, robotics, rock n roll, san francisco, sheep, sickle cell disease, sindh, sir isaac newton, standard deviation, tata young, thyroid gland, titration, treason, tundra, utilitarianism, vida guerra, volcanos, warwick, wastewater treatment, wellbutrin, western canada, wlan, yoghurt

Table 3. Correctness labels for the manual assessment of attributes

Label	Value	Examples of Attributes
vital	1.0	beans: calories, digoxin: side effects, maya angelou: age
okay	0.5	fredericton: heritage, library science: current events, robotics: three laws
wrong	0.0	alton towers: park view, kidney stones: pain, contemporary art: urban institute

it provides useful but non-essential information; and *wrong* if it is incorrect [11]. Thus, a correctness label is manually assigned to a total of 9,137 attributes extracted for the 100 target instances, in a process that confirms that evaluation of information extraction methods can be quite time consuming.

An analysis of the correctness labels assigned by two human judges to 500 extracted attributes indicates an inter-annotator agreement of 88.79%, resulting in a Kappa score of 0.85, indicating substantial agreement in this task.

To compute the precision score over a ranked list of attributes, the correctness labels are converted to numeric values (*vital* to 1, *okay* to 0.5 and *wrong* to 0), as shown in Table 3. Precision at some rank N in the list is thus measured as the sum of the assigned values of the first N attributes, divided by N.

5   Evaluation Results

Extracted Instance Attributes: The first row in Table 4 contains the total number of instances for which at least one attribute was extracted and the total number of <instance,attribute> pairs extracted for $Q_y S_n$. For those 483,344 instances, the second row shows how many have at least one attribute extracted by $Q_n S_y$, and the total number of <instance,attribute> pairs extracted for them. Note that the numbers are not comparable, given that the original data sources: a) are different in size (the input

Table 4. For Q_yS_n: number of instances with at least one attribute and total number of <instance,attribute> pairs extracted. For Q_nS_y, among the previous instances, how many have attributes extracted and total number of attributes.

Method	Instances	<Instance,Attribute> Pairs
Q_yS_n	483,344	8,974,433
Q_nS_y	462,701	14,823,701

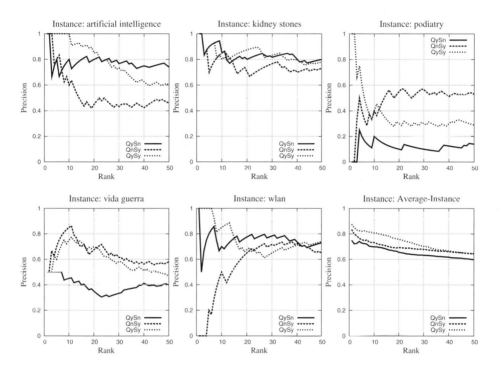

Fig. 3. Accuracy of the ranked lists of attributes extracted by various runs, for a few target instances and as an average over all target instances

for Q_yS_n is ten times smaller) and b) are different in nature (the input for Q_yS_n contains isolated queries, not sessions), and c) are affected by different restrictions (e.g., the vocabulary of instances in Q_yS_n is limited to the most frequent 5 million queries). However, as will be shown, precision improves when using session logs, without a drop in recall. Indeed, almost 96% of the instances for which Q_yS_n extracted some attributes also have attributes extracted by Q_nS_y, as illustrated in Table 4.

Session information may be more useful for extracting information about rare entities: in fact, Q_nS_y extracts attributes for many new instances whose attributes are not found with Q_yS_n. A possible reason is the fact that a user issuing two consecutive queries $\mathcal{I}$ and $\mathcal{I}\mathcal{A}$, where $\mathcal{A}$ is a common attribute name and $\mathcal{I}$ is an instance, may be a strong indicator that $\mathcal{I}$ has $\mathcal{A}$ as one of its attributes, and Q_nS_y is able to extract this

Table 5. Comparative accuracy of the attributes extracted in various experimental runs, for a few target instances and as an average over the entire set of target instances. Scores are expressed as *Abs*olute scores, *Rel*ative boosts (over Q_yS_n), and *Err*or reduction rates (over Q_yS_n). Scores are marked with $*$ and $\dagger$ if they are indistinguishable at 99% and 95% confidence respectively.

Instance	Precision											
	@5			@10			@20			@50		
	Q_yS_n	Q_nS_y	Q_yS_y	Q_yS_n	Q_nS_y	Q_yS_y	Q_yS_n	Q_nS_y	Q_yS_y	Q_yS_n	Q_nS_y	Q_yS_y
17th century (*Abs*)	0.50	0.80	0.70	0.40	0.70	0.60	0.35	0.68	0.53	0.40	0.76	0.54
artificial intelligence (*Abs*)	0.80	0.80	1.00	0.80	0.60	1.00	0.80	0.45	0.85	0.74	0.46	0.60
brazil (*Abs*)	0.90	1.00	0.90	0.85	0.95	0.95	0.78	0.97	0.95	0.79	0.92	0.87
communication skills (*Abs*)	0.70	0.90	1.00	0.85	0.80	1.00	0.68	0.62	0.82	0.67	0.56	0.52
electronic commerce (*Abs*)	1.00	1.00	1.00	0.60	0.90	0.90	0.75	0.75	0.95	0.64	0.60	0.62
gurgaon (*Abs*)	0.70	0.90	0.80	0.85	0.75	0.85	0.65	0.72	0.78	0.60	0.76	0.70
kidney stones (*Abs*)	0.90	0.70	0.90	0.85	0.85	0.80	0.80	0.70	0.88	0.80	0.73	0.78
medicaid (*Abs*)	0.60	1.00	0.60	0.60	0.80	0.70	0.62	0.70	0.75	0.49	0.64	0.61
podiatry (*Abs*)	0.20	0.40	0.60	0.20	0.40	0.40	0.10	0.55	0.30	0.14	0.53	0.29
robotics (*Abs*)	0.50	0.40	0.80	0.75	0.30	0.55	0.57	0.30	0.53	0.58	0.31	0.41
sickle cell disease (*Abs*)	0.90	0.90	0.90	0.90	0.85	0.95	0.78	0.82	0.88	0.73	0.73	0.76
vida guerra (*Abs*)	0.50	0.70	0.60	0.45	0.85	0.75	0.35	0.68	0.70	0.40	0.58	0.47
wlan (*Abs*)	0.80	0.20	1.00	0.70	0.50	0.85	0.78	0.65	0.68	0.73	0.65	0.74
Avg-Inst (*Abs*)	0.73$*$	0.76$*\dagger$	0.82	0.70	0.73$*$	0.81	0.66	0.69	0.76	0.60	0.64$*\dagger$	0.65
Avg-Inst (*Rel*)	-	+4%	+12%	-	+4%	+16%	-	+5%	+15%	-	+7%	+8%
Avg-Inst (*Err*)	-	-11%	-33%	-	-10%	-37%	-	-9%	-29%	-	-10%	-13%

pair. On the contrary, if $\mathcal{I}$ does not appear any more in the logs, just those two individual queries do not provide enough support for Q_yS_n to extract that information.

Accuracy of Instance Attributes: Figure 3 plots precision values for ranks 1 through 50, for each of the experimental runs. The first five graphs in the figure show the precision over individual target instances. Several conclusions can be drawn after inspecting the results. First, the quality of the attributes extracted by a given run varies among instances. For example, the attributes extracted for the instance *kidney stones* are better than for *vida guerra*. Second, the experimental runs have variable levels of accuracy. The last (i.e., lower right) graph in Figure 3 shows the precision as an average over all target instances. Although none of the runs outperforms the others on each and every target instance, on average, Q_yS_y performs the best and Q_yS_n (i.e., the baseline) the worst, with Q_nS_y placed in-between. In other words, attributes are more accurate when extracted from sessions (Q_nS_y) rather than from individual queries (Q_yS_n) - although

Table 6. Ranked lists of attributes extracted for a sample of the target instances

Run	Top Extracted Attributes
Instance: 17th century:	
Q_yS_n	timeline, pictures, politics, fashion, anatomy, french classicism art, accessories, austria, weapons, composers, era
Q_nS_y	fashion, clothing, art, costume, paintings, houses, weapons, names, timeline, music
Q_yS_y	fashion, timeline, composers, pictures, politics, authors, clothing, art, costume, anatomy
Instance: kidney stones:	
Q_yS_n	symptoms, causes, pictures, treatment, types, signs, prevention, signs and symptoms, removal, symptons
Q_nS_y	symptoms, treatment, pictures, causes, diet, symptoms in women, natural remedies, size, prevention, cure
Q_yS_y	symptoms, causes, pictures, treatment, prevention, types, size, symptons, signs, images
Instance: robotics:	
Q_yS_n	history, three laws, future, laws, 3 laws, basics, definition, applications, introduction, fundamentals
Q_nS_y	history, logo, basics, parts, career, competition, jobs, definition, technology, pictures
Q_yS_y	history, basics, definition, future, introduction, pictures, advantages, disadvantages, laws, types
Instance: sickle cell disease:	
Q_yS_n	symptoms, pictures, history, geographical distribution, causes, management, treatment, pathogenesis, pathology, new considerations in the treatment
Q_nS_y	symptoms, treatment, pictures, history, life expectancy, statistics, causes, cure, genetics, diagnosis
Q_yS_y	symptoms, pictures, history, treatment, causes, life expectancy, pathophysiology, effects, incidence, management

the attributes extracted in Q_yS_n across all instances do serve as a filtering mechanism for Q_nS_y, as explained earlier. The unsupervised merging of the extracted lists of attributes (Q_yS_y) gives an even larger improvement in accuracy relative to Q_yS_n.

For a more detailed analysis of qualitative performance, the upper part of Table 5 provides the precision scores for a sample of the target instances. For completeness, the scores in the table capture precision at the top of the extracted lists of attributes (rank 5) as well as over a wider range of those lists (ranks 10 and above). The table gives another view at how widely the quality of the extracted attributes may vary depending on the target instance. At the lower end, the precision of the attributes for the instances *podiatry* (with Q_yS_n) and *wlan* (with Q_nS_y) is as low as 0.20 at rank 5. At the higher end, the attributes for *sickle cell disease* are very good across all runs, with precision scores above 0.78 even at rank 20. The top attributes extracted for various instances are shown in Table 6.

When considering the comparative precision of the experimental runs in Table 5, run Q_yS_n extracts better attributes than Q_nS_y for some instances, e.g., at all ranks for *robotics* and especially for *wlan*. However, the opposite is true for most of the individual instances (e.g., *17th century, brazil, medicaid, podiatry, vida guerra*).

To better quantify the quality gap, the last rows of Table 5 show the precision computed as an average over all instances, rather than for each instance individually, and therefore they correspond to points on the curves from the last graph of Figure 3. Also shown in the table are the relative increases (*Rel*) and the reduction in the error rates (*Err*) at various ranks, for Q_nS_y and Q_yS_y, on one hand, relative to Q_yS_n, on the other hand. Consistently over all computed ranks, the precision is about 5% better on average when using sessions rather than individual queries, and about 12% better when merging attributes from sessions and individual queries in an unsupervised fashion. The results of P@20 shows that the combination is better than any of the standalone systems with 99%, and for P@50 the Q_yS_y system is better than the baseline Q_yS_n, also with 99% confidence. This is the most important result of the paper. It shows that query sessions represent a useful resource as a complement to individual queries, in instance attribute extraction.

6   Conclusions

This paper describes a procedure for extracting instance attributes from session query logs, by looking for pairs of consecutive queries such that the second one contains the first one as a prefix. A simple ranking function by query frequency produces results that improve (on aggregate) over a previous state-of-the-art system [10]. Compared to it, the main advantages are an increase in relative recall and precision scores, and the possibility of extracting attributes for less frequent queries.

More importantly, the combination of the two resources reduces the error rate between 13% and 37%, with respect to the precision of the extracted lists of attributes at various ranks. The improvement over the baseline system is statistically significant with 99% confidence for precision scores P@N with N higher or equal than 10.

Current work investigates ways of exploiting other information present in sessions, such as user clicks, to further improve the quality and coverage of the results.

References

1. Grishman, R., Sundheim, B.: Message Understanding Conference-6: a brief history. In: Proceedings of the 16th Conference on Computational Linguistics, vol. 1, pp. 466–471 (1996)
2. Chklovski, T., Gil, Y.: An analysis of knowledge collected from volunteer contributors. In: Proceedings of the National Conference on Artificial Intelligence, p. 564 (2005)
3. Etzioni, O., Banko, M., Soderland, S., Weld, S.: Open information extraction from the web. Communications of the ACM 51(12) (December 2008)
4. Sekine, S.: On-demand information extraction. In: Proceedings of the COLING/ACL on Main conference poster sessions, pp. 731–738 (2006)
5. Banko, M., Cafarella, M.J., Soderland, S., Broadhead, M., Etzioni, O.: Open information extraction from the Web. In: Proceedings of the 20th International Joint Conference on Artificial Intelligence (IJCAI 2007), pp. 2670–2676 (2007)
6. Tokunaga, K., Kazama, J., Torisawa, K.: Automatic discovery of attribute words from web documents. In: Proceedings of the 2nd International Joint Conference on Natural Language Processing (IJCNLP 2005), Jeju Island, Korea, pp. 106–118 (2005)

7. Yoshinaga, N., Torisawa, K.: Open-domain attribute-value acquisition from semi-structured texts. In: Proceedings of the Workshop on Ontolex, pp. 55–66 (2007)
8. Cafarella, M., Halevy, A., Wang, D., Zhang, Y.: Webtables: Exploring the power of tables on the web. Proceedings of the VLDB Endowment archive 1(1), 538–549 (2008)
9. Wu, F., Hoffmann, R., Weld, D.: Information extraction from Wikipedia: Moving down the long tail. In: Proceedings of the 14th ACM SIGKDD Conference on Knowledge Discovery and Data Mining (KDD 2008), pp. 731–739 (2008)
10. Paşca, M., Van Durme, B.: What you seek is what you get: Extraction of class attributes from query logs. In: Proceedings of the 20th International Joint Conference on Artificial Intelligence (IJCAI 2007), pp. 2832–2837 (2007)
11. Paşca, M.: Organizing and searching the World Wide Web of facts - step two: Harnessing the wisdom of the crowds. In: Proceedings of the 16th World Wide Web Conference (WWW 2007), Banff, Canada, pp. 101–110 (2007)
12. Pustejovsky, J.: The Generative Lexicon: a Theory of Computational Lexical Semantics. The MIT Press, Cambridge (1991)
13. Guarino, N.: Concepts, attributes and arbitrary relations. Data and Knowledge Engineering 8, 249–261 (1992)
14. Schubert, L.: Turing's dream and the knowledge challenge. In: Proceedings of the 21st National Conference on Artificial Intelligence (AAAI 2006), Boston, Massachusetts (2006)
15. Bellare, K., Talukdar, P., Kumaran, G., Pereira, F., Liberman, M., McCallum, A., Dredze, M.: Lightly-supervised attribute extraction. In: NIPS 2007 Workshop on Machine Learning for Web Search (2007)
16. Probst, K., Ghani, R., Krema, M., Fano, A., Liu, Y.: Semi-supervised learning of attribute-value pairs from product descriptions. In: Proceedings of the 20th International Joint Conference on Artificial Intelligence (IJCAI 2007), pp. 2838–2843 (2007)
17. Silverstein, C., Marais, H., Henzinger, M., Moricz, M.: Analysis of a very large Web search engine query log. In: ACM SIGIR Forum, pp. 6–12 (1999)
18. Jansen, B., Spink, A., Taksa, I.: Handbook of Research on Web Log Analysis. Information Science Reference (2008)
19. He, D., Goker, A.: Detecting session boundaries from web user logs. In: Proceedings of the BCS-IRSG 22nd Annual Colloquium on Information Retrieval Research, pp. 57–66 (2000)
20. Wen, J., Nie, J., Zhang, H.: Clustering user queries of a search engine. In: Proceedings of the International Conference on World Wide Web (2001)
21. Zhang, Z., Nasraoui, O.: Mining search engine query logs for query recommendation. In: Proceedings of the 15th International Conference on World Wide Web, pp. 1039–1040 (2006)
22. Church, K., Hanks, P.: Word association norms, mutual information, and lexicography. Computational Linguistics 16(1), 22–29 (1990)
23. Jones, R., Rey, B., Madani, O., Greiner, W.: Generating query substitutions. In: Proceedings of the 15th International Conference on World Wide Web, pp. 387–396 (2006)
24. Rey, B., Jhala, P.: Mining associations from Web query logs. In: Proceedings of the Web Mining Workshop, Berlin, Germany (2006)
25. Xue, G.R., Zeng, H.J., Chen, Z., Yu, Y., Ma, W.Y., Xi, W., Fan, W.: Optimizing Web search using Web click-through data. In: CIKM 2004: Proceedings of the 13th ACM International Conference on Information and Knowledge Management, pp. 118–126 (2004)
26. Ma, H., Yang, H., King, I., Lyu, M.R.: Learning latent semantic relations from clickthrough data for query suggestion. In: Proceeding of the 17th ACM Conference on Information and Knowledge Management, pp. 709–718 (2008)
27. Lau, T., Horvitz, E.: Patterns of search: Analyzing and modeling web query refinement. In: Proceedings of the International User Modelling Conference (1999)

28. Jones, R., Klinkner, K.L.: Beyond the session timeout: automatic hierarchical segmentation of search topics in query logs. In: CIKM 2008: Proceeding of the 17th ACM conference on Information and Knowledge Management, pp. 699–708 (2008)

29. Boldi, P., Bonchi, F., Castillo, C., Donato, D., Gionis, A., Vigna, S.: The query-flow graph: model and applications. In: CIKM 2008: Proceeding of the 17th ACM conference on Information and Knowledge Management, pp. 609–618 (2008)

30. Baeza-Yates, R., Tiberi, A.: Extracting semantic relations from query logs. In: Proceedings of the 13th ACM SIGKDD International Conference on Knowledge Discovery and Data Mining, pp. 76–85 (2007)

31. Shen, D., Qin, M., Chen, W., Yang, Q., Chen, Z.: Mining Web query hierarchies from click-through data. In: Proceedings of the National Conference on Artificial Intelligence (2007)

32. Paşca, M., Van Durme, B.: Weakly-supervised acquisition of open-domain classes and class attributes from web documents and query logs. In: Proceedings of the 46th Annual Meeting of the Association for Computational Linguistics (ACL 2008), Columbus, Ohio, pp. 19–27 (2008)

33. Komachi, M., Makimoto, S., Uchiumi, K., Sassano, M.: Learning semantic categories from clickthrough logs. In: Proceedings of the ACL-IJCNLP 2009 Conference, Short Papers, pp. 189–192 (2009)

34. Pennacchiotti, M., Pantel, P.: Entity extraction via ensemble semantics. In: Proceedings of the 2009 Conference on Empirical Methods in Natural Language Processing (EMNLP 2009), Singapore, pp. 238–247 (2009)

35. Wang, X., Chakrabarti, D., Punera, K.: Mining broad latent query aspects from search sessions. In: Proceedings of the 15th ACM SIGKDD International Conference on Knowledge Discovery and Data Mining, pp. 867–876 (2009)

36. Wong, T., Lam, W.: An unsupervised method for joint information extraction and feature mining across different web sites. Data & Knowledge Engineering 68(1), 107–125 (2009)

37. Ravi, S., Paşca, M.: Using structured text for large-scale attribute extraction. In: Proceedings of the 17th ACM Conference on Information and Knowledge Management (CIKM 2008), pp. 1183–1192 (2008)

38. Suchanek, F., Kasneci, G., Weikum, G.: Yago: a core of semantic knowledge unifying WordNet and Wikipedia. In: Proceedings of the 16th World Wide Web Conference (WWW 2007), Banff, Canada, pp. 697–706 (2007)

39. Nastase, V., Strube, M.: Decoding Wikipedia categories for knowledge acquisition. In: Proceedings of the 23rd National Conference on Artificial Intelligence (AAAI 2008), Chicago, Illinois, pp. 1219–1224 (2008)

40. Wu, F., Weld, D.: Automatically refining the Wikipedia infobox ontology. In: Proceedings of the 17th World Wide Web Conference (WWW 2008), Beijing, China, pp. 635–644 (2008)

41. Raju, S., Pingali, P., Varma, V.: An unsupervised approach to product attribute extraction. In: Proceedings of the 31th European Conference on IR Research on Advances in Information Retrieval, pp. 796–800 (2009)

42. Paşca, M., Van Durme, B., Garera, N.: The role of documents vs. queries in extracting class attributes from text. In: Proceedings of the 16th International Conference on Information and Knowledge Management (CIKM 2007), Lisbon, Portugal, pp. 485–494 (2007)

43. Spink, A., Jansen, B., Wolfram, D., Saracevic, T.: From e-sex to e-commerce: Web search changes. IEEE Computer 35(3), 107–109 (2002)

44. Hogan, K.: Interpreting hitwise statistics on longer queries. Technical report, Ask.com (2009)

45. Barr, C., Jones, R., Regelson, M.: The linguistic structure of english web-search queries. In: Proceedings of the 2008 Conference on Empirical Methods in Natural Language Processing, pp. 1021–1030 (2008)

Transliteration Equivalence Using Canonical Correlation Analysis

Raghavendra Udupa[1] and Mitesh M. Khapra[2]

[1] Microsoft Research India
[2] Indian Institute of Technology, Bombay

Abstract. We address the problem of Transliteration Equivalence, *i.e.* determining whether a pair of words in two different languages (*e.g.* **Auden**, ऑडेन) are name transliterations or not. This problem is at the heart of Mining Name Transliterations (MINT) from various sources of multilingual text data including parallel, comparable, and non-comparable corpora and multilingual news streams. MINT is useful in several cross-language tasks including Cross-Language Information Retrieval (CLIR), Machine Translation (MT), and Cross-Language Named Entity Retrieval. We propose a novel approach to Transliteration Equivalence using language-neutral representations of names. The key idea is to consider name transliterations in two languages as two views of the same semantic object and compute a low-dimensional common feature space using Canonical Correlation Analysis (CCA). Similarity of the names in the common feature space forms the basis for classifying a pair of names as transliterations. We show that our approach outperforms state-of-the-art baselines in the CLIR task for Hindi-English (3 collections) and Tamil-English (2 collections).

Keywords: Information Retrieval, Cross-Language Information Retrieval, Machine Translation, Cross-Language Named Entity Retrieval, Machine Transliteration, Mining, Canonical Correlation Analysis.

1 Introduction

Names play a very important role in several tasks including Document Retrieval, Question Answering, Entity Retrieval, Product Search, and Information Extraction. In Web Search, names are particularly important as they are both highly frequent[1] in queries and very helpful in understanding the intent of the query. Furthermore, names often determine the focus of informational queries. Take for example CLEF Hindi topic 447: पिम फोरत्यून की राजनीति [2]. In the LA-Times 2002 document collection, there is only one relevant document for this

---

[1] About 40% of the terms in web search queries are proper nouns (*e.g.* Texas) and about 30% of the query terms are common nouns (*e.g.* pictures) [1]. Furthermore, about 71% of web search queries contain at least one named entity (*e.g.* Harry Potter) [2].

[2] पिम फोरत्यून की राजनीति translates to English as Pim Fortuyn's Politics.

C. Gurrin et al. (Eds.): ECIR 2010, LNCS 5993, pp. 75–86, 2010.
© Springer-Verlag Berlin Heidelberg 2010

topic whereas there are innumerous documents that discuss politics. Therefore, it is absolutely essential that the name पिम फोरत्यून [3] is translated accurately to English in the Hindi-English CLIR task. Otherwise, the CLIR system will predictably fail in retrieving the relevant document.

Name translation is important not only to CLIR but also to several cross-language tasks including Machine Translation (MT), Cross-Language Speech Retrieval and Cross-Language Named Entity Retrieval. Unfortunately, most translation lexicons do not provide a good coverage of name translations as names form an open class of words in any language. New names are added to the language every day and hence it is impossible to keep the translation lexicons up-to-date with respect to name translations. In fact, many of the Out-Of-Vocabulary (OOV) terms in CLIR and MT are names [3,4,5]. As translation lexicons do not provide a good coverage of name translations, names are either transliterated using a Machine Transliteration system or left untranslated in CLIR and MT.

The effectiveness of a Machine Transliteration system is limited by its ability to produce correct and relevant transliterations. In CLIR, for instance, it is not sufficient if a Machine Transliteration system generates a correct transliteration of the name. It must generate the transliteration(s) present in the documents relevant to the query [6]. Recent works have shown that Mining Name Transliterations (MINT) from various sources of multilingual text data including parallel, comparable, and non-comparable corpora and multilingual news streams is a potent alternative to Machine Transliteration [7,8]. The mined name transliterations can be used to augment translation lexicons. In the specific case of CLIR, name transliterations can be mined from the top results of the first pass retrieval and used to improve substantially the retrieval performance of the second pass retrieval [6]. At the heart of MINT is the problem of Transliteration Equivalence: given a pair of names in two different languages (*e.g.* **Auden**, ऑडेन) determine whether the two names are transliterations of each other.

In this work, we propose a novel approach for identifying name transliterations based on multi-view learning [9]. Given two words, say **Auden** in English and ऑडेन in Hindi, we extract surface features from each of the words independently. For instance, the features extracted for **Auden** could be the character bigrams {∧A, Au, ud, de, en, n$} [4] and the features for ऑडेन could be the character bigrams {∧ऑ, ऑड, डे, ेन, न$ }. We view the resulting language/script specific feature vectors as two representations/views of the same semantic object, the entity whose name is written as **Auden** in English and as ऑडेन in Hindi. This enables us to compute a common, low-dimensional, and language/script neutral feature space for English and Hindi names from a training set of name pairs using the well-known multi-view learning technique Canonical Correlation Analysis (CCA). We map English and Hindi names to the common feature space using the projection matrices computed by CCA. In the common feature space, names that are mutual transliterations (*e.g.* {**Mary**, मॅरी}) are mapped to

---

[3] Pim Fortuyn.

[4] ∧ and $ are respectively the beginning-of-the-string and end-of-the-string markers.

similar feature vectors. On the contrary, names that are not mutual transliterations (*e.g.* {**Mary**, बिल}) are mapped to dissimilar feature vectors.

Given a new pair, say {**Wystan**, विस्टन}, we first determine the feature vectors for each of the words by projecting the surface features {∧W, Wy, ys, st, ta, an} and {∧व, वि, स्टि, स्ट, टन, न$ } to the common feature space. Transliteration equivalence of **Wystan** and विस्टन} (*i.e.* whether they form a transliteration pair or not) can now be decided using the cosine similarity of the two feature vectors.

Our approach has the following advantages over other methods for identifying name transliterations: 1) unlike generative models for name transliteration, we can employ any set of language/script dependent features for representing a name. For instance, we can use character and phoneme n-grams as features. Further, features need not be mutually independent. 2) unlike generative models for name transliteration, we do not require to compute hidden alignments between characters/phonemes of the two names. 3) unlike discriminative methods for name transliteration, we do not need negative examples for training the model.

In the remainder of this paper we provide a full exposition of our approach along with experimental results on several languages. We start by discussing relevant prior research in Section 2. Next we describe our approach in Section 3. We discuss the experimental setup and results of our experiments in Section 4. Finally, we discuss the results and propose some ideas for future investigation in Section 5.

2 Related Work

Machine Transliteration has been studied extensively in both CLIR and MT [3,4,10,5]. For a detailed bibliography of research in Machine Transliteration, please see [10]. As noted by Udupa *et. al.*, the effectiveness of a Machine Transliteration system is limited by its ability to produce correct and relevant transliterations [6].

Techniques for identifying cross-language spelling variants have been explored by Pirkola *et. al.* [11]. Similarly, techniques for identifying cognates in the vocabularies of related languages based on phonetic similarity have also been studied [12]. While such techniques are effective, their application is restricted to languages that share the same script and are closely related.

Several techniques for augmenting translation lexicons with translations mined from comparable corpora have been proposed in the literature [4,13,14,15]. Early work in this space focused on discovering translations using co-occurrence patterns of words in comparable corpora [4,15]. While these techniques have been moderately successful in mining translation lexicons, their application is restricted to the most frequent nouns. Recent work has explored sophisticated algorithms for mining parallel sentences and even parallel sub-sentential fragments from large comparable corpora [14]. Such methods rely on statistical alignment techniques for learning translations and unfortunately, name translations are

Let (A,B) be the first $K > 0$ basis vectors computed by CCA. Note that (A,B) maps the two views $x \in \mathbb{R}^m$ and $y \in \mathbb{R}^n$ to a common feature space:

$$x \rightarrow A^T x = u \in \mathbb{R}^K \tag{4}$$

$$y \rightarrow B^T y = v \in \mathbb{R}^K \tag{5}$$

Therefore, we can compare one view with the other directly in the common feature space:

$$sim(x, y) = cos(u, v) = \frac{<u, v>}{||u|| \cdot ||v||}$$
$$= \frac{x^T A B^T y}{\sqrt{x^T A A^T x}\sqrt{y^T B B^T y}} \tag{6}$$

3.3   CCA for Transliteration Equivalence

As mentioned in Section 3.1, we pose the problem of transliteration equivalence of a pair of words as one of determining the similarity of the words in the common feature space. We learn the common feature space by using CCA on a set of training examples $\{e_i, h_i\}_{i=1}^N$ where (e_i, h_i) is a pair of name transliterations. Each English (and resp. Hindi) name e_i (and resp. h_i) is represented by a feature vector x_i (and resp. y_i). Feature vectors x_i and y_i are the two views of the same name for the purpose of computing the common feature space. The basis vectors (A, B) of the common feature space are determined by solving the CCA optimization problem (Equation 1). Given a new pair of names (e, h), we check their equivalence by first projecting their feature vectors (x, y) to the common feature space (Equations 4 and 5) and then computing the cosine similarity of the feature vectors (Equation 6).

Features: In principle, we can use any set of features for representing a name. Furthermore, the features can be overlapping. A natural choice is character n-grams and phoneme n-grams observed in the name. These features can either be binary or weighted (tf-idf score of the feature in the training corpus of name transliterations). The character and phoneme n-gram features capture the orthographic and phonetic characteristics of the name. By adding the length of the name as a feature to the orthographic and phonetic features, we can ensure that name pairs that have very different lengths are mapped to dissimilar vectors in the common feature space.

Apart from the character and phoneme level features, we can also use higher-level features if additional resources such as comparable corpora are available. For example, the IDF of the name in the corpora can be a useful feature as name transliterations are likely to have similar IDF in the comparable corpora. If the documents in the comparable corpora have time stamps, then the time series of the name can be used as features. Prior work has shown that time series similarity is valuable in mining name transliterations from comparable corpora [8]. Our framework makes it possible to use time series along with character and phoneme level features in a principled way.

4   Experiments

In this Section, we describe the empirical studies that we conducted on Hindi-English and Tamil-English CLIR system to test the practicality of our approach.

Experimental Setup: We used the same experimental procedure as the one used in [6]. The input to our CCA-based classifier is OOV query terms and terms from the top results of the first pass retrieval. We remove stop-words from the query as well as the feedback documents. The output of the classifier is a set of name pairs which are added to the translation lexicon and used in the second pass retrieval.

Data: We conducted our experiments on three English language document collections: LA Times 2002 with topics 401-450 (CLEF 2007), LA Times 94 + Glasgow Herald 95 with topics 301-350 (CLEF 2006), Telegraph01-07 with topics 26-75 (FIRE 2008). The CLEF topics 401-450 are in Hindi and Tamil and 301-350 are in Hindi [20,21]. The FIRE 2008 topics 26-75 are in both Hindi and Tamil [22]. As the collections and topics are from past years, their relevance judgments are also available. We used all the three fields (title, description, and narrative) of the CLEF and FIRE topics in the first set of experiments and only the title in the second set of experiments.

Dictionaries: We used statistical dictionaries for both Hindi-English and Tamil-English CLIR. We generated the dictionaries by training statistical word alignment models on Hindi-English parallel corpora (about 55K parallel sentences) and Tamil-English parallel corpora (about 40 K parallel sentences) using the GIZA++ tool [23]. We used 5 iterations of IBM Model 1 and 5 iterations of HMM. We retained only the top 4 translations for every source word.

CLIR System: We used a KL-divergence based ranking approach for ranking the documents [24]. We used only the textual content of the documents for indexing and indexed only non-empty documents. We removed stop words from the text before indexing and stemmed the words using the Porter stemmer [25]. We did not stem the queries (in Hindi and Tamil) as we did not have a good stemmer for the query languages.

CLIR Baselines: We compared the retrieval performance of our system (KLD-CCA) with three baselines: 1) KLD: A CLIR system based on KL-divergence ranking and Jelinek-Mercer smoothing 2) KLD-MT: KLD + a state-of-the-art Machine Transliteration system [26] and 3) KLD-HMM: KLD + a state-of-the-art transliteration mining system [6]. The Machine Transliteration system used given query should depend not only on the individual ranked documents, but also on how they relate to each other. For example, it is questionable whether users will find a given document relevant to their information need after examining other similar documents [3]. The general problem of minimising the redundancy among the retrieved documents—or, conversely, of maximising their coverage with respect to different aspects of the original query—is NP-hard [4]. Most previous works on search result diversification are based on a greedy approximation to this problem [5]. In common, these works attempt to reduce the redundancy among the retrieved documents by comparing them with respect to their content or their estimated relevance to the original query. By doing so, they implicitly assume that similar documents will cover similar aspects underlying the query.

On the other hand, as queries often carry some ambiguity, the broader topic represented by a given query can be usually decomposed into distinct sub-topics. This, in turn, motivates an alternative approach to search result diversification, centred on explicitly modelling the possibly several aspects underlying a query. In this paper, we introduce a new framework for search result diversification that exploits this intuition in order to maximise the aspects covered in a document ranking by comparing the retrieved documents with respect to their estimated relevance to each of these aspects. In particular, we uncover different aspects underlying the original query in the form of *sub-queries*, which are then used as a central element for comparing a given pair of documents based on how well they satisfy each sub-query. By doing so, we can take into account both the diversity of aspects covered by a single document, as well as its novelty in face of the aspects already covered by other retrieved documents. Moreover, the relative importance of each identified sub-query can be directly incorporated within our framework, so as to bias the diversification process towards those sub-queries likely to represent more plausible aspects of the initial query.

We compare our proposed framework to both classical as well as some recent search result diversification approaches using a standard TREC collection with relevance assessments at the sub-topic level. Our results show that the explicit account of the possible aspects underlying the original query as sub-queries can produce substantial improvements over both implicit and explicit state-of-the-art diversification approaches. The remainder of this paper is organised as follows. Section 2 provides an overview of recent works on search result diversification. Section 3 describes the major components of our proposed diversification framework, built around the concept of sub-queries. Section 4 details our experimental settings, while Section 5 discusses our main findings. Lastly, Section 6 presents our conclusions and directions for future work.

2   Background and Related Work

The problem of diversifying search results can be stated as:

> Given a query q, retrieve a ranking of documents $R(q)$ with maximum relevance with respect to q and minimum redundancy with respect to its coverage of the possible aspects underlying q.

In its general form, this problem can be reduced from the maximum coverage problem [6], which makes it NP-hard [4]. Most previous approaches to search result diversification are based on a greedy approximation to this problem, namely, the so-called maximal marginal relevance (MMR) method [5]. The general idea of MMR is to iteratively re-rank an initial set of documents retrieved for a given query by selecting, at each iteration, the document not yet selected with the highest estimated relevance to the query and highest dissimilarity to the already selected documents. The various approaches based on MMR differ mostly by how the similarity between documents is computed. For example, Carbonell and Goldstein [5] suggested using a content-based similarity function, e.g., the cosine distance between the vectors representing the retrieved documents. Zhai and Lafferty [7] proposed to model relevance and novelty within the language modelling framework. They devised six different methods, based on either the KL divergence measure or a simple mixture model. More recently, Wang and Zhu [8] employed the correlation between documents as a measure of their similarity.

In common, all of these approaches consider the possible aspects associated to the original query only in an implicit way, namely, by directly comparing the retrieved documents against one another, under the assumption that similar documents will cover similar aspects. By demoting similar documents in the ranking, these approaches aim to reduce the ranking overall redundancy. An alternative approach is to explicitly consider the different aspects associated to a query by directly modelling these aspects. For instance, Agrawal et al. [4] investigated the diversification problem by employing a taxonomy for both queries and documents. In their work, documents retrieved for a query are considered similar if they are confidently classified into one or more common categories covered by the query. By doing so, documents covering well-represented categories in the ranking are penalised, as they would bring little novelty in face of the already selected documents. A related approach was investigated by Radlinski and Dumais [9]. In order to compose a diverse ranking, they proposed to filter the results retrieved for a given query so as to limit the number of those satisfying each of the aspects of this query, represented as different query reformulations, as obtained from a large query log from a commercial search engine.

Our approach also considers the possible aspects associated to a query explicitly. Differently from the approaches of Agrawal et al. [4] and Radlinski and Dumais [9], however, we do not rely on a predefined taxonomy or on classification schemes, nor do we reserve predetermined shares of the final ranking for results answering each of the identified aspects associated to a query. Instead, we uncover different aspects underlying a query as sub-queries, and estimate the similarity between any two documents based on their estimated relevance to common sub-queries. Moreover, we make use of the associations between documents and the identified sub-queries in order to perform a richer re-ranking of the results retrieved for the original query. By doing so, we can take into account not only the estimated relevance of a document to the original query, but also the relative importance of the different aspects underlying this query, their coverage in the ranking, and how well the given document satisfies each of them.

3   The xQuAD Diversification Framework

In this section, we describe our novel framework for search result diversification, centred around the concept of sub-queries. The *eXplicit Query Aspect Diversification* (xQuAD) framework is inspired by the greedy approximation approach to the general diversification problem, which is at the heart of most of the previous works on search result diversification, as described in Section 2. Differently from these approaches, however, our framework performs an explicit diversification of the documents retrieved for a given query, by exploiting the relationship between these documents and the aspects uncovered from this query. In particular, we aim to promote a diverse ranking of documents according to the following four components: *aspect importance*, based on the relevance of each identified aspect with respect to the initial query; *document coverage*, based on the estimated relevance of a given document to multiple aspects; *document novelty*, based on the estimated relevance of the document to aspects not well represented among the already selected documents; and *document relevance*, based on the estimated relevance of the document to the initial query.

Aspect importance is discussed in Section 3.2. The document coverage and novelty components follow from the intuitive definition of an ideal diverse ranking, which should provide a broad coverage of the aspects underlying the initial query, while reducing its overall redundancy with respect to aspects already well covered. Just as for aspect importance, the effectiveness of these components depend on the quality of the aspects uncovered from the initial query in the form of sub-queries, as discussed in Section 3.1. As for the last component, namely, document relevance, we argue that it can help cope with the uncertainty associated with the relevance estimations for multiple sub-queries. Indeed, it provides a common basis for comparing documents retrieved for different sub-queries, as the relevance scores based on these sub-queries may not be comparable.

Our proposed framework integrates all these components into Algorithm 1. The algorithm takes as input the initial query q, the set $R(q)$ of documents retrieved for q, a set $Q(q)$ of sub-queries q_i derived from q, a scoring function $r(d, q)$ that estimates the relevance of a document d to a query q (analogously, $r(d, q_i)$ estimates the relevance of d to the sub-query q_i), and a sub-query importance estimator $i_X(q_i, q)$ (see Section 3.2). Additionally, it has two parameters: the number τ of top-ranked results from $R(q)$ to be returned, and the weight ω, used for balancing the influence of the relevance and diversity estimations.

The algorithm constructs a result set $S(q)$ by iteratively selecting a document d^* which contributes the most relevant and novel information among the remaining documents from the initial ranking, $R(q)$. The core of the algorithm is the computation of $r(d, q, Q(q))$ (lines 3-5), which combines the relevance score of d with respect to the query q, and a diversity score, computed as a summation over each of the sub-queries $q_i \in Q(q)$ that are satisfied by this document. In particular, the contribution of a given sub-query q_i to the diversity of document d takes into account: (1) the relative importance $i_X(q_i, q)$ of q_i in light of the query q, (2) the estimated relevance of d to q_i, $r(d, q_i)$, and (3) a measure of the novelty of any document satisfying q_i. The latter is given by $m(q_i)^{-1}$, with

xQuAD$[q, R(q), Q(q), r, i_X, \tau, \omega]$

1   $S(q) \leftarrow \emptyset$
2   **while** $|S(q)| < \tau$ **do**
3   **for** $d \in R(q)$ **do**
4           $r(d, q, Q(q)) \leftarrow r(d, q) \times \left(\sum_{q_i \in Q(q)} i_X(q_i, q) r(d, q_i) / m(q_i) \right)^{\omega}$
5   **end for**
6       $d^* \leftarrow \arg\max_d \; r(d, q, Q(q))$
7   **for** $q_i \in Q(q)$ **do**
8           $m(q_i) \leftarrow m(q_i) + r(d^*, q_i)$
9   **end for**
10      $R(q) \leftarrow R(q) \setminus \{d^*\}$
11      $S(q) \leftarrow S(q) \cup \{d^*\}$
12 **end while**
13 **return** $S(q)$

Algorithm 1. The xQuAD framework

$m(q_i)$ defined as the "mass" of information satisfying q_i that is already included in the final ranking, $S(q)$. After the top scored document d^* is selected at the end of each iteration (line 6), the information mass $m(q_i)$ is updated to account for the selection of this document from all the sub-queries it satisfies (lines 7-9). The selected document is then removed from $R(q)$ (line 10) and included in the final document ranking, $S(q)$ (line 11). At the end of the process, $S(q)$ is the final diverse ranking to be presented to the user (line 13).

3.1   Uncovering Query Aspects

An important component of our proposed diversification framework is the generation of sub-queries, in the form of keyword-based representations of the possible aspects underlying the initial query. Several techniques can be used for this purpose. For instance, we could mine a query log for common reformulations of the initial query [9], or use a large external corpus, such as Wikipedia, in order to obtain possible disambiguation terms [10]. Alternatively, sub-queries can be generated from the target collection itself, e.g., by uncovering the most salient phrases from the top retrieved results for a given query [11].

To validate our approach, we use a test collection with relevance assessments at the level of sub-topics. These sub-topics can be seen as a simulation of ground-truth sub-queries, as discussed in Section 4. This allows us to investigate the full potential of our approach, by focusing on how to effectively exploit sub-queries within the xQuAD diversification framework. Additionally, we propose a technique inspired by traditional text clustering, in order to generate sub-queries from the baseline ranking. Given a ranking of documents retrieved for the original query, a clustering algorithm is applied to partition this ranking into a predefined number of clusters. In our experiments, we use the k-means algorithm [12]. From each generated cluster, we select the most informative terms as a sub-query, using the Bo1 information-theoretic query expansion model from the Divergence From Randomness (DFR) framework [13].

3.2   Estimating Aspect Relative Importance

The importance of the different aspects underlying a given query should ultimately reflect the interests of the user population—i.e., information consumers—
with respect to each of these aspects [3], e.g., based on the popularity of each
corresponding sub-query in a query log. In the absence of such data, an alternative is to rely on sub-query importance as conveyed by information producers. To
do so, we propose four different aspect importance estimators, which implement
the $i_X(q_i, q)$ component in the xQuAD framework, as presented in Algorithm 1.
The first one, $i_U(q_i, q)$, considers a uniform distribution of importance:

$$i_U(q_i, q) = \frac{1}{|Q(q)|}, \tag{1}$$

where $|Q(q)|$ is the total number of identified sub-queries. As a more refined
estimator, we introduce $i_N(q_i, q)$, which considers the total number of results
retrieved from the target collection for a particular sub-query q_i as an indication
of the importance of q_i:

$$i_N(q_i, q) = \frac{n(q_i)}{\sum_{q_j \in Q(q)} n(q_j)}, \tag{2}$$

where $n(q_i)$ is the total number of results retrieved for q_i, and $Q(q)$ is the set
of all sub-queries derived from the initial query q. Alternatively, inspired by
resource selection techniques in distributed information retrieval [14], we devise
richer importance estimators by considering the top retrieved documents for
each sub-query as a sample from the resource represented by all the documents
associated to that particular sub-query in the whole collection.

In particular, in this work, we investigate two effective resource selection algorithms as estimators of sub-query importance: Relevant Document Distribution Estimation (ReDDE) [15], and Central Rank-based Collection Selection
(CRCS) [16]. The ReDDE algorithm estimates the number of relevant documents contained in a given resource based on the estimated size of this resource
and on the number of its documents that are ranked above a certain threshold in a centralised ranking comprising samples from all resources. We devise a
ReDDE-inspired sub-query importance estimator $i_R(q_i, q)$ as:

$$i_R(q_i, q) = \sum_{d|r(d,q_i)>0} r(d, q) \times r(d, q_i) \times n(q_i), \tag{3}$$

where $r(d, q)$ is the estimated relevance of a document d with respect to the
query q. Analogously, $r(d, q_i)$ estimates the relevance of d to the sub-query q_i.
As above, $n(q_i)$ is the total number of results associated with q_i.

Besides $i_U(q_i, q)$, $i_N(q_i, q)$, and $i_R(q_i, q)$, we propose another way of estimating the relative importance of different sub-queries, inspired by the CRCS algorithm. CRCS ranks resources according to their estimated sizes, differing from
other approaches—including ReDDE—by also considering the position (or rank)

of each of the sampled documents in the centralised ranking of resource descriptions. The idea is that a document ranked higher should convey more importance of its resource than a document appearing towards the bottom of the ranking. Inspired by CRCS, we devise the $i_C(q_i, q)$ importance estimator as:

$$i_C(q_i, q) = \frac{n(q_i)}{\max_{q_j \in Q(q)} n(q_j)} \times \frac{1}{\hat{n}(q_i)} \sum_{d | r(d,q_i) > 0} \tau - j(d, q), \tag{4}$$

where $n(q_i)$ is as above, $\hat{n}(q_i)$ corresponds to the number of results associated to the sub-query q_i that are among the top τ ranked results for the query q, with $j(d, q)$ giving the ranking position of the document d with respect to q.

4   Experimental Setup

In this section, we describe our experimental setup, in order to support the investigation of the following research questions:

1. Is the explicit account of the aspects underlying a given query an effective approach for diversifying the results retrieved for this query?
2. Is our proposed framework an effective diversification approach when compared to other explicit diversification approaches?
3. Can we further improve the effectiveness of our framework by taking into account the relative importance of individual sub-queries?
4. Can we effectively derive sub-queries from the baseline ranking itself?

In the following, we detail the test collection, topics, and metrics used in our evaluation, as well as the diversification approaches to which ours is compared, including the procedure for training their parameters. The Terrier Information Retrieval platform[1] [17] is used for both indexing and retrieval.

4.1   Collection and Topics

In our experiments, we index the Financial Times of London 1991-1994, a standard test collection with 210,158 news articles. In particular, this collection was used in a diversity-oriented task investigated under the standard experimentation paradigm of the Text REtrieval Conference (TREC), as part of the TREC Interactive track in TREC-6, TREC-7, and TREC-8 [18]. The investigated task, then called "aspect retrieval", involved finding documents covering as many different aspects of a given query as possible. As part of this evaluation campaign, a total of 20 topics were adapted from the corresponding years of the TREC Adhoc track. Each topic includes from 7 to 56 sub-topics, as identified by TREC assessors, with relevance assessments provided at the sub-topic level. Figure 1 illustrates one of such topics, 353i, along with some of its identified aspects.

---

[1] http://www.terrier.org

```
<top>                                    353i-1 mining prospection
<num> Number: 353i                       353i-2 oil resources
<title> Antarctic exploration           353i-3 rhodium exploration
<desc>                                   353i-4 ozone hole / upper atmosphere
   Identify systematic explorations and  353i-5 greenhouse effect
   scientific investigations of Antarctica, 353i-6 measuring chemicals in the atmosphere
   current or planned.                   353i-7 analysis of toxic wast
</top>                                       ...
```

Fig. 1. TREC-7 Interactive track, topic 353i, and corresponding sub-topics

In the example, several sub-topics were identified by the assessors for topic 353i, as shown on the right-hand side of Figure 1. In order to test the full benefit of our explicit diversification framework, we follow Zhai et al. [7] and derive ground-truth sub-queries based on the official aspects associated to each of the TREC Interactive track topics. As discussed in Section 3, this experimental design choice allows us to simulate a best-possible sub-query generation mechanism in order to focus our attention to evaluating the diversification framework itself.

4.2 Retrieval Baselines

We evaluate the effectiveness of our framework at diversifying the rankings produced by two effective document ranking approaches as retrieval baselines, namely, BM25 [19] and the DPH hypergeometric, parameter-free model from the DFR framework [20]. On top of the initial ranking produced by either of these baselines, we compare our framework to several other diversification approaches, namely, the previously described approaches of Carbonell and Goldstein [5], Radlinski and Dumais [9], and Agrawal et al. [4]. In particular, as the last two make use of external resources or judgements, such as query logs or a classification taxonomy, which are not available for the test collection at hand, we simulate their best-case scenario, by considering the ground-truth sub-topics provided by the collection as input to their proposed diversification models.

4.3 Evaluation Metrics

Our analysis is based on two evaluation metrics that reward diversity, namely, α-normalised discounted cumulative gain (α-NDCG) [3], and intent-aware mean average precision (MAP-IA) [4], reported at two different cutoffs: 10 and 100. α-NDCG balances relevance and diversity through the tuning parameter α. The larger its value, the more diversity is rewarded. In the opposite end, when $\alpha = 0$, this metric is equivalent to the normal NDCG [21]. Following Wang and Zhu [8], we use $\alpha = 0.5$, in order to give equal weights to either of these dimensions.

Differently from other evaluation metrics in the literature, MAP-IA also takes into account how well a given document satisfies each aspect underlying the initial query, as well as the relative importance of each aspect, as given by a ground-truth importance distribution. In our evaluation, we devise two variants of MAP-IA. The first of these is a uniform variant, u-MAP-IA, which considers all aspects underlying a query as equally important, so as to provide a fair ground for the approaches that do not take the aspect importance into account. The second proposed variant, i-MAP-IA, estimates an ideal importance distribution over aspects as the ratio of relevant documents that cover each aspect when compared to all documents judged relevant for the initial query, as given by the provided ground-truth relevance assessments. Note, however, that although our framework can take the relative importance of different aspects into account, it relies on different estimation mechanisms, as proposed in Section 3.2.

4.4   Training Settings

To train the interpolation parameter of our framework, as well as the one used by the approach of Carbonell and Goldstein [5], we perform a 5-fold cross validation over the 20 topics, optimising for u-MAP-IA. The approaches of Radlinski and Dumais [9] and Agrawal et al. [4] do not require training under their simulated best-case scenario. As for our proposed clustering-based query expansion technique to uncover sub-queries from the baseline ranking, we apply the k-means algorithm on the top 1000 retrieved documents. In particular, we use $k = 20$ (the average number of sub-topics per considered topic), and extract the 10 most informative terms from each generated cluster as a sub-query.

5   Experimental Evaluation

In this section, we evaluate our framework with respect to the research questions stated in Section 4. Table 1 shows the performance of xQuAD and several baseline approaches in terms of α-NDCG, u-MAP-IA, and i-MAP-IA. In Table 1, MMR stands for the maximal marginal relevance method of Carbonell and Goldstein [5], whereas the simulated approaches of Agrawal et al. [4] and Radlinski and Dumais [9] using the official TREC Interactive track sub-topics as input are referred to as IA-Select and QFilter, respectively. It is worth noting that although IA-Select can also consider the relative importance of the different aspects underlying the initial query, our simulated version does not take this information into account, as it is not trivial to derive an analogy for their classification scheme in this case, without relying on relevance assessments. Nonetheless, to provide for a fairer comparison, we report the performance of xQuAD using the uniform aspect importance estimator given by Equation (1). This variant of our framework is denoted xQuAD$_U$. All approaches are applied over the top 1000 documents retrieved by the underlying baseline ranking. Significance with respect to this ranking is given by the Wilcoxon signed-rank test. In particular, the superscript symbols ▲ (▼) and △ (▽) denote a significant increase (decrease) at the $p < 0.01$ and $p < 0.05$ levels, respectively. A second such symbol (subscript) denotes significance with respect to the strongest among the considered baseline diversification approaches.

From Table 1, recalling our first and second research questions, we observe that xQuAD$_U$ markedly outperforms all other diversification approaches across all settings, except for the u-MAP-IA metric, when DPH is used as the baseline

Table 1. Comparative performance with a uniform aspect importance estimator

	α-NDCG		u-MAP-IA		i-MAP-IA	
	@10	@100	@10	@100	@10	@100
BM25	0.4505	0.5308	0.2286	0.1710	0.1416	0.1969
+MMR	0.4364	0.5102	0.2289	0.1700	0.1380	0.1841
+IA-Select	0.3392	0.4141	0.1592	0.1141	0.0868	0.1271
+QFilter	0.4509	0.5200	0.2300	0.1856	0.1416	0.1934
+xQuAD$_U$	**0.5727$^\blacktriangle_\blacktriangle$**	**0.6120$^\triangle_\blacktriangle$**	**0.2760**	**0.2240**	**0.1825**	**0.2235**
DPH	0.4633	0.5476	0.2464	0.1827	0.1620	0.2134
+MMR	0.4087$^\blacktriangledown$	0.4273$^\blacktriangledown$	**0.2876**	**0.2422**	0.1479	0.1805
+IA-Select	0.3585	0.4340	0.1765	0.1318	0.1029	0.1403
+QFilter	0.4634	0.5342	0.2466	0.1947	0.1620	0.2103
+xQuAD$_U$	**0.5935$^\blacktriangle_\blacktriangle$**	**0.6151$^\triangle_\triangle$**	0.2871	0.2371	**0.1998**	**0.2424**

ranking. This attests the effectiveness of our proposed framework when compared to both the implicit diversification performed by MMR and the explicit diversification provided by IA-Select and QFilter. Moreover, xQuAD$_U$ is the only approach to consistently improve over the baseline document rankings across all settings, with significant gains in terms of α-NDCG at both cutoffs. Indeed, all other approaches perform generally worse than the baseline rankings. In particular, the performance of IA-Select is disappointing, given its simulation with the ground-truth sub-topics. Nevertheless, these differences are not significant.

Next, we address our third research question, by assessing the impact of accounting for the relative importance of the different query aspects. Table 2 shows the performance of our framework using the different importance estimators introduced in Section 3.2. In particular, the subscript 'X' in xQuAD$_X$ reflects the use of the corresponding importance estimator $i_X(q_i, q)$, with $X \in \{U, N, R, C\}$ corresponding to Equations (1)-(4), respectively. Analogously to Table 1, a superscript symbol denotes statistical significance with respect to the baseline ranking, whereas a subscript symbol denotes significance with respect to xQuAD$_U$.

Table 2. Comparative performance using different aspect importance estimators

	α-NDCG		u-MAP-IA		i-MAP-IA	
	@10	@100	@10	@100	@10	@100
BM25	0.4505	0.5308	0.2286	0.1710	0.1416	0.1969
+xQuAD$_U$	**0.5727$^\blacktriangle$**	0.6120$^\triangle$	0.2760	0.2240	0.1825	0.2235
+xQuAD$_N$	0.4856$^\updownarrow$	0.5666$^\blacktriangle$	0.2484	0.1919	0.1597	0.2142
+xQuAD$_R$	0.4796	0.5716	0.2715	0.2132	0.1728	0.2274
+xQuAD$_C$	0.5204	**0.6238**	**0.3815$_\triangle$**	**0.2622**	**0.1871**	**0.2387**
DPH	0.4633	0.5476	0.2464	0.1827	0.1620	0.2134
+xQuAD$_U$	**0.5935$^\blacktriangle$**	**0.6151$^\triangle$**	0.2871	0.2371	**0.1998**	**0.2424**
+xQuAD$_N$	0.4878$_\blacktriangledown$	0.5281$_\triangledown$	0.2649	0.2171	0.1720$_\blacktriangledown$	0.2213
+xQuAD$_R$	0.4695$_\triangledown$	0.5664	0.2684	0.2131$_\triangledown$	0.1696	0.2240
+xQuAD$_C$	0.4894	0.5812	**0.3099**	**0.2409**	0.1708$_\triangledown$	0.2270

From Table 2, we first note that the variants of our framework improve over the baseline rankings in all but one case (DPH+xQuAD$_N$, α-NDCG@100). As for our stated research question, these results show that further improvements can be attained by taking into account the estimated relative importance of the different aspects underlying a query. In particular, the results using the importance estimators inspired by resource selection techniques are promising, notably for the xQuAD$_C$ variant, which outperforms the uniform estimation variant according to all but the α-NDCG@10 metric over BM25, with gains of up to 38% in terms of u-MAP-IA. The improvements, however, are less marked when the DPH baseline is considered, in which case the variant using the uniform aspect importance estimator is surprisingly the best one. This suggests that the performance of the different variants can be highly influenced by the performance of the baseline ranking. Indeed, as the same retrieval technique is used to estimate the relevance of the retrieved documents with respect to each different aspect, it can directly impact the estimation of the importance of this aspect.

Lastly, we address the question of whether sub-queries can be effectively generated from the baseline ranking itself. This can be particularly useful in cases when additional resources, such as a query log or a taxonomy of categories over queries and documents, are not available. As discussed in Section 3.1, we propose a clustering-based query expansion technique, in an attempt to uncover terms representative of different aspects underlying a query from a clustering of the top retrieved results for this query. Table 3 shows the results of xQuAD$_U$ using sub-queries generated by the DFR Bo1 query expansion model.

Table 3. Performance using sub-queries generated from the target collection

	α-NDCG		u-MAP-IA		i-MAP-IA	
	@10	@100	@10	@100	@10	@100
BM25	0.4505	**0.5308**	0.2286	0.1710	0.1416	**0.1969**
+xQuAD$_{U(Bo1)}$	**0.4509**	0.5193$^{\blacktriangledown}$	**0.2300**	**0.1742**	0.1416	0.1919$^{\blacktriangledown}$
DPH	0.4633	**0.5476**	0.2464	0.1827	0.1620	**0.2134**
+xQuAD$_{U(Bo1)}$	**0.4634**	0.5226$^{\blacktriangledown}$	**0.2466**	**0.1906**$^{\triangle}$	0.1620	0.2084$^{\blacktriangledown}$

From Table 3, as expected, we first observe that the obtained performances are much lower than those observed for the variants of our framework using the ground-truth sub-topics, as shown in Table 2. Nevertheless, they are comparable to the performances attained by our competing diversification approaches using the ground-truth sub-topics, as shown in Table 1. This suggests that investigating alternative methods for sub-query generation is a promising direction for further enhancing the performance of our framework.

6   Conclusions and Future Work

In this paper, we have proposed a novel framework for search result diversification. Given an initial set of documents retrieved for a query, the *eXplicit*

Interpreting User Inactivity on Search Results

Sofia Stamou[1] and Efthimis N. Efthimiadis[2]

[1] Computer Engineering and Informatics Department, Patras University
Patras, 26500 Greece, and
Department of Archives and Library Science, Ionian University
Corfu, 49100 Greece
stamou@ceid.upatras.gr
[2] Information School, University of Washington
Seattle, WA, USA
efthimis@u.washington.edu

Abstract. The lack of user activity on search results was until recently perceived as a sign of user dissatisfaction from retrieval performance, often, referring to such inactivity as a failed search (negative search abandonment). However, recent studies suggest that some search tasks can be achieved in the contents of the results displayed without the need to click through them (positive search abandonment); thus they emphasize the need to discriminate between successful and failed searches without follow-up clicks. In this paper, we study users' inactivity on search results in relation to their pursued search goals and investigate the impact of displayed results on user clicking decisions. Our study examines two types of post-query user inactivity: *pre-determined* and *post-determined* depending on whether the user started searching with a preset intention to look for answers only within the result snippets and did not intend to click through the results, or the user inactivity was decided after the user had reviewed the list of retrieved documents. Our findings indicate that 27% of web searches in our sample are conducted with a pre-determined intention to look for answers in the results' list and 75% of them can be satisfied in the contents of the displayed results. Moreover, in nearly half the queries that did not yield result visits, the desired information is found in the result snippets.

Keywords: Task-oriented search, queries without clickthrough, search abandonment, user study, interactive IR.

1 Introduction

Search logs are a convenient tool for capturing some aspects of the user interaction with retrieved results. Today, the study of query logs is a popular approach for identifying user querying trends and search patterns as well as for attempting to infer the search goals users are trying to accomplish via their queries [1, 7, 10, 11, 15, 12, 21, 22]. Although search logs provide valuable evidence about how and what users search for, they are less revealing about the criteria upon which information seekers base their search behavioral patterns. To fill this void, researchers have carried out user studies in which searchers explicitly indicate their satisfaction from retrieval performance in relation to

C. Gurrin et al. (Eds.): ECIR 2010, LNCS 5993, pp. 100–113, 2010.
© Springer-Verlag Berlin Heidelberg 2010

their underlying search goals. The commonality in these approaches is that they evaluate search tasks and retrieval effectiveness based entirely on the analysis of the user activity (i.e., clicks) on search results. This is because clickthrough data has been traditionally perceived by the information retrieval community as an indicator of implicit user feedback on the relevance of search results [7]. In this context, the users' inactivity on search results has been generally interpreted as a sign of dissatisfaction from retrieval performance [19].

However, researchers [20, 23] have recently reported that some queries might not be followed by user clicks on search results, simply because the user information needs were successfully addressed in the content of the snippets displayed in the search results page. Therefore, for those queries the lack of user activity on the returned documents should not be interpreted as a sign of decreased user satisfaction from retrieval performance. Despite the acknowledgment that some queries do not yield user visits on search results because the desired information is presented in the snippets (abstracts) of the retrieved documents [17], currently very few studies exist that investigate searches not followed by user visits on the results in relation to users' tasks [20, 12, 23]. To our knowledge, none of the reported research examines the impact of retrieved but unvisited results on both the user clicking decisions and the satisfaction of their information needs.

In this paper, we present a user study that investigates the intentions behind queries without clickthrough and examines the search tasks that can be successfully accomplished based entirely on the information provided on the results' page. Unlike previous work on abandoned searches, i.e., queries not followed by any click or any further query within a 24-hour period [20], our study investigates all queries without clickthrough, that is, both those abandoned and those followed by another search. This is because we seek to understand why users choose not to visit a single page for some of their queries rather than investigate why they abandon their searches altogether. Such understanding will help us interpret the user inactivity on retrieved results in a more precise manner so as to propose more user-centric retrieval evaluation methods. Moreover, we rely on explicit user feedback in order to interpret the user inactivity on search results in a precise manner.

To accomplish our study objective, we examine two types of user inactivity on search results: *pre-determined* and *post-determined* inactivity. We comparatively analyze the search tasks they pursue in order to understand why users decide not to visit results and how their decisions are influenced by the displayed result snippets. We define the user inactivity on the results as pre-determined when the user submits a query with a preset intention to look for answers in the result snippets and without following any link. On the other hand, the user inactivity is post-determined when the user issues a query with the intention to visit some results, but after reviewing the list of retrieved documents decides not to visit any of the results.

The paper is organized as follows. In section 2 we present a discussion of related work. Section 3 describes in detail the methodology of the user study. In section 4 we discuss the main findings, and in section 5 we present conclusions and plans for future work.

2 Related Work

The idea of utilizing the searcher activity on the returned results as an indicator of implicit relevance judgments is not new. Numerous studies exist on how the different post-query activities can be interpreted as implicit feedback signals (for an overview see [15]). The searchers' behavior that researchers attribute as implicit measures of interest are: time spent on a page combined with the amount of scrolling on a page [4], duration of search and number of result sets returned [7], click data on and beyond the search results [14], use of eye-tracking methods to capture the user's visual attention on the results [8], repetition of result clicks across user sessions [25]. Although, the above measures have been proposed for inferring the user satisfaction from the results visited for some query, they have not been explored for capturing the user satisfaction from the results viewed but not visited for a query. Based on the observation that what users did not click on might reveal interesting information about their perception of the results' usefulness in satisfying their search needs [13], a number of researchers studied the search tasks that can be achieved by viewing the result snippets and without clicking through the results [6, 20, 23]. The findings of those studies indicate that query abandonment can be *good* when the users find what they look for in the results' list, without requiring a click on the result contents.

In this paper, we build upon previous studies about the interpretation of searches without follow up clicks and we investigate via a user survey the intentions hidden behind queries lacking clickthrough, as well as the result snippets' impact on the user decisions not to visit retrieved results. In particular, we concentrate on queries that yield search results but none of which are accessed by the user who issued the query and investigate whether the user started searching with a pre-determined intention to visit results or not. Our investigation aims at capturing the result snippets' impact on the user post-query activities. Unlike previous work, that restricted their findings to abandoned searches [20], or examined the type of information needs that can be satisfied in the displayed snippets [6], our study relies on actual user feedback for identifying the search tasks that can be achieved without visiting search results. Moreover, we examine the results snippets' impact on the nature of post-query user activity instead of merely looking at the search tasks that can be fulfilled in the contents of the displayed snippets.

3 Methodology

The goals that lead people to engage in information seeking behavior affect their judgments of usefulness of the retrieved results [2]. This, coupled with the observation that up to 50% of the queries do not yield a single click on the results [5, pp.27], motivated our study on how to interpret searches not followed by user visits on the retrieved results. To that end, we carried out a user study in order to firstly identify the search tasks associated with queries lacking clickthrough and then examine for which tasks the user inactivity on results is pre-determined and for which tasks the user inactivity is post-determined (i.e., imposed) by the contents of the displayed results. Based on our findings and the feedback supplied by our participants, we try to capture the impact that the displayed results might have on both the post-query user behavior and the user satisfaction from search results.

3.1 The User Study

In this section, we discuss the details of the user study we carried out in order to understand why information seekers decide not to visit any result for some of their queries. Unlike prior work that attempts to interpret searches without clicks using query log analysis, we rely on explicit user feedback in order to analyze the lack of post-query user interaction with search results.

The study recruited six postgraduate computer science students (four male/ two female). Participants attended a training session where the study and its processes and procedures were explained and any questions were addressed. A browser plug-in was installed at each participant's workstation in order to collect their search trace for a time period of one week. In particular, we asked participants to conduct their web searches as they would normally do and we explained to them that we would record their HTTP requests from which we could obtain their queries and subsequent clickthrough. For every query recorded in our test set, we asked our participants to answer a set of questions (see Table 1) that we presented them via an online questionnaire. Before conducting our survey we familiarized our participants with the questions by giving them verbal explanations for every question and its candidate answers. The questions on the online questionnaire were presented incrementally to each participant and not all of them at once. We instructed our participants to open the questionnaire in a new browser window while conducting their searches and answer the first two questions before issuing their queries and the next two questions after their search was completed.

3.2 Questionnaire Design

The questions and the selected choices for answers were pilot-tested and refined prior to data collection. As Table 1 shows, the first question aims at capturing the participants' search goals and lists a number of tasks that web queries may pursue. The determination of the tasks presented to our participants relies on existing query classification schemes that have been proposed by many researches [3, 11, 18, 20, 23, 25]. In particular, the search tasks examined pertain to the following categories of queries: (i) informational (answer a in question #1), (ii) transactional (answer b in question #1), (iii) navigational (answer c in question #1), (iv) local (answer d in question #1), (v) person (answer e in question #1), (vi) product (answer f in question #1), (vii) quick answer (answer g in question #1), (viii) language-related (answer h in question #1), (ix) update (to answer i in question #1) and (x) repeat (answer j in question #1). Based on the above categories, we asked our participants to indicate for each of their searches the tasks that better reflected their information intentions. Note that we asked them to indicate their selections before actually submitting their queries and for every query to make a single selection depending on the task that they deemed the most suitable in describing their underlying intentions.

Following the selection of a search task, the second question appeared on the screen, which again had to be answered before the submission of the query. Question #2 aims at capturing the pre-query user intention of visiting or not search results. Again, our participants had to indicate a single pre-query intention for each of their searches by selecting an appropriate answer from the three provided. In particular, answer *yes* to question #2 indicates that the user initiated a search with a pre-determined intention not to click through results, answer *no* indicates that the user started searching with a pre-determined intention to click through results and answer *maybe* indicates that user started searching with an unclear intention about clicking through results.

After answering the first two questions, we asked our participants to conduct their searches, i.e. issue their queries and interact with search results as they preferred. When they completed their searches and before proceeding with the submission of another query or the termination of search, we asked them to answer question #3, which captures the post-query user activity on search results. In particular, question #3 provides a *yes/no* answer and records whether participants visited or not some result for each of their queries. In case participants indicated the answer *no* to question #3 a fourth question appeared on the screen otherwise the survey for the recorded search was terminated.

Table 1. Survey Questionnaire

1. What is the task of your search?
a. Find detailed information about a subject of interest
b. Obtain/download/ interact with a resource (e.g. game, song, greeting card)
c. Find a specific URL/ web site
d. Find local information (e.g. transportation timetable, movie showtimes)
e. Look for a person (including myself and fiction characters)
f. Find out about a product (e.g. price, features, retailers, discounts)
g. Get a quick answer to my query (e.g. currency exchange rates, sport scores, stock quote)
h. Look for linguistic information about my query terms (e.g. spelling, translation, definition)
i. See if there is any new result retrieved since the last time I issued the query
j. Re-find the information I got in one of my previous submissions of the query
2. Did you intend to visit some result(s) before issuing the query?
Yes
No
Maybe
3. Did you actually visit some result(s)?
Yes
No
4. What was the reason for not visiting any result?
a. I found what I was looking for in the result page
b. Results' seemed irrelevant
c. I had already seen these results for the query
d. Search was interrupted

Question #4 aims at collecting feedback on the reasons why searchers did not visit any result for their corresponding queries, i.e. those not followed by clickthrough events. To capture the reasons for not clicking, we provided our participants with four possible answers. The first reason for not clicking (answer *a* to question#4) suggests that the searcher found what she was looking for in the search results' page, therefore there was no need to click through the results. The second reason for not visiting results (answer *b*) suggests that the searcher did not receive the information she hoped for in the results' list, therefore she assumed that the retrieved results could not satisfy her search need and as such decided not to visit any of them. The third reason for not clicking (i.e. answer *c*) assumes that the searcher wanted to obtain new (i.e. unseen) information for her query, but none of the retrieved results was new. Therefore, even if results seemed relevant the searcher decided not to visit any of them, since she had already seen them and the quest for her search was to find new information. Finally the fourth reason for not clicking (answer *d*) suggests that for some unexpected reason (i.e. network connectivity problems, participant's need to take an urgent break) search was interrupted right after the submission of the query and the retrieval of query results.

3.3 Session Segmentation and Data De-identification

Having collected our participants' search traces and feedback, we firstly anonymized our data by replacing the user workstation IPs with random IDs and then we grouped the recorded queries and post-query activities into individual search sessions. To identify the distinct search sessions, we worked as follows. The first session is initiated the first time the participant issues a search request and records all her web transactions for as long as she demonstrates some search activity. We consider that a participant remains active in a search either when she views search results (i.e., she scrolls down the list of retrieved pages or moves to the 'next' page of results), or when she accesses the retrieved documents (i.e. clicks on/opens them), or when she continues to submit queries. Under this approach, we deem that a session expires if both of the following criteria apply: (i) the period of user inactivity in a search gets longer than 10 minutes, i.e., the participant does not perform any of the above tasks within a time interval of 10 minutes, and (ii) the participant's activity that is recorded right after an idle time of 10 minutes is not a click on a retrieved result for the same query. Both criteria are suggested by [16]. Upon session expiration, a new session starts. After applying the above criteria to our collected search trace, we ended up with a total number of 966 queries that span to 239 search sessions.

4 Results

4.1 Query Statistics

Having identified the individual search sessions of every participant and the set of queries issued in each session, the next step is to measure the fraction of queries for which participants did not demonstrate any activity on the search results. In this respect, we relied on the recorded post-query user activity and we identified the queries that were not followed by result clicks. Table 2 summarizes the statistics of our experimental data.

Table 2. Statistics on the experimental dataset

Collection period	1 week
# of unique user ID's	6
# of queries	966
# of sessions	239
# of queries without clickthrough	261
# of queries with clickthrough	705
# of clickthrough events	1,762
avg. # of sessions/user	39.83
avg. # of queries/session	4.04
avg. # of clicks/query with clickthrough	2.49

According to our results, in 27.02% (261) of the queries participants did not click on any retrieved result, while in 72.98% (705) of the queries, participants visited at least one retrieved document. Having collected and processed the experimental data, we proceed with the analysis of the feedback our participants supplied for each of their reported queries. Such analysis will assist the identification of search goals that are associated with queries without clicks. Furthermore, it will help determine for which search tasks the user inactivity on search results is pre-determined, and for which tasks the user post-query inactivity is imposed by the displayed result snippets. The answers to the above issues, which we present and discuss next, will help model the user search behavior and develop user-centric retrieval evaluation models.

4.2   Classification of Queries by Search Task

The distribution of the 966 queries by search tasks is given in Table 3 and generally conforms to the findings of previous work on what information seekers search for [11, 20, 25].

Table 3. Distribution of queries with and without clickthrough across search tasks

Search Goals	Queries without clickthrough	261	Queries with clickthrough	705
Informational		2.29%		33.47%
Transactional		3.84%		13.76%
Navigational		3.06%		16.74%
Local		21.45%		6.95%
Person		8.05%		8.65%
Product		6.51%		9.08%
Language-related		13.03%		1.71%
Update		9.97%		0.56%
Repeat		4.21%		8.55%
Quick answer		27.59%		0.44%

Our findings indicate that for a significant fraction (i.e., 27.59%) of the searches not followed by clicks on the results, searchers aim at finding a quick answer in the result snippets, while for 33.47% of the searches that yield visits to the retrieved documents, searches aim at finding information in the contents of the retrieved documents. Next, we examine the intended user activity on search results. This will improve our understanding on whether the displayed results affected in any way the user decisions of clicking or not clicking on the retrieved results.

4.3   Pre-query User Intentions across Search Tasks

Based on the participants' feedback, we grouped our test queries into three categories depending on whether participants issued a query with an initial intention to visit the search results, not to visit the results, or there was an unclear intention. In this respect, we relied on the answers our participants gave to question #2 and we classified the queries to the following categories: (i) searches of intended inactivity, i.e., queries that start with the intention to look for answers in the result snippets, (ii) searches of intended activity, i.e., queries that get started with the intention to look for answers in the retrieved documents, and (iii) searches with an unclear intention about where to look for answers. The results of the pre-query user intentions grouped by category are shown in Table 4.

Table 4. Distribution of pre-query user intentions across search tasks

Search Goals	Queries of intended inactivity	Queries of intended activity	Queries of unclear intention
	Total: 263	Total: 499	Total: 204
Informational	2.28%	39.08%	20.10%
Transactional	1.52%	12.43%	20.10%
Navigational	3.04%	17.64%	14.70%
Local	19.39%	8.81%	4.91%
Person	17.11%	3.81%	8.82%
Product	6.47%	4.00%	21.57%
Language-related	11.41%	2.00%	2.95%
Update	9.51%	0.20%	1.96%
Repeat	3.42%	11.83%	1.95%
Quick answer	25.85%	0.20%	2.94%

Of the 966 test queries, 499 (51.65%) were submitted with a pre-determined intention to look for answers in the contents of the retrieved documents. On the other hand, 263 (27.23%) of the queries were submitted with a pre-determined intention to seek for answers in the results' list, whereas 204 (21.12%) of the queries have no pre-determined intention about where to look for the desired information.

Moreover, Table 4 illustrates the distribution of our test queries across search tasks. According to our results most searches that seek the desired information in the retrieved result snippets pertain to the following search tasks: *quick answer, local, person* and *language-related* searches in that order. This practically indicates that users engaging in the above search tasks have most of the times a pre-determined intention to look for answers in the contents of the result lists and not to visit search results. On the other hand, most queries that look for information in the contents of the returned documents adhere to *informational, navigational, transactional* and *repeat* search tasks. This suggests that most of the time, users who engage in the above tasks intend to visit some of the retrieved results for satisfying their information needs. Finally, the majority of the searches for which users do not have a pre-determined intention about where to look for the desired information pertain to *product, transactional* and *informational* tasks. Results so far demonstrate that a considerable amount of searches are conducted with the intention to be satisfied in the contents of the result snippets. For those searches, we emphasize the need to evaluate retrieval effectiveness and user satisfaction from retrieval performance against the results the user views for a query. This is also shown in the work of [17, 20, 23], who suggest that the lack of click-through on search results should not be interpreted as a sign of decreased user satisfaction from retrieval performance.

4.4   Post-query User Activity across Search Tasks

Another aspect of the human search behavior we examined is the correlation between the initial user intention and the demonstrated user activity on search results. This is in order to assess whether the user pre-query intentions are maintained in the post-query activity or rather they are influenced based on what the user sees in the list of search results. For our assessment, we cross-examined the answers our participants gave to questions #2 and #3. Note that question #2 captures the pre-query participant intentions, while question #3 captures the demonstrated post-query participant activity.

Table 5 presents the fraction of queries across search tasks for which participants maintained initial intentions and the fraction of searches across tasks for which participants altered their initial intentions. Note that we have eliminated from our estimations queries of uncertain initial intentions (i.e. those for which our subjects selected the answer maybe to question #2). This is because for those queries participants lacked a specific initial intention about where to look for the desired information; therefore there is no evidence to judge whether their corresponding participant activity was intended or not. Under the above, the results reported in Table 5 concern the 762 queries submitted with a clear initial intention (i.e. 499 queries of intended activity and 263 queries of intended inactivity).

Table 5. Distribution of queries for which participants maintained or altered their initial intentions across search tasks

Search Goals	Queries for which subjects maintained initial intention	557	Queries for which subjects altered initial intention	205
	Distribution to search goals		Distribution to search goals	
Informational	33.93%		5.86%	
Transactional	10.41%		3.90%	
Navigational	14.36%		7.81%	
Local	5.56%		31.22%	
Person	3.60%		21.46%	
Product	3.41%		8.78%	
Language-related	4.31%		7.80%	
Update	3.77%		2.44%	
Repeat	8.98%		8.78%	
Quick answer	11.67%		1.95%	

Results indicate that from the 762 queries in 73.10% (557) of the examined searches participants maintained their pre-query intentions, whereas in the remaining 26.90% (205) of the examined searches participants altered their pre-query intentions. A possible interpretation for the discrepancies between the participants' pre-query intentions and post-query activity might be that the information displayed in the results' list (e.g. result titles and/or snippets) influenced the participant decisions of visiting or not the search results. This might also apply to searches conducted under no pre-determined intention to visit or not search searches, which according to our findings represent 21.12% of our test queries.

Given that the objective of our study is to understand the causes of user inactivity on search results, in the remainder of our analysis we concentrate on queries that were not followed by user clickthrough and investigate the impact of result snippets in satisfying their corresponding search goals. In this respect, we estimate the fraction of searches not followed by user activity on search results that were successfully addressed in the contents of the result snippets.

4.5   User Satisfaction from Searches without Clickthrough

To capture the satisfaction of the participants' goals pursued via searches characterized by intentional absence of clickthrough events, we relied on the answers our participants gave to question #4 (what was the reason for not visiting any result?) for their pre-determined inactive queries (i.e., the queries for which participants maintained their initial intention of not visiting search results). Table 6 reports the user perception from retrieval performance for searches that intentionally lacked user activity on the retrieved results.

Results indicate that users engaging in *informational*, *navigational*, *transactional* and *repeat* searches did not satisfy their information needs in the contents of the result snippets. Therefore, for such types of searches, the absence of clickthrough on search results is a sign of user dissatisfaction from retrieval performance.

On the other hand, *language-related* searches conducted with a pre-determined intention to look for answers in the result snippets can be successfully accomplished without the need to click through the results. Another interesting observation is that although for *update* searches people seek the desired information in the contents of the results' list, only a small fraction of them (i.e., 28.57%) are successfully addressed in the result snippets. This might suggest that in order to better serve *update* queries search engines could maintain a profile for searchers or local cached copies that links their queries to viewed results. The displayed results from a new search could then display both previously seen and new results and highlight the already seen results for easier inspection by the searcher.

Table 6. Post retrieval evaluation of results with respect to participants' pre-determined decision to not click

Distribution to search goals	Queries of pre-determined inactivity (with no clicks)	% of satisfied goals
Informational	0	—
Transactional	0	—
Navigational	0	—
Local	19	57.89%
Person	10	50.00%
Product	8	75.00%
Language-related	22	100.00%
Update	21	28.57%
Repeat	0	—
Quick answer	65	89.23%
Total: 145		**Avg: 74.48%**

post-determined searches and the distribution of queries without clickthrough in a search session in order to build models that can interpret the user search activity and inactivity on retrieved results. Another area for future research is how to explore the findings of our study in order to improve the search engine's effectiveness on serving diverse user needs. The development of predictive models of user activity could be used to implement search aids that would assist users in completing their search tasks successfully. Finally, we hope that this study would contribute towards the design of retrieval evaluation frameworks where search is seen holistically and incorporate multiple features for measuring user satisfaction from retrieval performance.

References

1. Agichtein, E., Brill, E., Dumais, S., Rango, R.: Learning user interaction models for predicting search result preferences. In: Proceedings of the 29th ACM SIGIR Conference (2006)
2. Belkin, N.: Some(what) grand challenges for information retrieval. ACM SIGIR Forum 42(1), 47–54 (2008)
3. Broder, A.: A taxonomy of web search. SIGIR Forum 36(2), 3–10 (2002)
4. Claypool, M., Le, P., Waseda, M., Brown, D.: Implicit interest indicators. In: Proceedings of the International Conference on Intelligent User Interfaces, pp. 33–40 (2001)
5. Callan, J., Allan, J., Clarke, C.L.A., Dumais, S., Evans, D.A., Sanderson, M., Zhai, C.: Meeting of the MINDS: an information retrieval research agenda. ACM SIGIR Forum 41(2), 25–34 (2007)
6. Cutrell, E., Guan, Z.: What are you looking for? an eye-tracking study of information usage in web search. In: Proceedings of the SIGCHI Conference on Human Factors in Computing Systems, pp. 407–416 (2007)
7. Fox, S., Karnawat, K., Mydland, M., Dumais, S., White, T.: Evaluating implicit measures to improve web search. ACM Transactions on Information Systems 23(2), 147–168 (2005)
8. Granka, L.A., Joachims, T., Gay, G.: Eye-tracking analysis of user behaviour in www results. In: Proceedings of the ACM SIGIR Conference, pp. 478–479 (2004)
9. Huang, J., Efthimiadis, E.N.: Analyzing and Evaluating Query Reformulation Strategies in Web Search Logs. In: Proceedings of the 18th ACM Conference on Information and Knowledge Management (CIKM), Hong Kong, November 2-6, pp. 77–86 (2009)
10. Jansen, B.J., Spink, A.: How are we searching the www: a comparison of nine search engine transaction logs. Information Processing & Management 42(1), 248–263 (2006)
11. Jansen, B.J., Booth, D.L., Spink, A.: Determining the informational, navigational and transactional intent of web queries. Information Processing & Management 44, 1251–1266 (2008)
12. Joachims, T., Granka, L., Pan, B., Hembrooke, H., Padlinski, F., Gay, G.: Evaluating the accuracy of implicit feedback from clicks and query reformulations in web search. ACM Transactions on Information Systems 25(2), 1–26 (2007)
13. Joachims, T., Radlinski, F.: Search engines that learn from implicit feedback. Computer 40, 34–40 (2007)
14. Jung, S., Herlocker, J.L., Webster, J.: Click data as implicit relevance feedback in web search. Information Processing & Management 43(3), 791–807 (2007)
15. Kelly, D., Teevan, J.: Implicit feedback for inferring user preference: a bibliography. ACM SIGIR Forum 37(2), 18–28 (2003)

16. Qiu, F., Liu, Z., Cho, J.: Analysis of user web traffic with a focus on search activities. In: Proceedings of the International Workshop on the Web and Databases, WebDB (2005)
17. Radlinski, F., Kurup, M., Joachims, T.: How does clickthrough data reflect retrieval quality. In: Proceedings of the CIKM Conference (2008)
18. Rose, D.E., Levinson, D.: Understanding user goals in web search. In: Proceedings of the 13th International Conference on World Wide Web (WWW 2004), pp. 13–19. ACM Press, New York (2004)
19. Sarma, A., Gollapudi, S., Ieong, S.: Bypass rates: reducing query abandonment using negative inferences. In: Proceedings of the 14th ACM SIGKDD International Conference on Knowledge Discovery and Data Mining, pp. 177–185 (2008)
20. Scott, J.L., Huffman, B., Tokuda, A.: Good abandonment in mobile and pc internet search. In: Proceedings of the 32nd Annual ACM SIGIR Conference, Boston, MA, pp. 43–50 (2009)
21. Sharma, H., Jansen, B.J.: Automated evaluation of search engine performance via implicit user feedback. In: Proceedings of the 28th ACM SIGIR Conference, pp. 649–650 (2005)
22. Spink, A.: A user centered approach to evaluating human interaction with web search engines: an exploratory study. Information Processing & Management 38(3), 401–426 (2002)
23. Stamou, S., Efthimiadis, E.N.: Queries without clicks: successful or failed searches? In: Proceedings of the SIGIR Workshop on the Future of Information Retrieval Evaluation, Boston, MA, USA (2009)
24. Taksa, I., Spink, A., Goldberg, R.: A task-oriented approach to search engine usability studies. Journal of Software 3(1), 63–73 (2008)
25. Teevan, J., Adar, E., Jones, R., Potts, M.: Information re-retrieval: repeat queries in Yahoo's logs. In: Proceedings of the 30th ACM SIGIR Conference (2007)

Learning to Select a Ranking Function

Jie Peng, Craig Macdonald, and Iadh Ounis

Department of Computing Science,
University of Glasgow, G12 8QQ, UK
{pj,craigm,ounis}@dcs.gla.ac.uk

Abstract. Learning To Rank (LTR) techniques aim to learn an effective document ranking function by combining several document features. While the function learned may be uniformly applied to all queries, many studies have shown that different ranking functions favour different queries, and the retrieval performance can be significantly enhanced if an appropriate ranking function is selected for each individual query. In this paper, we propose a novel Learning To Select framework that selectively applies an appropriate ranking function on a per-query basis. The approach employs a query feature to identify similar training queries for an unseen query. The ranking function which performs the best on this identified training query set is then chosen for the unseen query. In particular, we propose the use of divergence, which measures the extent that a document ranking function alters the scores of an initial ranking of documents for a given query, as a query feature. We evaluate our method using tasks from the TREC Web and Million Query tracks, in combination with the LETOR 3.0 and LETOR 4.0 feature sets. Our experimental results show that our proposed method is effective and robust for selecting an appropriate ranking function on a per-query basis. In particular, it always outperforms three state-of-the-art LTR techniques, namely Ranking SVM, AdaRank, and the automatic feature selection method.

1   Introduction

The effective ranking of documents in search engines is based on various document features, such as the frequency of query terms in each document, the length of each document, or link analysis. In order to obtain a better retrieval performance, instead of using a single or a few features, there is a growing trend to create a ranking function by applying a Learning To Rank (LTR) technique on a large set of features [1,2,3].

Typically, a LTR technique learns a ranking function by assigning a weight to each document feature, then uses this obtained ranking function to estimate the relevance scores for each document [1,4]. In recent years, many effective LTR techniques have been proposed to build such ranking functions, such as AdaRank [1], Ranking SVM [5,6] or the Automatic Feature Selection (AFS) method [2]. Most of the current LTR literature mainly focuses on how to efficiently and effectively learn such ranking functions, but simply equally applies

C. Gurrin et al. (Eds.): ECIR 2010, LNCS 5993, pp. 114–126, 2010.

the learned ranking function to all queries. However, many studies have shown that different queries benefit differently from each ranking function [7,8,9,10,11] and the retrieval performance can be significantly enhanced if an appropriate ranking function is used for each individual query.

In this paper, we propose the Learning To Select (LTS) framework for selectively applying an appropriate ranking function on a per-query basis. We believe that the effectiveness of a ranking function for an unseen query can be estimated based on similar training queries. A divergence measure can be used to determine the extent that a document ranking function alters the scores of an initial ranking of documents for a given query. We propose that this divergence can be used to identify similar training queries. In this case, a ranking function which performs well for training queries that have a similar divergence to the unseen query, will also perform well on the unseen query.

We conduct a comprehensive experimental investigation using query and relevance assessment sets from the TREC Web and Million Query tracks, in combination with the LETOR 3.0 and LETOR 4.0 feature sets [4]. Moreover, we use three state-of-the-art LTR techniques as our baselines, namely Ranking SVM, AdaRank, and the AFS method.

There are four key contributions from this work. First, we propose the novel LTS framework for selecting an appropriate ranking function on a per-query basis. This approach estimates the effectiveness of a ranking function for an unseen query based on the retrieval performance of this ranking function on already seen neighbour queries. Second, we propose the use of divergence, which measures the extent that a document ranking function alters the scores of an initial ranking of documents for a given query, as a query feature for identifying neighbour queries. Third, we show the effectiveness of our approach by comparing it to three state-of-the-art LTR techniques. Fourth, we show the robustness of our approach by selecting an appropriate ranking function from a large candidate set, which is created by using a LTR technique on different feature sets.

The remainder of this paper is organised as follows. Section 2 introduces the motivation of this paper. In Section 3, we describe several selective approaches presented in the literature. Section 4 describes our proposed LTS framework, which is used to select an appropriate ranking function for a given query. We present the experimental setup in Section 5, and analyse the experimental results in Section 6. Finally, we draw conclusions in Section 7.

2   Motivation

In this section, we provide an illustrative example showing the importance of the selective application of a ranking function. Table 1 shows the retrieval performance of three LTR techniques, namely Ranking SVM, AdaRank and the AFS method, on four different datasets[1]. Moreover, the upper bounds (denoted MAX) are achieved by manually selecting the most effective ranking function on a per-query basis. From this table, it is clear that the retrieval performance

---

[1] The detailed settings can be found in Section 5.

Table 1. MAX is the upper bound (100% correct per-query application). The highest score in each column is highlighted in bold and scores that are statistically better than $RankingSVM$, $AdaRank$, and AFS are marked with $\star$, $*$, and $\dagger$, respectively (Wilcoxon matched-pairs signed-ranks test, $p < 0.05$).

	MAP			
	TREC2003	TREC2004	TREC2007	TREC2008
Ranking SVM	0.5366	0.4193	0.4641	0.4752
AdaRank	0.5977	0.5062	0.4537	0.4766
AFS	0.6145	0.5079	0.4596	0.4784
MAX	**0.6933** $\star * \dagger$	**0.5744** $\star * \dagger$	**0.5057** $\star * \dagger$	**0.5226** $\star * \dagger$

can be significantly enhanced if we apply the most appropriate ranking function for each query. This observation suggests that different ranking functions do favour different queries and that the appropriate selective application of a ranking function could enhance the retrieval performance.

3   Related Works

Some selective application techniques have been previously proposed in Information Retrieval IR [3,8,9,10,11,14,15]. For example, in [3], Geng et al. proposed a query-dependent ranking approach. For each given query, they employ a specific ranking function, which is obtained by applying a LTR technique (e.g. Ranking SVM) on a training query set. This training query set is dependent on the given query, which can be identified by using a classification technique (K-nearest neighbour (KNN)) based on a query feature. The query feature used in their work is the mean of the document feature scores (e.g. $tf \cdot idf$) of the top retrieved documents, which can be obtained by a reference model (e.g. BM25), given as follows:

$$score(q, r_i) = \frac{\sum_{\tau=1}^{n} rel(d_\tau)}{n} \tag{1}$$

where n is the number of the top retrieved documents returned by ranking function r_i for a given query q. $rel(d_\tau)$ is the document relevance score of a document d at position τ of the document ranking list.

Extensive experiments on a large dataset, which was sampled from a commercial search engine, showed the effectiveness of the aforementioned approach [3]. However, [3] only investigated the selective application of a ranking function obtained from a single LTR technique and using a fixed set of document features. Hence, the effectiveness of the query-dependent ranking approach is not clear when there is more than one LTR technique and the number of features is varied. Moreover, the use of the mean of the feature scores from the top retrieved documents simply ignores the importance of the distribution of the feature scores, which has been effectively used in many retrieval applications [11,12]. For example, Manmatha et al. [12] use the relevance score distribution to estimate the effectiveness of a search engine.

In [8], a query performance predictor was used to decide whether to apply collection enrichment for a given query. Generally speaking, query performance prediction relies on the statistics of the collection for a given query, such as query term frequency in the collection and the number of documents containing the query term. Hence, query performance predictors may not be applicable to the selective application of ranking functions, as these statistics are invariant to changes in the ranking function.

Plachouras et al. [9,10] proposed a method to selectively apply an appropriate retrieval approach for a given query, which is based on a Bayesian decision mechanism. Features such as the link patterns in the retrieved document set and the occurrence of query terms in the documents were used to determine the applicability of the retrieval approaches. This method was shown to be effective when there were only two candidate retrieval approaches. However, the retrieval performance obtained using this method only improved slightly and actually decreased when more than two candidate retrieval approaches were used.

Peng et al. [11] select a single query-independent feature for a given query. Such a query-independent feature could be, for example, HostRank [13], PageRank or document length. However, current IR systems usually apply a large set of features in order to achieve a high retrieval performance, and it is not clear how to select multiple document features using this approach. Other approaches [14,15] attempt to predict the type of the query (e.g. known-item search query, information seeking query), and from this, apply different retrieval approaches. However, the accuracy of state-of-the-art query type prediction is not high [16]. Moreover, queries of the same type may benefit from having different retrieval approaches applied [11].

In order to selectively apply an appropriate ranking functions from a large set of candidate ranking functions, in this paper, we propose the LTS framework, which will be presented in the following section. The proposed method is agnostic to the number of ranking functions, as well as to the type of the queries.

4   The Learning to Select Framework

In this section, we present a novel LTS framework for selectively applying an appropriate ranking function on a per-query basis. We first introduce the general idea of this framework, then provide a more detailed algorithm.

4.1   General Idea

A document ranking function created by a LTR technique is based on the assumption that the training dataset is representative of unseen queries. However, some queries may benefit from applying different ranking functions. We believe that the effectiveness of a ranking function for an unseen query can be estimated based on similar training queries. A divergence measure can be used to determine the extent that a document ranking function alters the scores of an initial ranking of documents. We propose that this divergence can be used to identify similar training queries. In this case, a ranking function which performs well for training queries that have a similar divergence to the unseen query, will also perform well on the unseen query.

4.2   Algorithm

Divergence Estimation. In this work, the ranking function which is used to obtain the initial ranking of documents for a given query is called the *base ranking function* r_b. Other ranking functions which may be applied are called *candidate ranking functions* r_i. They assign different document relevance scores to the same documents as were retrieved by r_b.

There are several different ways to estimate the divergence between two document score distributions that are obtained by using a base ranking function and a candidate ranking function. Two commonly used divergence measures are studied in this work, namely Kullback-Leibler (KL) [17] and Jensen-Shannon (JS) [18], given as follows:

$$KL(r_b||r_i, q) = \sum_{d=1}^{n} r_b(d) \cdot \log_2 \frac{r_b(d)}{r_i(d)} \tag{2}$$

$$JS(r_b||r_i, q) = \frac{1}{2} \cdot KL(r_b||r_i, q) + \frac{1}{2} \cdot KL(r_i||r_b, q) \tag{3}$$

$$= \sum_{d=1}^{n} r_b(d) \cdot \log_2 \frac{r_b(d)}{\frac{1}{2} \cdot r_b(d) + \frac{1}{2} \cdot r_i(d)}$$

where for the top n retrieved documents of a given query q, $r_b(d)$ and $r_i(d)$ are the relevance scores of document d in the base ranking r_b and candidate ranking r_i, respectively.

It is easy to verify that adding a constant to r_i does not change the ranking position of each document in r_i, however, this does affect the divergence between r_b and r_i. In order to avoid the issue of translation invariance, we apply a score normalisation, which was proposed by Lee [19], on each document of the document rankings r_b and r_i:

$$r_N(d) = \frac{r(d) - r(min)}{r(max) - r(min)} \tag{4}$$

where $r(max)$ and $r(min)$ are the maximum and minimum document relevance scores that have been observed in the top retrieved documents from the input ranking r. $r(d)$ is the relevance score of document d in the input ranking.

Learning to Select. We have shown how to estimate a divergence score between two rankings of documents. In this work, we consider the divergence score to be an example of a query feature. Next, we present how to use the query feature, such as the divergence score, to selectively apply an appropriate ranking function for a given query.

Initially, on a training dataset, we have a set of queries $Q = \{q_1, q_2, ..., q_m\}$ and a set of candidate ranking functions $R = \{r_1, r_2, ..., r_n\}$. For each query q_j, we use one of the above described divergence measures to estimate the divergence score $d(r_b||r_i, q_j)$ of each ranking function r_i from the base ranking function

Table 2. Sample divergence scores and MAP evaluations for 5 training queries and 2 ranking functions

	MAP		divergence	
q_ϕ	$E(q_\phi, r_1)$	$E(q_\phi, r_2)$	$d(r_b\|\|r_1, q_\phi)$	$d(r_b\|\|r_2, q_\phi)$
q_1	0.1	0.2	0.5	0.3
q_2	0.5	0.3	0.7	0.6
q_3	0.3	0.2	0.4	0.5
q_4	0.4	0.5	0.2	0.4
q_5	0.2	0.1	0.8	0.7

r_b. Note that one divergence score will be estimated for each ranking function on each query. For all training queries Q, each ranking function r_i's divergence scores set is denoted $\mathbf{d}(r_b\|\|r_i, Q) = \{d(r_b\|\|r_i, q_1), ..., d(r_b\|\|r_i, q_m)\}$.

Next, in response to an unseen query q', for each ranking function r_i, we first estimate a divergence score $d(r_b\|\|r_i, q')$, then employ KNN to identify the k nearest queries from $\mathbf{d}(r_b\|\|r_i, Q)$ in a similar manner to [3] but using the divergence score. KNN is widely used for finding the closest objects in a metric space when there is little or no prior knowledge about the distribution of the objects. Each identified neighbour corresponds to a training query q_ϕ. Let $E(q_\phi, r_i)$ be the outcome of an evaluation measure calculated on the ranking function r_i for the query q_ϕ. The effectiveness of ranking function r_i on this test query q' is predicted based on the performance of r_i on the neighbours of q', denoted $\sum_{\phi=1}^{k} E(q_\phi, r_i)$.

We apply the ranking function r_i for the query q' that has the highest retrieval performance on the set of k nearest neighbouring queries:

$$r_i^*(q') = \arg\max_{r_i} \frac{\sum_{\phi=1}^{k} E(q_\phi, r_i)}{k} \qquad (5)$$

4.3 Example of LTS

Let us illustrate the LTS framework using an example. Assuming our training dataset has 5 queries, namely $Q = \{q_1, q_2, q_3, q_4, q_5\}$, and that we have two candidate ranking functions, namely $R = \{r_1, r_2\}$. The retrieval performance (e.g. MAP) of each ranking function and its divergence score for each training query are presented in Table 2 for a particular divergence measure.

Then, for an unseen query q', we estimate a divergence for each ranking function. For the purpose of our example, let $d(r_b\|\|r_1, q') = 0.3$ and $d(r_b\|\|r_2, q') = 0.6$, and let $k = 3$ in order to find the three nearest neighbouring queries. In this case, according to the difference between divergence scores, the nearest queries for ranking function r_1 are $\{q_1, q_3, q_4\}$, while for ranking function r_2 they are $\{q_2, q_3, q_5\}$. Therefore, for q', we apply r_1 as its mean retrieval performance for the nearest queries is higher than for r_2 ($\frac{0.1+0.3+0.4}{3} > \frac{0.3+0.2+0.1}{3}$).

5   Experimental Settings

In our experiments, we address three main research questions:

- Firstly, we test how effective our proposed LTS framework is for selecting an appropriate ranking function for a given query, by comparing it to three state-of-the-art LTR techniques.
- Secondly, as the number of candidate ranking functions increases, the selection becomes more challenging. To test how robust our proposed LTS framework is, we apply it on a larger number of candidate ranking functions.
- Thirdly, we test how important the query feature is for identifying neighbour queries, by investigating three different query features, namely KL divergence, JS divergence and the mean of the relevance scores, which has been shown to be effective in identifying neighbouring queries [3].

We conduct our experiments on two datasets, namely LETOR 3.0 and LETOR 4.0 [4]. The LETOR 3.0 dataset contains 64 different document features, including document length and HostRank [13], among others. The documents in LETOR 3.0 are sampled from the top retrieved documents by using BM25 on the .GOV corpus with the TREC 2003 and TREC 2004 Web track queries. In contrast, the LETOR 4.0 dataset contains 46 document features for documents similarly sampled from the .GOV2 corpus using the TREC 2007 and TREC 2008 Million Query track queries.

Many different LTR techniques have been proposed during the past few years. In this work, we employ three state-of-the-art LTR techniques, namely Ranking SVM [5,6], AdaRank [1], and the AFS method [2]. In addition, the average of the retrieval performance of the candidate ranking functions that were created by the three LTR techniques is used as an additional baseline (denoted AS).

In our experiments, we use a 5-fold cross-validation process by separating each LETOR dataset into 5 folds of equal size. We iteratively test our LTS framework on one fold after training on the remaining four folds.

BM25 is used as our base ranking function as the features included in the LETOR datasets are computed over the top retrieved documents, which are sampled using BM25 [4]. The feature weights that are related with each candidate ranking function by using the AFS method are set by optimising Mean Average Precision (MAP) on the training dataset, using a simulated annealing procedure [20]. The number of top retrieved documents and the number of neighbours, namely n and k in Section 4, are also set by optimising MAP over the training dataset, using a large range of different value settings. We evaluate our experimental results using MAP, Precision at N, and normalised Discounted Cumulative Gain (nDCG). We report the obtained results and their analysis in the following section.

6   Results and Discussion

6.1   Effectiveness of Our LTS Framework

In order to test the effectiveness of our proposed method for selectively applying an appropriate ranking function for a given query, we compare it with the AS baseline and three state-of-the-art LTR techniques, namely Ranking SVM, AdaRank, and the AFS method, which are systematically applied to all queries.

Tables 3 & 4 present the evaluation of the retrieval performances obtained by using the three state-of-the-art LTR techniques and by applying our proposed LTS framework in terms of MAP, Precision at N and nDCG on the LETOR 3.0 and LETOR 4.0 datasets, respectively. The best retrieval performances for each evaluation measure on each different dataset are emphasised in bold. The α, β, γ and δ symbols indicate that the retrieval performance obtained by the best of our proposed LTS framework is significantly better than the Ranking SVM, AdaRank, AFS and AS baselines, respectively, according to the Wilcoxon Matched-Pairs Signed-Ranks Test ($p < 0.05$). The best retrieval performance obtained by systematically applying a LTR technique on all queries is highlighted with underline. The $*$ symbol indicates that the retrieval performance obtained by using our proposed LTS framework is significantly better than the underlined score. $LTS-JS$, $LTS-KL$ and $LTS-Mean$ denote the application of the LTS framework by using JS divergence, KL divergence and the mean of the relevance scores as the query feature for identifying neighbour queries, respectively.

From the results in Tables 3 & 4, we observe that the best retrieval performance in each column is achieved by using our proposed LTS framework. The

Table 3. Comparison between LTS and state-of-the-art LTR techniques using different evaluation measures on the LETOR 3.0 dataset. Results are the mean over 5 folds.

TREC 2003				
	MAP	P@5	P@10	nDCG@5	nDCG@10
Ranking SVM	0.5366 α	0.1811	0.1063	0.6450	0.6615
AdaRank	0.5977 β	0.1731	0.0974	0.6612	0.6679
AFS	<u>0.6145</u> γ	0.1777	0.1023	0.6766	0.6914
AS	0.5829 δ	0.1773	0.1020	0.6610	0.6736
LTS-Mean	0.6305 $*$	0.1771	0.1043	0.6776	0.7047
LTS-KL	0.6446 $*$	0.1794	0.1040	**0.6908**	**0.7059**
LTS-JS	**0.6483** $*$	**0.1843**	**0.1080**	0.6884	0.6959
TREC 2004				
	MAP	P@5	P@10	nDCG@5	nDCG@10
Ranking SVM	0.4193 α	0.2116	0.1427	0.5527	0.5659
AdaRank	0.5062 β	0.2124	0.1364	0.6053	0.6110
AFS	<u>0.5079</u> γ	0.2169	**0.1436**	0.6075	0.6186
AS	0.4778 δ	0.2136	0.1409	0.5885	0.5985
LTS-Mean	0.5158	0.2178	0.1427	0.6054	0.6145
LTS-KL	**0.5423** $*$	0.2124	0.1400	0.6152	0.6345
LTS-JS	0.5397 $*$	**0.2204**	0.1413	**0.6286**	**0.6365**

Table 4. Comparison between LTS and state-of-the-art LTR techniques using different evaluation measures on the LETOR 4.0 dataset. Results are the mean over 5 folds.

	MAP	P@5	P@10	nDCG@5	nDCG@10
TREC 2007					
Ranking SVM	0.4641 α	0.4113	**0.3844**	0.4120	0.4421
AdaRank	0.4537 β	0.3932	0.3686	0.3916	0.4223
AFS	0.4603 γ	0.4080	0.3753	0.4118	0.4378
AS	0.4594 δ	0.4042	0.3761	0.4051	0.4341
LTS-Mean	0.4637	0.4060	0.3760	0.4076	0.4352
LTS-KL	**0.4692**	0.4116	0.3795	0.4149	0.4419
LTS-JS	0.4676	**0.4132**	0.3783	**0.4182**	**0.4422**
TREC 2008					
Ranking SVM	0.4752 α	0.3457	**0.2499**	0.4742	0.2296
AdaRank	0.4766 β	0.3419	0.2449	0.4701	0.2230
AFS	0.4784 γ	0.3480	0.2490	0.4761	0.2286
AS	0.4767 δ	0.3452	0.2480	0.4735	0.2281
LTS-Mean	0.4861 *	0.3465	0.2488	0.4747	0.2297
LTS-KL	0.4908 *	0.3485	0.2493	0.4794	0.2285
LTS-JS	**0.4911** *	**0.3488**	0.2494	**0.4813**	**0.2300**

only exceptions are for the P@10 evaluation measure. However, in each case, the performance of LTS is close to the highest P@10. Note that all training was conducted using the MAP evaluation measure.

In particular, from the MAP column, the best retrieval performance obtained by using our proposed LTS framework makes improvements over all the LTR techniques and the AS baseline. Moreover, the improvements are statistically significant, e.g. on the TREC 2003 dataset: $0.5366 \rightarrow 0.6483$; $0.5977 \rightarrow 0.6483$; $0.6145 \rightarrow 0.6483$; and $0.5829 \rightarrow 0.6483$. Furthermore, by comparing our LTS framework with the best LTR technique, we note that the results obtained by using the LTS framework are significantly better in 3 out of 4 cases, e.g., on the TREC 2003 dataset: $0.6145 \rightarrow 0.6305$; $0.6145 \rightarrow 0.6483$; and $0.6145 \rightarrow 0.6446$.

The above observations suggest that our proposed LTS framework is effective in applying an appropriate ranking function on a per-query basis.

6.2   Robustness of Our LTS Framework

The analysis in Section 6.1 demonstrates the effectiveness of the proposed LTS framework on a small set of candidate ranking functions. In this section, we investigate the robustness of our LTS framework when the number of candidate ranking functions increases. To achieve this, we simulate a number of candidate ranking functions by applying a single LTR technique on several different combinations of document features. In particular, in order to have a strong baseline, we use AFS, since it produces on average higher retrieval performance than the other two LTR techniques.

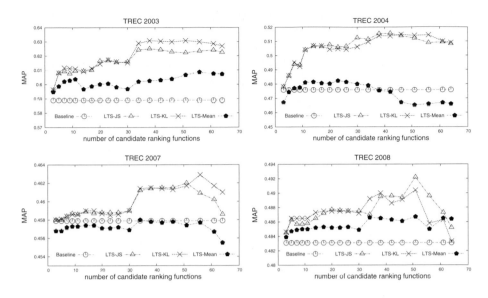

Fig. 1. MAP versus the number of candidate ranking functions on the LETOR 3.0 and LETOR 4.0 datasets

In this investigation, we choose the best 6 features from the LETOR feature set, based on the training dataset. There are then $2^6 - 1$ possible combinations (excluding the empty combination). Integrating each combination into the base ranking function (namely BM25) produces one candidate ranking function. In this case, our task becomes to select an appropriate ranking function from a set of candidate ranking functions, which contains as many as $2^6 - 1 = 63$ candidate ranking functions.

Figure 1 shows the effect on MAP as the number of candidate ranking functions is varied, on different TREC datasets. We order the 63 ranking functions according to their performance on the training dataset and the best performing ranking function is used as our baseline. The x axis denotes the number of top performing ranking functions applied. For example, for TREC 2003, 10 in the x axis means we use the top 10 performing ranking functions, which are assessed on the training dataset, as candidate ranking functions to selectively apply on the test dataset.

From Figure 1, we can observe that both LTS-JS and LTS-KL consistently outperform the baseline while the LTS-Mean sometimes underperforms the baseline when the number of candidate ranking functions is increased. This suggests that the effectiveness of our LTS framework is related to the query feature that is used for identifying the neighbouring queries. A more detailed analysis of this point is presented in Section 6.3. In the remaining of this section, we only report the MAP distributions based on LTS-JS and LTS-KL as both of these proposed query features always perform better than LTS-Mean.

Indeed, in contrast to LTS-Mean, both LTS-JS and LTS-KL are enhanced as the number of candidate ranking functions increases from 2 to 35. This is mainly because each newly added ranking function (from 2 to 35) has different behaviour and favours different queries. Hence, with more ranking functions added, the retrieval performance can be further improved if they can be selectively applied in an appropriate manner.

However, the retrieval performance only improves slightly when the number of candidate ranking functions increases to 50. The reason for this is that these ranking functions are created based on the combination of a small set of features, most of them with similar behaviour, which results in these newly added ranking functions (from 35 to 50) favouring the same queries as the previously chosen ranking functions.

We also observe that the best retrieval performance is obtained when there are around 50 candidate ranking functions. However, the performance starts to decrease after 50 — as the last added ranking functions perform poorly on the training dataset and bring noise to the LTS framework.

This investigation suggests that our method is a robust approach, which can increasingly improve the retrieval performance as more ranking functions are added. The only caveat is when the most poorly-performing ranking functions are added to the candidate set. This can be controlled by setting the number of top-performing candidate ranking functions from which to select.

6.3   Importance of Query Feature

The above two investigations have shown the effectiveness and robustness of our proposed LTS framework. It is of note that the key component of the proposed LTS framework is the query feature that is used to identify the neighbouring queries, namely JS divergence, KL divergence, and the mean of the relevance scores.

By using different query features for identifying similar queries in our proposed LTS framework, from Tables 3 & 4, we observe that the JS divergence measure (LTS-JS) and the KL divergence measure (LTS-KL) are producing very close retrieval performances and both of them consistently outperform the mean of the relevance scores (LTS-Mean). For example, on the TREC 2003 dataset: $0.6305 \rightarrow 0.6483$; and $0.6305 \rightarrow 0.6483$.

In addition, from the distribution of MAP versus the number of candidate ranking functions shown in Figure 1, we observe that LTS-JS, LTS-KL and LTS-Mean have a similar distribution when increasing the number of candidate ranking functions. The only exception is on the TREC 2004 dataset (right top), where the MAP distribution obtained by LTS-Mean goes down quickly just after the number of candidate ranking functions reaches 30. In addition, LTS-Mean always underperforms compared to both LTS-JS and LTS-KL, and can fail to improve the baseline. This is particularly noticeable for the TREC 2007 (left bottom), where LTS-Mean always fails to improve over the baseline. Finally, we note that KL divergence and JS divergence have comparable performance, which is explained in that they are mathematically related.

The above observations suggest that our proposed use of divergence measures as a query feature for identifying neighbour queries is very effective.

7   Conclusion

In this paper, we proposed a novel Learning To Select (LTS) framework for selecting an appropriate ranking function for a given query. In particular, for an unseen query, we identify similar training queries by using novel query features based on the divergence between document score distributions. Such similar queries are then used to select an appropriate highly performing ranking function to apply. We tested our framework on the LETOR 3.0 & LETOR 4.0 feature sets and their corresponding TREC tasks, by comparing it with three state-of-the-art Learning To Rank (LTR) techniques, namely Ranking SVM, AdaRank, and the AFS method.

Our experimental results showed that the retrieval performance obtained by using our proposed LTS framework could constantly outperform the three state-of-the-art techniques using different evaluation measures and on different datasets, the only exception being for the $P@10$ measure. In addition, improvements over all LTR techniques were statistically significant in most cases.

Moreover, we investigated the effectiveness of our framework when the number of candidate ranking functions increases. By plotting the distribution of MAP versus the number of candidate ranking functions, we found that by using our proposed framework, the retrieval performance was enhanced when increasing the number of candidate ranking functions.

Furthermore, our proposed use of divergence measures as query features to identify neighbouring queries was always more effective than the mean of the relevance scores measure, which ignores the distribution of relevance scores. For our future work, we plan to investigate other query features.

References

1. Xu, J., Li, H.: AdaRank: A Boosting Algorithm for Information Retrieval. In: Proceedings of SIGIR 2007, Amsterdam, The Netherlands (2007)
2. Metzler, D.: Automatic Feature Selection in the Markov Random Field Model for Information Retrieval. In: Proceedings of CIKM 2007, Lisbon, Portugal (2007)
3. Geng, X., Liu, T.Y., Qin, T., Arnold, A., Li, H., Shum, H.Y.: Query Dependent Ranking Using K-Nearest Neighbour. In: Proceedings of SIGIR 2008, Singapore (2008)
4. Liu, T.Y., Qin, T., Xu, J., Xiong, W.Y., Li, H.: LETOR: Benchmark Dataset for Research on Learning to Rank for Information Retrieval. In: Proceedings of SIGIR 2007 Learning to Rank workshop, Amsterdam, The Netherlands (2007)
5. Herbrich, R., Graepel, T., Obermayer, K.: Large Margin Rank Boundaries for Ordinal Regression. MIT Press, Cambridge (2000)
6. Joachims, T.: Optimizing Search Engines using Clickthrough Data. In: Proceedings of SIGKDD 2002, Alberta, Canada (2002)

7. Kamps, J., Mishne, G., de Rijke, M.: Language Models for Searching in Web Corpora. In: Proceedings of TREC 13, Gaithersburg, MD, USA (2004)
8. Peng, J., He, B., Ounis, I.: Predicting the Usefulness of Collection Enrichment for Enterprise Search. In: Azzopardi, L., Kazai, G., Robertson, S., Rüger, S., Shokouhi, M., Song, D., Yilmaz, E. (eds.) ICTIR 2009. LNCS, vol. 5766, pp. 366–370. Springer, Heidelberg (2009)
9. Plachouras, V., Ounis, I.: Usefulness of Hyperlink Structure for Query-Biased Topic Distillation. In: Proceedings of SIGIR 2004, Sheffield, UK (2004)
10. Plachouras, V.: Selective Web Information Retrieval. PhD thesis, University of Glasgow, UK (2006)
11. Peng, J., Ounis, I.: Selective Application of Query-Independent Features in Web Information Retrieval. In: Proceedings of ECIR 2009, Toulouse, France (2009)
12. Manmatha, R., Rath, T., Feng, F.: Modeling Score Distributions for Combining the Outputs of Search Engines. In: Proceedings of SIGIR 2001, New Orleans LA, USA (2001)
13. Xue, G.R., Yang, Q., Zeng, H.J., Yu, Y., Chen, Z.: Exploiting the Hierarchical Structure for Link Analysis. In: Proceedings of SIGIR 2005, Salvador, Brazil (2005)
14. Song, R., Wen, J.R., Shi, S., Xin, G., Liu, T.Y., Qin, T., Zheng, X., Zhang, J., Xue, G., Ma, W.Y.: Microsoft Research Asia at Web Track and Terabyte Track of TREC 2004. In: Proceedings of TREC 2004, Gaithersburg, MD, USA (2004)
15. Yang, K., Yu, N., Wead, A., La Rowe, G., Li, Y.H., Friend, C., Lee, Y.: WIDIT in TREC 2004 Genomics, Hard, Robust and Web Tracks. In: Proceedings of TREC 2004, Gaithersburg, MD, USA (2004)
16. Craswell, N., Hawking, D.: Overview of the TREC 2004 Web Track. In: Proceedings of TREC 2004, Gaithersburg, MD, USA (2004)
17. Kullback, S.: Information Theory and Statistics. John Wiley & Sons, New York (1959)
18. Lin, J.: Divergence Measures Based on the Shannon Entropy. IEEE Transactions on Information Theory 37(1), 145–151 (1991)
19. Lee, J.H.: Analyses of Multiple Evidence Combination. In: Proceedings of SIGIR 1997, Philadelphia, USA (1997)
20. Kirkpatrick, S., Gelatt, C., Vecchi, M.: Optimization by Simulated Annealing. Science 220(4598), 671–680 (1983)

Mining Anchor Text Trends for Retrieval

Na Dai and Brian D. Davison

Department of Computer Science & Engineering, Lehigh University, USA
{nad207,davison}@cse.lehigh.edu

Abstract. Anchor text has been considered as a useful resource to complement the representation of target pages and is broadly used in web search. However, previous research only uses anchor text of a single snapshot to improve web search. Historical trends of anchor text importance have not been well modeled in anchor text weighting strategies. In this paper, we propose a novel temporal anchor text weighting method to incorporate the trends of anchor text creation over time, which combines historical weights of anchor text by propagating the anchor text weights among snapshots over the time axis. We evaluate our method on a real-world web crawl from the Stanford WebBase. Our results demonstrate that the proposed method can produce a significant improvement in ranking quality.

1 Introduction

When a web page designer creates a link to another page, she will typically highlight a portion of the text of the current page, and embed it within a reference to the target page. This text is called the anchor text of the link, and usually forms a succinct description of the target page so that the reader of the current page can decide whether or not to follow the hyperlink.

Links to a target page are ostensibly created by people other than the author of the target, and thus the anchor texts likely include summaries and alternative representations of the target page content. Because these anchor texts are typically short and descriptive, they are potentially similar to queries [7] and can reflect user's information needs. Hence, anchor text has been widely utilized as an important part in ranking functions in general for commercial search engines.

Anchor text can also be important to other tasks, such as query intent classification [15], query refinement [14], query translation [17] and so on. Mining anchor text can help to better understand queries and users' information needs and enrich the representation of the linked page content.

When used for retrieval, one anchor text might not be as useful as another, and so recent work [19,7] has focused on how to determine the importance of anchor text for a given destination page. However, such work only considers one snapshot of the web graph (the current web), and so the influence from historical anchor text is "excluded".

More importantly, the creation of anchor text reflects how web content creators view the destination page. A historical trace of the variation in such viewpoints can help determine how to interpret the page. Consider a page which has 10 newly created in-links associated with a specific anchor text in the past 3 days. When compared with another page which only received ten in-links (with the same anchor text) within the past

C. Gurrin et al. (Eds.): ECIR 2010, LNCS 5993, pp. 127–139, 2010.

10 years, the importance of the anchor text on the former page should be emphasized, even if the absolute weights based on the current snapshot cannot differentiate them.

The primary goal of our work is to incorporate the trends of the creation of page in-links associated with anchor text into measuring the anchor text importance for representing page content in the retrieval task. We operate on the assumption that better anchor text representation of pages can improve retrieval quality. We incorporate the historical trends on anchor text by propagating the anchor text weights among historical and predicted future snapshots over the time axis. The significance of our work can be generalized onto other tasks, such as web page clustering and classification. It can also help to build time-sensitive document models. Furthermore, we propose a variety of ways to incorporate the trends from historical snapshots to better estimate the importance of anchor text in the current snapshot. Finally, we verify our models via empirical experiments, and our experiment shows the retrieval quality has significant improvement on a real-world web crawl from the Stanford WebBase.

Related work. Anchor text has been widely exploited in commercial search engines to improve web search. Brin and Page [3] recognized the importance of anchor text to be associated with the page to which a link points. Eiron and McCurley [8] investigated properties of anchor text in a large intranet and show its similarity with real user queries and consensus titles. It shed light on the evidence that better understanding of anchor text can help translate user queries into search results with high quality. Recent work by Fujii [9] investigated anchor text distributions on the web for query intent classification, which is used to construct an anchor text based retrieval model for web search. Anchor text can be modeled in different ways, incorporating diverse factors. Metzler et al. [19] focused on anchor text sparsity problem by incorporating anchor text associated with external links to enrich document representation. Dou et al. [7] took into consideration the relationship among anchor text which refers to the same destination page into anchor text based retrieval models to enhance the quality of search results.

Both Dou et al. [7] and Metzler et al. [19] used the retrieval model of BM25F [23], which is an extended BM25 model adaptive to fielded documents, to verify their models on anchor text empirically. The BM25F model can be further understood as smoothing destination page content by its surrogate anchor text, which is similar with the ideas of cluster-based smoothing and neighbor-based smoothing in language models. In this paper, we also use this model, but we aim to show the effectiveness of using historical information to model anchor text importance for retrieval.

Previous work has been focused on improving retrieval quality by incorporating temporal factors. Acharya et al. [1] proposed a series of temporal signals concealed in search context, which benefit web search quality. Work in [16,12,5] incorporated the bias from the temporal distribution of (relevant) search results into language models. Berberich et al. [2] took into account the temporal factors into link analysis algorithms. However, none of them considers to use link creation rate reflected on the anchor text weights to improve ranking quality. Our work catch the evidence of link occurrence at different time points such that we can infer the creation rate of every node and associated anchor text for every link. This enables us to analyze the anchor text importance change over time, and further use it to model anchor text weights for retrieval.

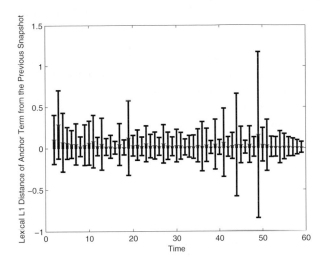

Fig. 1. Average and deviation of the lexical L_1 distance of anchor text term distribution over time for each of the top 2000 search result of the query *"paris hilton"*. X-axis is the time axis from early to late (from Jan. 2001 to Dec. 2005 with the time unit being 1 month). Y-axis records the average and deviation of the lexical L_1 distance of destination nodes' anchor term distribution between two successive time points.

The remainder of this paper is as follows. We introduce the temporal anchor text data used in this work. We then start to describe our methods in Section 3, which utilizes temporal anchor text to better evaluate the importance of anchor text for retrieval. The experiments in Section 4 and 5 show the effectiveness of our approaches. We conclude this work in Section 6.

2   Temporal Anchor Data

A destination page gets in-links from multiple source pages at different time points, each with distinct anchor text. We assign a timestamp to each pair of source and destination page, which represents the creation time of the associated link. Naturally, we consider the item <source page, destination page, anchor text, creation time> to be unique. If the anchor text on the link changes, we assume that the link associated with the old anchor text is removed and another link associated with the new anchor text is created.

Figure 1 demonstrates the variation of the similarity of anchor text terms from month to month over a five-year time period. Earlier months show somewhat larger changes, while the changes are more moderate in later time periods. This follows as many in-links were created during the time period from 2001 to 2002. However, we also found the change in 2004 has a larger deviation. We infer that in-links have sharp increase for some destination nodes, but not for others. To better understand the fine-grained variation of anchor text on links, we keep track of how the anchor text on each link change over time. The Jaccard coefficient of anchor terms on a specific link between two successive time points is 0.9954 ± 0.0514 on average. Based on these observations, we believe that the anchor text on links are relatively stable. Most anchor text does not change from the time point when the associated link was created to the time point when it was removed. The change of aggregated effects from anchor texts on destination nodes can be potentially used to benefit web search. Motivated by these observations, we propose our temporal anchor text based retrieval method.

3   Temporal Anchor Text Based Retrieval

In this section, we describe our proposed methods which incorporate historical trends of page in-links creation rate and smooth the anchor text weights for destination pages in anchor text based retrieval. Our method requires a web graph and the time point t_0 on which it is crawled. Here, we define t_0 to be the current time point, and assume the retrieval evaluation is based on the situation at t_0. We follow the approach proposed by Metzler et al. [19] to determine weights on anchor text at each time point. Metzler et al. aggregated a set of unique *anchor text lines* for each given destination page, and calculated weights on them individually. However, we add extra time points to the weights on anchor text lines. Such weights on anchor text lines represent their importance on a given destination page at a specific time point. The output of our method is a collection of anchor terms and the final smoothed weights on them for a destination page at time point t_0. Specifically, our approach can be divided into the following three steps:

- aggregate anchor text lines and calculate weights on them for destination pages at each time points before t_0;
- analyze the trend and use it to predict the possible weights on anchor text lines at the time points after t_0;
- propagate and diffuse the weights on anchor text lines through the time axis;

3.1   Aggregate Historical Anchor Text

In order to better understand how to collect and weight the aggregated historical anchor text, we first describe how we weight the anchor text of the current snapshot. We use the methods in [19] to collect and weight anchor text for a specific web snapshot. Indeed, there are other ways to weight anchor text beyond the methods presented in [19]. However, [19] is the only one to deal with *anchor text sparsity* problem. Considering the fact that historical link information may sometimes be unavailable, to better enrich the anchor text representation becomes a good choice. We now briefly review the way of collecting and weighting anchor text in that work.

Given a URL u, all in-link pages P that are within the same site(domain) as u are collected as *internal links*. Those in-link pages A that are in different domains from u are defined as *external links*. The anchor text on the external links are called *original anchor text*. For internal links, we further collect the external links of these internal links. The anchor text on the new collected external links are known as *aggregated anchor text* of u. The original anchor text are weighted as follows:

$$wt(a, u) = \sum_{s \in S(u)} \frac{\delta(a, u, s)}{|anchors(u, s)|}$$

where $S(u)$ is the set of external sites that links to u, $\delta(a, u, s)$ is 1 iff anchor text line a links to u from site s. The aggregated anchor text are weighted in multiple ways, we choose two of them which are shown to have best performance in general in [19], defined as follows:

$$wt_{Min}(a, u) = \min_{u' \in N(u)} wt(a, u') \qquad wt_{Max}(a, u) = \max_{u' \in N(u)} wt(a, u')$$

where $N(u)$ is the set of internal in-links and $wt(a, u')$ is the original weight of anchor text line a for URL u'.

Both original anchor text lines and external anchor text lines are used to enrich anchor text representation. We choose to use *combined* representation and *back off* representation to enrich destination pages' representation. Combined representation keeps the document structure and augments both original anchor text and aggregated anchor text, whereas back off representation exempts from the aggregated anchor text which have already appeared in original anchor text lines.

Once we have weights on anchor texts at the current time point t_0, we have actually known which links should contribute to anchor text weights. We keep track of these links by looking back to seek their creation time (see Section 3.4 for details). We define the difference of two successive time points as Δt, i.e., $\Delta t = t_i - t_{i-1}$. We map each link onto the time axis according to its creation time. If link l is created before t_i but after t_{i-1}, then for any given time point $t_j (0 > j \geq i)$, l is included in the snapshot at t_j. Given any time point $t_i (i < 0)$, we calculate the weight $w_i(a, u)$ of anchor text line a on the web page u based on all the links included at time point t_i.

3.2 Quantify Trends to Predict Future

Quantifying trends of weights on anchor text can help to predict how the weights change at future time points. Given a destination page, if the importance of its particular anchor text increases more greatly than its other anchor text, we may have higher confidence to believe such anchor text should be emphasized in some way since the trend shows it may get a higher weight in the near future. Here, we assume that: (1) newly created in-links prefer to associate with the anchor text used by recent in-links, according to their popularity, and (2) the weights on anchor text have reflected the number of page/site pointing to the target page, using that anchor text.

ARIMA (Auto-Regressive Integrated Moving Average) is a powerful way to predict time-series, but it is complex to use. Instead, we use the linear regression on moving average of order m to predict the value at the next time point. The reasons are as follows: (1) we observe that weights on anchor text have stable and monotone trends through time once the anchor text begins associating with the destination page; (2) we test the linear regression on weights (Max+combined) of individual anchor text line for a given destination page over time axis. The average mean square error (MSE) is only 0.0656. Based on these observations, we believe the linear model can well fit the trends of historical anchor text weights.

Given a URL u and one associated anchor text line a, we have a series of historical weights $w_{-n}(a, u), w_{-n+1}(a, u), \ldots, w_0(a, u)$. We first use a sliding window with size

$2k+1(k > 0)$ to smooth the time series. We calculate a moving average of order $2k+1$ as the following sequence of arithmetic means:

$$\frac{\sum_{i=-n}^{-n+2k} w_i(a, u)}{2k+1}, \frac{\sum_{i=-n+1}^{-n+2k+1} w_i(a, u)}{2k+1}, \ldots, \frac{\sum_{i=-2k}^{0} w_i(a, u)}{2k+1}$$

By using the sequence calculated above, we next use linear regression to predict the possible average at time point $(-k+2)$. The model assumes the moving average of order $2k+1$ has a linear relationship with the time points given a pair of anchor text and destination page, which is given by:

$$\overline{w}_i(a, u) = b + c \times i + \epsilon_i, i \geq (-k+2) \tag{1}$$

See [22] for deep analysis of the non-stationary correlated errors ϵ_i. We use existing evidence to estimate the parameters b and c. Once the weight $\widehat{\overline{w}}_{-k+2}(a, u)$ is achieved, $w_1(a, u)$ can be calculated by:

$$w_1(a, u) = \widehat{\overline{w}}_{-k+2}(a, u) \times (2k+1) - \sum_{i=-2k+1}^{0} w_i(a, u)$$

After we get the value of $w_1(a, u)$, we move the sliding window forward to calculate $w_i(a, u)(i > 1)$.

3.3 Diffuse Temporal Anchor Text Weights

Analyzing the trends of anchor text weights on a destination page allows us to predict the anchor text weights in the future. However, in order to better measure the importance of anchor text lines at t_0, we need to combine both the predicted future weights and the historical weights. As discussed in previous section, the predicted weights are extrapolated from historical trends, which help to differentiate two anchor text lines with the same weights at t_0. On the other hand, historical anchor text weights provide confirmation about what a destination page looks like. When we emphasize the predicted future weights, we actually give preference to the newly created destination pages, since the new pages tend to have higher anchor text creation rate, and the predicted anchor text weights are usually overemphasized. Whereas, when we combine some historical weights, we likely emphasize old pages which have stable anchor text distribution. By combining both the historical weights and predicted future weights, we can harmonize the influence from these two sides.

Specifically, we assume that the weights on an anchor text line at one time point can propagate through time axis to influence the weights of the same anchor text line at other time points for a given destination page. The intuition is that if an anchor text have a weight at a time point t_i, we hope it can influence the weights on the same anchor text at other time points by a decayed way which is proportional to a temporal distance. Thus, weights on two close time points would have more influence to each other than those on two far time points. Furthermore, we assume that the change ratio of the destination page content will also influence the weight propagation since huge change is likely to cause such propagation decayed more quickly. Given a time window, we calculate weights at the middle time point by aggregating the discount weights from all time points within it.

We now describe our method to propagate the weights formally. Let γ be the size of time window T, i.e., the number of time points within the time window. Let a be an anchor text line. Let u be a destination node, and u_i be a destination node at time point t_i. $w_1(u, a), w_2(u, a), \ldots, w_\gamma(u, a)$ are the weights of a on u at time points within the time window T. The weights at time point $t_{\frac{\gamma}{2}}$ after combining the propagated weights other time points within the time window is given by:

$$w'_{\frac{\gamma}{2}}(u, a) = \sum_{i=1}^{\gamma} f(u, \gamma, i) w_i(u, a) \tag{2}$$

where $f(u, \gamma, i)$ is the kernel function which determines the way of combining weight $w(u, a)$ at time point t_i.

Enlightened by previous work [6,13,21,18] which used proximity-based methods, we use five modified kernel functions originated from Gaussian kernel (equation 3), Triangle kernel (equation 4), Cosine kernel (equation 5), Circle kernel (equation 6), and Rectangle kernel (equation 7), which are defined by:

$$f_1(u, \gamma, i) = \exp[-\frac{1}{2}(\frac{i - \frac{\gamma}{2}}{\gamma(1 + \overline{B}_u(i \leftrightarrow \frac{\gamma}{2}))})^2] \tag{3}$$

$$f_2(u, \gamma, i) = 1 - \frac{|i - \frac{\gamma}{2}|}{\gamma(1 + \overline{B}_u(i \leftrightarrow \frac{\gamma}{2}))} \tag{4}$$

$$f_3(u, \gamma, i) = \frac{1}{2}[1 + \cos(\frac{\pi(i - \frac{\gamma}{2})}{\gamma(1 + \overline{B}_u(i \leftrightarrow \frac{\gamma}{2}))})] \tag{5}$$

$$f_4(u, \gamma, i) = \sqrt{1 - (\frac{|i - \gamma/2|}{\gamma(1 + \overline{B}_u(i \leftrightarrow \frac{\gamma}{2}))})^2} \tag{6}$$

$$f_5(u, \gamma, i) = 1 \tag{7}$$

where $\overline{B}_u(i \leftrightarrow \frac{\gamma}{2})$ is the average similarity between the destination page u's content at two successive time points within the scale $[i, \gamma/2]$ if $i < \gamma/2$ or $[\gamma/2, i]$ if $i \geq \gamma/2$. Without loss of generality, we assume $i < \gamma/2$. $\overline{B}_u(i \leftrightarrow \frac{\gamma}{2})$ is defined by:

$$\overline{B}_u(i \leftrightarrow \frac{\gamma}{2}) = \frac{1}{\frac{\gamma}{2} - i} \sum_{i'=i}^{\frac{\gamma}{2}-1} B_u(i', i' + 1) \tag{8}$$

We compare the similarity of two snapshots of page u's content by comparing their associated language models via the Bhattacharyya correlation:

$$B_u(i', i' + 1) = \sum_{v \in V} \sqrt{P(w|\theta_{u_{i'}}) P(w|\theta_{u_{i'+1}})} \tag{9}$$

This metric renders a similarity score between 0 and 1. Although this similarity is only based on $P(w|\theta_u)$, we can consider to combine other measures based on topic, timestamp, or out-link overlap so that all these measures can influence the probability of propagating the anchor text importance through the time axis.

3.4   Implementation

One key problem for utilizing temporal anchor text is that it is difficult to keep track of the information about when a link was created. Given a link appearing in the current snapshot, we looked back to the archival copies of the source page via the Wayback Machine portal of Internet Archive [10]. We parsed these copies to get all out-links within this page, and checked whether the given link was still in the out-link collection and whether the anchor text associated with the given link had any change. If either the anchor text has changed or the link did not exist, we utilized the timestamp of the next latest copy to be the time when the given link was created.

4   Experiment Setup

4.1   Data Set and Evaluation

Although many datasets, such as TREC .GOV collection [20], have been built for research purposes, they are usually small and biased, and cannot represent the characteristics of the real-world web graph. Hence, we choose to use a May 2005 crawl from the Stanford WebBase [4] as our dataset for ranking evaluation. This crawl has 58 million pages, and approximately 900 million links.

For ranking evaluation, 50 queries are selected from a set of consisting of those frequently used by previous researchers, ODP category names, and popular queries from Lycos and Google. For each query, we have judgment of 35 URLs on average. When human editors (members of our research lab) judge each pair of <query, URL>, they are asked to give a score based on how relevant the URL is to the given query. The rating results in the selection among excellent, good, not sure, bad, and worse. We use five-value scale which translates the ratings into the integers from 4 to 0. If the average score for this pair is more than 2.5, it is marked as relevant.

Based on the available relevance judgments, we evaluate the retrieval quality of our ranking algorithms over the Normalized Discounted Cumulative Gain (NDCG) [11] metric. NDCG credits ranking list with high precision at top ranks by weighting relevant documents according to their positions in the returned search results. Precision@10 is also used to show ranking performance, which calculates the average fraction of relevant URLs within the top ten results across all queries.

4.2   Ranking Function

Combining different fields of web pages has been shown highly effective for retrieval on the web in previous work [24]. BM25F is such a ranking model, which combines term frequencies in different fields linearly for BM25 score calculation. In this work, we test our anchor text weighting strategies by combining body text and anchor text in the BM25F model for retrieval. Suppose $w_{body}(i, j)$ is the weight of term i for page j in the body field, i.e., the term frequency of term i in page j. Let $w_{anchor}(i, j)$ be the weight of term i in the anchor text lines associated with page j, which is calculated by:

$$w_{anchor}(i, j) = \sum_{a \in A(j)} wt(a, j) \times tf_{anchor}(i, a)$$

Table 1. Performance comparison for different windows and different anchor text representations. The †and ‡symbols demonstrate the performance has statistically significant improvement when compared with the baseline (Latest anchors) at the level of p<0.1 and p<0.05 by one-tailed student t test.

Baseline								
	P@10	NDCG@3	NDCG@5	NDCG@10		P@10	NDCG@3	NDCG@5 NDCG@10
No anchors	1.6150	0.1860	0.1830	0.1749	Latest anchors	1.6170	0.1899	0.1846 0.1781
All historical anchors	1.6596	0.2023	0.1901	0.1856		–		

Backoff+Max				Combined+Max			
Window	P@10	NDCG@3	NDCG@5 NDCG@10	Window	P@10	NDCG@3	NDCG@5 NDCG@10
1	1.6383†	0.2019†	0.1911† 0.1822†	1	1.6383†	0.2019‡	0.1889 0.1841‡
2	1.6383†	0.2064‡	0.1945† 0.1858‡	2	1.6809‡	0.2064‡	0.1935‡ 0.1889‡
4	1.6809‡	0.2064‡	0.1945† 0.1879‡	4	1.7234‡	0.2064‡	0.1951‡ 0.1909‡
7	1.7234‡	0.2076‡	0.1984‡ 0.1915‡	7	1.7660‡	0.2094‡	0.1972‡ 0.1944‡
12	1.7234‡	0.2085‡	0.1990‡ 0.1916‡	12	1.7660‡	0.2105‡	0.1980‡ 0.1964‡
24	1.7660‡	0.2086‡	0.2002‡ 0.1950‡	24	1.8298‡	0.2129‡	0.2025‡ 0.2003‡

Backoff+Min				Combined+Min			
Window	P@10	NDCG@3	NDCG@5 NDCG@10	Window	P@10	NDCG@3	NDCG@5 NDCG@10
1	1.6170	0.1956†	0.1901† 0.1813†	1	1.6170	0.1956†	0.1875 0.1830‡
2	1.6170	0.2024‡	0.1913† 0.1829†	2	1.6809‡	0.1994†	0.1902† 0.1875‡
4	1.6596‡	0.2050‡	0.1921† 0.1853‡	4	1.7234‡	0.2033‡	0.1941‡ 0.1899‡
7	1.7021‡	0.2063‡	0.1979‡ 0.1892‡	7	1.7660‡	0.2081‡	0.1963‡ 0.1937‡
12	1.7234‡	0.2072‡	0.1975‡ 0.1909‡	12	1.7660‡	0.2092‡	0.1980‡ 0.1958‡
24	1.7660‡	0.2073‡	0.1990‡ 0.1943‡	24	1.8298‡	0.2115‡	0.2015‡ 0.1996‡

where $wt(a, j)$ is the weight on anchor text line a for the page j, and $tf_{anchor}(i, a)$ is the term frequency of i in the anchor text line a.

The aggregated term weights on i is a linear combination of weights i on anchor text and page body, which is given by:

$$w(i, j) = (1 - \alpha) \times w_{anchor}(i, j) + \alpha \times w_{body}(i, j)$$

where α is a combination parameter, which controls the balance between term weights on anchor text and page body used in BM25F ranking function. The document length is calculated by the same method.

5   Experimental Results

In this section, we describe the results of the experimental ranking evaluation. We start by showing how much the proposed ranking algorithms significantly improve the retrieval quality. We then render some deeper analysis about the characteristics of this ranking algorithms with respect to the improvement of ranking quality.

5.1   Performance Comparison

The first experiment shows the effectiveness of enlarging the window for propagating historical weights on anchor text lines over multiple aggregation functions and anchor text representation. The results are reported in Table 1. The baseline is the best performance by using the methods in [19] (on the latest snapshots). Since we remove all inlinks if there is no historical snapshots for the inlinked nodes in the Internet Archive, the improvement of using anchor text versus without using anchor text is not obvious. The performance of almost all combinations of window sizes, aggregation functions and document representation over all the metrics outperform the baseline significantly. Furthermore, the performance of all combinations of aggregation functions consistently increases with the window size, which indicates the temporal inlinks, especially long term historical inlink context is a good resource to reflect the link evolution that can be utilized in improving the ranking quality in terms of document relevance. Furthermore, the combined aggregation functions outperform the Backoff approaches, which suggests that the benefits from the "confirmation" influence brought by duplicate anchor text lines outweigh the noise they introduced.

5.2   Analysis

We now describe the effectiveness of kernel functions used in propagating anchor text line weights. The results are rendered in Table 2. The performance of Rectangle kernel is arguable the best in general among all combinations of aggregation functions. Gaussian and Circle kernels show comparable performance, which outperform Triangle and Cosine kernels. This observation demonstrates that search results benefit from emphasizing both historical and predicted future anchor weights without deemphasizing the influence of time points far away from the current point. We infer that the ranking quality will benefits from long-term temporal information rather than short-term one since long-term information tends to express more stable trends.

Table 2. Performance comparison for different kernels for propagating temporal anchor line weights when the window size is 12. The kernel 1, 2, 3, 4, and 5 represent Gaussian kernel, Triangle kernel, Cosine kernel, Circle kernel, and Rectangle kernel respectively.

Baseline								
	P@10	NDCG@3	NDCG@5	NDCG@10		P@10	NDCG@3	NDCG@5	NDCG@10
No anchors	1.6150	0.1860	0.1830	0.1749	Latest anchors	1.6170	0.1899	0.1846	0.1781

Backoff+Max					Combined+Max				
Kernel	P@10	NDCG@3	NDCG@5	NDCG@10	Kernel	P@10	NDCG@3	NDCG@5	NDCG@10
1	1.7022	**0.2085**	0.1962	0.1897	1	1.7457	**0.2105**	**0.1980**	0.1940
2	1.7021	0.2044	0.1955	0.1900	2	1.7447	0.2063	0.1955	0.1931
3	1.7020	0.2044	0.1955	0.1900	3	1.7447	0.2063	0.1955	0.1931
4	**1.7234**	0.2063	0.1985	0.1899	4	**1.7660**	0.2057	0.1966	0.1930
5	1.7023	0.2050	**0.1990**	**0.1916**	5	**1.7660**	0.2068	0.1977	**0.1964**

Backoff+Min					Combined+Min				
Kernel	P@10	NDCG@3	NDCG@5	NDCG@10	Kernel	P@10	NDCG@3	NDCG@5	NDCG@10
1	1.7021	**0.2072**	0.1953	0.1890	1	1.7447	0.2086	**0.1980**	0.1933
2	1.7019	0.2030	0.1950	0.1889	2	1.7438	0.2050	0.1946	0.1920
3	1.7019	0.2030	0.1950	0.1889	3	1.7438	0.2050	0.1946	0.1920
4	**1.7234**	0.2050	0.1955	0.1891	4	**1.7660**	0.2092	0.1957	0.1924
5	1.7021	0.2037	**0.1975**	**0.1909**	5	**1.7660**	0.2055	0.1967	**0.1958**

Next, we investigate the relationship between the average age of search results and the relative improvement of ranking quality in Table 3. We bucketize the queries according to the average age of their top 2000 search results. Bucket 0 is the one, in which query results have the short average age, and bucket 3 is the one in which query results have the longest average age. From Table 3, query results with longer ages benefit more

Table 3. Performance comparison for queries bucketized by the average age of search results. The weighting strategy is Combined+Max. P: Propagating weights on anchor text lines from past time points; F: Propagating predicted weights on anchor text lines from future points; T: Propagating weights on anchor text lines from both sides.

Window	Time	Bucket 0	Bucket 1	Bucket 2	Bucket 3	Window	Time	Bucket 0	Bucket 1	Bucket 2	Bucket 3
1	P	4.00%	2.72%	2.05%	2.92%	7	P	5.92%	7.11%	2.88%	4.58%
	F	4.00%	2.72%	2.05%	2.92%		F	7.24%	6.46%	3.04%	4.07%
	T	7.79%	2.67%	2.20%	3.54%		T	8.93%	7.18%	3.14%	4.27%
2	P	7.79%	2.75%	2.20%	3.54%	12	P	6.11%	4.78%	3.01%	4.64%
	F	7.79%	2.67%	2.20%	3.54%		F	8.08%	7.02%	3.03%	4.27%
	T	7.01%	6.44%	2.70%	4.27%		T	12.04%	6.18%	2.69%	4.27%
4	P	7.12%	6.32%	2.70%	4.27%	24	P	5.83%	2.88%	0.85%	5.05%
	F	8.82%	6.44%	2.63%	4.03%		F	12.76%	7.46%	2.75%	4.27%
	T	8.80%	7.12%	2.83%	4.27%		T	11.04%	5.30%	2.15%	3.92%

by propagating anchor text weights from past time points, whereas the query results with shorter ages have better improvements by propagating predicted weights from future time points over all window sizes. By combining the weights on both past time points and future time points, the relative improvement is greater than only combining weights in one direction for most buckets in different window sizes.

6   Conclusion

The dynamic page in-links and associated anchor text reflect how other pages view destination page changes over time. However, the ever-changing weights on anchor text, as an indicator of the change of anchor text importance, is seldom used for web search, partly because such information is not available. In this paper, we utilize the historical archival copies of web pages provided by the Internet Archive (a public resource) to investigate the way to benefit web search. We propose new methods to quantify anchor text importance, which is motivated by differentiating pages with different in-link creation rate over time and different historical in-link context. Evaluation experiments on one crawl of Stanford WebBase show the ranking performance of our proposed methods has more than 10% improvement over a baseline without considering historical information.

From this work, we realize that the current archival web pages only cover a small portion of the historical web, which causes a large amount of missing anchors (only 2.57% anchors have archival copies in our data set) and thus limits the application of the proposed method. Furthermore, the crawling policies used to collect these archival web page copies might not accurately records the trace of web activities. However, as an initial work, our results unraveled that with enough historical information for pages on the web, we can potentially give more accurate estimation about anchor text importance and page in-link importance to improve web search. As future work, we hope to find a way to mitigate the gap caused by the anchors having archival copies and those without in searching process, such that the method can be completely applicable.

Acknowledgments

This work was supported in part by a grant from the National Science Foundation under award IIS-0803605 and an equipment grant from Sun Microsystems.

References

1. Acharya, A., Cutts, M., Dean, J., Haahr, P., Henzinger, M., Hoelzle, U., Lawrence, S., Pfleger, K., Sercinoglu, O., Tong, S.: Information retrieval based on historical data. US Patent 7346839 (March 2008)
2. Berberich, K., Vazirgiannis, M., Weikum, G.: Time-aware authority ranking. In: Leonardi, S. (ed.) WAW 2004. LNCS, vol. 3243, pp. 131–142. Springer, Heidelberg (2004)
3. Brin, S., Page, L.: The anatomy of a large-scale hypertextual Web search engine. In: Proc. of the 7th Int'l. World Wide Web Conf. (April 1998)
4. Cho, J., Garcia-Molina, H., Haveliwala, T., Lam, W., Paepcke, A., Raghavan, S., Wesley, G.: Stanford WebBase components and applications. ACM Trans. on Internet Technology 6(2), 153–186 (2006)
5. Dakka, W., Gravano, L., Ipeirotis, P.G.: Answering general time sensitive queries. In: Proc. of the 17th ACM Conf. on Information and Knowledge Management (CIKM), pp. 1437–1438. ACM, New York (2008)
6. de Kretser, O., Moffat, A.: Effective document presentation with a locality-based similarity heuristic. In: Proc. 22nd Annual Int'l ACM SIGIR Conf. on Research and Dev. in Information Retrieval (July 1999)
7. Dou, Z., Song, R., Nie, J.-Y., Wen, J.-R.: Using anchor texts with their hyperlink structure for web search. In: Proc. 32nd Annual Int'l. ACM SIGIR Conf. on Research and Dev. in Information Retrieval (July 2009)
8. Eiron, N., McCurley, K.S.: Analysis of anchor text for web search. In: Proc. 26th Annual Intl. ACM SIGIR Conf. on Research and Dev. in Information Retrieval (July 2003)
9. Fujii, A.: Modeling anchor text and classifying queries to enhance web document retrieval. In: Proc. of the 17th Int'l. World Wide Web Conf. (April 2008)
10. Internet Archive. The Internet Archive (2009), http://www.archive.org/
11. Jarvelin, K., Kekalainen, J.: IR evaluation methods for retrieving highly relevant documents. In: Proc. of the 23rd Annual Int'l ACM SIGIR Conf. on Research and Development in Information Retrieval, July 2000, pp. 41–48 (2000)
12. Jones, R., Diaz, F.: Temporal profiles of queries. ACM Trans. Inf. Syst. 25(3), 14 (2007)
13. Kise, K., Junker, M., Dengel, A., Matsumoto, K.: Passage retrieval based on density distributions of terms and its applications to document retrieval and question answering. In: Dengel, A.R., Junker, M., Weisbecker, A. (eds.) Reading and Learning. LNCS, vol. 2956, pp. 306–327. Springer, Heidelberg (2004)
14. Kraft, R., Zien, J.: Mining anchor text for query refinement. In: Proc. of the 13th Int'l World Wide Web Conf., pp. 666–674. ACM, New York (2004)
15. Lee, U., Liu, Z., Cho, J.: Automatic identification of user goals in web search. In: Proc. of the 14th Int'l World Wide Web Conf., pp. 391–400. ACM Press, New York (2005)
16. Li, X., Croft, W.B.: Time-based language models. In: Proc. of the 20th Int'l. Conf. on Information and knowledge management, pp. 469–475. ACM, New York (2003)
17. Lu, W.-H., Chien, L.-F., Lee, H.-J.: Anchor text mining for translation of web queries: A transitive translation approach. ACM Trans. Inf. Syst. 22(2), 242–269 (2004)
18. Lv, Y., Zhai, C.: Positional language models for information retrieval. In: Proc. 32nd Annual Int'l. ACM SIGIR Conf. on Research and Dev. in Information Retrieval (July 2009)
19. Metzler, D., Novak, J., Cui, H., Reddy, S.: Building enriched document representations using aggregated anchor text. In: Proc. 32nd Annual Int'l ACM SIGIR Conf. on Research and Dev. in Information Retrieval (July 2009)
20. NIST. Text REtrieval Conference (TREC) home page (2008), http://trec.nist.gov/

21. Petkova, D., Croft, W.B.: Proximity-based document representation for named entity retrieval. In: Proc. of the 16th ACM Conf. on information and knowledge management, pp. 731–740. ACM, New York (2007)
22. Rao, S.S.: On multiple regression models with nonstationary correlated errors. Biometrika 91(3), 645–659 (2004)
23. Robertson, S., Zaragoza, H., Taylor, M.: Simple bm25 extension to multiple weighted fields. In: Proc. of the 13th ACM Int'l. Conf. on Information and knowledge management (CIKM), pp. 42–49. ACM, New York (2004)
24. Zaragoza, H., Craswell, N., Taylor, M., Saria, S., Robertson, S.: Microsoft cambridge at trec 13: Web and hard tracks. In: TREC '13: Proceedings of the thirteenth Text REtrieval Conference (2004)

Predicting Query Performance via Classification

Kevyn Collins-Thompson and Paul N. Bennett

Microsoft Research
1 Microsoft Way
Redmond, WA USA 98052
{kevynct,paul.n.bennett}@microsoft.com

Abstract. We investigate using topic prediction data, as a summary of document content, to compute measures of search result quality. Unlike existing quality measures such as query clarity that require the entire content of the top-ranked results, class-based statistics can be computed efficiently online, because class information is compact enough to pre-compute and store in the index. In an empirical study we compare the performance of class-based statistics to their language-model counterparts for two performance-related tasks: predicting query difficulty and expansion risk. Our findings suggest that using class predictions can offer comparable performance to full language models while reducing computation overhead.

1 Introduction

When the performance of an information retrieval system on a query can be accurately predicted, an informed decision can be made as to whether the query should be expanded, reformulated, biased toward a particular intent or altered in some other way. Increasing evidence points to the fact that valuable clues to a query's ambiguity and quality of corresponding results can be gleaned from query pre-retrieval features, and post-retrieval properties of the query's result set [9]. For example, the *query clarity* score [7] measures the divergence of a language model over the top-ranked pages from the generic language model of the collection. A separate but related performance prediction problem is to assess the *likely effect of query expansion* for a given query. Because query expansion is both inherently risky and adds further computational expense, methods for predicting the likely success of expansion and correctly scaling back expansion when it is unlikely to be effective are both valuable.

However, existing research in this area has been somewhat incomplete. Figure 1 gives a graphical summary of different pre- and post-retrieval models being compared, highlighting existing and missing work in the current body of research. First, properties of the top-ranked documents retrieved using an expanded query may not only be informative in relation to the original result set but also in relation to pre-retrieval features. Second, while a shift in word distribution between the collection, initial top-ranked results, and expansion results may be informative, because of vocabulary variation these comparisons are necessarily noisy.

C. Gurrin et al. (Eds.): ECIR 2010, LNCS 5993, pp. 140–152, 2010.

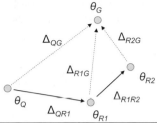

Symbol	Name	Study
Δ_{QG}	Simplified clarity	He & Ounis [10]
Δ_{QR1}	Query drift/coverage	Winaver & Kurland [21]
Δ_{R1G}	Clarity	Cronen-Townsend & Croft [7]
	Improved clarity	Hauff, Murdock, Baeza-Yates [9]
Δ_{R1R2}	Expansion drift	Zhou & Croft[23] (variant)
Δ_{R2G}	Expansion clarity	This study.

Fig. 1. Graphical depiction of model divergences. The general collection model is shown as θ_G. Inter-model KL-divergences of interest are shown as directed arrows, e.g. $\Delta_{QG} = KL(Q||G)$, *i.e.*, the distance between the query model and background model. A summary of related work on each model comparison is also shown.

To help alleviate this noise in the comparison and capture more of the underlying semantics of queries and documents, we investigate performing a precomputed classification of documents into a set of topics, such as defined by the Open Directory Project (ODP) [15], via models learned from labeled data. With this pre-computed classification, we later perform fast online comparisons in *topic space* to help restore this focus on semantic distance. Analogous to traditional clarity, we introduce *topic clarity* counterparts to the traditional language model components in Figure 1 and investigate their effectiveness.

A significant drawback of methods that analyze the result set is that they must incur the computational cost of performing an initial retrieval, as well as the cost of processing the full text of each top-ranked document. Since performance prediction is only one part of the entire retrieval process, adding computational load at intermediary steps is undesirable, especially in applications like Web search where speed is critical. Thus, we also examine whether the benefits of result-set analysis like query clarity can be approximated with less computational cost than using a full language model.

Throughout our analysis, we focus on how effective different model-divergence features are at predicting two types of query performance measure: *query difficulty*, which measures retrieval risk via the average precision (AP) of the top-ranked results; and *expansion risk* which estimates the likely magnitude of the relative gain or loss in AP obtained from using query expansion. Predicting the latter directly is an interesting problem since whether or not to do expansion may be the end goal. Furthermore predicting query difficulty and expansion risk are distinct problems that are only weakly correlated [3].

Our main contributions are as follows. We introduce new models and representations for estimating two important measures of query performance: query difficulty and expansion risk. Our work brings together features from previous studies on query difficulty based on divergences between language models of the query, collection and initial results. We extend this to include a model of *expansion results* from the expanded query. With these models and features, we compare the performance of two model representations: a low-dimensional pre-computed topic representation and a much larger unigram language model over two standard Web collections. We also develop a simple, effective method for deriving a topic representation, modeled as a distribution over ODP categories, of a query by estimating and combining pre-computed topic representations from the individual query terms.

2   Related Work

A number of previous models for query performance prediction can be viewed as special cases within a framework where various distances are calculated between a global background model of the collection, θ_G, a query model using pre-retrieval features, θ_Q, a language model based on the results of the original query, θ_{R1}, and a language model based on the results of the expanded query, θ_{R2}. A summary of related work comparing different query and expansion models is shown in Fig. 1.

In Section 3 we give analogues to each of these that can be computed using document topic prediction data. We focus on the change between pre- and post-retrieval models relative to the global background model, since this is where the majority of effects are observed. This comprises the following arcs. The divergence Δ_{QG}, which we call *simplified clarity*, is a pre-retrieval measure of query specificity and compares the query against the collection model. The post-retrieval divergence Δ_{QR1} measures *query drift* in the initial results. The divergence Δ_{R1G} is the analogue of traditional clarity, measuring the similarity of the results model to the generic collection model. The divergence Δ_{R2G} is a new additional measure that we call *expansion clarity* that estimates the specificity of the expanded results compared to the collection. We also include for completeness Δ_{R1R2}, the drift from the initial results to the expanded results.

Our examination of model drift extends recent studies that find *variance* to be an important facet of predicting query performance. More specifically, the sensitivity of some aspect of the retrieval process to variation in input or model parameters has been shown to be effective in varying degrees. This includes variance of results ranking (by varying document models) [23], query variation [22], query term *idf* weights [20] and document scores [8]. Aslam & Pavlu [2] introduced variation in the retrieval function instead of the model, by combining TREC runs from multiple systems for the same query.

While the above studies have looked at query difficulty, few have looked at predicting for expansion risk or difficulty. The significant downside risk of query expansion has been noted for decades [17] but has been largely neglected as an evaluation criterion in favor of average performance, with some recent exceptions [6] [14][1]. For query expansion algorithms to become more reliable, it will be important for them to correctly identify and manage risk for queries. We define *expansion risk* in this study to be the magnitude of the relative gain or loss in average precision from applying query expansion, relative to the unexpanded query. Thus, queries with small expansion risk are unlikely to be affected one way or the other by the application of expansion.

A variety of other work has examined query classification and use of class labels. Recently [16] quantified query ambiguity using ODP metadata for individual query terms, and [18] examined the category spread of top-ranked documents to identify ambiguous queries. In contrast to these studies our focus is on establishing and comparing analogues for query performance prediction based on class labels.

3   Methods

Statistics for predicting performance properties of a query can be categorized by the type of observations required to calculate them. Basic *pre-retrieval statistics* use features of the query alone, such as query length or query term *idf* values, without requiring document retrieval using the query [10]. *Post-retrieval statistics* require at least one retrieval step where documents are ranked. The content and/or meta-data of the resulting documents then give us additional information for estimation. The efficiency of post-retrieval statistics depends on the particular document representation used: topic predictions may be pre-computed and do not require fetching or analyzing potentially large documents at run-time. Although document language models may also be precomputed, they use a much larger representation proportional to the vocabulary size. In addition, any document similarity or distance computations for clustering or smoothing are also of correspondingly higher cost.

In the following sections we denote the collection by G, the query by Q, and assume that a set $R1$ of k documents is returned from G in response to Q. Furthermore, after applying query expansion to Q to obtain an expanded query Q', we obtain a set $R2$ of k documents in response to Q'.

Figure 1 shows the models and the relations between them that are of interest in this study. We use the notation Δ_{AB} to denote a divergence measure between two models A and B. For example, in the context of language-model based statistics Δ_{AB} denotes the KL-divergence $KL(A||B)$ between models A and B. Since KL-divergence is not symmetric, the ordering of A and B is important, and we use an arrow in Figure 1 to specify the direction of comparison.

3.1   Language-Model Based Statistics

As is standard, we use unigram language models as the representation basis for computing the language-model based statistics. This is a K-dimensional vector representing the parameters of a multinomial distribution over the K words in the vocabulary. Model similarity is computed using KL-divergence with Dirichlet smoothing, with KL-divergence defined as $\Delta(u, v) = \sum_i u_i \log \frac{u_i}{v_i}$ for language model distributions u and v.

3.2   Topic Based Statistics

We chose to use the ODP [15] for classification because of its broad, general-purpose topic coverage and availability of reasonably high-quality training data. Using a crawl of ODP from early 2008, we first split the data into a 70%/30% train/validation set, then identified the topic categories (some categories like "regional" are not topical and were discarded) that had at least 1K documents as good candidates for models that could be learned well and would be broadly applicable – resulting in 219 categories. We leave study of comparing distances in a hierarchy to future work and simply flattened the two levels to a m-of-T (where $T = 219$) prediction task. We then augmented the search index for every document with at least one and up to 3 predictions for each document, assuming the predictions surpass a minimal confidence threshold (approx. 0.05). Thus, minimal index bloat is incurred.

When aggregating the topic distribution for a result set, the topic representation θ is a T-dimensional vector, with one element per ODP class containing the average document class probability for that class. We computed the topic representations θ_G and θ_{R1} by aggregating the topic distribution for all documents in the collection and result set respectively. Model similarity between representations u and v is computed using the 'city block' (or Manhattan) metric

$$\Delta(u, v) = 1/2 \cdot \sum_{i=1}^{T} |u_i - v_i|. \tag{1}$$

We chose this standard symmetric similarity measure due to the nature of the class prediction vector, which unlike language models, is typically not normalized because documents can belong to more than one class, and because magnitude information is important to retain to assess topic prediction confidence.

Because the user's query is expressed in words and not topic categories, we must somehow compute a topic representation θ_Q of the query Q to obtain the pre-retrieval topic-based statistic. We do this in two steps. First, we pre-compute off-line a topic distribution θ_w for each word w in the corpus, by aggregating the predicted classes of the documents in which the word occurs. Then, for a given query we combine the topic representations for its individual terms using an operator of the form

$$\theta_Q[t] \propto \prod_{w \in Q} (\theta_w[t] + \epsilon) \tag{2}$$

which after expanding and collecting like terms in ϵ, can be written as

$$\theta_Q[t] \propto \prod_{w \in Q} \theta_w[t] + \epsilon \sum_{w_1 \in Q} \prod_{v \in Q \backslash w_1} \theta_v[t] + \epsilon^2 \sum_{w_1, w_2 \in Q} \prod_{v \in Q \backslash w_1, w_2} \theta_v[t] \tag{3}$$

$$+ \ldots + \epsilon^{N-1} \sum_{w \in Q} \theta_w[t] + \epsilon^N. \tag{4}$$

The parameter ϵ controls the conjunctive behavior of the operator: setting $\epsilon = 0$ gives a pure multiplicative AND operator, and increasing ϵ relaxes this condition, so that $\epsilon = 1$ gives all subsets of Q's terms equal weight. Large values of $\epsilon >> 1$ give increasing OR-like behavior that emphasize the sum over terms, rather than the product. In our experiments we focus on conservative AND-like behavior by using a value of $\epsilon = 0.001$. This approach is easily and efficiently generalized to an inference network, where the query terms are evidence nodes and a richer set of operators is possible, such as those in the Indri retrieval system [19].

Examples of the resulting topic distribution for three different queries are shown in Figure 2. The horizontal axis gives the (flattened) ODP level 1 and 2 categories, while the vertical axis gives $P(c|Q)$, the probability of category c given the query Q. We note that this new pre-retrieval topic-based query representation has many uses beyond performance prediction applications, such as providing an additional set of features for estimating query similarity.

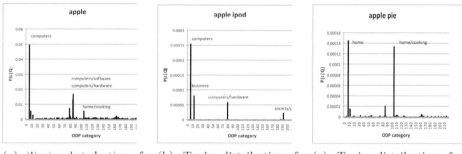

(a) Topic distribution for query 'apple'   (b) Topic distribution for query 'apple ipod'   (c) Topic distribution for query 'apple pie'

Fig. 2. Example showing how the ODP category distribution profiles for different queries can reflect ambiguity or clarity in topic. The ambiguous query 'apple' has two main senses of 'computer' and 'home/cooking', with the computer sense predominating. Refining the 'apple' query to 'apple ipod' (2b) focuses on the computer topic, while refining to 'apple pie' (2c) focuses on the 'cooking' sense.

4   Evaluation

Our evaluation is structured as follows. After describing our datasets and experimental setup, we first examine the cases where the topic (TP) representation produces features with comparable predictive power to their language model (LM) counterparts. We do this for both query difficulty and expansion risk prediction tasks. Second, we examine the predictive power of the information in the results from the *expanded* query via the resulting new expansion clarity feature, the divergence Δ_{R2G}, as well as the related expansion drift feature Δ_{R1R2}. Third, we examine combined models in which both TP and LM features are used to predict query difficulty and expansion risk.

4.1   Datasets and Experimental Setup

Our evaluation is based on two TREC Web datasets that have been widely used for query difficulty prediction: wt10g (1.7m pages, topics 451–550) and gov2 (25m pages, topics 701-850). Also, query performance prediction is known to be more difficult for these Web topics [9]. Indexing and retrieval were performed using the Indri system in the Lemur toolkit [13].

Our queries were derived from the title field of the TREC topics. Phrases were not used. We wrapped the initial query terms with Indri's `#combine` operator, performed Krovetz stemming, and used a stoplist of 419 common English words. To compute the query expansion baseline we used the default expansion method in Indri 2.2, which first selects terms using a log-odds calculation, then assigns final term weights using the Relevance Model [12]: document models were Dirichlet-smoothed with $\mu = 1000$. Indri's feedback model is linearly interpolated with the original query model weighted by a parameter α. By default we used the top 50 documents for feedback and the top 20 expansion terms, with the feedback interpolation parameter $\alpha = 0.5$ unless otherwise stated.

4.2   Comparing Topic and Language Model Representations

Our goal in this section is to compare the predictive power of TP and LM representations for the model divergence features shown in Figure 1 as well as some basic pairwise ratios of these features.

Query difficulty. The Kendall's tau correlations with average precision for each feature are shown in Table 1. We note that our relatively low query clarity Δ_{R1G} correlation is in line with published studies using similar methods [9] for the same collection. On both collections, the LM version of traditional query clarity Δ_{R1G} gave a higher correlation with AP than its TP counterpart. Performance for the post-expansion drift feature Δ_{R1R2}, however, was not only better than query clarity, but TP and LM performance was comparable: the TP improvement over LM for Δ_{R1R2} was significant for gov2 and statistically equivalent for wt10g. The best performing TP feature on both wt10g and gov2 was Δ_{R1R2} (correlation = 0.11 and 0.25 respectively). The best performing LM feature on wt10g was Δ_{QR1}/Δ_{QG} (correlation = 0.26) and for gov2 $\Delta_{R2G}/\Delta_{R1G}$ (correlation = 0.20).

Table 1. Query difficulty: Predictive power of different model divergence features according to Kendall-τ correlation with average precision. Document representation (DocRep) is either TP (topic prediction) or LM (language model). Superscripts $\star$ and $+$ denote significance of $p < 0.01$ and $p < 0.10$ respectively.

	DocRep	Δ_{QG}	Δ_{R1G}	Δ_{QR1}	Δ_{R1R2}	Δ_{R2G}	$\frac{\Delta_{QR1}}{\Delta_{QG}}$	$\frac{\Delta_{R1G}}{\Delta_{QG}}$	$\frac{\Delta_{R2G}}{\Delta_{R1G}}$
wt10g	TP	0.013	0.089	0.077	0.110	0.060	0.091	0.033	0.000
	LM	0.032	0.126	0.256	0.140	0.026	$0.260^\star$	$0.161^\star$	0.231^+
gov2	TP	0.108^+	0.047	0.069	$0.250^\star$	0.010	$0.130^\star$	0.001	0.100
	LM	0.001	$0.137^\star$	0.071	0.151	0.011	0.077	$0.141^\star$	0.204^+

Table 2. Expansion risk: Kendall-τ correlation of different model divergence features. Document representation (DocRep) is either TP (topic prediction) or LM (language model). Superscripts $\star$ and $+$ denote significance of $p < 0.01$ and $p < 0.10$ respectively.

	DocRep	Δ_{QG}	Δ_{R1G}	Δ_{QR1}	Δ_{R1R2}	Δ_{R2G}	$\frac{\Delta_{QR1}}{\Delta_{QC}}$	$\frac{\Delta_{R1G}}{\Delta_{QC}}$	$\frac{\Delta_{R2G}}{\Delta_{R1G}}$
wt10g	TP	$0.320^\star$	0.052	$0.322^\star$	$0.300^\star$	0.169^+	$0.355^\star$	$0.330^\star$	$0.280^\star$
	LM	$0.250^\star$	$0.240^\star$	0.071	$0.260^\star$	0.150^+	0.019	$0.225^\star$	0.110
gov2	TP	0.063	0.124^+	0.048	$0.260^\star$	0.040	0.070	0.090^+	0.188^+
	LM	0.001	0.100^+	0.060	$0.201^\star$	0.040	0.060	0.100^+	$0.281^\star$

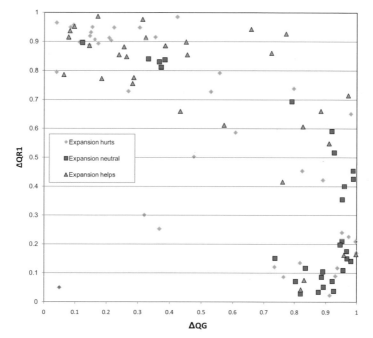

Fig. 3. Example showing how expansion-neutral ($< 15\%$ AP gain/loss) wt10g queries (dark squares) typically have high topic specificity (TP:Δ_{QG}) and low post-retrieval topic drift (TP:Δ_{QR1})

Expansion risk. Recall that expansion risk is defined as the magnitude of the relative gain or loss in average precision of applying the expansion algorithm, compared to the unexpanded query. Kendall's-tau correlations are shown in Table 2. Although LM-based features were more effective at predicting query difficulty, TP-based features were generally more effective at predicting expansion risk, especially when multiple features were combined (e.g. as in the ratio Δ_{QR1}/Δ_{QG}). Figure 3 shows how combining information from both Δ_{QR1} and Δ_{QG} helps isolate expansion-neutral queries – those queries for which expansion is unlikely to be effective. Queries with higher Δ_{QG} are more specific, being farther from the general collection model. At the same time, queries with low topic query drift Δ_{QR1} have results that match the expected topic profile based on the query terms alone. In this example, queries that are both topic-specific and with focused results are more unlikely to be affected by applying query expansion.

4.3   Predictive Power of Expansion Clarity (Δ_{R2G}) and Expansion Drift (Δ_{R1R2}) Features

The expansion clarity (Δ_{R2G}) and expansion drift (Δ_{R1R2}) features are interesting because they use additional new 'post-expansion' evidence: the results of the expanded query, not just the initial query. We find that such post-expansion features are indeed more effective and stable than features based only on pre-expansion models. For example, the expansion drift feature Δ_{R1R2}, which is dependent on both initial and expansion results models, is remarkably effective and stable across all tasks, representations, and collections compared to any pre-expansion feature. Looking at the Δ_{R1R2} column in Tables 1 and 2, we can see that the Δ_{R1R2} feature is consistently among the best-performing features for

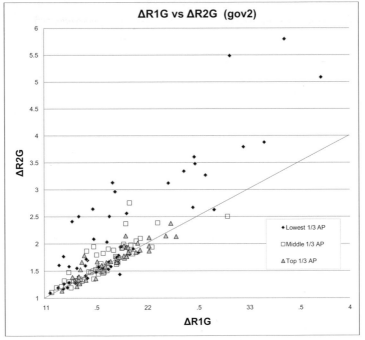

(a) Query clarity (Δ_{R1G}) vs. expansion clarity (Δ_{R2G})

Fig. 4. Queries whose initial results clarity (Δ_{R1G}) is hurt by expansion (higher Δ_{R2G}) appear as points above the line and are substantially more likely to have poor initial average precision. Query clarity is on the x-axis and expansion clarity on the y-axis. Shown are queries partitioned into the lowest-, mid-, and highest-scoring (AP) third for the gov2 corpus. Results for wt10g are similar and not shown for space reasons.

either TP or LM representations: it is the top-performing TP feature for predicting both wt10g and gov2 query difficulty and for gov2 expansion risk (with excellent performance on wt10g). For LM, in 3 out of 4 cases it is one of the top two best features, second only to the ratio $\frac{\Delta_{R2G}}{\Delta_{R1G}}$, which uses the additional information about the collection model. Figure 4 gives further insight into how adding expansion clarity Δ_{R2G} to the basic query clarity Δ_{R1G} feature helps discriminate the most difficult queries more effectively than query clarity alone.

4.4 Combining Topic- and LM-Based Features for Prediction

To analyze the interaction of the input variables, we used the WinMine v2.5 toolkit [5] to build a regression-based predictor of average precision using a 70/30

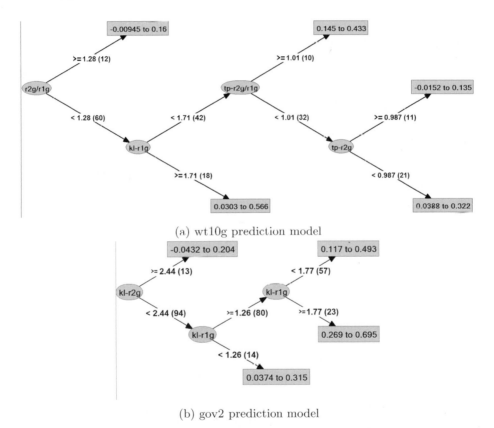

(a) wt10g prediction model

(b) gov2 prediction model

Fig. 5. Prediction models for query difficulty for the wt10g (top) and gov2 (bottom) collections, estimated using a Bayesian decision tree learning algorithm, using both topic and LM model divergences as input features. Inequalities at the branches indicate the threshold value for the input variable (shown in the ellipses). The resulting value of the target variable (average precision) is shown with a range of one standard deviation in the shaded rectangle. Both models have selected Δ_{R1G} (query clarity) and Δ_{R2G} (expansion clarity) together as the primary factors in predicting average precision.

train/test split. In particular, we used WinMine to build dependency networks – essentially a decision tree built using a Bayesian machine learning algorithm [4,11]. We chose this representation for its amenability to qualitative analysis. Note that a more direct comparison to the ranking correlations presented above would require training a ranking model. We defer that problem to future work and simply present the harder task of predicting the actual value of the dependent variable as a means of studying the interaction of the input variables.

The resulting decision trees for query difficulty on the gov2 and wt10g corpora are shown in Figure 5. The measure of accuracy was Root Mean Squared Error (RMSE), where the dependent variable to be predicted was average precision and the input variables were the model divergences from Figure 1 for both the topic and language model representations. Models for both corpora were able to attain better performance (lower RMSE) than a default baseline that simply predicted the distribution mean. The gov2 model using all the features attained an RMSE of 0.147, compared to a default baseline with a higher RMSE of 0.198. The wt10g model using all the features attained an RMSE of 0.163, compared to a default baseline RMSE of 0.175.

While the specific models estimated for each collection are different, they both rely exclusively on the Δ_{R1G} and Δ_{R2G} divergences (or the ratio between them) as the two primary prediction factors, ignoring Δ_{QG}, Δ_{QR1}, and Δ_{R1R2} in both topic and LM representations. This suggests that query clarity and expansion clarity together are most effective at summarizing the trajectory of topic drift that occurs when query expansion is applied, compared to other features, or either clarity feature alone. We also note that the model estimated using wt10g relies on a combination of both topic and language model features to achieve lower RMSE, making use of the complementary aspects of these representations.

5   Discussion and Conclusion

A significant amount of implicit pre-computation lies behind the topic-based representation of a document or query: from the training of the ODP category classifier from thousands of examples, to index-time labeling of topics for individual pages, and aggregating these for building a run-time mapping from terms to topic distributions. A similar effect might be accomplished for the language model representation by learning a global term-term translation matrix to smooth the query model, but with a corresponding increase in size and complexity, moving from a few hundred static ODP categories for the topic representation, to potentially hundreds of thousands of co-occurring terms per word for the language model representation.

Other algorithms for distilling a topic representation of a query are certainly possible: adding phrases, or making more subtle distinctions between morphological variants of terms, becomes important since typical Web queries are usually less than five words long. For example, the topic distributions for 'cat' and 'cats' (independent of other query terms) could be quite different (e.g. since 'cats' is a musical theatre title). This processing would take place at indexing time and thus could make use of large linguistic resources, and potentially increased computation that might not be practical at query run-time. Using latent topic models trained on the corpus could be another possible research direction, in cases where the corpus is of manageable size or effective methods of sampling representative content are available. Such probabilistic models would also have the advantage of giving another principled way to estimate the probability of a topic given a set of terms.

The evaluation shows that while the LM representation can sometimes give slightly better performance for query difficulty, using pre-computing topic predictions is not far behind for some features. In particular, the topic-based representation is more effective for pre-retrieval prediction (query classification) and superior for predicting expansion risk. This suggests that topic information may often serve as an acceptable, and much more efficient, proxy for predicting query properties and analyzing search results. In addition, our analysis also revealed the value of estimating expansion clarity – the divergence between the expansion top-ranked results and the collection – in either representation, with post-expansion features such as expansion drift being highly effective and stable.

Future applications of topic-based representations include more robust, efficient query similarity measures and measures of result diversity. Also interesting to consider are more sophisticated inference methods for estimating a topic distribution from a query based on the use of additional term dependency features.

Acknowledgments

We thank Sue Dumais for valuable discussions and feedback on this work.

References

1. Amati, G., Carpineto, C., Romano, G.: Query difficulty, robustness, and selective application of query expansion. In: McDonald, S., Tait, J.I. (eds.) ECIR 2004. LNCS, vol. 2997, pp. 127–137. Springer, Heidelberg (2004)
2. Aslam, J.A., Pavlu, V.: Query hardness estimation using Jensen-Shannon divergence among multiple scoring functions. In: Amati, G., Carpineto, C., Romano, G. (eds.) ECIR 2007. LNCS, vol. 4425, pp. 198–209. Springer, Heidelberg (2007)
3. Billerbeck, B.: Efficient Query Expansion. PhD thesis, RMIT University, Melbourne, Australia (2005)
4. Chickering, D., Heckerman, D., Meek, C.: A Bayesian approach to learning Bayesian networks with local structure. In: UAI 1997, pp. 80–89 (1997)
5. Chickering, D.M.: The WinMine toolkit. Technical Report MSR-TR-2002-103, Microsoft, Redmond, WA (2002)
6. Collins-Thompson, K., Callan, J.: Estimation and use of uncertainty in pseudo-relevance feedback. In: Proceedings of SIGIR 2007, pp. 303–310 (2007)
7. Cronen-Townsend, S., Croft, W.: Quantifying query ambiguity. In: Proceedings of HCL 2002, pp. 94–98 (2002)
8. Diaz, F.: Performance prediction using spatial autocorrelation. In: Proceedings of SIGIR 2007, pp. 583–590 (2007)

9. Hauff, C., Murdock, V., Baeza-Yates, R.: Improved query difficulty prediction for the web. In: Proceedings of CIKM 2008, pp. 439–448 (2008)
10. He, B., Ounis, I.: Query performance prediction. Information Systems 31, 585–594 (2006)
11. Heckerman, D., Chickering, D., Meek, C., Rounthwaite, R., Kadie, C.: Dependency networks for inference, collaborative filtering, and data visualization. Journal of Machine Learning Research 1, 49–75 (2000)
12. Lavrenko, V.: A Generative Theory of Relevance. PhD thesis, University of Massachusetts, Amherst (2004)
13. Lemur. Lemur toolkit for language modeling & retrieval (2002), http://www.lemurproject.org
14. Metzler, D., Croft, W.B.: Latent concept expansion using Markov Random Fields. In: Proceedings of SIGIR 2007, pp. 311–318 (2007)
15. Netscape Communication Corp. Open directory project, http://www.dmoz.org
16. Qiu, G., Liu, K., Bu, J., Chen, C., Kang, Z.: Quantify query ambiguity using ODP metadata. In: Proceedings of SIGIR 2007, pp. 697–698 (2007)
17. Smeaton, A., van Rijsbergen, C.J.: The retrieval effects of query expansion on a feedback document retrieval system. The Computer Journal 26(3), 239–246 (1983)
18. Song, R., Luo, Z., Wen, J.-R., Yu, Y., Hon, H.-W.: Identifying ambiguous queries in web search. In: Proceedings of WWW 2007, pp. 1169–1170 (2007)
19. Strohman, T., Metzler, D., Turtle, H., Croft, W.B.: Indri: A language model-based search engine for complex queries. In: Proceedings of the International Conference on Intelligence Analysis (2004)
20. Vinay, V., Cox, I.J., Milic-Frayling, N., Wood, K.: On ranking the effectiveness of searches. In: Proceedings of SIGIR 2005, pp. 398–404 (2005)
21. Winaver, M., Kurland, O., Domshlak, C.: Towards robust query expansion: model selection in the language modeling framework. In: Proceedings of SIGIR 2007, pp. 729–730 (2007)
22. YomTov, E., Fine, S., Carmel, D., Darlow, A.: Learning to estimate query difficulty. In: Proceedings of SIGIR 2005, pp. 512–519 (2005)
23. Zhou, Y., Croft, W.B.: Ranking robustness: a novel framework to predict query performance. In: Proceedings of CIKM 2006, pp. 567–574 (2006)

A Case for Automatic System Evaluation

Claudia Hauff[1], Djoerd Hiemstra[1], Leif Azzopardi[2], and Franciska de Jong[1]

[1] University of Twente, Enschede, The Netherlands
{c.hauff,hiemstra,f.m.g.dejong}@ewi.utwente.nl
[2] University of Glasgow, Glasgow, UK
leif@dcs.gla.ac.uk

Abstract. Ranking a set retrieval systems according to their retrieval effectiveness without relying on relevance judgments was first explored by Soboroff *et al.* [13]. Over the years, a number of alternative approaches have been proposed, all of which have been evaluated on early TREC test collections. In this work, we perform a wider analysis of *system ranking estimation* methods on sixteen TREC data sets which cover more tasks and corpora than previously. Our analysis reveals that the performance of system ranking estimation approaches varies across topics. This observation motivates the hypothesis that the performance of such methods can be improved by selecting the "right" subset of topics from a topic set. We show that using topic subsets improves the performance of automatic system ranking methods by 26% on average, with a maximum of 60%. We also observe that the commonly experienced problem of underestimating the performance of the best systems is data set dependent and not inherent to system ranking estimation. These findings support the case for automatic system evaluation and motivate further research.

1   Introduction

Ranking retrieval systems according to their retrieval effectiveness *without* relying on costly relevance judgments was first explored by Soboroff *et al.* [13]. The motivation for this research stems from the high costs involved in the creation of test collections and in particular human relevance assessments. If system evaluation without relevance assessments is achievable, then the cost of evaluation could be greatly reduced. Automatic system evaluation could also be useful in the development of methods for data fusion and source selection.

In recent years, a number of *system ranking estimation* approaches have been proposed [5,12,13,14,16], which attempt to rank a set of retrieval systems (for a given topic set and a test corpus) without human relevance judgments. All approaches estimate a performance-based ranking of systems by considering the relationship of the top retrieved documents across systems. While the initial results highlighted the promise of this new direction, the utility of these approaches remains unclear. This is mainly because they usually underestimate the performance of the best systems, which is attributed to the "tyranny of the masses" effect [5]. In the analysis presented in this paper, we will show this

C. Gurrin et al. (Eds.): ECIR 2010, LNCS 5993, pp. 153–165, 2010.

problem not to be inherent to system ranking estimation methods. In previous work [5,12,13,14,16], the evaluations were mostly performed on the TREC-{3,5,6,7,8} data sets[1]. In this work, perform a much wider analysis. We consider sixteen different TREC data sets, including a range of non-adhoc task data sets (such as expert search) and adhoc tasks on non-traditional corpora (such as the blog corpus). We find that the extent of mis-ranking the best systems varies considerably between data sets and is indeed strongly related to the degree of human intervention in the manual runs of a data set[2].

We also investigate the number of topics required to perform system ranking estimation. In all existing approaches, the retrieval results of the full TREC topic set are relied upon to form an estimate of system performance. However, in [11] it was found that some topics are better suited than others to differentiate the performance of retrieval systems. We hypothesize and verify experimentally, that with the right subset of topics, the current methods for estimating system rankings without relevance judgment can be significantly improved. These findings suggest that under certain conditions, automatic system evaluation is a viable alternative to human relevance-judgments based evaluations.

This paper is organized as follows: in Sec. 2 we provide an overview of related work and the motivation for relying on subsets of topics for system ranking. The research questions and the experimental setup are outlined in Sec. 3. The result section (Sec. 4) contains (i) a comparison of four ranking estimation approaches, (ii) motivational experiments to show the validity of using topic subsets, and, (iii) a first attempt to automatically find "good" subsets of topics. In Sec. 5 conclusions are drawn and directions for future work are discussed.

2   Related Work

Research aiming to reduce the cost of evaluation has been conducted along two lines: a number of approaches focus on *reducing* the amount of manual assessments required [2,6,8], while others rely on *fully automatic* evaluation. We only consider approaches of the second category, that is, we focus on algorithms that require no manual assessments at all.

The first work in this area is attributed to Soboroff *et al.* [13]. It was motivated by the fact that the relative ranking of retrieval systems remains largely unaffected by the assessor disagreement in the creation of relevance judgments [15]. This observation led to the proposal to use automatically created *pseudo* relevance judgments which are derived as follows: first, the top retrieved documents of all systems to rank for a topic are pooled together such that a document that

---

[1] When we speak of a data set, such as TREC-3, we mean all retrieval runs submitted to the Text REtrieval Conference (TREC, http://trec.nist.gov/) for the topics of that task. A retrieval run is the output of a retrieval system and thus by ranking retrieval runs, we rank retrieval systems.

[2] In the setting of TREC, a run is labelled automatic, if no human intervention was involved in its creation, otherwise it is considered to be manual (e.g. by providing explicit relevance judgments, manually re-ranking documents etc.).

is retrieved by x systems, appears x times in the pool. Then, a number of documents, the so called *pseudo relevant documents*, are drawn at random from the pool. This process is performed for each topic and the subsequent evaluation of each system is performed with pseudo relevance judgments in place of relevance judgments. A system's effectiveness is estimated by its pseudo mean average precision. To determine the accuracy of the estimated system ranking, it is compared against the ground truth ranking, that is the ranking of systems according to Mean Average Precision (MAP). The experiments in [13] were performed on TREC-{3,5,6,7,8}. The reported correlations were significant, however, one major drawback was discovered: whereas the ranking of the poorly and moderately performing systems was estimated quite accurately, the best performing systems were always ranked too low. It was suggested in [5] that this observation can be explained by the "tyranny of the masses" effect, where the best systems are estimated to perform poorly due to being different from the average system.

The exploitation of pseudo relevant documents has been further investigated by Nuray & Can [12], on very similar data sets (TREC-{3,5,6,7}). In contrast to [13], not all systems to be ranked participate in the derivation of pseudo relevance judgments. The authors experimented with different methods to find a good subset of $P\%$ of systems; overall, the best approach was to select those systems that were most different from the average system. Once a subset of non-average systems is determined, the top b retrieved documents of each selected system are merged and the top $s\%$ of the merged result list constitute the pseudo relevance judgments. The best performing result list merging mechanism was found to be Condorcet voting, where each document in the list is assigned a value according to its rank. This way, not only the frequency of occurrence of a document in various result lists is a factor as in [13], but also the rank the document is retrieved at. The reported correlations were generally higher than in [13]. However, our experiments will show that this is not always the case when evaluating a wider range of data sets.

In [16] it was proposed to rank the systems according to their reference count. The reference count of a system and its ranked list for a particular topic is the number of occurrences of documents in the ranked lists of the other retrieval systems. Experiments on TREC-{3,5,6,7,10} generally yielded lower correlations than in [5,12,13]. A somewhat similar strategy, the structure of overlap method, was proposed by Spoerri [14]. In contrast to [5,12,13,16], not all systems are ranked at once, instead random groupings of five systems are ranked repeatedly. For each grouping and for each of the topics, the percentage $S\%$ of documents in the ranked lists found by only one and the percentage $A\%$ of documents found by all five systems are determined. The three scores $S\%$, $A\%$ and $(S\% - A\%)$ were proposed as estimated system score. These scores are further averaged across the number of topics in the topic set. Since each system participates in a number of groupings, the scores across those groupings are again averaged, leading to the final system score. The reported correlations are significantly higher than in [5,12,13,16]. However, not all the systems that participated in TREC were used, only particular subsets (specifically the best automatic systems). In Sec. 4, we will show that this approach does not perform better than the originally proposed method by Soboroff *et al.* [13] when all available systems are included.

While each system ranking estimation method takes a different approach, a common underlying assumption is that all topics are equally useful when estimating the performance based ranking of systems. However, recent research on evaluation which relies on manual judgments to rank systems has found that only a subset of topics is needed [11]. In order to explore the relationship between a set of topics and a set of systems, Mizzaro & Robertson [11] took a network analysis based view. They proposed the construction of a complete bipartite *Systems-Topic graph* where systems and topics are nodes and a weighted edge between a system and a topic represents the retrieval effectiveness of the pair. While the study in [11] was theoretical in nature, an empirical study has been performed by Guiver *et al.* [9] yielding similar conclusions: that selecting the right subset of topics provides a similar indication of relative system performance to using the full topic set. If the right subset of topics could be selected, then the number of topics needed to compare systems could be greatly reduced. In this work, we examine whether topic subsets could also be used to improve the performance of system ranking estimation methods.

3   An Analysis of System Ranking Estimation Methods

Given the prior research on automatic system ranking estimation, the following analysis of these methods is undertaken to (i) evaluate the main approaches on a wide variety of data sets, (ii) to validate (or not) previous findings, and (iii) to improve the current approaches by using subsets of the full set of topics. Specifically, we examine the following research questions:

1. To what extent does the performance of system ranking estimation approaches depend on the set of systems to rank and the set of topics available?
2. By reducing the topic set size, can the performance of current system ranking estimation methods be improved?
3. Can topic subsets, that improve the performance of system ranking estimation methods, be selected automatically?

We focus the analysis on four system ranking estimation approaches [7,12,13,14] and evaluate them across sixteen different data sets. The following subsections detail the data sets used, the system ranking estimation approaches and the commonly employed evaluation measure of system ranking estimation.

3.1   Data Sets

As in [5,12,13,14,16], we rely on TREC tasks over different years in our experiments. However, whereas earlier studies focused mainly on TREC-{3,5,6,7,8}, we include a much wider variety of data sets. In particular, we evaluated **TREC-{6,7,8}** (adhoc tasks on the TREC Vol. 4+5 corpus), **TREC-{9,10}** (adhoc tasks on the WT10g corpus), **TB-{04,05,06}** (adhoc tasks on the GOV2 corpus with topics from the TeraByte tracks), **CLIR-01** (Cross-Language track 2001), **NP-02** (Named Page Finding track 2001), **EDISC-05** (Enterprise Discussion track 2005), **EEXP-05** (Enterprise Expert Search track 2005), **BLTR-06** (Blog Topical Relevance track 2006), **GEN-07** (Genomics track 2007), **LEGAL-07** (Legal track 2007) and **RELFB-08** (Relevance Feedback track 2008). All data sets can be downloaded from the TREC website.

The number of retrieval systems to rank for each data set varies between 37 and 129, while the number of topics, that are used to evaluate the systems, ranges from 25 to 237 (Tab. 1). Included in our experiments are all available runs, automatic as well as manual and short as well as long runs. We preprocessed the available corpora (TREC Vol. 4+5, WT10g and GOV2) by applying Krovetz stemming [10] and stopword removal.

3.2   Algorithms

Based on the results in the literature, we employ four different system ranking estimation methods: the data fusion (DF) approach by Nuray & Can [12], the random sampling (RS) approach by Soboroff *et al.* [13], the structure of overlap approach (SO) by Spoerri [14], and the document similarity auto-correlation ($ACSim$) approach by Diaz [7].

While DF, RS and SO were introduced in Sec. 2, $ACSim$ has not been applied to system ranking estimation yet. The motivation for evaluating these approaches is their mix of information sources. RS relies on document overlap as shown in [5], while DF takes the rank a system assigns to a document into account. $ACSim$ goes a step further and considers the content similarity of documents. Finally, SO ranks a large number of subsets of systems to achieve a ranking across all systems. Due to space constraints, we only briefly sketch the parameter settings of DF, RS and SO as these methods have already been described in Sec. 2.

Data fusion (DF). We evaluate the three parameters of the approach over a wide range of values: $s = \{1\%, 5\%, 10\%, 20\%, .., 50\%\}$, $P = \{10\%, 20\%, .., 100\%\}$ and $b = \{10, 20, .., 100, 125, .., 250\}$. To determine the parameter setting of a data set, we train on the remaining data sets available for that corpus, e.g. the parameters of TREC-6 are those that lead to the best performance on TREC-$\{7,8\}$. Data sets for training are only available for TREC-$\{6\text{-}10\}$ and TB-$\{04\text{-}06\}$ though. For the remaining data sets, we choose the parameter setting, that gives the best performance across those data sets: $s = 10\%$, $b = 50$ and $P = 100\%$, that is, the best results are achieved when *not* biasing the selection of systems ($P = 100\%$). Since the parameters are optimized on the test set, we expect DF to perform very well in those instances.

Random sampling (RS). We follow the methodology from [13] and rely on the 100 top retrieved documents per system. The percentage of documents to sample from the pool is sampled from a normal distribution with a mean according to the mean percentage of relevant documents in the relevance judgments and a standard deviation corresponding to the deviation between the different topics.

This requires some knowledge about the distribution of relevance judgments; this proves not to be problematic however, as fixing the percentage to a small value yields little variation in the results. As in [13], due to the randomness of the process, we perform 50 trials and average the results.

Structure of overlap (*SO*). As in [14], we rely on the top 50 retrieved documents per system and report the results of the $(S\% - A\%)$ score, as it gives the best results which is in accordance to [14].

Document similarity auto-correlation (*ACSim*). *ACSim* [7][3], is based on the notion that well performing systems are likely to fulfill the cluster hypothesis, while poorly performing systems are not. Based on a document's retrieval scores vector $\mathbf{y}$ of the top 75 retrieved documents, a perturbed score vector $\tilde{\mathbf{y}}$ is derived. Each element y_i is replaced in $\tilde{\mathbf{y}}$ by the weighted average of scores of the 5 most similar documents (based on TF.IDF) in $\mathbf{y}$. If the cluster hypothesis is fullfilled, we expect the most similar documents to also receive a similar score by the retrieval system, otherwise high document similarity is not expressed in similar scores and $\tilde{y}_i$ will be different from y_i. The average score vector $\mathbf{y}_\mu$ is formed by averaging $\mathbf{y}$ over all systems. To score each system, the linear correlation coefficient between $\mathbf{y}_\mu$ and $\tilde{\mathbf{y}}$ is determined. Here, a system is estimated to perform well, if it is similar to the average system. In [7] this approach has been used to rank systems according to single topics, while we use it to rank systems across a set of topics. Please note, that we can only report *ACSim* for the data sets, for which the corpora were available to us: TREC-{6-10} and TB-{04-06}.

3.3 Correlation of System Rankings

Each system ranking estimation method outputs a list of systems ordered by estimated system performance (i.e. system ranking). To provide an indication of the quality of an approach, this ranking is correlated against the ground truth ranking, which is determined by the retrieval effectiveness using relevance judgments. In all but two data sets, the effectiveness measure is MAP. Mean reciprocal rank and statistical AP [4] are the measures used for NP-02 and RELFB-08 respectively. Reported is the rank correlation coefficient Kendall's Tau $\tau \in [-1, 1]$, which measures the degree of correspondence between two rankings [1].

4 Empirical Findings

In Sec. 4.1 we compare the performances of *DF*, *RS*, *SO* and *ACSim* on the full set of topics. Then, in Sec. 4.2, we will show that: (i) system rankings cannot be estimated equally well for each topic in a topic set, (ii) relying on topic subsets can improve the accuracy of system ranking estimation, and (iii) automatically selecting subsets of topics can lead to improvements in system ranking.

---

[3] Referred to as $\rho(\tilde{\mathbf{y}}, \mathbf{y}_\mu)$ in [7].

4.1   System Ranking Estimation on the Full Set of Topics

In Tab. 1, the results of the evaluation are shown in terms of Kendall's τ. DF performs best on TREC-{6,7}, the poor result on TREC-8 is due to an extreme parameter setting learned from TREC-{6,7}. The highly data set dependent behavior of DF is due to its bias: the pseudo relevant documents are selected from non-average systems. A system that is dissimilar to the average system, can either perform very well or very poorly. RS outperforms DF on the remaining six data sets, where training data for DF is available (TREC-{9,10}, TB-{0,4-06}). On the eight data sets without training data, DF's parameters were optimized on the test set - despite this optimization RS outperforms DF in four instances.

Table 1. System ranking estimation on the full set of topics. All correlations reported are significant ($p < 0.005$). The highest τ per data set is bold. Column **#sys** contains the number of retrieval systems to rank for a data set with **#top** topics. The final three columns contain the name of the best system according to the ground truth which is expressed in mean reciprocal rank (NP-02), statistical AP [4] (RELFB-08) and MAP (all other data sets) respectively. **M/A** indicates whether the best system is manual (M) or automatic (A). **ER** shows the estimated rank of the best system by the RS approach. Rank 1 is the top rank.

	#sys	#top	Kendall's Tau τ				RS based rank estimate		
			DF	ACSim	SO	RS	best system	M/A	ER
TREC-6	73	50	**0.600**	0.425	0.470	0.443	*uwmt6a0*	M	57
TREC-7	103	50	**0.486**	0.417	0.463	0.466	*CLARIT98COMB*	M	74
TREC-8	129	50	0.395	0.467	0.532	**0.538**	*READWARE2*	M	113
TREC-9	105	50	0.527	0.639	0.634	**0.677**	*iit00m*	M	76
TREC-10	97	50	0.621	**0.649**	0.598	0.643	*iit01m*	M	83
TB-04	70	50	0.584	0.647	0.614	**0.708**	*uogTBQEL*	A	30
TB-05	58	50	0.606	0.574	0.604	**0.659**	*indri05AdmfL*	A	32
TB-06	80	50	0.513	0.458	0.447	**0.518**	*indri06AtdnD*	A	20
CLIR-01	47	25	0.697	-	0.650	**0.702**	*BBN10XLB*	A	2
NP-02	70	150	0.667	-	0.668	**0.693**	*thunp3*	A	17
EDISC-05	57	59	**0.668**	-	0.614	0.666	*TITLETRANS*	A	1
EEXP-05	37	50	**0.589**	-	0.502	0.483	*THUENT0505*	A	10
BLTR-06	56	50	0.482	-	0.357	**0.523**	*wxoqf2*	A	5
GEN-07	66	36	**0.578**	-	0.362	0.563	*NLMinter*	M	1
LEGAL-07	68	43	**0.754**	-	0.749	0.741	*otL07frw*	M	4
RELFB-08	117	237	0.537	-	0.544	**0.559**	*Brown.E1*	M	65

Relying on TF.IDF based document similarity does not aid, shown by $ACSim$ performing worse than RS in seven out of eight data sets. SO performs slightly better than RS on three data sets, on the remaining twelve its performance is worse.

As discussed, the commonly cited problem of automatic system evaluation is the mis-ranking of the best systems. To give a better impression of the ranking accuracy, in Fig. 1 scatter plots of the estimated system ranks versus the MAP based ground truth system ranks are shown for two data sets. Apart from the best systems, which are severely mis-ranked in Fig. 1(a), the estimated ranking shows a good correspondence to the true ranking. In contrast, in Fig. 1(b), the best systems are estimated more accurately.

As in previous work evaluations have mostly been carried out on early TREC data sets, where the problem of underestimating the best systems occurs

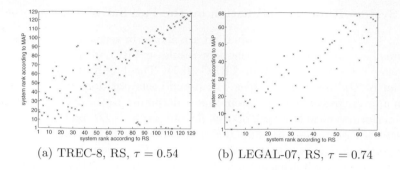

(a) TREC-8, RS, $\tau = 0.54$ (b) LEGAL-07, RS, $\tau = 0.74$

Fig. 1. Scatter plots of estimated system ranks versus ground truth (MAP) based system ranks. Each marker indicates one system. Rank 1 is assigned to the best system.

consistently, it has been assumed to be a general issue. When considering more recent and diverse data sets, we find this problem to be dependent on the set of systems to rank. Tab. 1 also contains the estimated rank (ER) of the best system by the *RS* approach, which in the ground truth is assigned rank 1. For instance, while for TREC-8, the best system is estimated to be ranked at rank 113 ($\tau = 0.54$), the corresponding estimated ranking on GEN-07 is 1 ($\tau = 0.56$), that is, the best system is correctly identified. The fact that both τ values are similar, suggests that both, τ and ER, should be reported, as sometimes we are most interested in identifying the best systems correctly. A look at the best systems for each data set shows, that in TREC-{6-10} the best systems are manual and derived with a great deal of human intervention[4]. The best systems of TB-{04-06} on the other hand, are automatic. A similar observation can be made for the more recent data sets, the best systems are mostly automatic. In the case of GEN-07 and LEGAL-07, where the best systems are classified as manual, the human intervention is less pronounced[5]. The severe mis-ranking of RELFB-08's best system is a result of the task, which is to exploit manually judged documents for ranking, that is the systems are based on considerable human intervention.

We conclude, that in contrast to previous work, *RS* is the most consistent and overall the best performing approach. Furthermore, in contrast to common belief, automatic system evaluation methods are capable of identifying the best runs, when they are automatic or involve little human intervention. The extent of the mis-ranking problem is largely influenced by the amount of human intervention in the best manual runs.

4.2   System Ranking Estimation on Topic Subsets

In this section, we will show that the ability of system ranking estimation methods to rank systems correctly, differs significantly between the topics of a topic

---

[4] E.g. the best system of TREC-6, *uwmt6a0*, was created by letting four human assessors judge documents for their relevance for more than 100 hours.

[5] E.g. the best system of GEN-07, *NLMinter*, relied on manually reformulated queries.

set. While for a number of topics the estimated rankings are highly accurate and close to the ground truth rankings, for other topics they fail. Based on this observation, we hypothesize that the performance of system ranking estimation approaches can be improved, if the "right" subset of topics is used.

Single Topic Performance. For each topic, we evaluate the estimated ranking of systems by correlating it against the ground truth ranking which is based on average precision. Here, we are not interested in how well a single topic can be used to approximate the ranking of systems over the entire topic set. We are interested in how well the system ranking estimation approach performs for each individual topic. Due to space constraints, in Tab. 2, we only present results for four data sets. Across the data sets and system ranking estimation methods, the spread in correlation between the best and worst case is very wide; in the worst case, there is no significant correlation between the ground truth and the estimated ranking, in the best case the estimated ranking is highly accurate. These findings form our motivation: if we can determine a subset of topics for which the system ranking estimation algorithms perform well, we hypothesize that this will enable us to achieve a higher estimation accuracy of the true ranking across the full set of topics.

Table 2. Single topic dependent ranking performance: minimum and maximum estimation ranking accuracy in terms of Kendall's τ. Significant correlations ($p < 0.005$) are marked with †.

	DF		ACSim		SO		RS	
	min. τ	max. τ	min. τ	max. τ	min. τ	max. τ	min. τ	max. τ
TREC-6	0.008	0.849†	−0.134	0.777†	−0.147	0.752†	−0.106	0.823†
TB-04	0.002	0.906†	0.038	0.704†	0.056	0.784†	0.025	0.882†
CLIR-01	0.268	0.862†	-	-	0.221	0.876†	0.248	0.839†
LEGAL-07	0.027	0.690†	-	-	0.058	0.691†	−0.008	0.690†

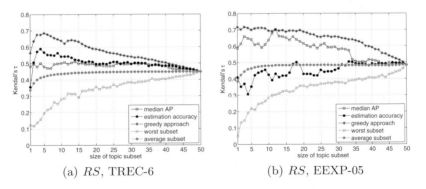

(a) *RS*, TREC-6   (b) *RS*, EEXP-05

Fig. 2. Topic subset selection experiments

Topic Subset Performance. Having shown that the quality of ranking estimation varies between topics, we now investigate if selecting a subset of topics from the full topic set is useful in the context of system ranking estimation

Table 3. Topic subset selection experiments: comparison of Kendall's τ achieved by *RS* when relying on the full topic set and the best greedy subset, $\pm$ indicates the percentage of change. All τ values are significant ($p < 0.005$). The *JSD* columns show the results when performing automatic topic subset selection with the *JSD* approach. Reported are the results of topic subset sizes $c = 10$ and $c = 20$. In bold, improvements of *JSD* over the full topic set.

	RS			JSD	
	full set	greedy	±%	c=10	c=20
	τ	τ		τ	τ
TREC-6	0.443	0.654	+47.6%	**0.455 0.485**
TREC-7	0.466	0.584	+25.3%	**0.489 0.505**
TREC-8	0.538	0.648	+20.4%	**0.585 0.588**
TREC-9	0.677	0.779	+15.1%	0.649 0.644
TREC-10	0.643	0.734	+14.2%	0.634 0.635
TB-04	0.708	0.846	+19.5%	**0.760 0.733**
TB-05	0.659	0.812	+23.2%	**0.670** 0.612
TB-06	0.518	0.704	+35.9%	0.495 0.508
CLIR-01	0.702	0.808	+15.1%	**0.706** 0.698
NP-02	0.693	0.853	+23.1%	0.623 0.597
EDISC-05	0.666	0.801	+20.3%	**0.709 0.729**
EEXP-05	0.483	0.718	+48.7%	**0.616 0.616**
BLTR-06	0.523	0.601	+14.9%	0.501 **0.528**
GEN-07	0.563	0.678	+20.4%	0.530 0.556
LEGAL-07	0.741	0.865	+16.7%	0.695 0.728
RELFB-08	0.559	0.872	+60.0%	**0.589 0.638**

algorithms. That is, we try to determine whether we can improve the accuracy of the algorithms over the results reported in Sec. 4.1 on the full set of topics. To investigate this point, we experiment with selecting subsets of topics according to different strategies.

A topic set consists of m topics. We thus test topic subsets of cardinality $c = \{1, ..., m\}$. Ideally, for each c we would test all possible subsets. As this is not feasible[6], for each c, we randomly sample 10000 subsets of topics, run a system ranking estimation algorithm, evaluate it and record the (i) the worst τ and (ii) the average τ achieved across all samples[7]. We also include three iterative subset selection strategies: (iii) the greedy strategy, where a subset of topics is greedily built by adding one topic at a time such that τ is maximized, (iv) the median AP strategy where a subset of topics is built by each time adding the easiest topic, which is the topic that exhibits the highest median average precision across all systems and (v) the estimation accuracy strategy where a subset of topics is built by each time adding the topic with the highest estimation accuracy (single topic performance). As strategies (iii)-(v) require knowledge of the relevance judgments, these experiments should only be seen as motivational: the goal is to determine whether it is beneficial at all to rely on subsets instead of the full topic set.

For the topic subsets of each cardinality, we determine the correlation between the estimated ranking of systems and the ground truth ranking of systems across the full set of topics. Now we are indeed interested in how well a subset of one or

---

[6] For each cardinality c, a total of $\binom{m}{c}$ different subsets exist.

[7] Recording the best τ of the sampled subsets does not perform better than the greedy approach and is therefore not reported separately.

more topics can be used to approximate the ranking of systems over the entire topic set.

Indicatively for two data sets, the results are shown in Fig. 2. The results of the other data sets and algorithms not shown are similar. The greedy strategy, especially at low subset sizes, yields significantly higher correlations than the baseline, which is the correlation at the full topic set size. The worst subset strategy shows the potential danger of choosing the wrong subset of topics - τ is significantly lower than the baseline for small c. When averaging τ across all sampled subsets (the average subset strategy) of a cardinality, at subset sizes of about $m/3$ topics, the correlation is only slightly worse than the baseline. For the median AP strategy, where first the easiest topics are added to the subset of topics, the gains in correlation over the baseline are visible, though less pronounced than for the greedy strategy. Less consistent is the estimation accuracy strategy, where first those topics are added to the topic subset, for whom the ranking of systems is estimated most accurately. While in Fig. 2(a) this strategy comes closest to the greedy approach, in Fig. 2(b) this strategy most of the time performs worse than the average τ strategy.

In Tab. 3, a summary of the results of the RS approach across all data sets is given (the results of DF, SO, $ACSim$ are similar but not shown due to lack of space): shown is Kendall's τ achieved on the full set of topics and the best performing τ of the greedy approach. The percentage of change varies between 14.2% and 60% with a maximum τ of 0.872. Thus, across all data sets, subsets of topics indeed exist that can significantly improve the accuracy of system ranking estimation.

Automatic Topic Subset Selection. Topic subset selection will only be useful in practice, if it becomes possible to automatically identify useful subsets. As RS proved overall to perform best, we focus on it. RS is popularity based, that is, the most often retrieved documents have the highest chance of being sampled from the pool and thus being declared pseudo relevant. This approach assumes that *popularity* $\approx$ *relevance*. This assumption is not realistic, but we can imagine cases of topics where it holds: in the case of *easy* topics. Easy topics are those, where all or most systems do reasonably well, that is, they retrieve the truly relevant document towards the top of the ranking and then, relevance can be approximated by popularity. The results of the median AP strategy in Fig. 2 confirms this reasoning.

This leads to the following strategy: adding topics to the subset of topics according to their estimated difficulty. As we do not have access to relevance judgments, we rely on an estimate of topic difficulty, as provided by the Jensen-Shannon Divergence (JSD) approach [3]. The JSD approach estimates a topic's difficulty with respect to the collection and in the process also relies on different retrieval systems: the more diverse the result lists of different systems as measured by the JSD, the more difficult the topic is. Thus, we perform an estimation task on two levels: first, a ranking of topics according to their difficulty is estimated, then we rely on the topics estimated to be easiest for system ranking estimation.

The results of this automatic subset selection approach are shown in the last two columns of Tab. 3: listed are the τ values achieved for subsets of 10 and 20 topics estimated to be easiest. The particular size of the topic subset is of no great importance as seen in the small variation in τ. It is also visible that this approach can help, though the improvements are small and inconsistent. As reason we suspect the fact, that the accuracy of the JSD approach itself is also limited [3].

5   Summary and Future Work

In this work, we have compared four system ranking estimation approaches on a wide variety of data sets and, contrary to earlier findings, we have shown that the first approach proposed (Soboroff *et al.* [13]) is still the best performing method. We also observed that the ability of system ranking estimation methods varies widely for each individual topic. A number of experiments with different topic subset selection strategies confirmed the hypothesis that some subsets of topics are better suited for the system ranking estimation algorithms than others.

We proposed a strategy to automatically identify good subsets of topics by relying on topics that have been estimated to be easy. This strategy yielded some improvement, though not consistently across all data sets. Considering the potential for improvement, this result should be considered as a first attempt at topic subset selection. In the future, we will focus on identifying topic features that distinguish between topics appearing in subsets which improve system ranking estimation and topics that occur mostly in poorly performing subsets.

Finally, we made the important finding, that the problem of mis-ranking the best systems encountered by previous authors, which has so far hampered the take-up of automatic system ranking estimation, is strongly data set dependent. In particular we found that the smaller the amount is of human intervention in the best systems, the smaller the problem of underestimating their performance. In fact, for some TREC data sets we were indeed able to identify the best system correctly, thus making the case for automatic system evaluation.

References

1. Rank Correlation Methods. Hafner Publishing Co., New York (1955)
2. Amitay, E., Carmel, D., Lempel, R., Soffer, A.: Scaling ir-system evaluation using term relevance sets. In: SIGIR 2004, pp. 10–17 (2004)
3. Aslam, J.A., Pavlu, V.: Query hardness estimation using Jensen-Shannon divergence among multiple scoring functions. In: Amati, G., Carpineto, C., Romano, G. (eds.) ECiR 2007. LNCS, vol. 4425, pp. 198–209. Springer, Heidelberg (2007)
4. Aslam, J.A., Pavlu, V., Yilmaz, E.: A statistical method for system evaluation using incomplete judgments. In: SIGIR 2006, pp. 541–548 (2006)
5. Aslam, J.A., Savell, R.: On the effectiveness of evaluating retrieval systems in the absence of relevance judgments. In: SIGIR 2003, pp. 361–362 (2003)
6. Carterette, B., Allan, J.: Incremental test collections. In: CIKM 2005, pp. 680–687 (2005)

7. Diaz, F.: Performance prediction using spatial autocorrelation. In: SIGIR 2007, pp. 583–590 (2007)
8. Efron, M.: Using multiple query aspects to build test collections without human relevance judgments. In: ECIR 2009, pp. 276–287 (2009)
9. Guiver, J., Mizzaro, S., Robertson, S.: A few good topics: Experiments in topic set reduction for retrieval evaluation. To appear in TOIS
10. Krovetz, R.: Viewing morphology as an inference process. In: SIGIR 1993, pp. 191–202 (1993)
11. Mizzaro, S., Robertson, S.: Hits hits trec: exploring ir evaluation results with network analysis. In: SIGIR 2007, pp. 479–486 (2007)
12. Nuray, R., Can, F.: Automatic ranking of information retrieval systems using data fusion. Information Processing and Management 42(3), 595–614 (2006)
13. Soboroff, I., Nicholas, C., Cahan, P.: Ranking retrieval systems without relevance judgments. In: SIGIR 2001, pp. 66–73 (2001)
14. Spoerri, A.: Using the structure of overlap between search results to rank retrieval systems without relevance judgments. Information Processing and Management 43(4), 1059–1070 (2007)
15. Voorhees, E.M.: Variations in relevance judgments and the measurement of retrieval effectiveness. Information Processing and Management 36, 697–716 (2000)
16. Wu, S., Crestani, F.: Methods for ranking information retrieval systems without relevance judgments. In: Matsui, M., Zuccherato, R.J. (eds.) SAC 2003. LNCS, vol. 3006, pp. 811–816. Springer, Heidelberg (2004)

Aggregation of Multiple Judgments for Evaluating Ordered Lists

Hyun Duk Kim, ChengXiang Zhai, and Jiawei Han

Department of Computer Science, University of Illinois at Urbana-Champaign,
201 N Goodwin Ave, Urbana, IL 61801, USA
hkim277@illinois.edu, czhai@cs.uiuc.edu, hanj@cs.uiuc.edu

Abstract. Many tasks (e.g., search and summarization) result in an ordered list of items. In order to evaluate such an ordered list of items, we need to compare it with an ideal ordered list created by a human expert for the same set of items. To reduce any bias, multiple human experts are often used to create multiple ideal ordered lists. An interesting challenge in such an evaluation method is thus how to aggregate these different ideal lists to compute a single score for an ordered list to be evaluated. In this paper, we propose three new methods for aggregating multiple order judgments to evaluate ordered lists: weighted correlation aggregation, rank-based aggregation, and frequent sequential pattern-based aggregation. Experiment results on ordering sentences for text summarization show that all the three new methods outperform the state of the art average correlation methods in terms of discriminativeness and robustness against noise. Among the three proposed methods, the frequent sequential pattern-based method performs the best due to the flexible modeling of agreements and disagreements among human experts at various levels of granularity.

Keywords: Evaluation, Sentence ordering, Judgment aggregation, Frequent sequential pattern mining.

1 Introduction

How to aggregate different human evaluators' judgments is a difficult problem in evaluation with multiple human annotations. When we evaluate the performance of a system, we often compare the output of system with a "gold standard output" created by a human evaluator; the more similar the system output is to the human-created gold standard, the better the performance of the system would be.

Unfortunately, when a task is difficult or inherently subjective to judge (as in the case of many information retrieval problems such as search and summarization), human experts may not agree with each other on the gold standard. Thus using only one single human expert to create the gold standard can be biased, and it would be necessary to have multiple experts to create a gold standard, leading naturally to multiple (gold standard) judgments, each created by a different human expert.

C. Gurrin et al. (Eds.): ECIR 2010, LNCS 5993, pp. 166–178, 2010.
© Springer-Verlag Berlin Heidelberg 2010

The research question we study in this paper is how to aggregate these multiple judgments created by multiple experts to evaluate ordered lists of items. Evaluation of ordered lists is quite important in information retrieval because in many tasks, the output of a system is an ordered list. For example, a search engine generates a ranked list of documents, while a text summarizer generates a ranked list of extracted sentences from documents. Existing approaches to evaluation of these tasks tend to simplify the task of evaluation by not requiring human experts to create a *complete* ideal ranked list and only asking them to distinguish items that should be ranked high from those that should be ranked low. For example, in evaluating retrieval results, we often ask a human assessor to judge which document is relevant and which is non-relevant. While this alleviates the problem of disagreement among human assessors, it does not allow us to distinguish finer granularity differences in ranking, as, e.g., changing the relative order of relevant documents (or non-relevant documents) would generally not affect performance. Moreover, such a coarse judgment would not be sufficient for evaluating a task where ranking is the primary goal. For example, a common last step in most text summarization approaches is to order the extracted representative sentences from documents appropriately to generate a coherent summary. To distinguish a good ordering from a poor one, it would be better to have human experts to generate ideal orderings of the extracted sentences. Furthermore, since it is unlikely that human experts would agree on a single way of ordering a set of sentences, it is necessary to have multiple human experts to create ideal orderings, raising the challenge to aggregate these potentially different ideal orderings to evaluate any ordered list given by a system.

The current way of solving this aggregation problem is to evaluate a given ordered list from a system with each gold standard separately and then take the average of the results of individual gold standard evaluation as the overall score of the system list (see, e.g., [1]). However, this simple method does not explicitly consider and effectively reflect agreements and disagreements among different gold standards since it trusts equally every part of an ideal ordering created by every expert. Intuitively, an expert agreeing with most other experts can be trusted more, and for the same expert, different parts of an ordering may not be equally trustable. Thus ideally, we should use only the "agreed" judgments for evaluation or put different weights on different parts depending on their "degree of agreement" by all the experts.

Based on these insights, we propose three new methods for aggregating multiple order judgments to evaluate ordered lists.

The first method, called weighted correlation aggregation (WCA), models the overall agreement among the assessors and is a direct extension of the current average correlation method. In this method, we would take a *weighted* average of the correlations between the list to be evaluated and all the gold standard ordered lists, where the weight on each expert is computed based on the degree of overall agreement between this expert and all other experts.

The second method, called rank-based aggregation (RBA), models the agreement among the assessors in the rank of each item, and generates a consensus ordered list that summarizes the judgments by all the assessors. Evaluation can then be based on this single consensus ordering.

The third method, called frequent sequential pattern-based aggregation (FreSPA), models the agreement among the assessors at various levels of granularity as represented by consensus sequential patterns of various lengths. We would first identify sequential patterns and their frequencies from all the human judgments of ordering, and then score an ordered list based on how well the list matches the frequent sequential patterns.

We compared the three proposed methods with two baseline methods on a publicly available data set for evaluating sentence ordering in text summarization. The two baseline methods are the uniform average correlation methods, using Kendall's τ and Spearman correlation, respectively; they represent the current state of the art. Experiment results show that our methods outperform both baseline methods in terms of discriminativeness and robustness against noise.

The rest of the paper is organized as follows. In Section 2, we discuss related work. We define our problem in Section 3, and present the proposed methods in Section 4. In Section 5, we evaluate our methods and discuss experiment results. We conclude our paper and discuss future work in Section 6.

2   Related Work

Most previous work on evaluating sentence ordering did not consider aggregation of human judgments [2, 3, 4, 5]. In many cases, sentence ordering was evaluated qualitatively by human [3, 4]. For quantitative evaluation, Kendall's τ [2] and Spearman's rank correlation coefficient are often used (e.g., [2, 1, 5]). However, these measures have not considered aggregation of multiple ordering judgments made by different human evaluators or have simply taken a uniform average of the correlation values computed using multiple judgments without considering the variation of trustworthiness of human assessors, which we address in our methods.

Barzilay et al. [6] looked into the agreement among human assessors and revealed the difficulty in evaluating the task of ordering sentences in text summarization. They created multiple judgments using multiple assessors, but did not propose a method for aggregating human judgments for evaluating sentence ordering. We propose methods for automatic evaluation based on aggregation of those human generated orderings and use their data set to evaluate the proposed method.

Subjectivity in human evaluators' annotations has been studied in [7, 8, 9, 10, 11]. Aggregation of judgments and votes have been studied in other study fields such as logic, economy and mathematics [12, 13, 14, 15]. These studies focus on finding the best item based on different voting results, whereas our work is to develop a method for evaluating a new ordered list based on multiple judgments.

Data fusion in information retrieval is also related to our work. The aim of data fusion is to combine many ranked results to generate one potentially better ranking. Various methods were proposed to combine ranked search lists, including, e.g., CombSUM/CombMNZ [16], ProbFuse [17], and generative model-based method [18]. Despite of its similarity to the ordering judgment aggregation, data fusion is mainly for making a new ranking and not for evaluation. In one of our proposed methods, we use a similar aggregation strategy to the one used in [16] to generate a single consensus judgment so as to convert multiple judgments into one single judgment.

Perhaps the most similar work to ours is the pyramid method for summary content evaluation [19], where agreements on content units are considered in designing a scoring method for evaluating the content of a summary, and the strategy is similar to our pattern-based aggregation. However, the pyramid method cannot be applied to a complex task such as sentence ordering. Moreover, the pyramid method relies on manual matching of the content units, whereas our method is automatic and can be regarded as adapting the philosophy of the pyramid method for evaluating ordered lists in an automatic way.

Frequent sequential pattern mining is one of the most popular topics in data mining. Many algorithms have been developed, including the generalized sequential pattern mining algorithm (GSP) [20], sequential pattern discovery using equivalent class (SPADE) [21], PrefixSpan [22], and CloSpan [23]. In this paper, we used PrefixSpan algorithm for finding frequent sequential patterns.

3   Problem Definition

We consider the general problem of evaluating an ordered list based on multiple gold standard ordered lists created by different human assessors.

Formally, let $X = \{x_1, ..., x_k\}$ be a set of k items to be ordered. Suppose n human assessors created n (generally different) orderings of the k items in X, denoted as O_1, ..., O_n. $O_i = (x_{i_1}, ..., x_{i_k})$ is an ordering, where i_1, ..., i_k are a permutation of integers $1, ..., k$. Given a new ordered list to be evaluated, $O = (y_1, ..., y_k)$, our goal is to find a scoring function $s(O; O_1, ..., O_n)$ that can be used to score the ordered list O based on its consistency with the gold standard orderings $O_1, ..., O_n$.

This problem setup is general and should cover many interesting application instances. A particular case that we will use to evaluate the proposed scoring function is sentence ordering in text summarization. Virtually all the methods for automatic summarization would first extract a set of most representative sentences from documents and then order them to form a coherent summary document. To evaluate the quality of the generated summary, one aspect we must measure is the optimality of the ordering of the sentences, which presumably would directly affect the readability of the summary.

4   Methods for Ordered List Evaluation

In this section, we first present two baseline methods that represent the state of the art and then present the three new methods that we proposed.

4.1   Baseline Methods

For general ordering evaluation, Kendall's τ [2] and Spearman's rank correlation coefficient are often used. Let π and σ be two different orders, and N the length of the order. Kendall's τ is defined as $\tau = 1 - \frac{2S(\pi,\sigma)}{N(N-1)/2}$, where $S(\pi,\sigma)$ means number of discordant pairs between π and σ. For example, the two ordered lists (ABCD) and (ACDB) have two discordant pairs since the orders of BC and BD are reversed.

Spearman's rank correlation coefficient is defined as

$$Spearman = 1 - \frac{6\sum_{i=1}^{N}(\pi(i)-\sigma(i))^2}{N(N^2-1)}$$

where $\pi(i)$ and $\sigma(i)$ mean the rank of item i in π and σ, respectively. That is, this measure uses rank difference of items in the two ordered lists. The range of both Kendall's τ and Spearman's rank correlation coefficient is [-1,1].

These two baseline methods are only able to compare a target list to one ideal ordered list. In order to use it for evaluation with multiple ideal ordered lists, in the existing work (e.g., [1]), the average of the correlation values for all the ideal ordered lists is taken as the overall score of a target list w.r.t. the multiple judgments. We thus call this baseline method *average correlation* (AC). Formally, $S_{AC}(O; O_1, ..., O_n) = \frac{1}{n}\sum_{i=1}^{n} C(O, O_i)$, where $C(O, O_i)$ is the correlation between two orderings as measured by either Kendall's τ or Spearman's rank correlation (denoted by AC-τ and AC-Sp, respectively).

The AC method gives each ideal ordered list an equal weight, thus does not address the issue of variation of trustworthiness of different assessors. Below we propose three methods to address this issue.

4.2   Weighted Correlation Aggregation (WCA)

Our first method is weighted correlation aggregation, which is a direct extension of AC by assigning different weights to different judgments, where the weight of each judgment (i.e., each assessor) is computed based on the degree of overall agreement of the assessor with other assessors. Formally, WCA is defined as:

$$S_{WCA}(O; O_1, ..., O_n) = \frac{\sum_{i}^{n} w_i C(O, O_i)}{\sum_{i}^{n} w_i},$$

where the weight $w_i = \frac{1}{n-1}\sum_{j=1, j\neq i}^{n} C(O_i, O_j)$.

The two-list correlation measure $C(O, O_i)$ can be either Kendall's τ or Spearman's rank correlation, leading to two corresponding variations of the WCA method, denoted by WCA-τ and WCA-Sp, respectively.

In WCA, the weight on each assessor is based on the *overall* agreement between the assessor and other assessors. However, it may be desirable to model the agreement of assessors at a finer granularity level, which motivates us to propose the following rank-based aggregation method.

4.3   Rank-Based Aggregation (RBA)

In this method, we model the agreement of assessors at the level of the rank of each item and aggregate all the ranks to generate a consensus ordered list which can summarize the overall judgments. Specifically, the consensus ordered list is generated by ordering items based on their combined ranking scores:

Combined Ranking Score of $x_i = \sum_{j=1}^{n} Rank_j(x_i)$,
where $Rank_j(x_i)$ is the rank of x_i in O_j.

A combined ranking score is lower (thus the item would be ranked higher in the consensus list) when it was highly ranked in more ideal ordered lists. Since small errors can be compensated by other experts in the summing process, this method can be more robust against noise.

After obtaining the consensus list, we can use either Kendall's τ or Spearman's rank correlation to compute the correlation of a target list with the consensus list, leading again to two variations of the RBA method, denoted by RBA-τ and RBA-Sp, respectively.

4.4   Frequent Sequential Pattern-Based Aggregation (FreSPA)

In this method, we attempt to model the agreement at various levels of granularity based on frequent sequential patterns. A sequential pattern is a subsequence of items in an ordering O_i possibly with gaps. For example, if an ordering is ABCD, both "BC" and "BD" are potential sequential patterns. Intuitively, a sequential pattern of an ordering captures the desired ordering of a subset of items. Thus if a sequential pattern has high frequency in all the judgments $O_1, ..., O_n$, it would mean that it has captured an ordering of a subset of items agreed by most assessors. We can thus use these frequent sequential patterns to score a target ordered list.

Formally, we denote a sequential pattern by pat_i, its length by $seqLen_i$, and its support (i.e., frequency) by sup_i. We would be especially interested in frequent sequential patterns (with a support at least $minSup$) in the ideal ordered lists $O_1, ..., O_n$, as they capture the agreed partial ordering of a subset of items.

Furthermore, there will be a trade-off in the length of the patterns. On the one hand, longer patterns are more useful as they give ordering information for more items, but on the other hand, they are also less likely agreed by more assessors. To retain the flexibility, we introduce two parameters, $minLen$ and $maxLen$, to restrict the lengths of patterns to be considered. If we set $minLen = 2, maxLen = 2$, we will only find length-2 pair-wise sequential patterns. If we restrict the $maxLen$ to 3, we only find frequent sequential patterns whose length is 2 or 3. Naturally, both $minLen$ and $maxLen$ must be between 2 and k.

Let P denote all the sequential patterns in $O_1, ..., O_n$ that satisfy our constraints. The FreSPA score for a target order which we want to evaluate, $O = y_1, y_2, ..., y_k$, is defined as follows:

$$S_{FreSPA}(O; O_1, ... O_n) = \frac{\sum_{pat_i \in O}(1 + wLen*(seqLen_i - 1))*(1 + wSup*(sup_i - 1))}{\sum_{pat_i \in P}(1 + wLen*(seqLen_i - 1))*(1 + wSup*(sup_i - 1))}$$

where $wLen$ and $wSup$ are two weighting parameters on the length and support (i.e., frequency) of a pattern. Clearly, $S_{FreSPA} \in [0, 1]$.

$wLen$ decides how much more weight we will put on a longer pattern. When $wLen = 0$, patterns with different lengths would have the same score contribution, whereas when $wLen = 1$, patterns m times longer would have m times more score contributions. $wSup$ decides how much more weight we will put on patterns with higher supports. When $wSup = 0$, all patterns have the same score contribution regardless of support, and when $wSup = 1$, patterns with m times higher support would have m times higher score contributions.

These parameters allow us to tune FreSPA based on the characteristics of any given set of judgments (e.g., level of consensus, amount of noisy judgments). For example, if all the judgments mostly agree with each other, we may trust longer patterns more. These parameters can be set empirically based on optimizing some effectiveness measure on the given set of judgments. Note that since such tuning can always be done before we apply FreSPA to evaluate any given ordered lists, the parameter tuning is not a problem; indeed, they provide the needed flexibility to adjust to different evaluation data set.

A main computational challenge is to find all the frequent patterns quickly. This can be done by using an efficient frequent sequential pattern mining algorithm. Many such algorithms have been proposed in the data mining community (see Section 2 for a discussion about them). In our experiments, we used the PrefixSpan algorithm [22] because it is very efficient and generates all sub-patterns. Since the number of human assessors is usually not very large, in general, the scalability of FreSPA is not a concern.

5   Experiment Results

In this section, we evaluate the proposed methods using the task of ordering sentences for text summarization. We chose this task because of the availability of a data set with multiple human judgments for evaluating our methods.

5.1   Data Set

We use the data set created by researchers for studying sentence ordering in summarization [6, 24]. The data set is publicly available on the web[1]. The data is created by asking human assessors to generate an ideal ordering of a set of sentences extracted from documents to form a coherent summary. The detailed instructions for the assessors are also published on the web [2]. There are 10 sentence sets. On average, each set of sentences has 8.8 sentences and is judged by 10.4 assessors.

5.2   Metric

Given a set of multiple judgments of ordering, how do we know whether one aggregation method is effective than another? We propose to measure the effectiveness of an aggregation method based on its "Evaluation Discriminativeness" (ED), which is defined as the score difference between a good order and a bad order of items.

---

[1] http://www1.cs.columbia.edu/~noemie/ordering/
[2] http://www1.cs.columbia.edu/~noemie/ordering/experiments/

Intuitively, a good aggregation method should give a high score to a good order and a low score to a bad order. Thus, the difference of the scores given to a good order and a bad order can indicate how well a method can distinguish a good order from a bad order; the higher the ED value is, the more discriminative the method is.

The remaining question is: how do we know which is a good order and which is a bad order? The fact that we have multiple gold standard judgments makes it tricky to answer this question. To avoid bias, we take the judgment from each assessor as a good order and take its reverse order as a bad order. We then compute the average ED over all the assessors. Since a gold standard order created by a human assessor can be assumed to be a good order, it would be reasonable to assume that its reverse is a bad order.

Formally, the ED value of an aggregation-based scoring method S is defined as:

$$ED = \frac{1}{n}\sum_{i}^{n}(S(O_i; O_1, ..., O_{i-1}, O_{i+1}, ..., O_n) - S(O_{iR}; O_1, ..., O_{i-1}, O_{i+1}, ..., O_n))$$

where O_{iR} is reverse order of O_i. When computing the ED value on the judgment created by one assessor, we use all the other assessors' judgments as gold standards. This is essentially similar to leave-one-out cross-validation. The final overall ED score of method S is the average of the ED scores on all the assessors. Through optimizing the ED value, we can tune any parameters of method S to maximize its discriminativeness; the tuned parameter setting can then be used to evaluate any new ordered lists generated by a system.

Since the ED measure requires all the measures to have the same score range, we normalize both Kendall's τ and Spearman's rank correlation using the min-max normalization method, $(x + 1)/2$, so that the scores would be in $[0,1]$, the range of the FreSPA score.

In addition to the discriminativeness, we also examine the robustness to noisy judgments of an aggregation-based evaluation method. Since the judgment of an optimal order in our problem setup is inherently subjective, there may be noisy judgments in our gold standard in the sense that some judgments may not be reliable, which can be caused by, e.g., careless judgment or biased judgment by an assessor. To compare their robustness, we also compare different methods by adding different amounts of random judgments into our data set.

5.3   Basic Comparison

We first present results from comparing all the variants of the proposed three methods (i.e., WCA-τ, WCA-Sp, RBA-τ, RBA-Sp, and FreSPA) with the two versions of the baseline method (i.e., AC-τ and AC-Sp) on the original data set.

The FreSPA method has several parameters to set. For this comparison, we set $wLen = 1.0$, $wSup = 1.0$, and $minSup = 0.75$ (a pattern must occur in at least 75% of the judgments), and used patterns of all lengths (i.e., $minLen = 2$ and $maxLen = k$).

The results are shown in the first row of Table 1. We see that all the three proposed methods outperform the corresponding baseline methods. Among the three proposed methods, FreSPA is the best, followed by RBA, which outperforms WCA. Since both FreSPA and RBA model agreement of assessors at a finer granularity level than WCA which models the overall agreement, this suggests that it is more effective to aggregate judgments at finer granularity level. Furthermore, the fact that FreSPA outperforms RBA suggests that it may be beneficial to model agreement at various levels of granularity. Note that the parameter setting we used here for FreSPA is not the best configuration for this method, and it is possible to further tune these parameters to achieve better ED as will be shown later when we analyze the influence of the parameters on the performance of FreSPA. We also see that Spearman generally performs better than Kendall's τ.

Table 1. ED of different methods on original and noisy data sets

Noise ratio	AC-τ	AC-Sp	WCA-τ	WCA-Sp	RBA-τ	RBA-Sp	FreSPA
0	0.454	0.542	0.459	0.549	0.564	0.676	0.722
0.25	0.300	0.354	0.357	0.435	0.460	0.564	0.656
0.5	0.220	0.264	0.236	0.292	0.380	0.468	0.551
0.75	0.167	0.202	0.173	0.215	0.358	0.433	0.510
1	0.113	0.136	0.133	0.159	0.307	0.379	0.395
Max-Min	0.341	0.406	0.326	0.389	0.257	0.297	0.327
% Degradation	75.1%	74.9%	71.0%	70.9%	45.6%	43.9%	45.3%

5.4   Comparison on Noisy Data

To test the robustness of these methods to noise, we repeated the comparison presented earlier by systematically adding some random orders into the gold standard judgment set. Specifically, we use a "noise ratio" parameter r to control the amount of noise to be added and add nr random orders to a data set with n original judgments. For example, if there are 10 human orderings and r is 0.5, we would add 5 random orders to the data set.

Table1 shows the ED values of all the methods with different noise ratios. Overall, the conclusions we made earlier on the original data set are all confirmed here. In particular, for all levels of noise ratios, the proposed methods outperform the baselines, FreSPA performs better than RBA which outperforms WCA, and Spearman performs better than Kendall's τ.

In the last two rows, we show the absolute degradation and relative percentage of degradation of ED from the original data set when we add maximum amount of noise ($r = 1.0$). We see that our proposed methods have less degradation than their corresponding baseline methods because of modeling agreement among the assessors. This is most clearly seen from the improvement of the two WCA variants over their corresponding baselines as their only difference is that in WCA, we will be able to assign lower weights to noisy orders as they are less likely correlated with other judgments. Moreover, RBA and FreSPA have significantly less degradation than WCA, indicating again the benefit of modeling agreement at finer granularity level. Indeed, when we use units such as ranks of items and frequent sequential patterns to model agreement, we can expect to eliminate most noise automatically due to their lack of agreement from other assessors. In particular, the minimum support threshold in FreSPA helps ensure that we only use reliable judgments.

5.5   Parameter Setting of FreSPA

While the WCA and RBA have no parameter to tune, FreSPA has several parameters to set, allowing it to adapt to special characteristics of a set of judgments. We now look into the influence of the setting of these parameters on the ED of FreSPA by varying the parameter values on both the original and noisy data sets. Unless otherwise stated, the parameters that are not varied are set to their default values, which we used for comparing different methods (i.e., $wLen = wSup = 1.0$, $minLen = 2$, $maxLen = k$, $minSupp = 0.75$).

Table 2. ED with different wLen and wSup

wLen	ED (original)	ED (noisy)	wSup	ED (original)	ED (noisy)
0	**0.7301**	0.5255	0	0.7106	0.5704
0.25	0.7261	0.5517	0.25	0.7186	0.5145
0.5	0.7239	0.4736	0.5	0.7203	0.5504
0.75	0.7225	0.5327	0.75	0.7211	0.5141
1	0.7215	0.5482	1	0.7215	**0.6019**
5	0.7177	0.5642	5	0.7227	0.5694
10	0.7169	0.5828	10	0.7229	0.5616
20	0.7165	0.5874	20	0.7230	0.5275
50	0.7163	**0.6120**	50	0.7230	0.5361
100	0.7162	0.5403	100	**0.7230**	0.5130

We first show the results from varying $wLen$ and $wSup$ in Table2. We see that the optimal settings of these two parameters differ between the original data set and the noisy data set. In particular, for the original (clean) data set, it is better to set $wLen = 0$, meaning that all patterns are equally important. However, for the noisy data, it is important to set $wLen$ to a large value so as to give more weight to longer patterns, which makes sense as when there is noise, longer patterns would be more trustable (since the high frequency of a short pattern is likely due to chance). Similarly, for the clean data, it is better to set $wSup$ to a large value, which means we can trust the support computed over all the judgments, whereas on the noisy data, setting it to a moderate value (around 1.0) performs the best, which means that we cannot entirely trust the voting from all the judgments.

Next, in Table3, we show the results from varying the pattern length parameters $minLen$ and $maxLen$ on both the original and noisy data sets. Since

Table 3. ED with different maxLen and minLen ratios

maxLen/k (minLen=2)	ED (original)	ED (noisy)	minLen/k (maxLen=k)	ED (original)	ED (noisy)
0	0.7493	0.5265	0	0.7215	0.5841
0.25	0.5818	0.4362	0.25	0.7215	0.5361
0.5	0.7267	0.5331	0.5	0.4137	0.0707
0.75	0.7215	0.5320	0.75	0.0824	0.0000
1	0.7215	0.5303	1	0.0000	0.0000

Table 4. ED with different minimum support

minSup	ED (original)	ED (noisy)
0	0.1741	0.0815
0.25	0.3678	0.2639
0.5	0.5328	0.4040
0.75	0.7215	0.5855
1	0.8985	0.0230

different sentence sets have different numbers of sentences (i.e., different k), we vary these two parameters based on their ratio to the total number of items to order k, which is also the maximum value that $minLen$ and $maxLen$ can take. A ratio of 0 means that $minLen$ (or $maxLen$) is set to 2 since their minimum value is 2. This time, the difference between the original and noisy data sets appears to be less clear, and it appears that using all the patterns (i.e., $maxLen = k$ and $minLen = 2$) is desirable. Also, it is clear that if we use only long patterns, as we would expect, the performance is very poor. However, $maxLen = 2$ seems to be an "outlier" when the performance is actually very good; further analysis would be needed to understand why.

Finally, we look at $minSup$, which controls the threshold for support, i.e., the frequency of a pattern. Intuitively, FreSPA would be more reliable with a higher $minSup$, which would demand more consensus by the assessors unless there are noisy judgments. To verify this intuition, we varied this parameter on both the original and noisy data sets. The results are shown in Table 4. The results confirm our intuition as on the original data set, the optimal value of $minSup$ is 1.0, meaning that it is important to require a pattern to be agreed by every assessor. When $minSup = 1$, we probably would use many short patterns as they are more likely to be agreed by all assessors. However, on the noisy data set, the optimal value is 0.75, which means that we should require only 75% assessors to agree on a pattern, which is reasonable given that some judgments are noise.

6   Conclusions

How to aggregate multiple human judgments to evaluate an ordered list is an important challenge in evaluating many information retrieval tasks. Previous work has not addressed well the variation of trustworthiness of different assessors. In this paper, we proposed three new methods to better address this issue, including the weighted correlation aggregation (WCA), rank-based aggregation (RBA), and frequent sequential pattern-based aggregation (FreSPA). Evaluation using a sentence ordering data set shows that all the three new methods outperform the state of the art average correlation methods in terms of discriminativeness and robustness against noise. Among the three proposed methods, FreSPA performs the best due to the flexible modeling of agreements and disagreements among human experts at various levels of granularity. Moreover, RBA and FreSPA are more effective and more robust than WCA due to modeling of agreement at a finer granularity level.

All the proposed methods are quite general and can be applied to any task where ordering needs to be evaluated with multiple human judgments. The general idea of the FreSPA method can also be extended to evaluate other non-ordering tasks. For example, by finding frequent patterns instead of finding frequent *sequential* patterns, we can adapt FreSPA to evaluate the content of a summary for summarization evaluation. The idea would be very similar to the pyramid method [19], but the modified FreSPA can be expected to give more flexibility and better performance with minimum support. These will be interesting directions for future work.

Acknowledgments. This material is based upon work supported by the National Science Foundation under Grant Numbers IIS-0347933, IIS-0713581, and IIS-0713571.

References

1. Lapata, M.: Probabilistic text structuring: experiments with sentence ordering. In: Proceedings of ACL 2003, pp. 545–552. Association for Computational Linguistics (2003)
2. Lapata, M.: Automatic evaluation of information ordering: Kendall's tau. Comput. Linguist. 32(4), 471–484 (2006)
3. Okazaki, N., Matsuo, Y., Ishizuka, M.: Improving chronological sentence ordering by precedence relation. In: Proceedings of COLING 2004, Morristown, NJ, USA, p. 750. Association for Computational Linguistics (2004)
4. Bollegala, D., Okazaki, N., Ishizuka, M.: A bottom-up approach to sentence ordering for multi-document summarization. In: Proceedings of ACL 2006, Morristown, NJ, USA, pp. 385–392. Association for Computational Linguistics (2006)
5. Bollegala, D., Okazaki, N., Ishizuka, M.: A machine learning approach to sentence ordering for multidocument summarization and its evaluation. In: Dale, R., Wong, K.-F., Su, J., Kwong, O.Y. (eds.) IJCNLP 2005. LNCS (LNAI), vol. 3651, pp. 624–635. Springer, Heidelberg (2005)
6. Barzilay, R., Elhadad, N., McKeown, K.R.: Inferring strategies for sentence ordering in multidocument news summarization. Journal of Artificial Intelligence Research 17, 35–55 (2002)
7. Reidsma, D., op den Akker, R.: Exploiting 'subjective' annotations. In: Proceedings of HumanJudge 2008, Morristown, NJ, USA, pp. 8–16. Association for Computational Linguistics (2008)

8. Wilson, T.: Annotating subjective content in meetings. In: Proceedings of LREC 2008, Marrakech, Morocco, European Language Resources Association, ELRA (2008), http://www.lrec-conf.org/proceedings/lrec2008/

9. Beigman Klebanov, B., Beigman, E., Diermeier, D.: Analyzing disagreements. In: Proceedings of HumanJudge 2008, Manchester, UK, pp. 2–7. International Committee on Computational Linguistics (2008)

10. Passonneu, R., Lippincott, T., Yano, T., Klavans, J.: Relation between agreement measures on human labeling and machine learning performance: Results from an art history domain. In: Proceedings of LREC 2008, Marrakech, Morocco (2008)

11. Wiebe, J.M., Bruce, R.F., O'Hara, T.P.: Development and use of a gold-standard data set for subjectivity classifications. In: Proceedings of ACL 1999, Morristown, NJ, USA, pp. 246–253. Association for Computational Linguistics (1999)

12. Lang, J.: Vote and aggregation in combinatorial domains with structured preferences. In: Proceedings of IJCAI 2007, pp. 1366–1371. Morgan Kaufmann Publishers Inc, San Francisco (2007)

13. Dietrich, F., List, C.: Judgment aggregation by quota rules. Public Economics 0501005, EconWPA (2005)

14. Hartmann, S., Sprenger, J.: Judgment aggregation and the problem of tracking the truth (2008)

15. Drissi, M., Truchon, M.: Maximum likelihood approach to vote aggregation with variable probabilities. Technical report (2002)

16. Fox, E.A., Shaw, J.A.: Combination of multiple searches. In: TREC, pp. 243–252 (1993)

17. Lillis, D., Toolan, F., Collier, R., Dunnion, J.: Probfuse: a probabilistic approach to data fusion. In: Proceedings of SIGIR 2006, pp. 139–146. ACM, New York (2006)

18. Efron, M.: Generative model-based metasearch for data fusion in information retrieval. In: Proceedings of JCDL 2009, pp. 153–162. ACM, New York (2009)

19. Nenkova, A., Passonneau, R., McKeown, K.: The pyramid method: Incorporating human content selection variation in summarization evaluation. ACM Trans. Speech Lang. Process. 4(2), 4 (2007)

20. Srikant, R., Agrawal, R.: Mining sequential patterns: Generalizations and performance improvements. In: Apers, P.M.G., Bouzeghoub, M., Gardarin, G. (eds.) EDBT 1996. LNCS, vol. 1057, pp. 3–17. Springer, Heidelberg (1996)

21. Zaki, M.J.: Spade: An efficient algorithm for mining frequent sequences. Mach. Learn. 42(1-2), 31–60 (2001)

22. Pei, J., Han, J., Mortazavi-asl, B., Pinto, H., Chen, Q., Dayal, U.: Prefixspan: Mining sequential patterns efficiently by prefix-projected pattern growth. In: Proceedings of ICDE 2001, p. 215. IEEE Computer Society, Washington (2001)

23. Yan, X., Han, J., Afshar, R.: Clospan: Mining closed sequential patterns in large datasets. In: Proceedings of SDM 2003, pp. 166–177 (2003)

24. Barzilay, R., Elhadad, N., McKeown, K.R.: Sentence ordering in multidocument summarization. In: Proceedings of HLT 2001, Morristown, NJ, USA, pp. 1–7. Association for Computational Linguistics (2001)

Evaluation and User Preference Study on Spatial Diversity

Jiayu Tang and Mark Sanderson

Department of Information Studies, University of Sheffield, UK
{j.tang,m.sanderson}@sheffield.ac.uk

Abstract. Spatial diversity is a relatively new branch of research in the context of spatial information retrieval. Although the assumption that spatially diversified results may meet users' needs better seems reasonable, there has been little hard evidence in the literature indicating so. In this paper, we will show the potentials of spatial diversity by not only the traditional evaluation metrics (precision and cluster recall), but also through a user preference study using Amazon Mechanical Turk. The encouraging results from the latter prove that users do have strong preference on spatially diversified results.

1 Introduction

In information retrieval, users' needs seem to be multi-dimensional. For example, a potential buyer searching for images of a particular camera is not only interested in image of "this" camera (first dimension, relevance), but also interested in images taken from different directions (second dimension, visual diversity [7]). A person who submitted the query "Castles in the UK" is perhaps interested in all kinds of castles from the whole UK. If the search system only shows a few well known castles repeatedly, although highly relevant, the user may never find out about castles from other places that would also interest him. Thus, a search result that is not only relevant (first dimension, relevance) but also covers many different places (second dimension, spatial diversity [9]) will perhaps meet users' needs better. The research in this paper will be focused on spatial diversity.

Although in spatial information retrieval, the assumption that users would prefer spatially diversified results seems to be reasonable, to the best of our knowledge there has been little hard evidence in the literature indicating so. Therefore, We have conducted a user experiment in Amazon Mechanical Turk (MTurk), to investigate whether such an assumption stands. The results are encouraging. Details of the experiment will be presented in this paper, as well as evaluation results on precision and cluster recall.

2 Background

2.1 Related Work on Diversity

By avoiding duplicated and near duplicated documents, diversity featured search systems try to meet different users' information needs by generating a more

C. Gurrin et al. (Eds.): ECIR 2010, LNCS 5993, pp. 179–190, 2010.
© Springer-Verlag Berlin Heidelberg 2010

diverse list of results. Carbonell and Goldstein's work [3] is probably the first to identify the need for diversity in IR. Their Maximal Marginal Relevance (MMR) algorithm realized diversity by selecting documents that are both relevant and the least similar to the documents already chosen. More recently, Zhai *et al.* [14] adopted language models to measure the novelty contained in documents and then tried to increase diversity by bringing more novelty. Zhang *et al.* [15] invented Affinity Ranking to improve the diversity in web search. Information richness scores are calculated from an affinity graph which models the asymmetric similarities between every pair of documents, and then used for re-ranking. Deselaers *et al.* [7] proposed a method to diversify image search results so that the top images are more diverse visually.

While most of the research on diversity in IR is concerned with the semantics of documents or the visual content of images, some other researchers raised the importance of another dimension of diversity - spatial diversity [13,5,9]. The work in [13] is perhaps the first to bring diversity to spatial information retrieval. Their *geographical distributed ranking* algorithms are effectively algorithms for diversifying results spatially. Later on, Paramita *et al.* [9] proposed two spatial diversity algorithms and evaluated them using a formal test data-set. However, both studies did not give a theoretical explanation on spatial diversity, rather they tackled the problem by selecting documents that are not only relevant but also geographically the furthest from previous ones.

2.2   Diversity Evaluation

The most popular evaluation metric for diversity is probably sub-topic recall (also known as S-recall or cluster recall) [14]. Sub-topic is analogous to the idea of "information nugget" [4] that is becoming popular in the summarization and question answering community. Clarke *et al.* [4] modeled the user's information needs as a set of nuggets $u \subseteq N$, here $N = \{n_1, ..., n_m\}$ is the space of possbile nuggets. A document is regarded as relevant if it has one or more nuggets that also exist in user's information need. Following this perspective, it is probably fair to say that the goal of a diversity approach is to generate a list of documents that cover as many different information nuggets, which exist in the user information needs, as possible. Sub-topic recall [14], which calculates the percentage of sub-topics covered, is effectively a measure of how well different information nuggets are covered. S-recall at rank K is computed as follows

$$\text{S-recall at K} = \frac{|\cup_{i=1}^{K} subtopics(d_i)|}{n_A} \qquad (1)$$

where n_A represents the total number of sub-topic in the given topic, and $subtopics(d_i)$ denotes the set of sub-topics to which document d_i is relevant. Therefore, S-recall at K represents the percentage of retrieved sub-topics in the top K documents.

In the recent campaign, ImageCLEFPhoto 2008, participants were encouraged to implement diversity in the retrieval process [2]. The relevant images of each test collection topic were manually grouped into topic clusters (i.e. sub-topics) by the organizers. For example, in geographic query, different clusters were represented by different places of where the images were captured. The organisers also used S-recall [14] for their diversity evaluation.

2.3   Amazon Mechanical Turk (MTurk)

Amazon Mechanical Turk (also known as MTurk) is an increasingly popular crowdsourcing Internet service. It provides a platform for collecting human intelligence for tasks which can not be done by computers. A task is called a HIT (Human Intelligence Task), in which two parties are involved, the requester and the worker. Requesters are people who submit the HIT, and workers (also called providers) are people who complete the HIT and receive a monetary payment from the requester. MTurk has been used by researchers for many different tasks, such as image region annotation [11], retrieval relevance assessment [1] and extracting document facets [6].

However, the main problem for using MTurk is how to ensure the quality of results, because it is argued that some workers will give random answers in order to complete the tasks quickly. MTurk provides a number of quality control mechanisms, such as a minimal approval rate which allows only workers who have an overall approval rate that is above the minimal to work on the task, and pre-task assessment questions which can be applied to filter out workers who cannot pass them. Unfortunately, these methods still can not solve the problem completely. A user who passed the pre-task assessment questions can still give random answers when he actually did the HIT. Researchers have proposed some other approaches to tackle the problem. [11] and [1] suggest to collect multiple results from each HIT and remove those deviating from the majority of workers. Furthermore, [1] proposed to calculate a weighted overall score for a HIT from multiple results, which may be calculated by giving more weight to workers are considered as more reputable (e.g. a high approval rate). Finally, [11] suggest to insert some "gold standard data" into the tasks to identify malicious workers. Those who gave too many wrong answers to the gold standard are more likely to add noise to the overall results and thus can be filtered out.

3   What Is Spatial Diversity?

Spatial diversity refers to diversity in the geographical dimension of spatial information retrieval, which normally contains other dimensions such as the semantics of text. A good result should be relevant not only semantically, but also geographically. For example, for the query "castles in UK", relevant documents are those describing "castles" and meanwhile located in "UK". In this work, a document takes the form of a photo annotated with text and the GPS location of where it was taken. Whilst diversity can also be applied to other dimensions, our research is focused on the geographical/spatial dimension.

In the work of [13], the phrase "high spreading" is used to express the notion of spatial diversity. The explanation of high spreading was "points further away from already ranked points are favored". Here, "points" denote geographical locations, and "further away" refers to distances. Paramita *et al.* [9], on the other hand, described spatial diversity as "a way to present diverse results to users by presenting documents from as many different locations as possible". However, neither of the two studies have given a clear definition.

[12] attempted to define spatial diversity from a more theoretical perspective. They used the information nugget idea [4] to analyze users' *spatial information needs* in queries that use "in" as the spatial relationship, e.g. "castles in UK". It is assumed that a user who issued an "in" query is generally interested in semantically relevant documents from all or a subset of the locations within the area of interest (e.g. UK). Thus, every location is regarded as a spatial information nugget and the whole collection of nuggets (i.e. locations) constitute the user's spatial information needs. When the user is shown a document from location A, his spatial information needs on A is met to some extent. It is therefore better to show him documents from other locations. However, documents from very close locations are intuitively unfavorable. For example, having seen a photo taken from the left side of a castle, despite that they are all from different locations, the user may not be as interested in a photo taken from the right side as in a totally different castle. Or in a more upper level, having seen a photo of a castle in England, the user may now become more interested in Scottish castles than English ones. This was explained by [12]'s assumption that "in spatial information retrieval, each document has a certain level of coverage of its neighboring locations". The word "coverage" refers to the assumption that every document has some influence on not only the exact location it comes from, but also on its neighboring ones. In other words, having seen a document from a location A, the user's spatial information needs on the neighboring locations of A are also *partially* met. The intensity of coverage generally decreases as the distance increases. Then, [12] defined spatial diversity as "a measure of location coverage". The more locations that are covered and more intense the coverage is, the better spatial diversity a list of documents achieves, which may satisfy users better.

4   Spatial Diversity Techniques

In this section, we will briefly introduce the three spatial diversity algorithms used in our experiments, namely Geometric Mean (GM) from [9], and SC-1 and SC-2 from [12]. They all work by re-ranking the standard search results so that documents from different locations are promoted to the top of the list, while trying to maintain the precision. Thus, two factors need to be taken into account - semantic relevance and spatial diversity.

4.1   The Geometric Mean (GM) Algorithm

The GM diversity algorithm [9] is an iterative re-ranking method inspired by the GMAP algorithm of [10] used for aggregating scores across topics. In each iteration, it chooses the document with the highest *diversity score* from the remaining list and appends it to the re-ranked list. Diversity score is updated in each iteration and takes into account both the relevance score of the document and it's distances from already chosen documents. A document that is not only relevant but also far away from previous ones will get a high diversity score. See [9] for more details.

4.2   The SC-1 and SC-2 Algorithms

The SC-1 and SC-2 algorithms [12] are built on the notion of *spatial coverage* (SC), or *location coverage* as referred in Section 3. The two have the same underlying statistical model, but differ in how spatial coverage is approximated. The underlying model is a greedy iterative algorithm. In each iteration, it chooses the document that ranks highest according to the sum of its semantic relevance rank and spatial diversity rank. The former is simply generated by a standard search engine, while the latter is measured by how well the location of a new document is covered by those already in the re-ranked list. If a document is from a location that is not covered or only slightly covered by previous documents, it will get a high spatial diversity rank, because the user is likely to have higher interest in such locations than those already covered. See [12] for more details.

Table 1. 22 Geographical Topics

ID	Query
0	church with more than two towers
1	straight road in the USA
2	destinations in Venezuela
3	black and white photos of Russia
4	exterior view of school building
5	night shots of cathedrals
6	lighthouse at the sea
7	sport stadium outside Australia
8	exterior view of sport stadium
9	accommodation provided by host families
10	snowcapped building in Europe
11	cathedral in Ecuador
12	group picture on a beach
13	tourist destinations in bad weather
14	winter landscape in South America
15	sunset over water
16	mountains on mainland Australia
17	indoor photos of a church or cathedral
18	views of walls with unsymmetric stones
19	famous television (and telecommunication) towers
20	drawings in Peruvian deserts
21	seals near water

5 Experiments

5.1 Data-Set

We have chosen the image collection that has been used in the ImageCLEFPhoto 2008 campaign [2], for our experiment. The data-set, known as the IAPR TC-12 image collection [8], is comprised of 20,000 images. Each image comes with a XML meta-data file which contains annotations describing the image content and the location name where the image was taken. Gazetteers were used for relating a location name to a pair of latitude and longitude, following the procedure described in [9] and [12].

Table 2. Cluster Recall at Different Ranks ($^*p < 0.05,^{**}p < 0.01$)

	Std	GM	SC-1	SC-2
CR@5	0.172	0.274*	0.262*	0.288**
CR@10	0.268	0.395**	0.345*	0.404**
CR@20	0.417	0.496*	0.469	0.486**
CR@30	0.480	0.523	0.533**	0.538**

Table 3. Precision at Different Ranks ($^*p < 0.05,^{**}p < 0.01$)

	Std	GM	SC-1	SC-2
P@5	0.482	0.427	0.436	0.446
P@10	0.496	0.423*	0.409*	0.450
P@20	0.468	0.418	0.414	0.396
P@30	0.452	0.391	0.408	0.394

5.2 Results on Precision and S-Recall

Firstly, we use Lucene as the full text search engine to generate the document list for each of the 22 topics listed in Table 1. Secondly, the three diversity algorithms are applied to re-rank each document list. Finally, average precision and S-recall over the 22 topics are calculated for each algorithm, using the "golden standard" assessment file [2]. The number of top documents used for re-ranking is varied, from 20 to 100 with an increment of 20.

Precision and cluster recall (S-recall), as used in ImageCLEFPhoto 2008, have been calculated. Precision calculates the percentage of relevant documents in the top N, while S-recall is calculated as in Equation 1. Due to limited space, only the results (averaged over the 22 topics) from re-ranking the top 60 documents are presented (Table 2 and 3). As can be seen, all the three spatial diversity algorithms achieved better diversity than the standard search, but slightly worse precision. Significance test using paired t-test has also been conducted to see if the the difference is significant. As shown in Table 2 and 3, overall the improvement in diversity is significant but the decrease in precision is not. Among the three algorithms, SC-2 seems to achieve the best performance.

5.3 User Preference Study on Spatial Diversity

In order to investigate if users actually favor spatially diversified results over standard results, we have run a user preference study in MTurk. The idea is simple - by showing two maps (one is the standard and the other is the spatially

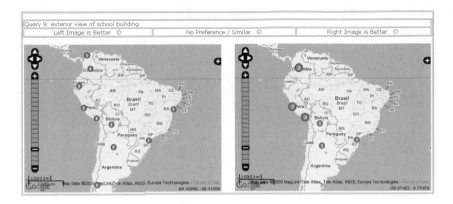

Fig. 1. One example of preference judgement unit (PJU) in spatial diversity user study. See the HIT description for more details.

diversified) in which search results are depicted, we ask the workers to decide which one they prefer based on the corresponding query.

Specifically, we compared each of the three spatial diversity algorithms described in Section 4 with the standard search. For each of the 22 topics, we generated two maps. One map was plotted with the top 10 results from the standard search, and the other was plotted with the top 10 results from one of the spatial diversity algorithms. Search results were plotted on the locations where the photos were taken. It should be noted that documents were represented as red circles rather than the actual images. There are several reasons for such a choice: 1) to make the task as simple as possible for the workers; 2) to focus on only the spatial dimension, i.e. spatial diversity and spatial relevance.

By placing the two maps side by side and showing the query based on which the maps were generated, we gave the worker three options to choose from: 1) the left map is better; 2) no preference/similar; 3) the right map is better. We call this a preference judgement unit (PJU). Figure 1 shows the PJU for the query "exterior view of school building". The left map is generated by SC-2, and the right one is from standard search (This information is of course hidden from the workers).

For quality control purpose, we used a combination of the methods mentioned in Section 2.3. Firstly, we required the workers to have a minimal approval rate of 90% if they want to work on our HITs. Secondly, we randomly inserted some "trap" PJUs into each HIT. A "trap" is a manually constructed PJU in which the preference of two maps is so obvious that every genuine worker should be able to tell easily and correctly. For example, two maps are exactly the same, or one is apparently better than the other (e.g. one shows 10 results and the other shows only 1 result). Each HIT is comprised of 10 PJUs (or topics), including 7 or 8 real PJUs, and 3 or 2 trap ones. In other words, we divided the 22 topics into 3 groups, with 7, 7 and 8 topics respectively, and then added 2 or 3 trap PJUs to each group so that the total number of PJU per group is 10. Every comparison between one spatial diversity algorithm and the standard search generated 3 HITs, resulting in 9 HITs in total.

In MTurk, one worker can only work on the same HIT once, so no duplicate work is allowed. We have requested 50 workers for each HIT, meaning that each topic is set to be judged by 50 different workers. Therefore, the total number of HITs submitted to MTurk is 450. Each HIT was paid \$0.02, and the task was available for 7 days. However, by the time when the pre-set deadline is reached, only 301 HITs (about 2/3 of the total requested) have been finished. As for quality control, we only use submissions from workers who have correctly answered all the trap PJUs of a HIT. If a worker gave wrong answer to any of the trap PJUs, his work on this HIT will be filtered out. As a result, 186 HITs (about 20 judgements per topic) were accepted and used for further analysis.

It is also worth mentioning that in order to conduct a fair experiment, we have randomized not only the order of topics, but also the order of the two maps in each PJU. Therefore, the position of traps in each HIT, and the order of the standard search map and the spatially diversified map, are all unpredictable. The following two paragraphs are the descriptions we provided in each HIT about what the workers are expected to do. As can be seen, in order to void biasing the workers' judgement, we did not give any mention of the notion or the word "diversity" in the task description.

> **INTRODUCTION** *This is a searching results preference survey. There are 10 queries. Each query is a short sentence issued by a user who wants to find the relevant photos via an image search engine. The results (i.e. photos) are returned to the user as small red circles on a geographical map and on the locations where the photos were taken. Inside every circle, a number (from 1 to 10) is shown to indicate how many photos are included the circle. Photos that belong to the same circle are taken from the same or very close geographical locations. For every query, two different maps (the left map and the right map) are shown, each containing a maximum of 10 results (photos).*
>
> **TASK** *Now, imagine you were the user who issued the query, as the searching result, which map do you prefer? the left one, the right one, or there's no apparent difference?*
>
> ---
> HIT Description Shown to the Worker.

5.4   Results on User Preference Study

A three level preference scheme has been applied. For each topic, if the diversified map is favored, a preference value of 1 is assigned to this topic. If the standard one is favored, -1 is assigned. If the worker decided there's no difference between the two, 0 is assigned. All the workers' choices on each topic is averaged to give an overall preference value, which has been shown in Figure 2. Error bars indicate the standard error of mean.

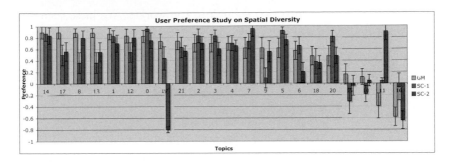

Fig. 2. The results of user preference study on spatial diversity. Topics ordered by the preference value of the GM method. Preference value equals to 1 if the users prefer spatially diversified results, -1 if they prefer the standard results.

As can be seen, users have shown a strong preference to spatially diversified results on most of the topics (17 out of the 22). The rest of the topics can be categorized into three groups:

1. Diversity algorithms failed to achieve apparent improvement in diversity. Topics fall into this kind are topic 10 and 15. As shown in Figure 2, there is no apparent preference (either to the standard or to the diverse result) on these two topics. Manual inspection implies one reason is that the standard one is diverse already.
2. Diversity algorithms caused apparent damage to precision. Topics fall into this kind is topic 16. Users exhibited high preference to results with higher precision. For topic 16, "mountains on mainland Australia", the diversified map did not achieve higher diversity but has one less document in "mainland Australia" than the standard one (spatial relevance). As a result, users favored the standard results. This however is more of a problem of the search engine, because it did not manage to restrict the returned documents to be actually in "mainland Australia".
3. Diversity algorithms performed differently. Some improved the standard result without apparent reduction in precision, but some compromised precision too much. Topics fall into this kind are topic 11 and 19.

Intuitively, topics with higher diversity scores (i.e. cluster recall) will receive stronger user preferences. In order to verify this, we have chosen to use scatter plots. The horizontal axis is the improvement (as in absolute value of difference) in cluster recall by each diversity algorithm over the standard results. The vertical axis is the average preference value of all workers on a particular topic. Each point represents a topic. In Figure 3(a), we have depicted all the topics, but the correlation is not clear. This is due to the fact that our user study is only concerned with spatial diversity, without any relevance judgement. Therefore, the workers probably assumed all documents (shown as circles) are relevant. As a result, diversified results with a very low precision (e.g. 0) will still get a high

preference as long as it is "diverse" enough to the worker. We have then filtered
out topics with a precision at rank 10 (denoted as P@10) that is less than 0.5.
The new scatter plot is shown in Figure 3(b). As can be seen, the correlation
becomes more clear. In general, the more improvement in diversity, the higher
preference score it achieves. Figure 3(c) and 3(d) used precision threshold value
of 0.7 and 0.9. It can probably be claimed that cluster recall is a reasonable
metric for evaluating users' preference in diversity.

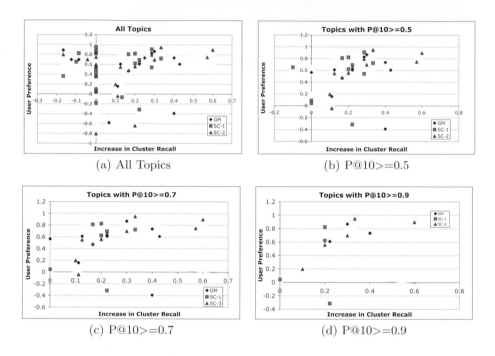

Fig. 3. Scatter Plot of User Preference vs. Increase in Cluster Recall

To compare the three diversity algorithms, we have also generated a chart
of the average preference value for each algorithm at different cut-off precision
thresholds. The horizontal axis is the cut-off precision value. For example, at a
threshold of 0.6, topics with P@10 less than 0.6 are filtered out. The vertical
axis is the average preference of the remaining topics (e.g. P@10>=0.6) for a
particular algorithm. Overall, SC-1 gained the least average preference. GM and
SC-2, however, are comparable. SC-2 is better in the middle range, but worse on
both ends. When the threshold is reduced from 1 to 0.7, the average preference
of GM decreased rapidly. This is probably because GM brought more spatially
irrelevant documents in, which is not surprise considering that GM works by
choosing documents as far away as possible from existing ones. It can probably
be concluded that GM tends to produce more spatially diverse results than SC-
2, but risk bringing in more spatially irrelevant ones. SC-2, on the other hand,

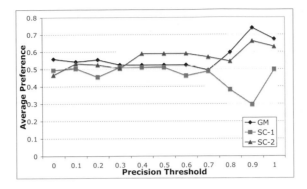

Fig. 4. Comparison of average preference at different cut-off precision thresholds

tends to perform more stably over all topics. However, another reason for the dramatic changes at high thresholds might be due to the small number of topics with a high precision.

To sum up, our study via MTurk has shown that users do have apparent preference to spatially diversified results. Although 5 of the topics exhibited different behaviours, they all seem to be caused by either no apparent diversity improvement or an apparent precision decrease. Given that the precision can be maintained, we believe spatially diversified results will meet users' needs better.

6   Conclusions and Future Work

Three spatial diversity algorithms have been utilized to re-rank the document list generated by a standard search engine. According to evaluations based on precision, cluster recall and significance test, the three algorithms seem to be able to improve diversity significantly while keeping an insignificant decrease in precision. Moreover, a user preference study has been conducted in MTurk to investigate whether users do favor spatially diversified results. Statistics calculated from the user study submissions, which had been filtered by several quality control methods, showed users' strong preference to spatial diversity. On the other hand, the correlation experiment on user preference and cluster recall showed that there is a positive correlation between them. The higher the cluster recall score is, the stronger preference exhibited by users.

We plan to do the experiments on another data-set with more documents and topics. In addition, using other diversity metrics and investigating the correlation to user preference will be interesting.

Acknowledgements

Work undertaken in this paper is supported by the EU-funded Tripod project (Contract No. 045335).

References

1. Alonso, O., Rose, D.E., Stewart, B.: Crowdsourcing for relevance evaluation. SIGIR Forum 42(2), 9–15 (2008)
2. Arni, T., Tang, J., Sanderson, M., Clough, P.: Creating a test collection to evaluate diversity in image retrieval. In: Proceedings of the Workshop on Beyond Binary Relevance: Preferences, Diversity, and Set-Level Judgments, held at SIGIR (2008)
3. Carbonell, J.G., Goldstein, J.: The use of mmr, diversity-based reranking for re-ordering documents and producing summaries. In: SIGIR 1998, pp. 335–336. ACM Press, Melbourne (1998)
4. Clarke, C.L.A., Kolla, M., Cormack, G.V., Vechtomova, O., Ashkan, A., Büttcher, S., MacKinnon, I.: Novelty and diversity in information retrieval evaluation. In: SIGIR 2008: Proceedings of the 31st annual international ACM SIGIR conference on Research and development in information retrieval, pp. 659–666. ACM Press, Singapore (2008)
5. Clough, P., Joho, H., Purves, R.: Judging the spatial relevance of documents for GIR. In: Lalmas, M., MacFarlane, A., Rüger, S.M., Tombros, A., Tsikrika, T., Yavlinsky, A. (eds.) ECIR 2006. LNCS, vol. 3936, pp. 548–552. Springer, Heidelberg (2006)
6. Dakka, W., Ipeirotis, P.G.: Automatic extraction of useful facet hierarchies from text databases. In: IEEE 24th International Conference on Data Engineering (ICDE 2008), April 2008, pp. 466–475. IEEE, Los Alamitos (2008)
7. Deselaers, T., Gass, T., Dreuw, P., Ney, H.: Jointly optimising relevance and diversity in image retrieval. In: ACM International Conference on Image and Video Retrieval 2009 (CIVR 2009), Santorini, Greece, 08/07/2009, ACM (2009)
8. Grubinger, M., Clough, P., Mller, H., Deselaers, T.: The iapr tc-12 benchmark: A new evaluation resource for visual information systems. In: International Workshop OntoImage 2006 Language Resources for Content-Based Image Retrieval held in conjunction with LREC 2006, Genoa, Italy, pp. 13–23 (2006)
9. Paramita, M.L., Tang, J., Sanderson, M.: Generic and spatial approaches to image search results diversification. In: 31st European Conference on Information Retrieval (ECIR), pp. 603–610 (2009)
10. Robertson, S.: On gmap: and other transformations. In: Conference on Information and Knowledge Management, Virginia, USA, pp. 78–83 (2006)
11. Sorokin, A., Forsyth, D.: Utility data annotation with amazon mechanical turk. In: Proceedings of the First IEEE Workshop on Internet Vision at CVPR 2008, pp. 1–8 (2008)
12. Tang, J., Sanderson, M.: The tripod project technical report - modeling diversity in spatial information retrieval using spatial coverage. Technical report, University of Sheffield (2009)
13. van Kreveld, M., Reinbacher, I., Arampatzis, A., van Zwol, R.: Distributed ranking methods for geographic information retrieval. In: Proceedings of the 20th European Workshop on Computational Geometry, pp. 231–243 (2004)
14. Zhai, C.X., Cohen, W.W., Lafferty, J.: Beyond independent relevance: Methods and evaluation metrics for subtopic retrieval. In: SIGIR 2006, Toronto, Canada, pp. 10–17 (2003)
15. Zhang, B., Li, H., Liu, Y., Ji, L., Xi, W., Fan, W., Chen, Z., Ma, W.-Y.: Improving web search results using affinity graph. In: SIGIR 2005, pp. 504–511. ACM, New York (2005)

News Comments:
Exploring, Modeling, and Online Prediction

Manos Tsagkias, Wouter Weerkamp, and Maarten de Rijke

ISLA, University of Amsterdam, Science Park 107, 1098 XG Amsterdam, The Netherlands
{m.tsagkias,w.weerkamp,derijke}@uva.nl

Abstract. Online news agents provide commenting facilities for their readers to express their opinions or sentiments with regards to news stories. The number of user supplied comments on a news article may be indicative of its importance, interestingness, or impact. We explore the news comments space, and compare the log-normal and the negative binomial distributions for modeling comments from various news agents. These estimated models can be used to normalize raw comment counts and enable comparison across different news sites. We also examine the feasibility of online prediction of the number of comments, based on the volume observed shortly after publication. We report on solid performance for predicting news comment volume in the long run, after short observation. This prediction can be useful for identifying news stories with the potential to "take off," and can be used to support front page optimization for news sites.

1 Introduction

As we increasingly live our lives online, huge amounts of content are being generated, and stored in new data types like blogs, discussion forums, mailing lists, commenting facilities, and wikis. In this environment of new data types, online news is an especially interesting type for mining and analysis purposes. Much of what goes on in social media is a response to, or comment on, news events, reflected by the large amount of news-related queries users ask to blog search engines [9]. Tracking news events and their impact as reflected in social media has become an important activity of media analysts [1] and there is a growing body of research on developing algorithms and tools to support this type of analysis (see the related work section below). In this paper, we focus on online news articles plus the comments they generate, and attempt to uncover the factors underlying the commenting behavior on these news articles. We explore the dynamics of user generated comments on news articles, and undertake the challenge to model and predict news article comment volume shortly after publication.

To make things more tangible, consider a striking example of unexpected commenting behavior in response to news stories: March 13, 2009, a busy day for one of the biggest news papers in the Netherlands, *De Telegraaf*. In less than 24 hours, more than 1,500 people commented on *Telegraaf*'s article regarding the latest governmental policy on child benefit abuse. One month later, the Dutch news reported a potential pandemic swine flu, first located in Mexico, but less than five hundred comments were posted to related articles across different news sites, even a week after the first publication. Given that both news events are important to the Dutch society, their numbers of comments

C. Gurrin et al. (Eds.): ECIR 2010, LNCS 5993, pp. 191–203, 2010.

differ greatly. What causes the first story to receive over three times as many comments as the second? What factors contribute to the impact of a news story?

Let us take a step back and ask why we should be interested in commenting behavior and the factors contributing to it in the first place? We briefly mention two types of application for predicting the number of comments shortly after publication. First, in *reputation analysis* one should be able to quickly respond to stories that "take off" and real-time observation and prediction of the impact of news articles is required. Second, the *lay-out decisions* of online news agents often depend on the expected impact of articles, giving more emphasis to articles that are likely to generate more comments, both in their online news papers (e.g., larger headline, picture included) and in their RSS feeds (e.g., placed on top, capitalized).

To come to these applications and answer the questions raised by the example, we need more insight in comments and commenting behavior on online news articles. Our aim is to gain this insight, and use these insights to predict comment volume of news articles shortly after publication. To this end, we seek to answer the following questions:

1. What are the dynamics of user generated comments on news articles? Do they follow a temporal cycle? The answers provide useful features for modeling and predicting news comments.
2. Can we fit a distribution model on the volume of news comments? Modeling the distribution allows for normalizing comment counts across diverse news sources.
3. Does the correlation between number of responses at early time and at later time found in social media such as Digg and Youtube hold for news comments? I.e., are patterns for online responses potentially "universal"? And can we use this to predict the number of comments an article will receive, having seen an initial number?

This paper makes several contributions. First, it explores the dynamics and the temporal cycles of user generated comments in online Dutch media. Second, it provides a model for news comment distribution based on data analysis from eight news sources. And third, it tries to predict comment volume once an initial number of comments is known, using a linear model. In §2 we discuss related work. §3 explores the dataset, and we use insights gained here to try to fit distribution models in §4. Finally, we try to predict comment volume in §5 and conclude in §6.

2   Related Work

Different aspects of the comment space dynamics have been explored in the past. Mishne and Glance [10] looked at weblog comments and revealed their usefulness for improving retrieval and for identifying blog post controversy. Duarte et al. [4] engaged in describing blogosphere access patterns from the blog server point, and identified three groups of blogs using the ratio of posts over comments. Kaltenbrunner et al. [6] measured community response time in terms of comment activity on Slashdot stories, and discovered regular temporal patterns in people's commenting behavior. Lee and Salamatian [7] report that the amount of comments in a discussion thread is inversely proportional to its lifespan after experimenting with clustering threads for two online discussion fora, and for a social networking site. Schuth et al. [12] explore the news comments space of four online Dutch media. They describe the commenters and derive a method for extracting discussion threads from comments. De Choudhury et al. [3] characterize conversations in online media through their interestingness.

We explore the comment space of online news articles, and model the commenting patterns for multiple news sources. Previous work finds that the distribution of comments over blog posts is governed by Zipf's law [8, 10, 12]. Lee and Salamatian [7] use a Weibull distribution for modeling comments in discussion threads. Kaltenbrunner et al. [5] point to discussions in the literature for selecting the log-normal over the Zipf distribution for modeling; they use four log-normal variants to model response times on Slashdot stories. Ogilvie [11] models the distribution of comment counts in RSS feeds using the negative binomial distribution; a similar approach is taken by Tsagkias et al. [15] to model news comments for prediction prior to publication. Finally, Wu and Huberman [16] find that diggs can be modeled with the log-normal distribution, and Szabó and Huberman [14] model popularity growth of online content using a linear model.

3   Exploring News Comments

In this section we describe our data, comments to online news articles, compare commenting behavior to that in the blogosphere, and discover temporal cycles.

The dataset consists of aggregated content from seven online news agents: *Algemeen Dagblad* (*AD*), *De Pers*, *Financieel Dagblad* (*FD*), *Spits*, *Telegraaf*, *Trouw*, and *WaarMaarRaar* (*WMR*), and one collaborative news platform, *NUjij*. We have chosen to include sources that provide commenting facilities for news stories, but differ in coverage, political views, subject, and type. Six of the selected news agents publish daily newspapers and two, *WMR* and *NUjij*, are present only on the web. *WMR* publishes "oddly-enough" news and *NUjij* is a collaborative news platform, similar to Digg, where people submit links to news stories for others to vote for or initiate discussion. We focus only on the user interaction reflected by user generated comments, but other interaction features may play a role on a user's decision to leave a comment.

For the period November 2008–April 2009 we collected news articles and their comments. Our dataset consists of 290,375 articles, and 1,894,925 comments. The content is mainly written in Dutch. However, since our approach is language independent and we believe that the observed patterns and lessons learned apply to news comments in other countries, we could apply our approach to other languages as well.

3.1   News Comments vs. Blog Post Comments

The commenting feature in online news is inspired by the possibility for blog readers to leave behind their comments. Here, we look at general statistics of our news sources and comments, and compare these to commenting statistics in blogs as reported in [10]; the numerical summary can be found in Table 1. News comments are found to follow trends similar to blog post comments. The total number of comments is an order of magnitude larger than the total number of articles, which is positively correlated with the case of influential blogs. In general, about 15% of the blog posts in the dataset in [10] receives comments, a number that increases for the news domain: the average percentage of commented articles across all sources in our dataset is 23%. *Spits* and *WMR* display the

Table 1. Dataset statistics of seven online news agents, and one collaborative news platform (*NUjij*) for the period November 2008–April 2009

News agent	Total articles (commented)		Total comments	Comments per article w/ comments			Time (hrs) 0–1 com. 1–last com.	
				mean	median	st.dev		
AD	41 740	(40%)	90 084	5.5	3	5.0	9.4	4.6
De Pers	61 079	(27%)	80 72	5.0	2	7.5	5.9	8.4
FD	9 911	(15%)	4 413	3.0	2	3.8	10.	9.3
NUjij	94 983	(43%)	602 144	14.7	5	32.3	3.1	6.3
Spits	9 281	(96%)	427 268	47.7	38	44.7	1.1	13.7
Telegraaaf	40 287	(21%)	584 191	69.9	37	101.6	2.5	30.2
Trouw	30 652	(8%)	19 339	7.9	4	10.3	11.7	8.1
WMR	2 442	(100%)	86 762	35.6	34	13.08	1.1	54.2

interesting characteristic of receiving comments on almost every article they publish. This can be explained by the two sites having very simple commenting facilities. In contrast, *Trouw* has the lowest ratio of commented articles: commenting is enabled only for some of the articles, partially explaining the low ratio of commented articles. Another reason can be the content's nature: WMR's oddly-enough news items are more accessible and require less understanding increasing the chance to be commented.

Half of the news sources receive the same number of comments as blogs (mean 6.3), whereas the other half enjoys an order of magnitude more comments than blogs. Looking at reaction time, the time required for readers to leave a comment, it is on average slower for news ($\sim$ 6 hrs) than for blogs ($\sim$ 2 hrs), although this differs significantly per news source. A speculation on the reason underlying these differences can be the news source's readers demographics, e.g., tech savvies or youngsters are rather quick to react, whilst older people, less acquainted with the internet, access the online version of the news papers less frequently.

3.2 Temporal Cycles of News Comments

We perform an exploration of temporal cycles governing the news comment space. We look at three levels of temporal granularity: monthly, daily, and hourly. In our dataset,

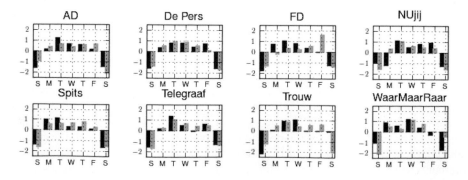

Fig. 1. Comments (black) and articles (grey) per day of the week and per source

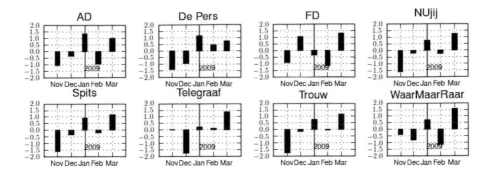

Fig. 2. Comments per month and per source. Vertical line is a year separator.

the volume of comments ranges two orders of magnitude making the comparison of raw comment counts difficult. We therefore report comments in z-scores: z-scores represent how many σ's (standard deviations) the score differs from the mean, and allows for comparison across sources.

Looking at comment volume per month in Fig. 2, we observe months with high and low comment volume, either reflecting the importance of published news, or the seasonal user behavior. For example, March shows the highest comment volume across the board, and November shows the least for most sources.

We explore the comment volume per day of the week in Fig. 1: weekdays receive more comments compared to weekends, with Wednesday being, on average, the most active day and Sunday the least active day across the board. These results are in agreement with the activity observed in social networks such as Delicious, Digg, and Reddit.[1] Comparing the number of comments to the number of articles published per day, most sources show an insignificant, negative correlation ($p \gg 0.05$). Three sources, however, have articles and comments highly correlated, but differ in polarity: *FD* and *Trouw* show a negative correlation and *NUjij* shows a positive correlation. The variety found in correlation polarity likely indicate the commenting behavior of a source's audience.

Finally, we look at the distribution of comments throughout the day. Fig. 3 reveals a correlation between posted comments, sleep and awake time, as well as working, lunch and dinner time. The comment volume peaks around noon, starts decreasing in the afternoon, and becomes minimal late at night. Interesting exceptions are *NUjij*, the collaborative news platform, and *FD*, a financial newspaper: comment volume in *NUjij* matches with blog post publishing [8], which has a slow start and gradually peaks late in the evening. *FD* on the other hand receives most of its comments early in the morning, and then drops quickly. This is in line with the business oriented audience of *FD*.

Overall, the commenting statistics in online news sources show similarities to those in the blogosphere, but are nevertheless inherent characteristics of each news source. The same goes for the temporal cycles, where we see similar patterns for most sources, but also striking differences. These differences in general and temporal characteristics

---

[1] http://3.rdrail.net/blog/thurday-at-noon-is-the-best-time-post-and-be-noticed-pst/

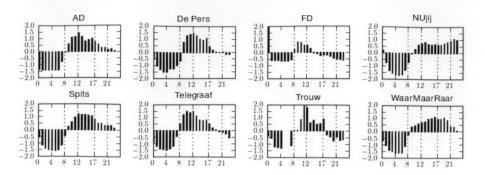

Fig. 3. Comments per hour and per source

possibly reflect the credibility of the news organisation, the interactive features they provide on their web sites, and their readers' demographics [2].

4   Modeling News Comments

In this section we seek to identify models (i.e., distributions) that underly the volume of comments per news source. We do so (1) to understand our data, and (2) to define "volume" across sources. If two articles from two sources receive the same number of comments, do they expose the same volume? Ten comments may signify a high volume for an article in one source, but a low volume in another. Expressing comment volume as a normalized score offers a common ground for comparing and analyzing articles between sources. Our approach is to express a news article's comment volume as the probability for an article from a news source to receive x many comments. We consider two types of distribution to model comment volume: log-normal and negative binomial.

Recall that the log-normal distribution is a continuous distribution, with probability density function defined for $x > 0$, cf. (1), and the two parameters μ (the mean) and σ (the standard deviation of the variable's natural logarithm) affect the distribution's shape. For a given source we estimate the parameters using maximum likelihood estimation.

$$LN_{pdf}(x; \mu, \sigma) = \frac{1}{x\sigma\sqrt{2\pi}}e^{-\frac{(\ln x - \mu)^2}{2\sigma^2}} \qquad (1)$$

The negative binomial distribution is a discrete distribution with probability mass function defined for $x \geq 0$, with two parameters r ($r-1$ is the number of times an outcome occurs) and p (the probability of observing the desired outcome), cf. (2). There is no analytical solution for estimating p and r, but they can be estimated numerically.

$$BN_{pmf}(k; r, p) = \binom{k + r - 1}{r - 1}p^r(1 - p)^k \qquad (2)$$

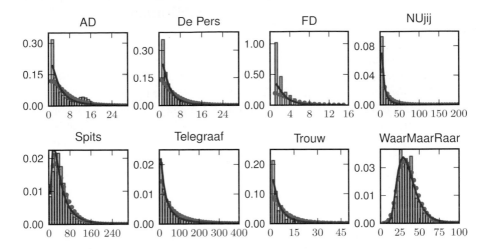

Fig. 4. Modeling comment volume distribution per source using the continuous log-normal (blue line), and the discrete negative binomial distribution (red dots). Grey bars represent observed data. Probability density is on y-axis, and number of comments (binned) is on x-axis.

For evaluating the models' goodness of fit we choose the χ^2 test. χ^2 is a good alternative to the widely used Kolmogorov-Smirnov goodness of fit test due to its applicability to both continuous and discrete distributions [13]. The metric tests whether a sample of observed data belongs to a population with a specific distribution. Note that, the test requires binned data, and as such is sensitive to the number of chosen bins.

For each news source we estimate the parameters for the log-normal and the negative binomial distributions over the entire period of our dataset (see Fig. 4), and report χ^2 goodness of fit results in Table 2. Both distributions fit our dataset well, with low χ^2 scores denoting strong belief that the underlying distribution of the data matches that of log-normal and negative binomial. Log-normal is rejected for *WaarMaarRaar* possibly because it failed to reach close enough the peak observed at 25 comments. We stress that the results should be taken as indicative, mainly due to the sensitivity of χ^2 to the number of bins (here 30). We experimented with different bin sizes, and observed that for different number of bins either the log-normal, or the negative-binomial failed to describe all sources. Although searching for the optimal number of bins for both distributions to fit all sources could be interesting, we did not exhaust the entire potential. An example of the test's sensitivity is shown in Table 3 where log-normal displays very similar results to negative-binomial even for the source that failed the χ^2 test.

The final decision on which distribution to favor, depends on the data to be modeled and task at hand. From a theoretical point of view, negative binomial seem better suited to the task of modeling comments: comments are not a continuous but discrete variable. From a practical point of view, for the same task, log-normal parameters are less expensive to estimate and the results match closely those of negative binomial.

The results of our data exploration and modeling efforts are put to the test in the next section, in which we explore the correlation between comment volume shortly and longer after publication.

5 Predicting Comment Volume After Publication

Predicting the number of news comments *prior* to publication in the long term has proved to be very challenging [15]. Szabó and Huberman [14] published promising work on predicting the long term popularity of Digg stories (measured in diggs), and Youtube videos (measured in views) *after* observing how their popularity evolves in the first hours of publication. First, we are interested in finding out whether the correlation between early and late popularity found by Szabó and Huberman also holds for the news comments space. Then, assuming such a relation has been confirmed, it can be employed for predicting the comment volume of a news story.

We begin to explore the relation between early and late comment volume by look-ing at the similarities of news comments and other online user generated content. In Section 3.2 we reported on the circadian pattern underlying news comment generation, which is found to be similar to blog posts [10], Diggs and Youtube video views [14]. The existence of a circadian pattern implies that a story's comment volume depends on the publication time, and therefore not all stories share the same prospect of being commented; stories published during daytime—when people comment the most—have a higher prior probability of receiving a comment.

Taking into account the above, publication time adds another dimension of com-plexity in finding temporal correlations. To simplify our task, we introduce a temporal transformation from real-time to *source-time*, following [14], a function of the com-ment volume entering a news site within a certain time unit. I.e., *source-time* is defined as the time required for $\bar{x}_i$ comments to enter a news agent system i, where $\bar{x}_i$ stands for the average comments per hour cast to a particular source, and is the division of a source's total comments by the total number of hours that we have data for. Conse-quently, *source-time* has the property of expanding or condensing the real-time scale in order to keep the ratio of incoming comments per hour fixed. Once the number of comments per time unit has been fixed, all stories share the same probability to be commented independently of their publication time. In the rest of this section, story comments are measured in their news agent specific *source-time*, e.g., for *Trouw* we measure in *trouw-time*, for *WMR* in *wmr-time*, etc. Once the temporal transformation is

Table 2. χ^2 goodness of fit for log-normal and negative binomial distributions at 0.10 significance level. Boldface indicates rejection of the null hypothesis: observed and expected data belong to the same distribution.

News site	Log-normal		Negative binomial	
	χ^2 score	p-value	χ^2 score	p-value
AD	0.08	1.00	0.08	1.00
De Pers	0.59	1.00	0.64	1.00
FD	0.18	1.00	0.26	1.00
NUjij	0.06	1.00	0.06	1.00
Spits	0.67	1.00	1.42	1.00
Telegraaf	0.04	1.00	0.04	1.00
Trouw	0.56	1.00	0.98	1.00
WaarMaarRaar	**236.89**	0.00	0.15	1.00

Table 3. Number of comments, per source corresponding at 0.5 of the inverse cumulative distribution function (ICDF)

Distribution	Comments for ICDF @ 0.5							
	AD	De Pers	FD	NUjij	Spits	Telegraaf	Trouw	WMR
Log-normal (LN)	3	3	2	6	36	32	4	34
Negative binomial (NB)	3	3	1	8	39	43	5	33

in place, we need a definition for *early* and *late* time, between which we are interested in discovering a correlation. We introduce *reference time* t_r as "late" time, and we set it at 30 source-days after the story has been published. For "early" time, we define *indicator time* t_r to range from 0 to t_r in hourly intervals [14]. Some news agents disable comments after a certain period. As a result, there are articles that constantly reach their maximum comments before t_r, however we have not marked them separately.

We choose Pearson's correlation coefficient ρ to measure the correlation strength between reference and indicator times. Using articles with more than one comment, we compute ρ in hourly intervals from publication time to reference time for all sources over the entire period of the dataset. Fig. 5 shows that the number of comments per source increases exponentially, yet with different rates, reflecting the commenting rules of each site: the time a story remains visible on the front page, for how long comments are enabled, etc. In the same figure we show a positive correlation that grows stronger as t_i approaches t_r due to stories that saturate to their maximum number of comments. The curve slope indicates how fast stories reach their maximum number of comments, e.g., *Spits* displays a very steep comment volume curve meaning that most stories stop receiving comments short after publication. Looking at when sources reach strong correlation ($\rho > 0.9$) we find that the corresponding indicator times reflect the average

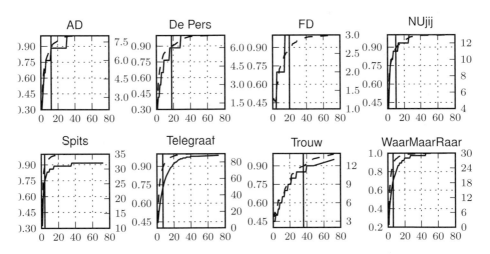

Fig. 5. Comment counts averaged over all stories (right y-axis, solid line), and ρ between indicator, and reference time (left y-axis, dashed line). Indicator time shown at x-axis. Vertical line shows the indicator time with $\rho \geq 0.9$.

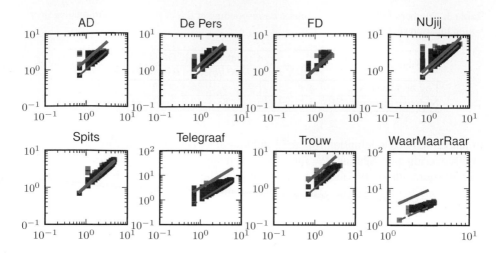

Fig. 6. Correlation of news stories comment volume per source between 2 hours, and 30 days after publication. Number of comments at $t_i(2)$ is x-axis, and comments at t_r is y-axis. K-means separates stories in two clusters depending on their initial comments. Green line shows a fitted model using only the upper stories, with slope fixed at 1. Red dashed line marks the boundary where no stories can fall below.

commenting lifespan of each source (see Table 1). In contrast to our expectations that *NUjij*, the collaborative news platform, follows a fast correlation pattern similar to Digg (0.98 after the 5th digg-hour), our findings suggest that a strong correlation is achieved much later (ρ at 0.90 after 11 source-hours). Although, nujij-time and digg-time are not directly comparable, we can compare the average user submissions entering each system per hour: 5.478 diggs vs. 140 comments. The difference in order of magnitude can be explained by the different popularity levels enjoyed by the two sites. One could argue that digg-ing and commenting are different tasks: on the one hand, commenting, similarly to writing, asks for some reflection on how to verbalize one's thoughts regardless of the size or the quality of the comment. On the other hand, digg-ing requires the click of a button, rendering the task easier, and hence more attractive to participate.

Given the exponential accumulation of comments over time, a logarithmic scale is appropriate for plotting. In contrast to Diggs or YouTube views, comments do not scale more than two orders of magnitude (compare $10^0 - 10^2$ for comments to $10^1 - 10^4$ for Diggs and Youtube views). Despite the difference in scale, our data shows an emerging pattern similar to Youtube, where a bump is observed in the middle range of early comments. From Fig. 6 two groups of stories emerge, both resulting in many comments: one with stories that begin with too few comments in early indicator times, and one with stories that begin with many comments. This pattern is different from Digg or Youtube where a linear correlation is evident in similar graphs [14].

For our prediction experiments, we are interested in minimizing noise to improve performance, and hence could exploit the emerging clusters by eliminating stories with too few comments at early indicator times. Since these stories exhibit a rather random

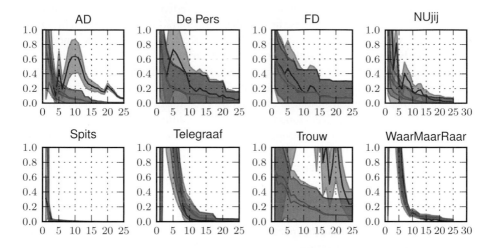

Fig. 7. Relative square error using Model 1 (blue line), Model 2 (green line), and Model 3 (red line). Standard deviation is shown in the shaded areas around the lines. QRE on y-axis, indicator time on x-axis.

pattern with regards to their final number of comments, we employ k-means clustering in an attempt to separate them from stories that show a more consistent pattern.

We follow [14] and estimate a linear model on a logarithmic scale for each source in our dataset. The linear scale estimate $\hat{N}_s$ for a story s at indicator time t_i given t_r is defined as $\hat{N}_s(t_i, t_r) = exp[ln(\alpha_0 N_s(t_i)) + \beta_0(t_i) + \sigma^2/2]$, where $N_s(t_i)$ is the observed comment counts, α_0 is the slope, β_0 is the intercept, and σ^2 is the variance of the residuals from the parameter estimation.

For evaluating our model we choose the relative squared error metric averaged over all stories from a certain source at t_i given t_r.

$$QRE(s, t_i, t_r) = \sum_c \left[\frac{\hat{N}_s(t_i, t_r) - N_s(t_r)}{N_s(t_r)} \right]^2 \tag{3}$$

For our experiments, we split our dataset in training and testing for each source. The training sets span from November 2008—January 2009, and the test sets cover February 2009. Model parameters are estimated on the training set, and QREs are calculated on the test set using the fitted models.

We define three experimental conditions based on which we estimate model parameters using our training set: (M1) using in the upper end stories as clustered by k-means, and fixing the slope at 1, (M2) using all stories, and fixing the slope at 1, and (M3) using all stories. Fig. 7 illustrates QREs for the three experimental conditions up to 25 hours after observation; we choose not to include all indicator times up to reference time to increase readability of the details at early times. From the three experimental conditions, M1 proved to underperform in most cases. M2 and M3 demonstrate similar performance across the board with one slightly outperforming the other depending on the source. QREs decrease to 0 as we move to reference time, followed by a similar decrease in standard error. M2 demonstrates strong predictive performance indicated by low QRE < 0.2 for all sources, in less than 10 hours of observation. The QREs converge to 0 faster for some sources and slower for others, exposing the underlying commenting dynamics of each source as discussed earlier.

In this section we looked at natural patterns emerging from news comments, such as the possible correlation of comment counts on news stories between early and later publication time. A relation similar to the one observed for Digg and Youtube has been confirmed, allowing us to predict long term comment volume with very small error. We observed that different news sources ask for different observation times before a robust prediction can be made. Using QRE curves one can find the optimum observation time per source, that balances between short observation period and low error.

6 Conclusion and Outlook

We studied the news comments space from seven online news agents, and one collaborative news platform. Commenting behavior in the news comments space follows similar trends as the behavior in the blogosphere. Our news sources show quite similar temporal cycles and commenting behavior, but that mainly the differences herein reflect differences in readers' demographics and could prove useful in future research.

As to modeling comments from various news agents, we compared the log-normal and negative binomial distributions. These estimated models can be used to normalize raw comment counts and enable comparison, and processing of articles from different news sites. According to χ^2 goodness of fit test, the underlying distribution of news comments matches with either log-normal or negative binomial. The latter is a discrete distribution and suits the task better, yet in our our setup log-normal showed similar results and parameter estimation for log-normal is computationally less expensive.

Finally, we looked at the feasibility of predicting the number of comments at a late time, based on the number of comments shortly after publication. Our goal was to find patterns similar to other online content such as Digg, and Youtube. We confirmed this relation, and exploited its potential using linear models. Our results showed that prediction of the long term comment volume is possible with small error after 10 source-hours observation. This prediction can be useful for identifying news stories with the potential to "take off," and can for example be used to support front page optimization for news sites.

Acknowledgments. This research was supported by the DAESO and DuOMAn projects carried out within the STEVIN programme which is funded by the Dutch and Flemish Governments under project numbers STE-05-24 and STE-09-12, and by the Netherlands Organisation for Scientific Research (NWO) under project numbers 640.-001.501, 640.002.501, 612.066.512, 612.061.814, 612.061.815, 640.004.802.

References

[1] Altheide, D.L.: Qualitative Media Analysis (Qualitative Research Methods). Sage Pubn. Inc., Thousand Oaks (1996)
[2] Chung, D.S.: Interactive features of online newspapers: Identifying patterns and predicting use of engaged readers. J. Computer-Mediated Communication 13(3), 658–679 (2008)

[3] De Choudhury, M., Sundaram, H., John, A., Seligmann, D.D.: What makes conversations interesting? In: WWW 2009, April 2009, p. 331 (2009)

[4] Duarte, F., Mattos, B., Bestravras, A., Almedia, V., Almedia, J.: Traffic characteristics and communication patterns in blogosphere. In: ICWSM 2006 (March 2007)

[5] Kaltenbrunner, A., Gomez, V., Lopez, V.: Description and prediction of slashdot activity. In: LA-WEB 2007, pp. 57–66 (2007)

[6] Kaltenbrunner, A., Gómez, V., Moghnieh, A., Meza, R., Blat, J., López, V.: Homogeneous temporal activity patterns in a large online communication space. CoRR (2007)

[7] Lee, J.G., Salamatian, K.: Understanding the characteristics of online commenting. In: CONEXT 2008, pp. 1–2 (2008)

[8] Mishne, G., de Rijke, M.: Capturing global mood levels using blog posts. In: AAAICAAW, pp. 145–152 (2006)

[9] Mishne, G., de Rijke, M.: A study of blog search. In: Lalmas, M., MacFarlane, A., Rüger, S.M., Tombros, A., Tsikrika, T., Yavlinsky, A. (eds.) ECIR 2006. LNCS, vol. 3936, pp. 289–301. Springer, Heidelberg (2006)

[10] Mishne, G., Glance, N.: Leave a reply: An analysis of weblog comments. In: Third annual workshop on the Weblogging ecosystem (2006)

[11] Ogilvie, P.: Modeling blog post comment counts (July 2008),
http://livewebir.com/blog/2008/07/
modeling-blog-post-comment-counts/

[12] Schuth, A., Marx, M., de Rijke, M.: Extracting the discussion structure in comments on news-articles. In: WIDM 2007, pp. 97–104 (2007)

[13] Sheskin, D.J.: Handbook of Parametric and Nonparametric Statistical Procedures. Chapman & Hall/CRC, USA (2000)

[14] Szabó, G., Huberman, B.A.: Predicting the popularity of online content. CoRR, abs/0811.0405 (2008)

[15] Tsagkias, M., Weerkamp, W., de Rijke, M.: Predicting the Volume of Comments on Online News Stories. In: CIKM 2009, pp. 1765–1768 (2009)

[16] Wu, F., Huberman, B.A.: Novelty and collective attention pnas,
http://www.pnas.org/content/104/45/17599.abstract

Query Performance Prediction: Evaluation Contrasted with Effectiveness

Claudia Hauff[1], Leif Azzopardi[2], Djoerd Hiemstra[1], and Franciska de Jong[1]

[1] University of Twente, Enschede, The Netherlands
{c.hauff,hiemstra,f.m.g.dejong}@ewi.utwente.nl
[2] University of Glasgow, Glasgow, UK
leif@dcs.gla.ac.uk

Abstract. Query performance predictors are commonly evaluated by reporting correlation coefficients to denote how well the methods perform at predicting the retrieval performance of a set of queries. Despite the amount of research dedicated to this area, one aspect remains neglected: how strong does the correlation need to be in order to realize an improvement in retrieval effectiveness in an operational setting? We address this issue in the context of two settings: Selective Query Expansion and Meta-Search. In an empirical study, we control the quality of a predictor in order to examine how the strength of the correlation achieved, affects the effectiveness of an adaptive retrieval system. The results of this study show that many existing predictors fail to achieve a correlation strong enough to reliably improve the retrieval effectiveness in the Selective Query Expansion as well as the Meta-Search setting.

1 Introduction

Predicting the performance, i.e. the retrieval effectiveness, of queries has become a very active area of research in recent years [1,4,6,8,13,15,18,19,20,21]. Accurately predicting a query's effectiveness would enable the development of adaptive components in retrieval systems. For instance, if the performance of a query is predicted to be poor, the system may ask the user for a refinement of the query or divert it to a specialized corpus. Conversely, if the performance of a query appears sufficiently good, the query's performance can be further improved by an affirmative action such as automatic query expansion (AQE) [1].

While the perceived benefits of query performance prediction (QPP) methods are clear, current evaluations often do not consider whether those benefits are indeed realized. Commonly, the correlation between the ground truth (the retrieval effectiveness in average precision for instance) and the predicted performance of queries is reported. The subsequent application of QPP methods in an operational setting is often missing and thus it remains unclear, if QPP methods perform well enough to be useful in practice.

In this work, we attempt to bridge this knowledge gap by investigating the relationship between the correlation that QPP methods achieve, and their effect on retrieval effectiveness in two operational settings: Meta-Search (MS) [15,18]

C. Gurrin et al. (Eds.): ECIR 2010, LNCS 5993, pp. 204–216, 2010.

and Selective Query Expansion (SQE) [1,5,18]. Our goal is to determine at what levels of correlation a QPP method can be considered good enough to be employable in practice. If we can determine such thresholds, we would be able to infer from a correlation-based evaluation, whether the quality of a QPP method is sufficient for an operational setting.

We conduct two empirical studies based on several TREC[1] data sets. In previous work [7], we performed preliminary experiments with respect to SQE. Here, we extend these experiments and add a second setting: Meta-Search. We will show, that the correlation a QPP method needs to achieve on average to be useful in practice, is dependent on the operational setting. In the SQE setting, moderate to high correlations result in reliable improvements. In the MS setting, low to moderate correlations are already sufficient, however for these improvements to be statistically significant, we find moderate to high correlations to be required.

The paper is organized as follows: in Sec. 2 we outline related work and the motivation for our work. The data sets and our general approach are described in Sec. 3. Then, the experiments on SQE (Sec. 4) and MS (Sec. 5) are presented, followed by the conclusions and directions for future work (Sec. 6).

2   Related Work and Motivation

First, we briefly describe the two main types (pre- and post-retrieval) of QPP methods. To give an indication of the success QPP methods have achieved in SQE and MS respectively, we also provide an overview of literature that employed QPP methods in either setting.

Query Performance Prediction. Pre-retrieval QPP methods predict the performance of a query without considering the ranked list of results returned by a retrieval system in response to a query. Approaches in this category are usually based on the query terms' corpus-statistics, such as the standard deviation of the query terms' IDF [8] or TF.IDF [19] values. Post-retrieval predictors base their predictions on the ranked list of results. The strategies employed are manifold, they include a comparison between the ranked list and the corpus [1,4], the perturbation of query terms and a subsequent comparison of the generated ranked lists [13,18,21], the perturbation of documents in the ranked list to determine the list's stability [13,20], and the reliance on different retrieval approaches to form predictions based on the diversity of the returned documents [6]. The two most commonly reported correlation coefficients in QPP evaluations are the rank correlation coefficient Kendall's Tau $\tau \in [-1, 1]$ and the linear correlation coefficient $r \in [-1, 1]$. Current state of the art QPP methods achieve up to $\tau \approx 0.55$ and $r \approx 0.65$, depending on the test corpus and query set.

Applications of SQE. The two SQE scenarios evaluated in [18] are based on the notion that easy queries (queries with a high retrieval effectiveness) improve with the application of AQE, while difficult queries (queries with a low

---

[1] Text REtrieval Conference (TREC), http://trec.nist.gov/

retrieval effectiveness) degrade. The reasoning is as follows: easy queries will have relevant documents among the top ranked results, and thus an AQE algorithm [3,12,16,17], which derives additional query terms from the top ranked documents returned for the initial query, is likely to pick terms related to the information need. The ranked list retrieved for the expanded query thus further improves the quality of the results. Difficult queries have few or no relevant documents in the top ranks and an AQE algorithm will add irrelevant terms to the query, degrading the result quality. In the first scenario in [18], a support vector machine is trained on features derived from the ranked list of the original query to classify queries as either to be expanded or not. In the second scenario [18], a QPP method is used to rank the queries according to their predicted performance. The 85% best performing queries are derived from TREC topic descriptions[2] (simulating AQE), while the bottom 15% of queries are derived from TREC topic titles (simulating no AQE). In both experiments, selectively expanding the queries based on a QPP method proves slightly better than uniformly expanding all queries, with a change in Mean Average Precision (MAP) of +0.001.

An analogous scenario with a different QPP method is evaluated in [1]; the greatest improvement reported is from a MAP of 0.252 (AQE of all queries) to 0.256 (selective AQE). Better results are reported in [5], where the threshold of when (not) to expand a query is learned: in the best case, MAP increases from 0.201 (AQE of all queries) to 0.212 (selective AQE). AQE is combined with collection enrichment in [9]: depending on how the QPP method predicts a query to perform, it is either not changed, expanded based on the results of the local corpus or expanded based on the results of the external corpus. The evaluation yields mixed results, while for one data set the MAP increases from 0.220 (AQE of all queries) to 0.236 (selective AQE), no change in effectiveness is observed for a second data set. These results indicate, that successfully applying a QPP method in the SQE setting is a challenging task.

Applications of MS. In [18], the following meta-search setup is evaluated: a corpus is partitioned into four parts, each query is submitted to each partition, and the result lists of each partition are merged with weights according to their predicted performance. In this experiment, MAP increases from 0.305 (merging without weights) to 0.315 (merging with QPP based weights).

In [15], a variety of retrieval algorithms are applied on one corpus. For each query and retrieval algorithm, a result list is derived and its predicted performance score is determined. Heuristic thresholds are used to classify each result list as either poor, medium or good. The result lists are then merged with weights according to the performance classification. The best weighted data fusion method performs 2.12% better than the unweighted baseline. Finally, in [14] it is proposed to generate a number of relevance models [11] for each query and pick the model that is predicted to perform best. The results indicate the

---

[2] A TREC topic usually consists of a title (a small number of keywords), a description (a sentence) and a narrative (a long description of the information need).

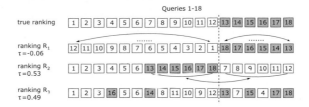

Fig. 1. Examples of predicted query rankings (R_1 to R_3) and their effect on SQE. Of the 18 queries, those ranked 1-12 (white) improve when AQE is employed, the results of queries ranked 13-18 (grey) degrade with the application of AQE.

feasibility of the approach, the QPP-based model selection strategy significantly outperforms the baseline.

Overall, few works have employed QPP in practice. Based on the results, we hypothesize, that it is easier to be successful in the MS than in the SQE setup.

Evaluation by Correlation. Evaluating QPP methods by reporting correlation coefficients is the current standard. The correlation based evaluation though can lead to problems: (i) very different predictions can lead to similar correlations, (ii) a QPP method for which a high (low) correlation is reported, may not lead to an increase (decrease) in retrieval effectiveness in an operational setting, and, (iii) following from (i) and (ii) is the observation that a single QPP method is not a reliable indicator of the level of correlation required, at which the application of a QPP method in practice is likely to lead to a consistent gain in retrieval effectiveness.

Fig. 1 contains a conrete example for Kendalls τ, the correlation coefficient we investigate in this work. The true ranking is the ranking of queries based on their retrieval effectiveness. Let us assume a SQE setup according to [18]: the queries ranked 1-12 benefit from AQE, the remaining queries do not. R_1, R_2 and R_3 are examples of predicted rankings of query performance. R_1 predicts all *ranks* incorrectly, thus $\tau_{R_1} = -0.06$ with respect to the true ranking. All AQE decisions though are correct, which leads to an optimal increase in retrieval effectiveness in SQE. The opposite case is R_2 with $\tau_{R_2} = 0.53$. The AQE decision is wrong for 12 queries, degrading the retrieval effectiveness. Based on the predicted ranking R_3, the wrong AQE decision is made for 4 queries. Although the retrieval effectiveness will be better than based on R_2, τ is similar: $\tau_{R_3} = 0.49$.

Thus, predictions resulting in similar correlations can have very different impacts on retrieval effectiveness, a problem which motivates our work.

3 Method and Materials

Ideally, we would like to perform the following experiment: given a large number of QPP methods, a large number of retrieval algorithms and a set of queries, let each QPP method predict the queries' quality. Evaluate the QPP methods in terms of τ, use the predictions in an operational setting and perform retrieval experiments to derive baseline and QPP-based results. Finally, check at what level of τ the QPP-based results generally improve over the baseline. In practice, this is not feasible as state of the art QPP methods only reach up to $\tau \approx 0.55$.

Data Sets. To perform experiments on SQE and MS, test collections, sets of queries and their respective retrieval performances are required. To make our results independent of a particular retrieval approach, we utilize TREC data sets, in particular the runs submitted by the participants to different adhoc retrieval tasks. All submitted runs with a MAP greater than 0.15 are included in this study. The data sets used are as follows (in brackets the number of runs): TREC6 (49), TREC7 (77), TREC8 (103), TREC9 (53), TREC10 (59), Terabyte04 (35), Terabyte05 (50) and Terabyte06 (71).

QPP Method of Arbitrary Accuracy. As state of the art QPP methods achieve a limited τ, they are unsuitable for our investigation. Since τ is rank based, it is possible to construct predicted rankings of any level of τ, simply by randomly permutating the true performance ranking of queries. The smaller the number of permutations, the closer τ is to 1. The larger the number of permutations, the closer τ is to 0. From the full range of $\tau \in [-1, 1]$, sixteen intervals $\{c_{0.1}, ..., c_{0.85}\}$ of size 0.05 each were investigated, starting a $c_{0.1} = [0.1, 0.15)$ and ending at $c_{0.85} = [0.85, 0.9)^3$. For each interval c_i, 1000 rankings were randomly generated with $\tau \in c_i$ with respect to the true ranking. We rely on such a large number of rankings due to the issues outlined in Fig. 1: a single predicted ranking can be misleading when applied in practice. By considering the results of 1000 rankings for each c_i, we can consider the change in retrieval effectiveness *on average*. Each predicted ranking is used in the SQE and MS experiments, which allows us to analyze the impact of varying levels of correlation against retrieval effectiveness.

4   Selective Query Expansion

We analyze the relationship between τ as evaluation measure of QPP methods and the change in retrieval performance when queries are expanded selectively in a setup analogous to [18]. The effect AQE has on retrieval effectiveness varies, depending on the AQE approach, the retrieval algorithm and the set of queries evaluated. While across the whole query set, AQE aids retrieval effectiveness (improvements range between $3 - 40\%$ [12,16,17]), not all queries benefit. The percentage of queries from a query set performing worse when AQE is applied varies between $20 - 40\%$ [1,10,12,16]. In [1,10] it is reported that the worst and the very best queries are hurt by AQE[4], whereas in [3,16] only the worst queries are reported to be hurt by AQE.

---

[3] This is sufficient, as negative correlations can be transformed into positive correlations by reversing the ranking and $\tau = 1$ indicates two perfectly aligned rankings.

[4] Further adding terms dilutes the results of the already highly performing queries.

4.1 Experimental Details

Let us for now assume that all well performing queries improve with AQE, while the worst performing queries all degrade with AQE. Let θ be a rank threshold. Our SQE setup is as follows: given a set of m queries, they are ranked according to their predicted performance. AQE is applied to the best $(\theta \times m - 1)$ performing queries, the remaining queries are not expanded. As this setup only requires predicted rankings, we can use our generated predicted rankings of arbitrary accuracy. To evaluate the retrieval effectiveness of SQE, we require pairs of baseline (no AQE) and AQE runs. Then, we perform SQE based on the predicted rankings and consider SQE successful if it improves over the retrieval effectiveness of the AQE run. We derive baseline and AQE run pairs from the runs in our data sets. As we are not interested in the result lists themselves, but in the effectiveness of each run on each query Q, we consider a run to consist of a list of average precision (AP) values, thus $run = (ap^{Q_1}, ap^{Q_2}, .., ap^{Q_m})$.

Run Pairs. Each run of our data sets is considered as a baseline run run_{base} (no AQE) for which a corresponding AQE run run_{aqe} is generated. Recall, that we work with the assumption that AQE improves the effectiveness of the well performing queries, while degrading the effectiveness of poorly performing queries. Thus, for each $ap_{base}^{Q_i}$ in run_{basc}, a respective $ap_{aqe}^{Q_i}$ value in run_{aqe} is generated such that $ap_{aqe}^{Q_i} > ap_{base}^{Q_i}$ when $ap_{base}^{Q_i}$ is among the top $(\theta \times m - 1)$ performing queries in run_{base}, otherwise $ap_{aqe}^{Q_i} < ap_{base}^{Q_i}$. The $ap_{aqe}^{Q_i}$ values are randomly sampled (with the outlined restrictions) from the other runs in the data sets, a strategy supported by [2], where no correlation between $ap_{base}^{Q_i}$ and the amount of improvement, i.e. $\Delta = ap_{aqe}^{Q_i} - ap_{base}^{Q_i}$, $\Delta > 0$, was found. The optimal SQE run run_{opt} is the run where the correct AQE decision is made for every query: $ap_{opt}^{Q_i} = max(ap_{base}^{Q_i}, ap_{aqe}^{Q_i})$. We only include run pairs where the MAP of run_{aqe} improves by between $15 - 30\%$ over run_{base} and run_{opt} improves by at least 3% over run_{aqe}. Due to the random component in the process, 500 run pairs are created for each setting of $\theta = \{1/2, 2/3, 3/4\}$ and $m = \{50, 150\}$. The choice of θ is based on [12,16], the settings of m are typical TREC topic set sizes.

Experiment. Given the 1000 rankings per c_i and the 500 run pairs (run_{basc}/run_{aqe}) for each setting of θ and m, SQE is thus performed $500,000$ times for each c_i. From each run pair and predicted ranking in c_i a selective AQE run run_{sqe} is formed: if according to the predicted ranking $ap_{base}^{Q_i}$ is among the top $(\theta \times m - 1)$ scores in run_{base}, then $ap_{sqe}^{Q_i} = ap_{aqe}^{Q_i}$, that is the AQE result is used. The remaining queries are not expanded and $ap_{sqe}^{Q_i} = ap_{base}^{Q_i}$. Recorded are the MAP of run_{base}, run_{aqe}, run_{opt} and run_{sqe}. We consider SQE to be successful if the MAP of run_{sqe} is higher than the MAP of run_{aqe}. Since the (run_{base}/run_{aqe}) pairs lead to different absolute changes in retrieval performance, we report a normalized value: $v_{sqe} = 100(MAP_{sqe} - MAP_{base})/(MAP_{opt} - MAP_{base})$. When the correct AQE decision is made for each query, $v_{sqe} = 100$. In contrast, $v_{sqe} < 0$ if the MAP of run_{sqe} is below the baseline's run_{base} MAP.

We present the results, derived for each c_i, in the form of box plots. Every box marks the lower quartile, the median and the upper quartile of $500,000$ v_{sqe} values. The whiskers mark the 1.5 inter-quartile range, the remaining separately marked points are outliers. We also plot the median normalized value of AQE runs as horizontal line, as it is the value v_{sqe} should improve upon.

4.2   Experimental Results

Best-Case Scenario. First, we evaluate the best-case scenario, where our assumption that AQE only hurts the worst performing queries holds for all run pairs. The results for the different settings of θ and m are presented in Fig. 2.

Independent of m and θ, for all correlation intervals c_i a number of positive outliers exist, where the MAP of run_{sqe} improves over run_{aqe}'s MAP. Thus, even if $\tau = 0.1$, a QPP method can be successful by chance. This supports the view that a single experiment and QPP method are inadequate indicators to show a QPP's method utility in practice.

When the correlation of the predicted ranking with respect to the ground truth is low, run_{sqe} may perform worse than run_{base} as observed for $\theta = 1/2$ and $m = 50$ (negative v_{sqe} values). This means that a poor QPP method may

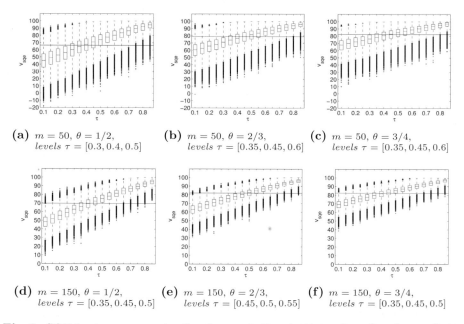

(a) $m = 50$, $\theta = 1/2$, *levels* $\tau = [0.3, 0.4, 0.5]$

(b) $m = 50$, $\theta = 2/3$, *levels* $\tau = [0.35, 0.45, 0.6]$

(c) $m = 50$, $\theta = 3/4$, *levels* $\tau = [0.35, 0.45, 0.6]$

(d) $m = 150$, $\theta = 1/2$, *levels* $\tau = [0.35, 0.45, 0.5]$

(e) $m = 150$, $\theta = 2/3$, *levels* $\tau = [0.45, 0.5, 0.55]$

(f) $m = 150$, $\theta = 3/4$, *levels* $\tau = [0.35, 0.45, 0.5]$

Fig. 2. SQE best-case scenario. On the x-axis the starting value of each correlation interval c_i is listed, that is the results for $c_{0.2} = [0.2, 0.25)$ are shown at position 0.2. The horizontal lines mark the normalized median value of the performance of the AQE runs, which v_{sqe} must improve upon for SQE to be successful. "*levels* τ" lists the lower bound of c_i where the SQE runs outperform the AQE runs in at least 25%, 50% and 75% of all $500,000$ samples (these levels can be read from the box plots).

significantly degrade the effectiveness of a system. An increase in τ generally leads to a smaller spread in performance (the height of the boxes in the plot) of v_{sqe}, that is, outliers are rarer and the performance drop is not as sharp. Finally, the number m of queries also influences the outcome - with increased m the spread of the results decreases and the results can be considered to be more stable. This suggests that the more queries are used in the evaluation, the better the correspondence between the evaluation measure Kendalls τ and the performance in an operational setting.

Random Perturbations. We now investigate what happens when the assumption that all well performing queries improve with AQE is violated. This experiment is motivated by the divergent observations [1,10,12,16] about the change in effectiveness of the best performing queries when AQE is applied.

To simulate such violation, we perturb run_{aqe}. Given a run pair (run_{base}/run_{aqe}), we randomly select a query Q_i from the top $(\theta \times m - 1)$ performing queries of run_{aqe} and perturb its score $ap_{aqe}^{Q_i}$ to $\hat{ap}_{aqe}^{Q_i}$, which is a random value below $ap_{base}^{Q_i}$. To keep the MAP of run_{aqe} constant, the difference $(ap_{aqe}^{Q_i} - \hat{ap}_{aqe}^{Q_i})$ is randomly distributed among the other queries' ap scores. This procedure is performed for $p = \{10\%, 20\%, 30\%\}$ of the queries. The results of this experiment with fixed $\theta = 1/2$ are shown in Fig. 3. Evidently, already a small number of perturbed queries has a considerable influence on the minimum τ which is required to improve run_{sqe}

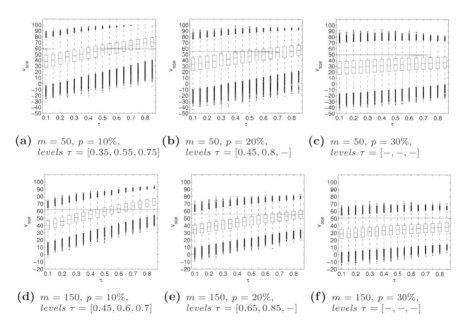

(a) $m = 50$, $p = 10\%$, *levels* $\tau = [0.35, 0.55, 0.75]$

(b) $m = 50$, $p = 20\%$, *levels* $\tau = [0.45, 0.8, -]$

(c) $m = 50$, $p = 30\%$, *levels* $\tau = [-, -, -]$

(d) $m = 150$, $p = 10\%$, *levels* $\tau = [0.45, 0.6, 0.7]$

(e) $m = 150$, $p = 20\%$, *levels* $\tau = [0.65, 0.85, -]$

(f) $m = 150$, $p = 30\%$, *levels* $\tau = [-, -, -]$

Fig. 3. SQE random perturbation scenario. $\theta = 1/2$ is fixed. *"levels* τ*"* lists the lower bound of c_i where the SQE runs outperform the AQE runs in at least 25%, 50% and 75% of all $500,000$ samples.

over run_{aqe}. When $p = 10\%$, $\tau \geq 0.55$ is required to ensure that for more than 50% of the samples run_{sqe} improves over run_{aqe}. A further increase in the number of perturbed queries leads to a situation as visible for $p = 30\%$, where independent of the accuracy of the predicted ranking, in less than 25% of all instances run_{sqe} improves over run_{aqe}.

To summarize, in the best-case scenario, we have shown that a QPP method should evaluate to at least $\tau = 0.4$, for 50% of the samples to improve with the application of SQE. This threshold increases to $\tau = 0.55$, when the assumption we make about for what queries AQE is successful, is slightly violated. These results explain the low levels of success in [1,5,9,18], where QPP methods were applied in the SQE setting. Only the best performing QPP methods reach the levels of τ required to be beneficial. Moreover, making the wrong assumption about when AQE will lead to increased system effectiveness quickly leads to a situation where it does not matter anymore, how well a QPP method performs.

5   Meta-search

The second operational setting we examine is Meta-Search where we adopt a setup analogous to [14]: given a query and a set of t runs, we select the run that is predicted to perform best for the query. We chose this setup over the data fusion setup [18], where the merging algorithm introduces an additional dimension, as it allows us to better control the experimental setting.

5.1   Experimental Details

In preliminary experiments we found two parameters influencing the retrieval effectiveness of QPP-based meta-search: (i) the number t of runs, and (ii) the percentage γ of improvement in MAP between the worst and the best performing run in the set of t runs. The experimental setup reflects these findings. We derived 500 sets of t runs from our data sets for each setting of γ: $0 - 5\%$, $15 - 20\%$, $30 - 35\%$ and $50 - 55\%$. A range $0 - 5\%$ means, that all t runs perform very similar in terms of MAP, while in the extreme setting of γ, the MAP of the best run is $50 - 55\%$ better than the MAP of the worst run. To generate each set of runs, t runs are randomly selected from the data sets. A set is valid if the maximum percentage of retrieval improvement lies in the specified interval of γ. Recall, that 1000 predicted rankings exist per correlation interval c_i. As we require t predicted rankings per set of runs, t rankings are randomly chosen from all rankings of a given c_i. To avoid result artifacts due to relying on predicted ranks instead of scores[5], the ith predicted rank is replaced by the ith highest ap score of the run. The meta-search run run_{meta} is then created by selecting for each query the result of the run with the highest predicted ap score. The optimal run run_{opt} is derived by $ap_{opt}^{Q_i} = max(ap_{run_1}^{Q_i}, .., ap_{run_t}^{Q_i})$.

---

[5] An extreme example is the case where the worst performing query of run A has a higher average precision than the best performing query of run B.

Analogous to the SQE experiments, we report the normalized performance of run_{meta}, that is, $v_{meta} = 100(MAP_{meta} - MAP_{worst})/(MAP_{opt} - MAP_{worst})$, where MAP_{worst} is the MAP of the worst of the t runs. When $v_{meta} < 0$, run_{meta} performs worse than the worst run of the set, a value of $v_{meta} = 100$ implies that run_{meta} is optimal. For MS to be successful, run_{meta} needs to perform better than the best run of the set. This threshold is indicated by the horizontal line in the plots (the normalized median of the best runs' MAP across all sets).

5.2   Experimental Results

We experimented with $t = \{2, 3, 4, 5\}$. Due to space constraints though, in Fig. 4 we only report the results for $t = 4$ and $\gamma = \{0 - 5\%, 30 - 35\%, 50 - 55\%\}$.

Evidently, low correlations are sufficient to improve over the effectiveness of the best run. Note though, that for low correlations outliers exist that perform worse than the worst run (negative v_{meta}). As the performance difference between the best and the worst run increases (γ), higher correlations are required to improve run_{meta} over the best individual run. For $m = 50$ and $\gamma = 50 - 55\%$ for example, $\tau \geq 0.4$ is required to improve 50% of the samples. As in the case of the SQE experiments, an increase in m leads to more stable results.

The table in Fig. 5 contains basic statistics of the MAP values of the worst and best run of the sets of t runs as well as run_{opt}. The results were averaged over the 500 sets of runs for each m and γ. A different view on the sets of runs offers the percentage of queries in run_{opt} that are drawn from the worst performing

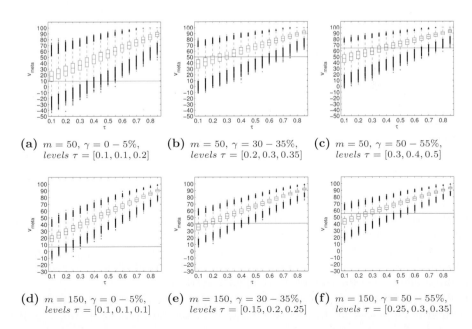

(a) $m = 50$, $\gamma = 0 - 5\%$, levels $\tau = [0.1, 0.1, 0.2]$

(b) $m = 50$, $\gamma = 30 - 35\%$, levels $\tau = [0.2, 0.3, 0.35]$

(c) $m = 50$, $\gamma = 50 - 55\%$, levels $\tau = [0.3, 0.4, 0.5]$

(d) $m = 150$, $\gamma = 0 - 5\%$, levels $\tau = [0.1, 0.1, 0.1]$

(e) $m = 150$, $\gamma = 30 - 35\%$, levels $\tau = [0.15, 0.2, 0.25]$

(f) $m = 150$, $\gamma = 50 - 55\%$, levels $\tau = [0.25, 0.3, 0.35]$

Fig. 4. MS experiments with $t = 4$ systems, m queries and varying γ

m	γ	% worst in opt.	MAP worst	MAP best	MAP opt
50	$0 - 5\%$	23.4%	0.236	0.245	0.328
	$15 - 20\%$	18.4%	0.220	0.259	0.334
	$30 - 35\%$	14.5%	0.209	0.277	0.344
	$50 - 55\%$	11.5%	0.196	0.298	0.356
150	$0 - 5\%$	23.5%	0.222	0.231	0.345
	$15 - 20\%$	20.0%	0.211	0.249	0.347
	$30 - 35\%$	17.5%	0.195	0.259	0.348
	$50 - 55\%$	14.5%	0.182	0.277	0.354

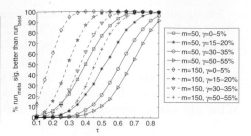

Fig. 5. The table contains the average MAP values over the 500 sets of $t = 4$ runs for each setting of m and γ. Listed are the MAP of the worst, best and optimal meta-search run.

The figure shows the development of the percentage of samples in which run_{meta} statistically significantly outperforms the best individual run (run_{best}) across the correlation intervals c_i.

run (column **% worst in opt.**). At $t = 4$, if the run selection is random, we expect 25% of the results to come from each run. As expected, with increasing γ, less results from the worst performing run are utilized in the optimal run. Though even for large differences in retrieval effectiveness ($\gamma = 50 - 55\%$), the worst performing run still contributes towards the optimal run.

We also investigate if the improvements of the MAP of run_{meta} are statistically significant[6]. We performed a paired t-test with significance level 0.05. Shown in Fig. 5 is the percentage of samples in which run_{meta} significantly outperforms the best individual run in the set of t runs. At $m = 50$, the threshold of τ to improve 50% of the samples significantly, ranges from 0.4 ($\gamma = 0 - 5\%$) to 0.7 ($\gamma = 50 - 55\%$). The thresholds are lower for $m = 150$: $\tau = 0.2$ ($\gamma = 0 - 5\%$) and $\tau = 0.5$ ($\gamma = 50 - 55\%$) respectively.

Thus, in the MS setting, QPP methods that result in low correlations can already lead to a gain in retrieval effectiveness. For these improvements to be statistically significant though, QPP methods with moderate to high correlations are needed.

6   Conclusions

We have investigated the relationship of one standard evaluation measure of QPP methods (Kendalls τ) and the change in retrieval effectiveness when QPP methods are employed in two operational settings: Selective Query Expansion and Meta-Search. We aimed to give a first answer to the question: when are QPP methods good enough to be usable in practice? To this end, we performed a comprehensive evaluation based on TREC data sets.

We found that the required level of τ depends on the particular setting a QPP method is employed in. In the case of SQE, in the best-case scenario, $\tau \geq 0.4$

---

[6] This analysis was not required in the SQE setup, due to the run construction.

was found to be the minimum level of τ for the SQE runs to outperform the AQE runs in 50% of the samples. In a second experiment we showed the danger of assuming AQE to behave in a certain way - slightly violating an assumption already requires QPP methods with $\tau \geq 0.75$ for them to be viable in practice.

The outcome is different for the MS experiments. Here, the level of τ is dependent on the performance differences of the participating runs. If the participating runs are similar, a QPP method with $\tau = 0.1$ is sufficient to improve 50% of the runs over the baseline. If the performance differences are great, $\tau = 0.3$ is required. To achieve statistically significant improvements for 50% of runs under large system differences, $\tau = 0.7$ ($m = 50$) and $\tau = 0.5$ ($m = 150$) are required.

These results indicate that (i) QPP methods need further improvement to become viable in practice, in particular for the SQE setting and (ii) that with increasing query set sizes m the evaluation in terms of τ relates better to the change in effectiveness in an operational setting.

We have restricted ourselves to an analysis of Kendalls τ as it is widely used in QPP evaluations. We plan to perform a similar analysis for the linear correlation coefficient r. In contrast to the rank-based τ, r is based on raw scores, which adds another dimension - the distribution of raw scores - to the experiments.

References

1. Amati, G., Carpineto, C., Romano, G.: Query difficulty, robustness and selective application of query expansion. In: McDonald, S., Tait, J.I. (eds.) ECIR 2004. LNCS, vol. 2997, pp. 127–137. Springer, Heidelberg (2004)
2. Billerbeck, B., Zobel, J.: When query expansion fails. In: SIGIR 2003, pp. 387–388 (2003)
3. Carmel, D., Farchi, E., Petruschka, Y., Soffer, A.: Automatic query refinement using lexical affinities with maximal information gain. In: SIGIR 2002, pp. 283–290 (2002)
4. Cronen-Townsend, S., Zhou, Y., Croft, W.B.: Predicting query performance. In: SIGIR 2002, pp. 299–306 (2002)
5. Cronen-Townsend, S., Zhou, Y., Croft, W.B.: A framework for selective query expansion. In: CIKM 2004, pp. 236–237 (2004)
6. Diaz, F.: Performance prediction using spatial autocorrelation. In: SIGIR 2007, pp. 583–590 (2007)
7. Hauff, C., Azzopardi, L.: When is query performance prediction effective? In: SIGIR 2009, pp. 829–830 (2009)
8. He, B., Ounis, I.: Inferring query performance using pre-retrieval predictors. In: Apostolico, A., Melucci, M. (eds.) SPIRE 2004. LNCS, vol. 3246, pp. 43–54. Springer, Heidelberg (2004)
9. He, B., Ounis, I.: Combining fields for query expansion and adaptive query expansion. Inf. Process. Manag. 43(5), 1294–1307 (2007)
10. Kwok, K.: An attempt to identify weakest and strongest queries. In: SIGIR 2005 Query Prediction Workshop (2005)
11. Lavrenko, V., Croft, W.B.: Relevance based language models. In: SIGIR 2001, pp. 120–127 (2001)
12. Mitra, M., Singhal, A., Buckley, C.: Improving automatic query expansion. In: SIGIR 1998, pp. 206–214 (1998)

13. Vinay, V., Cox, I.J., Milic-Frayling, N., Wood, K.: On ranking the effectiveness of searches. In: SIGIR 2006, pp. 398–404 (2006)
14. Winaver, M., Kurland, O., Domshlak, C.: Towards robust query expansion: model selection in the language modeling framework. In: SIGIR 2007, pp. 729–730 (2007)
15. Wu, S., Crestani, F.: Data fusion with estimated weights. In: CIKM 2002, pp. 648–651 (2002)
16. Xu, J., Croft, W.B.: Query expansion using local and global document analysis. In: SIGIR 1996, pp. 4–11 (1996)
17. Xu, J., Croft, W.B.: Improving the effectiveness of information retrieval with local context analysis. ACM Trans. Inf. Syst. 18(1), 79–112 (2000)
18. Yom-Tov, E., Fine, S., Carmel, D., Darlow, A.: Learning to estimate query difficulty: including applications to missing content detection and distributed information retrieval. In: SIGIR 2005, pp. 512–519 (2005)
19. Zhao, Y., Scholer, F., Tsegay, Y.: Effective pre-retrieval query performance prediction using similarity and variability evidence. In: Macdonald, C., Ounis, I., Plachouras, V., Ruthven, I., White, R.W. (eds.) ECIR 2008. LNCS, vol. 4956, pp. 52–64. Springer, Heidelberg (2008)
20. Zhou, Y., Croft, W.B.: Ranking robustness: a novel framework to predict query performance. In: CIKM 2006, pp. 567–574 (2006)
21. Zhou, Y., Croft, W.B.: Query performance prediction in web search environments. In: SIGIR 2007, pp. 543–550 (2007)

A Framework for Evaluating Automatic Image Annotation Algorithms

Konstantinos Athanasakos, Vassilios Stathopoulos, and Joemon M. Jose

Department of Computing Science, University of Glasgow,
Sir Alwyn Williams Building, Lilybank Gardens, Glasgow G12 8QQ, UK
{athanask,stathv,jj}@dcs.gla.ac.uk

Abstract. Several Automatic Image Annotation (AIA) algorithms have been introduced recently, which have been found to outperform previous models. However, each one of them has been evaluated using either different descriptors, collections or parts of collections, or "easy" settings. This fact renders their results non-comparable, while we show that collection-specific properties are responsible for the high reported performance measures, and not the actual models. In this paper we introduce a framework for the evaluation of image annotation models, which we use to evaluate two state-of-the-art AIA algorithms. Our findings reveal that a simple Support Vector Machine (SVM) approach using Global MPEG-7 Features outperforms state-of-the-art AIA models across several collection settings. It seems that these models heavily depend on the set of features and the data used, while it is easy to exploit collection-specific properties, such as tag popularity especially in the commonly used Corel 5K dataset and still achieve good performance.

1 Introduction

During the last decade, we have witnessed a major transformation of the digital multimedia information field. A lot of effort has been invested in identifying modern and efficient ways of browsing, navigating and retrieving multimedia data, while the traditional challenge of bridging the semantic gap [1] remains unsolved. The ultimate goal of understanding multimedia content requires us to identify a way to effectively combine low-level features in order to reach a high-level understanding of objects and semantics portrayed in an image. A question arises however, as to whether a correlation between these two levels actually exists.

Automatic Image Annotation (AIA) attempts to learn the afore-mentioned correlation and build a dictionary between low-level features and high-level semantics [2]. The idea is to use a manually annotated set of multimedia data in order to train a system to be able to identify the joint or conditional probability of an annotation occurring together with a certain distribution of multimedia content feature vectors. Two ways have been suggested. The first one is using supervised machine learning techniques in order to classify an image into pre-defined categories. In this approach, the class and the non-class model have to

C. Gurrin et al. (Eds.): ECIR 2010, LNCS 5993, pp. 217–228, 2010.

[13], the Corel dataset is far too "easy", while the TrecVid datasets essentially comprise an effort to build more realistic collections. Nevertheless, none of these models were directly compared to other models using these enhanced collections, since this would be an expensive and time-consuming procedure requiring the implementation of other models as well and carrying out more experiments. On the other hand, some models, such as the Corr-LDA [7] were not directly compared to any previous models. Moreover, although the SML has achieved the best performance so far on the Corel 5K dataset, we do not have enough evidence to support that this is due to the model and that SML would outperform previous models in other settings as well. Especially with the Corel 5K dataset, it would be easy to exploit collection-specific properties and still get good results. As such, we cannot be certain as to whether models are robust and independent of their setting and whether some perform better because of their descriptive ability or because of the discriminating ability of the features sets used.

3   Evaluation Framework

In this section, we describe the Evaluation Framework which we propose to be used for the evaluation of already introduced and future Automatic Image Annotation algorithms. It essentially defines a set of test collections, a sampling method which attempts to extract normalised and self-contained samples, a variable-size block segmentation technique with varying degrees of overlapping and a set of multimedia content descriptors. These are all discussed in details in the following sections.

3.1   Multimedia Collections

A very common challenge related to image classification algorithms and machine learning methods in general is the fact that these are usually dependent on the data on which they are applied. This actually means that their performance and discriminating ability varies significantly depending on the test collection which is used each time. In the case of image classification algorithms, the setting on which such an algorithm might be evaluated consists of a multimedia collection and the kind of features that will be used to represent its images.

Regarding multimedia collections, facts such as whether images depict single or multiple objects, and whether an annotation implies dominance of an object or simply its presence are some examples of these factors. Moreover a collection could be strongly or weakly labelled, depending on whether all instances of an object are annotated or not, while the existence of object hierarchies having tags such as "cat" and "tiger", "car" and "exotic car" or "water" and "ocean" might not only affect the performance of the algorithm, but also the results that one would expect. Collections also define the level of semantics that an algorithm should target for. Searching for objects is a totally different task than searching for scene categories or emotional states. It would perhaps require a different way of treating images, namely segmenting and representing, thus again modifying the overall setting on which the algorithm would have to operate.

As such, an evaluation of a set of image classification algorithms would simply be incomplete, if it did not involve testing these algorithms on various settings in order to prove their robustness, namely whether they perform equally well under various settings. Therefore, a set of three multimedia collections was selected to be incorporated in our evaluation procedure. These are the Corel 5K [2], TrecVid 2007 [14] and Caltech 101 [15] collections.

Corel 5K is considered a rather easy setting, since Global Colour Features alone are considered to provide enough discriminative power for this collection. It was first used by Duygulu et al. [2] in the field of automatic image annotation algorithms, while since then, it has been used by each new model in the literature, in order for the results to be comparable to previously proposed models. The TrecVid 2007 dataset on the other hand comprises an extremely challenging setting. Since it is intended to be used for several high level tasks such as shot boundary detection and high level feature extraction, one can appreciate that using this dataset in the AIA domain will be equally difficult and unpredictable. Caltech 101 has a major advantage over other multimedia datasets, in that each image depicts a single object, thus removing any confusion associated with the multiple-labels paradigm. As such, it can be employed to learn precisely the class and non-class model of certain categories and objects. Although the categories are not described by the same number of images, the fact that images belong to only one category each allows for a sample which is fair towards all categories, namely it has the same number of images describing each category, while still being consistent and self-contained.

It is obvious that these collections present various settings ranging from controlled, "laboratory" ones to more realistic collections incorporating issues such as statistically unbalanced tag distributions, weak labelling and so on. Ideally, an AIA algorithm should be able to cope with all of the various challenges present in the afore-mentioned collections. However, no algorithm has been found and proved to meet this condition. In addition, as suggested by Westerveld and de Vries in [13], we might have to consider different performance measures in terms of granularity depending on the difficulty level of a collection.

3.2   Sampling Procedure

In this paper, the afore-mentioned collections were not used as a whole, rather we used a sampling procedure to extract a smoother and self-contained representative sample of each collection. By smoother, we mean that most of the tags would contain approximately the same number of images, and only a few, if any, would be described by significantly more example images. By self-contained, we mean that we would not discard any instances of the sampled classes which were included in the sampled images, as this would harm their class and non-class models. This sampling process was performed for two reasons. First, using the whole collections would require an immense amount of time to complete evaluating these algorithms, as in the case of memory-based models like MBRM which require examining the whole training set each time a test image is being classified, while at the same time, it would not add significant value to the validity of our experiments. Second and more importantly, all of these collections have a highly unbalanced distribution of images over classes. There are a lot of classes which are inadequately described, a set of classes with a reasonable number of images belonging to them and a few which are very popular and frequent within each collection. Using the whole collections would probably create an easier setting for all of the algorithms for two reasons. When evaluating such an algorithm, popular tags would be more likely to be selected to be tested, while on the other hand, when classifying an image it would be more likely to annotate it with a more frequent tag. Moreover, we did not want to allow models to exploit attributes of collections which were unrelated to visual information, such as tag popularity. Hence, a sampling procedure was applied on all of the collections, which attempted to smooth these settings removing extreme conditions, namely classes which were either inadequately or very precisely described, while at the same time preserving the rest of the attributes of these collections.

The collections were first analysed, plotting the distribution of all N_{total} images over all of the C_{total} tags of each collection. In that way, it would be feasible to empirically determine on a reasonable number of images N_{min} with which each one of the classes should at least be described. This parameter N_{min} was set on a per collection basis. The second step was to remove any classes which were inadequately described, namely being described by a number of example images $N_C < N_{min}$. The result thus far would be having identified a part of the collection which contains only classes for which we have enough images (N_{min}) at our disposal. Next, we would randomly select a number of C_{sample} classes to form our sample. However, in order to also remove tags which were very frequent, we did not select the C_{sample} classes from the whole range of the remaining classes ($C_{remaining}$), but from the first $C_{sample-from}$ classes after sorting them based on the number of images belonging to them. As such, the result now would be having a sample of N_{sample} images from each collection which contained only medium-frequency classes. However, when selecting an image, we would consider all of the tags which belonged to the sampled C_{sample} classes regardless of the fact that this would make some tags appear as more popular than others. Discarding some of the instances of a class, might have a negative impact on the performance and the overall operation of an image classification algorithm, as it would be very difficult to define the class's class and non-class models.

In Figure 1, the reader is provided with the distributions of images over classes for the Corel 5K and the Caltech 101 collections. In the left column, the distribution of images over tags for the whole collection is plotted. In the middle, we have removed the inadequately described tags, which enables us to empirically determine, how many tags should be sampled (C_{sample}) and how many of the most popular tags, which appear on the right side of the graph should be discarded. Finally, at the right column, the distribution of images over classes for our sample of each collection is plotted. Moreover, in Table 1 the reader is provided with some statistics and the values of the afore-mentioned parameters for each collection.

Table 1. Statistics and parameters' values regarding the sampling procedure

	Corel 5K	TrecVid 2007	Caltech 101
N_{total}	4079	17675	8242
C_{total}	374	36	100
N_{min}	40	40	40
$C_{remaining}$	75	30	75
$C_{sample-from}$	60	30	70
C_{sample}	50	30	50
N_{sample}	1195	527	2009

However, in order to be fair with an algorithm and remove any chance of the results being based on luck, such an evaluation should be cross-validated. In our evaluation, we decided to evaluate the algorithms using $N = 10$ folds. However, it would be extremely difficult and even impossible to be able to split our samples of these collections into $N = 10$ totally separated, self-contained and statistically-balanced parts. Certain parts might not include any train images for some classes, or certain tags might not be tested in some folds. Therefore, we modified the afore-mentioned sampling procedure, executing it N times for each collection sampling each time $N_{fold} = N_{min}/N$ images for each one of the predetermined C_{sample} set of classes. In that way, the result of this process would be having N consistent, separated and self-contained samples from each collection.

3.3   Image Segmentation

Since fixed-size block-segmentation is an essential part of one of the models which were chosen to be evaluated, namely the MBRM model, block segmentation was also used while extracting local features. However, dividing images into equally-sized regions might be misleading, since even small-size objects may be split into two or more regions while large-sized ones are always seen in part and never in whole. Hence, the optimal size of blocks is dependent on the images and the collection itself. In order to overcome the two afore-mentioned obstacles of fixed-size block segmentation, we used variable-size block segmentation with varying degrees of overlapping. First, all images were resized to fit in a 512×512 window. Then, we empirically identified a set of block sizes $S = \{32, 64, 128, 256\}$ which would be meaningful when used in block segmentation given the average size of the images and the average size of the objects depicted in them. Depending on the size s_i of the square blocks, we would determine on the degree of overlapping. The step d between two neighbouring blocks was set to $d = s_i$ when $s_i \leq 32$, $d = s_i/3$ when $32 < s_i \leq 64$, and finally $d = s_i/4$ when $s_i > 64$. By considering overlapping multi-resolution block-segmentation, we ensure that objects and classes will be seen both in part and as a whole during the annotation process, which is a very desired property in object class recognition.

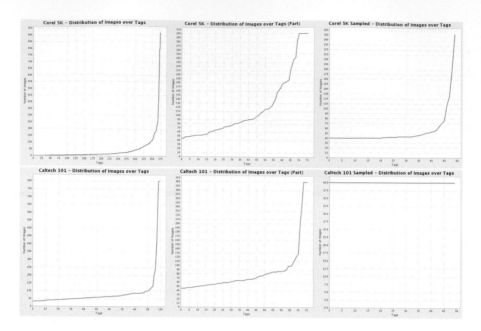

Fig. 1. Distributions of images over classes in the Corel 5K (first row) and Caltech 101 (second row) collections

3.4   Content Descriptors

Image representation and feature extraction is an important and definitive step when attempting to use an automatic image annotation algorithm. It is important to identify the appropriate set of features, one which would provide not only the appropriate level of discrimination among images, but also enough compactness, so that the algorithm itself will not suffer from the challenging problems of computational complexity, immense resource requirements and the curse of dimensionality. In addition, it is not unusual for a multimedia collection to be known to yield better results when used in combination with a specific set of features, while on the other hand, certain image classification algorithms also perform better when used with certain sets of features. Hence an evaluation of image classification algorithms incorporating various features sets representing different attributes and characteristics of the same images from the same collections might shed some light into the operation of these algorithms through their variation in performance when applied on various such settings of collections and features sets.

When deciding on the features sets which would be incorporated in the evaluation process, the objective was to use standardised features sets, no matter how well they would actually perform. The goal of the present work was not to get better results, but to investigate patterns in the relative performance and the presence of any consistency between certain image classification algorithms.

As such, by using colour and texture features defined in the MPEG-7 Standard [16], such as Colour Histogram (CH), Edge Histogram (EH), and Homogeneous Texture (HT), as well as SIFT features introduced in 2004 by Lowe [17], it would be clear that we did not act in favour of a specific algorithm, while the results of this work would still be meaningful in the future, as it would be straightforward to implement a new algorithm, run experiments on the same collections using these standardised features sets and get comparable results.

4   Results

In this section, results showing mean per-word precision and recall for each setting individually are presented.

In Table 2, results of experiments with our implementation of MBRM and SML using MPEG-7 and SIFT Features respectively are presented for the three collections. Our results are significantly lower than the ones reported in the original papers [4,5]. The reason for this is that we used normalised parts of the collections, as well as other sets of features. On the other hand, in Table 3, the MBRM is contrasted to the simpler Support Vector Machines (SVM) approach using the SVM-light implementation [18].

First of all, with respect to the collections, we would say that Corel was the most "extreme" setting, followed by that of TrecVid 2007, and then the completely normalised sample of Caltech 101. By "extreme", we mean that only a few tags were more popular than others, while these had significantly more example images. Moreover, we would assume that, as TrecVid 2007 is supposed to be used for high level video tasks, it would be extremely difficult to detect similarity between frames using common image descriptors.

From Table 2, we can see that the variance of both Precision and Recall around the means was significantly high. We also see that only a small percentage of tags has $Recall > 0$ and most of these tags are popular tags in the collection. This is similar to previously reported results [4,5,11] on the Corel 5K collection. However, since we have removed most of the popular tags the numbers tend to be significantly smaller. This shows that previous optimistic results on Corel 5k are actually due to the tag distribution rather than the descriptive ability of the models. Interestingly, MBRM would always return the most popular words when evaluated on Corel 5K and TrecVid 2007. On the contrary, in Caltech 101, in which tag frequencies were completely normalised, more words were returned and the diversity among them was high. Also, regarding the TrecVid dataset, we see that MBRM had exactly the same response across all descriptors, meaning that similarity across images was not taken into account by the model. On the other hand, SML achieved the best performance on TrecVid 2007, followed by Corel 5K and Caltech 101. The bad performance on Caltech might be due to the fact that it is a single-label environment, and the actual number of classes depicted in an image was considered during the annotation process. The difference in performance between Corel 5K and TrecVid 2007 might be either due to the visual content of the images, or due to collection-specific properties. Nevertheless,

Table 2. Mean Precision and Recall of MBRM (MPEG-7) and SML (SIFT)

Collections	Corel 5K				TrecVid 2007				Caltech 101			
Models	MBRM			SML	MBRM			SML	MBRM			SML
Descriptors	CH	EH	HT	SIFT	CH	EH	HT	SIFT	CH	EH	HT	SIFT
# of words in total	70				30				50			
# of words with Recall>0	4	4	4	6	8	8	8	9	18	14	13	2
Precision and Recall on all words												
Mean Per-word Recall	0.034	0.045	0.045	0.046	0.194	0.194	0.194	0.130	0.125	0.265	0.270	0.015
Variance in Recall	0.151	0.193	0.193	0.175	0.356	0.356	0.356	0.269	0.207	0.275	0.286	0.077
Mean Per-word Precision	0.020	0.010	0.010	0.003	0.163	0.163	0.163	0.073	0.127	0.286	0.251	0.0009
Variance in Precision	0.121	0.044	0.044	0.011	0.296	0.296	0.296	0.160	0.214	0.316	0.268	0.005
Precision and Recall on words with Recall > 0												
Mean Per-word Recall	0.569	0.750	0.750	0.495	0.534	0.534	0.534	0.397	0.222	0.377	0.422	0.360
Variance in Recall	0.284	0.238	0.238	0.303	0.295	0.295	0.295	0.320	0.206	0.182	0.174	0.125
Mean Per-word Precision	0.334	0.172	0.174	0.036	0.449	0.449	0.449	0.188	0.227	0.360	0.284	0.021
Variance in Precision	0.374	0.054	0.054	0.012	0.232	0.232	0.232	0.235	0.218	0.244	0.217	0.015

overall in all collections, our results are not as optimistic as previously reported ones, and this seems to be related to the normalised tag distributions of our samples. However, although different categories of features were used with each model, the results between them are still comparable and can be interpreted in a generic way.

Moreover, we applied a Support Vector Machine using global MPEG-7 features on the Corel 5K collection and compared it with MBRM and SML. Results are presented in table 3, where we can see that a simple SVM with global features achieves better results than MBRM and SML, which are considered state-of-the-art methods. We have also implemented a SVM with local MPEG-7 features by using k-means to cluster local features and create visual terms. The local features are associated to their closest visual term (cluster centroid) and images are represented by the frequency of the visual terms they contain, similarly to a bag of word model used in Information Retrieval. Despite the quantisation errors introduced by the k-means algorithm, results are still better than MBRM and SML although not as good as using the SVM directly on the global MPEG-7 features.

Finally, with respect to SML, it was not feasible to combine it with local MPEG-7 Features. The image segmentation procedure which was used for extracting MPEG-7 local features led to a quite homogeneous representation of each image individually. The MBRM was not affected by this homogeneity since features were homogeneous only at the image level. SML however was not able to cluster the feature vectors representing each image with a mixture model of a reasonable number of components. As SML uses a mixture of Gaussians, it essentially makes strong assumptions about the nature and the properties of the features, thus making it feature-dependent. Hence, the SML would require a significantly larger dataset, and a descriptor which would provide an appropriate degree of heterogeneity at the image level.

Table 3. Comparison between MBRM and SVM using MPEG-7 Descriptors

Collections	Corel 5K						Caltech 101				TrecVid 2007			
Models	MBRM	SVM	MBRM	SVM	MBRM	SVM	MBRM	SVM	MBRM	SVM	MBRM	SVM	MBRM	SVM
Descriptors	CH	GCH	CH	EH	GEH	EH	CH	GCH	EH	GEH	CH	GCH	EH	GEH
# of words in total	70						50				30			
# words (Recall>0)	4	32	26	4	37	29	18	20	14	23	8	15	8	16
Precision and Recall on all words														
Mean Recall	0.034	0.204	0.102	0.045	0.402	0.314	0.125	0.327	0.265	0.580	0.194	0.405	0.194	0.611
Recall Variance	0.151	0.236	0.159	0.193	0.430	0.250	0.207	0.221	0.275	0.325	0.356	0.265	0.356	0.374
Mean Precision	0.020	0.131	0.051	0.010	0.242	0.193	0.127	0.372	0.286	0.740	0.163	0.454	0.163	0.564
Precision Variance	0.121	0.226	0.087	0.044	0.301	0.183	0.214	0.158	0.316	0.363	0.296	0.347	0.296	0.336
Precision and Recall on words with Recall > 0														
Mean Recall	0.569	0.149	0.173	0.750	0.188	0.245	0.222	0.173	0.377	0.300	0.534	0.227	0.534	0.265
Recall Variance	0.284	0.095	0.144	0.238	0.195	0.169	0.206	0.080	0.182	0.124	0.295	0.140	0.295	0.144
Mean Precision	0.334	0.173	0.087	0.172	0.193	0.139	0.227	0.142	0.360	0.057	0.449	0.009	0.449	0.028
Precision Variance	0.374	0.269	0.092	0.054	0.294	0.182	0.218	0.234	0.244	0.288	0.232	0.258	0.232	0.288

5   Conclusion

In this paper, we considered the lack of proper evaluation in the domain of Automatic Image Annotation. We found that the evaluation methodologies followed by AIA researchers are insufficient and do not support and prove the models' initial assumptions. Hence, we defined an Evaluation Framework, which is comprised by more than one multimedia collections and standardised descriptors, uses a sampling method to extract smoother, self-contained and representative samples and a multi-resolution block-segmentation method. We used this framework to evaluate and compare two state-of-the-art AIA models and we found that they heavily depend on the underlying test set. MBRM was found to return the most popular tags, while the SML was found to be extremely feature-dependent, and could not be integrated with standardised MPEG-7 Features. Thus, the high reported performance measures could be artifacts of the collections and not due to the descriptive power of the models. Finally, we have demonstrated that a simple SVM approach performs better than state-of-the-art models across several collections and descriptors.

We argue that as the number of experimental settings increases and as we keep their diversity high, we get more insight on a model's functionality, while strong and weak points emerge. As such, this study sets forward an evaluation paradigm for future annotation models, while the proposed framework should be integrated in the whole process of the development of a model, from the conceptualisation and the development phases until the validation and evaluation.

Acknowledgements

The research leading to this paper was supported by European Commission under contract FP6-027122 (Salero).

References

1. Smeulders, A.W.M., Worring, M., Santini, S., Gupta, A., Jain, R.: Content-based image retrieval at the end of the early years. IEEE Trans. Pattern Anal. Mach. Intell. 22(12), 1349–1380 (2000)
2. Duygulu, P., Barnard, K., de Freitas, J.F.G., Forsyth, D.A.: Object recognition as machine translation: Learning a lexicon for a fixed image vocabulary. In: Heyden, A., Sparr, G., Nielsen, M., Johansen, P. (eds.) ECCV 2002. LNCS, vol. 2353, pp. 97–112. Springer, Heidelberg (2002)
3. Lavrenko, V., Manmatha, R., Jeon, J.: A model for learning the semantics of pictures. In: Thrun, S., Saul, L., Schölkopf, B. (eds.) Advances in Neural Information Processing Systems, vol. 16. MIT Press, Cambridge (2004)
4. Feng, S., Manmatha, R., Lavrenko, V.: Multiple bernoulli relevance models for image and video annotation. In: Proceedings of the 2004 IEEE Conference on Computer Vision and Pattern Recognition, June-2 July 2004, vol. 2, pp. II–1002–II–1009 (2004)
5. Carneiro, G., Chan, A.B., Moreno, P.J., Vasconcelos, N.: Supervised learning of semantic classes for image annotation and retrieval. IEEE Trans. Pattern Anal. Mach. Intell. 29(3), 394–410 (2007)
6. Blei, D.M., Ng, A.Y., Jordan, M.I.: Latent dirichlet allocation. J. Mach. Learn. Res. 3, 993–1022 (2003)
7. Blei, D.M., Jordan, M.I.: Modeling annotated data. In: Proceedings of the 26th annual international ACM SIGIR, pp. 127–134. ACM, New York (2003)
8. Boiman, O., Shechtman, E., Irani, M.: In defense of nearest-neighbor based image classification. In: IEEE CVPR 2008, June 2008, pp. 1–8 (2008)
9. Vasconcelos, N., Lippman, A.: Learning mixture hierarchies. In: Advances in Neural Information Processing Systems, vol. II, pp. 606–612. MIT Press, Cambridge (1999)
10. Li, J., Wang, J.Z.: Automatic linguistic indexing of pictures by a statistical modeling approach. IEEE Trans. Pattern Anal. Mach. Intell. 25(9), 1075–1088 (2003)
11. Jeon, J., Lavrenko, V., Manmatha, R.: Automatic image annotation and retrieval using cross-media relevance models. In: Proceedings of the 26th annual international ACM SIGIR Conference, pp. 119–126. ACM, New York (2003)
12. Kwasnicka, H., Paradowski, M.: On evaluation of image auto-annotation methods. In: Proceedings of the 6th Int. Conf. on Intelligent Systems Design and Applications, pp. 353–358. IEEE Computer Society, Washington (2006)
13. Westerveld, T., de Vries, A.P.: Experimental evaluation of a generative probabilistic image retrieval model on 'easy' data. In: Proceedings of the SIGIR Multimedia Information Retrieval Workshop (August 2003)
14. Ayache, S., Quénot, G.: Trecvid 2007 collaborative annotation using active learning. In: TRECVID 2007 Workshop (November 2007)
15. Fei-Fei, L., Fergus, R., Perona, P.: One-shot learning of object categories. IEEE Trans. Pattern Anal. Mach. Intell. 28(4), 594–611 (2006)
16. Manjunath, B.S., Salembier, P., Sikora, T.: Introduction To Mpeg-7: Multimedia Content Description Interface. John Wiley & Sons, Chichester (2002)
17. Lowe, D.G.: Distinctive image features from scale-invariant keypoints. Int. J. Comput. Vision 60, 91–110 (2004)
18. Joachims, T.: Making large-scale support vector machine learning practical. In: Advances in kernel methods: support vector learning, pp. 169–184 (1999)

BASIL: Effective Near-Duplicate Image Detection Using Gene Sequence Alignment

Hung-sik Kim[1], Hau-Wen Chang[1], Jeongkyu Lee[2], and Dongwon Lee[3,*]

[1] Computer Science and Engineering, Penn State University, USA
[2] Computer Science and Engineering, University of Bridgeport, USA
[3] College of Information Sciences and Technology, Penn State University, USA
{hungsik,hzc120,dongwon}@psu.edu, jelee@bridgeport.edu

Abstract. Finding near-duplicate images is a task often found in Multimedia Information Retrieval (MIR). Toward this effort, we propose a novel idea by bridging two seemingly unrelated fields – *MIR* and *Biology*. That is, we propose to use the popular gene sequence alignment algorithm in Biology, i.e., BLAST, in detecting near-duplicate images. Under the new idea, we study how various image features and gene sequence generation methods (using gene alphabets such as A, C, G, and T in DNA sequences) affect the accuracy and performance of detecting near-duplicate images. Our proposal, termed as BLASTed Image Linkage (BASIL), is empirically validated using various real data sets. This work can be viewed as the "first" step toward bridging MIR and Biology fields in the well-studied near-duplicate image detection problem.

Keywords: Image Matching, CBIR, NDID, BLAST, Copy detection.

1 Introduction

Determining if two images are *similar* or not is a frequently studied task in the Contents-Based Image Retrieval (CBIR) problem. In particular, the task of detecting *near-duplicate* images becomes increasingly important in many applications of Multimedia Information Retrieval (MIR) – e.g., detecting illegally copied images on the Web [6] or detecting near-duplicate keyframe retrieval from videos [14]. We refer to such a problem as the *Near-Duplicate (ND)* problem, informally defined as follows:

Near-Duplicate Problem. Given a set of query images I_q and a collection of source images I_s, for each query image i_q ($\in I_q$), find all images, I_r ($\subseteq I_s$) that are "near-duplicate" to i_q.

Depending on the types of duplicate images, ND problem can be classified into two folds: (1) Near-Duplicate Keyframes (NDK) [8,13,14], and (2) Near-Duplicate Image Detection (NDID) problems. Generally, NDK is defined as a

---

* Partially supported by NSF DUE-0817376 and DUE-0937891 awards.

C. Gurrin et al. (Eds.): ECIR 2010, LNCS 5993, pp. 229–240, 2010.

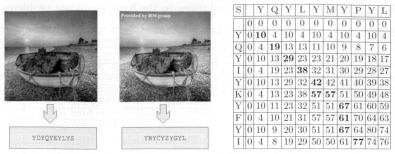

S		Y	Q	Y	L	Y	M	Y	P	Y	L
	0	0	0	0	0	0	0	0	0	0	0
Y	0	**10**	4	10	4	10	4	10	4	10	4
Q	0	4	**19**	13	13	11	10	9	8	7	6
Y	0	10	13	**29**	23	23	21	20	19	18	17
I	0	4	19	23	**38**	32	31	30	29	28	27
Y	0	10	13	29	32	**42**	42	41	40	39	38
K	0	4	13	23	38	**57**	**57**	51	50	49	48
Y	0	10	11	23	32	51	51	**67**	61	60	59
F	0	4	10	21	31	57	57	**61**	70	64	63
Y	0	10	9	20	30	51	51	**67**	64	80	74
I	0	4	8	19	29	50	50	61	**77**	74	76

(a) Example images and sequences   (b) Example of Sequence Alignment

Fig. 1. Sequence Alignment Example

NDID problem well for many reasons. In general, near-duplicate images tend to have near-identical characteristics which in turn are mapped to a long gene sequence of identical alphabetical "hits."

Figure 1(a) illustrates an example of two ND images. The image on the right is modified from the one on the left via operations such as changing contrast, compression and adding logo. The protein sequences below images are generated by BASIL using Y component in YUV color domain. The similarity of the two sequences i_s and i_q can be evaluated by means of a local alignment (e.g., Smith-Waterman) algorithm. In the algorithm, the alignment is operated on two-dimensional matrix S in which each cell $S(i, j)$ keeps a score of the current matching. S is initialized with $S(i, 0) = 0, 0 \le i \le |i_q|$ and $S(0, j) = 0, 0 \le j \le |i_s|$, and is built as follows:

$$S(i, j) = \max \begin{cases} S(i-1, j-1) + s(i_q(i), i_s(j)) \\ \max_{0 \le k \le i-1} \{S(k, j) - \sigma(i-k)\} \\ \max_{0 \le k \le j-1} \{S(i, k) - \sigma(j-k)\} \\ 0 \end{cases}, 1 \le i \le |i_q| \text{ and } 1 \le j \le |i_s|,$$

where $s(i_q(i), i_s(j))$ is the pairwise score of i-th letter of i_q and j-th letter of i_s in scoring matrix, $\sigma(k)$ is the gap penalty of a gap of length k. Figure 1(b) shows the result of the alignment of the two sequences. By utilizing BLAST, alignments can be done much faster than the dynamic programming algorithms. Moreover, single BLAST query can match a sequence against the whole database of sequences, and find the similar sequences instead of pairwise matching in such algorithms.

3.1   Overview of BASIL

Figure 2 shows the overview of the proposed BASIL framework. First, for each image i_s ($\subseteq I_s$, source image set), we extract a set of features, $\mathcal{F}$, and transform $\mathcal{F}$ to a (either DNA or protein) gene sequence, s_s. All the generated sequences are stored in the BLAST database D. Similarly, a query image i_q is also transformed to a corresponding gene sequence s_q. Then, using the BLAST algorithm and an appropriate scoring matrix, s_q is compared to sequences in D and top-k near-duplicate sequences (and their corresponding images) are returned as an answer.

When we generate gene sequences from images, depending on *how* we translate *which* of the extracted image features, we end up with different gene representations. In particular, since it is difficult to find a set of image features

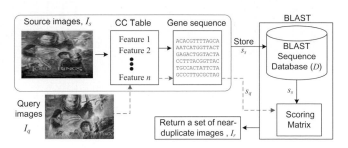

Fig. 2. Overview of BASIL

that work universally well for all data sets, it is important to devise a solution orthogonal to the choice of image features. Toward this first challenge, we propose the *Composite Conversion* (CC) table that contains both pre-defined conversion rules and candidate image features so that users can select desirable features and gene sequences depending on a given data set (see Section 3.2). In addition, the second challenge is to devise solutions in BASIL such that the kernel of BLAST algorithm and implementation should *not* be changed to make existing tools remain useful. Instead, our proposal sits atop BLAST algorithm and manipulates query and source image sequences. For instance, the scoring matrix (that reflects the similarity between different gene alphabets) used in BLAST is originally adjusted to the Biology domain. Therefore, we propose variations of new scoring matrices that reflect the characteristics of near-duplicate image matching scenarios (see Section 3.3).

3.2   The Composite Conversion (CC) Table

ND images are often created by deliberate editing methods (e.g., changing colors/contrasts or cropping images), involuntary distortion (e.g., changing format/size) and variations of capturing conditions (e.g., different angle/time). To find appropriate features for BASIL, therefore, we have tested and selected a variety of features of three groups: color-based ($\mathcal{F}_C$, Y in YC_bC_r and H in HSV), texture-based ($\mathcal{F}_T$, edge density by Law's texture energy), and semantic ($\mathcal{F}_S$, keywords and annotations) features. Each image, i, will be divided to some blocks, say 16×16 macro blocks, and both color- and texture-based features are computed within a macro block while semantic feature is computed from associated keywords or annotations of i. Then, the feature set, $\mathcal{F}$, is the union of $\mathcal{F}_C$, $\mathcal{F}_T$, and $\mathcal{F}_S$.

In order to generate the gene sequences from $\mathcal{F}$, we consider two types of sequences used in BLAST: (1) a protein sequence is made of 23 alphabets (i.e., A, B, C, D, E, F, G, H, I, K, L, M, N, P, Q, R, S, T, V, W, X, Y, and Z), while (2) a

n-Value	Pro.	DNA	n-Value	Pro.	DNA
0 ~ δ	A	AAC	~ 13δ	L	ATT
~ 2δ	R	CCT	~ 14δ	K	ATG
~ 3δ	N	CAG	~ 15δ	M	CAC
~ 4δ	B	AAG	~ 16δ	F	ACT
~ 5δ	D	ACC	~ 17δ	P	CCC
~ 6δ	C	AAT	~ 18δ	S	CGC
~ 7δ	Q	CCG	~ 19δ	T	CGG
~ 8δ	Z	GAG	~ 20δ	W	CTG
~ 9δ	E	ACG	~ 21δ	Y	GAC
~ 10δ	G	AGC	~ 22δ	V	CTC
~ 11δ	H	AGG	~ 23δ	X	CTT
~ 12δ	I	AGT	~ 24δ		

letter	Pro.	DNA	letter	Pro.	DNA
A	A	AAC	N	N	CAG
B	B	AAG	O	Y	CAT
C	C	AAT	P	P	CCC
D	D	ACC	Q	Q	CCG
E	E	ACG	R	R	CCT
F	F	ACT	S	S	CGC
G	G	AGC	T	T	CGG
H	H	AGG	U	Z	CGT
I	I	AGT	V	V	CTC
J	X	ATC	W	W	CTG
K	K	ATG	X	X	CTT
L	L	ATT	Y	Y	GAC
M	M	CAC	Z	Z	GAG

(a) Mapping chart for $\mathcal{F}_C$ and $\mathcal{F}_T$ ($\sigma = \frac{1}{23}$)   (b) Mapping chart for $\mathcal{F}_S$

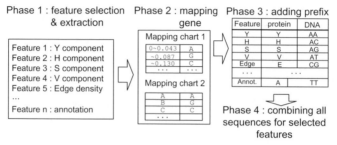

(c) The Composite Conversion Table

Fig. 3. The CC table with two mapping charts

DNA sequence is made of four gene alphabets (i.e., A, C, G, and T). BASIL can take both protein and DNA sequences.

The *Composite Conversion (CC)* table, as illustrated in Figure 3(c), contains various image features and two mapping charts. A mapping chart in Figure 3(a) is used for mapping numeric values obtained from image contents, while another in Figure 3(b) is for literal words obtained from descriptive annotations. For $\mathcal{F}_C$ and $\mathcal{F}_T$, we use the normalized values to use the same mapping chart in Figure 3(a). Since we have 23 gene alphabets for protein, for the best transformation of feature values, we place the normalized real values to 23 bins, as shown in Figure 3(a). For DNA gene sequences, since 4 gene letters are not enough to express 23 bins, 3-bit combination of 4 letters is used for each bin. For $\mathcal{F}_S$, similarly, each literal alphabet is mapped to gene alphabet(s) by pre-defined rules, as shown in Figure 3(b). For protein sequences with 23 protein letters, we add 3 more artificial letters (X, Y, and Z) to have 1-to-1 mapping to 26 literal alphabets. For DNA sequences, we use 3-bit combination letters with A, G, C, T. Figure 3(c) shows the four phases of the CC table to generate the final gene sequences:

- **Phase 1 (Feature selection & extraction).** Among all available image features, a set of features are selected (by users) and normalized. The selection of features depends on the availability of features as well as the characteristic of the given image sets. In addition, the size of a macro block that determines the length of gene sequences is fixed.

– **Phase 2 (Mapping to gene letters).** According to the mapping tables in
Figure 3(a), the normalized feature values from Phase 1 are mapped to appropriate gene letters. If semantic features are used in Phase 1, for instance,
they are also mapped to gene letters using Figure 3(b). At this phase, one
can decide whether to use DNA or protein genes as the final representation.
– **Phase 3 (Adding prefix).** Because of the limitation of gene alphabets,
the same gene letters can be used in different features. For ensuring stronger
connection within the same features, therefore, each letter from phase 2 is
combined with corresponding letters representing a specific feature, as shown
in Figure 3(c). This phase can be skipped if only one image feature is selected
in phase 1.
– **Phase 4 (Combining all features).** All gene sequences from different features are combined. The final output sequence of the CC table thus captures
all features of an image holistically. This phase is also skipped if only one
image feature is selected in phase 1.

Since an individual feature in a CC table is very independent, the features in a
CC table can be obtained by separating homogeneous components of an image
such as color components. With the same reason, the features in a CC table can
be acquired very heterogeneously. For example, all of image color components,
texture information, meta data (such as resolution, format, and date), and annotations can be included as features in a CC table. The final gene sequence
of an image captures all selected homogeneous and heterogeneous components,
and is passed through BLAST to compare all features at once.

3.3   The Scoring Matrix

When two sequences are compared in BLAST, a similarity score is computed to
quantify the quality of the pair-wise alignments. For this task, BLAST uses a scoring matrix that includes all possible pair-wise scores of letters in 2-dimensional
matrix. For the scoring matrix, Percent Accepted Mutation (PAM), and BLOcks
SUbstitution Matrix (BLOSUM) derived from theoretical or empirical frequencies
of amino acid substitutions are popular.

Since both matrices are originally created for biological data in mind, they are
not suitable for BASIL with image data. Therefore, we propose to use new scoring
matrices: (1) **Uniform matrix** assigns uniform score for each identity and substitution. For example, "1" is assigned for all identities (i.e., diagonal), and "-1"
is assigned for the others of the matrix; and (2) **Gaussian distributed matrix**:
The uniform matrix cannot capture the diverse characteristics of features used
in BASIL. For example, red and orange colors are more similar than red and blue
in terms of hue (H) color domain. In general, the gaussian distributed matrix is
good for numeric features, such as $\mathcal{F}_C$ and $\mathcal{F}_T$.

There are several important advantages to employ characterized scoring matrices into BASIL: (1) The semantics of image features can be represented using
the matrix; (2) The different weights can be applied for image matching using
identities' values in the matrix; (3) Positive credits and negative penalties can be

adjusted for exact/fuzzy matched and unmatched letters, respectively; (4) The more sophisticated scoring matrix than Uniform or Gaussian (e.g., Probabilistic, Linguistic, or Trained matrices) can be easily added to the CC table. We will leave this as future work.

4   Experimental Validation

4.1   Set-Up

The CC table is implemented by Matlab 7.0 on Intel Core 2 Duo (1.8GHz, 2GB RAM, Windows XP Home), and both BLAST DB generation and gene sequence matching (near-duplicate image matching) was done by WU-BLAST 2.0[1] on IBM Z60t (Intel Pentium-M 1.6GHz, 1.5GB RAM, Ubuntu 7.10).

Real world data set					Modified data set	
Dark Knight (DK)		The Lord of The Ring (LR)			Flickr (FK)	
Category	# of images	Category	# of images		Category	# of images
Back	19					1
Batman	17	Poster with annotations	20		Each original image (240 original images)	original image + 12 edited images
Face	27					
Fire	42					
Joker	9					
Wsos	41	Others with annotations	200			
Others	1108					
Sub-total	1263	Sub-total	220		Sub-total	3120

Total number of images : 4603

Fig. 4. Image data sets

As summarized in Figure 4, two real-world data sets and one edited data set are used in our experiments: Dark Knight (DK), The Lord of The Rings (LR)[2], and Flickr (FK)[3]. The DK data set is manually classified into 6 categories (9~41 images in each category), and further augmented by 1,108 irrelevant images for each category. For LR data set, one category (LR poster, 20 images) is selected with additional semantic annotations such as title, file name, and descriptions of images, while the other category is of 200 irrelevant images with their annotations. In the FK data set, for each original image, 12 near-duplicate images are generated by 12 typical editing methods, i.e., *blur, changing brightness, changing format, changing color, color enhancement, changing contrast, compression, crop, adding logo, changing resolution, changing size, and multi-editing (e.g. crop+compression+logo)*.

As an evaluation metric, we mainly use the average precision (P) and recall (R) in a PR graph using top-k model. In DK and LR image sets, for each category except others, 9-15 query images are randomly chosen to achieve 95% confidence levels with 6.5-9.9 confidence intervals on average precision and recall. With FK image set, for each category (total 240 categories), 10 query images are randomly chosen to achieve 99% confidence levels with 2.9 confidence interval.

4.2   Comparison within BASIL

DNA vs. Protein and Scoring Matrix. BASIL's CC table is flexible to take different mapping and scoring matrix – e.g., DNA/Protein for mapping and

---

[1] http://www.advbiocomp.com/blast/obsolete/

[2] Both DK and LR are gathered from the Web.

[3] FK is gathered from http://www.flickr.com/

Uniform/Gaussian for scoring matrix. Therefore, we first examined the performance under different selections of mapping and scoring matrix. **While we omit details in the interest of space**, we found that in general: (1) Protein yields better accuracy than DNA does, because protein utilizes finer granularity than DNA such that the gap between two numerical values are less ambiguous; and (2) The Gaussian matrix provides better accuracy than Uniform or BLOSUM62 does. This is because the Gaussian matrix compensates the strict difference between letters. Based on above observations, in subsequent experiments, we use Protein and Gaussian as the default mapping and scoring matrix.

Comparison among Image Features. Since gene sequence is generated by extracted features, the performance of BASIL depends on the quality of $\mathcal{F}$. In this experiment, we use 16 macro blocks per image for $\mathcal{F}_C$ and $\mathcal{F}_T$, collected meta data for $\mathcal{F}_S$, and 23 gene letters in Figure 3 (a) and (b).

We used 6 popular features in the CC table: Y component from YC_bC_r color domain, H, S, V components from HSV color domain, Law's edge energy component, and semantic feature. Y, H, S, V, E (energy), and A (semantic annotation) stand for each component, respectively. For the evaluation of the effect of selected features, among these 6 features, one can choose any combination of them. In Figures 5(a) and (b), we evaluate the performance of various combination of features including 1 feature (i.e., H, V, and E), 2 features (i.e., SE and VE), 3 features (i.e., HSE and YVE), 5 features (i.e., $YHSVE$), and all 6 image features in the CC table. Note that feature A is only available in the LR image set.

For the FK set, in Figure 5(a), all of H, V, and E features have a high precision until recall becomes 0.5. However, afterward, H feature becomes the best. In the real-world data set (DK and LR), in Figure 5(b), both V and E give the best result overall in terms of both precision and recall, while H yields the worst accuracy. Since color feature is more sensitive to H, in the real-world data set (DF and LR), people often copy and modify images with color change/enhancement functions before images are uploaded to the Web. On the other hand, the FK data set is generated by 12 editing methods. However, only 2 of them are related to the color in FK data set. Therefore, the results show that V and E for DK and LR are better features than H.

When multiple features are selected, one can usually gain the average performances of different features. For instance, in the real-world image set (DK and LR), the accuracy with multiple features is always between those with an individual feature. However, note that the combination of features from image contents usually outperforms the average of the accuracies from individual feature selection. This is because the accuracy of BASIL system follows the top-k model. That is, even though the similarity between genes are averaged from multiple features, the similarity ranking from BLAST can be changed when features are combined. Another benefit of using multiple features combined is the improved robustness of BASIL for unknown image sets. In this paper, note that all sets are set to be unknown since we do not analyze the characteristic of data sets by sample or whole images in data sets. As a result, by combining all six features, $YHSVEA$,

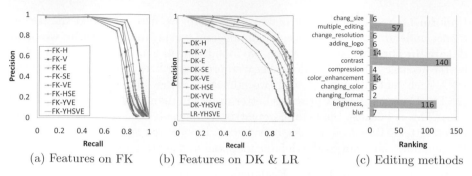

(a) Features on FK   (b) Features on DK & LR   (c) Editing methods

Fig. 5. Comparison among BASIL insides

in LR image set, we achieve the highest accuracy from BASIL system shown in Figure 5 (b).

Comparison among Editing methods. Using FK image set, we compare the impact of different editing methods on the accuracy of BASIL. For this evaluation, we select 5 features ($YHSVE$) and 23 protein letters. Since we use 12 editing methods, when an original image is queried, ideally, all 12 edited images must be returned at high ranking. The worst ranking of a returned image is 3,120, since we have 3,120 images in FK set, i.e., 3,120 gene sequences in BLAST DB. Since BLAST DB also contains original images, note that the best ranking of edited images always starts from 2.

Figure 5(c) shows the average rank of edited images in the returned list from BASIL with 12 different editing methods. BASIL system reveals that the average ranking of the expected duplicate images is about 2 for the best case and about 140 for the worst case. While some editing methods (e.g., contrast and brightness) make the detection of ND images really challenging for BASIL, in general, majority of editing methods are well covered by BASIL framework. Overall, BASIL is robust on various editing methods that are typically used by image editing tools.

4.3 Comparison against Other Methods

Due to the difficulty in obtaining the implementations of other NDID methods (summarized in Table 1), instead, we compare the performance of BASIL against two publicly available non-NDID solutions – Ferret for CBIR and ND_PE for NDK.

Comparison with Ferret. Here we first evaluate BASIL against one of the state-of-the-art CBIR alternatives, Ferret, from the CASS project at Princeton[4]. Ferret is a toolkit for content-based similarity search for various data types including digital image. The result using the FK image set is shown in Figure 6(a), where Ferret and HSE exhibit the best results while the balanced $YHSVE$ is behind them after the recall of 0.5. With the DK set, BASIL achieves the best

---

[4] http://www.cs.princeton.edu/cass/

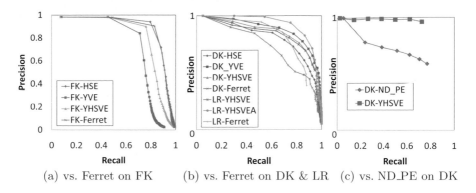

Fig. 6. Comparison BASIL to Ferret and ND_PE

accuracy using the YVE feature selection as shown in 6(b). Overall, both BASIL with $YSHVE$ features and Ferret show similar accuracy. One of the benefits of the CC table in BASIL is that it enables to combine any heterogenous features to the final gene sequences. For instance, heterogeneous features such as semantic or content-based one can be uniformly represented in gene sequences. As a result, Figure 6(b) shows that the line of LR-YHSVEA (6 features including a *semantic information*) significantly outperforms Ferret.

Comparison with ND_PE. The ND_PE is a near-duplicate keyframe (NDK) detection toolkit based on local features of images, developed by Video Retrieval Group (VIREO) from City University of Hong Kong[5]. In ND_PE, a set of local interest points of images are extracted and represented in PCA-SIFT descriptor. The similarity of two images is then determined on the degree of matches between two sets of keypoints such as a bipartite graph matching. We compare the accuracy of ND_PE and BASIL with $YSHVE$ features on DK data set in Figure 6(c). In this test, 9–10 images in each category are selected to measure the similarity against all images in the data set. The top-30 returned images per query are used to generate the average PR graph[6]. Figure 6(c) shows that overall the accuracy of BASIL outperforms that of ND_PE for the real near-duplicate data set, DK. Note that ND_PE was originally designed to solve the NDK problem, not the NDID problem. Since both the NDK and NDID problems are slightly different, therefore, direct comparison between the results of BASIL and ND_PE should be interpreted with much care.

5   Conclusion

In this paper, we proposed a novel solution, named as BLASTed Image Linkage (BASIL), to the near-duplicate image detection (NDID) problem by bridging

---

[5] http://vireo.cs.cityu.edu.hk/research/NDK/ndk.html

[6] The implementation of ND_PE crashed for a few pairs of images in testing. In preparing the PR graph of Figure 6(c), such images were ignored.

two seemingly unrelated fields – *Multimedia* and *Biology*. In BASIL, we use the popular gene sequence alignment algorithm in Biology, BLAST, to determine the similarity between two images. To be able to handle flexible transformation from diverse image features to gene sequences, we also proposed the Composite Conversion (CC) table that hosts different images features and pre-fixed transformation rules. The validity of BASIL is positively measured using three real image sets on various aspects. In future, we plan to extend BASIL to apply it to different mediums such as video, audio, or time series. In addition, the structural characteristics of multi-media inputs will be studied to achieve structural alignment matching algorithms. BASIL implementations and data sets used in this paper are available at: `http://pike.psu.edu/download/ecir10/basil/`

References

1. Altschul, S., Gish, W., Miller, W., Myers, E., Lipman, D.: Basic Local Alignment Search Tool. J. Mol. Biology 215(3), 403–410 (1990)
2. Cai, D., He, X., Han, J.: Spectral Regression: A Unified Subspace Learning Framework for Content-Based Image Retrieval. In: ACM Multimedia (2007)
3. Dong, W., Wang, Z., Charikar, M., Li, K.: Efficiently Matching Sets of Features with Random Histograms. In: ACM Multimedia (2008)
4. Falchi, F., Lucchese, C., Orlando, S., Perego, R., Rabitti, F.: Caching Contentbased Queries for Robust and Efficient Image Retrieval. In: EDBT (2009)
5. Foo, J.J., Zobel, J., Sinha, R.: Clustering near-duplicate images in large collections. In: ACM MIR (2007)
6. Foo, J.J., Zobel, J., Sinha, R., Tahaghoghi, S.M.M.: Detection of Near-Duplicate Images for Web Search. In: ACM CIVR (2007)
7. Howarth, P., Rüger, S.M.: Evaluation of Texture Features for Content-Based Image Retrieval. In: ACM CIVR (2004)
8. Ke, Y., Sukthankar, R., Huston, L.: An Efficient Parts-based Near-Duplicate and Sub-Image Retrieval System. In: ACM Multimedia (2004)
9. Kim, H., Chang, H., Liu, H., Lee, J., Lee, D.: BIM: Image Matching using Biological Gene Sequence Alignment. In: IEEE Int'l Conf. on Image Processing (ICIP) (November 2009)
10. Mehta, B., Nangia, S., Gupta, M., Nejdl, W.: Detecting Image Spam using Visual Features and Near Duplicate Detection. In: WWW (2008)
11. Valle, E., Cord, M., Philipp-Foliguet, S.: High-dimensional Descriptor Indexing for Large Multimedia Databases. In: ACM CIKM (2008)
12. Wu, X., Hauptmann, A.G., Ngo, C.-W.: Practical Elimination of Near-Duplicates from Web Video Search. In: ACM Multimedia (2007)
13. Zhang, D.-Q., Chang, S.-F.: Detecting Image Near-Duplicate by Stochastic Attributed Relational Graph Matching with Learning. In: ACM Multimedia, October 2004, pp. 877–884 (2004)
14. Zhao, W.-L., Ngo, C.-W., Tan, H.-K., Wu, X.: Near-Duplicate Keyframe Identification with Interest Point Matching and Pattern Learning. IEEE Trans. On Multimedia 9, 1037–1048 (2007)
15. Zheng, Y.-T., Neo, S.-Y., Chua, T.-S., Tian, Q.: The Use of Temporal, Semantic and Visual Partitioning Model for Efficient Near-Duplicate Keyframe Detection in Large Scale News Corpus. In: ACM CIVR (2007)

Beyond Shot Retrieval: Searching for Broadcast News Items Using Language Models of Concepts

Robin Aly[1], Aiden Doherty[2], Djoerd Hiemstra[1], and Alan Smeaton[2]

[1] Datbase Systems, University Twente, 7522AE Enschede, The Netherlands
{r.aly,hiemstra}@ewi.utwente.nl
[2] CLARITY: Center for Sensor Web Technology, Dublin City University, Ireland
{aiden.doherty,alan.smeaton}@dcu.ie

Abstract. Current video search systems commonly return video shots as results. We believe that users may better relate to longer, semantic video units and propose a retrieval framework for news story items, which consist of multiple shots. The framework is divided into two parts: (1) A concept based language model which ranks news items with known occurrences of semantic concepts by the probability that an important concept is produced from the concept distribution of the news item and (2) a probabilistic model of the uncertain presence, or risk, of these concepts. In this paper we use a method to evaluate the performance of story retrieval, based on the TRECVID shot-based retrieval groundtruth. Our experiments on the TRECVID 2005 collection show a significant performance improvement against four standard methods.

1 Introduction

Video search systems have usually concentrated on retrieval at the shot level, with a shot being the smallest unit of a video which still contains temporal information [16]. However not as much attention has been focused on searching for larger/semantic units of retrieval, generally referred to as stories. We believe that users may relate better to these retrieval units. However, current retrieval models for video search are difficult to adapt to these story retrieval units, since they are tailored to find shots, which are most often represented by a single keyframe. Therefore, the main contribution of this paper is a retrieval framework (based on language modelling of semantic concepts, for example a "Person", "Outdoor" or "Grass"), which can be applied to longer video segments, such as a news item.

Throughout this paper we assume that the user's information need is specified by a textual query. Given the growing prominence and attention afforded to lifelog data from wearable cameras such as the SenseCam, where audio data isn't recorded [3], we want our model to be also applicable to search in video data without considering the audio stream. As a result we focus on working with concepts extracted from the content of images. Current concept based video retrieval systems normally operate on a fixed-number of features per retrieval unit, for example the confidence scores of detectors for a number of concepts [6,18].

C. Gurrin et al. (Eds.): ECIR 2010, LNCS 5993, pp. 241–252, 2010.

Results") we see the rank and confidence scores $o_d = (o_1, o_2, o_3)$ for each concept occurrence. On the right ("Possible Representations"), we see the possible concept occurrences $C_1 - C_3$ in document 5.

We identify four different classes of how systems rank these documents, two for unit ranking and two for unit representation. Class (5) represents our proposed approach which combines the possible representations to the expected score and additionally considers the risk of using this score.

Confidence Score Based (1). Many approaches from this class are variations of the score functions $CombMNZ$ and $CombSUM$, which originate from meta search, see [2]. In parallel to class (1) score based approaches are only applicable to a fixed number of concepts. Another problem is the normalization and weighting of the confidence scores.

Component Rank Based (2). Systems of this class use methods such as the Borda Count method, which originates from election theory in politics. See Donald *et al.* [6] for the application in concept based video retrieval. However, this method is not directly applicable to longer video segments since it relies on the fixed number of concepts to rank.

Top-1 Representation (3). Systems in this class consider the most probable representation. Therefore, the system completely ignores other possible representations. While this technique was found suitable for retrieval tasks with high detection quality [21], Momou *et al.* report in [10] that under high word error rate, the performance deteriorates quickly.

Expected Component Value (4). More recent approaches in spoken document retrieval use the lattice output of the speech recognizer to obtain more variability [4,5]. For example, Chia [5] calculates the expected term frequencies of the top-N documents, weighted by their probability. For high N and a linear score function this value approximates the expected score, which is one part of our ranking framework. However, there is no notion of the risk a system takes, if it uses this score.

2.3   Uncertainty in Text IR - Mean-Variance Analysis

Markowitz proposes in [11] the Portfolio Selection theory in economics. Wang [22] successfully transferred this theory into the Mean-Variance Analysis framework for uncertain scores in text information retrieval. According to Wang, a retrieval system should rank a news item at position m which optimizes the following expression, which considers "risk" of the item:

$$d^* = \operatorname*{argmax}_d E[r_d] - b\, w_m \operatorname{var}[r_d] - 2b \sum_{i=1}^{m-1} w_i \operatorname{cov}[r_d, r_{d_i}] \tag{1}$$

Here, r_d is the uncertain score of the news item d and the three ingredients of the framework are: (1) The expected uncertain score $E[r_d]$; (2) the variance $\operatorname{var}[r_d]$

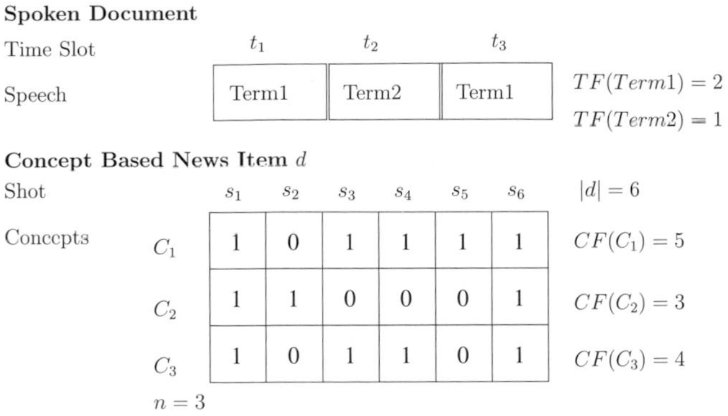

Spoken Document

	t_1	t_2	t_3	
Time Slot				
Speech	Term1	Term2	Term1	$TF(Term1) = 2$
				$TF(Term2) = 1$

Concept Based News Item d

		s_1	s_2	s_3	s_4	s_5	s_6	
Shot								$|d| = 6$
Concepts	C_1	1	0	1	1	1	1	$CF(C_1) = 5$
	C_2	1	1	0	0	0	1	$CF(C_2) = 3$
	C_3	1	0	1	1	0	1	$CF(C_3) = 4$
$n = 3$								

Fig. 2. A concept based news item representation and its analogy to a spoken document

of the possible scores nearby $E[r_d]$; and (3) the covariance $\mathrm{cov}[r_d, r_{d_i}]$, which specifies whether the scores of news item d and d_i are correlated. The parameter b represents the different risk affinities of different users. For $b > 0$ is the risk averse user who prefers to have a stable ranking, while for $b < 0$ he is willing to take more risk to have the chance of getting relevant news items to the top of the ranking. For $b = 0$ the system ranks by the expected score of each news item, which is the same as ranking by the original score function.

Essentially given the problem of the semantic gap in image retrieval, and hence the difficulty in producing highly accurate semantic concept detectors, we believe there may be merit in boosting the rank of more risky items to the top of the ranked list (given that the safe items may be proposed by mildly accurate concept detectors). Indeed this approach builds upon that proposed by Varian [20].

3   News Item Search

3.1   Concept Based News Item Representation

A news broadcast video can naturally be segmented into news items. Furthermore, these items can be subdivided into shots. Until now, this unit was used to present results to the user. Figure 2 shows how concept-based and spoken document-based representations would approach the problem of representing a news story. The spoken document consists of three spoken words at time position $t_1 - t_3$ and the news item of six shots $s_1 - s_6$. The concept lexicon consists of three concepts $C_1 - C_3$. We denote the presence of concept i in shot j as $C_{ij} \in \{0, 1\}$ where 1 stands for the presence of the concept. On the right, we see the term and concept frequencies as the count of the values on the left, as the analogy to spoken documents. We can express the frequency as a sum: $CF(C_i) = \sum_j C_{ij}$. For a given information need, we then select n important concepts $C_1, ..., C_n$ according to our prior work [1] and the vector of the concept frequencies $rep = (CF(C_1), ..., CF(C_n))$ is our document representation.

3.2   Concept Language Models

We now describe our ranking function for concept based news item retrieval. To our knowledge, this is the first proposal of a concept retrieval function for this domain. The basic idea behind our approach is to consider the occurrence and absence of a concept as two words of the language of this concept. Therefore a word is either "present" or "absent" and instead of a single stream of terms we have multiple "concept streams". As mentioned before, by simply counting we can get the concept frequency in a news item.

Because the concept frequencies between news items are difficult to compare we consider, in parallel to language modelling [8,15], the probability that a concept is present in a news item. However, the extracted concepts will not fully reflect the content of the video. For example, since they are normally extracted at discrete points in time a concept detector may miss the occurrence of a concept in the news story as a whole. To solve this, we apply Dirichlet smoothing [23] to the probability and obtain the language model score for concept C_i in news item d :

$$P(C_i|d) = \frac{CF(C_i) + \mu P(C_i)}{|d| + \mu} \tag{2}$$

where $P(C_i)$ is the prior of encountering a concept C_i in the collection, $|d|$ is the length (in shots), finally μ is the scale parameter of the Dirichlet prior distribution. We now can rank news items by the probability of drawing a list of concepts independently from their "concept stream":

$$score(rep) = P(C_1, ..., C_n|d) = \prod_i^n P(C_i|d) \tag{3}$$

Here, the right part calculates the probability of sampling these concepts independently from the news item d.

3.3   Uncertain Concept Occurrences

Until now we have considered concept based news item search for the case of known concept occurrences. However in reality we will only have uncertain knowledge about their presence through the output of detectors. Let o_{ij} be the detector's confidence score that concept C_i occurs in shot s_j and $\boldsymbol{o_d}$ is now the combination of all confidence scores of an news item. For each concept in each shot of the news item this output can be transformed into a probability: $P(C_{ij} = 1|o_{ij})$. This probability can for example be estimated by a method described by Platt [14], which considers each concept occurrence independently. Work in the somewhat related domain of lifelogging has shown that the occurrences of many concepts within a shot and within adjacent shots is statistically dependent on each other [3]. However in this work we concentrate on more generic representations and leave the investigations of these dependencies for future work.

With this knowledge, we can determine the probability distribution over the possible frequency values $CF(C_i)$. For example, the probabiltiy that concept C_i has a frequency of 1 in a news item with $|d| = 3$ is

$$P(CF(C_i) = 1|\boldsymbol{o_d}) = P(\boldsymbol{C_i} = 1, 0, 0|\boldsymbol{o_d}) + P(\boldsymbol{C_i} - 0, 1, 0|\boldsymbol{o_d}) + P(\boldsymbol{C_i} = 0, 0, 1|\boldsymbol{o_d})$$

where $\boldsymbol{C_i}$ is a short form for (C_{i1}, C_{i2}, C_{i3}) and the first probability is calculated as follows:

$$P(\boldsymbol{C_i} = 1, 0, 0|\boldsymbol{o_d}) = P(C_{i1} = 1|o_{i1})(1 - P(C_{i2} = 1|o_{i2}))(1 - P(C_{i2} = 1|o_{i3}))$$

The probability of a whole representation is calculated as follows $P(rep|\boldsymbol{o_d}) = \prod_i^n P(CF(C_i)|\boldsymbol{o_d})$. Furthermore, the expected concept frequency of a concept, which is for example used by Chia et al. [5], can be determined as $E[CF(C_i)|\boldsymbol{o_d}] = \sum_{j=1}^n P(C_{ij}|o_{ij})$.

3.4   Retrieval under Uncertainty

We now describe how we combine the concept language score and the uncertainty of the concept occurrences into one ranking function. Because of the representation uncertainty, the document score is a random variable S_d.

Similar to the Mean-Variance Analysis framework from Wang [22], our framework now ranks by a combination of the expected score and the standard deviation of the score:

$$RSV(d) = E[S_d|\boldsymbol{o_d}] - b\sqrt{\text{var}[S_d|\boldsymbol{o_d}]} \tag{4}$$

Here, RSV(d) is the final score under which a document is ranked, $E[S_d|\boldsymbol{o_d}]$ is the score we expect from the distribution of concept occurrences and $\text{var}[S_d|\boldsymbol{o_d}]$ is the variance of the score around the expected value and specifies how dispersed the scores are. the risk factor was easier to control when considering the standard deviation[1] rather than the variance which was used in the Mean-Variance Analysis framework. The constant b specifies the risk perception of a system in the same way as in the Mean-Variance Analysis framework. We now define the expected score, the first component of our ranking framework:

$$E[S_d|\boldsymbol{o_d}] = \sum_{rep} score(rep)P(rep|\boldsymbol{o_d}) \tag{5}$$

That is, we iterate over all possible concept frequency combinations, calculate the resulting score and merge these scores according to their probability of being the right representation. Additionally the variance of the score, which represents the risk can be defined as:

$$\text{var}[S_d|\boldsymbol{o_d}] = E[S_d^2|\boldsymbol{o_d}] - E[S_d|\boldsymbol{o_d}]^2 \tag{6}$$

---

[1] Standard deviation = square root of variance.

$$\text{with } E[S_d^2|\mathbf{o_d}] = \sum_{rep} score(rep)^2 P(rep|\mathbf{o_d}) \tag{7}$$

While there is a very large number of possible representations ($2^{n|d|}$) in fully calculating Equations 5 and 7, we apply the Monte Carlo estimation method which samples from the given distribution. The method is defined as follows: Let $rep_{d_1}, ..., rep_{d_N}$ be random samples from $P(Rep|\mathbf{o_d})$. The expectations from Equation 5 and Equation 7 can then be approximated by:

$$E[S_d|\mathbf{o_d}] \simeq \frac{1}{N}\sum_{l=1}^{N} score(rep_{d_l}) \qquad E[S_d^2|\mathbf{o_d}] \simeq \frac{1}{N}\sum_{l=1}^{N} score(rep_{d_l})^2$$

To attain a random sample rep_{d_1} of a news item we iterate over each shot j and flip for each concept i a coin with the probability $P(C_{ij}|o_{ij})$ for head, the output of the concept detector. If we observe head (i.e. the probability is greater than a random number from the interval $[0:1]$), we add one to the concept frequency $CF(C_i)$ of this concept. After processing all concepts for all shots we calculate the score of the sample according to Equation 3. Because the standard error of the Monte Carlo estimate is in the order of $\sqrt{N}$ we achieve a relatively good estimate already with few samples.

4   Experiments

4.1   Experiment Setup

Our experiments are based on the TRECVID 2005 dataset which comprises 180 hours of Chinese, Arabic and English broadcast news [17]. NIST announced the automatic shot segmentation from Peterson [13] as the official shot boundary reference, defining a total of $45,765$ shots. For the segmentation of the videos into news items, we used the results from [9], which essentially looked for the anchor person in the video, to determine an item change. This segmentation resulted in $2,451$ news items of an average length of 118 seconds. We associated a shot with a news item, if it began within the time interval of the aforementioned news item. This resulted on average in 17.7 shots per news item.

Because of the novelty of our approach no standard set of queries with relevance judgments existed for this search task. Therefore, we decided on using the 24 official, existing queries from TRECVID 2005, replacing the *"Find shot of ..."* with *"Find news items about ..."*. Furthermore, we assumed that a news item is relevant to a given query, if it contained a relevant shot (which can be determined from the standard TRECVID groundtruth set). We argue that for most topics this is realistic since the user is likely searching for the news item as a whole, rather than shot segments within it.

We used the lexicon of 101 concepts and the corresponding detector set from the MediaMill challenge experiment for our experiments [19]. The reason for this is that it is an often referenced stable detector set with good performance on the mentioned data set. As detailed above, we use a concept selection method to select important concepts for a query [1]. Therefore we executed the query with a standard text retrieval engine on a textual representation of the development set and assume the top-N documents to be relevant. We then used the first n most frequent concepts for this query, as we have used before [1].

We compared our approach to four other approaches (classes 1-4 discussed in Section 2). As the approaches from concept based shot retrieval only work on a fixed number of features we used the average probability of each considered concept as the score for this concept $s(C_i) = \sum_j P(C_{ij}|o))/|d|$. To quickly recap, the considered approaches are: (1) Borda Count which considers the rank of the average concept occurrence probability [6], (2) CombMNZ which multiplies the scores as long as they are not zero [2]. (3) Top-1, which ranks the news items by the concept language model score of the most probable representation. To be more concrete, a concept occurrence was counted if the probability of the concept was above 0.5. The resulting concept frequencies were then used to calculate the concept language model score described in Equation 3. Finally, (4) we used an approach similar to that of Chia [5], which uses the expected concept frequency as the concept frequency in Equation 3.

4.2   Comparison to Other Methods

Table 1 shows the result of the comparison of the described methods with our expected score method. The first row, n, beneath the class names indicates the number of concepts under which this class performed the best. We see that classes (1)-(3) perform much worse than the two methods which include multiple possible concept frequencies. Among them, there is only a small difference. For our method we used $N = 200$ samples, a Dirichlet prior of $\mu = 60$, and a risk factor $b = -2$. Since these parameters returned the best results while using few samples. To rule out random effects, we repeated the run ten times and report the average. Our method is significantly better that the expected frequency method and has a mean average precision of 0.214.

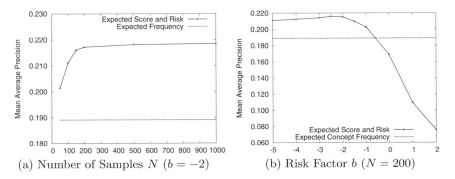

(a) Number of Samples N $(b = -2)$ (b) Risk Factor b $(N = 200)$

Fig. 3. Robustness study of parameter settings

Table 1. Results of the comparison of our method (5) against four other methods described in related work. The actual methods are (1) Borda Count fusion, (2) $CombMNZ$, (3) Top-1, (4) Considering the expected frequency and (5) Our method of taking the expected score plus a risk expression. The MAP of our method has been successfully tested for significant improvement with a paired t-test with significance level 0.05, marked by *.

	(5) Expected Score + Risk	(1) Rank	(2) Score	(3) Top-1	(4) Expected Frequency
n	10	1	10	5	10
MAP	0.214*	0.090	0.105	0.094	0.192
P10	0.291	0.000	0.045	0.245	0.287

4.3   Study of Parameter Values

In Figure 3 we summarise the result of a study over the two most important parameters in our model. Here we also set $\mu = 60$ and repeated each run ten times, to rule out random effects. In Figure 3 (a) the sensitivity of our method over the number of samples is shown. We see that already with few samples ($N = 50$) the performance is better than the expected concept frequency. As usual for a Monte Carlo estimator, the precision increases in line with the square root of the number of samples. After $N = 250$ samples we barely see any improvement.

Figure 3 (b) shows the behaviour of our model for changes of the risk parameter b. We see, with values of $b > -1$ our method performs worse than the expected concept frequency. The reason for this is that the concept detectors still show a low performance and therefore the variance of the concept frequencies can be quite high. A risk averse system will, in extreme cases, value the item with a certain score of practically zero higher than an item with still slightly lower score but a higher variance. This also makes clear why a positive risk perception increases performance.

5   Conclusions

In this work we proposed a ranking method for longer video segments than the commonly assumed retrieval unit of a video shot. Because of the novelty of the task we focused on the search for news items, a particular segment type. After identifying four major classes of ranking methods for general multimedia data, we found that current shot based methods are hard to adapt to longer video segments. Therefore, we proposed a new ranking function, a concept based language model, which ranks a news item with known concept occurrences by the probability that important concepts are produced by the item. However, since we only have probabilistic knowledge about the concept occurrence we included this uncertainty in our ranking framework by considering the expected language model score plus the associated risk, represented by the standard deviation of the score.

We also proposed a means to creating a groundtruth for future story retrieval tasks, whereby we infer relevance from existing TRECVID judgements on shot-based retrieval. In the experiment which we performed on the TRECVID 2005 test collection, our proposed concept based language modelling retrieval method was able to show significant improvements over all representative methods from the identified classes.

We have shown that models which consider all possible concept frequencies perform better than systems that take only one of the scores, ranks, or most probable representations of documents into account. We have also shown that our method, which considers the expected score of a concept based language model, performs significantly better than an adapted method from spoken document retrieval (which takes the expected concept frequency and only then applies the concept based language model).

Acknowledgments

CLARITY authors AD & AS wish to acknowledge the support of Science Foundation Ireland (grant number 07/CE/I1147).

References

1. Aly, R., Hiemstra, D., de Vries, A.P.: Reusing annotation labor for concept selection. In: CIVR 2009: Proceedings of the International Conference on Content-Based Image and Video Retrieval 2009. ACM, New York (2009)
2. Aslam, J.A., Montague, M.: Models for metasearch. In: SIGIR 2001: Proceedings of the 24th annual international ACM SIGIR conference on Research and development in information retrieval, pp. 276–284. ACM, New York (2001)
3. Byrne, D., Doherty, A., Snoek, C.G.M., Jones, G., Smeaton, A.F.: Everyday concept detection in visual lifelogs: Validation. relationships and trends. Multimedia Tools and Applications Journal (2009)
4. Chelba, C., Acero, A.: Position specific posterior lattices for indexing speech. In: ACL 2005: Proceedings of the 43rd Annual Meeting on Association for Computational Linguistics, Morristown, NJ, USA, pp. 443–450. Association for Computational Linguistics (2005)
5. Chia, T.K., Sim, K.C., Li, H., Ng, H.T.: A lattice-based approach to query-by example spoken document retrieval. In: SIGIR 2008. Proceedings of the 31st annual international ACM SIGIR conference on Research and development in information retrieval, pp. 363–370. ACM, New York (2008)
6. Donald, K.M., Smeaton, A.F.: A comparison of score, rank and probabilitybased fusion methods for video shot retrieval. In: Leow, W.-K., Lew, M., Chua, T.-S., Ma, W.-Y., Chaisorn, L., Bakker, E.M. (eds.) CIVR 2005. LNCS, vol. 3568, pp. 61–70. Springer, Heidelberg (2005)
7. Hauff, C., Aly, R., Hiemstra, D.: The effectiveness of concept based search for video retrieval. In: Workshop Information Retrieval (FGIR 2007), Halle-Wittenberg. LWA 2007: Lernen - Wissen - Adaption, vol. 2007, pp. 205–212. Gesellschaft fuer Informatik (2007)

8. Hiemstra, D.: Using Language Models for Information Retrieval. PhD thesis, University of Twente, Enschede (January 2001)
9. Hsu, W.H., Kennedy, L.S., Chang, S.-F.: Video search reranking via information bottleneck principle. In: MULTIMEDIA 2006: Proceedings of the 14th annual ACM international conference on Multimedia, pp. 35–44. ACM Press, New York (2006)
10. Mamou, J., Carmel, D., Hoory, R.: Spoken document retrieval from call-center conversations. In: SIGIR 2006: Proceedings of the 29th annual international ACM SIGIR conference on Research and development in information retrieval, pp. 51–58. ACM Press, New York (2006)
11. Markowitz, H.: Portfolio selection. The Journal of Finance 7(1), 77–91 (1952)
12. Natsev, A.P., Haubold, A., Tešić, J., Xie, L., Yan, R.: Semantic concept-based query expansion and re-ranking for multimedia retrieval. In: MULTIMEDIA 2007: Proceedings of the 15th international conference on Multimedia, pp. 991–1000. ACM Press, New York (2007)
13. Petersohn, C.: Fraunhofer hhi at trecvid 2004: Shot boundary detection system. In: TREC Video Retrieval Evaluation Online Proceedings, TRECVID (2004)
14. Platt, J.: Advances in Large Margin Classifiers. In: Probabilistic outputs for support vector machines and comparison to regularized likelihood methods, pp. 61–74. MIT Press, Cambridge (2000)
15. Ponte, J.M.: A language modeling approach to information retrieval. PhD thesis, University of Massachusetts Amherst (1998)
16. Smeaton, A.F., Over, P., Doherty, A.R.: Video shot boundary detection: Seven years of trecvid activity. Computer Vision and Image Understanding (2009)
17. Smeaton, A.F., Over, P., Kraaij, W.: Evaluation campaigns and trecvid. In: MIR 2006: Proceedings of the 8th ACM International Workshop on Multimedia Information Retrieval, pp. 321–330. ACM Press, New York (2006)
18. Snoek, C.G.M., Worring, M.: Concept-based video retrieval. Foundations and Trends in Information Retrieval 4(2), 215–322 (2009)
19. Snoek, C.G.M., Worring, M., van Gemert, J.C., Geusebroek, J.-M., Smeulders, A.W.M.: The challenge problem for automated detection of 101 semantic concepts in multimedia. In: MULTIMEDIA 2006: Proceedings of the 14th annual ACM international conference on Multimedia, pp. 421–430. ACM Press, New York (2006)
20. Varian, H.R.: Economics and search. SIGIR Forum 33(1), 1–5 (1999)
21. Voorhees, E.M., Harman, D.: Overview of the ninth text retrieval conference (trec-9). In: Proceedings of the Ninth Text REtrieval Conference TREC-9, pp. 1–14 (2000)
22. Wang, J.: Mean-variance analysis: A new document ranking theory in information retrieval. In: ECIR 2009: Proceedings of the 31th European Conference on IR Research on Advances in Information Retrieval, pp. 4–16. Springer, Heidelberg (2009)
23. Zhai, C., Lafferty, J.: A study of smoothing methods for language models applied to information retrieval. ACM Trans. Inf. Syst. 22(2), 179–214 (2004)

Ranking Fusion Methods Applied to On-Line Handwriting Information Retrieval

Sebastián Peña Saldarriaga[1], Emmanuel Morin[1], and Christian Viard-Gaudin[2]

[1] LINA UMR CNRS 6241, Université de Nantes, France
[2] IRCCyN UMR CNRS 6597, École Polytechnique de l'Université de Nantes, France

Abstract. This paper presents an empirical study on the application of ranking fusion methods in the context of handwriting information retrieval. Several works in the electronic text-domain suggest that significant improvements in retrieval performance can be achieved by combining different approaches to IR. In the handwritten-domain, two quite different families of retrieval approaches are encountered. The first family is based on standard approaches carried out on texts obtained through handwriting recognition, therefore regarded as noisy texts, while the second one is recognition-free using word spotting algorithms. Given the large differences that exist between these two families of approaches (document and query representations, matching methods, etc.), we hypothesize that fusion methods applied to the handwritten-domain can also bring significant effectiveness improvements. Results show that for texts having a word error rate (WER) lower than 23%, the performances achieved with the combined system are close to the performances obtained with clean digital texts, i.e. without transcription errors. In addition, for poorly recognized texts (WER > 52%), improvements can also be obtained with standard fusion methods. Furthermore, we present a detailed analysis of the fusion performances, and show that existing indicators of expected improvements are not accurate in our context.

1 Introduction

The use of ranking, or data fusion to combine document retrieval results has been addressed for several years by the Information Retrieval (IR) community [1,2,3,4,5,6,7,8,9]. The underlying assumption is that different retrieval techniques for the same query, retrieve different sets of documents [10]. By merging results from multiple systems, better retrieval effectiveness should be achieved for the same information need. However, as shown in Figure 1, there are several scenarios in which data fusion is ineffective or even more, harmful for retrieval performances. In particular, ranking fusion is not likely to improve results when the rankings involved are highly correlated [8].

Lee [11] claimed that different systems retrieve similar sets of relevant documents (+) but retrieve different sets of non-relevant documents (∘), however when the overlap between relevant documents is too high, as in Figure 1(a), little or no improvement is to be expected [12]. In scenario 1(b), the common

C. Gurrin et al. (Eds.): ECIR 2010, LNCS 5993, pp. 253–264, 2010.

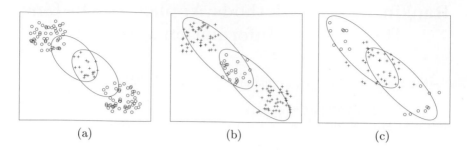

Fig. 1. Several scenarios in data fusion for a given query. (a) high overlap between relevant documents (b) high overlap between non-relevant documents (c) ideal scenario.

non-relevant documents will dominate the fusion process and be promoted to higher positions, thus affecting the performances of the merged result. In the ideal scenario, shown in Figure 1(c), the two sets share relevant documents but some documents appear only in one of them. In such a scenario, performance improvements are more likely to be observed when the non-common documents are merged into the fused set at a high rank [12].

To the best of our knowledge, the use of data fusion in handwriting retrieval has not yet been investigated. This study aims to investigate the effect of ranking fusion methods on handwriting retrieval. Current state-of-the-art in handwriting retrieval distinguishes two families of methods: recognition-based and recognition-free approaches. Since these methods use different document representations, query representations and retrieval algorithms, we hypothesize that the data fusion assumption holds in the handwritten domain. Thus, by combining the results of different handwriting retrieval methods, we can expect to improve retrieval effectiveness and leverage the strength of both method families.

The rest of this paper is organized as follows: Specificities and current methods in handwriting retrieval are reviewed in Section 2. Section 3 outlines the rank aggregation methods used in our experiments and the experimental methodology and data are described in Section 4. Experimental results are presented and discussed in Section 5, then conclusions are drawn in the final section.

2   Previous Works on Handwriting Retrieval

For several years, on-line handwriting was confined to the role of a convenient input method for PDAs, Tablet PCs, etc. With the recent evolutions of pen computers and digital pens, the production of on-line documents has become commonplace. These devices generate a series of two-dimensional coordinates corresponding to the writing trajectory as a function of time [13], called *on-line handwriting* or *digital ink*. As a result, algorithms for efficient storing and retrieval of on-line data are being increasingly demanded.

While the task of efficient retrieval of text documents has been addressed by researchers for many years, the retrieval of handwritten documents has been addressed only recently [14,15,16,17,18,19,20,21]. Existing methods for handwritten document retrieval can be divided into recognition-based (noisy IR) [14,18] and recognition-free (word spotting) approaches [15,16,17,19,20,21].

Word spotting aims to detect words in a document by comparing a query word with the individual words in the document without explicit recognition, the query itself being either a handwritten text or an electronic string. The challenges with word spotting approaches is to deal with segmentation of handwritten texts into words and to cope with arbitrary writing styles.

In the case of noisy IR, a handwriting recognition engine is in charge of processing handwriting before the retrieval process. Then standard IR methods are applied to the output text. When carried out on the noisy texts obtained through handwriting recognition, IR techniques will be penalized by recognition errors. On the other hand, the critical point in word spotting is the proper selection of image features and similarity measures. Retrieval errors in word spotting are expected for words with similar shapes [16].

While being robust in determining which documents of a collection contain the keywords in the user query, word spotting robustness in satisfying information needs is not well established. In contrast, IR techniques carried out on transcribed texts should perform as well as standard methods as long as recognition is not very noisy [18]. Actually, we cannot tell *a priori* which method performs better than the others under all circumstances.

The interest of data fusion for handwriting retrieval is twofold. At first, an obvious reason is that fusion methods might improve retrieval effectiveness, since it is already the case in the text-domain. The second reason is that the combination involved two different levels of representation of handwritten texts: digital ink and transcribed texts.

According to the previous observations, we argue that by combining the results of different handwriting retrieval methods, we can improve retrieval effectiveness, while taking advantage of strengths of different techniques. A corpus of more than 2,000 on-line documents is used for experimental validation. Several fusion models were applied to rankings returned by different baseline retrieval algorithms, including both, recognition-free and recognition-based approaches. The fusion methods used are outlined in the section below.

3   Ranking Fusion Methods

Several ranking fusion methods have been proposed in the past in the IR literature [1,2,3,4,5,6,7,8,9]. Early works by Fox and Shaw [1] introduced a group of result merging operators such as CombMAX, CombSUM, and CombMNZ. Vogt and Cotrell [3] proposed a weighted linear combination method, in which training data is needed to determine the appropriate weight given to each input system. Wu & McLean [8] used correlation weights that do not require training data.

Data fusion can also be seen as a voting procedure where a consensus ranking can be found using Borda count [4], Condorcet method [6], Markov chains [7] and methods inspired from the social choice theory [9]. Manmatha et al. [5] proposed to fit a mixture model consisting of an exponential and a Gaussian to the score distributions, then to average probabilities, thus minimizing the Bayes' error if the different systems are considered as independent classifiers; Bayesian fusion has also been explored [4]. Logistic regression has been employed successfully on one TREC collection [2].

As pointed out by Aslam and Montague [4], all of these methods can be characterized by the data they require: relevance scores or ordinal ranks, and whether they need training or not. In the present work we choose to use simple methods that require no training data, i.e. explicit user feedback. In particular the CombMNZ method has become a high-performance standard method in ranking fusion literature, whereas more complicated methods and weighting techniques exhibit mixed results.

In the following we will describe the methods used in our experiments. We adopt notational conventions from previous work [7]. These conventions are reviewed in Table 1.

Table 1. Notational conventions

Symbol	Definition
i	a document
τ	ranking of documents
$\tau(i)$	rank of document i
$\omega^\tau(i)$	normalized score of document i
R	$\{\tau_1, \tau_2, \ldots \tau_{|R|}\}$; set of rankings to fuse
$h(i, R)$	$|\{\tau \in R : i \in \tau\}|$; number of rankings containing i
$s^{\hat\tau}(i)$	fused score of document i

The CombSUM operator corresponds to the sum of the normalized scores for i given by each input system, while CombMNZ is the same score multiplied by $h(i, R)$. Besides these two methods, a simple approach to combine estimated scores from each input system is to take the harmonic mean (CombHMEAN, Equation 1). Since the harmonic mean will tend towards the smallest score assigned to a document, any agreement between systems that is not supported by score agreement, i.e. similar scores are assigned to the same document, is thus minimized.

$$s^{\hat\tau}(i) = \frac{|R|}{\sum_{\tau \in R} \frac{1}{\omega^\tau(i)}} \tag{1}$$

Since the scores from the different systems are normalized $[0, 1]$, we propose to average the log odds, a method that has been shown to be effective in combining document filtering approaches [22] (CombODDS).

$$s^{\hat\tau}(i) = \frac{1}{|R|} \sum_{\tau \in R} \log \frac{\omega^\tau(i)}{1 - \omega^\tau(i)} \tag{2}$$

Since CombHMEAN and CombODDS are only defined for $\omega^\tau > 0$, i.e. they cannot deal with partial rankings, we assign an extremely small score to documents missing in one of the rankings to combine.

As suggested by Lee [11], using scores is equivalent to doing an independent weighting, i.e. without considering the whole result list. For this reason, the last two methods used in our experiments are based on ordinal ranks, i.e. considering the whole ranking. First we define a rank-derived score as follows:

$$r^\tau(i) = 1 - \frac{\tau(i) - 1}{|\tau|} \tag{3}$$

Then we use $r^\tau(i)$ with the CombSUM and CombMNZ operators. For convenience, the resulting methods will be called *RankCombSUM* and *RankCombMNZ*.

4   Methodology

The test collection used in our experiments is a handwritten subset of the Reuters-21578 corpus for text categorization (TC). An obvious difference between TC and IR test collections is that TC collections do not have a standard set of queries with their corresponding relevant judgments. However, we can use category codes to generate queries using relevance feedback techniques. Previous works reported IR experiments with the Reuters-21578 collection [23] using an approach similar to the one that is described here.

In the following, we consider the ground truth texts of our test collection randomly partitioned into two subsets of nearly-equal sizes:

- Q is the set used for query generation (1016 documents)
- T is the set used for retrieval (1013 documents)

Ranking Relevant Terms
In order to provide relevant terms for query generation, we used an adaptation of the basic formula for the binary independence retrieval model [24], which is the log odds ratio between the probability of term t occurring in a document labeled with category c ($p_{t,c}$); and the probability of term t occurring in a document not belonging to c ($q_{t,c}$).

$$score_{t,c} = \log \frac{p_{t,c} \times (1 - q_{t,c})}{(1 - p_{t,c}) \times q_{t,c}} \tag{4}$$

In practice, the probability estimates are $p_{t,c} = (x + 0.5)/(X + 1.0)$ and $q_{t,c} = (n - x + 0.5)/(N - X + 1.0)$, where a correction is applied to avoid zero denominators, and where x is the number of documents of c containing term t, X is the number of documents of c in Q, n is the number of documents containing t, and N is the number of documents in Q. Stopwords are not considered and words are stemmed. We keep the top 100 scoring terms as a basis for query generation as explained below.

Query Generation

The characteristic query q_c of a category c is generated using the Ide dec-hi [25] relevance feedback formula as follows:

$$q_c = \sum_{i=1}^{X} C_i - \sum_{j=1}^{X} S_j \tag{5}$$

Where C_i is the vector for the i-th document of c, S_j is the vector for the j-th document not belonging to c, and X is the number of documents of c. It is worth noting that the vector space dimension is 100, and that documents are indexed using unnormalized term frequencies. All the documents of c are used but only $|c|$ random negative samples. The queries generated are 5 terms long as suggested by Sanderson's results [23].

For each category in our corpus, the relevance feedback query is given in Table 2. By choosing a category, we can now perform a retrieval on T using the generated query. The documents tagged with the chosen category are considered as relevant.

Table 2. Generated queries for each of the 10 categories represented in the test collection. Query terms are stemmed.

Category	Query terms
Earnings	vs ct net shr loss
Acquisitions	acquir stake acquisit complet merger
Grain	tonn wheat grain corn agricultur
Foreign Exchange	stg monei dollar band bill
Crude	oil crude barrel post well
Interest	rate prime lend citibank percentag
Trade	surplu deficit narrow trade tariff
Shipping	port strike vessel hr worker
Sugar	sugar raw beet cargo kain
Coffee	coffe bag ico registr ibc

It is worth noting that for 6 categories, the name of the category is part of the query. Even though the generated queries are likely to be representative of their corresponding category, it is not clear if they make sense from a human perspective. Nevertheless, within the scope of our experiments, what is important is that all the retrieval methods perform the same task. Moreover, the query generation process can be seen as a single iteration of relevance feedback during an interactive retrieval session [23].

5   Results and Discussion

In this section, we present the experimental results for the ranking fusion methods described in Section 3 with the corpus and queries presented above.

5.1   Baseline Methods

Several existing retrieval methods were used as baseline systems to be combined. On the noisy IR side, three models, namely, **cosine**, **okapi** and **language modeling** (LM)[1] are used. Okapi parameters are set as they are usually set in the literature, and documents are weighted using the $tf \times idf$ measure in cosine retrieval.

On the word spotting side, we use **InkSearch®** (IS)[2], which is a stable and out-of-the-box system. It enables the searching of text in handwriting, and does not require training. Documents are scored by adding up the confidence scores of query word occurrences.

For the recognition-based methods, the recognition engine of MyScript Builder[3] is used. Recognition can be performed on a character-level basis (termed as *free*) or on a word-level basis (termed as *text*). The word recognition error rates for each of these two recognition strategies are reported in Table 3.

Table 3. Recognition error rates for *free* and *text* strategies on the handwritten dataset

Recognition type	Word error rate
text	22.19%
free	52.47%

Unsurprisingly word-level recognition clearly outperforms character-level recognition [26]. Half the information is lost with the latter, while the former achieves low WER considering that no prior linguistic knowledge specific to this kind of documents is used. Results for baseline IR methods are reported in Table 4. In our work, we report the mean average precision (MAP) which provides a standard and stable evaluation metric [27].

The impact of recognition errors is obvious. In the case of documents recognized with the *text* strategy, a performance loss of roughly 5% is observed for every method, whereas this loss ranges from 15% to 20% with the *free* strategy.

5.2   Ranking Fusion Experiments

Table 5 shows the MAP obtained after fusion of the text-based approaches and IS. When using the *text* documents, significant improvements in performance are observed for *IS/Cosine* and *IS/Okapi* pairs with every fusion method. Combining IS and LM retrieval systematically degrades performance but never significantly.

---

[1] LM retrieval is based on the Kullback-Leibler divergence as implemented in the lemur toolkit, www.lemurproject.org

[2] InkSearch® is part of MyScript Builder SDK.

[3] MyScript Builder SDK can be found at http://www.visionobjects.com/products /software-development-kits/myscript-builder

Table 4. MAP obtained with the baseline methods. Columns indicate the type of recognition performed, the MAP obtained with the ground truth texts is also reported.

	Truth	Text	Free
Cosine	0.6887	0.6385	0.4980
Okapi	0.6989	0.6546	0.5005
LM	0.5589	0.4960	0.4101
IS	-	0.6547	0.6547

CombODDS and CombHMEAN perform as well as the standard methods. When the *text* documents are used, CombODDS and CombHMEAN achieves the best performance with *IS/Cosine* and *IS/Okapi* pairs respectively.

Rank-based methods are slightly outperformed by their score-based counterparts. Attempting to derive a cardinal score from an ordinal rank does not have a positive impact on the final results. Furthermore, this is highly questionable especially when the input rankings have different lengths [9], contain ties, and when the score distribution has a small statistical dispersion.

Table 5. Retrieval scores after fusion. Bold numbers indicate improvements with respect to IS and italic numbers indicate degradations in performances. An asterisk (*) indicates that the performance difference between IS and the combined method is statistically significant according to the Wilcoxon signed rank test at the 95% confidence interval.

(a) text

	Cosine	Okapi	LM
CombSUM	**0.6826***	**0.6933***	*0.6361*
CombMNZ	**0.6857***	**0.6933***	*0.6346*
CombODDS	**0.6871***	**0.6935***	*0.6346*
CombHMEAN	**0.6852***	**0.6940***	*0.6393*
RankCombSUM	**0.6775**	**0.6785**	*0.6351*
RankCombMNZ	**0.6808**	**0.6795**	*0.6347*

(b) free

	Cosine	Okapi	LM
CombSUM	**0.6782**	**0.6741**	*0.6451*
CombMNZ	**0.6760**	**0.6721**	*0.6428*
CombODDS	**0.6737**	**0.6719**	*0.6408*
CombHMEAN	**0.6710**	**0.6729**	*0.6410*
RankCombSUM	**0.6734**	**0.6692**	*0.6411*
RankCombMNZ	**0.6691**	**0.6644**	*0.6385*

Concerning the *free* documents, similar behaviour is observed. Once again, *IS/Cosine* and *IS/Okapi* pairs always produce improvements with every method, but these are not statistically significant. CombSUM achieves the best performance for every column in Table 5(b). It is worth noting that the performance difference between all the combining methods for *text* and *free* documents is only of 1% or 2%, while the difference between the individual methods ranges from 9% to 15%. Also, performances are slightly less degraded when the *free* documents are used in fusions with LM results.

5.3   Further Observations

In order to better understand why fusion techniques fail or succeed in bringing effectiveness improvements, we will try to examine different properties of the rankings. Lee [11] stated that improvements are expected when ranks have a greater overlap of relevant documents than of non-relevant documents. Later, Beitzel et al. [12] claimed that overlap rates were a poor indicator and proposed to relate fusion improvements to the Spearman's rank correlation coefficient.

Table 6 confirms that the unequal overlap property is not a good indicator of expected improvements. In the case of *text* documents, relevant/non-relevant overlap difference is about 64% for *IS/Cosine* and *IS/Okapi* pairs, and 56% for *IS/LM*. With *free* documents, a difference of 60% can be observed for *IS/Cosine* and *IS/Okapi* pairs, and 55% for *IS/LM*.

Table 6. Overlap and Spearman's correlation coefficient of recognition based systems with respect to IS

	(a) text				(b) free		
	IS/Cosine	IS/Okapi	IS/LM		IS/Cosine	IS/Okapi	IS/LM
R Overlap	95.71%	95.81%	78.20%	R Overlap	78.08%	76.90%	66.83%
NR Overlap	31.73%	30.84%	22.04%	NR Overlap	17.48%	16.85%	11.42%
SR Correlation	0.7259	0.7366	0.7378	SR Correlation	0.6439	0.6424	0.6753

Next we examined the predictive quality of Spearman's rank correlation coefficient in our context. For each query, we computed the correlation coefficient between the rankings returned by IS and the text-based systems, and the average precision after fusion. The average correlation coefficient across queries is shown in Table 6.

Figure 2 shows the performance of CombMNZ as a function of the correlation coefficient. Disregarding the type of recognized documents, there is a positive Spearman correlation between all ranking pairs. Following Beitzel et al. [12], these correlations can be considered "moderate" to "strong" positive correlations, and any positive effects due to fusion should be minimized. However, we can observe in Figure 2 that for several queries, substantial improvement is achieved.

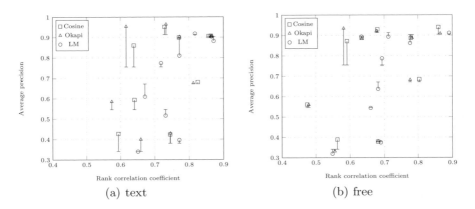

(a) text   (b) free

Fig. 2. Performance of fusion with CombMNZ as a function of Spearman's rank correlation coefficient. Each point represent a query, a point above the bar indicates an improvement with respect to IS, and a point below indicate degradations.

Furthermore, referring back to Tables 5 and 6, we can see that the best performances are obtained by combining ranks that exhibit higher correlation. Note that combining IS with LM retrieval always lead to performance degradation regardless of the correlation. Note also that the rank correlation coefficient does not take into account the relevance of documents in the rankings. This can explain why it cannot accurately predict expected improvements.

6   Conclusion

In this contribution, we have presented an empirical study on the combination of different approaches to on-line handwriting retrieval using data fusion methods. Experiments have been conducted with a handwritten subset of the Reuters-21578 corpus. Since this test collection does not contain a set of standard queries, we used relevance feedback techniques to generate queries, then we used category labels as relevance judgments. The corpus was divided into two subsets of equal sizes, the first one was used to generate the queries, and the second one for retrieval experiments.

The experimental results for individual baseline methods showed that recognition errors have an important impact on retrieval performances, losses ranging from 5 to 20% were observed when passing from clean digital texts to recognized documents. We also observed that score-based combining operators produces better retrieval performance than rank-based ones. The proposed methods, CombODD and CombHMEAN, perform as well as standard methods in the literature. Despite performance losses induced by recognition errors, data fusion improves retrieval performances in most configurations.

Further analysis of ranking fusion results showed that overlap rates [11] and Spearman's rank correlation coefficient [12] are not accurate indicators of expected performances after fusion. The baseline methods fused satisfy the unequal overlap property, yet degradations in performance can be observed. On the other hand, runs that have higher correlation coefficients achieve the best results. It has also been argued that ideally ranks that are being combined should be close in performance [8], however results obtained with the *free* documents suggest that this is not necessarily true.

The conclusions presented in this paper are not to be quoted out of the context of handwriting retrieval, and the systemic differences between the families of methods combined. We are aware that our experiments have several limitations, in particular those related to the size of the test collection, the number of queries, and their impact on the reliability of significance tests. The number of systems to combine can be seen as another limitation. To the best of our knowledge there is no available test collection for IR in the on-line handwriting domain, however we think that further experimental validation using bigger databases and human-prepared queries and relevance judgments should be conducted.

Another direction that needs to be explored is the relationship between recognition errors, degradations in baseline scores and expected fusion improvements, models and measures are needed to describe more precisely their inextricable relationships. This will be the subject of future work.

Acknowledgments

This research was partially supported by the French National Research Agency grant ANR-06-TLOG-009.

References

1. Shaw, J.A., Fox, E.A.: Combination of Multiple Searches. In: TREC-2, Proceedings of the 2nd Text REtrieval Conference, pp. 243–252 (1994)
2. Savoy, J., Le Calvé, A., Vrajitoru, D.: Report on the TREC-5 Experiment: Data fusion and Collection Fusion. In: TREC-5, Proceedings of the 5th Text Retrieval Conference, pp. 489–502 (1997)
3. Vogt, C.C., Cottrell, G.W.: Fusion via a Linear Combination of Scores. Information Retrieval 1(3), 151–173 (1999)
4. Aslam, J.A., Montague, M.: Models for Metasearch. In: SIGIR 2001, Proceedings of the 24th Annual ACM SIGIR Conference on Research & Development in Information Retrieval, pp. 276–284 (2001)
5. Manmatha, R., Rath, T.M., Feng, F.: Modeling Score Distributions for Combining the Outputs of Search Engines. In: SIGIR 2001, Proceedings of the 24th Annual ACM SIGIR Conference on Research & Development in Information Retrieval, pp. 267–275 (2001)
6. Montague, M., Aslam, J.A.: Condorcet Fusion for Improved Retrieval. In: CIKM 2001, Proceedings of the 11th International Conference on Information & Knowledge Management, pp. 538–548 (2002)
7. Renda, M.E., Straccia, U.: Web Metasearch: Rank vs. Score Based Rank Aggregation Methods. In: SAC 2003, Proceedings of the 18th Annual ACM Symposium on Applied Computing, pp. 841–846 (2003)
8. Wu, S., McClean, S.: Data Fusion with Correlation Weights. In: Losada, D.E., Fernández-Luna, J.M. (eds.) ECIR 2005. LNCS, vol. 3408, pp. 275–286. Springer, Heidelberg (2005)
9. Farah, M., Vanderpooten, D.: An Outranking Approach for Rank Aggregation in Information Retrieval. In: SIGIR 2007, Proceedings of the 30th Annual ACM SIGIR Conference on Research & Development in Information Retrieval, pp. 591–598 (2007)
10. Belkin, N.J., Cool, C., Croft, W.B., Callan, J.P.: The E_ect of Multiple Query Representations on Information Retrieval System Performance. In: SIGIR 1993, Proceedings of the 16th Annual ACM SIGIR Conference on Research & Development in Information Retrieval, pp. 339–346 (1993)
11. Lee, J.H.: Analysis of Multiple Evidence Combination. In: SIGIR 1997, Proceedings of the 20th Annual ACM SIGIR Conference on Research & Development in Information Retrieval, pp. 267–276 (1997)
12. Beitzel, S.M., Jensen, E.C., Chowdury, A., Grossman, D., Frieder, O., Goharian, N.: On Fusion of Effective Retrieval Strategies in the Same Information Retrieval System. Journal of the American Society of Information Science & Technology 50(10), 859–868 (2004)
13. Plamondon, R., Srihari, S.N.: On-line and o_-line handwriting recognition: a comprehensive survey. IEEE Transactions on Pattern Analysis & Machine Intelligence 22(1), 63–84 (2000)

14. Russell, G., Perrone, M., Chee, Y.M.: Handwritten document retrieval. In: Proceedings of the 8th International Workshop on Frontiers in Handwriting Recognition, pp. 233–238 (2002)
15. Rath, T.M., Manmatha, R.: Word image matching using dynamic time warping. In: CVPR 2003, Proceedings of the IEEE Conference on Computer Vision & Pattern Recognition, pp. 521–527 (2003)
16. Jain, A.K., Namboodiri, A.M.: Indexing and retrieval of on-line handwritten documents. In: ICDAR 2003, Proceedings of the 10th International Conference on Document Analysis & Recognition, pp. 655–659 (2003)
17. Rath, T.M., Manmatha, R., Lavrenko, V.: A search engine for historical manuscript images. In: SIGIR 2004, Proceedings of the 27th Annual ACM SIGIR Conference on Research & Development in Information Retrieval, pp. 369–376 (2004)
18. Vinciarelli, A.: Application of information retrieval techniques to single writer documents. Pattern Recognition Letters 26(14), 2262–2271 (2005)
19. Jawahar, C.V., Balasubramanian, A., Meshesha, M., Namboodiri, A.M.: Retrieval of online handwriting by synthesis and matching. Pattern Recognition 42(7), 1445–1457 (2009)
20. Terasawa, K., Tanaka, Y.: Slit style HOG feature for document image word spotting. In: ICDAR 2009, Proceedings of 10th International Conference on Document Analysis & Recognition, pp. 116–120 (2009)
21. Cheng, C., Zhu, B., Chen, X., Nakagawa, M.: Improvements in keyword search japanese characters within handwritten digital ink. In: ICDAR 2009, Proceedings of 10th International Conference on Document Analysis & Recognition, pp. 863–866 (2009)
22. Hull, D.A., Pedersen, J.O., Schütze, H.: Method Combination For Document Filtering. In: SIGIR 1996, Proceedings of the 19th Annual ACM SIGIR Conference on Research & Development in Information Retrieval, pp. 279–287 (1996)
23. Sanderson, M.: Word sense disambiguation and information retrieval. In: SIGIR 1994, Proceedings of the 17th Annual ACM SIGIR Conference on Research & Development in Information Retrieval, pp. 142–151 (1994)
24. Robertson, S.E., Spärck Jones, K.: Relevance weighting of search terms. Journal of the American Society for Information Science 27(3), 129–146 (1976)
25. Ide, E.: New Experiments in Relevance Feedback. In: The Smart Retrieval System, pp. 337–354. Prentice-Hall, Inc., Englewood Cliffs (1971)
26. Perraud, F., Viard-Gaudin, C., Morin, E., Lallican, P.M.: Statistical language models for on-line handwriting recognition. IEICE Transactions on Information & Systems E88-D(8), 1807–1814 (2005)
27. Buckley, C., Voorhees, E.M.: Evaluating evaluation measure stability. In: SIGIR 2000, Proceedings of the 23rd Annual ACM SIGIR Conference on Research & Development in Information Retrieval, pp. 33–40 (2000)

Improving Query Correctness Using Centralized Probably Approximately Correct (PAC) Search

Ingemar Cox[1], Jianhan Zhu[1], Ruoxun Fu[1], and Lars Kai Hansen[2]

[1] University College London
[2] Technical University of Denmark
{i.cox,j.zhu,r.fu}@cs.ucl.ac.uk, lkh@imm.dtu.dk

Abstract. A non-deterministic architecture for information retrieval, known as probably approximately correct (PAC) search, has recently been proposed. However, for equivalent storage and computational resources, the performance of PAC is only 63% of a deterministic system. We propose a modification to the PAC architecture, introducing a centralized query coordination node. To respond to a query, random sampling of computers is replaced with pseudo-random sampling using the query as a seed. Then, for queries that occur frequently, this pseudo-random sample is iteratively refined so that performance improves with each iteration. A theoretical analysis is presented that provides an upper bound on the performance of any iterative algorithm. Two heuristic algorithms are then proposed to iteratively improve the performance of PAC search. Experiments on the TREC-8 dataset demonstrate that performance can improve from 67% to 96% in just 10 iterations, and continues to improve with each iteration. Thus, for queries that occur 10 or more times, the performance of a non-deterministic PAC architecture can closely match that of a deterministic system.

1   Introduction

High query rates together with a very large collection size combine to make web search computationally challenging. To meet this challenge commercial search engines use a centralized distributed architecture in which the index is disjointly partitioned across a number of clusters [2]. Within each cluster, the partial index is then replicated across all machines in the cluster. When a query is received, the query is forwarded to a single machine in each partition/cluster, and the results are then consolidated before transmitting the retrieved results to the user.

This centralized distributed architecture guarantees that the entire index is searched. Moreover, for a fixed number of machines (fixed computational and storage budget), the number of partitions and the number of computers per partition (replication) can be altered in order to ensure that queries are responded to with low latency. This distributed architecture works well when the data set and the query rate do not change frequently. Nevertheless, repartitioning and replication cannot be avoided, and the procedure can be time consuming and expensive. For example, in [9], it is reported that Google must take half of their

C. Gurrin et al. (Eds.): ECIR 2010, LNCS 5993, pp. 265–280, 2010.

machines offline during this process and that terabytes of data must be copied between machines.

A key characteristic of a centralized distributed architecture is that it is deterministic, i.e. every query is compared to all documents in the collection and multiple instances of the same query generate the same result set. Recently a non-deterministic architecture has been proposed [4], called probably approximately correct (PAC) search. In the PAC search architecture, it is assumed that (i) the computers are independent, i.e. there is no communication between computers, (ii) each computer locally stores a random subset or partition of the index, (iii) the partitions are *not* disjoint, i.e. documents indexed on one machine may also be indexed on other machines, but there is no replication of documents within a single machine, (iv) a query is sent to a random subset of computers and the results from each machine are then consolidated before being displayed to the user.

A PAC search architecture is non-deterministic because (i) only a random subset of the index is searched in response to a query, and (ii) multiple instances of the same query may generate different result sets. The correctness of a PAC search is defined with respect to the result set achieved using a deterministic search. That is, it is assumed that both architectures implement the same retrieval model, and that, for each query, it is desired to approximate the result set that would be returned with a deterministic implementation of this retrieval model. Given the random nature of PAC search, the results are approximate and probabilistic, hence probably approximately correct search. Note that this definition of correctness does *not* incorporate traditional measure of precision and recall. Precision and recall are considered to be *only* a function of the retrieval model. The purpose of the PAC search is to approximate the results of this retrieval model as closely as possible using a non-deterministic search architecture. The correctness of a PAC search is measured by retrieval accuracy, which is defined as the ratio of the results returned by a non-deterministic system to the results returned by a deterministic system.

The PAC architecture has several advantages. For example, it does not need to be repartitioned. Rather, should the rate of queries increase to a point where the latency becomes too high, the PAC architecture can gracefully degrade performance in order to maintain responsiveness. In addition, it is designed so that there is very little communication overhead within the system. It is also inherently scalable and fault tolerant. However, for the same computational and storage budgets, the expected performance of a PAC system is less than that for a deterministic system.

In [4] theoretical bounds on the expected performance of PAC search were derived and verified by simulation. In particular, the performance of PAC search was compared to the centralized architecture used by commercial search engines such as Google. Using the same number of computers to store partial copies of the index, and querying a random subset of these computers (equivalent to one computer in each disjoint partition), it was shown that the retrieval accuracy of PAC search is 63%. And if we are interested in the top-10 documents, then there is over an 88% chance of finding 5 or more documents in common with the deterministic solution.

These percentages seem surprisingly high given the random nature of PAC search. However, to be practical, the retrieval accuracy is expected to be much closer to 100%. Of course, this can be achieved by querying a larger number of computers. In the example above, 1,000 computers are randomly chosen from a set of 300,000 available computers. If instead, the query is sent to 5,000 computers then the retrieval accuracy is 99%. Unfortunately, this accuracy requires fives times the computational resources expended for an equivalent deterministic search, and is therefore not economically feasible. In this paper, we examine a number of ways in which PAC search can be modified to improve its accuracy, while utilizing computational resources comparable to a deterministic search.

In Section 3 we briefly review some basics of a PAC search system. In Section 4 we then propose a novel approach which utilizes pseudo-random query node selection to improve PAC search performance. In particular, we consider those queries that frequently occur, and propose an iterative algorithm whereby PAC accuracy quickly increases to close to 100%. Section 5 presents both simulation results, as well as results based on a TREC data set. Finally, we conclude and discuss future works in Section 6.

2   Related Work

Two broad classes of distributed computer architectures have been proposed for information retrieval. These are centralized distributed architectures, used by commercial search engines such as Google, and decentralized distributed architectures used by a variety of peer-to-peer systems.

In the centralized architecture, the index is distributed over a number of disjoint partitions [2]. And within each partition, the partial index is replicated across a number of machines. A query must be sent to one machine in each partition and their partial responses are then consolidated before being returned to the user. The number of partitions and the number of machines per partition is a careful balance between throughput and latency [9]. Changes to the collection or to the query distribution may necessitate that the index be repartitioned, a process than can be complex and time consuming. Note that while the index is distributed across machines, the machines themselves are typically housed within a central server facility.

The problem of repartitioning was addressed in, ROAR (Rendezvous On a Ring) [9]. However, the proposed system is still deterministic.

Peer-to-peer networks offer a more geographically dispersed arrangement of machines that are not centrally managed. This has the benefit of not requiring an expensive centralized server facility. However, the lack of a centralized management can complicate the communication process. And the storage and computational capabilities of peers may be much less than for nodes in a commercial search engine. Li *et al.* [8] provide an overview of the feasibility of peer-to-peer web indexing and search. Their analysis assumes a deterministic system in which, if necessary, a query is sent to all peers in the network. The authors do comment on the possibility of compromising result quality by streaming the results to the users based on incremental intersection. However such a compromise is quite different from the non-deterministic search proposed here.

Terpstra *et al.* proposed a non-deterministic architecture called Bubblestorm [16]. They viewed the problem from a different perspective than PAC, assuming queries and documents are randomly replicated to machines and estimating the chance of a query meeting a document on a same machine. Bubblestorm faces the same problem as PAC search: under fixed communication, computation and storage budgets, the low storage capacity of individual machines results in unacceptable retrieval performance.

A variety of peer-to-peer decentralized architectures [14,15,6,10,18,17,13] have also been proposed and deployed previously, with a variety of search capabilities. They all have deterministic indexing and retrieval processes, and can be classified as deterministic systems following the definition in [4]. A problem with these peer-to-peer decentralized architectures is that their search depends on the network structure. Thus, they are relatively fragile to network structure changes, i.e. nodes entering, nodes leaving and nodes failing.

In this paper, we focus our attention on queries that occur more than once, our goal being to improve our performance with each new instance of the query. There has been considerable work investigating the distribution of queries. Analysis of a search engine query log [12] shows that query occurrences roughly follow a power law distribution, where 63.7% of unique queries only occur once (forming the so-called "long tail"). An hourly analysis of a query log [3] highlighted the temporal behavior of query traffic during a day. The temporal and spatial features of queries have been taken into account in Web caching [5].

3   PAC Search

Here we review the concept of PAC search and provide some results derived in [4] that will be needed subsequently. Figure 1 illustrates the basic elements of the PAC architecture. It is assumed that there are N unique documents in the collection to be indexed. For Web search, N is the unique number of web pages, currently estimated to be of the order of 65 billion documents [1]. It is further assumed that K computers are available and each computer indexes a random sample of n documents. It is reported in [4] that Google utilizes 300,000 computers and we therefore set $K = 300,000$. The n documents on each computer are assumed to be unique, i.e. no document appears more than once. However, each computer's sample of n documents is *not* disjoint with other computers, so a document may, and very likely will, be indexed by more than one computer. In [9] it was reported that the fraction of the collection indexed by each machine in the Google search architecture, $\frac{n}{N}$ is 0.001, and we therefore use this ratio in our subsequent analysis.

The union of the K samples is referred to as the collection sample, C_s, and is the union of all the individual samples. The size of the collection sample,

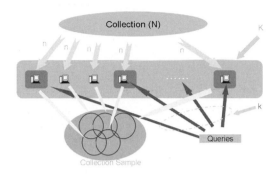

Fig. 1. Basic elements of the PAC architecture

$|C_s| = Kn$. Note that in a deterministic centralized distributed architecture the same storage capacity is need. However, in this case, each document is replicated on each machine within a partition.

It was shown in [4] that if each computer randomly samples the Web to acquire its n documents, then the expected coverage, $E(\text{coverage})$, i.e. the ratio of the expected number of unique documents in the collection sample to the number of unique documents in the collection, N, is

$$E(\text{coverage}) = \hat{N}/N = 1 - \epsilon, \qquad (1)$$

where ϵ is given by

$$\epsilon = (1 - n/N)^K \qquad (2)$$

Since the value inside the parentheses is less than one and $K = 300,000$, then ϵ is effectively zero and our expected coverage $E(\text{coverage})$ is effectively one. Thus, it is (nearly) certain that all documents in the collection will exist within the collection sample.

During retrieval, only a subset, k, of computers are queried, where $k = 1,000$ based on the number of disjoint partitions attributed to Google [9]. The union of n documents on each of the k computers is referred to as the sample index. Since k is much smaller than the total number of computers, K, the ratio of the expected number of unique documents in the sample index to the number of unique documents in the collection, (equivalent to the probability of any document being present in the retrieval index), given by

$$P(d) = E(\text{coverage}) = 1 - (1 - n/N)^k \qquad (3)$$

is 0.63 when $k = 1,000$ and $\frac{n}{N} = 0.001$. Thus, the coverage during retrieval is much smaller than the coverage during acquisition, and there is therefore a finite chance that the retrieval set provided by PAC will not be the same as for a deterministic search.

If r represents the set of documents retrieved by a deterministic system, and r' represents the number of documents retrieved by PAC search that are contained

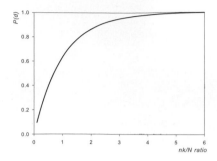

Fig. 2. The probability of retrieving any given relevant document of PAC search for different ratios of the sample index to collection size, $\frac{nk}{N}$

in r, i.e. $|r'| \leq |r|$, then in [4] it was shown that the probability of retrieving $|r'|$ documents, $P(|r'|)$ is

$$P(|r'|) = \binom{r}{r'} P(d)^{|r'|}(1 - P(d))^{|r|-|r'|}, \tag{4}$$

For the specific case where $\frac{n}{N} = 0.001$, $k = 1000$, and $r=10$, the probability of retrieving 5, 6, 7, 8, 9, and 10 documents in r is 17.1%, 24.5%, 24.1%, 15.5%, 5.9%, and 1%, respectively, and the probability of retrieving 5 or more documents in the top-10 is therefore over 88%.

The expectation of $|r'|$ is $E(|r'|) = |r|P(d)$. Therefore, Equation (3) indicates that acceptable performance using PAC search can be achieved provided the probability, $P(d)$, is sufficiently high.

The performance of PAC search is determined by the size of the sample index, kn in comparison with the collection size, N. Following the same setting as Google where $n/N = 0.001$, Figure 2 shows the accuracy of PAC search for various ratios of $\frac{nk}{N}$. Here we see that the accuracy rapidly increases from 63% when the sample index is equal to collection size to 86% when the sample index is twice as big as the collection, and to 95% when the sample index is three times as big as the collection. If the sample index is 5 times the size of the collection, then the accuracy is 99%.

4   Centralized PAC Search

The previous discussion of PAC search assumed a decentralized architecture in which a client device randomly selects k computers to issue a query to. As discussed in [4] a non-deterministic search may be disconcerting to users, since the same query, issued more than once, is likely to retrieve different result sets, albeit with significant overlap. In [4], it was proposed to ameliorate this problem, by issuing the query to a set of k pseudo-randomly selected computers, where the pseudo-random function was seeded with the query. As a result, a user issuing

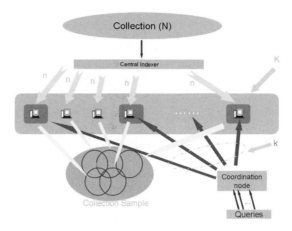

Fig. 3. A centralized PAC search architecture

the same query more than once will always see the same result set, since the set of pseudo-random computers will always be the same for a given query.

If this pseudo-random selection is performed independently by each client/ user, then different users will still see different results when issuing the same query, again albeit with significant overlap. To resolve this, we can construct a centralized non-deterministic PAC search architecture in which a centralized computer or computers receives queries from users, performs the pseudo-random selection centrally, and then forwards the query to the chosen set of k machines. This is illustrated in Figure 3.

Of course, such a configuration reduces the fault tolerance of the system. Reliability is now a function of single point of failure, the coordination node. However, this is no worse than for deterministic centralized distributed architectures.

Given a centralized PAC search architecture, we now investigate how such an arrangement can be used to improve the correctness of PAC search for frequently issued queries.

4.1   Adaptive Pseudo-random Selection of the Sample Index

It is observed that queries on the Web follow a power law distribution, where a small proportion of popular queries have a large number of occurrences, while a large proportion of less popular queries only have a small number of occurrences [3,5]. For frequently issued queries, we consider whether it is possible to improve upon our initial selection of k randomly selected computers in order to improve the correctness of the PAC search. The fundamental idea is the following. Given the first instance of a frequently occurring query, q, we select k pseudo-random computers. These computers form the sample index for this query. After receiving the results from the k machines, consolidating the results, and transmitting the retrieval results back to the user, we record the subset of computers that provided the highest ranked documents. The choice and size, k_0, of this subset will be discussed shortly. When we receive the same query a second time, the query is issued to this subset k_0 together with a new set, $k - k_0$ of pseudo-randomly chosen machines. Once again, after receiving the results from the k machines, consolidating the results, and transmitting the retrieval results back to the user, we record the subset, k_1, of computers that provided the highest ranked documents. And the process repeats. At each iteration, i, the identifiers of the retained computers, k_{i-1}, are cached. We refer to this as node caching. Note that nodes, rather than the documents retrieved from these nodes, are cached i.e. we are not caching queries in the traditional sense [5].

The purpose of query caching is to reduce the computational requirements of the information retrieval system by saving and caching the results of common searches. In contrast, the purpose of node caching is to iteratively improve the accuracy of the PAC search. There is no saving in computation. Of course, node and query caching could be combined, but this is outside the scope of this paper.

After responding to the initial query, we have examined k computers, and the expected accuracy is given by Equation (3). After responding to the query a second time, we have examined $k+k-k_0$ computers. From Equation (3) we know that if we had looked at all $2k - k_0$ computers simultaneously, then our expected accuracy would increase. This provides an upper limit on our performance. In practice, we did *not* examine all $2k - k_0$ computers simultaneously. Rather, at each iteration we examined k computers, this being the computation resource available to each query. Based on the adaptive pseudo-random selection of the sample index outlined above, can we design an algorithm that closely follows the upper limit provided by Equation (3)?

Before answering this, we consider a more fundamental question. What is the probability that there exists k computers from the set of all computers, K, such that, for a particular query, q, the accuracy of the k computers is 100%? That is to say, given a set of r documents retrieved deterministically in response to the query, q, what is the probability of these r documents existing on at least one configuration of k computers?

For a specific set of k computers, the expected number of unique documents in the sample index, $\hat{N}$, is, from Equation (1)

$$\hat{N} = (1 - \epsilon)N = \left(1 - (1 - n/N)^k\right)N \tag{5}$$

The probability of finding a specific document within the sample index is simply $\frac{\hat{N}}{N}$. The probability, $P(|r|)$, of finding a specific set of r documents in the sample index is

$$P(|r|) = \prod_{\delta=0}^{r-1} \left(\frac{\hat{N} - \delta}{N - \delta}\right) \tag{6}$$

and the probability of all r documents not being in the sample index is

$$P(|\bar{r}|) = \left[1 - \prod_{\delta=0}^{r-1} \left(\frac{\hat{N} - \delta}{N - \delta}\right)\right] \tag{7}$$

The probability of *not* finding the r documents in any sample of k computers is

$$P_{all}(|\bar{r}|) = \left[1 - \prod_{\delta=0}^{r-1} \left(\frac{\hat{N} - \delta}{N - \delta} \right) \right]^{\binom{K}{k}} \tag{8}$$

where $\binom{K}{k}$ is the number of ways of choosing k computers from K. Thus, the probability that all r documents will be present in at least one sample of k computers is

$$P = 1 - \left[1 - \prod_{\delta=0}^{r-1} \left(\frac{\hat{N} - \delta}{N - \delta} \right) \right]^{\binom{K}{k}} \tag{9}$$

Clearly the quantity in the square brackets is less than one, and this is raised to the power of $\binom{K}{k}$, which, for $K = 300,000$ and $k = 1,000$, is an extremely large number. Thus, there is near certainty that there exists a set of k computers on which all r relevant documents (as defined by deterministic search) are present in the sample index.

Now, let us consider the number of relevant documents we expect to see on a single computer. For a given query, q, let r denote the number of relevant documents in the collection (of size N). remembering that each computer samples n documents, the probability that a sampled document will be relevant is simply $\frac{r}{N}$. And the probability, $P(r')$ of sampling $r' \leq r$ documents is given by

$$P(r') = \binom{n}{r'} \left(\frac{r}{N} \right)^{r'} \left(1 - \frac{r}{N} \right)^{n-r'} \tag{10}$$

This is a standard binomial distribution, so the expected number of relevant documents to be found on a single machine, $E(r')$ is

$$E(r') = nr/N \tag{11}$$

The preceding analysis indicates that (i) there are configurations of k computers on which all relevant documents will be present in the sample index, and (ii) for a known number of relevant documents, Equation (11) provides the number of relevant documents expected on each computer. This number can be used to guide an heuristic search to find a k-configuration that includes all relevant documents for a specific query. We begin with a description of an algorithm that assumes knowledge of the number of relevant documents, see Section 4.2. Of course, in practice this is not the case, and the algorithm is modified in Section 4.3 to account for this.

4.2 Known Number of Relevant Documents

Before we consider the practical problem in which the number of relevant documents is unknown, we first consider the ideal case in which, for a given query, the number of relevant documents is known. The purpose of this exercise is to investigate the performance of an algorithm under idealized conditions. By so doing, we are better able to understand the upper limit on the performance of our algorithm prior to introducing heuristic assumptions. Experimental results are provided based on a simulation described in Section (5.1).

At iteration, i, each of our k computers has a retrieved r_j relevant documents, where $1 \leq j \leq k$. We order the computers based on the number of relevant documents retrieved, r_j. For simplicity, and without loss of generality, we assume that $r_1 \geq r_2 \cdots \geq r_k$.

At each iteration, we need to decide which computers to keep, and which to discard. Intuitively, we expect to retain more computers with each iteration, as we converge to a "optimum" configuration of k computers for a specific query. Therefore, we initially retain $x\%$ of computers (i.e. the $x\%$ that retrieve that largest number of relevant documents), and for each subsequent iteration, i, retain $(x + (i - 1)y)\%$ of computers. We refer to this percentage as the "keep" ratio.

Furthermore, at each iteration, any computers in the keep ratio that retrieved less than the expected number of documents given by Equation (11) are discarded.

4.3   Unknown Number of Relevant Documents

In practice, the number of relevant documents for a particular query is unknown. The assumption of the iterative method proposed in Section 4.2 is that the set of relevant documents is known beforehand. However, this is not realistic in a real retrieval environment. Instead, we assume that the top ranked r documents, based on a retrieval model such as the BM25 model [11], are relevant to the query. Indeed, in a deterministic system such as Google, the top ranked r documents are presented to users.

At iteration, i, our k computers produce a merged ranked list of documents, where each of our k computers has retrieved r_j documents, $1 \leq j \leq k$, in the top r rank positions. Intuitively, we favor a computer with many documents ranked highly in the top r positions.

As before, we can simply rank the k computers based on the number of documents each computer has in the top r positions, r_j. However, this measure gives equal weight to documents at the top and bottom of the result set. In order to also consider the rank of documents, we propose an NDCG-like score [7] for judging how well each computer is responding to a query. For computer, j, its score, s_j is defined as

$$s_j = \sum_{m=1,\ldots,r} \frac{\delta_m}{log_2(1 + m)} \tag{12}$$

where $\delta_m = 1$ is an indicator variable that is one if the document at rank position m is one of the documents on computer j, and 0 otherwise.

We set r=1,000 in our TREC data based experiments in Section 5.2, and the results show that our NDCG-like score performs better than simply counting the number of documents in the top rank positions.

Once again, we retain the top x% of the ordered list of computers in the first iteration. And in iteration, i, we retain the top $(x + (i - 1) * y)$% of the order list.

5   Experimental Results

In order to investigate the effectiveness of our two methods we performed both simulations and experiments based on a TREC dataset.

For both simulations and TREC experiments, we used the same computation and storage requirements as assumed of Google. There are K=300,000 computers which independently sample N documents. We kept the ratio between the number of documents indexed by each computer, n, to N as 0.001, and fix the number of computers queried at each step $k = 1,000$.

We fixed the initial keep ratio at $x = 20$% in the first iteration. This is incremented by $y = 3$% in each subsequent iteration. A less heuristic pruning strategy is a topic of future work.

5.1   Simulation

Due to computational cost, we set N=100,000 and $n = 100$. Note that the expected accuracy is only affected by the *ratio*, $\frac{n}{N}$. Therefore, our results generalize to much larger N and n, provided the ratio remains fixed (0.001). We randomly sample r documents from the entire N and treat these as the relevant set to a query. The measured accuracy is the ratio of relevant documents, r', found on k computers, to the total number of relevant documents r.

We consider the cases where the number of relevant documents are r=1,000, 2,000, and 4,000. The expected number of relevant documents on a computer is $\frac{nr}{N} = 1$, 2, and 4, respectively. For r=1,000, 2,000, and 4,000, we ran ten

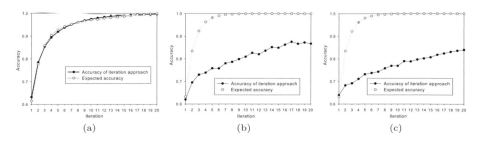

(a)   (b)   (c)

Fig. 4. Accuracy for our iteration approach and comparison with expected accuracy. The data points are based on the performance at each step. Under different number of relevant documents, $r = $ (a) 1,000 (b) 2,000 (c) 4,000.

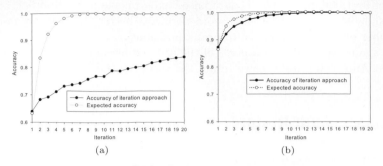

Fig. 5. When the number of relevant document r is 4,000, when we adjust the number of sampled computers k, accuracy for our iteration approach and comparison with expected accuracy. The data points are based on the performance at each step. $k=$ (a) 1000 (b) 2000.

trials of our simulations, and average the accuracy scores at each iteration step over the ten trials. The results are shown in Fig. (4). Note that the variances of the accuracy scores at each step are small, and do not affect the overall trend. Fig. (4) shows that when the number of relevant documents $r = 1,000$, the iterative algorithm closely follows the upper bound on performance, and an accuracy of 99% is reached in 15 iterations. As the number of relevant documents increases ($r = 2,000$ and $r = 4,000$), the iterative algorithm performs less well. The reason for this degradation in performance as r increases is unclear, and a topic of ongoing research.

Next, we study how the number of sampled computers, k, affects the performance of the algorithm when r is large. We fix the number of relevant documents to $r = 4,000$, and considered two values of k, namely 1,000 and 2,000. The results are shown in Fig. (5), where we observe that for the larger k the simulation results much more closely track the upper bound on performance. By querying 2000 rather than 1000 machines, our initial accuracy increase from 63% to 86% and this appears to allow us to perform a much better pruning at subsequent iterations. Conversely, it appears that when we query $k = 1000$ machines and our initial accuracy is 63%, our heuristic pruning strategy is retaining poor machines. The reason for this is unknown and the subject of ongoing work.

5.2   TREC Experimental Results

In Section 5.1, we investigated PAC performance in a simulated retrieval environment in which we assumed that all the relevant documents for a query are known. However, in a real retrieval environment, the number of relevant documents is unlikely to be known. Thus we need to evaluate our approach based on document retrieval models. Our proposed approach in Section 4.3 allows us to carry out real world retrieval.

We tested our approach using the TREC-8 50 topics, using only the title part. The dataset consists of approximately half a million documents. We used

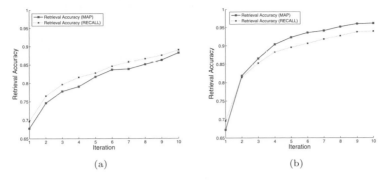

(a)   (b)

Fig. 6. Retrieval accuracy of our iterative approach on MAP and Recall-1000. Results are based on the TREC8 50 queries. (a) when nodes are pruned based on the number of documents they retrieve in the ranked list, and (b) when nodes are pruned based on the NDCG-like metric.

the same settings as previous, i.e., K=300,000, and $\frac{n}{N}$=0.001. We used the BM25 retrieval model [11] for ranking documents on each computer. In order to eliminate any unreliability from a single trial, we ran 10 trails, and in each trial, we performed 10 iterations for each query. In the first iteration, each query is issued to 1,000 randomly chosen computers. Then each computer's performance is evaluated based on the metric proposed in Section 4.3. At each iteration we query 1,000 computers, a portion retained from the previous iterations together with a randomly selected set of computers. We used standard IR metrics including MAP (mean average precision) and Recall at 1,000 to evaluate the performance at each iteration. The MAP and Recall values at each iteration are averaged over the 10 trials and the result is reported here. To compare our iterative approach with a deterministic system, following the definition of the correctness of the PAC search, we define the retrieval accuracy for MAP and recall as:

$$RetrievalAccuracy(Metric) = Metric_{PAC}/Metric_{det}, \qquad (13)$$

where $Metric_{PAC}$ is the MAP or Recall-1000 of the PAC system, and $Metric_{det}$ is the MAP or Recall-1000 of the deterministic system.

Figure 6 shows the performance of the system when pruning is based on (a) simple counting and (b) an nDCG-based metric. It is clear that the nDCG-like metric outperforms the simple counting. The retrieval accuracy when pruning based on simple counting increases from 67% to 81% in the first 5 iterations. In comparison, the retrieval accuracy using the nDCG-like metric increases from 67% to 92% in the first 5 iterations. The retrieval accuracy when pruning based on the simple counting method only reaches 88% at iteration 10, while the retrieval accuracy of the nDCG-like metric reaches 96% at iteration 10.

Figure 6 (b), also shows that both MAP and Recall-1000 follow a similar trend, i.e., performance improves quickly from iteration 1 to 5, and then improves more slowly from iteration 5 to 10. It is worth noting that only 4 iterations are required for our approach to improve from around 67% to 90% of the performance of a deterministic approach. The final retrieval accuracy reaches 96% at iteration 10. Note that the the standard deviations across trials are very small, and do not affect our observations here. This indicates that our approach has stable performance.

These experiments suggest that, for frequently occurring queries, a centralized PAC architecture can perform at a similar level as a deterministic system.

6   Conclusions and Future Work

The non-deterministic probably approximately correct search architecture is an interesting alternative search architecture. However, the performance of the PAC architecture must be improved if it is to become competitive with traditional deterministic systems.

In this paper, we proposed adding a centralized query coordination node to the architecture. This configuration is not as distributed as the original PAC proposal. However, it retains much of its benefits, at the expense of introducing a single point of failure. Of course, in practice such a node could itself be replicated for fault tolerance and load balancing.

Using a centralized PAC architecture, and in response to a query, the random selection of nodes is replaced by a pseudo-random selection, seeded with the query. This has the advantage, as noted in [4] of eliminating the variation in results sets in response to the same query. More importantly, for frequently occurring queries, we considered the problem of iteratively refining the pseudo-random choice of k computers in order to improve the performance.

A theoretical analysis provided a proof that there exists (with near certainty) a configuration of k nodes on which all relevant documents will be found. The analysis also provided an upper bound on the performance of any iterative algorithm. The analysis also allowed us to estimate the expected number of relevant documents that should be present on any single machine. This information was then used to partly guide a heuristic search.

Two heuristic search algorithms were proposed. The first scored computers based simply on the number of relevant documents they retrieved. The second scored computers based on a nDCG-like score that gives higher weight to higher ranked documents.

Simulations showed that the search algorithm very closely followed the theoretical upper bound when the number of relevant documents was less than $r = 1000$. However, as the number increased to 2000 or 4000, deviation from the upper bound increased. Nevertheless, in all cases, retrieval performance continued to improve with each iteration. For $r = 4000$, the case where twice as many machines ($k = 2000$) were queried per iteration was examined. In this case, the initial accuracy is 86% and the search once again closely follows the theoretical upper bound. This suggests that when $k = 1000$, the iterative algorithm is retaining a sub-optimal choice of machines. Analysis to the relationship among the number of relevant documents, the number of queried machines and the retrieval performance remains a source of on-going research.

Experiments on the TREC-8 dataset showed that the nDCG-based algorithm provided superior performance. In particular, the MAP score relative to a deterministic system increased from an initial 67% to 90% in just 4 iterations. And after 10 iterations performance is 96% of a deterministic system. These experiments suggests that, for frequently occurring queries, a centralized PAC architecture can perform at a similar level as a deterministic system.

This iterative approach is only applicable for queries that occur frequently (e.g. more than 10 times). For less frequently occurring queries, alternative approaches must be developed, which are the subject of future work.

References

1. http://www.worldwidewebsize.com (2009)
2. Barroso, L.A., Dean, J., HÄolzle, U.: Web search for a planet: The google cluster architecture. IEEE Micro 23(2), 22–28 (2003)
3. Beitzel, S.M., Jensen, E.C., Chowdhury, A., Grossman, D.A., Frieder, O.: Hourly analysis of a very large topically categorized web query log. In: SIGIR, pp. 321–328 (2004)
4. Cox, I., Fu, R., Harsen, L.K.: Probably approximately correct search. In: Proc. of the Internationla Conference on Theoretical Information Retrieval, ICTIR (2009)
5. Fagni, T., Perego, R., Silvestri, F., Orlando, S.: Boosting the performance of web search engines: Caching and prefetching query results by exploiting historical usage data. ACM Trans. Inf. Syst. 24(1), 51–78 (2006)
6. Harren, M., Hellerstein, J.M., Huebsch, R., Loo, B.T., Shenker, S., Stoica, I.: Complex queries in dht-based peer-to-peer networks. In: Druschel, P., Kaashoek, M.F., Rowstron, A. (eds.) IPTPS 2002. LNCS, vol. 2429, p. 242. Springer, Heidelberg (2002)
7. JÄarvelin, K., KekÄalÄainen, J.: Cumulated gain-based evaluation of ir techniques. ACM Trans. Inf. Syst. 20(4), 422–446 (2002)
8. Li, J., Loo, B.T., Hellerstein, J.M., Kaashoek, M.F., Krager, D.R., Morris, R.: On the feasibility of peer-to-peer web indexing and search. In: Kaashoek, M.F., Stoica, I. (eds.) IPTPS 2003. LNCS, vol. 2735, pp. 207–215. Springer, Heidelberg (2003)
9. Raiciu, C., Huici, F., Handley, M., Rosenblum, D.S.: Roar: increasing the flexibility and performance of distributed search. SIGCOMM Comput. Commun. Rev. 39(4), 291–302 (2009)
10. Reynolds, P., Vahdat, A.: Efficient peer-to-peer keyword searching. In: Proceedings of the International Middleware Conference (2003)
11. Robertson, S., Walker, S., Jones, S., Hancock-Beaulieu, M., Gatford, M.: Okapi at trec-3. In: Proc. of the Third Text REtrieval Conference (TREC 1994), pp. 109–126 (1996)
12. Silverstein, C., Henzinger, M.R., Marais, H., Moricz, M.: Analysis of a very large web search engine query log. SIGIR Forum 33(1), 6–12 (1999)
13. Skobeltsyn, G., Luu, T., Zarko, I.P., Rajman, M., Aberer, K.: Web text retrieval with a p2p query-driven index. In: SIGIR, pp. 679–686 (2007)
14. Stoica, I., Morris, R., Karger, D., Kaashoek, F., Balakrishnan, H.: Chord: Scalable peer-to-peer lookup service for internet applications. In: Proceedings of the 2001 ACM SIGCOMM Conference, pp. 149–160 (2001)

15. Tang, C., Xu, Z., Mahalingam, M.: psearch: Information retrieval in structured overlays. In: HotNets-I (2002)
16. Terpstra, W.W., Kangasharju, J., Leng, C., Buchmann, A.P.: Bubblestorm: resilient, probabilistic, and exhaustive peer-to-peer search. In: SIGCOMM, pp. 49–60 (2007)
17. Yang, K.-H., Ho, J.-M.: Proof: A dht-based peer-to-peer search engine. In: Conference on Web Intelligence, pp. 702–708 (2006)
18. Yang, Y., Dunlap, R., Rexroad, M., Cooper, B.F.: Performance of full text search in structured and unstructured peer-to-peer systems. In: INFOCOM (2006)

Learning to Distribute Queries into Web Search Nodes

Marcelo Mendoza, Mauricio Marín, Flavio Ferrarotti, and Barbara Poblete

Yahoo! Research Latin America
Av. Blanco Encalada 2120, 4th floor, Santiago, Chile
{mendozam,mmarin,flaviof,bpoblete}@yahoo-inc.com

Abstract. Web search engines are composed of a large set of search nodes and a broker machine that feeds them with queries. A location cache keeps minimal information in the broker to register the search nodes capable of producing the top-N results for frequent queries. In this paper we show that it is possible to use the location cache as a training dataset for a standard machine learning algorithm and build a predictive model of the search nodes expected to produce the best approximated results for queries. This can be used to prevent the broker from sending queries to all search nodes under situations of sudden peaks in query traffic and, as a result, avoid search node saturation. This paper proposes a logistic regression model to quickly predict the most pertinent search nodes for a given query.

1 Introduction

Data centers supporting Web search engines are composed of large sets of processors which form clusters of computers. These systems can be seen as a collection of slave search nodes (processors) which are fed with queries by master query-receptionist/answer-provider machines called brokers. The scale of these systems – along with issues of energy consumption, hardware/software investment and maintenance costs – make it relevant to devise efficient and scalable query solution strategies. These strategies include: data structures which are suitable for distributed indexing, different levels of caching and parallel query processing algorithms. All of these factors, when operating in combination, enable search engines to cope efficiently with heavy and highly dynamic query traffic. Certainly an important design goal is to be able to reduce the level of hardware redundancy in order to allow the search engine to operate at a high utilization level in steady state query traffic, while at the same time respond in a stable manner to sudden peaks in traffic. This paper describes a method that allows search engines to achieve this last objective.

Broker machines contain a results cache which stores the answers to the most frequent user queries. We call this cache RCache. When a given query is not found in the RCache it is sent to P search nodes which respond with their top-N results for the query and then the broker merges these local top results to get the global top-N results and produce the answer set for the query. The document

C. Gurrin et al. (Eds.): ECIR 2010, LNCS 5993, pp. 281–292, 2010.
© Springer-Verlag Berlin Heidelberg 2010

collection is evenly partitioned into the P search nodes and each sub-collection is indexed with the well-known inverted file which is usually kept in secondary memory. A search node can also contain additional levels of caching in its main memory. The most basic one is a cache of inverted lists, but there can also exist additional caches such as a cache of pre-computed inverted list intersections between pairs of terms frequently cooccurring in queries. The method presented in this paper assumes the existence of an additional cache – which we call LCache – in the broker machine and, possibly, another cache in each search node – which we call global top-N cache. The LCache is a *small location cache* which stores, for each cached query, the set of search node IDs that produce the global top-N results for the query.

The relationship between the RCache and LCache is as follows. As pointed out in [10], the objective of caching is, to improve the query throughput of search engines. Therefore, frequent queries which require large amounts of resources be processed, are better candidates to be hosted by the RCache, than frequent queries which require fewer resources. The goal of the LCache is to prevent these last more *inexpensive* queries from being sent to all search nodes, including also unfrequent queries that obtain their top-N from very few processors. Under heavy query traffic conditions it is critical to reduce the number of processors hit by each query. This significantly reduces the overhead and therefore increases scalability. This occurs because processor utilization is better achieved for a large number of different queries, than for fewer queries that require more processors and producing a similar work-load.

The LCache admission policy caches queries containing the largest values for the product $f \cdot L \cdot P/m$ where f is the frequency of the query, L the average cost of processing the query in a search node, and m the number of search nodes producing the documents within the global top-N results. In contrast, the queries cached in the RCache are those queries with the largest values for the product $f \cdot L \cdot m$. For the purpose of the method proposed in this paper we also keep in the LCache the search node IDs for the results of the queries stored in the RCache.

Typically a single entry in the RCache is of the order of KBs whereas a single entry in the LCache is of the order of a few bytes. Wherever the answers of the RCache are stored, in secondary memory or in another set of processors, the LCache can also store its entry contents there. The entries in the LCache can be efficiently compressed by performing document clustering and distributing co-related clusters in the same or consecutive processor IDs to balance the load properly. In addition, document IDs can be re-labeled in order to make them consecutive within each cluster so that (doc_ID, proc_ID) pairs can also be compressed efficiently. So, alternatively each entry in the LCache could be composed of a pair (doc_ID, proc_ID) which allows the query processing algorithm to go to the respective processors and directly get the snippets and other meta-data to build the result set for the query. Certainly this occupies more space than simply compressing processors IDs and thereby we prefer to keep the doc_ID values distributed onto the processors in what we call global top-N caches. The objective is to allow the broker store as much location data as possible for queries that require few processors to be completely solved.

Apart from improving overall query throughput, which has been shown in [16], the work in [9] shows that the LCache can also be used to improve the search engine ability to cope efficiently with sudden peaks in query traffic. This work resembles the idea in [19] which proposes that upon sudden peaks in traffic, the queries not found in the RCache must be sent to less than P search nodes as would have been the case when traffic is in steady state. The aim is to avoid processor saturation and temporarily respond to users with approximated answers to queries. The queries are routed to the search nodes capable of providing good approximations to the exact results.

However, this is where the similarity between [9] and [19] ends. To determine the set of processors to which the queries that are not found in the RCache must be sent, the method in [19] proposes a technique based on representation of search node contents. This method is used to dynamically rank search nodes in accordance with specific query terms. Whereas the method proposed in [9] views the query terms and search node IDs stored in the LCache as a semantic cache [4,3]. The approach presented in [9] determines which processors are most likely to contain good approximated results for queries that are not found both in the RCache *and* in the LCache. We emphasize "and" to note that the LCache by itself contributes to reduce the work-load into processors which is critical in this case. The space used by the representation of search node contents can be used to host LCache entries with the advantage that hits in the LCache can be used to respond exact answers to queries without affecting significantly the overall work-load of the processors.

Nevertheless, ignoring the fact that the LCache is able to deliver exact answers at low processor overload cost in high query traffic situations, the LCache semantic method proposed in [9] is not as effective as the method proposed in [19]. On the average and for our experimental datasets, the method in [19] is able to rank search nodes for more than 80% of the queries, whereas in the LCache semantic method if is below 40%. In this paper we propose a new LCache-based method that matches the performance of the method proposed in [19] for this metric. Given the equivalent performance, the advantages of our method over the methods proposed in [9] and [19] are evident in practical terms since it is well-known that large Web search engines are daily faced with drastic variations in query traffic. The method proposed in this paper uses the LCache as a training dataset for a standard machine learning method. This makes it possible to build a predictive model for the search nodes capable of producing the best results for queries not found in the RCache and the LCache. The method uses a logistic regression model to predict those search nodes.

The remaining sections of the paper are organized as follows. In Section 2 we review related work. In Section 3 we introduce the learning to cache issue, modelling the location cache problem in the machine learning domain. Experimental results are shown in Section 4. Finally, we conclude in Section 5.

2   Related Work

Regarding caching strategies, one of the first ideas studied in literature was having a static cache of results (RCache abridged) which stores queries identified as frequent from an analysis of a query log file. Markatos et al. [17] showed that the performance of the static caches is generally poor mainly due to the fact that the majority of the queries put into a search engine are not frequent ones and therefore, the static cache reaches a low number of hits. In this sense, dynamic caching techniques based on replacement policies like LRU or LFU achieved a better performance. In another research, Lempel and Moran [13] calculated a score for the queries that allows for an estimation of how probable it is that a query will be made in the future, a technique called Probability Driven Caching (PDC). Lately, Fagni et al. [6] proposed a structure for caching where they maintain a static collection and a dynamic one, achieving good results, called Static-Dynamic Caching (SDC). In SDC, the static part stores frequent queries and the dynamic part handles replacement techniques like LRU or LFU. With this, SDC achieved a hit ratio higher than 30% in experiments conducted on a log from Altavista. Long and Suel [15] showed that upon storing pairs of frequent terms determined by the co-occurrence in the query logs, it is possible to increase the hit ratio. For those, the authors proposed putting the pairs of frequent terms at an intermediate level of caching between the broker cache and the end-server caches. Baeza-Yates et al. [2] have shown that caching posting lists is also possible, obtaining higher hit rates than when just doing results and/or terms. Recently, Gan and Suel [10] have studied weighted result caching in which the cost of processing queries is considered at the time of deciding the admission of a given query to the RCache.

A common factor that is found in the previous techniques is that they attempt to give an exact answer to the evaluated queries through the cache exactly. The literature shows that the use of approximation techniques like semantic caching can be very useful for distributed databases. A seminal work in this area is authored by Godfrey and Gryz [11]. They provide a framework for identifying the distinct cases that are presented to evaluate queries in semantic caches. Fundamentally, they identify cases which they call semantic overlap, for which it is possible for the new query to obtain a list of answers from the cache with a good precision. In the context of Web search engines, Chidlovskii et al. [4,3] poses semantic methods for query processing. The proposed methods store clusters of co-occurring answers identified as regions, each of these associated to a signature. New frequent queries are also associated to signatures, where regions with similar signatures are able to develop their answer. Experimental results show that the performance of the semantic methods differs depending on the type of semantic overlap analyzed. Amiri et al. [1] deals with the problem of the scalability of the semantic caching methods for a specific semantic overlap case known as query containment.

A related work to our objective is authored by Puppin et al. [19]. The authors proposed a method where a large query log is used to form P clusters of documents and Q clusters of queries by using a co-clustering algorithm proposed in

[5]. This allows defining a matrix where each entry contains a measure of how pertinent a query cluster is to a document cluster. In addition, for each query cluster a text file containing all the terms found in the queries of the cluster is maintained. Upon reception of a query q the broker computes how pertinent the query q is to a query cluster by using the BM25 similarity measure. These values are used in combination with the matrix co-clustering entries to compute a ranking of document clusters. The method is used as a collection selection strategy achieving good precision results. At the moment this method is considered as state-of-the-art in the area. This method, namely PCAP, will be evaluated against our method in the experimental results section.

The another similar work to our objective is our prior work [9]. Here we propose proposed using an LCache as a semantic cache. The proposed method allows reducing the visited processors for queries not found in an LCache. The semantic method for evaluating new queries uses the inverse frequency of the terms in the queries stored in the cache (Idf) to determine when the results recovered from the cache are a good approximation to the exact answer set. The precision of the results differs depending on the semantic overlap case analyzed. In the context of this paper we call this method SEMCACHE and it will be evaluated against our method in the experimental results section. Recently [16], we show that the combination of an LCache and an RCache using dynamic caching strategies is able to achieve efficient performance under severe peaks in query traffic, as opposite to the strategy that only uses the RCache that gets saturated as they are not able to follow the variations in traffic.

To the best of our knowledge, there is no previous work that uses machine learning methods to predict the most likely search nodes for queries that are not cached in the broker machine.

3   Learning to Rank Search Nodes

3.1   Modelling the Location Cache Problem in the Machine Learning Domain

A classification task is based on the appropriate processing of training and evaluation datasets. A training dataset consists of data instances, each of which are composed of a collection of features and one or more tags that indicates which category the instance belongs to. A training dataset is used to construct a classification model that allows it to categorize new data instances. An evaluation dataset is used to measure the classifier's performance. Each data instance from the evaluation dataset is categorized using the classifier. Each of these data instances contains one or more labels that indicate which category they belong to. They are compared to the categories predicted by the classifier.

The LCache's problem could be modeled in the machine learning domain like a classification task. Let LC_i be an entry in an LCache. An entry LC_i is formed by the query terms q_i stored in that position, and by the list of machines that allows the top-N results to be obtained for q_i. Given a new query q (not stored in the cache) we want to determine the list of machines that allows the top-N results of

q to be obtained. In the machine learning domain we can interpret each machine as a category. We can also interpret each LC_i entry from the compact cache as an instance of training data. From these, we prepare a collection of features (the terms of q_i) and the categories to which it (the list of machines) belongs. Given that the distributed search engine typically considers multiple machines, the problem in the machine learning domain is multi-class. In addition, given that the responses to the query could be distributed into several machines, we need the predictive model to give us a list of likely machines. Due to this factor, the problem is also multi-labeled.

Given an LC_i entry in the LCache, we will call the list of terms of q_i $\mathbf{x}_i$ and the list of machines $\mathbf{y}_i$. Given an LCache of l entries, we arrange a training dataset formed by pairs $(\mathbf{x}_i, \mathbf{y}_i)$, $i = 1, \ldots, l$. We also consider that the LCache vocabulary (the group of terms made from the queries stored in the LCache) contains n elements, and that the distributed document system considers m machines. Later, $\mathbf{x}_i \in \{0, 1\}^n$, where 1 indicates that the term was used to form the query, 0 in another case, and $\mathbf{y}_i \in \{0, 1\}^m$, where 1 indicates that the machine contains one of the top-N responses for q, 0 in another case.

The predictive model generally consists of a $y(\mathbf{x}, \mathbf{w})$ function where $\mathbf{w}$ corresponds to the parameters of a combination (generally convex) over the vector of features $\mathbf{x}$ of the entry. In the simplest case, the model is linear to the parameters $\mathbf{w}$ and so $y = \mathbf{w}^T \mathbf{x} + w_0$. The coefficient w_0 corresponds to the model's bias. To decide each category, an activation function is applied to y, which we will call $f(\cdot)$. Frequently, the activation function corresponds to the sign function which makes it possible to evaluate the categorization based on a Boolean variable.

3.2   Logistic Regression Model

For the first approximation we will consider the classification problem restricted to two classes. The construction of a classification model consists of the estimation of the following conditional probability:

$$\mathbf{P}(y = \pm 1 \mid \mathbf{x}, \mathbf{w}).$$

Notice that the probability is conditioned to $\mathbf{x}$, meaning the feature vector of the entry instance that we want to classify. It is also conditioned for the predictive model represented by the parameter vector $\mathbf{w}$.

The logistic regression model assumes that the conditional probability can be calculated in the following way:

$$\mathbf{P}(y = \pm 1 \mid \mathbf{x}, \mathbf{w}) = \frac{1}{1 + e^{-y(\mathbf{w}^T \mathbf{x} + w_0)}},$$

that corresponds to an activation function on the predictive model known as a logistic function with trajectory on $[0, 1]$. The classification model could be estimated minimizing the negative log-likelihood function (loss function) as follows:

$$\min_{\mathbf{w}} \frac{1}{2} \mathbf{w}^T \mathbf{w} + C \sum_{i=1}^{l} \log(1 + e^{-y_i(\mathbf{w}^T x_i + w_0)}),$$

where $C > 0$ corresponds to the penalization parameter of the classification errors and the pairs $(\mathbf{x}_i, \mathbf{y}_i)$, $i = 1, \ldots, l$ correspond to the training dataset. The factor $\frac{1}{2}\mathbf{w}^T\mathbf{w}$ is known as the regularization factor. It allows the classification model to have good generalization properties.

The logistic regression model can be successfully applied to many text categorization problems due to the fact that it is scalable on high dimensional data. In the compact cache problem domain we will consider the probability $\mathbf{P}(y_m = \pm 1 \mid \mathbf{x}_q, \mathbf{w})$ where $\mathbf{x}_q$ represents the features of the new query q and y_m represents the event *to process the query q in the m machine*. In the following section we will show how we can use the logistic regression model for the multi-class multi-label case.

3.3   Multi-class Multi-label Logistic Regression

One form of extending the logistic regression model for multi-class problems is to use the method one-vs.-rest (OVR). This method consists of developing a binary classifier for each class that allows the objective class to be separated from the rest of the classes. Later we provide a probability $\mathbf{P}(y_m = \pm 1 \mid \mathbf{x}_q, \mathbf{w})$ for each category.

In the compact cache domain problem using OVR approximation for each new query, the predictive model gives a list of probabilities, one for each machine. This is done by modelling the new query in the feature space defined by the LCache (the terms that constitute the vocabulary for all the queries stored in the LCache) and calculating the logistic activation function for each machine in the distributed system. For each machine, we have a predictive model that allows us to indicate how promissory the machine is for the new query. The classification model then determines a list of probabilities associated with each machine, which indicates the order in which the machines must be visited in order to retrieve relevant results.

One of the strong points of the OVR method is the fact that the multi-class methods existing in literature today are generally more costly and complex to implement, leading to a slow training phase. The OVR methods have shown comparable precision to the multi-class methods with faster training times.

Since a binary classification method will be used but the training data is multi-label, we must utilize a data transformation method. Following Tsoumakas *et al.* [20] each instance of training data $(\mathbf{x}_i, \mathbf{y})$, will be decomposed of z uni-label instances $(\mathbf{x}_i, \mathbf{y}_m)$, $m = 1, \ldots, z$, where z corresponds to the number of categories that the query q_i belongs to (the length of the list of machines that allows the top-N results for q_i to be obtained).

4   Experimental Results

4.1   Dataset and Document Distribution

The dataset we will use in this paper corresponds to a query log file of Yahoo! from the year 2009. The query log contains 2,109,198 distinct queries and

3,991,719 query instances. 1,629,113 distinct queries were considered as training data, leaving the 602,862 remaining queries for evaluation. The testing set considers 999,573 query instances in the query log trace. The training set considers 2,992,146 query instances. The vocabulary of the 3.9 million query instances is compound by 239,274 distinct query terms.

We initialize the LCache with the 150,000 most frequent queries in the training query log trace. The LCache contains 204,485 different terms. Out of the 602,862 distinct queries in the testing dataset, there are 453,354 queries that have at least one term in the set of terms of the LCache.

For each query considered in the query log data, we obtained the top-50 results retrieved using the Yahoo! Search BOSS API [21]. The generated document collection corresponds to 50,864,625 documents. The API uses the Yahoo! services to get the same answers that are presented to actual users by the production Yahoo! search engine.

The document collection was distributed using a query-document co-clustering algorithm. The co-clustering algorithm allows us to find clusters of related documents. We choose to use this clustering algorithm because we wanted to compare our results with those of PCAP - the strategy of collection selection that is considered as state-of-the-art in the area [19]. A **C++** implementation of the algorithm was used for the experiments. The source code of this algorithm is available in [18].

The document collection was distributed over two arrays of processors considering 16 and 128 machines. Each entry in the LCache stores the set of processors that have the top-20 results for the corresponding query.

4.2   Performance Evaluation

Performance measures. To evaluate the performance of our method we compare the results coming from the predictive model with the results coming from the Yahoo! Search Engine. This is an effective approach to evaluate the performance in our distributed search engine because the document collection was built using the Yahoo! Search BOSS API. Thus, the best case for our method is to predict the machines where the top-N results for the testing queries are.

Following Puppin *et al.* [19], we will measure the effectiveness of our approach using the *Intersection* measure defined as follows:

Intersection: Let's call G_q^N the top-N results returned for a query q by the Yahoo! Search Engine, and H_q^N the top-N results returned for q by our method. The intersection at N, $INTER_N(q)$, is the fraction of results retrieved by the the proposed method that appear among the top-N documents in the Yahoo! Search Engine BOSS API:

$$INTER_N(q) = \frac{|H_q^N \cap G_q^N|}{|G_q^N|}.$$

The $INTER_N$ measure for a testing query set is obtained as the average of the $INTER_N(q)$ measured for each testing query.

Implementation of the methods. For the evaluation we will consider the state-of-the-art approximated methods PCAP and SEMCACHE, described in Sections 2 and 3, respectively. We tested the correctness of our PCAP implementation by using the query training set over the array of 16 processors. In this setting, PCAP obtained a $INTER_{20}$ measure of 31.40%, 43.25%, 59.19%, and 77.68%, when we visit 1, 2, 4, and 8 processors, respectively. These $INTER_{20}$ values are very similar to those obtained by Puppin *et al.* [19] in their experiments. Regarding the SEMCACHE method, we use an Idf threshold equals to 10, as it is recommended in [9].

Our method (LOGRES abridged) was implemented using a trust region Newton method that has been proved on high dimensional data [14]. The method is based on a sequential dual method suitable for large scale multi-class problems [12] that reduces the time involved in the training process. The logistic regression implementation used in these experiments is available on [8].

For the training phase of LOGRES, we perform a 5-fold cross validation process to determine the value of the C parameter. We choose the value of C measuring the accuracy of LOGRES over an array of 16 and 128 processors. Results are shown in Table 1.

Table 1. LOGRES 5-folds cross-validation training accuracy for differents penalty factors (percentages). Bold fonts indicate best performance values.

Penalty factor	16 Machines	128 Machines
C = 0.00001	15.2067	7.9573
C = 0.0001	15.2067	7.9573
C = 0.001	15.2067	7.9573
C = 0.01	**15.2067**	**7.9573**
C = 0.1	14.9742	7.9411
C = 1	10.0769	6.5423
C = 10	9.7191	5.2433

Results. We use the 602,862 distinct queries reserved for testing to evaluate the *INTER* measure. As it is shown in Table 1, SEMCACHE approximates a fraction of these queries (the 39.75% of the testing set match a semantic overlap case). On the other hand, PCAP approximates the 87.56% of the testing set (the remaining testing queries do not match terms with the query-document clusters). LOGRES approximates to the 75.27% of the testing set (we need at least one term of the query shared with the LCache feature space to build the query vector representation). Notice that for the SEMCACHE and LOGRES methods we are not considering hits in the LCache (app 4.6% of the testing dataset) because we are evaluating only the predictive capacity of each method.

As Table 1 shows, the coverage over the testing set differs depending on the method evaluated. To perform a fair comparison between these methods, we calculate the *INTER* measure over the whole query testing dataset, sending the queries not covered by the methods to all processors in a random order.

Table 2. Testing set coverage for the methods

	Number of Queries	Percentage
PCAP	527,918	87.56
SEMCACHE	239,682	39.75
LOGRES	453,354	75.27

For the 16 processors array, we calculate the *INTER* measure for 1, 2, 4 and 8 processors, at N = 5, 10, and 20. For the 128 processors array, we calculate the measure for 8, 16, 32 and 64 processors, at N = 5, 10, and 20. As a baseline we consider a random order generator (RANDOM abridged). Table 2 shows the results obtained for 16 and 128 processors.

Table 3. Performance evaluation of the methods over a distributed search engine of 16 and 128 processors

$INTER_5\%$

	1	2	4	8	16
RANDOM	6.30	12.55	25.08	49.98	100
PCAP	22.38	33.12	48.31	69.61	100
SEMCACHE	10.26	17.49	30.54	54.50	100
LOGRES	13.29	20.16	32.46	55.09	100

$INTER_5\%$

	8	16	32	64	128
RANDOM	6.21	12.51	24.81	49.81	100
PCAP	16.48	23.58	36.64	64.03	100
SEMCACHE	11.68	19.17	32.55	57.66	100
LOGRES	16.81	24.46	37.72	65.88	100

$INTER_{10}\%$

	1	2	4	8	16
RANDOM	6.29	12.55	25.08	49.99	100
PCAP	22.85	33.82	49.05	70.01	100
SEMCACHE	10.27	17.46	30.48	54.43	100
LOGRES	12.72	19.31	31.32	54.06	100

$INTER_{10}\%$

	8	16	32	64	128
RANDOM	6.20	12.51	24.85	49.88	100
PCAP	14.56	21.36	34.34	62.31	100
SEMCACHE	11.24	18.65	31.97	57.24	100
LOGRES	15.11	22.27	34.82	63.90	100

$INTER_{20}\%$

	1	2	4	8	16
RANDOM	6.28	12.53	25.05	49.99	100
PCAP	23.77	35.11	50.51	71.01	100
SEMCACHE	10.44	17.66	30.67	54.59	100
LOGRES	12.21	18.58	30.29	53.10	100

$INTER_{20}\%$

	8	16	32	64	128
RANDOM	6.19	12.48	24.84	49.89	100
PCAP	13.21	19.83	32.81	61.06	100
SEMCACHE	11.01	18.37	31.65	56.97	100
LOGRES	13.56	20.16	31.94	61.90	100

PCAP outperforms LOGRES in the 16 processors array by approximately 10% when we visit the first and second machine. This difference increases when we visit 4 and 8 machines. When we tested the methods over an array of 128 processors, we can see that PCAP degrades by a considerable margin. PCAP degrades its performance by approximately 6% when we visit the first 8 processors. This difference increases when we visit more machines. On the other hand, LOGRES improves its performance by approximately 3% when we visit the first 8 and 16 processors. This improvement increases when we visit more processors.

These results show that LOGRES scales well with regards to the number of processors.

Notice that in this comparison we are using a very small LCache and we are not considering hits in the LCache. The comparison was performed by only considering the predictive capacity of each method, but in the cases of SEMCACHE and LOGRES, this means that we discard 27,934 testing queries that made hit in the LCache (42,833 query instances in the query log trace).

5   Conclusions

We have presented the compact cache problem by dealing with it from the machine learning perspective. We have shown that each entry in an LCache can be used as a training instance for a standard machine learning technique in which we use the query terms as features and each machine is represented by a category. The problem corresponds to a multi-class problem (m machines are considered, $m > 2$, in which the documents in the collection are distributed) and multi-label (given a query q, the top-N responses for q can be distributed in several machines).

As more entries are added to the compact cache, the size of its vocabulary, (i.e., the number n of different terms needed to formulate all queries stored in the cache) also grows, which in turn, increases the dimensionality of the feature space of the corresponding learning machine. On the contrary, for the problem studied in this work, the number of machines m among which the document collection is partitioned, keeps constant. Furthermore, the number of different terms stored in the cache is usually of the order of 10^5 while the number of machines is of much smaller order, usually 10^2. The literature shows that for this problem setting linear classifiers offer good results. This is due to the high dimensionality of the data allowing linear separability which is in general the case of the text. In this scenario, the state of the art shows that there are several linear classifiers that could be useful for learning cache. Among these we find the linear Support Vector Machine (linear SVMs) and the recently presented Logistic Regression. We decide to target the problem by using logistic regression, because logistic activation function allows a probability to be estimated for each machine. This favors the definition of an order in which visiting the machines is recommended, unlike the rest of the linear classification models which frequently use the sign function for the activation.

An advantage of the LOGRES method in regard to the other evaluated methods is the small amount of space that it occupies. Strictly speaking, LOGRES could even do without the LCache, storing only the parameters of the predictive model for each machine like we showed in the experiment section. This obtained a predictive capacity comparable to the capacity of PCAP. When the LCache is added to the broker, the performance of the method would be even better because a large proportion of queries would hit the LCache. Given that the vector representation of each query is sparse, the method involves low computational costs.

References

1. Amiri, K., Park, S., Tewari, R., Padmanabhan, S.: Scalable template-based query containment checking for web semantic caches. In: ICDE (2003)
2. Baeza-Yates, R., Gionis, A., Junqueira, F., Murdock, V., Plachouras, V., Silvestri, F.: Design trade-offs for search engine caching. ACM TWEB 2(4) (2008)
3. Chidlovskii, B., Roncancio, C., Schneider, M.: Semantic Cache Mechanism for Heterogeneous Web Querying. Computer Networks 31(11-16), 1347–1360 (1999)
4. Chidlovskii, B., Borghoff, U.: Semantic Caching of Web Queries. VLDB Journal 9(1), 2–17 (2000)
5. Dhillon, I., Mallela, S., Modha, D.: Information-theoretic co-clustering. In: KDD (2003)
6. Fagni, T., Perego, R., Silvestri, F., Orlando, S.: Boosting the performance of Web search engines: Caching and prefetching query results by exploiting historical usage data. ACM TOIS 24(1), 51–78 (2006)
7. Falchi, F., Lucchese, C., Orlando, S., Perego, R., Rabitti, F.: A Metric Cache for Similarity Search. In: LSDS-IR (2008)
8. Fan, R., Chang, K., Hsieh, C., Wang, X., Lin, C.: LIBLINEAR: A library for large linear classification. Journal of Machine Learning Research 9, 1871–1874 (2008)
9. Ferrarotti, F., Marin, M., Mendoza, M.: A Last-Resort Semantic Cache for Web Queries. In: SPIRE (2009)
10. Gan, Q., Suel, T.: Improved Techniques for Result Caching in Web Search Engines. In: WWW (2009)
11. Godfrey, P., Gryz, J.: Answering Queries by Semantic Caches. In: Bench-Capon, T.J.M., Soda, G., Tjoa, A.M. (eds.) DEXA 1999. LNCS, vol. 1677, pp. 485–498. Springer, Heidelberg (1999)
12. Keerthi, S., Sundararajan, S., Chang, K., Hsieh, C., Lin, C.: A sequential dual method for large scale multi-class linear SVMs. In: SIGKDD (2008)
13. Lempel, R., Moran, S.: Predictive caching and prefetching of query results in search engines. In: WWW (2003)
14. Lin, C., Weng, R., Keerthi, S.: Trust region Newton method for large-scale logistic regression. Journal of Machine Learning Research 9, 627–650 (2008)
15. Long, X., Suel, T.: Three-level caching for efficient query processing in large Web search engines. In: WWW (2005)
16. Marin, M., Ferrarotti, F., Mendoza, M., Gomez, C., Gil-Costa, V.: Location Cache for Web Queries. In: CIKM (2009)
17. Markatos, E.: On caching search engine query results. Computer Communications 24(7), 137–143 (2000)
18. Puppin, D., Silvestri, F.: C++ implementation of the co-cluster algorithm by Dhillon, Mallela, and Modha, http://hpc.isti.cnr.it
19. Puppin, D., Silvestri, F., Perego, R., Baeza-Yates, R.: Load-balancing and caching for collection selection architectures. In: INFOSCALE (2007)
20. Tsoumakas, G., Katakis, I.: Multi-label Classification: An Overview. International Journal of Data Warehousing and Mining 3(3), 1–13 (2007)
21. Yahoo! Search BOSS API (2009), http://developer.yahoo.com/search/boss/
22. Yan, H., Ding, S., Suel, T.: Inverted index compression and query processing with optimized document ordering. In: WWW (2009)

Text Clustering for Peer-to-Peer Networks with Probabilistic Guarantees

Odysseas Papapetrou[1], Wolf Siberski[1], and Norbert Fuhr[2]

[1] L3S Research Center
{papapetrou,siberski}@l3s.de
[2] Universität Duisburg-Essen
norbert.fuhr@uni-due.de

Abstract. Text clustering is an established technique for improving quality in information retrieval, for both centralized and distributed environments. However, for highly distributed environments, such as peer-to-peer networks, current clustering algorithms fail to scale. Our algorithm for peer-to-peer clustering achieves high scalability by using a probabilistic approach for assigning documents to clusters. It enables a peer to compare each of its documents only with very few selected clusters, without significant loss of clustering quality. The algorithm offers probabilistic guarantees for the correctness of each document assignment to a cluster. Extensive experimental evaluation with up to 100000 peers and 1 million documents demonstrates the scalability and effectiveness of the algorithm.

1 Introduction

Text clustering is widely employed for automatically structuring large document collections and enabling cluster-based information browsing, which alleviates the problem of information overflow. It is especially useful on large-scale distributed environments such as distributed digital libraries [8] and peer-to-peer (P2P) information management systems [2], since these environments operate on significantly larger document collections. Existing P2P systems also employ text clustering to enhance information retrieval efficiency and effectiveness [11,17]. However, these systems do not address the problem of efficient *distributed* clustering computation; they assume that clustering is performed on a dedicated node, and are therefore not scalable. Existing distributed and P2P clustering approaches, such as [3,5,7,9], are also limited to a small number of nodes, or to low dimensional data. Hence, a distributed clustering approach that scales to large networks and large text collections is required.

In this work we focus on text clustering for large P2P networks. We are particularly interested in systems in which the content distribution is imposed by the nature of the system, e.g., P2P desktop sharing systems [2]. We require a P2P clustering algorithm which can cluster such distributed collections effectively and efficiently, without overloading any of the participating peers, and without requiring central coordination.

A key factor to reduce network traffic for distributed clustering in these systems is to reduce the number of required comparisons between documents

C. Gurrin et al. (Eds.): ECIR 2010, LNCS 5993, pp. 293–305, 2010.
© Springer-Verlag Berlin Heidelberg 2010

and clusters. Our approach achieves this by applying probabilistic pruning: Instead of considering all clusters for comparison with each document, only a few most relevant ones are taken into account. We apply this core idea to K-Means, one of the frequently used text clustering algorithms. The proposed algorithm, called Probabilistic Clustering for P2P (PCP2P), reduces the number of required comparisons by an order of magnitude, with negligible influence on clustering quality.

In the following section, we present current distributed and P2P clustering algorithms, and explain why they are not applicable for P2P text clustering. In Section 3 we introduce PCP2P, an efficient and effective P2P text clustering algorithm. We present the probabilistic analysis for PCP2P in Section 4, and show how it is parameterized to achieve a desired correctness probability. In Section 5 we verify scalability and quality of PCP2P in large experimental setups, with up to 100000 peers and 1 million documents, using real and synthetic data.

2   Related Work

Distributed Hash Tables. PCP2P relies on a Distributed Hash Table (DHT) infrastructure. DHTs provide efficient hash table capabilities in a P2P environment by arranging peers in the network according to a specific graph structure, usually a hypercube. DHTs offer the same functionality as their standard hash table counterparts, a `put(key,value)` method, which associates `value` with `key`, and a `get(key)` method, which returns the value associated with `key`. Our implementation uses Chord [16], but any other DHT could be used as well.

Distributed Clustering. Several algorithms for parallelizing K-Means have been proposed, e.g., [4,6]. These algorithms focus on harnessing the power of multiple nodes to speed up the clustering of large datasets. They assume a controlled network or a shared memory architecture, and therefore are not applicable for P2P, where these assumptions do not apply.

Eisenhardt et al. [5] proposed one of the first P2P clustering algorithms. The algorithm distributes K-Means computation by broadcasting the centroid information to all peers. Due to this centroid broadcasting, it cannot scale to large networks. Hsiao and King [9] avoid broadcasting by employing a DHT. Clusters are indexed in the DHT using manually selected terms. This requires extensive human interaction, and the algorithm cannot adapt to new topics.

Datta et al. [3] proposed LSP2P and USP2P, two P2P approximations of K-Means. LSP2P uses gossiping to distribute the centroids. In an evaluation with 10-dimensional data, LSP2P achieves an average misclassification error of less than 3%. However, as we show in Section 5, LSP2P fails for text clustering, because it is based on the assumption that data is uniformly distributed among the peers, i.e., each peer has at least some documents from each cluster. This assumption clearly does not hold for text collections in P2P networks. The second algorithm, USP2P, uses sampling to provide probabilistic guarantees. However, the probabilistic guarantees are also based on the assumption that data is uniformly distributed among the peers. Also, USP2P requires a coordinating peer

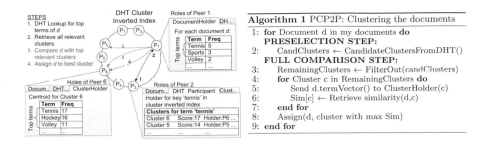

Fig. 1. PCP2P: a. System architecture, b. Algorithm for document assignment

which gets easily overloaded, since it is responsible for exchanging centroids with a significant number of peers, for sampling, e.g., 500 peers out of 5500 peers.

Hammouda et al. [7] use a hierarchical topology for the coordination of K-Means computation. Clustering starts at the lowest level, and the local solutions are aggregated until the root peer of the hierarchy is reached. This algorithm has the disadvantage that clustering quality decreases noticeably for each aggregation level, because of the random grouping of peers at each level. Therefore, quality decreases significantly for large networks. The authors report a quality of less than 20% of the quality of K-Means already for 65 nodes.

3   PCP2P: Probabilistic Clustering for P2P

In PCP2P, a peer has up to three different roles (Fig. 1.a.). First, it serves as document holder, i.e., it keeps its own document collection, and it assigns its documents to clusters. Second, it participates in the underlying DHT by holding part of the distributed index. Third, a peer can be a *cluster holder*, i.e., maintain the centroid and document assignments for one cluster.

PCP2P consists of two parallel activities, *cluster indexing* and *document assignment*. Both activities are repeated periodically to compensate churn, and to maintain an up-to-date clustering solution. *Cluster indexing* is performed by the cluster holders. In regular intervals, these peers create compact cluster summaries and index them in the underlying DHT, using the most frequent cluster terms as keys. We describe this activity in Section 3.1. The second activity, *document assignment*, consists of two steps, preselection and full comparison. In the *preselection step*, the peer holding d retrieves selected cluster summaries from the DHT index, to identify the most relevant clusters (Fig. 1.b, Line 2). Preselection already filters out most of the clusters. In the *full comparison step*, the peer computes similarity score estimates for d using the retrieved cluster summaries. Clusters with low similarity estimates are filtered out (Line 3, see Section 3.2 for details), and the document is sent to the few remaining cluster holders for full similarity computation (Lines 4-7). Finally, d is assigned to the cluster with highest similarity (Line 8). This two-stage filtering algorithm reduces drastically the number of full comparisons (usually less than five comparisons per document, independent of the number of clusters). At the same time, it provides probabilistic guarantees that the resulting clustering solution exhibits nearly the same quality as centralized clustering (Section 4).

3.1 Indexing of Cluster Summaries

Cluster holders are responsible for indexing summaries of the clusters in the DHT. Particularly, each cluster holder periodically recomputes its cluster centroid, using the documents assigned to the cluster at the time. It also recomputes a *cluster summary* and publishes it to the DHT index, using selected cluster terms as keys. As we explain later, this enables peers to identify relevant clusters for their documents efficiently. For this identification, it is sufficient to consider the most frequent terms of a cluster c as keys, i.e., all terms t with $TF(t, c) \geq CluTF_{min}(c)$, where $CluTF_{min}(c)$ denotes the frequency threshold for c. We use $TopTerms(c)$ to denote the set of these terms. Note that $TopTerms(c)$ does not include stopwords; these are already removed when building the document vectors. For the rest of this section we assume that $CluTF_{min}(c)$ is given. Section 4 shows how a value for this threshold can be derived that satisfies the desired probabilistic guarantees.

The cluster summary includes (1) all cluster terms in $TopTerms(c)$ and their corresponding TF values, (2) $CluTF_{min}(c)$, and (3) the sum of all term frequencies (the L1 norm), cluster length (the L2 norm), and dictionary size.

Load Balancing. To avoid overloading, each cluster holder selects random peers to serve as *helper cluster holders*, and replicates the cluster centroid to them. Their IP addresses are also included in the cluster summaries, so that peers can randomly choose a replica without going through the cluster holder. Communication between the master and helper cluster holders only occurs for updating the centroids, by exchanging the respective local centroids as in [4]. Since only one centroid needs to be transferred per helper cluster holder and only a small number of peers is involved, load balancing does not affect scalability.

3.2 Document Assignment to Clusters

Each peer is responsible of clustering its documents periodically. Clustering of a document consists of two steps: (a) the preselection step, where the most promising clusters for the document are detected, and, (b) the full comparison step, where the document is fully compared with the most promising clusters and assigned to the best one.

Preselection step. Consider a peer p which wants to cluster a document d. Let $TopTerms(d)$ denote all terms in d with $TF(t, d) \geq DocTF_{min}(d)$, where $DocTF_{min}(d)$ denotes a frequency threshold for d (we explain how $DocTF_{min}$ is derived in Section 4). For each term t in $TopTerms(d)$, peer p performs a DHT lookup and finds the peer that holds the cluster summaries for t (Fig. 1.a, Step 1). It then contacts that peer directly to retrieve all summaries published using t as a key (Step 2). To avoid duplicate retrieval of summaries, p executes these requests sequentially, and includes in each request the cluster ids of all summaries already retrieved. We refer to the list of all retrieved summaries as the *preselection list*, denoted with $\mathcal{C}_{pre}$. The summary of the optimal cluster for d is included in $\mathcal{C}_{pre}$ with high probability, as shown in Section 4.

Full comparison step. After constructing $\mathcal{C}_{pre}$, peer p progressively filters out the clusters not appropriate for the document at hand, using one of two alternative filtering strategies, as follows. Using the retrieved cluster summaries, p estimates the cosine similarities for all clusters in $\mathcal{C}_{pre}$. For the cluster with the highest similarity estimate, p sends the compressed term vector of d to the respective cluster holder for a full cosine similarity, and retrieves the similarity score (Fig. 1.a, Step 3). Based on this score and the employed filtering strategy, more clusters are filtered out. The process is repeated until $\mathcal{C}_{pre}$ is empty. Finally, p assigns d to the cluster with the highest similarity score, and notifies the respective cluster holder (Step 4).

Filtering Strategies. We propose two different strategies to filter out clusters from $\mathcal{C}_{pre}$, (a) *conservative*, and (b) *Zipf-based filtering*. Both strategies employ the information contained in the cluster summaries to estimate the cosine similarity between the document and each candidate cluster. Let $t_1, t_2, \ldots, t_n$ denote the terms of d sorted *descending* by their frequency, i.e., $TF(t_i, d) \geq TF(t_j, d)$ for all $j > i$. Cosine similarity is estimated as follows.

$$ECos(d, c) = \sum_{i=1}^{n} \frac{TF(t_i, d) \times f(t_i, c)}{|d| \times |c|} \tag{1}$$

$|d|$ and $|c|$ denote the L2-Norm of the document and cluster. The function $f(t_i, c)$ denotes an estimation for $TF(t_i, c)$, which is specific to the filtering strategy. We will address this function in detail in the next paragraphs.

Having estimated the cosine similarities for all clusters in $\mathcal{C}_{pre}$, PCP2P proceeds as follows. Let $c_{\overline{max}}$ denote the cluster in $\mathcal{C}_{pre}$ with the maximum estimated similarity. Peer p removes $c_{\overline{max}}$ from the list and sends the compressed term vector of d to the cluster holder of $c_{\overline{max}}$, for cosine comparison. After retrieving the real cosine value $Cos(d, c_{\overline{max}})$, it removes from $\mathcal{C}_{pre}$ all clusters c with $ECos(d, c) < Cos(d, c_{\overline{max}})$. This process is repeated, until $\mathcal{C}_{pre}$ is empty. Finally, d is assigned to the cluster with the highest cosine similarity.

The key distinction between the two filtering strategies is the way they compute $ECos(d, c)$, and, in particular, their definition of $f(t_i, c)$.

Conservative filtering. The conservative strategy computes an upper bound for cosine similarity. For the terms included in the cluster's summary, conservative strategy uses the actual cluster frequency, included in the summary. For all other terms, it progressively computes an upper bound for the term frequency in the cluster. Formally, $f(t_i, c)$ is defined as follows.

$$f(t_i, c) = \begin{cases} TF(t_i, c) & \text{if } t_i \in TopTerms(c) \\ min(CluTF_{min}(c) - 1, SC - ST - SE) & \text{otherwise} \end{cases}$$

where SC is the sum of cluster frequencies for all terms included in the cluster, and ST is the sum of cluster frequencies for all terms included in $TopTerms(c)$. SE holds the sum of all term frequencies estimated up to now by the algorithm for c. By definition, the conservative strategy never underestimates the cosine similarity value. Therefore, this strategy always detects the best cluster.

Zipf-based filtering. A more accurate similarity estimation can be derived based on the assumption that term frequencies in the cluster follow a Zipf distribution. Recall that document terms $t_1, t_2, \ldots, t_n$ are ordered descending on their frequency. We use this order to estimate the rank of terms not included in the cluster summary. Zipf-based filtering defines $f(t_i, c)$ as follows:

$$f(t_i, c) = \begin{cases} TF(t_i, c) & \text{if } t_i \in TopTerms(c) \\ min(SC/(r^s \times \sum_{k=1}^{DT} 1/k^s), SC - ST - SE) & \text{otherwise} \end{cases}$$

SC, ST, and SE are defined as in the conservative strategy. DT denotes the number of distinct terms in c, and with r we represent the estimated rank for the missing term. $SC/(r^s \times \sum_{k=1}^{DT} 1/k^s)$ gives the expected term frequency of t_i in c, assuming that term frequencies follow a Zipf distribution with exponent s. Ranks of missing terms are estimated as follows: the i-th document term that is not included in the $TopTerms(c)$ is assumed to exist in the cluster centroid, with rank $r = |TopTerms(c)| + i$.

3.3   Cost Analysis

We express cost in number of messages and transfer volume. For a cluster c, the cost of indexing the cluster summary (both in number of messages and transfer volume) is $Cost_{ind} = O(|TopTerms(c)| \times \log(n))$, where n is the number of peers. The cost of the preselection step for each document d is $Cost_{pre} = O(|TopTerms(d)| \times \log(n))$. The full comparison step incurs a cost of $Cost_{fcs} = O(|\mathcal{C}_{fcs}|)$, where $\mathcal{C}_{fcs}$ denotes the set of clusters fully compared with d.

The dominating cost is the one incurred for assigning documents to clusters, namely $Cost_{pre} + Cost_{fcs}$. Per document, this cost has the following properties: (a) it grows logarithmically with the number of peers, because DHT access cost grows logarithmically, and (b) it is independent of the size of the document collection. It also depends on $|\mathcal{C}_{fcs}|$. Our experimental evaluation (Section 5) shows that $|\mathcal{C}_{fcs}|$ is on average very small, and independent of the total number of clusters k. This means that PCP2P scales to networks of large sizes, and with large numbers of documents and clusters.

4   Probabilistic Analysis

In the previous section, we assumed that the optimal values for $CluTF_{min}(c)$ and $DocTF_{min}(d)$ are given. We now describe how PCP2P computes these values dynamically for each cluster and document to satisfy the desired clustering quality requirements. Due to space limitations, we only provide the final results of the analysis here. The reader can find the full proofs in [14].

Our analysis uses a probabilistic document generation model [13,15]. Briefly, the model assumes that each document belongs to a topic $\mathcal{T}$, and each topic $\mathcal{T}_i$ is described by a term probability distribution ϕ_i (a language model). A document of length l that belongs to $\mathcal{T}_i$ is created by randomly selecting l terms with replacement from ϕ_i. The probability of selecting a term t is given by ϕ_i.

Notations. $Pr_{correct}$ denotes the desired correctness probability, i.e., the probability of each document to be assigned to the correct cluster. We use $\mathcal{C}_{sol} := \{c_1, \ldots, c_k\}$ to denote a snapshot of clusters on an ongoing clustering. Each cluster $c_i \in \mathcal{C}_{sol}$ follows the language model ϕ_i. We use $t_1[\phi_i], \ldots, t_n[\phi_i]$ to denote the terms of ϕ_i sorted by descending probabilities. Also, $TopDistr(\alpha, \phi_i)$ denotes the set of α terms with highest probability in ϕ_i, i.e., $t_1[\phi_i], \ldots, t_\alpha[\phi_i]$.

For the case of conservative filtering, a document d is assigned correctly to its optimal cluster, denoted with c_{opt}, if c_{opt} is detected in the preselection step and included in $\mathcal{C}_{pre}$. The probability that c_{opt} is included in $\mathcal{C}_{pre}$ is denoted with Pr_{pre}. Clearly, by setting $Pr_{pre} = Pr_{correct}$ we satisfy the desired correctness probability. The purpose of the analysis is to find the values of $CluTF_{min}$ and $DocTF_{min}$ that satisfy Pr_{pre}.

PCP2P computes these values automatically, as follows. c_{opt} is retrieved in the preselection step if there exists at least one term $t \in TopDistr(\alpha, \phi_i)$ with $TF(t, c_{opt}) \geq CluTF_{min}$, and at the same time $TF(t, d) \geq DocTF_{min}$. Let Pr_{find} denote the probability that each of the terms from $TopDistr(\alpha, \phi_i)$ has a frequency in d of at least $DocTF_{min}$. With Pr_{ind} we denote the probability that each term from $TopDistr(\alpha, \phi_i)$ has a frequency in c_{opt} of at least $CluTF_{min}$. With Eqn. 4, peers compute the proper values for Pr_{find} and Pr_{ind}, such that clustering succeeds with probability Pr_{pre}. Given these probabilities, each peer computes the proper values for $DocTF_{min}$ and $CluTF_{min}$ per document and cluster respectively, according to Eqns. 2 and 3.

The probabilistic guarantees are based on the following theorem.

Theorem 1. *Given a document d which follows language model ϕ_i. The expected frequency of term t in d according to ϕ_i is denoted with $\hat{TF}(t, d)$. For any term t with $\hat{TF}(t, d) > DocTF_{min}$, the probability of the actual term frequency $TF(t, d)$ exceeding $DocTF_{min}$ is at least $1 - \exp(-\hat{TF}(t, d) \times (1 - DocTF_{min}/\hat{TF}(t, d))^2/2)$. Furthermore, with a probability Pr_{find} the frequency of term t in d is at least*

$$\hat{TF}(t, d) - \sqrt{2 \times \hat{TF}(t, d) \times \log\left(\frac{1}{1 - Pr_{find}}\right)}.$$

Sketch. The proof uses the lower-tail Chernoff bound ([12], p. 72) to compute a lower bound for the term frequency of a term t in d according to ϕ_i. In particular, we model the generation of d as independent Poisson trials, using the term probabilities of ϕ_i, as is standard in language generation models. Then, we apply Chernoff bounds to find $Pr[TF(t, c_i) < DocTF_{min}]$. Finally:

$$Pr[TF(t, d) \geq DocTF_{min}] = 1 - Pr[TF(t, c_i) < DocTF_{min}]$$
$$\geq 1 - \exp(-\hat{TF}(t, d) \times (1 - DocTF_{min}/\hat{TF}(t, d))^2/2) \quad (2)$$

The second part of the theorem is derived by solving Eqn. 2 for $DocTF_{min}$. □

With a similar theorem (see [14]), we can find a probabilistic lower bound for the frequency of a term t in c_i:

$$CluTF_{min} = \hat{TF}(t, c_i) - \sqrt{2 \times \hat{TF}(t, c_i) \times \log\left(\frac{1}{1 - Pr_{ind}}\right)} \tag{3}$$

To compute the expected term frequencies for $t_x[\phi_i]$, i.e., the x'th most frequent term in ϕ_i, peers use the Zipf distribution (validated, for example, in [1] for text).

Using Eqn. 3, the cluster holder of c_i selects $CluTF_{min}$ such that $Pr[TF(t_\alpha[\phi_i], c_i) \geq CluTF_{min}] \geq Pr_{ind}$. The probability that both d and c_{opt} have a common term with corresponding frequencies at least $DocTF_{min}$ and $CluTF_{min}$, is at least:

$$Pr_{pre} \geq 1 - \prod_{j=1}^{\alpha} (1 - Pr[TF(t_j, d) \geq DocTF_{min}] \times Pr_{ind}) \tag{4}$$

Using Eqn. 4, peers set the value of $DocTF_{min}$ per document, s.t. $Pr_{pre} \geq Pr_{correct}$.

Algorithm Configuration. As in standard K-Means, the number of clusters k can be freely chosen. In addition, PCP2P allows to set the desired correctness probability $Pr_{correct}$. All other parameters for satisfying the required probabilistic guarantees are derived using the results of our analysis.

First, a few sampled documents are collected from the network and are used to estimate the Zipf distribution skew. The algorithm then computes the remaining parameters. By default, α is set to 5, and Pr_{ind} and Pr_{pre} are set to $Pr_{correct}$. As shown earlier, these probability values satisfy the desired correctness probability for conservative filtering. Each peer then computes the proper $DocTF_{min}$ and $CluTF_{min}$ per document and cluster which satisfy these probabilities. With respect to Zipf-based filtering, we offer probabilistic guarantees only for the preselection step. The above values also satisfy the probabilistic guarantees for the preselection step of Zipf-based filtering.

5   Experimental Evaluation

The purpose of the experiments was to evaluate PCP2P with respect to effectiveness, efficiency, and scalability. Effectiveness was evaluated based on a human-generated classification of documents, using the two standard quality measures, entropy and purity. Additionally, we measured how well PCP2P approximates K-Means. With respect to efficiency, we measured number of messages, transfer volume, and document-cluster comparisons, for different correctness probabilities and collection characteristics. Finally, we examined scalability of PCP2P by varying the network and collection size, and the number of clusters. In the following we report average results after 10 repetitions of each experiment. Unless otherwise mentioned, we report results for 50 clusters.

As a real-world dataset, we have used the REUTERS Corpus Volume 1 (RCV1) [10]. We chose RCV1 because it is the largest collection with a classification of articles, which is necessary for evaluation of clustering. To be able

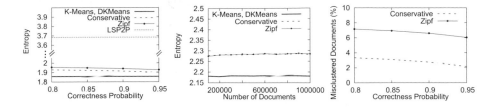

Fig. 2. Quality: a. Entropy, b. Entropy for different dataset sizes, c. Approximation

to apply standard quality measures, we restricted RCV1 to all articles which belonged to exactly one class (approx. 140000 articles). To systematically examine the effect of the collection's characteristics on the algorithm, and to evaluate it with a significantly larger dataset, we also used synthetic document collections (SYNTH) with a size of 1.4 million documents each. These collections were created according to the well-accepted Probabilistic Topic Model [15] from 200 composite language models, with different term distribution skews.

We compared PCP2P with two other P2P clustering algorithms: (a) LSP2P [3], the state-of-the-art for P2P clustering, and, (b) DKMeans, a P2P implementation of K-Means. DKMeans works like PCP2P, but without preselection and filtering, i.e., each compressed document vector is sent to all cluster holders for comparison purposes before assigned to a cluster. DKMeans was included in our experiments because it accurately simulates K-Means, i.e., it produces exactly the same results in a distributed fashion.

5.1 Evaluation Results

Clustering Quality. Figure 2.a plots entropy of all algorithms for the RCV1 dataset (lower entropy denotes better clustering quality). The X-axis is only relevant to the two PCP2P approaches, but we also include K-Means and LSP2P results for comparison. For clarity, the Y-axis is discontinuous. For the reported experiments we have used a network of 10000 peers, with 20% churn. Note that network size has no effect on the quality of PCP2P and DKMeans.

We see that both PCP2P filtering strategies achieve nearly the same clustering quality as K-Means. On the other hand, LSP2P, the current state-of-the-art in P2P clustering, converges to a significantly worse clustering solution, even for a moderate networks of 10000 peers. LSP2P fails because it is based on the assumption that each peer has at least some documents from each cluster, which is unsatisfiable with respect to text clustering[1]. Concerning the PCP2P variants, conservative filtering yields the best (lowest) entropy and the best (highest) purity, as expected. Zipf-based filtering shows comparable results. The same outcome is observed with respect to purity: K-Means has a purity of 0.645, and the maximum relative difference between PCP2P and K-Means is less than 1%.

---

[1] Note that we reproduced the good results of LSP2P on a smaller network with a 10-dimensional dataset, as reported in [3].

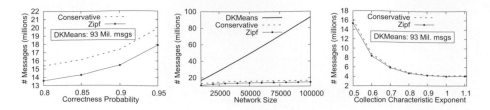

Fig. 3. Efficiency for varying: a. $Pr_{correct}$, b. Network size, c. Term Distribution Skew

Table 1. Cost and Quality of PCP2P and DKMeans for varying number of clusters

	Messages		Transfer vol.(Mb)		Comparisons		Entropy	
Clusters	25	100	25	100	25	100	25	100
DKMeans	4.66E7	1.86E8	2176	8706	25E5	100E5	2.03	1.61
Conservative	1.45E7	1.70E7	995	2490	568078	1704332	2.08	1.66
Zipf	1.34E7	1.36E7	603	1228	10673	31907	2.12	1.69

LSP2P has an average purity of 0.29, which shows insufficient clustering quality. Since LSP2P fails to produce a clustering solution of comparable quality with PCP2P and K-Means, we do not include it in our further experiments.

To evaluate the effect of document collection size on clustering quality, we repeated the experiment with the SYNTH collection, varying the number of documents between 100000 and 1 million. The SYNTH collection was generated as explained earlier, with a term distribution skew of 1.0 [18]. Figure 2.b plots entropy in correlation to the number of documents, using correctness probability $Pr_{correct} = 0.9$. We see that for PCP2P, entropy remains very close to the entropy achieved by K-Means, even for the largest collection with one million documents. The same applies for purity. This confirms that quality of PCP2P is not affected by collection size. We also see that for collections with high term distribution skews (1.0 in this experiment, compared to 0.55 for the RCV1 collection), Zipf-based filtering is nearly as effective as conservative filtering.

Approximation quality. In addition to the standard quality measures, we also counted the number of documents that PCP2P assigned to a different cluster than K-Means after each clustering round. Fig. 2.c plots the percentage of misclustered documents for different values of $Pr_{correct}$. As expected, conservative yields the best approximation, with less than 4% misclusterings even for $Pr_{correct} = 0.8$. Zipf-based PCP2P is also very accurate. We also see that the actual number of misclustered documents for PCP2P is always better than the probabilistic guarantees, since guarantees refer to upper bounds for number of errors.

Efficiency. We used PCP2P and DKMeans to cluster the RCV1 collection in a network of 100000 peers, with 20% churn. We did not include LSP2P here, since it fails with respect to quality, as already shown. Figure 3.a shows the number of messages required to perform one clustering iteration in correlation to $Pr_{correct}$. The plot also includes DKMeans cost as reference. We see that the two PCP2P variants generate significantly fewer messages than DKMeans. Also, as expected, Zipf-based filtering is more efficient than conservative filtering. For $Pr_{correct} < 0.9$, Zipf-based filtering requires an order of magnitude less messages than DKMeans, and conservative filtering requires less than 20% of the messages. The reason for the significantly better performance of PCP2P compared to DKMeans is that it reduces the number of document-cluster comparisons. Conservative filtering requires less than 20% of the comparisons of DKMeans in all setups, whereas Zipf requires less than 1%. Therefore, documents are sent over the network fewer times. Regarding transfer volume, conservative filtering requires less than 22% of the respective transfer volume of DKMeans, whereas Zipf-based filtering requires around 5%.

Scalability. We evaluated scalability of PCP2P with respect to network size, number of documents, and number of clusters. Figure 3.b shows the cost for different network sizes. The cost for PCP2P increases only logarithmically with network size, while cost for DKMeans increases linearly. This behavior is expected, because the only factor changing with network size for PCP2P is the DHT access cost, which grows logarithmically. The same behavior is observed with respect to transfer volume. We do not show quality measures, because they are independent of the network size. Our experimental results also confirmed that PCP2P scales linearly with collection size, as shown in Section 3.3. We repeated the clustering of RCV1 with 25 and 100 clusters, on a network of 100000 peers. Table 1 summarizes the results. Both PCP2P filtering strategies scale well with the number of clusters. In fact, network savings of PCP2P grow with the number of clusters, compared to DKMeans. Regarding entropy, the relative difference between PCP2P variants and DKMeans is stable. The same is observed regarding purity (not included in the table). Summarizing, PCP2P approximation quality is independent of the number of clusters, but PCP2P cost savings become even higher for a larger number of clusters.

Influence of Term Distribution Skew. PCP2P relies on the fact that term frequencies follow a Zipf distribution. Although it is accepted that document collections follow Zipf distribution, different document collections exhibit different distribution skews [1]. For example, the RCV1 collection used in our experiments has a skew of 0.55, while values reported in the literature for other text collections are around 1.0 [1,18]. To evaluate the influence of the skew on PCP2P, we used SYNTH collections generated with different Zipf skew factors, between 0.5 and 1.1. We do not present details with respect to quality, because the quality of PCP2P was always high and unaffected by the skew factor.

Figure 3.c displays the execution costs in number of messages, for a varying distribution skew, and for $Pr_{correct} = 0.9$. To achieve the same quality level, PCP2P cost is significantly lower for higher skews. This behavior is expected: with higher skews PCP2P needs to perform fewer document-cluster comparisons for satisfying the probabilistic guarantees. For commonly reported skew values (around 1.0), the number of messages is reduced by an order of magnitude. But even for a skew as low as 0.5, the cost of both PCP2P variants is significantly lower than the cost of DKMeans.

6 Conclusions

We presented PCP2P, the first scalable P2P text clustering algorithm. PCP2P achieves a clustering quality comparable to standard K-Means, while reducing communication costs by an order of magnitude. We provided a probabilistic analysis for the correctness of the algorithm, and showed how PCP2P automatically adapts to satisfy the required probabilistic guarantees. Extensive experimental evaluation with real and synthetic data confirm the efficiency, effectiveness and scalability of the algorithm, and its appropriateness for text collections with a wide range of characteristics.

Our future work focuses on applying the core idea of PCP2P, i.e., probabilistic filtering, to other clustering algorithms, both for distributed and centralized settings. Furthermore, we work towards a P2P IR method based on clustering, similar to [11,17], but now based on PCP2P, a truly distributed clustering infrastructure.

References

1. Blake, C.: A comparison of document, sentence, and term event spaces. In: ACL (2006)
2. Cudré-Mauroux, P., Agarwal, S., Aberer, K.: Gridvine: An infrastructure for peer information management. IEEE Internet Computing 11(5) (2007)
3. Datta, S., Giannella, C.R., Kargupta, H.: Approximate distributed K-Means clustering over a peer-to-peer network. IEEE TKDE 21(10), 1372–1388 (2009)
4. Dhillon, I.S., Modha, D.S.: A data-clustering algorithm on distributed memory multiprocessors. In: Workshop on Large-Scale Parallel KDD Systems (1999)
5. Eisenhardt, M., Müller, W., Henrich, A.: Classifying documents by distributed P2P clustering. In: INFORMATIK (2003)
6. Forman, G., Zhang, B.: Distributed data clustering can be efficient and exact. SIGKDD Explor. Newsl. 2(2), 34–38 (2000)
7. Hammouda, K., Kamel, M.: HP2PC: Scalable hierarchically-distributed peer to peer clustering. In: SDM (2007)
8. Haslhofer, B., Knezevié, P.: The BRICKS digital library infrastructure. In: Semantic Digital Libraries, pp. 151–161 (2009)
9. Hsiao, H.-C., King, C.-T.: Similarity discovery in structured P2P overlays. In: ICPP (2003)
10. Lewis, D.D., Yang, Y., Rose, T.G., Li, F.: RCV1: a new benchmark collection for text categorization research. J. Mach. Learn. Res. 5, 361–397 (2004)
11. Lu, J., Callan, J.: Content-based retrieval in hybrid peer-to-peer networks. In: CIKM 2003, pp. 199–206. ACM, New York (2003)
12. Motwani, R., Raghavan, P.: Randomized Algorithms. Cambridge University Press, Cambridge (1995)

13. Papadimitriou, C.H., Tamaki, H., Raghavan, P., Vempala, S.: Latent semantic indexing: a probabilistic analysis. In: PODS (1998)
14. Papapetrou, O., Siberski, W., Fuhr, N.: Text clustering for P2P networks with probabilistic guarantees. Extended version (2009),
    http://www.l3s.de/~papapetrou/publications/pcp2p-ecir-ext.pdf
15. Steyvers, M., Griffiths, T.: Handbook of Latent Semantic Analysis. In: Probabilistic Topic Models, pp. 427–448. Lawrence Erlbaum, Mahwah (2007)
16. Stoica, I., Morris, R., Karger, D., Kaashoek, F., Balakrishnan, H.: Chord: A scalable peer-to-peer lookup service for internet applications. In: SIGCOMM (2001)
17. Xu, J., Croft, W.B.: Cluster-based language models for distributed retrieval. In: SIGIR (1999)
18. Zipf, G.K.: Human Behavior and the Principle of Least-Effort. Addison-Wesley, Cambridge (1949)

XML Retrieval Using Pruned Element-Index Files

Ismail Sengor Altingovde, Duygu Atilgan, and Özgür Ulusoy

Department of Computer Engineering, Bilkent University, Ankara, Turkey
{ismaila,atilgan,oulusoy}@cs.bilkent.edu.tr

Abstract. An element-index is a crucial mechanism for supporting content-only (CO) queries over XML collections. A full element-index that indexes each element along with the content of its descendants involves a high redundancy and reduces query processing efficiency. A direct index, on the other hand, only indexes the content that is directly under each element and disregards the descendants. This results in a smaller index, but possibly in return to some reduction in system effectiveness. In this paper, we propose using static index pruning techniques for obtaining more compact index files that can still result in comparable retrieval performance to that of a full index. We also compare the retrieval performance of these pruning based approaches to some other strategies that make use of a direct element-index. Our experiments conducted along with the lines of INEX evaluation framework reveal that pruned index files yield comparable to or even better retrieval performance than the full index and direct index, for several tasks in the ad hoc track.

1 Introduction

Classical information retrieval (IR) is the quest for identifying the documents in a collection that are most relevant to a user's information need, usually expressed as a keyword query. Although there have been efforts for passage retrieval, most of the previous works in the IR field presume a document as the typical unit of retrieval. In contrary, XML documents, which started to emerge since late 90s, have a logical structure and may allow finer-grain retrieval, i.e., at the level of elements. Given that XML is used for the representation of lengthy documents, such as e-books, manuals, legal transcripts, etc., the "focused" retrieval is expected to provide further gains for the end users in locating the specific relevant information [16].

In the last decade, especially under the INitiative for the Evaluation of XML retrieval (INEX) [13] campaigns, several indexing, ranking and presentation strategies for XML collections have been proposed and evaluated. Given the freshness of this area, there exists a number of issues that are still under debate. One such fundamental problem is indexing the XML documents. As mentioned above, the focused retrieval aims to identify the most relevant parts of an XML document to a query, rather than retrieving the entire document. This requires constructing an index at a lower granularity, say, at the level of elements, which is not a trivial issue given the nested structure of XML documents.

In this paper, we essentially focus on the strategies for constructing space-efficient element-index files to support content-only (CO) queries. In the literature, the most

C. Gurrin et al. (Eds.): ECIR 2010, LNCS 5993, pp. 306–318, 2010.
© Springer-Verlag Berlin Heidelberg 2010

straight-forward element-indexing method considers each XML element as a separate document, which is formed of the text directly contained in it and the textual content of all of its descendants. We call this structure a *full* element-index. Clearly, this approach yields significant redundancy in terms of the index size, as elements in the XML documents are highly nested. To remedy this, a number of approaches, as reviewed in detail in the next section, are proposed in the literature. One such method is restricting the set of elements indexed, based on the size or type of the elements [18]. Such an approach may still involve redundancy for the elements that are selected to be indexed. Furthermore, determining what elements to index would be collection and scenario dependent. There are also other indexing strategies that can fully eliminate the redundancy. For instance, a *direct* element-index is constructed by only considering the text that is directly contained under an element (i.e., disregarding the content of the element's descendents). In the literature, propagation based mechanisms, in which the score of each element is propagated upwards in the XML structure, are coupled with the direct index for effective retrieval (e.g., [11]). In this case, the redundancy in the index is somewhat minimized, but query processing efficiency would be degraded.

For an IR system, it is crucial to optimize the underlying inverted index size for the efficiency purposes. A lossless method for reducing the index size is using compression methods (see [22] for an exhaustive survey). On the other hand, many lossy static index pruning methods are also proposed in the last decade. All of these methods aim to reduce the storage space for the index and, subsequently, query execution time, while keeping the quality of the search results unaffected. While it is straight-forward to apply index compression methods to (most of) the indexing methods proposed for XML, it is still unexplored how those pruning techniques serve for XML collections, and how they compare to the XML-specific indexing methods proposed in the literature.

In this paper, we propose to employ static index pruning techniques for XML indexing. We envision that these techniques may serve as a compromise between a full element-index and a direct element-index. In particular, we first model each element as the concatenation of the textual content in its descendants, as typical in a full index. Then, the redundancy in the index is eliminated by pruning this initial index. In this way, an element is allowed to contain some terms, say, the most important ones, belonging to its descendants; and this decision is given based on the full content of the element in an adaptive manner.

For the purposes of index pruning, we apply two major methods from the IR literature, namely, term-centric [5] and document-centric pruning [4] to prune the full element-index. We evaluate the performance for various retrieval tasks as described in the latest INEX campaigns. More specifically, we show that retrieval using pruned index files is comparable or even superior to that of the full index up to very high levels of pruning. Furthermore, we compare these pruning-based approaches to a retrieval strategy coupled with a direct index (as in [11]) and show that pruning-based approaches are also superior to that strategy. As another advantage, the pruning-based approaches are more flexible and can reduce an index to a required size.

In the next section, we first review a number of indexing strategies for XML collections as well as the associated retrieval techniques to support content-only queries. Next, we summarize some of the static index pruning strategies that are proposed for large-scale IR systems and search engines. In Section 3, we describe the pruning techniques that are adapted for reducing the size of a full element-index. Section 4 is devoted to the experimental evaluations. Finally, we conclude and point to future work directions in Section 5.

2 Related Work

2.1 Indexing Techniques for XML Retrieval

In the literature, several techniques are proposed for indexing the XML collections and for query processing on top of these indexes. In a recent study, Lalmas [16] provides an exhaustive survey of indexing techniques—essentially from the perspective of IR discipline—that we briefly summarize in the rest of this section.

The most straight-forward approach for XML indexing is creating a "full" element-index, in which each element is considered along with the content of its descendants. In this case, how to compute inverse document frequency (IDF), a major component used in many similarity metrics, is an open question. Basically, IDF can be computed across all elements, which also happens to be the approach taken in our work. As a more crucial problem [16], a full element-index is highly redundant because the terms are repeated for each nested element and the number of elements is typically far larger than the number of documents.

To cope with the latter problem, an indexing strategy can only consider the direct textual content of each element, so that redundancy due to nesting of the elements would be totally removed. In [9, 10], only leaf nodes are indexed, and the scores of the leaf elements are propagated upwards to contribute to the scores of the interior (ancestor) elements. In a follow-up work [11], the direct content of each element (either leaf or interior) is indexed, and again a similar propagation mechanism is employed. Another alternative is propagating the representations of elements, e.g., term statistics, instead of the scores. However, the propagation stage, which has to be executed during the query processing time, can degrade the overall system efficiency.

In the database field, where XML is essentially considered from a data-centric rather than a document-centric point of view, a number of labeling schemes are proposed especially to support structural queries (see [20] for a survey). In XRANK system [12], postings are again created only for the textual content directly under an element; however document identifiers are encoded using the Dewey ids so that the scores for the ancestor elements can also be computed without a propagation mechanism. This indexing strategy allows computing the same scores as a full index while the size of the index can be in the order of a direct index. However, this scheme may suffer from other problems, such as the excessive Dewey id length for very deeply located elements. We provide a further discussion of this strategy in Section 4.

An in-between approach to remedy the redundancy in a full element-index is indexing only certain elements of the documents in the collection. Element selection can be based upon several heuristics (see [16] for details). For instance, shorter elements (i.e., with only few terms) can be discarded. Another possibility is selecting elements based on their popularity of being assessed as relevant in INEX framework. The semantics of the elements can also be considered while deciding which elements to index by a system designer. Yet another indexing technique that is also related is distributed indexing, which proposes to create separate indexes for each element type, possibly selected with one of the heuristics discussed above. This latter technique may be especially useful for computing the term statistics in a specific manner to each element type.

In this paper, our main concern is reducing the index size to essentially support content-only queries. Thus, we attempt to make an estimation of how the index sizes for the above approaches can be ordered. Of course, I_{full}, i.e., full element-index, would have the largest size. Selective (and/or distributed) index, denoted as I_{sel}, would possibly be smaller; but it can still involve some degree of redundancy for those elements that are selected for indexing. Thus, a direct index (I_{direct}) that indexes only the text under each element would be smaller than the former. Finally, the lower bound for the index size can be obtained by discarding all the structuring in an XML document and creating an index only on the document basis (i.e., I_{doc}). Thus, a rough ordering can be like size(I_{full})$\geq$ size(I_{sel}) $\geq$ size(I_{direct}) $\geq$ size(I_{doc}). In this paper, we employ some pruning methods that can yield indexes of sizes comparable to size(I_{direct}) or size(I_{sel}). We envision that such methods can prune an index up to a given level in a robust and adaptive way, without requiring a priori knowledge on the collection (e.g., semantics or popularity of elements). Furthermore, the redundancy that remains in the index can even help improving the retrieval performance. These index pruning methods are reviewed in the next section.

2.2 Static Pruning Strategies for Inverted Indexes

In the last decade, a number of different approaches are proposed for the static index pruning. The underlying goal of static index pruning is to reduce the file size and query processing time, while keeping the search result quality and subsequently, the effectiveness of the system unaffected, or only slightly affected. In this paper, as in [4], we use the expressions *term-centric* and *document-centric* to indicate whether the pruning process iterates over the terms (or, equivalently, the posting lists) or the documents at the first place, respectively.

In one of the earliest works in this field, Carmel et al. proposed term-centric approaches with uniform and adaptive versions [5]. Roughly, adaptive top-k algorithm sorts the posting list of each term according to some scoring function (Smart's TF-IDF in [5]) and removes those postings that have scores under a threshold determined for that particular term. The algorithm is reported to provide substantial pruning of the index and exhibit excellent performance at keeping the top-ranked results intact in comparison to the original index. In our study, this algorithm (which is referred to as TCP strategy hereafter) is employed for pruning full element-index files, and it is further discussed in Section 3.

As an alternative to term-centric pruning, Büttcher et al. proposed a document-centric pruning (referred to as DCP hereafter) approach with uniform and adaptive versions [4]. In the DCP approach, only the most important terms are left in a document, and the rest are discarded. The importance of a term for a document is determined by its contribution to the document's Kullback-Leibler divergence (KLD) from the entire collection. In a more recent study [1], a comparison of TCP and DCP for pruning the entire index is provided in a uniform framework. It is reported that for disjunctive query processing TCP essentially outperforms DCP for various parameter selections. In this paper, we also use the DCP strategy to prune the full element-index, and further discuss DCP in Section 3.

There are several other proposals for static index pruning in the literature. A locality based approach is proposed in [6] for the purposes of supporting conjunctive and phrase queries. In a number of other works, search engine query logs are exploited to guide the static index pruning [2, 8, 17, 19].

3 Pruning the Element-Index for XML Retrieval

The size of the full element-index for an XML collection may be prohibitively large due to the nested structure of the documents. A large index file does not only consume storage space but also degrades the performance of the actual query processing (as longer posting lists should be traversed). The large index size would also undermine the other optimization mechanisms, such as the list caching (as longer list should be stored in the main memory).

Previous techniques in the literature attempt to reduce the size of a full element-index by either totally discarding the overlapping content, or only indexing a subset of the elements in a collection. In contrast, we envision that some of the terms that appear in an element's descendants may be crucial for the retrieval performance and should be repeated at the upper levels; whereas some other terms can be safely discarded. Thus, instead of a crude mechanism, for each element, the decision for indexing the terms from the element's descendants should be given adaptively, considering the element's textual content and search system's ranking function. To this end, we employ two major static index pruning techniques, namely term-centric pruning (TCP) [5] and document-centric pruning (DCP) [4] for indexing the XML collections. Below, we outline these strategies as used in our study.

- *TCP(I, k, ε)*: As it is mentioned in the previous section, TCP, the adaptive version of the top-k algorithm proposed in [5], is reported to be very successful in static pruning. In this strategy, for each term t in the index I, first the postings in t's posting list are sorted by a scoring function (e.g, TF-IDF). Next, the kth highest score, z_t, is determined and all postings that have scores less than $z_t * \varepsilon$ are removed, where ε is a user defined parameter to govern the pruning level. Following the practice in [3], we disregard any theoretical guarantees and determine ε values according to the desired pruning level.

    A recent study shows that the performance of the TCP strategy can be further boosted by carefully selecting and tuning the scoring function used in the pruning stage [3]. Following the recommendations of that work, we employ BM25 as the scoring function for TCP.

- *DCP(D, λ)*: In this paper, we apply the DCP strategy for the entire index, which is slightly different from pruning only the most frequent terms as originally proposed by [4]. For each document d in the collection D, its terms are sorted by the scoring function. Next, the top $|d|*\lambda$ terms are kept in the document and the rest are discarded, where λ specifies the pruning level. Then, the inverted index is created over these pruned documents.

KLD has been employed as the scoring function in [4]. However, in a more recent work [1], it is reported that BM25 performs better when it is used during both pruning and retrieval. Thus, we also use BM25 with DCP algorithm.

In this paper, we compare the retrieval performance of four different XML indexing approaches:

- I_{full}: Full element-index (as described before)
- I_{direct}: An index created by using only the text directly under each element
- $I_{TCP, \varepsilon}$: Index files created from I_{full} by using TCP algorithm at a pruning level ε
- $I_{DCP, \lambda}$: Index files created from I_{full} by using DCP algorithm at a pruning level λ.

4   Experiments

4.1   Experimental Setup

Collection and queries. In this paper, we use English Wikipedia XML collection [7] employed in INEX campaigns between 2006 and 2008. The dataset includes 659,388 articles obtained from Wikipedia. After conversion to XML, the collection includes 52 million elements. The textual content is 1.6 GB whereas the entire collection (i.e., with element tags) takes 4.5 GB.

Our main focus in this paper is content-only (CO) queries whereas content-and-structure queries (CAS) are left as a future work. In the majority of the experiments reported below, we use 70 query topics with relevance assessments provided for the Wikipedia collection in INEX 2008 (see [15; p. 8] for the exact list of the queries). The actual query set is obtained from the title field of these topics after eliminating the negated terms and stopwords. No stemming is applied.

Indexing. As we essentially focus on CO queries, the index files are built upon only using the textual content of the documents in the collection; i.e., tag names and/or paths are not indexed. In the best performing system in all three tasks of INEX 2008 ad hoc retrieval track, only a subset of elements in the collection are used for scoring [14]. Following the same practice, we only index the following elements: <p>, <section>, <normallist>, <article>, <body>, <td>, <numberlist>, <tr>, <table>, <definitionlist>, <th>, <blockquote>, <div>, <li>, <u>. Each of these elements in an XML document is treated as a separate document and assigned a unique global identifier. Thus, the number of elements to be indexed is found to be 7.4 million out of 52 million elements in Wikipedia collection.

During indexing, we use the open-source Zettair search engine [21] to parse the documents in the collection and obtain a list of terms per element. Then, an element-level index is constructed by using each of the strategies described in this paper. The posting lists in the resulting index files include <element-id, frequency> pairs for each term in the collection, as this is adequate to support the CO queries. Of course, the index can be extended to include, say, term positions, if the system is asked to support phrase or proximity queries, as well. Posting lists are typically stored in a binary file format where each posting takes 8 bytes (i.e., a 4 byte integer is used per each field).

The resulting element-index takes 4 GB disk space. In the below discussions, all index sizes are considered in terms of their raw (uncompressed) sizes.

Retrieval tasks and evaluation. In this study, we concentrate on three ad-hoc retrieval tasks, namely, Focused, Relevant-in-Context (RiC) and Best-in-Context (BiC), as described in recent INEX campaigns (e.g., see [13, 15]). In short, the Focused task is designed to retrieve the most focused results for a query without returning overlapping elements. The underlying motivation for this task is retrieving the relevant information at the correct granularity. Relevant-in-Context task requires returning a ranked list of documents and a set of relevant (non-overlapping) elements listed for each article. Finally, Best-in-Context task is designed to find the best-entry-point (BEP) for starting to read the relevant articles. Thus, the retrieval system should return a ranked list of documents along with a BEP for each document.

We evaluate the performance of different XML indexing strategies for all these three tasks along with the lines of INEX 2008 framework. That is, we use INEXeval software provided in [13] which computes a number of measures for each task, which is essentially based on the amount of retrieved text that overlaps with the relevant text in assessments. In all experiments, we return up to 1500 highest scoring results.

4.2 Performance Comparison of Indexing Strategies: Focused Task

For the focused retrieval task, we return the highest scoring 1500 elements after eliminating the overlaps. The overlap elimination is simply achieved by choosing the highest scoring element on a path in the XML document.

In Figure 1a, we plot the performance of TCP and DCP based indexing strategies with respect to the full element-index, I_{full}. The evaluation measure is interpolated precision at 1% recall level, iP[0.01], which happens to be the official measure of INEX 2008. For this experiment, we use BM25 function as described in [4] to rank the elements using each of the index files. For the pruned index files, the element length, i.e., number of terms in an element, reduces after pruning. In earlier studies [1, 3], it is reported that using the updated element lengths results better in terms of effectiveness. We observed the same situation also for XML retrieval case, and thus, use the updated element lengths for each pruning level of TCP and DCP.

To start with, we emphasize that the system performance with I_{full} is reasonable in comparison to INEX 2008 results. That is, focused retrieval based on I_{full} yields an iP figure of 0.643 at 1% recall level. The best official result in INEX 2008 for this task is 0.689 and our result is within the top-10 results of this task (see Table 6 in [15]). This is also the case for RiC and BiC results that will be discussed in the upcoming sections, proving that we have a reasonable baseline for drawing conclusions in our experimental framework.

Figure 1a reveals that DCP based indexing is as effective as I_{full} up to 50% pruning and indeed, at some pruning levels, it can even outperform I_{full}. In other words, it is possible to halve the index and still obtain the same or even better effectiveness than the full index. TCP is also comparable to I_{full} up to 40%. For this setup, DCP seems to be better than TCP, an interesting finding given that just the reverse is observed for typical document retrieval in previous works [1, 2]. However, the situation changes for higher levels of recall (as shown in Table 1), and, say, for iP[.10], TCP performs better than DCP up to 70% pruning.

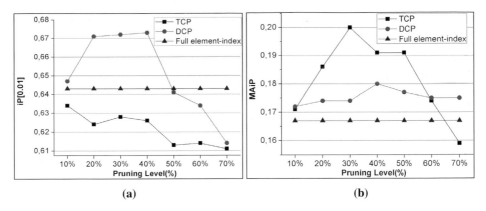

Fig. 1. Effectiveness comparison of I_{full}, I_{TCP} and I_{DCP} in terms of (a) iP[0.01], and (b) MAiP

In Table 1, we report the interpolated precision at different recall levels and mean average interpolated precision (MAiP) computed for 101 recall levels (from 0.00 to 1.00). For these experiments, due to lack of space, we only show three pruning levels (30%, 50% and 70%) for both TCP and DCP. The results reveal that, up to 70% pruning, both indexing approaches lead higher MAiP figures than I_{full} (also see Figure 1b). The same trend also applies for iP at higher recall levels, namely, 5% and 10%.

In Table 1, we further compare the pruning based approaches to other retrieval strategies using I_{direct}. Recall that, as discussed in Section 3, I_{direct} is constructed by considering only the textual content immediately under each element, disregarding the element's descendants. For this collection, the size of the index turns out to be almost 35% of the I_{full}, i.e., corresponding to 65% pruning level. In our first experiment, we evaluate focused retrieval using I_{direct} and BM25 as in the above. In this case, I_{direct} also performs well and yields 0.611 for iP[.01] measure, almost the same effectiveness for slightly smaller indexes created by TCP and DCP (see the case for 70% pruning for TCP and DCP in Table 1). However, in terms of iP at higher levels and MAiP, I_{direct} is clearly inferior to the pruning based approaches, as shown in Table 1.

As another experiment, we decided to implement the propagation mechanism used in the GPX system that participated in INEX between 2004 and 2006 [9, 10, 11]. In these campaigns, GPX is shown to yield very competitive results and ranked among the top systems for various retrieval tasks. Furthermore, GPX is designed to work with an index as I_{direct}, i.e., without indexing the content of descendants for the elements. In this system, first, the score of every element is computed as in the typical case. However, before obtaining the final query output, the scores of all elements in an XML document are propagated upwards so that they can also contribute to their ancestors' scores.

In this paper, we implemented this propagation mechanism (denoted as PROP) of GPX and used Equation 2 given in [11]. In accordance with the INEX official run setup described in that work, we set the parameters $N1$=0.11 and $N2$=0.31 in our implementation. Their work also reports that another scoring function (denoted as SCORE here) performs quite well (see Equation 1 in [11]) when coupled with the propagation. Thus, we obtained results for the propagation mechanism using both scoring functions,

Table 1. Effectiveness comparison of indexing strategies for Focused task. Prune (%) field denotes the percentage of pruning with respect to full element-index (I_{full}). Shaded measures are official evaluation measure of INEX 2008. Best results for each measure are shown in bold.

Indexing Strategy	Prune (%)	iP[.00]	iP[.01]	iP[.05]	iP[.10]	MAiP
I_{full}	0%	0.725	0.643	0.507	0.446	0.167
$I_{TCP, 0.3}$	30%	0.700	0.628	**0.560**	0.511	**0.200**
$I_{DCP, 0.3}$	30%	**0.750**	**0.672**	0.529	0.469	0.174
$I_{TCP, 0.5}$	50%	0.666	0.613	0.549	**0.518**	0.191
$I_{DCP, 0.5}$	50%	0.708	0.641	0.518	0.473	0.177
$I_{TCP, 0.7}$	70%	0.680	0.611	0.511	0.446	0.159
$I_{DCP, 0.7}$	70%	0.681	0.614	0.534	0.477	0.175
I_{direct}	65%	0.731	0.611	0.448	0.362	0.126
$I_{direct}+PROP_{SCORE}$	65%	0.519	0.473	0.341	0.302	0.110
$I_{direct}+PROP_{BM25}$	65%	0.450	0.435	0.384	0.302	0.116

namely, BM25 and SCORE. In Table 1, corresponding experiments are denoted as $I_{direct}+PROP_{BM25}$ and $I_{direct}+PROP_{SCORE}$, respectively. The results reveal that, SCORE function performs better at early recall levels, but for both cases the effectiveness figures are considerably lower than the corresponding results (i.e., 70% pruning level) based on TCP and DCP. We attribute the lower performance of PROP mechanism to the following observation. For the Wikipedia dataset, 78% of the data assessed as relevant resides in the leaf nodes. This means that, returning leafs in the result set would improve effectiveness, and vice versa. In contrast, PROP propagates element scores to the upper levels in the document, which may increase the number of interior nodes in the final result and thus reduce the effectiveness.

Note that, we also attempted to verify the reliability of our implementation of propagation mechanism by using INEX 2006 topics and evaluation software, and to see how our results compare to the GPX results reported in [11]. We observed that the results slightly differ at early ranks but then match for higher rank percentages.

We can summarize our findings as follows: In terms of the official INEX measure, which considers the performance at the first results most, the index files constructed by the static pruning techniques lead to comparable to or even superior results than I_{full} up to 70% pruning level. A direct element-index also takes almost 35% of the full index. Its performance is as good as the pruned index files for iP[.01], but it falls rapidly at higher recall levels. Score propagating retrieval systems, similar to GPX, perform even worse with the I_{direct} and do not seem to be a strong competitor.

Finally, there is another indexing approach proposed in the XRANK system [12] that can serve as a natural competitor of the strategies discussed here. In the XRANK's approach, as reviewed in Section 2, element ids are represented with Dewey encoding. This representation can yield an index of size I_{direct}, while the exact element scores that could be obtained from a full element-index can also be computed (with some overhead during query processing). Furthermore, it can support structure based constraints for CAS queries. However, this indexing technique may also cause some problems. For instance, since element ids are Dewey encoded, it may be hard to represent some elements in deeply nested XML documents. Another issue may be updating the Dewey codes when an XML document is updated (see [20] for a general discussion). Also, to our best knowledge, the performance of typical inverted index

Table 2. Effectiveness comparison of indexing strategies for RiC task

Indexing Strategy	Prune (%)	gP[5]	gP[10]	gP[25]	gP[50]	MAgP
I_{full}	0%	0.364	**0.321**	0.246	0.198	0.190
$I_{TCP,\,0.3}$	30%	0.380	0.321	0.248	0.199	0.193
$I_{DCP,\,0.3}$	30%	0.381	0.321	**0.256**	**0.202**	**0.196**
$I_{TCP,\,0.5}$	50%	**0.385**	0.321	0.247	0.196	0.185
$I_{DCP,\,0.5}$	50%	0.366	0.321	0.252	0.196	0.185
$I_{TCP,\,0.7}$	70%	0.355	0.297	0.223	0.170	0.152
$I_{DCP,\,0.7}$	70%	0.352	0.305	0.232	0.172	0.157
I_{direct}	65%	0.312	0.281	0.210	0.168	0.140
$I_{direct}+PROP_{SCORE}$	65%	0.275	0.242	0.201	0.158	0.142
$I_{direct}+PROP_{BM25}$	65%	0.223	0.199	0.184	0.145	0.108

compression techniques for Dewey encoded index files is not evaluated yet. On the other hand, the index files created by the static pruning techniques can be processed by typical IR systems without requiring any modifications in the query processing, indexing and compression modules. Additionally, as shown in above results, these techniques may yield better effectiveness than a full index in several cases. Nevertheless, we leave comparison with the XRANK's approach as a future work.

4.3   Performance Comparison of Indexing Strategies: Relevant-in-Context Task

For the Relevant-in-Context task, after scoring the elements, we again eliminate the overlaps and determine the top-1500 results as in the Focused task. Then, those elements from the same document are grouped together. The result is a ranked list of documents along with a set of elements. While ranking the documents, we use the score of the highest scoring retrieved element per document as the score of this particular document. We also experimented with another technique in the literature; i.e. using the average score of elements in a document for ranking the documents, which performs worse than the former approach and is not investigated further.

In Table 2, we present the results in terms of the generalized precision (gP) metric at early ranks and mean average generalized precision (MAgP), i.e., the official measure of INEX 2008 for both RiC and BiC tasks. The results show that approaches using I_{direct} are inferior to those using the pruned index files based on either TCP or DCP at 70% pruning level; however with a relatively smaller margin with respect to the previous task. In comparison of TCP and DCP based approaches to I_{full}, we observe that the former cases still yield comparable or better performance, however up to 50% pruning, again a more conservative result than that reported for the previous task.

4.4   Performance Comparison of Indexing Strategies: Best-in-Context Task

For the Best-in-Context task, we obtain the relevant documents exactly in the same way as in RiC. However, while ranking the documents, if the article node of the document is within these retrieved elements, we use its score as the document score. Otherwise, we use the score of the highest scoring retrieved element as the score of

Table 3. Effectiveness comparison of indexing strategies for BiC task

Indexing Strategy	Prune (%)	gP[0.05]	gP[0.10]	gP[0.25]	gP[0.50]	MAgP
I_{full}	0%	0.367	0.314	0.237	0.186	0.178
$I_{TCP,\,0.3}$	30%	0.369	0.318	0.237	0.187	0.178
$I_{DCP,\,0.3}$	30%	**0.388**	**0.332**	**0.246**	**0.187**	**0.184**
$I_{TCP,\,0.5}$	50%	0.364	0.319	0.232	0.179	0.165
$I_{DCP,\,0.5}$	50%	0.363	0.310	0.234	0.178	0.166
$I_{TCP,\,0.7}$	70%	0.335	0.287	0.198	0.154	0.138
$I_{DCP,\,0.7}$	70%	0.340	0.280	0.214	0.157	0.143
I_{direct}	65%	0.215	0.183	0.141	0.116	0.087
I_{direct}+PROP$_{SCORE}$	65%	0.127	0.132	0.120	0.100	0.086
I_{direct}+PROP$_{BM25}$	65%	0.156	0.151	0.136	0.115	0.074

this particular document. Then, we identify a best-entry-point (BEP) per document. In INEX 2008, a simple approach of setting the BEP as 1 is found to be very effective and ranked second among all participants [15]. Note that, this suggests starting to read each ranked document from the beginning. For our work, we also experimented with providing the offset of the highest scoring element per document as BEP [15], which yielded inferior results to the former approach. Thus, we only report the results where BEP is set to 1.

In Table 3, we compare indexing strategies in terms of the same evaluation metrics used in RiC task. As in RiC case, the performance obtained by using pruned index files with TCP and DCP is comparable to that of using the full element-index up to 50% pruning. At the 70% pruning level, both pruning approaches have losses in effectiveness with respect to I_{full}, but they are still considerably better than using I_{direct} with the (approximately) same index size. For instance, while MAgP for DCP is 0.143, the retrieval strategies using I_{direct} (with BM25), I_{direct}+PROP$_{SCORE}$ and I_{direct}+PROP$_{BM25}$ yield the MAgP figures of 0.087, 0.086 and 0.074, respectively. Again, basic retrieval using I_{direct} outperforms propagation based approaches, especially at the earlier ranks for generalized precision metric.

5   Conclusion

Previous experiences with XML collections suggest that element indexing is important for high performance in ad hoc retrieval tasks. In this study, we propose to use static index pruning techniques for reducing the size of a full element-index, which would otherwise be very large due to the nested structure of XML documents. We also compare the performance of term and document based pruning strategies to those approaches that use a direct element index that avoids indexing nested content more than once. Our experiments are conducted along the lines of previous INEX campaigns. The results reveal that pruned index files are comparable or even superior to the full element-index up to very high pruning levels for various ad hoc tasks (e.g., up to 70% pruning for Focused task and 50% pruning for RiC and BiC tasks) in terms of the retrieval effectiveness. Moreover, the performance of pruned index files is also better than that of the approaches using the direct index file at the same index size.

Future work directions involve extending our framework with some other static index pruning techniques and investigating the performance for the other query types (e.g., conjunctive and phrase queries, CAS queries, etc.).

Acknowledgments. This work is supported by The Scientific and Technological Research Council of Turkey (TÜBİTAK) by the grant number 108E008.

References

1. Altingovde, I.S., Ozcan, R., Ulusoy, Ö.: A practitioner's guide for static index pruning. In: Proc. of ECIR 2009, pp. 675–679 (2009)
2. Altingovde, I.S., Ozcan, R., Ulusoy, Ö.: Exploiting Query Views for Static Index Pruning in Web Search Engines. In: Proc. of CIKM 2009, pp. 1951–1954 (2009)
3. Blanco, R., Barreiro, A.: Boosting static pruning of inverted files. In: Proc. of SIGIR 2007, pp. 777–778 (2007)
4. Büttcher, S., Clarke, C.L.: A document-centric approach to static index pruning in text retrieval systems. In: Proc. of CIKM 2006, pp. 182–189 (2006)
5. Carmel, D., Cohen, D., Fagin, R., Farchi, E., Herscovici, M., Maarek, Y.S., Soffer, A.: Static index pruning for information retrieval systems. In: Proc of SIGIR 2001, pp. 43–50 (2001)
6. de Moura, E.S., Santos, C.F., Araujo, B.D., Silva, A.S., Calado, P., Nascimento, M.A.: Locality-Based pruning methods for web search. ACM TOIS 26(2), 1–28 (2008)
7. Denoyer, L., Gallinari, P.: The Wikipedia XML Corpus. SIGIR Forum 40(1), 64–69 (2006)
8. Garcia, S.: Search Engine Optimization Using Past Queries. Doctoral Thesis, RMIT (2007)
9. Geva, S.: GPX – Gardens Point XML Information Retrieval at INEX 2004. In: Fuhr, N., Lalmas, M., Malik, S., Szlávik, Z. (eds.) INEX 2004. LNCS, vol. 3493, pp. 211–223. Springer, Heidelberg (2005)
10. Geva, S.: GPX – Gardens Point XML IR at INEX 2005. In: Fuhr, N., Lalmas, M., Malik, S., Kazai, G. (eds.) INEX 2005. LNCS, vol. 3977, pp. 240–253. Springer, Heidelberg (2006)
11. Geva, S.: GPX – Gardens Point XML IR at INEX 2006. In: Proc. of INEX 2006 Workshop, pp. 137–150 (2006)
12. Guo, L., Shao, F., Botev, C., Shanmugasundaram, J.: XRANK: Ranked Keyword Search Over XML Documents. In: Proc. of the ACM SIGMOD 2003, pp. 16–27 (2003)
13. Initiative for the Evaluation of XML Retrieval (2009), http://www.inex.otago.ac.nz/
14. Itakura, K.Y., Clarke, C.L.A.: University of Waterloo at INEX 2008: Adhoc, Book, and Link-the-Wiki Tracks. In: Geva, S., Kamps, J., Trotman, A. (eds.) INEX 2008. LNCS, vol. 5631, pp. 132–139. Springer, Heidelberg (2009)
15. Kamps, J., Geva, S., Trotman, A., Woodley, A., Koolen, M.: Overview of the 2008 Ad Hoc Track. In: Geva, S., Kamps, J., Trotman, A. (eds.) INEX 2008. LNCS, vol. 5631, pp. 1–28. Springer, Heidelberg (2009)
16. Lalmas, M.: XML Retrieval. Morgan & Claypool, San Francisco (2009)
17. Ntoulas, A., Cho, J.: Pruning policies for two-tiered inverted index with correctness guarantee. In: Proc. of SIGIR 2007, pp. 191–198 (2007)

18. Sigurbjörnsson, B., Kamps, J.: The effect of structured queries and selective indexing on XML retrieval. In: Fuhr, N., Lalmas, M., Malik, S., Kazai, G. (eds.) INEX 2005. LNCS, vol. 3977, pp. 104–118. Springer, Heidelberg (2006)
19. Skobeltsyn, G., Junqueira, F., Plachouras, V., Baeza-Yates, R.: ResIn: a combination of results caching and index pruning for high-performance web search engines. In: Proc. of SIGIR 2008, pp. 131–138 (2008)
20. Su-Cheng, H., Chien-Sing, L.: Node Labeling Schemes in XML Query Optimization: A Survey and Trends. IETE Tech Rev 26, 88–100 (2009)
21. Zettair search engine (2009), http://www.seg.rmit.edu.au/zettair/
22. Zobel, J., Moffat, A.: Inverted files for text search engines. ACM Computing Surveys 38(2), 1–56 (2006)

Category-Based Query Modeling for Entity Search

Krisztian Balog, Marc Bron, and Maarten de Rijke

ISLA, University of Amsterdam, Science Park 107, 1098 XG Amsterdam, The Netherlands
k.balog@uva.nl, m.m.bron@uva.nl, derijke@uva.nl

Abstract. Users often search for entities instead of documents and in this setting are willing to provide extra input, in addition to a query, such as category
information and example entities. We propose a general probabilistic framework
for entity search to evaluate and provide insight in the many ways of using these
types of input for query modeling. We focus on the use of category information
and show the advantage of a category-based representation over a term-based
representation, and also demonstrate the effectiveness of category-based expansion using example entities. Our best performing model shows very competitive
performance on the INEX-XER entity ranking and list completion tasks.

1 Introduction

Users often search for specific entities instead of documents mentioning them [8, 20].
Example information needs include "Impressionist art museums in The Netherlands"
or "Experts on authoring tools," where answers to be returned are museums and experts
and not just articles discussing them. In such scenarios, users may be assumed to be
willing to express their information need more elaborately than with a few keywords [3].
These additional means may include categories to which target entities should belong
or example entities. We focus on abstractions of these scenarios, as they are evaluated in
the context of INEX, the INitiative for the evaluation of XML retrieval. In 2007, INEX
launched an *entity ranking* track [8, 9]. Here, entities are represented by their Wikipedia
page and queries asking for an entity are typed (asking for entities belonging to certain
categories) and may come with examples. Two tasks are being considered at INEX:
(1) *entity ranking*, where a query and target categories are given, and (2) *list completion*,
where a query, example entities, and (optionally) target categories are given.

Given that information needs involving entities can be formulated in so many ways,
with so many ingredients (query, categories, examples), the obvious system-oriented
question to ask is how to map these ingredients into the query component of a retrieval
model. In this paper, we focus on effectively capturing and exploiting category-based
information. Several approaches to incorporating such information have been proposed
(see §2 below), but there is no systematic account of approaches yet.

We introduce a probabilistic framework for entity retrieval that models category information in a theoretically transparent manner. Information needs and entities are represented as a tuple: a term-based model plus a category-based model, both characterized
by probability distributions. Ranking of entities is then based on similarity to the query,
measured in terms of similarities between probability distributions. Our framework is

C. Gurrin et al. (Eds.): ECIR 2010, LNCS 5993, pp. 319–331, 2010.

capable of synthesizing all previous approaches proposed for exploiting category information in the context of the INEX Entity Ranking task. Our focus is on two core steps: query modeling and query model expansion.

We seek to answer the following questions. Does our two-component query model improve over single component approaches (either term-based or category-based)? What are effective ways of modeling (blind) relevance feedback in this setting, using either or both of the term-based and category-based components?

Our main contribution is the introduction of a probabilistic retrieval model for entity search, in which we are able to effectively integrate term-based and category-based representations of queries and entities. We provide extensive evaluations of our query models and approaches to query expansion. Category-based feedback is found to be more beneficial than term-based feedback, and category-based feedback using example entities brings in the biggest improvements, bigger than combinations of blind feedback and information derived from example entities.

In §2 we discuss related work. We introduce our retrieval model in §3. In §4 we zoom in on query modeling and query model expansion; §5 is devoted to an experimental evaluation. An analysis and conclusion complete the paper.

2   Related Work

We review work on entity-oriented search tasks and then consider work on entity ranking at INEX.

2.1   Entity Retrieval

A range of commercial providers now support entity-oriented search, dealing with a broad range of entity types: people, companies, services and locations. Examples include TextMap, ZoomInfo, Evri, and the Yahoo! correlator demo.[1] They differ in their data sources, supported entity types, functionality, and user interface. Common to them, however, is their ability to rank entities with respect to a topic or another entity.

Conrad and Utt [6] introduce techniques for extracting entities and identifying relationships between entities in large, free-text databases. The degree of association between entities is based on the co-occurrence within a fixed window size. A more general approach is also proposed, where all paragraphs containing a mention of an entity are collapsed into a single pseudo document. Raghavan et al. [21] re-state this approach in a language modeling framework. Sayyadian et al. [23] introduce the problem of finding missing information about a real-world entity from text and structured data; entity retrieval over text documents can be significantly aided by structured data.

The TREC Question Answering track recognized the importance of search focused on entities with factoid questions and list questions (asking for entities that meet certain constraints) [29]. To answer list questions [22], systems have to return instances of

---

[1] See http://www.textmap.com/, http://www.zoominfo.com/,
http://www.evri.com/, and http://sandbox.yahoo.com/correlator,
respectively.

the class of entities that match the description in the question. List questions are often treated as (repeated) factoids, but special strategies are called for as answers may need to be collected from multiple documents [5]. Google Sets allows users to enter some instances of a concept and retrieve others that closely match the examples provided [14]. Ghahramani and Heller [13] developed an algorithm for completing a list based on examples using machine learning techniques.

Entity search generalizes *expert finding*: given a topic, return a ranked list of experts on the topic. TREC 2005–2008 featured an expert finding track [4]. Lessons learned involve models, algorithms, and evaluation methodology [1, 2].

Zaragoza et al. [32] retrieve entities in Wikipedia, where instances are not necessarily represented by textual content other than their descriptive label. In 2007, the INEX Entity Ranking track (INEX-XER) [8, 9] introduced tasks where candidate items are restricted to have their own Wikipedia page. Fissaha Adafre et al. [10] addressed an early version of the entity ranking and list completion tasks and explored different representations of list descriptions and example entities.

2.2   Entity Ranking at INEX

INEX has traditionally focused on the use of document structure to improve retrieval effectiveness. While the initial focus was on document and element retrieval, over the years, INEX has expanded. In recent years, INEX has mainly been using Wikipedia as its document collection. The main lesson learned is that exploiting the rich structure of the collection (text plus category information, associations between entities, and query-dependent link structure) helps improve retrieval performance [8].

As to query formulation for entity retrieval, usually stemming and stopwording is performed. Craswell et al. [7] go beyond this and modify the query with NLP techniques, removing verbs while focussing on adjectives, nouns and named entities; others use the query to retrieve a set of documents and use co-occurrence statistics between retrieved article titles and query terms as a component in their ranking score [34].

INEX participants have considered expanding queries using category information. E.g., the latter has been used for expanding the query with labels of the target category [7]. Others have used set-based metrics to measure the overlap between target categories and categories of candidate entities [27, 30]. Target category information provided as part of the query need not be complete, as manual assignment of categories to documents is imperfect. Some teams expand the target categories, e.g., by using the category structure to expand with categories [15, 25, 30]. Others expand the target categories using lexical similarity between category labels and query terms [17, 26].

A standard way of combining category and term-based components is to use a language modeling approach and to estimate the probability of an entity given the query and category information [16, 30, 34]. Vercoustre et al. [28] integrate query difficulty prediction and Zhu et al. [34] treat categories as metadata fields and apply a multi-field retrieval model. Several participants in the list completion task use the categories of example entities for category expansion, using various expansion techniques [7, 16, 26, 30, 34]; some use category information to expand the term-based model [30].

3   Modeling Entity Ranking

We present a general retrieval scheme for two entity ranking tasks. In the *entity ranking* task one is given a "standard" keyword query (Q) along with a set of target categories (C) and has to return entities. For *list completion* we need to return entities given a keyword query (Q), a set of target categories (C) and a set of similar entities (E).

3.1   A General Scheme for Entity Ranking

Fig. 1 depicts our scheme for ranking entities. The process goes as follows. We are given the user's input, consisting of a query, a set of input categories, and, optionally, example entities. This input is translated into a query model, with a term-based and/or a category-based component (I). During retrieval (Step III) this model is compared against models created for indexed entities (derived in Step II). In Step IV a ranked list of entities is produced (based on Step III), which, in turn, may (optionally) give rise to an expanded query model (Step V)—from that point onwards Steps III, IV, V can be repeated.

Our focus is limited to the problem of modeling the query: (i) How can the user's input be translated into an initial query model (Step I)? And (ii) how can this—often sparse—representation be refined or extended to better express the underlying information need (Step V)? Specifically, we are interested in sources and components that play a role in estimating the term- and category-based representations of query models. Some models may involve additional components or steps for entity modeling (e.g., priors) or for ranking (e.g., links between entities); this does not affect our approach, as this concerns the ranking part (II, III and IV), and this information is not (explicitly) present on the query side (I and V).

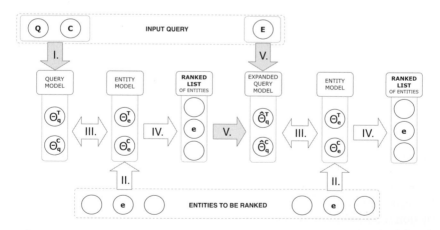

Fig. 1. A general scheme for entity ranking. The steps on which we focus are indicated with grey arrows; all steps are explained in §3.1.

3.2   A Probabilistic Model for Entity Ranking (Steps III and IV)

We introduce a probabilistic framework that implements the entity ranking approach depicted in Fig. 1. We rank entities according to their probability of being relevant given the query: $P(e|q)$. We apply Bayes' rule and rewrite this probability to:

$$P(e|q) \propto P(q|e) \cdot P(e), \tag{1}$$

where $P(q|e)$ expresses the probability that query q is generated by entity e, and $P(e)$ is the *a priori* probability of e being relevant, i.e., the entity prior. We assume that $P(e)$ is uniform, thus, does not affect the ranking.

Each entity is represented as a pair $\theta_e = (\theta_e^T, \theta_e^C)$, where θ_e^T is a distribution over terms and θ_e^C is a distribution over categories. The query is also represented as a pair: $\theta_q = (\theta_q^{\prime T}, \theta_q^C)$, which is then (optionally) further refined, resulting in an expanded query model $\hat{\theta}_q = (\hat{\theta}_q^T, \hat{\theta}_q^C)$ that is used for ranking entities.

The probability of an entity generating the query is estimated using a mixture model:

$$P(q|e) = \lambda \cdot P(\theta_q^T|\theta_e^T) + (1 - \lambda) \cdot P(\theta_q^C|\theta_e^C), \tag{2}$$

where λ controls the interpolation between the term-based and category-based representations. The estimation of $P(\theta_q^T|\theta_e^T)$ and $P(\theta_q^C|\theta_e^C)$ requires a measure of the difference between two distributions. Here, we opt for the Kullback-Leibler divergence. The term-based similarity is estimated as follows:

$$KL(\theta_q^T||\theta_e^T) = \sum_t P(t|\theta_q^T) \cdot \log \frac{P(t|\theta_q^T)}{P(t|\theta_e^T)}, \tag{3}$$

where $P(t|\theta_e^T)$ and $P(t|\theta_q^T)$ remain to be defined. Since KL divergence is a score (which is lower when two distributions are more similar), we turn it into a probability using Eq. 4:

$$P(\theta_q^T|\theta_e^T) \propto \max KL(\theta_q^T||\cdot) - KL(\theta_q^T||\theta_e^T). \tag{4}$$

The category-based component of the mixture in Eq. 2 is calculated analogously to the term-based case. Consequently, $P(c|\theta_e^C)$ and $P(c|\theta_q^C)$ remain to be defined.

This completes Step III. Next, we describe the entity model component, i.e., Step IV. Steps I and V are discussed in §4.

3.3   Entity Modeling (Step IV)

Term-based representation. To estimate $P(t|\theta_e^T)$ we smooth the empirical entity model with the background collection to prevent zero probabilities. We employ Bayesian smoothing using Dirichlet priors which has been shown to achieve superior performance on a variety of tasks and collections [33, 19] and set:

$$P(t|\theta_e^T) = \frac{n(t,e) + \mu^T \cdot P(t)}{\sum_t n(t,e) + \mu^T}, \tag{5}$$

where $n(t, e)$ is the number of occurences of t in e, $\sum_t n(t, e)$ is the total number of term occurrences, i.e., the document length, and $P(t)$ is the background model (the relative frequency of t in the collection). Since entities correspond to Wikipedia articles, this representation of an entity is identical to constructing a smoothed document model for each Wikipedia page, in a standard language modeling approach [24, 18]. Alternatively, the entity model can be expanded with terms from related entities [10].

Category-based representation. Analogously to the term-based case, we smooth the maximum-likelihood estimate with a background model. We employ Dirichlet smoothing, and use the parameter μ^C to avoid confusion with μ^T:

$$P(c|\theta_e^C) = \frac{n(c,e)+\mu^C \cdot P(c)}{\sum_c n(c,e)+\mu^C}. \tag{6}$$

In Eq. 6, $n(c,e)$ is 1 if entity e is assigned to category c, and 0 otherwise; $\sum_c n(c,e)$ is the total number of categories e is assigned to; $P(c)$ is the background category model and is set using a maximum-likelihood estimate:

$$P(c) = \frac{\sum_e n(c,e)}{\sum_c \sum_e n(c,e)}, \tag{7}$$

where $\sum_c \sum_e n(c,e)$ is the number of category-entity assignments in the collection.

4   Estimating and Expanding Query Models

In this section we introduce methods for estimating and expanding query models: Steps I and V in Figure 1. We construct initial (θ_q) and expanded ($\hat{\theta}_q$) query models, which boils down to estimating the probabilities $P(t|\theta_q^T)$, $P(c|\theta_q^C)$, $P(t|\hat{\theta}_q^T)$, and $P(c|\hat{\theta}_q^C)$) as listed in Fig. 1 and discussed in §3.

4.1   Query Models (Step I)

We define a series of query models, each consisting of a term-based component and/or a category-based component; graphical depictions of the models are given in Figure 2.

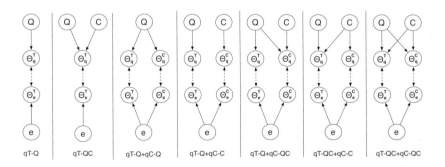

Fig. 2. Query models without expansion; Q stands for the topic title, E for example entities and C for the target categories; solid arrows from input to query model indicate the input is used to create the model; dashed arrows indicate a comparison between models; the models and acronyms are explained in §4.1.

qT-Q This query model only has a term-based component and uses no category information (i.e., it amounts to standard language modeling for document retrieval). Writing $n(t, Q)$ for the number of times term t is present in query Q, we put

$$P(t|\theta_q^T) = P_{bl}(t|\theta_q^T) = \frac{n(t,Q)}{\sum_t n(t,Q)}. \tag{8}$$

qT-QC This model uses the names of input categories added to the term-based query model (θ_q^T); for the sake of simplicity, original terms and terms from category names contribute with the same weight to the total probability mass ($\alpha^T = 0.5$).

$$P(t|\theta_q^T) = \alpha^T \cdot P_{bl}(t|\theta_q^T) + (1 - \alpha^T) \cdot \frac{\sum_{c \in C} n(t,c)}{\sum_{c \in C} \sum_t n(t,c)}. \tag{9}$$

qT-Q+qC-Q This model uses the keyword query (Q) to infer the category-component of the query model (θ_q^C), by considering the top N_c most relevant categories given the query; relevance of a category is estimated based on matching between the name of the category and the query, i.e., a standard language modeling approach on top of an index of category names, where $P(Q|c)$ is the probability of category c generating query Q.

$$P(c|\theta_q^C) = P_q(c|\theta_q^C) = \begin{cases} P(Q|c)/\sum_{c \in C} P(Q|c), & \text{if } c \in \text{top } N_c \\ 0, & \text{otherwise.} \end{cases} \tag{10}$$

qT-Q+qC-C This model uses input categories to form a category-based query model. Setting $n(c, q)$ to 1 if c is a target category, and $\sum_c n(c, q)$ to the total number of target categories provided with the topic statement, we put

$$P(c|\theta_q^C) = P_{bl}(c|\theta_q^C) = \frac{n(c,q)}{\sum_c n(c,q)}. \tag{11}$$

qT-Q+qC-QC This model combines qT-Q+qC-C with categories relevant to the query added to the category-based query model; again, to keep things simple we allocate the probability mass equally between the two components ($\alpha^C = 0.5$).

$$P(c|\theta_q^C) = \alpha^C \cdot P_{bl}(c|\theta_q^C) + (1 - \alpha^C) \cdot P_q(c|\theta_q^C). \tag{12}$$

qT-QC+qC-C This model combines qT-Q+qC-C with names of input categories added to qT-Q (and contributes half of the probability mass).

qT-QC+qC-QC This model combines qT-Q+qC-QC and qT-QC+qC-C.

4.2   Expanded Query Models (Step V)

Expansions of the basic query model can take place on either (or both) of the two components. The general form we use for expansion is a mixture of the baselines defined in §4.1 (subscripted with bl) and an expansion (subscripted with ex). We set

$$P(t|\hat{\theta}_q^T) = \lambda^T \cdot P_{ex}(t|\theta_q^T) + (1 - \lambda^T) \cdot P_{bl}(t|\theta_q^T) \tag{13}$$

for the term-based component. And for the category-based component we set:

$$P(c|\hat{\theta}_q^C) = \lambda^C \cdot P_{ex}(c|\theta_q^C) + (1 - \lambda^C) \cdot P_{bl}(c|\theta_q^C). \tag{14}$$

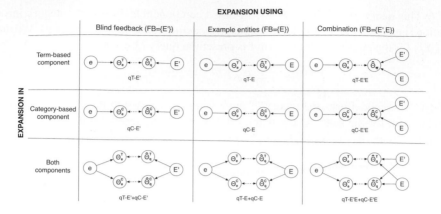

Fig. 3. Models with expansion; same graphical and notational conventions as in Fig. 2; acronyms explained in §4.2

We present a general method for estimating the expansions $P_{ex}(t|\theta_q^T)$ and $P_{ex}(c|\theta_q^C)$, using a set of feedback entities, FB. This feedback set may be obtained by taking the top N relevant entities according to a ranking obtained using the initial query. We use E' to denote this set of blind feedback entities. Alternatively, one might assume explicit feedback, such as the example entities (denoted by E) in our scenario. Constructing the feedback set by using either blind feedback ($FB = E'$), example entities ($FB = E$), or a combination of both ($FB = E' \cup E$) yields three query expansion methods. Depending on where feedback takes place we have 9 variations, shown in Fig. 3. Next, we construct expanded query models from a set of feedback entities FB.

Term-based expansion. Given a set of feedback entities FB, the expanded query model is constructed as follows:

$$P(t|\hat{\theta}_q^T) = \frac{P_{K_T}(t|FB)}{\sum_{t'} P_{K_T}(t'|FB)}, \tag{15}$$

where $P_{K_T}(t|FB)$ denotes the top K_T terms with the highest $P(t|FB)$ value, calculated according to Eq. 16.

$$P(t|FB) = \frac{1}{|FB|}\sum_{e\in FB} \frac{n(t,e)}{\sum_t n(t,e)} \tag{16}$$

where $\sum_t n(t,e)$ is the total number of terms, i.e., the length of the document corresponding to e. (This is [3]'s best performing query model generation method using example documents, with all feedback documents assumed to be equally important.)

Category-based expansion. Here we put

$$P(c|\hat{\theta}_q^C) = \frac{P_{K_C}(c|FB)}{\sum_{c'} P_{K_C}(c'|FB)}, \tag{17}$$

where $P_{K_C}(c|FB)$ denotes the top K_C categories with the highest $P(c|FB)$ value, calculated according to Eq. 18, (where, as before, $n(c,e)$ is 1 if e belongs to c):

$$P(c|FB) = \frac{1}{|FB|}\sum_{e\in FB} \frac{n(c,e)}{\sum_t n(c,e)}. \tag{18}$$

5   Experimental Evaluation

In order to answer the research questions listed in §1, we run a set of experiments. We detail our experimental setup, present the results and formulate answers.

5.1   Experimental Setup

Test Collection. We use test sets of the INEX Entity Ranking track (INEX-XER) [8, 9], that use (a dump of the English) Wikipedia as document collection from which (articles corresponding to) entities are to be returned. The collection consists of over 650,000 documents plus a hierarchy of (over 115,000) categories; this is not a strict hierarchy: assignments of categories to articles are not always consistent [8].

Tasks. INEX-XER has two tasks: *entity ranking* and *list completion*. An entity ranking topic specifies a keyword query (Q) and target categories (C). In the list completion task, the topic also specifies example entities (E) in addition. We also consider a variation of the task where all three input sources (Q, C, and E) are provided by the user.

Topics and judgments. Two sets of topics are available for INEX-XER. For XER2007 a test set of 46 topics was created for the entity ranking track, 25 of which were specifically developed and assessed by track participants. For XER2008, 35 topics were developed and assessed by track participants [8]. We report on Mean Average Precision (MAP) and Mean Reciprocal Rank (MRR) for XER2007. For XER2008, xinfAP replaces MAP [31] as a better estimate of AP in the case of incomplete assessments [9].

Parameter settings. Our models involve a number of parameters. In this section we apply our baseline settings for these parameters and we use them for all models. We use the average document length for the term-smoothing parameter ($\mu^T = 411$) and the average number of categories assigned to an entity for the category-based smoothing parameter ($\mu^C = 2.2$). Our mixture models involve two components, to which we assign equal importance, i.e., $\lambda = \lambda^T = \lambda^C = 0.5$. In §5.4 we briefly analyze the sensitivity of our models w.r.t. these parameters.

5.2   The Performance of Query Models

We examine the effectiveness of our query models and of the use of two components—for terms and categories—for the *entity ranking* task. In the experiments that involve the keyword query in the construction of the category-component of the query model, we set $N_c = 10$ (see Eq. 10). Table 1 list the results for the query models defined in §4.1, using the default parameter settings detailed in §5.1; in §5.4 we report on optimized runs and compare them against the best scores obtained at INEX-XER.

 We compare the performance of the models using a two-tailed t-test at a significance level of $p = 0.05$. Simply flattening the target category information and adding category names as terms to the term component is not an effective strategy; see (1) vs. (2), (4) vs. (6), and (5) vs. (7). When we consider category-based information provided with the input query as a separate component, we see improvements across the test sets: see (1) vs. (3) (significant for 2007 and 2008) and (2) vs. (6) (significant for 2008). The switch

Table 1. Entity ranking results, no expansion. Best results per collection in boldface.

Model	λ	θ_q^T	θ_q^C	XER2007		XER2008	
				MAP	MRR	xinfAP	MRR
(1) qT-Q	1.0	Eq. 8	-	0.1798	0.2906	0.1348	0.2543
(2) qT-QC	1.0	Eq. 9	-	0.1706	0.3029	0.1259	0.2931
(3) qT-Q+qC-Q	0.5	Eq. 8	Eq. 10	0.2410	0.3830	0.1977	0.3190
(4) qT-Q+qC-C	0.5	Eq. 8	Eq. 11	0.2162	0.4168	0.3099	0.4783
(5) qT-Q+qC-QC	0.5	Eq. 8	Eq. 12	**0.2554**	**0.4531**	**0.3124**	**0.5024**
(6) qT-QC+qC-C	0.5	Eq. 9	Eq. 11	0.1881	0.2948	0.2911	0.4439
(7) qT-QC+qC-QC	0.5	Eq. 9	Eq. 12	0.2255	0.3346	0.2950	0.4357

from using only the keyword query for the construction of the category component to using target categories defined explicitly by the user ((3) vs. (4)) does not lead to consistent improvements (although the improvement is significant on the 2008 set); the move from the latter to a combination of both ((4) vs. (5) (significant on the 2007 set) and (6) vs. (7)) leads to consistent improvements for all tasks and measures.

5.3   The Performance of Expanded Query Models

We report on the effectiveness of the expansion models depicted in Fig. 3. When we only use Q and C, results are evaluated on the *entity ranking* task. When E is also used we evaluate results on the *list completion task*. Some notation: E' denotes a pseudo-relevant set of entities, i.e., the top N obtained using methods detailed §5.1; and E denotes a set of example entities. When we use example entities for expansion, we need to remove them from the runs, i.e., use the list completion qrels: in order to have a fair comparison between approaches reported in this subsection we need to do that for the pseudo-feedback runs as well, i.e., when we use only E'. We use the following settings; number of feedback entities (N): 5; number of feedback categories (K_C): 10; number of feedback terms (K_T): 15; default values for λ, λ^T, and λ^C (0.5).

Table 2 presents the results of query expansion, applied on top of the best performing run from §5.1 (qT-Q+qC-QC). Category-based feedback helps more than term-based feedback; term-based blind feedback does not lead to significant improvements and hurts in one case (entity ranking, 2007, MAP); category-based expansion improves in terms of MAP scores on both tasks, in both years (significantly in 2008). For list completion, category-based feedback using examples leads to the biggest improvements (both years, both measures); relative improvements can be up to +47% in MAP (2007) and +50% in MRR (2008). Blind feedback plus examples improves over blind feedback alone, but is outperformed by category-based feedback using examples.

5.4   Analysis

We briefly analyze the sensitivity of our models w.r.t. their parameters. For each parameter we perform a sweep,[2] while using the default settings for all others. The best

---

[2] The sweep for mixture model parameters is performed in 0.1 steps in the $[0 \ldots 1]$ range; for the number feedback entities/terms/categories we use values $\{3, 5, 10, 15, 20, 25\}$.

Table 2. Results, with expansion. Best results in boldface. Baseline corresponds to model (5) in Table 1. Significance differences with baseline denoted with $^\triangle$ and $^\triangledown$.

Model	FB	$\hat{\theta}_q^T$	$\hat{\theta}_q^C$	XER2007		XER2008	
				MAP	MRR	xinfAP	MRR
Entity ranking (blind feedback only)							
BASELINE (no expansion)	-	-	-	0.2554	**0.4531**	0.3124	0.5024
qT-E'	$\{E'\}$	Eq. 15	-	0.2511	$0.3654^\triangledown$	0.3214	0.4694
qC-E'	$\{E'\}$	-	Eq. 17	**0.2590**	0.4516	$0.3317^\triangle$	**0.5042**
qT-E'+qC-E'	$\{E'\}$	Eq. 15	Eq. 17	0.2536	0.4144	$\mathbf{0.3369}^\triangle$	0.4984
List completion (blind feedback and/or examples)							
BASELINE (no expansion)	-	-	-	0.2202	0.4042	0.2729	0.4339
Blind feedback							
qT-E'	$\{E'\}$	Eq. 15	-	0.2449	0.3818	0.2814	0.4139
qC-E'	$\{E'\}$	-	Eq. 17	0.2251	0.3858	$0.2970^\triangle$	0.4777
qT-E'+qC-E'	$\{E'\}$	Eq. 15	Eq. 17	0.2188	0.3576	$0.3022^\triangle$	0.4768
Examples							
qT-E	$\{E\}$	Eq. 15	-	$0.2376^\triangle$	0.3875	0.2886	0.4274
qC-E	$\{E\}$	-	Eq. 17	$0.3139^\triangle$	$\mathbf{0.5380}^\triangle$	$0.3750^\triangle$	$0.6127^\triangle$
qT-E+qC-E	$\{E\}$	Eq. 15	Eq. 17	$\mathbf{0.3254}^\triangle$	$0.5357^\triangle$	$\mathbf{0.3827}^\triangle$	$\mathbf{0.6526}^\triangle$
Blind feedback plus examples							
qT-E'E	$\{E', E\}$	Eq. 15	-	0.2200	$0.3193^\triangledown$	0.2843	0.4036
qC-E'E	$\{E', E\}$	-	Eq. 17	$0.2563^\triangle$	0.4421	$0.3299^\triangle$	0.5048
qT-E'E+qC-E'E	$\{E', E\}$	Eq. 15	Eq. 17	$0.2474^\triangle$	0.3961	$0.3319^\triangle$	0.4701

Table 3. Results, using default parameters vs. parameters optimized for MAP. Significance tested against default parameter setting. Best results for each are in boldface.

Model	XER2007		XER2008	
	MAP	MRR	xinfAP	MRR
Entity ranking				
No expansion, default parameters	0.2554	0.4531	0.3124	0.5024
No expansion, λ optimized	$0.2873^\triangle$	0.4648	0.3156	0.5023
Expansion (blind feedback), default parameters	0.2590	0.4516	0.3369	0.4984
Expansion (blind feedback), optimized parameters	$\mathbf{0.3863}^\triangle$	$\mathbf{0.6509}^\triangle$	**0.3703**	**0.5849**
Best performing INEX run	0.306	-	0.341	-
List completion				
No expansion, default parameters	0.2202	0.4042	0.2729	0.4339
No expansion, λ optimized	0.2410	0.3997	0.2784	0.4693
Expansion (using examples), default parameters	0.3254	0.5357	0.3827	0.6526
Expansion (using examples), optimized parameters	**0.3384**	**0.5909**	$\mathbf{0.4182}^\triangle$	**0.7041**
Best performing INEX run	0.309	-	0.402	-

individually found values are then put together and used in the optimized run, reported in Table 3. This method may not result in the overall best possible parameter settings, however, it is not our aim here to tweak and fine-tune parameters. Runs without query expansion involve only one parameter, λ; the best empirically found value is 0.7. The feedback runs are insensitive to the number of feedback terms; we use $K_T = 15$, as before, however, more weight ($\lambda^T = 0.7$) is given to the expanded (term-based) query model. For the blind feedback runs, using a small number of feedback entities ($N = 3$) performs best. When example entities are given, relying heavily ($\lambda^C = 0.8$) on a small number of feedback categories ($K_C = 3$) performs best; for blind feedback a conservative strategy pays off ($K_C = 10, \lambda^C = 0.6$).

6   Conclusions

We have introduced a probabilistic framework for entity search. The framework allows us to systematically explore combinations of query and category information as well as example entities to create query models. It also allows us to integrate term-based and category-based feedback information. We explored our models along many dimensions; experimental evaluation was performed using the 2007 and 2008 editions of the INEX Entity Ranking track. We demonstrated the advantage of a category-based representation over a term-based one for query modeling. We also showed the effectiveness of category-based feedback, which was found to outperform term-based feedback. The biggest improvements over a competitive baseline based on term and category-based information were achieved when category-based feedback is used with example entities (provided along with the keyword query). State-of-the-art performance was achieved on the entity ranking and list completion tasks on all available test sets. In future work we plan a more detailed result analysis than we were able to include here and to examine ways of automatically estimating parameters that are topic dependent (i.e., dependent on the query terms, and/or target categories and/or example entities).

Acknowledgments. This research was supported by the DAESO and DuOMAn projects carried out within the STEVIN programme which is funded by the Dutch and Flemish Governments under project numbers STE-05-24 and STE-09-12, and by the Netherlands Organisation for Scientific Research (NWO) under project numbers 640.001.501, 640.002.501, 612.066.512, 612.061.814, 612.061.815, 640.004.802.

References

[1] Balog, K.: People Search in the Enterprise. PhD thesis, University of Amsterdam (2008)
[2] Balog, K., Azzopardi, L., de Rijke, M.: Formal models for expert finding in enterprise corpora. In: SIGIR 2006, pp. 43–50 (2006)
[3] Balog, K., Weerkamp, W., de Rijke, M.: A few examples go a long way. In: SIGIR 2008, pp. 371–378 (2008)
[4] Balog, K., Soboroff, I., Thomas, P., Craswell, N., de Vries, A.P., Bailey, P.: Overview of the TREC 2008 enterprise track. In: TREC 2008, NIST (2009)
[5] Chu-Carroll, J., Czuba, K., Prager, J., Ittycheriah, A., Blair-Goldensohn, S.: IBM's PIQUANT II in TREC 2004. In: Proceedings TREC 2004 (2004)
[6] Conrad, J., Utt, M.: A system for discovering relationships by feature extraction from text databases. In: SIGIR 1994, pp. 260–270 (1994)
[7] Craswell, N., Demartini, G., Gaugaz, J., Iofciu, T.: L3S at INEX2008: retrieving entities using structured information. In: Geva, et al. (eds.) [12], pp. 253–263
[8] de Vries, A., Vercoustre, A.-M., Thom, J.A., Craswell, N., Lalmas, M.: Overview of the INEX 2007 entity ranking track. In: Fuhr, et al. (eds.) [11], pp. 245–251

[9] Demartini, G., de Vries, A., Iofciu, T., Zhu, J.: Overview of the INEX 2008 entity ranking track. In: Geva, et al. (eds.) [12], pp. 243–252

[10] Fissaha Adafre, S., de Rijke, M., Tjong Kim Sang, E.: Entity retrieval. In: Recent Advances in Natural Language Processing (RANLP 2007) (September 2007)

[11] Fuhr, N., Kamps, J., Lalmas, M., Trotman, A. (eds.): INEX 2007. LNCS, vol. 4862. Springer, Heidelberg (2008)

[12] Geva, S., Kamps, J., Trotman, A. (eds.): INEX 2008. LNCS, vol. 5631. Springer, Heidelberg (2009)

[13] Ghahramani, Z., Heller, K.A.: Bayesian sets. In: NIPS 2005 (2005)

[14] GoogleSets (2009), http://labs.google.com/sets (accessed January 2009)

[15] Jämsen, J., Näppilä, T., Arvola, P.: Entity ranking based on category expansion. In: Fuhr, et al. (eds.) [11], pp. 264–278

[16] Jiang, J., Liu, W., Rong, X., Gao, Y.: Adapting language modeling methods for expert search to rank wikipedia entities. In: Geva, et al. (eds.) [12], pp. 264–272

[17] Kaptein, R., Kamps, J.: Finding entities in wikipedia using links and categories. In: Geva, et al. (eds.) [12], pp. 273–279

[18] Lafferty, J., Zhai, C.: Document language models, query models, and risk minimization for information retrieval. In: SIGIR 2001, pp. 111–119 (2001)

[19] Losada, D., Azzopardi, L.: An analysis on document length retrieval trends in language modeling smoothing. Information Retrieval 11(2), 109–138 (2008)

[20] Mishne, G., de Rijke, M.: A study of blog search. In: Lalmas, M., MacFarlane, A., Rüger, S.M., Tombros, A., Tsikrika, T., Yavlinsky, A. (eds.) ECIR 2006. LNCS, vol. 3936, pp. 289–301. Springer, Heidelberg (2006)

[21] Raghavan, H., Allan, J., Mccallum, A.: An exploration of entity models, collective classification and relation description. In: Link KDD 2004 (2004)

[22] Rose, D.E., Levinson, D.: Understanding user goals in web search. In: WWW 2004, pp. 13–19 (2004)

[23] Sayyadian, M., Shakery, A., Doan, A., Zhai, C.: Toward entity retrieval over structured and text data. In: WIRD 2004 (2004)

[24] Song, F., Croft, W.B.: A general language model for information retrieval. In: CIKM 1999, pp. 316–321 (1999)

[25] Tsikrika, T., Serdyukov, P., Rode, H., Westerveld, T., Aly, R., Hiemstra, D., de Vries, A.P.: Structured document retrieval, multimedia retrieval, and entity ranking using PF/Tijah. In: Fuhr, et al. (eds.) [11], pp. 306–320

[26] Vercoustre, A.-M., Pehcevski, J., Thom, J.A.: Using wikipedia categories and links in entity ranking. In: Fuhr, et al. (eds.) [11], pp. 321–335

[27] Vercoustre, A.-M., Thom, J.A., Pehcevski, J.: Entity ranking in wikipedia. In: SAC 2008, pp. 1101–1106 (2008)

[28] Vercoustre, A.-M., Pehcevski, J., Naumovski, V.: Topic difficulty prediction in entity ranking. In: Geva, et al. (eds.) [12], pp. 280–291

[29] Voorhees, E.: Overview of the TREC 2004 question answering track. In: Proceedings of TREC 2004 (2005) NIST Special Publication: SP 500–261

[30] Weerkamp, W., He, J., Balog, K., Meij, E.: A generative language modeling approach for ranking entities. In: Geva, et al. (eds.) [12], pp. 292–299

[31] Yilmaz, E., Kanoulas, E., Aslam, J.A.: A simple and efficient sampling method for estimating AP and NDCG. In: SIGIR 2008, pp. 603–610 (2008)

[32] Zaragoza, H., Rode, H., Mika, P., Atserias, J., Ciaramita, M., Attardi, G.: Ranking very many typed entities on wikipedia. In: CIKM 2007, pp. 1015–1018 (2007)

[33] Zhai, C., Lafferty, J.: A study of smoothing methods for language models applied to information retrieval. ACM Trans. Inf. Syst. 22(2), 179–214 (2004)

[34] Zhu, J., Song, D., Rüger, S.: Integrating document features for entity ranking. In: Fuhr, et al. (eds.) [11], pp. 336–347

Maximum Margin Ranking Algorithms for Information Retrieval

Shivani Agarwal and Michael Collins

Massachusetts Institute of Technology, Cambridge MA 02139, USA
{shivani,mcollins}@csail.mit.edu

Abstract. Machine learning ranking methods are increasingly applied to ranking tasks in information retrieval (IR). However ranking tasks in IR often differ from standard ranking tasks in machine learning, both in terms of problem
structure and in terms of the evaluation criteria used to measure performance.
Consequently, there has been much interest in recent years in developing ranking
algorithms that directly optimize IR ranking measures. Here we propose a family
of ranking algorithms that preserve the simplicity of standard pair-wise ranking
methods in machine learning, yet show performance comparable to state-of-the-
art IR ranking algorithms. Our algorithms optimize variations of the hinge loss
used in support vector machines (SVMs); we discuss three variations, and in each
case, give simple and efficient stochastic gradient algorithms to solve the resulting optimization problems. Two of these are stochastic gradient projection algorithms, one of which relies on a recent method for $l_{1,\infty}$-norm projections; the
third is a stochastic exponentiated gradient algorithm. The algorithms are simple and efficient, have provable convergence properties, and in our preliminary
experiments, show performance close to state-of-the-art algorithms that directly
optimize IR ranking measures.

1 Introduction

Ranking methods in machine learning have gained considerable popularity in information retrieval (IR) in recent years [1, 2, 3, 4, 5, 6, 7, 8, 9, 10, 11, 12]. Although the benefit
of using such methods is rarely in question, there has been much debate recently about
what types of ranking algorithms are best suited for the domain. In particular, ranking
tasks in IR often differ from standard ranking tasks in machine learning in a variety
of ways: for example, often in IR, one does not wish to learn a single ranking over all
objects (in this case documents), but rather wishes to learn a ranking function that can
rank different sets of documents with respect to different queries. Moreover, the ranking
performance measures used in IR are usually different from standard pair-wise ranking
measures, often focusing on the ranking quality at the top of a retrieved list.

These differences have led to several questions about how to best design ranking
algorithms for IR, as well as several worthwhile adaptations and improvements of existing ranking algorithms [4,7,13]. For example, Qin et al. [13] argue that loss functions
in IR should be defined at the level of queries rather than individual documents or document pairs. There has also been much discussion on pair-wise vs. list-wise ranking
algorithms, where the latter employ loss functions that directly take into account the
total order in a ranked list [14].

C. Gurrin et al. (Eds.): ECIR 2010, LNCS 5993, pp. 332–343, 2010.

More recently, there has been much interest in algorithms that attempt to directly optimize ranking measures that are popular in IR, such as the normalized discounted cumulative gain (NDCG), mean average precision (MAP), and others [15, 5, 8, 7, 16, 10, 11, 12]. For example, Joachims [15] proposed a general method, inspired by large margin methods for structured prediction [17, 18], for optimizing multivariate performance measures that include as special cases the area under the ROC curve and measures related to recall and precision; this was extended by Yue et al. [8] to a support vector method for optimizing the MAP. Large margin structured prediction methods have also been used by Chapelle et al. [16] to optimize the NDCG. Other algorithms that attempt to optimize the NDCG include the LambdaRank algorithm of Burges et al. [5], the AdaRank algorithm of Xu and Li [7], the SoftRank algorithm of Taylor et al. [9], the regression-based algorithm of Cossock and Zhang [11], and the recent algorithm of Chapelle and Wu [12]. These algorithms have shown considerable promise, often resulting in significant improvement in performance over standard pair-wise ranking algorithms, such as RankSVM [19, 2] and RankBoost [20], applied to IR ranking tasks.

Here we propose a family of algorithms that preserve the simplicity of the pair-wise approach, yet exhibit performance comparable to state-of-the-art IR ranking algorithms. Our algorithms are based on optimizing variations of the hinge loss used in support vector machines (SVMs). We start with a pair-wise ranking loss that takes into account the degree of difference in the relevance (with respect to a query) of a pair of documents, and then use this to construct a query-level loss function. We discuss three variations of the query-level loss; in each case, we provide a stochastic gradient algorithm for solving the dual of the corresponding optimization problem. Two of these are stochastic gradient projection algorithms, one of which relies on a recent algorithm for projections onto $l_{1,\infty}$-norm constraints [21]; the third is a stochastic exponentiated gradient algorithm, similar to exponentiated gradient algorithms developed for structured prediction [22]. The resulting algorithms are simple and efficient, have provable convergence properties, and in our preliminary experiments, show performance close to state-of-the-art algorithms that optimize the MAP or NDCG.

The rest of the paper is organized as follows. In Section 2, we describe more formally the problem setting we consider. Section 3 gives our algorithms; this is followed by our experimental results in Section 4. We conclude with a discussion in Section 5.

2   Preliminaries

The problem we consider can be described as follows. There is a query space $\mathcal{Q}$ and a document space $\mathcal{D}$. The learner is given as training examples m queries $q^1, \ldots, q^m \in \mathcal{Q}$, the ith query q^i being associated with n_i documents $d_1^i, \ldots, d_{n_i}^i \in \mathcal{D}$, together with real-valued relevance labels $y_j^i \in \mathbb{R}$ denoting the (human-judged) relevance of document d_j^i to query q^i, and the goal is to learn from these examples a ranking function which, given a new query q, can rank the documents associated with the query such that more relevant documents are ranked higher than less relevant ones.

More formally, as is standard in the use of machine learning methods in IR, we shall assume a query-document feature mapping $\phi : \mathcal{Q} \times \mathcal{D} \rightarrow \mathbb{R}^d$ that maps each query-document pair to a d-dimensional feature vector. The learner then receives labeled training examples of the form $S = (S^1, \ldots, S^m)$, where $S^i = ((\phi_1^i, y_1^i), \ldots, (\phi_{n_i}^i, y_{n_i}^i))$ is the training sample associated with the ith query; here $\phi_j^i \equiv \phi(q^i, d_j^i)$. The goal is to learn a real-valued ranking function $f : \mathbb{R}^d \to \mathbb{R}$ that ranks accurately documents associated with future queries; f is taken to rank a document d_j associated with a query q higher than a document d_k if $f(\phi(q, d_j)) > f(\phi(q, d_k))$, and lower than d_k otherwise. In this paper, we shall be interested in linear ranking functions $f_\mathbf{w} : \mathbb{R}^d \to \mathbb{R}$ given by

$$f_\mathbf{w}(\phi) = \mathbf{w} \cdot \phi$$

for some weight vector $\mathbf{w} \in \mathbb{R}^d$.

Let us consider first the loss of such a function $f_\mathbf{w}$ on a pair of documents d_j^i, d_k^i associated with query q^i. In earlier SVM algorithms for ranking [2, 19], the following pair-wise hinge loss was often used:

$$\ell_\mathsf{H}(\mathbf{w}, (\phi_j^i, y_j^i), (\phi_k^i, y_k^i)) = \left[1 - z_{jk}^i \mathbf{w} \cdot (\phi_j^i - \phi_k^i)\right]_+ ,$$

where $[a]_+ = \max(0, a)$ and

$$z_{jk}^i = \mathrm{sign}(y_j^i - y_k^i) .$$

These algorithms then consisted of minimizing an l_2-regularized version of the average pair-wise hinge-loss across all document pairs and all queries. In particular, if

$$R_i = \left\{(j, k) \mid y_j^i > y_k^i\right\}$$

denotes the set of 'preference pairs' for the ith query, then these early algorithms learned $\mathbf{w}$ by solving the following optimization problem:

$$\min_\mathbf{w} \left[\frac{1}{2}\|\mathbf{w}\|^2 + \frac{C}{\left(\sum_{i=1}^m |R_i|\right)} \sum_{i=1}^m \sum_{(j,k) \in R_i} \ell_\mathsf{H}(\mathbf{w}, (\phi_j^i, y_j^i), (\phi_k^i, y_k^i))\right] , \quad (1)$$

where $C > 0$ denotes an appropriate regularization parameter. The pair-wise hinge loss ℓ_H can be seen as a convex upper bound on the following binary mis-ranking error, which simply assigns a constant penalty of 1 to each mis-ranked pair of documents:

$$\ell_\mathsf{0\text{-}1}(\mathbf{w}, (\phi_j^i, y_j^i), (\phi_k^i, y_k^i)) = \mathbf{1}\left(z_{jk}^i \mathbf{w} \cdot (\phi_j^i - \phi_k^i) < 0\right) ,$$

where $\mathbf{1}(\psi)$ is the indicator function that takes the value 1 if the predicate ψ is true, and 0 otherwise. Thus, early SVM ranking algorithms ignored the possibility of a need to assign different penalties to different mis-ranked pairs, which arises when documents can have multiple relevance levels y_j^i. Recently, Cao et al. [4] addressed this issue by suggesting a modification to the above pair-wise hinge loss; however, the loss they propose relies on certain heuristics in order to set some parameters. Here we use the following simple and intuitive variation of the pair-wise hinge loss, which takes into account the different relevance levels of different documents; this loss was also used recently in the context of standard ranking algorithms in [23]:

$$\ell_\mathsf{H,rel}(\mathbf{w}, (\phi_j^i, y_j^i), (\phi_k^i, y_k^i)) = \left[|y_j^i - y_k^i| - z_{jk}^i \mathbf{w} \cdot (\phi_j^i - \phi_k^i)\right]_+ .$$

This can be viewed as a convex upper bound on the following relevance-weighted mis-ranking error:

$$\ell_{\text{rel}}(\mathbf{w}, (\boldsymbol{\phi}_j^i, y_j^i), (\boldsymbol{\phi}_k^i, y_k^i)) = |y_j^i - y_k^i| \, \mathbf{1}\left(z_{jk}^i \mathbf{w} \cdot (\boldsymbol{\phi}_j^i - \boldsymbol{\phi}_k^i) < 0\right) \; .$$

Thus, under ℓ_{rel} (and therefore $\ell_{\text{H,rel}}$), mis-ranking a pair of documents with relevance labels 1 and 5 incurs a larger penalty than mis-ranking a pair of documents with relevance labels 1 and 2.

Our main interest will be not in the above pair-wise loss $\ell_{\text{H,rel}}$ itself, but rather in query-level loss functions derived from it. In particular, we shall be interested in learning algorithms that select $\mathbf{w} \in \mathbb{R}^d$ as follows:

$$\min_{\mathbf{w}} \left[\frac{1}{2} \|\mathbf{w}\|^2 + \frac{C}{m} \sum_{i=1}^{m} L(\mathbf{w}, S^i) \right] \; ,$$

where $L(\mathbf{w}, S^i)$ is an appropriate loss term that measures the ranking loss incurred by $f_{\mathbf{w}}$ on the training sample S^i associated with the ith query, and $C > 0$ acts as a regularization parameter.

We note that our approach is different from recent large margin methods aimed at directly optimizing measures such as the MAP or NDCG [8, 16], which also use query-level loss functions based on the hinge loss. In particular, the query-level loss L in these approaches takes the form

$$L(\mathbf{w}, S^i) = \max_{\pi} \left[\Delta(\pi, \pi_y) - \left(g_{\mathbf{w}}((\boldsymbol{\phi}_1^i, \dots, \boldsymbol{\phi}_{n_i}^i), \pi_y) - g_{\mathbf{w}}((\boldsymbol{\phi}_1^i, \dots, \boldsymbol{\phi}_{n_i}^i), \pi)) \right]_+ \; ,$$

where π denotes either a permutation of the n_i documents (for NDCG) or a vector of binary assignments to the documents (for MAP, in the case of binary relevance labels y_j^i); π_y denotes a 'true' permutation or binary assignment vector induced by the relevance labels y_j^i; $\Delta(\pi, \pi_y)$ measures the loss incurred in predicting π instead of π_y (which can be taken to be one minus the NDCG or MAP of π relative to π_y); and $g_{\mathbf{w}}$ is an appropriately defined function that assigns a score to each permutation or binary assignment vector over the documents, and is used to predict such a permutation or assignment vector (rather than directly rank the documents based on the scores $\mathbf{w} \cdot \boldsymbol{\phi}_j^i$ as in our case) via

$$\widehat{\pi} = \arg\max_{\pi} \left[g_{\mathbf{w}}((\boldsymbol{\phi}_1^i, \dots, \boldsymbol{\phi}_{n_i}^i), \pi) \right] \; .$$

See [8, 16] for further details of such approaches.

In contrast, the query-level loss L in our case will be constructed from the simple and intuitive relevance-weighted pair-wise hinge loss $\ell_{\text{H,rel}}$ described above. In the following, we describe three different constructions for L, and in each case give efficient stochastic gradient algorithms to solve the resulting optimization problems.

3   Algorithms

As discussed above, the ranking algorithms we consider learn a linear ranking function $f_{\mathbf{w}} : \mathbb{R}^d \to \mathbb{R}$ by solving an optimization problem of the following form:

$$\min_{\mathbf{w}} \left[\frac{1}{2} \|\mathbf{w}\|^2 + \frac{C}{m} \sum_{i=1}^{m} L(\mathbf{w}, S^i) \right] \; ,$$

where $L(\mathbf{w}, S^i)$ denotes a query-level loss function that will be constructed from the relevance-weighted pair-wise hinge loss $\ell_{H,rel}$ described above.

3.1 Stochastic Gradient Projection Algorithm for Average Pair-Wise Loss

The first construction we consider for the query-level loss L is the following **average pair-wise loss:**

$$L_1^{H,rel}(\mathbf{w}, S^i) = \frac{1}{|R_i|} \sum_{(j,k) \in R_i} \ell_{H,rel}(\mathbf{w}, (\phi_j^i, y_j^i), (\phi_k^i, y_k^i)).$$

Notice that in addition to the relevance weighting in $\ell_{H,rel}$, this differs from the early SVM ranking algorithms described in Section 2 (see Eq. (1)) in that the loss is normalized by query, taking into account different numbers of document pairs for different queries. While one could in principle use stochastic subgradient methods (such as those of [24]) to directly solve the resulting optimization problem, we focus here on a dual version as this will facilitate the development of similar algorithms for the other two loss formulations. Using standard techniques involving the introduction of slack variables, we can write the minimization problem corresponding to the above loss as

$$\min_{\mathbf{w}, \boldsymbol{\xi}} \left[\frac{1}{2} \|\mathbf{w}\|^2 + \frac{C}{m} \sum_{i=1}^m \frac{1}{|R_i|} \sum_{(j,k) \in R_i} \xi_{jk}^i \right]$$

subject to

$$\xi_{jk}^i \geq 0 \qquad\qquad \forall\, i, j, k$$
$$\xi_{jk}^i \geq (y_j^i - y_k^i) - \mathbf{w} \cdot (\phi_j^i - \phi_k^i) \quad \forall\, i, j, k.$$

Introducing Lagrange multipliers and taking the Lagrangian dual then results in the following convex quadratic program (QP):

$$\min_{\boldsymbol{\alpha}} \left[\frac{1}{2} \sum_{i=1}^m \sum_{(j,k) \in R_i} \sum_{i'=1}^m \sum_{(j',k') \in R_{i'}} \alpha_{jk}^i \alpha_{j'k'}^{i'} Q_{jk,j'k'}^{i,i'} - \sum_{i=1}^m \sum_{(j,k) \in R_i} \alpha_{jk}^i (y_j^i - y_k^i) \right]$$

subject to

$$0 \leq \alpha_{jk}^i \leq \frac{C}{m|R_i|} \qquad \forall\, i, j, k\,,$$

where

$$Q_{jk,j'k'}^{i,i'} = (\phi_j^i - \phi_k^i) \cdot (\phi_{j'}^{i'} - \phi_{k'}^{i'}).$$

Solving the above QP using standard QP solvers would take $O\big(\big(\sum_{i=1}^m |R_i| \big)^3 \big)$ time, which is $O(m^3 n^6)$ if $n_i = n$ and $|R_i| = O(n^2)$ for all i. Instead, we use a simple stochastic gradient projection method which starts with some initial values $\boldsymbol{\alpha}^{(1)}$ for $\boldsymbol{\alpha}$, and on each iteration t, randomly selects a single query i and updates the corresponding $|R_i|$ variables $\alpha_{jk}^i{}^{(t)}$ using a gradient and projection step:

$$\boldsymbol{\alpha}^{i(t+1)} \leftarrow \mathcal{P}_{\Omega_i}\Big(\boldsymbol{\alpha}^{i(t)} - \eta_t \nabla^{i(t)} \Big),$$

$$\boldsymbol{\alpha}^{i'(t+1)} \leftarrow \boldsymbol{\alpha}^{i'(t)} \quad \text{for } i' \neq i,$$

where $\eta_t > 0$ is a learning rate; $\nabla^{i(t)} \in \mathbb{R}^{|R_i|}$ is the partial gradient of the objective function in the above QP with respect to $\boldsymbol{\alpha}^i$, evaluated at $\boldsymbol{\alpha}^{(t)}$; $\Omega_i = \{\boldsymbol{\alpha}^i \in \mathbb{R}^{|R_i|} : 0 \leq \alpha_{jk}^i \leq \frac{C}{m|R_i|} \ \forall (j,k) \in R_i\}$ is constraint set for $\boldsymbol{\alpha}^i$ in the above QP; and $\mathcal{P}_{\Omega_i}$ denotes Euclidean projection onto Ω_i. The projection onto the box constraints in Ω_i is straightforward: values of α_{jk}^i outside the interval $[0, \frac{C}{m|R_i|}]$ are simply clipped to the interval. The convergence proof for standard gradient projection methods can be extended to show that if $\eta_t = \frac{\eta_0}{\sqrt{t}}$ for some constant $\eta_0 > 0$, then the above stochastic gradient projection algorithm converges (in expectation) to an optimal solution, and moreover, the number of iterations required to reach a solution whose objective value (in expectation) is within ϵ of the optimal is $O(m^2/\epsilon^2)$; we omit the details for lack of space. An iteration that updates the variables associated with the ith query takes $O(|R_i|)$ time; this leads to a total of $O(m^2 n^2/\epsilon^2)$ time if $n_i = n$ and $|R_i| = O(n^2)$ for all i. On solving the above QP for $\boldsymbol{\alpha}$, the weight vector $\mathbf{w}$ can be recovered from $\boldsymbol{\alpha}$ in a standard manner.

3.2 Stochastic Exponentiated Gradient Algorithm for Maximum Pair-Wise Loss

The next construction we consider for L is the following **maximum pair-wise loss**:

$$L_2^{\mathsf{H,rel}}(\mathbf{w}, S^i) = \max_{(j,k) \in R_i} \left[\ell_{\mathsf{H,rel}}(\mathbf{w}, (\boldsymbol{\phi}_j^i, y_j^i), (\boldsymbol{\phi}_k^i, y_k^i)) \right] .$$

Define the ranking margin of $\mathbf{w}$ on a pair of documents $(j,k) \in R_i$ as $\mathbf{w} \cdot (\boldsymbol{\phi}_j^i - \boldsymbol{\phi}_k^i)$ if $\mathbf{w} \cdot (\boldsymbol{\phi}_j^i - \boldsymbol{\phi}_k^i) < (y_j^i - y_k^i)$, and $(y_j^i - y_k^i)$ otherwise. Then the resulting algorithm in this case will maximize not the average ranking margin over document pairs associated with a query, but rather the minimum ranking margin across all document pairs associated with each query. Again, we can write the corresponding minimization problem as

$$\min_{\mathbf{w}, \boldsymbol{\xi}} \left[\frac{1}{2} \|\mathbf{w}\|^2 + \frac{C}{m} \sum_{i=1}^m \xi^i \right]$$

subject to

$$\xi^i \geq 0 \qquad\qquad\qquad\qquad\qquad \forall i$$
$$\xi^i \geq (y_j^i - y_k^i) - \mathbf{w} \cdot (\boldsymbol{\phi}_j^i - \boldsymbol{\phi}_k^i) \quad \forall i, j, k.$$

Introducing Lagrange multipliers and taking the Lagrangian dual then results in the following convex QP (after an appropriate scaling of variables, and introduction of an additional variable α_0^i for each i):

$$\min_{\boldsymbol{\alpha}} \left[\frac{1}{2} \sum_{i=1}^m \sum_{(j,k) \in R_i} \sum_{i'=1}^m \sum_{(j',k') \in R_{i'}} \alpha_{jk}^i \alpha_{j'k'}^{i'} \frac{C}{m} Q_{jk,j'k'}^{i,i'} - \sum_{i=1}^m \sum_{(j,k) \in R_i} \alpha_{jk}^i (y_j^i - y_k^i) \right]$$

subject to

$$\alpha_0^i + \sum_{(j,k) \in R_i} \alpha_{jk}^i = 1 \quad \forall i$$
$$\alpha_{jk}^i, \ \alpha_0^i \geq 0 \qquad\qquad \forall i, j, k.$$

The constraints in the above problem force $\boldsymbol{\alpha}^i$ to lie in the simplex Δ_i of distributions over $|R_i| + 1$ elements for each i. This allows us to derive an efficient exponentiated gradient (EG) algorithm in this case which starts with an initial set of distributions $\boldsymbol{\alpha}^{(1)} \in \Delta_1 \times \ldots \times \Delta_m$, and on each iteration t, updates the distribution associated with a single randomly chosen query i using an exponentiated gradient step:

$$\alpha_{jk}^{i\ (t+1)} \leftarrow \frac{\alpha_{jk}^{i\ (t)} \exp(-\eta_0 \nabla_{jk}^{i\ (t)})}{Z^{(t)}} ; \quad \alpha_0^{i\ (t+1)} \leftarrow \frac{\alpha_0^{i\ (t)}}{Z^{(t)}} ,$$

$$\boldsymbol{\alpha}^{i'\ (t+1)} \leftarrow \boldsymbol{\alpha}^{i'\ (t)} \text{ for } i' \neq i ,$$

where $\eta_0 > 0$ is a constant learning rate; $\nabla_{jk}^{i\ (t)}$ is the partial derivative of the objective in the above QP with respect to α_{jk}^i, evaluated at $\boldsymbol{\alpha}^{(t)}$; and $Z^{(t)}$ is chosen to ensure $\boldsymbol{\alpha}^{i\ (t+1)} \in \Delta_i$. The algorithm can actually be implemented in a way that requires only $O(n_i)$ time per iteration rather than $O(|R_i|)$ time using ideas developed for structured prediction problems [22]; we omit the details for lack of space. As in [22], the resulting algorithm can be shown to converge (in expectation) to an optimal solution; if $n_i = n$ for all i, then the time required to reach a solution whose objective value (in expectation) is within ϵ of the optimal is $O\left(\frac{mn}{\epsilon}\left(|A|_\infty D[\boldsymbol{\alpha}^* \| \boldsymbol{\alpha}^{(1)}] + Q(\boldsymbol{\alpha}^{(1)}) - Q(\boldsymbol{\alpha}^*)\right)\right)$, where $\boldsymbol{\alpha}^{(1)}, \boldsymbol{\alpha}^*$ denote the initial and optimal sets of distributions, respectively; $Q(\boldsymbol{\alpha})$ denotes the objective function in the above problem; $D[\boldsymbol{\alpha}^* \| \boldsymbol{\alpha}^{(1)}]$ denotes the sum of the Kullback-Leibler divergences between $\boldsymbol{\alpha}^{i*}$ and $\boldsymbol{\alpha}^{i\ (1)}$ over all i, and $|A|_\infty$ is the largest entry in the matrix whose entries are $A_{(i,j,k),(i',j',k')} = \frac{C}{m} Q_{jk,j'k'}^{i,i'}$.

3.3   Stochastic Gradient Projection Algorithm for Maximum-Average Pair-Wise Loss

In taking the average or maximum pair-wise ranking loss as above, one does not distinguish ranking errors at the top of the list from ranking errors at the bottom. However, in practice, and in IR in particular, ranking errors at the top of the list are often more costly (for example, in web search, the accuracy of the first few web pages returned by a search engine is paramount). To this end, we consider a hybrid **maximum-average pair-wise loss** construction for L:

$$L_3^{\mathsf{H,rel}}(\mathbf{w}, S^i) = \max_{k:|P_{ik}|>0} \frac{1}{|P_{ik}|} \sum_{j \in P_{ik}} \ell_{\mathsf{H,rel}}(\mathbf{w}, (\phi_j^i, y_j^i), (\phi_k^i, y_k^i)) ,$$

where $P_{ik} = \{j \mid y_j^i > y_k^i\}$ denotes the set of documents that are preferred to d_k^i. To see why this loss term might penalize ranking errors at the top more heavily than ranking errors at the bottom, note that the cost of each 'mis-ranking up' of a document is inversely weighted by the number of documents preferred to it; therefore, 'mis-ranking up' a lower-relevance document by a few positions is less costly than 'mis-ranking up' a higher-relevance document. By minimizing the largest such 'mis-ranking up' cost over all documents, the resulting algorithm should therefore discourage mis-ranking of higher-relevance documents, resulting in good accuracy at the top of the returned ranking. Indeed, the above loss is reminiscent of the l_p-norm based loss studied in [25], where a greater value of p corresponds to a greater push toward more accurate ranking performance at the top of the returned list; the above loss can be viewed as an l_∞ extreme, using the relevance-weighted hinge loss instead of the (binary) exponential loss used in [25]. Introducing slack variables as before, the corresponding minimization problem can be written as

$$\min_{\mathbf{w},\boldsymbol{\xi}} \left[\frac{1}{2} \|\mathbf{w}\|^2 + \frac{C}{m} \sum_{i=1}^{m} \xi^i \right]$$

subject to

$$\xi^i \geq \frac{1}{|P_{ik}|} \sum_{j \in P_{ik}} \xi^i_{jk} \qquad \forall\, i, k$$

$$\xi^i_{jk} \geq 0 \qquad \forall\, i, j, k$$

$$\xi^i_{jk} \geq (y^i_j - y^i_k) - \mathbf{w} \cdot (\phi^i_j - \phi^i_k) \qquad \forall\, i, j, k.$$

Introducing Lagrange multipliers and taking the Lagrangian dual then results in the following convex optimization problem (after an appropriate scaling of variables):

$$\min_{\boldsymbol{\alpha},\boldsymbol{\gamma}} \left[\frac{1}{2} \sum_{i=1}^{m} \sum_{(j,k) \in R_i} \sum_{i'=1}^{m} \sum_{(j',k') \in R_{i'}} \alpha^i_{jk} \alpha^{i'}_{j'k'} \frac{Q^{i,i'}_{jk,j'k'}}{|P_{ik}||P_{i'k'}|} - \sum_{i=1}^{m} \sum_{(j,k) \in R_i} \alpha^i_{jk} \frac{(y^i_j - y^i_k)}{|P_{ik}|} \right]$$

subject to

$$0 \leq \alpha^i_{jk} \leq \gamma^i_k \qquad \forall\, i, j, k$$

$$\sum_{k:|P_{ik}|>0} \gamma^i_k = \frac{C}{m} \qquad \forall\, i.$$

The constraints in this case can be interpreted as a set of constraints on the $l_{1,\infty}$-norm of $\boldsymbol{\alpha}^i$ for each i, together with non-negativity constraints. Quattoni et al. [21] recently developed an efficient algorithm for $l_{1,\infty}$-norm projections; this allows us to use a stochastic gradient projection method similar to that discussed for the average pair-wise loss in Section 3.1. In this case, each projection step consists of a projection onto the $l_{1,\infty}$ constraints using the method of [21], followed by a projection onto the non-negativity constraints (which simply involves setting negative values of α^i_{jk} to zero). The number of iterations to reach an ϵ-optimal solution (in expectation) is $O(m^2/\epsilon^2)$ as before; each iteration now takes $O(|R_i| \log |R_i|)$ time owing to the projection, thus leading to a total of $O(m^2 n^2 \log n/\epsilon^2)$ time if $n_i = n$ and $|R_i| = O(n^2)$ for all i.

4   Experiments

In preliminary experiments, we evaluated our algorithms on the OHSUMED data set, a benchmark data set for IR ranking algorithms available publicly as part of the LETOR distribution[1] [26] (we used LETOR 3.0). The data set consists of 106 medical queries.

---

[1] Available from http://research.microsoft.com/en-us/um/beijing/projects/letor/

Each query is associated with a number of documents, each of which has been judged by human experts as being either definitely relevant to the query (label 2), partially relevant (label 1), or not relevant (label 0); there are a total of 16,140 such query-document pairs with relevance judgments (an average of roughly 152 judged documents per query). Each query-document pair is represented as a vector of 45 features.

There are five folds provided in the data set; each fold consists of a split of the queries into roughly 60% for training, 20% for validation, and 20% for testing. We evaluated our algorithms on these five folds and compared their performance with several other ranking algorithms for which results are available as baselines in the LETOR distribution. The following performance measures were used to evaluate the algorithms:

1. **NDCG@k**: The NDCG@k of a ranking function f on a query q with n documents $d_1, \ldots,$ d_n, associated with relevance labels $y_1, \ldots, y_n$, is the NDCG (which is simply NDCG@n) [27] truncated to the top k documents returned by f:

$$\text{NDCG@}k[f] = \frac{1}{Z_k} \sum_{j=1}^{k} \frac{2^{y_{\sigma(j)}} - 1}{\text{discount}(j)},$$

where $\sigma(j)$ denotes the index of the document ranked at the jth position by f; discount(j) is a discounting factor that discounts the contribution of documents ranked lower in the list; and Z_k is a constant that ensures the maximum NDCG@k over all rankings is 1. In the LETOR evaluation tool, discount(j) is defined to be 1 if $j = 1$, and $\log_2(j)$ otherwise.

2. **Prec@k**: For binary labels, where a label of 1 is considered as relevant and 0 as irrelevant, the Prec@k of a ranking function f on a query q as above is the proportion of relevant documents in the top k documents returned by f:

$$\text{Prec@}k[f] = \frac{1}{k} \sum_{j=1}^{k} \mathbf{1}(y_{\sigma(j)} = 1),$$

where $\mathbf{1}(\cdot)$ is an indicator function whose value is 1 if its argument is true and 0 otherwise. For the OHSUMED data, relevance labels of 1 and 2 are considered relevant, and 0 irrelevant.

3. **MAP**: The average precision (AP) of a ranking function f on a query q as above is the average Prec@k over all positions k occupied by relevant documents:

$$\text{AP}[f] = \frac{\sum_{k=1}^{n} \text{Prec@}k[f] \cdot \mathbf{1}(y_{\sigma(j)} = 1)}{\sum_{k=1}^{n} \mathbf{1}(y_{\sigma(j)} = 1)}.$$

The mean average precision (MAP) refers to the mean AP across a set of queries.

The results are shown in Figure 1; here the algorithms of Sections 3.1, 3.2, and 3.3 are referred to as RankMM-1, RankMM-2, and RankMM-3, respectively. For comparison, we also show results for the following algorithms (all obtained from the LETOR website): regression, RankSVM [2, 19], RankBoost [20], ListNet [14], two versions of AdaRank [7], and SVMMAP [8]. All results shown are averages across the five folds. For each fold, the regularization parameter C, learning rate η_0, and number of iterations T in our algorithms were selected from the ranges $\{0.1, 1, 10, 100, 1000\}$, $\{10^{-2}, 10^{-3}, 10^{-4}, 10^{-5}, 10^{-6}\}$, and $\{100, 250, 500, 750, 1000\}$, respectively; in particular, for consistency with the other results reported in LETOR, the parameters that gave the highest MAP on the validation set were used in each case for training.

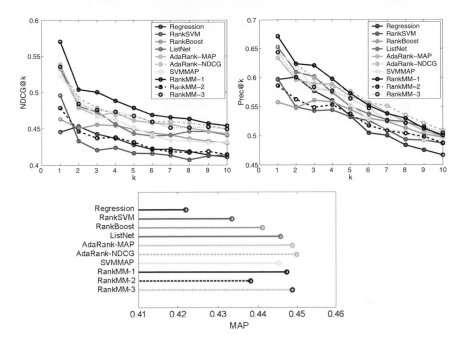

Fig. 1. Results on the OHSUMED data set in terms of (**top left**) NDCG@k; (**top right**) Prec@k; and (**bottom**) MAP. See text for details.

Of the algorithms shown for comparison, regression is a point-wise algorithm that predicts labels of individual documents; RankSVM and RankBoost are pair-wise ranking algorithms; and the remainder are list-wise ranking algorithms, with the last three directly optimizing the MAP or NDCG. Other than RankBoost, which uses thresholded features as weak rankers, all algorithms learn a linear ranking function. As can be seen, the performance of our algorithms, particularly RankMM-1 and RankMM-3, is considerably superior to the standard pair-wise (and point-wise) ranking algorithms[2], and indeed, in many cases, is comparable to the performance of the best algorithms that directly optimize the MAP or NDCG. We note that RankMM-2 appears not to be as suited to IR performance measures; RankMM-3 appears to be particularly suited to MAP. The best overall performance (for this data set) is obtained using RankMM-1.

5   Conclusion and Open Questions

We have proposed a family of ranking algorithms for IR that employ loss functions derived from a pair-wise ranking loss, yet show performance comparable to a number of algorithms that have been proposed recently for optimizing IR ranking measures such as the MAP and NDCG. Our two best performing algorithms are both stochastic

---

[2] It was recently found that RankSVM, with proper training, yields better results than those reported in LETOR [28]. Our algorithms show improvement over these new results as well.

gradient projection algorithms, one of which requires $l_{1,\infty}$-norm projections, for which we use a method of [21]; the third is a stochastic exponentiated gradient (EG) algorithm.

There are several open questions regarding ranking algorithms in IR. The relationships between different ranking methods and performance measures are still not clearly understood; for example, it would be of interest to study statistical convergence properties of these algorithms, as has been done for example in [11]. Another practical issue, given the scale of many IR applications, is efficiency. We have obtained an $O(1/\epsilon^2)$ rate of convergence to an ϵ-optimal solution for our stochastic gradient projection algorithms, and an $O(1/\epsilon)$ rate for the EG algorithm. We expect it may be possible to obtain an $O(1/\epsilon)$ rate for the projection algorithms as well; we leave this to future work.

Finally, a larger scale comparison of different ranking algorithms is required to better understand their respective merits and shortcomings. Unfortunately obtaining appropriate data sets for this purpose that include more than two relevance levels has been a challenge due to their mostly proprietary nature. We hope to evaluate these algorithms on larger scale web search data in the near future.

Acknowledgments. This work was supported in part by the National Science Foundation (NSF) under Grant No. DMS-0732334. Any opinions, findings, and conclusions or recommendations expressed in this article are those of the authors and do not necessarily reflect the views of the NSF.

References

1. Herbrich, R., Graepel, T., Bollmann-Sdorra, P., Obermayer, K.: Learning preference relations or information retrieval. In: Proceedings of the ICML-1998 Workshop on Text Categorization and Machine Learning (1998)
2. Joachims, T.: Optimizing search engines using clickthrough data. In: Proceedings of the 8th ACM Conference on Knowledge Discovery and Data Mining (2002)
3. Burges, C.J.C., Shaked, T., Renshaw, E., Lazier, A., Deeds, M., Hamilton, N., Hullender, G.: Learning to rank using gradient descent. In: Proceedings of the 22nd International Conference on Machine Learning (2005)
4. Cao, Y., Xu, J., Liu, T.Y., Li, H., Hunag, Y., Hon, H.W.: Adapting ranking SVM to document retrieval. In: Proceedings of the 29th ACM SIGIR Conference on Research and Development in Information Retrieval (2006)
5. Burges, C.J.C., Ragno, R., Le, Q.V.: Learning to rank with non-smooth cost functions. In: Advances in Neural Information Processing Systems, vol. 19. MIT Press, Cambridge (2007)
6. Tsai, M.F., Liu, T.Y., Qin, T., Chen, H.H., Ma, W.Y.: FRank: A ranking method with fidelity loss. In: Proceedings of the 30th ACM SIGIR Conference on Research and Development in Information Retrieval (2007)
7. Xu, J., Li, H.: AdaRank: A boosting algorithm for information retrieval. In: Proceedings of the 30th ACM SIGIR Conference on Research and Development in Information Retrieval (2007)
8. Yue, Y., Finley, T., Radlinski, F., Joachims, T.: A support vector method for optimizing average precision. In: Proceedings of the 30th ACM SIGIR Conference on Research and Development in Information Retrieval (2007)
9. Taylor, M., Guiver, J., Robertson, S., Minka, T.: Softrank: optimizing non-smooth rank metrics. In: Proceedings of the 1st ACM International Conference on Web Search and Data Mining (2008)

10. Chakrabarti, S., Khanna, R., Sawant, U., Bhattacharyya, C.: Structured learning for nonsmooth ranking losses. In: Proceedings of the 14th ACM Conference on Knowledge Discovery and Data Mining (2008)
11. Cossock, D., Zhang, T.: Statistical analysis of bayes optimal subset ranking. IEEE Transactions on Information Theory 54(11), 5140–5154 (2008)
12. Chapelle, O., Wu, M.: Gradient descent optimization of smoothed information retrieval metrics. Information Retrieval Journal (to appear, 2010)
13. Qin, T., Zhang, X.D., Tsai, M.F., Wang, D.S., Liu, T.Y., Li, H.: Query-level loss functions for information retrieval. Information Processing and Management 44(2), 838–855 (2008)
14. Cao, Z., Qin, T., Liu, T.Y., Tsai, M.F., Li, H.: Learning to rank: From pairwise approach to listwise approach. In: Proceedings of the 24th International Conference on Machine Learning (2007)
15. Joachims, T.: A support vector method for multivariate performance measures. In: Proceedings of the 22nd International Conference on Machine Learning (2005)
16. Chapelle, O., Le, Q., Smola, A.: Large margin optimization of ranking measures. In: Proceedings of the NIPS-2007 Workshop on Machine Learning for Web Search (2007)
17. Taskar, B., Guestrin, C., Koller, D.: Max-margin markov networks. In: Advances in Neural Information Processing Systems, vol. 16. MIT Press, Cambridge (2004)
18. Tsochantaridis, I., Joachims, T., Hofmann, T., Altun, Y.: Large margin methods for structured and interdependent output variables. Journal of Machine Learning Research (JMLR) 6, 1453–1484 (2005)
19. Herbrich, R., Graepel, T., Obermayer, K.: Large margin rank boundaries for ordinal regression. Advances in Large Margin Classifiers, 115–132 (2000)
20. Freund, Y., Iyer, R., Schapire, R.E., Singer, Y.: An efficient boosting algorithm for combining preferences. Journal of Machine Learning Research 4, 933–969 (2003)
21. Quattoni, A., Carreras, X., Collins, M., Darrell, T.: An efficient projection for $l_1 \infty$ regularization. In: Proceedings of the 26th International Conference on Machine Learning (2009)
22. Collins, M., Globerson, A., Koo, T., Carreras, X., Bartlett, P.: Exponentiated gradient algorithms for conditional random fields and max-margin Markov networks. Journal of Machine Learning Research 9, 1775–1822 (2008)
23. Agarwal, S., Niyogi, P.: Generalization bounds for ranking algorithms via algorithmic stability. Journal of Machine Learning Research 10, 441–474 (2009)
24. Shalev-Shwartz, S., Singer, Y., Srebro, N.: Pegasos: Primal estimated sub-gradient solver for SVM. In: Proceedings of the 24th International Conference on Machine Learning (2007)
25. Rudin, C.: Ranking with a p-norm push. In: Proceedings of the 19th Annual Conference on Learning Theory (2006)
26. Liu, T.Y., Xu, J., Qin, T., Xiong, W., Li, H.: LETOR: Benchmark dataset for research on learning to rank for information retrieval. In: Proceedings of the SIGIR-2007 Workshop on Learning to Rank for Information Retrieval (2007)
27. Järvelin, K., Kekäläinen, J.: Cumulated gain-based evaluation of IR techniques. ACM Transactions on Information Systems 20(4), 422–446 (2002)
28. Chapelle, O., Keerthi, S.S.: Efficient algorithms for ranking with SVMs. Information Retrieval Journal (to appear, 2010)

Query Aspect Based Term Weighting Regularization in Information Retrieval

Wei Zheng and Hui Fang

Department of Electrical and Computer Engineering
University of Delaware, USA
{zwei,hfang}@ece.udel.edu

Abstract. Traditional retrieval models assume that query terms are independent and rank documents primarily based on various term weighting strategies including TF-IDF and document length normalization. However, query terms are related, and groups of semantically related query terms may form query aspects. Intuitively, the relations among query terms could be utilized to identify hidden query aspects and promote the ranking of documents covering more query aspects. Despite its importance, the use of semantic relations among query terms for term weighting regularization has been under-explored in information retrieval. In this paper, we study the incorporation of query term relations into existing retrieval models and focus on addressing the challenge, i.e., how to regularize the weights of terms in different query aspects to improve retrieval performance. Specifically, we first develop a general strategy that can systematically integrate a term weighting regularization function into existing retrieval functions, and then propose two specific regularization functions based on the guidance provided by constraint analysis. Experiments on eight standard TREC data sets show that the proposed methods are effective to improve retrieval accuracy.

1 Introduction

It has been a long standing challenge to develop robust and effective retrieval models. Many retrieval models have been proposed and studied including vector space models [19], classic probabilistic models [18,23,7], language models [15,25] and recently proposed axiomatic models [5]. These retrieval models rank documents based on the use of various term weighting strategies including term frequency, inverse document frequency and document length normalization [4].

Although these retrieval models rank documents differently, they may fail to return relevant documents for the same reasons. In the previous studies [8,2], researchers conducted failure analysis for the state-of-the-art retrieval models and showed that one of the common failures is that the retrieval models fail to return documents covering all the query aspects. This failure, in a way, is caused by the underlying assumption that query terms are independent to each other. Traditional retrieval models often ignore query term relations in term

C. Gurrin et al. (Eds.): ECIR 2010, LNCS 5993, pp. 344–356, 2010.

weighting and treat every query term as a query aspect. However, such an assumption is not always true. Query terms could be related to each other, and multiple semantically related query terms may form a query aspect. Intuitively, query term relations are useful to identify different aspects in the query and can provide guidance on the term weighting. For example, consider query "stolen or forged art" (i.e., topic 422 in TREC8). The query contains two aspects, i.e., "stolen or forged" and "art". Intuitively, documents covering both aspects should be ranked higher than those covering only one aspect. Thus, a document talking about "stolen or forged money" should not be ranked higher than the one talking about "stolen art". Unfortunately, a query aspect may contain one or multiple terms. Since existing retrieval models treat query terms independently, they may assign lower relevance scores to the documents covering more query aspects. In particular, Buckley [2] reported that all the analyzed retrieval models over-emphasized one aspect of the query, i.e., "stolen or forged", while missing the other aspect, i.e., "art". Clearly, it is important to exploit query aspect information to regularize term weighting and incorporate the term regularization into existing retrieval functions. Despite its importance, the use of query term relations for term weighting regularization has been under-explored in the IR literature. It remains unclear how to regularize term weighting based on query term relations and how to systematically incorporate the term weighting regularization functions into existing retrieval functions.

In this paper, we study the problem of incorporating query term relations into existing retrieval functions. Specifically, we discuss how to utilize term semantic similarities to identify query aspects and how to systematically exploit the query aspect information to regularize term weighting in existing retrieval functions. We first present a general strategy based on the recently proposed inductive definition scheme [5]. We show that the inductive definition provides a natural way of extending an existing retrieval function with the aspect based term weighting regularization - all we need to do is to generalize the query growth function of a retrieval function to incorporate an aspect-based term regularization function. We then propose two term weighting regularization functions that can utilize the query term relations such as query aspects in order to avoid favoring documents that cover fewer query aspects. To evaluate the effectiveness of the proposed methods, we integrate them into four representative retrieval functions (i.e., pivoted normalization retrieval function [21], Okapi BM25 retrieval function [18], Dirichlet prior retrieval function [25] and axiomatic retrieval function [5]), and conduct experiments over eight representative TREC data sets. Experiment results show that, for verbose queries, the proposed methods can significantly and consistently improve the retrieval accuracy on almost all the data sets we experimented with. The rest of the paper is organized as follows. We discuss related work in Section 2 and briefly review the basic ideas of inductive definition and axiomatic approaches in Section 3. We then present our work on aspect-based term weighting regularization in Section 4, and discuss experiment results in Section 5. Finally, we conclude in Section 6.

2   Related Work

Most traditional retrieval models assume that query terms are independent. To improve retrieval accuracy, many studies have recently tried to exploit the relations among query terms. They range from the early studies on the use of phrases in document retrieval [3,14,12] to the recent work on query segmentation [9,16,10], term proximity [22], and term dependencies [13]. Previous studies on the use of phrases in retrieval models [3,14,12] often identified phrases using either statistical or syntactic methods, scored documents with matched phrases, and then heuristically combined the term-based and phrase-based relevance scores. Recent studies [1,11] focused on using supervised learning techniques to support verbose queries. In particular, Bendersky and Croft [1] proposed a probabilistic model for combining the weighted key concepts with the original queries. Query segmentation refers to the problem of segmenting a query into several query concepts. The commonly used methods are based on term co-occurrences [9,16]. Similar to previous work [9,16,6], we assume that term co-occurrences such as mutual information can be used to compute term semantic similarity. But our work focuses on aspect-based term weighting regularization instead of query aspect identification. Kumaran and Allan [10] proposed to interact with users and allow them to extract the best sub-queries from a long query. However, they did not study how to utilize the sub-queries or segmented queries to regularize term weighting. Our work is also related to the studies of term dependencies and term proximity [13,22]. For example, Metzler and Bruce [13] proposed a term dependence model, which can model different dependencies between query terms. However, the proposed model affects the retrieval efficiency, and it remains unclear how to incorporate term dependencies into other retrieval models. Tao and Zhai [22] studied how to exploit term proximity measures, but our work focuses on the semantic relations among query terms.

Although the motivation is similar, our work differs from the previous work in that (1) we attempt to systematically integrate aspect based term weighting regularization into a variety of existing retrieval models; (2) we propose to use constraint analysis to provide guidance on the implementation of term weighting regularization functions; (3) our methods do not rely on the use of external resources and are less computational expensive than the method proposed in the previous study [1]. Moreover, as shown in Section 5, the performance of our methods are comparable to the performance reported in the previous study [1].

3   Axiomatic Approaches to IR

Axiomatic approaches have recently been proposed as a new way of analyzing and developing retrieval functions [4,5]. The basic idea is to search in a space of candidate retrieval functions for the ones that can satisfy a set of desirable retrieval constraints. Retrieval constraints are often defined by formalizing various retrieval heuristics that any reasonable retrieval functions should satisfy. Previous studies proposed several retrieval constraints for TF-IDF weighting, document length normalization, semantic term matching and term proximity [4,5,6,22]. These constraints are shown to be effective to provide guidance on how to improve the performance of an existing retrieval function and how to develop new retrieval functions.

To constrain search space of retrieval functions, an inductive definition of retrieval functions was proposed [5]. The inductive definition decomposes a retrieval function into three component functions: (1)primitive weighting function, which gives the relevance score of a one-term document for a given one-term query; (2)document growth function, which captures the change of relevance scores when a term is added to a document; and (3)query growth function, which captures the score change when a term is added to a query. There are multiple ways of instantiating each of these component functions. In general, different instantiations of the three component functions would lead to different retrieval functions.

Previous study [5] showed that most existing "bag of words" representation based retrieval functions can be decomposed with the proposed inductive definition, and they have similar instantiations of query growth function as follows.

$$S(Q \cup \{q\}, D) = S(Q, D) + S(\{q\}, D) \times \Delta(c(q, Q)) \tag{1}$$

where D denotes a document, Q denotes a query, and $Q \cup \{q\}$ denotes a new query generated by adding a term q to query Q, $S(Q, D)$ denotes the relevance score and $c(q, Q)$ is the term occurrence of q in query Q. Four existing retrieval functions differ in the implementation of $\Delta(c(q, Q))$. In particular, Okapi implements it as $\Delta(x) = \frac{(k_3+1) \times (x+1)}{k_3+x+1} - \frac{(k_3+1) \times x}{k_3+x}$, where k_3 is the parameter in the Okapi BM25 retrieval function, and other functions including Pivoted, Dirichlet and axiomatic retrieval functions implement it as $\Delta(x) = 1$. Note that these query growth functions are only related to query term frequency and do not consider the semantic relations among query terms.

4   Aspect-Based Query Term Regularization

4.1   Problem Formulation

It is known that terms are semantically related. For example, the occurrence of a term in a document may indicate the occurrences of its related terms in the document. Within a query, groups of semantically related terms may form different query aspects. In general, we define a *query aspect* as a group of query terms that are semantically similar to each other. A query may contain one or more query aspects, and a query aspect may contain one or more query terms. For example, query "ocean remote sensing" has two aspects, i.e., "ocean" and "remote sensing". Formally, let $Q = \{q_1, q_2, ..., q_n\}$ be a query with n terms. $\mathcal{A}(q) \subseteq Q$ denotes the aspect of query term q. If the aspect of term q_1 has two terms, i.e., q_1 and q_2, then $\mathcal{A}(q_1) = \{q_1, q_2\}$. $s(t_1, t_2) \in [0, +\infty]$ denotes the semantic similarity between two terms t_1 and t_2. If $\mathcal{A}(q_1) = \mathcal{A}(q_2)$ and $\mathcal{A}(q_1) \neq \mathcal{A}(q_3)$, then $s(q_1, q_2) > s(q_1, q_3)$ and $s(q_1, q_2) > s(q_2, q_3)$. The underlying assumption is that terms within a query aspect should be more semantically similar than those from different query aspects.

Indeed, the definition of query aspects suggests that one possible way of identifying query aspects is to cluster query terms based on a term semantic similarity function. We explore a single-link hierarchical clustering algorithm in the paper. Specifically, we start with each term in a query as a cluster, and then keep combining two clusters when there exist two terms, one from each cluster, whose similarity is higher than a threshold. The threshold can be set as the average similarity of all term pairs for the query. The algorithm stops when no clusters can be further combined. As a result, every cluster can be regarded as a query aspect.

$s(t_1, t_2)$ may be any given term semantic similarity function. Following the previous studies [20,24,6], we assume that co-occurrences of terms reflect underlying semantic relations among query terms, and adopt the expected mutual information measure (EMIM) [23,24] as the term semantic similarity function. Formally, the term semantic similarity function is defined as follows.

$$s(t_1, t_2) = I(X_{t_1}, X_{t_2}) = \sum_{X_{t_1}, X_{t_2} \in \{0,1\}} p(X_{t_1}, X_{t_2}) \log \frac{p(X_{t_1}, X_{t_2})}{p(X_{t_1})p(X_{t_2})}. \tag{2}$$

X_t is a binary random variable corresponding to the presence/absence of term t in each document. We compute the mutual information for query term pairs using the test collection itself and leave other possible term semantic similarity functions and other aspect identification methods as our future work.

Note that most traditional retrieval models [19,23,7,15,25,5] assume that query terms are independent, and each query term corresponds to a query aspect, i.e., $\forall q \in Q, \mathcal{A}(q) = \{q\}$. As shown in the previous studies [2,8], the assumption often leads to non-optimal retrieval performance because the retrieval models may incorrectly assign higher relevance scores to the documents that cover fewer query aspects. For example, for the query "ocean remote sensing" mentioned earlier, all the analyzed retrieval models over-emphasized one aspect "remote sensing" and failed to return documents covering both aspects.

In this paper, we aim to study how to utilize the semantic relations among query terms, such as query aspect information, to regularize term weighting in order to improve the retrieval performance of an existing retrieval function.

4.2 General Strategy

The occurrence of a query term often indicates the occurrences of its semantically related terms. If a query term has many semantically related terms in a query, this term and its query aspect might be over-emphasized because of the matching of these related terms. To solve this problem, we now propose a general strategy that can regularize term weighting based on semantic relations among query terms for existing retrieval functions. Specifically, we first define a constraint for term weighting regularization based on query term relations and integrate the regularization function into existing retrieval functions through the inductive definition scheme under the guidance of constraint analysis [4,5].

The basic idea is to adjust the weights of a query term based on its semantic relations with other query terms so that the documents covering more query aspects would be ranked higher than those covering fewer aspects. We can formalize this idea as a retricval constraint. Let us first introduce some notations. Q denotes a query and $\mathcal{A}(q) \subseteq Q$ denotes the query aspect of query term q. Query terms q_1 and q_2 belong to different query aspects if $\mathcal{A}(q_1) \neq \mathcal{A}(q_2)$. Let $td(t)$ denote any reasonable measure of term discrimination value of term t (usually based on term popularity in a collection), such as IDF. The term weighting regularization constraint can be defined formally as follows.

Regularization Constraint: Let $Q = \{q_1, q_2, q_3\}$ be a query with three query terms q_1, q_2 and q_3, where $td(q_2) = td(q_3)$. We assume that $\mathcal{A}(q_1) = \mathcal{A}(q_2)$ and $\mathcal{A}(q_1) \neq \mathcal{A}(q_3)$, or equivalently $s(q_1, q_2) > s(q_1, q_3)$ and $s(q_1, q_2) > s(q_2, q_3)$ based on the definition of query aspects. Let D_1 and D_2 be two documents, and $c(t, D)$ denotes the count of term t in document D. If $c(q_1, D_1) = c(q_1, D_2) > 0$, $c(q_2, D_1) = c(q_3, D_2) > 0$, $c(q_3, D_1) = c(q_2, D_2) = 0$, and $|D_1| = |D_2|$, then $S(Q, D_1) < S(Q, D_2)$.

The constraint requires a retrieval function to assign a higher relevance score to the document that covers more query aspects. Thus, even though both D_1 and D_2 match two query terms with the same term discrimination values, we would like D_2 to have a higher scorc because the matched query terms in D_1 (i.e., q_1 and q_2) are from the same aspect while the matched query terms in D_2 (i.e., q_1 and q_3) are from different aspects.

We analyze four representative retrieval functions with the constraint. The functions are *pivoted normalization function* derived from vector space models [19,21]), *Okapi BM25* derived from classical probabilistic models [23,7,18], *Dirichlet prior* derived from language models [15,25] and *F2-EXP* derived from axiomatic retrieval models [5]. The constraint analysis results show that none of the functions satisfies the constraint because they ignore the query term relations and would assign the same scores to both documents.

To make the retrieval functions satisfy the constraint, we need to incorporate semantic relations among query terms into retrieval functions. As reviewed in Section 3, the inductive definition makes it possible to decompose a retrieval function into three component functions. Clearly, a natural way of incorporating semantic relations among query terms into rctricval functions is to generalize the query growth function so that it is related to not only the query term frequency but also the semantic relations between a query term and other terms in the query. Thus, we propose to define the following *generalized query growth function* by extending Equation (1) with a function $f(q, Q, s(\cdot))$ that regularizes the query term weighting based on the semantic relations between term q and query Q.

$$S(Q \cup \{q\}, D) = S(Q, D) + S(\{q\}, D) \times \Delta(c(q, Q)) \times f(q, Q, s(\cdot)) \tag{3}$$

where $s(\cdot)$ is a term semantic similarity function such as Equation (2).

To integrate term regularization function $f(q, Q, s(\cdot))$ into a retrieval function, we can first decompose the retrieval function into three component functions, and then combine its original primitive weighting function and original document growth function with the generalized query growth function as shown in Equation (3). Thus, the new retrieval function is an extension of the original retrieval function, and it uses the regularization function $f(\cdot)$ to regularize term weighting based on the semantic relations among query terms. The extended versions for the analyzed four retrieval functions are shown as follows.

– Extended Pivoted Normalization:

$$S(Q, D) = \sum_{t \in D \cap Q} \frac{1 + \log(1 + \log(c(t, D)))}{1 - b + b \frac{|D|}{avdl}} \times c(t, Q) \times \log \frac{N + 1}{df(t)} \times f(t, Q, s(\cdot)) \qquad (4)$$

– Extended Okapi BM25:

$$S(Q, D) = \sum_{t \in Q \cap D} \log \frac{N - df(t) + 0.5}{df(t) + 0.5} \cdot \frac{(k_3 + 1) \cdot c(t, Q)}{k_3 + c(t, Q)} \cdot \frac{(k_1 + 1) \cdot c(t, D)}{k_1((1 - b) + b \frac{|D|}{avdl}) + c(t, D)} \cdot$$
$$f(t, Q, s(\cdot)) \qquad (5)$$

– Extended Dirichlet Prior:

$$S(Q, D) = \sum_{t \in Q \cap D} c(t, Q) \cdot \log(1 + \frac{c(t, D)}{\mu \times p(t|C)}) \cdot f(t, Q, s(\cdot)) + \log \frac{\mu}{|D| + \mu} \cdot \sum_{q \in Q} f(q, Q, s(\cdot)) \quad (6)$$

– Extended Axiomatic:

$$S(Q, D) = \sum_{t \in Q \cap D} c(t, Q) \times (\frac{N}{df(t)})^{0.35} \times f(t, Q, s(\cdot)) \times \frac{c(t, D)}{c(t, D) + b + \frac{b \times |D|}{avdl}} \qquad (7)$$

$c(t, D)$ is the count of term t in document D, $c(t, Q)$ is the count of term t in query Q, $df(t)$ is the number of documents with term t, $|D|$ is the length of document d, $avdl$ is the average document length of the document collection, and $p(t|C)$ is the probability of term t in collection C. k_1 is set to 1.2 and k_3 is set to 1000. μ and b are parameters in the original retrieval functions. Note that the original retrieval functions are the special cases of their corresponding extensions when $f(q, Q, s(\cdot)) = 1, \forall q \in Q$.

The proposed general strategy provides a systematical way of incorporating query term relations into any retrieval functions that can be decomposed with the inductive definition scheme. Moreover, although we discussed how to compute term semantic similarity and how to identify aspect, the proposed strategy is generally applicable for any other reasonable term semantic similarity functions and aspect identification methods.

The remaining challenge of the proposed general strategy is to select appropriate implementation for function $f(q, Q, s(\cdot))$, which regularizes the term weighting based on the semantic relations among query terms so that the extended retrieval functions satisfy the defined constraint. We discuss two term weighting regularization functions in the next subsection.

4.3   Term Weighting Regularization Functions

To make extended retrieval functions satisfy the constraint, we need to implement the regularization function f in a way so that it would demote the weights of terms that either belong to large query aspects or have many semantically related terms. Thus, we propose the following two term weighting regularization functions, with the first one explicitly capturing the query aspect information and the second one implicitly capturing aspect information through the semantic relations among query terms.

Aspect size based regularization: The proposed regularization constraint is to avoid over-favoring documents covering fewer aspects. One possible solution is to penalize terms from larger aspects. The rationale is that a larger query aspect needs to be penalized more harshly because the aspect is more likely to be over-favored due to the larger number of query terms in the aspect. Thus, we propose an aspect size based regularization function, f_{size}, which explicitly uses the query aspect information and regularizes the term weighting based on the size of its aspect. Formally,

$$f_{size}(q, Q, s(\cdot)) = 1 - \alpha + \alpha \cdot \left(\frac{|\mathcal{A}_s(q)|}{|Q|} \right)^{-\beta} \tag{8}$$

where $|\mathcal{A}_s(q)|$ is the number of terms in query aspect $\mathcal{A}_s(q)$ identified based on term similarity function $s(\cdot)$ and $|Q|$ is the number of terms in query Q. Clearly, the value of f_{size} for a query term q is inversely correlated with the size of its query aspect. It means that if a query term is in a larger query aspect, the weights of the query term should be penalized more because the matching of its semantic related terms may also contribute to the relevance score. There are two parameters in the regularization function: β controls the curve shape of the regularized function, and α balances the original and the regularized term weighting. We will examine the parameter sensitivity in the experiment section.

Semantic similarity based regularization: The aspect size based regularization function requires us to explicitly identify query aspects. However, the accuracy of aspect identification may greatly affect the retrieval performance for the regularization based retrieval function. To overcome the limitation, we propose a semantic similarity based regularization function, i.e., f_{sim}, which does not require the explicit aspect identification. Specifically, f_{sim} exploits the semantic similarity between term q and its query Q, which is computed by taking the average of the semantic similarity between q and other query terms in Q. Formally, we have

$$f_{sim}(q, Q, s(\cdot)) = 1 - \alpha + \alpha \times (-\log(\frac{\sum_{q' \in Q - \{q\}} s(q, q')}{|Q| - 1})) \tag{9}$$

where $|Q|$ is the number of terms in query Q and α is a parameter that balances the original term weighting and the regularized term weighting. If a query term is more semantically related to the query, f_{sim} would decrease the term weighting so that the term and its related terms would not be over-emphasized by the retrieval functions. f_{sim} does not require the identification of query aspects. Instead, it implicitly assumes that the relations between query aspects can be approximated by the relations between query terms.

Summary: We incorporate the proposed regularization functions, $f_{size}(\cdot)$ or $f_{sim}(\cdot)$, into the four extended retrieval functions shown in Equation (4)-(7). After analyzing the extended retrieval functions, we can show that all of them satisfy the defined retrieval constraint because both regularization functions satisfy

$$f(q_2, Q, s(\cdot)) < f(q_3, Q, s(\cdot)),$$

which leads to $S(Q, D_1) < S(Q, D_2)$ for all extended retrieval functions. It means that the proposed regularization functions can demote the weights of terms that are from a larger query aspect or have more semantically related terms in the query.

5   Experiments

5.1   Experiment Setup

We evaluate the proposed methods on eight representative TREC data sets: the ad hoc data used in the ROBUST track of TREC 2004 (Robust04), the ad hoc data used in the ROBUST track of TREC 2005 (Robust05), the ad hoc data used in TREC7 (Trec7), the ad hoc data used in TREC8 (Trec8), the Web data used in TREC8 (Web), news articles (Ap88-89), technical reports(Doe) and government documents (Fr88-89). We use two types of queries: keyword queries (i.e., title-only) and verbose queries (i.e., description-only). Table 1 shows some statistics of the test sets, including the collection size, the number of documents, the number of queries and average number of terms per keyword query, and average number of terms per verbose query.

The preprocessing only involves stemming with Porter's stemmer. No stop words is removed for two reasons: (1) A robust retrieval model should be able to discount the stop words appropriately; (2) Removing stop words would introduce at least one extra parameter, i.e., the number of stop words into the experiments. The performance is measured in terms of MAP (mean average precision).

We now explain the notations for different methods. BL is the original retrieval function without regularization. f_{size} and f_{sim} denote the aspect size based and semantic similarity based regularization functions respectively. As shown in Equation (4)-(7), we can integrate the proposed functions into four retrieval functions, i.e., Pivoted (**Piv.**), Okapi BM25 (**Okapi**), Dirichlet Prior (**Dir.**) and axiomatic function (**AX**). Okapi is known to perform poorly for verbose queries [17,4], so instead we report the performance of the modified Okapi (**Mod. Okapi**) [4], which is a stronger baseline for verbose queries. Each of

Table 1. Statistics of Test Collections

Collection	Size	#d	#q	#t/kq	#t/vq	Collection	Size	#d	#q	#t/kq	#t/vq
Robust04	2GB	528K	249	2.75	15.5	Robust05	3GB	1,033K	50	2.7	17.5
Trec7	2GB	528K	50	2.5	14.3	Trec8	2GB	528K	50	2.42	15.8
Web	2GB	247K	50	2.42	15.8	Ap88-89	491MB	165K	146	3.76	18.1
Doe	184MB	226K	35	3.69	18.8	Fr88-89	469MB	204K	42	3.5	19.6

these retrieval functions has one retrieval parameter, and we tune the retrieval parameters as well as the new parameters introduced in the proposed methods and report the optimal performance. In all the tables, ‡ and † indicate that the improvement is statistically significant according to the Wilcoxin signed rank test at the level of 0.05 and 0.1 respectively.

5.2 Comparison of Proposed Methods

Table 2 shows the optimal performance comparison of the proposed methods for verbose queries. Clearly, both proposed term regularization functions can significantly and consistently improve the retrieval performance for all four retrieval functions over almost all of the eight data collections. Moreover, f_{sim} performs better than f_{size}. After analyzing the results, we find that the performance difference might be caused by the fact that f_{sim} is not as dependent to the accuracy of aspect identification as f_{size}. The aspect identification method used in this paper can correctly identify aspects for some queries but not for

Table 2. Performance Comparison for Verbose Queries (MAP)

Function		Robust04	Robust05	Trec7	Trec8	Web	Ap88-89	Doe	Fr88-89
Piv.	BL	0.2145	0.1406	0.1461	0.2032	0.2122	0.1931	0.1031	0.1424
	f_{size}	**0.2277‡**	0.1408	**0.1706‡**	**0.2208‡**	**0.2481‡**	0.1931	**0.1237‡**	**0.1854‡**
	f_{sim}	**0.2422‡**	**0.1517‡**	**0.1782‡**	**0.2336‡**	**0.2555‡**	**0.2047‡**	**0.1216‡**	**0.1637‡**
		(+13.1%)	(+7.8%)	(+21.9%)	(+15.3%)	(+20.8%)	(+6.22%)	(+17.3%)	(+15.5%)
Mod. Okapi	BL	0.2114	0.1391	0.1527	0.2014	0.2371	0.1812	0.1037	0.1526
	f_{size}	**0.2404‡**	0.1428	**0.1785‡**	**0.2311‡**	**0.2758‡**	0.1821	**0.1173**	**0.1942‡**
	f_{sim}	**0.2532‡**	**0.1547 ‡**	**0.1843‡**	**0.2408‡**	**0.2814†**	**0.1936†**	**0.1141†**	**0.1772†**
		(+12.9%)	(+8.4%)	(+17.2%)	(+12.0%)	(+17.1%)	(+5.91%)	(+10.2%)	(+13.3%)
Dir.	BL	0.2326	0.1598	0.1811	0.2279	0.2693	0.1990	0.1253	0.1522
	f_{size}	**0.2448‡**	0.1598	**0.1902‡**	**0.2393‡**	**0.2938‡**	0.1990	0.1253	**0.1856‡**
	f_{sim}	**0.2578‡**	**0.1623**	**0.1971‡**	**0.2507‡**	**0.2946‡**	0.1990	**0.1270**	**0.1801‡**
		(+10.3%)	(+1.25%)	(+7.64%)	(+10.1%)	(+8.39%)	(+0%)	(+1.6%)	(+13.9%)
AX	BL	0.2421	0.1612	0.1864	0.2357	0.2715	0.2016	0.1161	0.1674
	f_{size}	**0.2531‡**	0.1612	**0.1881**	**0.2434‡**	**0.2896‡**	0.2016	**0.1245**	**0.2013‡**
	f_{sim}	**0.2534‡**	0.1620	**0.1904†**	**0.2446‡**	**0.2866‡**	**0.2071‡**	0.1232	**0.1958‡**
		(+4.5%)	(+1.24%)	(+2.15%)	(+3.81%)	(+5.51%)	(+4.55%)	(+6.03%)	(+17.4%)

Table 3. Performance Comparison for Keyword Queries (MAP)

Function		Robust04	Robust05	Trec7	Trec8	Web	Ap88-89	Doe	Fr88-89
Piv.	BL	0.2406	0.1999	0.1762	0.2438	0.2883	0.2267	0.1788	0.2183
	f_{sim}	**0.2432‡**	0.1999	**0.1780†**	**0.2442**	**0.2892**	0.2267	0.1788	**0.2205**
Okapi	BL	0.2477	0.2013	0.1857	0.2512	0.3105	0.2255	0.1847	0.2247
	f_{sim}	**0.2507‡**	0.2013	**0.1892‡**	**0.2518**	0.3105	0.2255	0.1847	**0.2267**

Table 4. Performance Comparison for Verbose Queries on Robust04

Function		MAP	Prec@5	Prec@10	Function		MAP	Prec@5	Prec@10
Piv.	BL	0.2145	0.4610	0.3964	Dir.	BL	0.2326	0.4554	0.4032
	f_{sim}	**0.2422‡**	**0.4956‡**	**0.4241 ‡**		f_{sim}	**0.2578‡**	**0.4859‡**	**0.4237**
Mod. Okapi	BL	0.2114	0.4827	0.4068	AX	BL	0.2421	0.4859	0.4253
	f_{sim}	**0.2532‡**	**0.5044‡**	**0.4293‡**		f_{sim}	**0.2534 ‡**	**0.4956**	**0.4281**

all queries. Thus, the performance of f_{size} may be affected more by the inaccurate aspect identification, which leads to the relatively worse performance of f_{size}. Table 3 shows the results of two functions for keyword queries, and the results for the other two functions are similar and not included due to the space limit. Clearly, the performance improvement for keyword queries is not as significant and consistent as for verbose queries. Our result analysis suggests that the smaller improvement is caused by the smaller number of terms in keyword queries, because the keyword queries often lead to the smaller query aspects and incorrect aspect identifications for some short queries.

Table 4 shows the performance of the proposed f_{sim} method for verbose queries on ROBUST04. In addition to MAP, we also report the performance measured with Prec@5 and Prec@10. We compare our results with the results of another recently proposed retrieval method for verbose queries [1]. The MAP of their baseline method for verbose queries is 0.2450, and the MAP of their proposed method is 0.2620. Prec@5 of their baseline method is 0.4726 and the Prec@5 of their proposed method is 0.4854. Due to the different pre-processing strategies and different baseline functions, these numbers cannot be directly compared with the results reported in this paper. However, it is quite encouraging to see that our proposed methods can achieve comparable performance with much less computational cost and without the use of external resources.

Finally, we conduct an additional set of experiments when stop words are removed in the pre-processing stage. With the stop word removal, the performance (MAP) of BL for the representative retrieval functions on Robust04 are 0.2196 (Piv.), 0.2351 (Mod. Okapi), 0.2323 (Dir.) and 0.2473 (AX) respectively. Clearly, the results show that stopword removal can improve retrieval performance a little bit but not that much, which suggests that the baseline method we used in the performance comparison is a strong baseline.

5.3   Parameter Sensitivity

As indicated in Equation (8)-(9), f_{sim} has one parameter α and f_{size} has two parameters α and β. Figure 1 shows the sensitivity curve for these parameters on Robust04. $\alpha = 0$ means that no regularization is used. The better performance

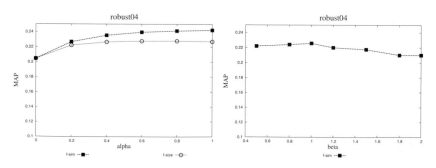

Fig. 1. Performance Sensitivity: α in f_{sim} and f_{size} (Left) and β in f_{size} (Right)

at larger values of α indicates that the proposed term regularization methods are effective. We also observe that the performance is not very sensitive to the values of β. We only show results on one data set due to the space limit. However, the results are similar for other data sets. In general, setting $\alpha = 0.6$ and $\beta = 1$ would lead to good performance for most data collections.

6   Conclusions and Future Work

In this paper, we study the problem of exploiting semantic relations among query terms to regularize term weighting in retrieval functions. Assuming that groups of semantically related query terms form query aspects, we present a general strategy that can systematically incorporate a term weighting regularization function into a variety of existing retrieval functions. Specifically, we propose two term regularization functions based on term semantic similarity, and then discuss how to integrate the regularization functions into existing retrieval functions through the inductive definition scheme. The proposed methods are incorporated into four representative retrieval functions and evaluated on eight representative TREC retrieval collections. Experiment results show that, for verbose queries, the proposed methods can significantly improve the retrieval performance of all the four retrieval functions across almost all the eight test collections. Note that the proposed methods do not require training data and external resources, and the computational cost is low when the MI values are stored in the index offline.

There are several interesting future research directions. First, we will explore other aspect identification methods and compute term semantic similarity using other resources, such as WordNet, query logs and external collections. Second, the semantic relations among query terms could be different for different domains. It would be interesting to explore whether the proposed methods work well in some specific domains, such as biomedical literature search or legal document search. Finally, we focus on only semantic relations among query terms in this work. It would be interesting to study the combination of query term semantic relations, semantic term matching between query and documents, the proximity of different query terms and any other reasonable properties of terms.

References

1. Bendersky, M., Croft, W.B.: Discovering key concepts in verbose queries. In: Proceedings of SIGIR 2008 (2008)
2. Buckley, C.: Why current ir engines fail. In: Proceedings of SIGIR 2004 (2004)
3. Croft, W., Turtle, H., Lewis, D.: The use of phrases and structured queries in information retrieval. In: Proceedings of SIGIR 1991 (1991)
4. Fang, H., Tao, T., Zhai, C.: A formal study of information retrieval heuristics. In: Proceedings of SIGIR 2004 (2004)
5. Fang, H., Zhai, C.: An exploration of axiomatic approaches to information retrieval. In: Proceedings of SIGIR 2005 (2005)
6. Fang, H., Zhai, C.: Semantic term matching in axiomatic approaches to information retrieval. In: Proceedings of SIGIR 2006 (2006)

7. Fuhr, N.: Probabilistic models in information retrieval. The Computer Journal 35(3), 243–255 (1992)
8. Harman, D., Buckley, C.: Sigir 2004 workshop: Ria and where can ir go from here? SIGIR Forum 38(2) (2004)
9. Jones, R., Rey, B., Madani, O., Greiner, W.: Generating query substitutions. In: Proceedings of WWW 2006 (2006)
10. Kumaran, G., Allan, J.: A case for shorter queries, and helping users create them. In: Proceedings of HLT 2006 (2006)
11. Lease, M.: An improved markov rndom field model for supporting verbose queries. In: Proceedings of SIGIR 2009 (2009)
12. Liu, S., Liu, F., Yu, C., Meng, W.: An effective approach to document retrieval via utilizing wordnet and recognizing phrases. In: Proceedings of SIGIR 2004 (2004)
13. Metzler, D., Croft, W.B.: A markov random field model for term dependencies. In: Proceedings of SIGIR 2005 (2005)
14. Mitra, M., Buckley, C., Singhal, A., Cardie, C.: An analysis of statistical and syntactic phrases. In: Proceedings of RIAO 1997 (1997)
15. Ponte, J., Croft, W.B.: A language modeling approach to information retrieval. In: Proceedings of the ACM SIGIR 1998, pp. 275–281 (1998)
16. Risvik, K.M., Mikolajewski, T., Boros, P.: Query segmentation for web search. In: Proceedings of the 2003 World Wide Web Conference (2003)
17. Robertson, S., Walker, S.: On relevance weights with little relevance information. In: Proceedings of SIGIR 1997, pp. 16–24 (1997)
18. Robertson, S.E., Walker, S., Jones, S., Hancock-Beaulieu, M.M., Gatford, M.: Okapi at TREC-3. In: Proceedings of TREC-3 (1995)
19. Salton, G.: Automatic Text Processing: The Transformation, Analysis and Retrieval of Information by Computer. Addison-Wesley, Reading (1989)
20. Schutze, H., Pedersen, J.O.: A co-occurrence based thesaurus and two applications to information retrieval. Information Processing and Management 33(3), 307–318 (1997)
21. Singhal, A., Buckley, C., Mitra, M.: Pivoted document length normalization. In: Proceedings of SIGIR 1996 (1996)
22. Tao, T., Zhai, C.: An exploration of proximity measures in information retrieval. In: Proceedings of SIGIR 2007 (2007)
23. van Rijbergen, C.J.: A theoretical basis for theuse of co-occurrence data in information retrieval. Journal of Documentation, 106–119 (1977)
24. van Rijsbergen, C.J.: Information Retrieval. Butterworths (1979)
25. Zhai, C., Lafferty, J.: A study of smoothing methods for language models applied to ad hoc information retrieval. In: Proceedings of SIGIR 2001 (2001)

Using the Quantum Probability Ranking Principle to Rank Interdependent Documents

Guido Zuccon and Leif Azzopardi

University of Glasgow, Department of Computing Science,
Glasgow G12 8QQ, UK
{guido,leif}@dcs.gla.ac.uk

Abstract. A known limitation of the Probability Ranking Principle (PRP) is that it does not cater for dependence between documents. Recently, the Quantum Probability Ranking Principle (QPRP) has been proposed, which implicitly captures dependencies between documents through "quantum interference". This paper explores whether this new ranking principle leads to improved performance for subtopic retrieval, where novelty and diversity is required. In a thorough empirical investigation, models based on the PRP, as well as other recently proposed ranking strategies for subtopic retrieval (i.e. Maximal Marginal Relevance (MMR) and Portfolio Theory(PT)), are compared against the QPRP. On the given task, it is shown that the QPRP outperforms these other ranking strategies. And unlike MMR and PT, one of the main advantages of the QPRP is that no parameter estimation/tuning is required; making the QPRP both simple and effective. This research demonstrates that the application of quantum theory to problems within information retrieval can lead to significant improvements.

1 Introduction

The Probability Ranking Principle (PRP) is the most widely used and accepted ranking criteria for the retrieval of documents [1]. Although this ranking principle has been shown to be generally applicable, it makes some assumptions which do not always hold [2,3]. When results need to be relevant but also diverse, as is the case in subtopic retrieval, the independence assumption made by the PRP is unrealistic. This is because the PRP neglects relationships between documents at relevance level ignoring the fact that a previous document may already contain similar relevant material [4,5,6,7].

To address this problem, different attempts have been made to formulate a better ranking principle by accounting for the similarity between documents in the ranking process. One such approach is called Maximal Marginal Relevance (MMR) [8]. This was used within a risk minimisation framework to compensate for the PRP's limitations [4]. While recently, Wang and Zhu [5] investigated the PRP's assumption with respect to the certainty in the estimation of a document's probability of relevance. These approaches have been motivated from empirical observations or heuristically adapted to the specific retrieval task, and require a

C. Gurrin et al. (Eds.): ECIR 2010, LNCS 5993, pp. 357–369, 2010.

significant amount of parameter tuning to be effective. However, a new ranking principle has been proposed for coping with interdependent document relevance; the Quantum Probability Ranking Principle (QPRP) [9].

The principle is derived from Quantum Probability Theory [10] where interdependent document relevance is captured by "quantum interference". While the theoretical foundations for the principle have been outlined in [9], it has not been empirically tested or validated. Thus, the aim of this paper is to conduct an empirical study comparing models based on the QPRP against models based on the PRP as well as against state-of-the-art non-PRP based models such as MMR [8] and PT [5]. On the subtopic retrieval task of the Interactive TREC Track, the results of this study show that the QPRP consistently outperforms these other models/principles. A distinct advantage that the QPRP has over the other more sophisticated approaches, is that no explicit parameter tuning is required. This work shows for the first time that Quantum Theory can be successfully applied within information retrieval.

The paper continues as follows: Section 2 provides an overview of the subtopic task, and explains how the assumptions of the PRP are too restrictive in this context, before describing the main approaches which aim to account for interdependence between documents. Then, Section 3 presents the QPRP and how it can be applied in order to account for document dependence through "quantum interference". To show the QPRP in action, Section 4 presents the extensive empirical study performed on the subtopic retrieval task, which explicitly focuses on novelty and diversity ranking. Finally, the paper concludes in Section 6, summarising the contribution of this work along with directions of future research.

2   Background and Related Works

The subtopic retrieval task stems from the need of providing a document ranking which covers all the possible different facets (subtopics) relevant to the user's information need. Thus, an IR system aims to maximise the user's satisfaction by retrieving documents which cover all the relevant subtopics in the ranking. Given a test collection, where the subtopics of the relevant documents have been identified, the effectiveness of the retrieval system can be measured in several ways [4,6,7,11]. The three main measures employed are S-recall, S-Mean Reciprocal Rank and S-precision. Subtopics coverage is measured by s-recall at rank k [4], formally defined as

$$s - recall(k) = \frac{|\cup_{i=1}^{k} subtopics(d_i)|}{n_s} \tag{1}$$

where $subtopics(d_i)$ returns the set of subtopics relevant to the topic that are contained in document d_i, and n_s is the number of possible subtopics relevant to the general topic. Intuitively, the fewer documents that have to be examined in order to retrieve all subtopics, the more effective the system. This intuition is also captured by the S-MRR measure, which is defined as the inverse of the rank at which full subtopic coverage is achieved.

For a precision oriented measure, S-precision at r is calculated by taking the ratio between the minimum rank that an optimal system $\mathcal{S}_{opt}$ achieves a S-recall value of r over the corresponding minimum rank the system $\mathcal{S}$ achieves the same S-recall value r [4]. The optimal system reaches recall value of r at the smallest rank k. Formally,

$$s - precision(r) = \frac{minRank(\mathcal{S}_{opt}, r)}{minRank(\mathcal{S}, r)} \qquad (2)$$

From the evaluation measures used for subtopic retrieval, it is clear that the intrinsic dependencies between documents need to be considered by the system when ranking. Consequently, models which adhere to the PRP have been shown to result in sub-optimal performance [12].

2.1   The Probability Ranking Principle and Its Limitations

The Probability Ranking Principle was first detailed by Robertson [1] in 1977, and its origins stem from initial work performed by Cooper [13] in 1971. Since its inception the PRP has played a vital part in shaping the development of IR models, methods and systems. The intuition underlying the PRP is as follows: in order to obtain the best overall effectiveness, an IR system should rank documents in descending order of their probability of relevance to the user's information need. It has been shown that adhering to this principle guarantees an optimal ranking [1]. In [14], Gordon also shows that the PRP maximises a suitably defined utility function.

The PRP assumes a probability distribution over the documents, which represents the space of events. In [14], the probabilities associated to documents represent the chances that a user is satisfied by observing those documents in response to his/her information need. Such probabilities are approximated by the probability of a document being relevant to the query; and, in practice, they are estimated from statistics extracted from the document and the collection. However, a key assumption made by the PRP is that the probability of relevance of a document is independent from the relevance of other documents (i.e. independence assumption). Consequently, the PRP dictates that at each rank the IR system should select document d such that:

$$d = \underset{d_i \in \mathcal{RE} \backslash RA}{\arg\max} \; P(d_i) \qquad (3)$$

with RA being the list of documents that have been ranked, d_i a document belonging to the set of retrieved documents ($\mathcal{RE}$) for a query but not ranked yet, and $P(.)$ being the probability of a document being relevant to the information need. However, when the independence assumption is not upheld, the PRP provides a suboptimal ranking [12].

In the case of subtopic retrieval, there is the explicit requirement of preferring relevant and novel information over redundant. Thus, the documents previously retrieved will influence what documents should be retrieved next. Under the PRP, however, if two documents have a high probability of relevance, but cover the same topics, they will both ranked in high positions. This has to be avoided in subtopic retrieval, since the two documents have the same relevant content and thus no novel information is conveyed to the user if both are retrieved at high ranks.

2.2   Beyond PRP: Attempts to Include Document Dependence

In the last decade, several attempts have been made to either model or include interdependent document relevance in the ranking process, in particular to cope with interactive information retrieval and subtopic retrieval. For example, the PRP is extended to interactive IR and framed within a situation-based framework in [15]. Under the **Interactive PRP**, users move between situations and the independent relevance assumption is substituted by a weaker condition within each situation. The ranking principle is then derived by the optimum ordering of the choices presented in each situation. As we shall see the QPRP differs from this approach because the quantum probability framework naturally encodes dependent relevance in the interference term[1]; and so, it considers dependence at the document level not at the situation level.

Maximal Marginal Relevance & Risk Minimization: The problem of a document ranking exploiting diversity amongst documents has been heuristically tackled in [8], where a technique called Maximal Marginal Relevance (MMR) was proposed. Document ranking is obtained by balancing the score of similarity between document and query, e.g. the probability of relevance, and a diversity score between the candidate document and all the documents ranked at earlier positions. A successful framework for coping with subtopics retrieval is based on risk minimization [4], where documents are ranked in increasing value of expected risk. In particular, language models are employed to represent documents and queries, while a loss function is used to model users preferences. The preference for retrieving documents that are both relevant and novel is encoded in the MMR function. Documents are selected following the objective function:

$$d = \arg\max \left(value_R(\theta_i; \theta_Q)(1 - c - value_N(\theta_I; \theta_1, \ldots, \theta_{i-1})) \right) \qquad (4)$$

where $value_R(\theta_i; \theta_Q)$ is the query likelihood estimated using language models, c represents the relative cost of seeing a non -relevant document compared with seeing a relevant but non-novel document, and $value_N(\theta_I; \theta_1, \ldots, \theta_{i-1})$ is the estimated novelty coefficient.

Portfolio Theory: Risk is also combined with the document relevance estimation in the Portfolio Theory (PT) approach recently proposed in [5]. The intuition behind the PT model for IR is that a measure of uncertainty (variance) is associated to each estimation of document relevance; when ranking documents the

---

[1] Interference is discussed in Section 3.

IR system should maximize relevance in the ranking while minimizing variance. The ranking criterion proposed by the Portfolio paradigm differs from the PRP because the rank of a document is not just function of the estimated probability of relevance of the document itself. Instead, the document's probability of relevance is combined with an additive term which synthesises the risk inclination of the user, and the uncertainty (variance) associated with the probability estimation, along with the correlation between the candidate document and documents ranked previously. In particular, documents are ranked according to:

$$d = \arg\max \left(P(d_i) - b w_{d_i} \delta_{d_i}^2 - 2b \sum_{d_k \in RA} w_{d_k} \delta_{d_i} \delta_{d_k} \rho_{d_i,d_k} \right) \tag{5}$$

where b encodes the risk propensity of the user, RA is the list of documents already ranked, $\delta_{d_i}^2$ is the variance associated to the probability estimation of document d_i, w_{d_i} is a weight inversely proportional to the rank position which express the importance of the rank position itself, and ρ_{d_i,d_k} is the correlation between document d_i and document d_k. Intuitively, the PT's ranking function is affected by the probability of relevance, the variance associated to the probability estimation and the correlation between candidate documents and documents already ranked.

In summary, the ranking functions suggested by the approaches for subtopic retrieval considered in this paper, i.e. MMR and PT, have two components. The first is the probability of relevance of a document with respect to user's information need, and it is in common with the PRP approach. However, the role of the second component in the ranking functions is to encode the degree of novelty/diversity of the candidate document with respect to the ones already ranked. In the QPRP, instead, although apparently it reflects the same schema as MMR and PT, relevance and novelty/diversity estimation are mixed together in the interference term. Moreover, MMR and PT have been inspired by empirical observations and require significant effort in parameter estimation. Conversely, the QPRP is derived from Quantum Probability Theory and does not contain any parameters which need explicit tuning.

3   The Quantum Probability Ranking Principle

Using Quantum Theory within IR was originally proposed by van Rijsbergen [16], and has been subsequently developed in a number of ways [17,18,19,20,21]. Here we consider the Quantum Probability Ranking Principle, that has been recently proposed by Zuccon et al. [9]. The QPRP is derived through the application of quantum probability to the problem of document ranking[2]. The resultant of this work was the following formulation: when ranking documents, the IR system has to maximise the total satisfaction of the user given the document ranking, achievable by maximising the total probability of the ranking. Using

---

[2] For a full derivation of the principle, we refer the reader to [9].

the quantum law of total probability, the resultant ranking strategy impose to select at each rank position a document d such that:

$$d = \arg\max \left(P(d_i) + \sum_{d_x \in RA} I_{d_x, d_i} \right) \tag{6}$$

where RA is the list of documents already ranked and I_{d_x, d_i} is the "quantum interference" between documents d_x and d_i. The pseudo-code of the QPRP algorithm is sketched in Algorithm 1. The intuition underlying this paradigm is that documents in a ranking share relationships at relevance level, i.e. they interfere with each other. For example, [3,4] showed that the user is more likely to be satisfied by documents addressing his information need in different aspects than documents with the same content. Then, it might be sensible to model documents expressing diverse information as having higher degree of interference than documents that are similar. For the same reason, documents containing novel information might highly interfere with documents ranked in previous positions. Even contrary information might be captured by the interference term: documents containing content contrary to the one presented at the previous rank position might trigger a revision of user's beliefs about the topic.

In summary, interference appears to capture the dependencies in documents' relevance judgements. The QPRP suggests that documents ranked until position $n-1$ interfere with the degree of relevance of the document ranked at position n. However, while the QPRP has been proposed, no experimental work has been performed which validates whether the Quantum based principle provides a better ranking or not. It is the aim of this paper to empirically explore the QPRP in the context of subtopic retrieval. In the following we detail how the QPRP formally differs from the PRP and we provide an outline of the estimation of quantum interference for subtopic retrieval.

Algorithm 1. The ranking strategy of QPRP

$\mathcal{RE} = \{\text{retrieved documents}\}$
$RA[k] = 0, \forall k : 1 \leq k \leq |\mathcal{RE}|$
comment: vector RA will contain the document ranking
$p = 1$
while $\mathcal{RE} \neq \varnothing$ **do**
$\quad d = \arg\max \left(P(d_i) + \sum_{d_x \in \mathcal{RA}} I_{d_x, d_i} \right)$, with $d_i \in \mathcal{RE}$
$\quad \mathcal{RE} = \mathcal{RE} \setminus \{d\}$
$\quad RA[p] = d$
$\quad p = p + 1$
end while
return RA

3.1 Estimating Probabilities and Interference

Ranking according to the QPRP is quite simple. Firstly, the QPRP uses the same probability estimates as the PRP (i.e the probability of relevance of a document $P(d_i)$). Next, is the interference component, which we shall explain in detail.

In a Quantum Probability Theory the law of total probability is different from standard probability theory (i.e. Kolmogorovian). In particular, the interference component might manifest. While in the PRP the law of total probability, i.e. the total probability associated with the ranking composed by document A and document B, is treated as the sum of the probabilities of the single, independent events (assumption of independent document relevance), the situation in the Quantum framework is different.

The QPRP assumes that underlying the relevance probability distribution there is a primitive concept of a complex amplitude distribution. This assumption follows from Quantum Probability Theory, and is the key point which differentiates the QPRP approach from the traditional PRP. What it means is that no assumption of independence between document relevance is made. The intuition is that a complex number $\phi_i = x_i + j y_i$ (called the amplitude) is associated to each event (i.e. document). Complex probability amplitudes ϕ_i and real[3] probabilities p_i are linked by the relationship $p_i = |\phi_i|^2 = (\sqrt{x_i^2 + y_i^2})^2$ where $|.|$ is the modulus of a complex number.

In this case, the total amplitude of a set of events relates to the sum of amplitudes associated to such events, similarly to what happens in the PRP with probabilities. When deriving the total probability of these events from the amplitudes, the amplitudes themselves are first summed and then their modulus is calculated, leading mathematically to the presence of an additional component, the interference, other than the square of the modulus of each amplitude. This is intuitive if we consider the polar form a complex number, $\phi_i = |\phi_i|(cos\psi_i + jsin\psi_i)$, where $j = \sqrt{-1}$ is the imaginary unit. Then,

$$p_{AB} = |\phi_{AB}|^2 = |\phi_A + \phi_B|^2 = \overline{(\phi_A + \phi_B)}(\phi_A + \phi_B) =$$
$$|\phi_A|^2 + |\phi_B|^2 + \overline{\phi_A}\phi_B + \phi_A\overline{\phi_B} =$$
$$|\phi_A|^2 + |\phi_B|^2 + 2|\phi_A||\phi_B|\cos(\psi_A - \psi_B) =$$
$$p_A + p_B + 2\sqrt{p_A}\sqrt{p_B}\cos\theta_{AB}. \tag{7}$$

where $\overline{\phi_A}$ indicates the complex conjugate of ϕ_A and θ_{AB} is the difference between the phases ψ_A and ψ_B. The interference is a real number, since both modulus and cosine are real-valued functions. However, the interference might not be zero: this is the case when the amplitudes do not have orthogonal phases. The difference between PRP and QPRP then resides in the interference component. If the difference of amplitudes's phases is an odd multiple of $\frac{\pi}{2}$, then the probability of joint events is simply the sum of the probabilities of the events, i.e. the sum of the squared amplitudes. When this is not true, the interference term is different from zero. In which case, the total probability obtained using Kolmogorovian probability theory assuming independence between events differs from the total probability employing Quantum Probability Theory. The difference is given by the additive interference term. The interference term might assume a positive value (i.e. constructive interference) or a negative one (i.e. destructive interference). In fact the interference is a function of the modulus of

---

[3] Meaning belonging to the field of the real numbers.

the amplitudes, which is always a positive real number, and of the amplitude's phase. In particular, the interference depends upon the cosine of the amplitude's phase difference.

In the QPRP, the total probability at each cutoff of the ranking is a function of the probabilities associated to the single documents and the interference between each pair of documents contained in the ranking. The maximization of the total probability depends upon the document's probabilities and their interference (see eq. 6). It has been shown that finding the solution of this maximisation problem can be reduced to a minimum set covering problem, which is NP-hard [4]. However, by using a greedy algorithm the complexity is significantly reduced, as it is the case for the MMR and PT approaches.

3.1.1   The QPRP in Action: Interference in Subtopic Retrieval

While the QPRP has been proposed, no concrete instantiation has been developed and tested. The main issue is the estimation of the interference term or, equivalently, the estimation of the amplitudes's phase. In this subsection, we outline how to estimate the interference for the task of subtopic retrieval, and then evaluate the instantiation in the remainder of the paper.

The main idea is to capture document interdependence through the interference component. In particular, relationships can be encoded in the phase, while the square roots of the estimated probabilities of relevance act as modulation component. The presence of relevance probabilities guarantees that the interference for documents that are diverse from the ones previously ranked but not (estimated) relevant, i.e. their probability is ~ 0, results null. Vice versa, if the documents involved in the interference have high probability of being relevant, then their contribution will be high. However, the interference also depends, both in sign and modulo, from the cosine of the phase difference between the amplitudes. Being able to derive this component directly from the amplitude distribution would mean being able to generate a complex amplitude distribution from real text statistics: the feasibility of this idea is still under investigation. However, we can try to estimate the phase difference between the amplitudes associated with documents. The estimation of this component depends upon the particular retrieval task. For example, in the subtopic retrieval task constructive interference (positive) might be used to model the interaction between documents covering different facets of the topic, while vice versa destructive interference might occur between documents covering the same subtopics. The converse situation, i.e. constructive interference to model coverage of the same topic while destructive models topical dissimilarity, seems feasible for encoding interdependent document relevance in ad-hoc retrieval task, acting similarly to an iterative implicit feedback mechanism: this is not covered in the present work, but will be topic of further investigations. In the subtopic retrieval scenario, we assume that redundant relevant documents destructively interfere, while documents conveying relevant but novel information generate constructive interference. The implementation details for the estimation of interference are discussed in Section 4.cterized by the correspondent Okapi weight.

4   Empirical Study

The aim of this empirical investigation is two fold:

1. test wether accounting for interdependent document relevance delivers a better document ranking for the subtopic retrieval task (i.e. PRP versus non-PRP), and,
2. compare the ranking strategy based on the QPRP against the classical (MMR) and state of the art (PT) techniques for subtopic retrieval.

To this aim, we conducted the following empirical investigation using the TREC subtopic retrieval track. This uses the documents from the Financial Times of London contained in TREC 6,7 and 8 collections and 20 ad-hoc retrieval topics from the TREC interactive tracks which are composed of subtopics, sometimes referred to as aspects. The collection was indexed using Lemur[4] where standard stop words were removed and Porter stemming applied. For each of the TREC Topics, we used the title of the topic to generate queries. We applied stopping and stemming to both kinds of queries.

The baseline method for the experiments was BM25 as it upholds the Probability Ranking Principle. The more competitive baselines employed were MMR and PT, and these were compared against the QPRP (see below) . For MMR, PT and QPRP methods, the top n documents retrieved by the BM25 baselines were re-ranked accordingly, where we tried n of 100, 200 and 1000. The normalised BM25 score was then used by these methods as the probability of relevance. While other ranking function may have been used, BM25 is a robust baseline previously used and delivers similar performance to Language Models [5]. For each of these methods, a kernel is required to compute the degree of dissimilarity/interdependence between documents. In the experiments reported here we used Pearson's correlation between the weighted term vectors associated to the documents as the kernel. The weighting schema was BM25. This kernel was previously used in [5]. While there are other choices of kernels, like the cosine similarity measures, in a set of preliminary experiments we found that the Pearson's correlation achieved the best results, across all strategies.

Maximal Marginal Relevance: We tested the MMR approach varying the value of the hyper-parameter c in the range $[0, 1]$ by decimal steps. For c equal one, the MMR strategies delivers the same rank as the PRP approach, while for a value of c equal zero the relevance score is discarded in favour of the dissimilarity score. The best results found are reported in the following section.

Portfolio Theory: To compute PT we need the variance associated to the probability estimated provided by the normalised BM25 scores. The variance (indicated with δ^2) becomes an adjunctive parameter of the PT ranking strategy. In [5], they suggest using a constant variance. We investigated the optimal value of the variance in combination with the value of the parameter b that encodes the risk propensity of a user. We considered values of b in the range $[1, 10]$ with

---
[4] http://www.lemurproject.org

unitary increments and values of δ^2 in the range $[10^{-10}, 10^{-1}]$. Here, we report the best results obtained by the possible combinations of parameters given the grid search of b by δ^2. Finally, the correlation ρ between pairs of documents is computed employing Pearson's correlation as described above, while the weight w associated to each rank position r is given by $\frac{1}{log_2 r}$, as in [5].

Quantum Probability Ranking Principle: The implementation of the QPRP ranking strategy does not require any parameter setting/tuning procedure. Without a method to estimate the complex probability amplitudes, we resort to an approximation of the phase by using Pearson's correlation. By using Pearson's correlation, it also enables us to fairly compare PT against QPRP. Interference between document d_i and d_x is then approximated using $-\sqrt{P(d_i)}\sqrt{P(d_x)}\rho_{d_i,d_x}$, where $P(.)$ is the estimation of the probability of relevance of a document and ρ is the Pearson's correlation between the two documents' term vectors.

5   Results

In Tables 1, 2 and 3 we report the results obtained by the different ranking strategies. In each table, the best results for MMR and PT are reported with respect to S-r@5 for Table 1, S-p@0.1 for Table 2 and S-MRR@100% for Table 3; and the the best performance overall for each measure is highlighted in bold. Note that we only report some of the results for $n = 100$ and $n = 200$ documents, but similar trends were witnessed at different levels on n. From these tables of results, the following points are of interest:

1. first note that in Tables 1, 2 and 3 the results for PT and MMR are generally higher than the equivalent listed in [5], both in absolute terms and percentage increase over the PRP. This is mainly due to the fact that we optimised each method specifically, making them very strong and competitive baselines. But in practice the performance of these methods is likely to be slightly lower;

2. MMR and PT improves upon the PRP at early levels of S-precision and S-recall, but fail to consistently outperform the PRP across all levels;

3. the QPRP improves upon PRP baselines for all levels of S-precision and S-recall. Furthermore, the QPRP outperforms MMR and PT across most levels;

4. the QPRP consistently outperforms other strategies across all topics when considering S-MRR@100%. This means on each topic the QPRP returns complete coverage of all subtopics at a rank lower than all the other strategies;

5. since the topics set is small, performing significance tests would not be appropriate [22, pages 178–180]. However, the QPRP delivers consistently better performance over the PRP, and also outperforms that state of the art methods. Also, as no parameter tuning is required for the QPRP, it represents a very attractive alternative to PT and MMR.

Table 1. Subtopic retrieval performance when $n = 200$: where PT ($b = 4$, $\delta^2 = 10^{-5}$) and MMR ($c = 0.5$) are optimized for S-r@5 measure

Models	S-r@5	S-r@10	S-r@20	S-r@50	S-MRR@100	S-p@.1	S-p@.2	S-p@.5
PRP	0.2466	0.3900	0.4962	0.6034	0.0086	0.3968	0.3062	0.1941
MMR	0.2697	0.3540	0.4795	0.6032	0.097	0.4203	0.2876	0.1964
	(+8.56%)	(-10.19%)	(-3.49%)	(-0.02%)	(+11.39%)	(+5.59%)	(-6.48%)	(+1.17%)
PT	0.2791	0.3654	0.4444	0.5494	0.0130	0.4587	0.2915	0.1769
	(+11.63%)	(-6.75%)	(-11.66%)	(-9.82%)	(+33.47%)	(+13.48%)	(-5.05%)	(-9.73%)
QPRP	0.3093	0.4063	0.5026	0.6186	0.0177	0.4237	0.3446	0.2362
	(+20.25%)	(+3.99%)	(+1.27%)	(+2.45%)	(+51.31%)	(+6.33%)	(+11.14%)	(+17.81%)

Table 2. Subtopic retrieval performance when $n = 100$: where PT ($b = 3$, $\delta^2 = 10^{-5}$) and MMR ($c = 0.9$) are optimised for S-p@.1 measure

Models	S-p@.1	S-p@.2	S-p@.5	S-p@1	S-MRR@100	S-r@5	S-r@10	S-r@50
PRP	0.3968	0.3062	0.1920	0.0101	0.0071	0.2466	0.3900	0.6034
MMR	0.4501	0.3162	0.1948	0.0106	0.0073	0.2690	0.3784	0.6077
	(+11.46%)	(+3.15%)	(+1.46%)	(+5.27%)	(+1.89%)	(+8.29%)	(-3.08%)	(+0.70%)
PT	0.4807	0.2992	0.1857	0.0086	0.0108	0.2791	0.3622	0.5929
	(+17.45%)	(-2.32%)	(-3.38%)	(-17.10%)	(+33.80%)	(+11.63%)	(-7.69%)	(-1.76%)
QPRP	0.4237	0.3452	0.2338	0.01167	0.01621	0.3093	0.4063	0.6150
	(+6.33%)	(+11.29%)	(+17.90%)	(+13.43%)	(+55.61%)	(+20.25%)	(+3.99%)	(+1.89%)

Table 3. The Subtopic Minimum Reciprocal Rank for various levels of coverage (25%-100%) for $n = 200$. The results for PT ($b = 6$, $\delta^2 = 10^{-5}$) and MMR ($c = 0.5$) are optimised on S-MRR@100%.

Models	S-MRR@25%	S-MRR@50%	S-MRR@75%	S-MRR@100%
PRP	0.2316	0.1056	0.0707	0.0086
MMR	0.2201	0.1135	0.0705	0.0097
	(-5.22%)	(+6.96%)	(-0.28%)	(+11.39%)
PT	0.2131	0.1098	0.0674	0.0154
	(-8.68%)	(+3.82%)	(-4.89%)	(+43.92%)
QPRP	0.2322	0.1355	0.0716	0.0177
	(+0.25%)	(+22.06%)	(+1.25%)	(+51.31%)

6   Conclusions and Future Works

In this paper we have explored how the QPRP can be applied in the setting of subtopic retrieval; which specifically requires models to account for interdependent document relevance. The QPRP naturally encodes the interdependence through quantum interference. The new ranking strategy has been empirically compared against the PRP and state-of-the-art ranking approaches. We have shown that accounting for documents dependencies at relevance level delivers a better ranking for subtopic retrieval. Also, the results of our empirical investigation have shown that the QPRP consistently outperforms previous approaches, i.e. MMR and PT, but with the additional advantage that no tedious parameter tuning is required. This research demonstrates that the use of Quantum

Probability Theory to model processes within information retrieval can lead to substantial improvements. Future investigations will consider:

1. alternative estimations of the interference;
2. how to derive a complex amplitude distribution from the document corpus;
3. the relationships between interference in the quantum probability framework and conditional probabilities in Kolmogorovian probability theory;
4. test the QPRP employing alternative collections for subtopic retrieval, and;
5. how to apply the QPRP paradigm to other retrieval tasks, e.g. ad-hoc retrieval.

Acknowledgments. We would like to thank Keith van Rijsbergen for his support and mentoring. This work has been conducted as part of the EPSRC Renaissance project (EP/F014384/1).

References

1. Robertson, S.E.: The probability ranking principle in IR. Journal of Documentation 33, 294–304 (1977)
2. Eisenberg, M., Berry, C.: Order effects: A study of the possible influence of presentation order on user judgments of document relevance. JASIS 39(5), 293–300 (2007)
3. Chen, H., Karger, D.R.: Less is more: probabilistic models for retrieving fewer relevant documents. In: SIGIR 2006, pp. 429–436 (2006)
4. Zhai, C.X., Cohen, W.W., Lafferty, J.: Beyond independent relevance: methods and evaluation metrics for subtopic retrieval. In: SIGIR 2003, pp. 10–17 (2003)
5. Wang, J., Zhu, J.: Portfolio theory of information retrieval. In: SIGIR 2009, pp. 115–122 (2009)
6. Clarke, C.L., Kolla, M., Vechtomova, O.: An effectiveness measure for ambiguous and underspecified queries. In: Azzopardi, L., Kazai, G., Robertson, S., Rüger, S., Shokouhi, M., Song, D., Yilmaz, E. (eds.) ICTIR 2009. LNCS, vol. 5766, pp. 188–199. Springer, Heidelberg (2009)
7. Agrawal, R., Gollapudi, S., Halverson, A., Ieong, S.: Diversifying search results. In: WSDM 2009, pp. 5–14 (2009)
8. Carbonell, J., Goldstein, J.: The use of MMR, diversity-based reranking for reordering documents and producing summaries. In: SIGIR 1998, pp. 335–336 (1998)
9. Zuccon, G., Azzopardi, L., van Rijsbergen, K.: The quantum probability ranking principle for information retrieval. In: Azzopardi, L., Kazai, G., Robertson, S., Rüger, S., Shokouhi, M., Song, D., Yilmaz, E. (eds.) ICTIR 2009. LNCS, vol. 5766, pp. 232–240. Springer, Heidelberg (2009)
10. Khrennikov, A.: Interpretations of probabilities. Walter de Gruyter, Berlin (2009)
11. Clarke, C.L., Kolla, M., Cormack, G.V., Vechtomova, O., Ashkan, A., Büttcher, S., MacKinnon, I.: Novelty and diversity in information retrieval evaluation. In: SIGIR 2008, pp. 659–666 (2008)
12. Gordon, M.D., Lenk, P.: When is the probability ranking principle suboptimal. JASIS 43(1), 1–14 (1999)
13. Cooper, W.S.: The inadequacy of probability of usefulness as a ranking criterion for retrieval system output. Unpublished working paper (1971)

14. Gordon, M.D., Lenk, P.: A utility theoretic examination of the probability ranking principle in information retrieval. JASIS 42(10), 703–714 (1999)
15. Fuhr, N.: A probability ranking principle for interactive information retrieval. JIR 12(3), 251–265 (2008)
16. van Rijsbergen, C.J.: The Geometry of Information Retrieval. Cambridge University Press, Cambridge (2004)
17. Melucci, M.: A basis for information retrieval in context. ACM TOIS 26(3), 1–41 (2008)
18. Piwowarski, B., Lalmas, M.: A quantum-based model for interactive information retrieval. In: Azzopardi, L., Kazai, G., Robertson, S., Rüger, S., Shokouhi, M., Song, D., Yilmaz, E. (eds.) ICTIR 2009. LNCS, vol. 5766, pp. 224–231. Springer, Heidelberg (2009)
19. Hou, Y., Song, D.: Characterizing pure high-order entanglements in lexical semantic spaces via information geometry. In: Bruza, P., Sofge, D., Lawless, W., van Rijsbergen, K., Klusch, M. (eds.) QI 2009. LNCS, vol. 5494, pp. 237–250. Springer, Heidelberg (2009)
20. Flender, C., Kitto, K., Bruza, P.: Beyond ontology in information systems. In: Bruza, P., Sofge, D., Lawless, W., van Rijsbergen, K., Klusch, M. (eds.) QI 2009. LNCS, vol. 5494, pp. 276–288. Springer, Heidelberg (2009)
21. Huertas-Rosero, A.F., Azzopardi, L., van Rijsbergen, C.J.: Eraser lattices and semantic contents. In: Bruza, P., Sofge, D., Lawless, W., van Rijsbergen, K., Klusch, M. (eds.) QI 2009. LNCS, vol. 5494, pp. 266–275. Springer, Heidelberg (2009)
22. van Rijsbergen, C.J.: Information Retrieval, 2nd edn. Butterworth (1979)

Wikipedia-Based Semantic Smoothing for the Language Modeling Approach to Information Retrieval

Xinhui Tu[1,3], Tingting He[1], Long Chen[2], Jing Luo[3], and Maoyuan Zhang[1]

[1] Engineering & Research Center For Information Technology On Education,
Huazhong Normal University. Wuhan, China
[2] Birkbeck, University of London
[3] Department of Computer Science and Technology,
Wuhan University of Science and Technology.Wuhan, China
tuxinhui@mail.wust.edu.cn, tthe@mail.ccnu.edu.cn,
long@dcs.bbk.ac.uk, luoluocat@mail.wust.edu.cn,
zhangmy@mail.ccnu.edu.cn

Abstract. Semantic smoothing for the language modeling approach to information retrieval is significant and effective to improve retrieval performance. In previous methods such as the translation model, individual terms or phrases are used to do semantic mapping. These models are not very efficient when faced with ambiguous words and phrases because they are unable to incorporate contextual information. To overcome this limitation, we propose a novel Wikipedia-based semantic smoothing method that decomposes a document into a set of weighted Wikipedia concepts and then maps those unambiguous Wikipedia concepts into query terms. The mapping probabilities from each Wikipedia concept to individual terms are estimated through the EM algorithm. Document models based on Wikipedia concept mapping are then derived. The new smoothing method is evaluated on the TREC Ad Hoc Track (Disks 1, 2, and 3) collections. Experiments show significant improvements over the two-stage language model, as well as the language model with translation-based semantic smoothing.

Keywords: Information Retrieval, Language Model, Wikipedia.

1 Introduction

Language modeling (LM) for Information Retrieval (IR) has been a promising area of research over the last ten years [10]. It provides an elegant mathematical model for ad-hoc text retrieval with excellent empirical results reported in the literature. An important problem in LM-based IR is the query and document models estimation. According to [10], this is essential in achieving optimal IR performance. However, it is very difficult to estimate an accurate document model due to the sparsity of training data. When a term does not occur in a document, the maximum likelihood estimator would give it a zero probability. This is unreasonable because the zero count is often due to insufficient sampling, and a larger sample of the data would likely contain the term.

C. Gurrin et al. (Eds.): ECIR 2010, LNCS 5993, pp. 370–381, 2010.
© Springer-Verlag Berlin Heidelberg 2010

Therefore, some effective smoothing approaches, which combine the document model with the background collection model, have been proposed by Zhai and Lafferty [14], [15].

However, it will be more effective if semantic relationship information can be incorporated into the semantic smoothing method for the language model [8]. Berger and Lafferty [1] subsequently integrate semantic smoothing into the language model where document terms are mapped onto query terms by taking advantage of translation model which coming from synthetic document-query pairs. However, this method judge semantic relationship between terms merely by using purely statistical techniques that don't make use of human's background knowledge. For example, without human's background knowledge, the term "mouse" may be translated to both "computer" and "cat" with high probabilities. Previous methods [2], [7], [13] have present other ways to train the translation probabilities between individual terms or phrase, but their approaches still suffer from the same problem as [1]. Thus, it is urgent to develop a framework to semantically smooth document models within the LM retrieval framework.

When computer programs face tasks that require human level intelligence, it is only natural to use an encyclopedia to endow the machine with the breadth of knowledge available to humans. Wikipedia is the largest encyclopedia in the world. The English version, as of May 10, 2009, contains 2,890,577 articles with about 100 million internal hyperlinks. And each article in Wikipedia describes a single topic; its title is a succinct, well-formed phrase that resembles a term in a conventional thesaurus.

In this paper, we propose a novel semantic smoothing method. In our method, a document is decomposed into a set of weighted Wikipedia concepts and, then, those Wikipedia concepts are mapped into individual terms for the purpose of semantic smoothing.

We define a Wikipedia concept as the topic described in a Wikipedia article that meets certain conditions. Because of the ambiguity in natural language, several concepts may share one title term. The ambiguous term in a document will be map into one concept with additional contextual constraints. For example, "mouse" in the document about computer technology could be map into Wikipedia concept "mouse (computing)", however, "mouse" in the document about mice could be map into Wikipedia concept "mouse (animal)". Because a Wikipedia concept is usually unambiguous, the mapping from Wikipedia concept to individual terms should have higher accuracy and result in better retrieval performance as compared to the semantic translations between single words or phrases.

We adopt an existing algorithm proposed by Wang [11], [12] to identify Wikipedia concepts occur in documents. Furthermore, we develop an expectation maximization (EM) based algorithm to estimate probabilities of mapping each Wikipedia concept into individual terms. The new smoothing method is tested on several collections from TREC Disks 1, 2, and 3, which contain news articles from several sources, including the Associated Press (AP), J.M.Smucker (SJM), and Wall Street Journal (WSJ). The experimental results show that significant improvements are obtained over the two-stage language model (TSLM) [15] and the language model with translation-based semantic smoothing (TSS) [1].

The contribution of this paper is threefold. First, it proposes a new LM-based document representation using a set of Wikipedia concepts. The new scheme also explores the relationship between individual terms and more complicated Wikipedia concepts. Second, it develops an EM-based algorithm to estimate the semantic relationships between Wikipedia concepts and individual terms and further uses those semantic relationships to smooth the document model, which is referred to as WSS (Wikipedia-based semantic smoothing) in this paper. The WSS model will be not only applicable to the area of text retrieval but also to many other text mining areas such as text categorization. Third, it empirically proves the effectiveness of the WSS for LM IR.

The remainder of this paper is organized as follows: In Section 2, we review previous work related to our method. In Section 3, we first formally define Wikipedia concept and present the approaches to extract Wikipedia concepts from documents and, then, we describe in detail the method of the WSS. Section 4 shows the experimental results on TREC Disks 1, 2, and 3. Section 5 concludes our paper.

2   Related Work

2.1   Semantic Smoothing Approaches

Semantic smoothing is potentially a more significant and effective method that incorporate semantic information into the language model. There have been various semantic smoothing techniques for language model.

Berger and Lafferty propose the translation-based semantic smoothing model [1] on the basis of the ideas and methods derive from statistical machine learning. Specifically, each term in a document of the collection is mapped into query terms, described as follows:

$$p(q|d) = \sum_w t(q|w)l(w|d) \tag{1}$$

where $t(q|w)$ is the translation probability which estimate the likelihood from document term w to query term q, and $l(w|d)$ is the traditional BOW probability. Even though it works well when cooperates with the simple language model, this model nevertheless, only captures part of semantic information in documents, namely the semantic relationship between individual terms, while the contextual information of the terms is fully ignored.

In order to address this problem, Liu and Croft, subsequently, propose the cluster language model [9] :

$$p(w|d) = \frac{N_d}{N_d+\mu} p_{ML}(w|d) + \left(1 - \frac{N_d}{N_d+\mu}\right) p(w|cluster) \tag{2}$$

where N_d is the document length and μ is a smoothing parameter. The document clusters in this model share a resemblance to our Wikipedia concepts in that we both integrate some context information by making use of a set of documents with similar background, instead of a single document estimation only. The fundamental problem for this model is that a document can only be classified into a single cluster, which is unreasonable for those long documents where they might actually belong to several different clusters, whereas a document can have multiple Wikipedia concepts in our model.

Zhou et al. [13] propose a semantic smoothing method by decomposing a document into a set of weighted topic signatures. In their model, the topic signatures are ontology-based concepts or multiword phrases. As previously mentioned, this model is not very efficient when confronted with ambiguous words and phrases because it is unable to disambiguate the sense of topic signatures.

2.2 Wikipedia-Based Text Mining

Wikipedia is the largest encyclopedia in the world. Many researchers have applied this encyclopedia to various text mining applications such as text categorization.

Gabrilovich et al. [4], [5] propose and evaluate a method to render text classification systems with encyclopedic knowledge – Wikipedia. They first build an auxiliary text classifier that can match documents with the most relevant articles of Wikipedia, and then augment the conventional BOW representation with new features which are the concepts (mainly the titles) represented by the relevant Wikipedia articles. Empirical results show that this representation improve text categorization performance across a diverse collection of datasets.

Wang et al [11], [12] propose another text classification model which overcomes the shortcoming of the BOW approach by embedding background knowledge derived from Wikipedia into a semantic kernel, which is then used to enrich the representation of documents. The semantic kernel in their model is able to keep multi-word concepts unbroken; it captures the semantic closeness of synonyms, and performs word sense disambiguation for polysemous terms. The empirical evaluation with real data sets demonstrates that the model successfully achieves improved classification accuracy with respect to the BOW technique, and to other recently developed methods.

Gabrilovich et al. [6] propose Explicit Semantic Analysis (ESA), a novel method to compute semantic relatedness of natural language texts by representing the meaning of texts in a high-dimensional space of concepts derived from Wikipedia. They use machine learning techniques to explicitly represent the meaning of any text as a weighted vector of Wikipedia-based concepts. Assessing the relatedness of texts in this space amounts to comparing the corresponding vectors using conventional metrics (e.g., cosine). In comparison to the previous methods, the ESA model results in substantial improvements in correlation of computed relatedness scores with human judgments. Importantly, due to the use of natural concepts, the ESA model is easy to explain to human users.

3 Wikipedia-Based Semantic Smoothing Approach

In this section, we describe Wikipedia-based semantic smoothing approach in detail. First, we define Wikipedia concept and estimate the mapping probabilities from each Wikipedia concept to individual terms through the EM algorithm. Second, we adopt the method proposed by Wang [11], [12] to automatically extract Wikipedia concepts in documents and disambiguate the concept senses. Last, the estimated mapping probabilities are used for document smoothing.

3.1 Wikipedia Concepts

As we regard each Wikipedia article as a single Wikipedia concept, which, in the same time, correspond to that of a Wikipedia article. So it is necessary to conduct a pre-processing procedure to abridge some useless or improper titles before constructing the Wikipedia concept index.

Firstly, we simply remove all the titles that have much to do with the chronology, such as "Years", "Decades" and "Centuries". The concept recognition from the title is then decided by the heuristic rules below:

1. If the title contain multiword, skip all the prepositions, determiners, conjunctions, etc while keep those capitalized words. A title is regarded as a concept directly when all the words of it are capitalized.

2. If the title consist of one word, it is regarded as a concept only if it appears in this article more than 3 times.

We then build a concept index, which will be able to search out the corresponding concept when a word or phrase is given, on the basis of above procedure.

Simultaneously, we build up an ambiguous title set to collect all the polysemous titles, which is the one bearing multiple meanings. For instance, the title "Mouse" is an ambiguous one, since it may refer to a kind of animal, a computing input device or something else. The following are some meanings of the title "Mouse".

Table 1. Example of Wikipedia Concept Models

Mouse (computing)		Mouse (animal)	
Term	Prob.	Term	Prob.
mouse	0.112	mice	0.093
mice	0.083	mouse	0.090
button	0.051	food	0.038
optical	0.046	pet	0.036
computer	0.042	animal	0.036
user	0.037	species	0.032
software	0.037	female	0.028
laser	0.035	male	0.026
wheel	0.029	human	0.025
system	0.024	rat	0.020

(1) Mouse (animal): a type of rodent, particularly of the genus Mus.
(2) Mouse (computing), plural mice, is the hardware item that controls the on-screen cursor.
(3) Mouse (programming language): a computer programming language
(4) Mouse (set theory): a small model of set theory with nice properties
(5) Mouse (musician), best known as the leader of the garage rock band Mouse and the Traps.

3.2 Wikipedia Concept Model Estimates

Each Wikipedia article whose title is defined as a Wikipedia concept will be indexed with individual terms. For each Wikipedia concept c_k, we can obtain a corresponding Wikipedia article d_k. Intuitively, we can use the document d_k to approximate the semantic profile for c_k, that is, to determine the probability of mapping the Wikipedia concept to terms in the vocabulary. If all terms appearing in the document d_k center on the Wikipedia concept c_k, then we can simply use a maximum likelihood estimator and the problem is as simple as frequency counting. However, some terms address the issue of other concept, whereas some are background terms of the whole collection. We use the generative model proposed in [16] to remove noise. Assume that the document corresponding to c_k is generated by a mixture model (that is, interpolating the topic model with the background collection model p(w|C):

$$p\big(w|\theta_{c_k}, C\big) = (1 - \alpha)p\big(w|\theta_{c_k}\big) + \alpha p(w|C) \tag{4}$$

Here, the coefficient is accounting for the background noise and c_k refers to the parameter set of the topic model associated with the Wikipedia concept c_k. In the experiments in this paper, the background coefficient is set to 0.5. Under this mixture language model, the log likelihood of generating the document set d_k is

$$\log p(d_k|\theta_{c_k}, C) = \sum_w c(w, d_k) \log p(w|\theta_{c_k}, C) \tag{5}$$

Here, $c(w,d_k)$ is the frequency of term w in d_k. The topic model for c_k can be estimated using the EM algorithm [3]. The EM update formulas are

$$p^{(n)}(w) = \frac{(1-\alpha)p^{(n)}p(w|\theta_{c_k})}{(1-\alpha)p^{(n)}p\big(w|0_{c_k}\big) + \alpha p(w|C)} \tag{6}$$

$$p^{(n+1)}\big(w|\theta_{c_k}\big) = \frac{c(w,d_k)p^{(n)}(w)}{\sum_w c(w_i,d_k)p^{(n)}(w_i)} \tag{7}$$

Our Wikipedia concept model is significantly different from previous ones described in [1], [2], [7], [9], [13]. In previous models, individual terms or phrases are   used to do semantic mapping. These models are not very efficient when faced with ambiguous words and phrases because they are unable to incorporate contextual information. Our model use unambiguous Wikipedia concepts to do semantic mapping. Thus, the resulting mapping will be more accurate.

From the two examples shown in Table 1, we can see that the concept-word mapping is quite accurate and unambiguous. For example, if we estimate the topic models for terms "mouse" without contextual information, then the models will contain mixed topics and are fairly general.

3.3 Search Wikipedia Concepts in Documents

For each document in the test collections, we search for the Wikipedia concepts that meet the criteria of 3.1. Such concepts appear in a document refer to as the candidate concepts. (If an m-gram concept is contained in an n-gram concept (with n > m), only the last one becomes a candidate concept.) We then construct a vector representation of a document, which consists of two parts: terms and candidate concepts.

For example, consider the text fragment "Machine Learning, Statistical Learning, and Data Mining are related subjects". Table 2 shows the traditional BOW term vector for this text fragment (after stemming), where feature values correspond to term frequencies. Table 2 also shows the concept vector representation, where boldface entries are candidate concepts.

Table 2. Traditional BOW term vector and Wikipedia concept vector

Traditional BOW term vector		Vector of Wikipedia concepts	
entry	tf	entry	tf
machine	1	machine learning	1
learn	2	statistical learning	1
statistic	1	data mining	1
data	1		
mine	1		
relate	1		
subject	1		

3.4 Disambiguation of Concept Senses

If a candidate concept is polysemous, i.e. it has multiple meanings, it is necessary to perform word sense disambiguation to find its most proper meaning in the context where it appears, prior to calculating its proximity to other related concepts. We utilize text similarity to do explicit word sense disambiguation. This method computes document similarity by measuring the overlapping of terms. For instance, a document #1 talks about computer mouse device, and the term "mouse" in Wikipedia refers to several different meanings, as listed in section 3.1. To pinpoint the correct meaning of a polysemous concept, we can calculate the cosine similarities between the text document (where the concept appears) and each of Wikipedia's articles(with a unique meaning of the polysemous concept), both of which are expressed as tf-idf vectors. The larger the cosine similarity between two tf-idf term vectors is, the higher the resemblance of the two corresponding text documents shares. Thus, the meaning with the largest cosine similarity of the article is selected as the best choice. From Table 3, the Wikipedia article describing "mouse" (computing) has the largest similarity with the document #1, and this is indeed confirmed to be the case by manual examination of the document.

Table 3. Cosine similarity between the document #1 and the Wikipedia's articles corresponding to the different meanings of the term "Mouse"

Meaning of "Mouse"	Similarity with the document #1
Mouse (computing)	0.2154
Mouse (programming language)	0.1811
Mouse (animal)	0.1309
Mouse (set theory)	0.0504
Mouse (musician)	0.0376

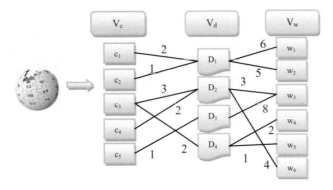

Fig. 1. Illustration of document indexing. V_c, V_d, and V_w are the Wikipedia concept set, document set, and word set, respectively. The number on each line denotes the frequency of the corresponding topic signature or word in the document.

3.5 Document Model Smoothing

Suppose we have indexed all documents in the test collection C with terms (individual words) and Wikipedia concepts (see Fig. 1). The probability of mapping a Wikipedia concept c_k to any individual term w, denoted as $p(w|c_k)$, is also given. Then, we can easily obtain a document model as follows.

$$p_t(w|d) = \sum_k p(w|c_k)p_{ml}(c_k|d) \tag{8}$$

The likelihood of a given document generating the Wikipedia concept c_k can be estimated with

$$p_{ml}(c_k|d) = \frac{c(c_k,d)}{\sum_i c(c_i,d)} \tag{9}$$

where $c(c_i,d)$ is the frequency of the Wikipedia concept c_k in a given document d.

However, If only the translation model is used, then there will be serious information loss. A natural extension is to interpolate the translation model with a unigram language model. The final document model for retrieval use is described as follows: It is a mixture model with two components, namely, a simple language model and a translation model:

$$p_{ml}(w|d) = (1 - \lambda)p_b(w|d) + \lambda p_t(w|d) \tag{10}$$

The translation coefficient λ controls the influence of the two components in the mixture model.

4 Experiments

4.1 Processing Wikipedia Data

As an open source project, the entire content of Wikipedia is available in the form of database dumps that are released periodically, from several days to several weeks apart. The version used in this study was released on May. 24, 2008. The full content and revision history at this point occupy about 120 GB of compressed data. We consider only the basic statistics for articles, which consume 2.3 GB (compressed).

These are about three million distinct entities in English Wikipedia. After filtered Wikipedia concepts as described in Section 3.1, we got 531,439 concepts.

4.2   Test Collections

We evaluate our method on six TREC ad hoc collections from Disks 1, 2,and 3. Table 4 shows the statistics of the six collections.

Table 4. The description of six testing collections

Collections	Word	Rel./Doc	Q.Len/Q.#
AP89/1-50	145,349	3,301/84,678	3.4/47
AP88&89/51-100	204,970	6,101/164,597	3.4/49
AP88&89/101-150	204,970	4,822/164,597	4.0/50
WSJ90-92/101-150	135,864	2,049/74,520	3.8/48
WSJ90-92/151-200	135,864	2,041/74,520	4.6/49
SJMN91/51-100	173,727	2,322/90,257	3.4/48

For each collection, we build two separate indices, the word index and the Wikipedia concept index. For Wikipedia concept indexing, we select all Wikipedia concepts in the documents as index terms. For word indexing, each document is processed in a standard way. Words are stemmed (using porter-stemmer), and stop words are removed. We use a 319-word stop list compiled by van Rijsbergen [17]. TREC topics are often described in multiple sections, including title, description, narrative, and concept. We only use the title for each TREC topic as our query.

4.3   Comparison with Previous Method

The comparison of our method (WSS) to the Two-Stage Language Model (TSLM) is shown in Table 5. We set the parameters γ and μ in the TSLM to 0.5 and 750, respectively, in the experiment because almost all collections achieve the optimal MAP at this configuration. For the translation coefficient λ in the WSS, all collections achieved the best performance at the setting point of $\lambda = 0.4$. In order to validate the significance of the improvement, we also run the paired-sample t-test. The incorporation of the Wikipedia concept translation improves both MAP and the overall recall over the baseline model on all six collections. Except for the recall on the collection of WSJ 90-92 topics 151-200, the improvements over the TSLM are all statistically significant at the level of $p < 0.05$ or even $p < 0.01$.

*(In Table 5, the signs ** and * indicate that the improvement is statistically significant according to the paired-sample t-test at the levels of p< 0:01 and p < 0:05, respectively.)*

Table 5. Comparison of the WSS to the TSLM and the TSS

Collections/Topics		TSLM	TSS	WSS	Change over TSLM	Change over TSS
AP89	MAP	0.187	0.195	0.214	+14.4%**	+9.7%**
1-50	Recall	1621	1730	1751	+8.0%**	+1.2%
AP88&89	MAP	0.252	0.272	0.287	+13.9%**	+5.5%*
51-100	Recall	3428	3735	3770	+10.0%**	+0.9%
AP88&89	MAP	0.219	0.235	0.251	+14.6%**	+6.8%*
101-150	Recall	3055	3237	3462	+13.3%**	+7.0%*
WSJ90-92	MAP	0.239	0.244	0.268	+12.1%**	+9.8%*
101-150	Recall	1510	1568	1591	+5.4%*	+1.5%
WSJ90-92	MAP	0.314	0.324	0.329	+4.8%*	+1.5%
151-200	Recall	1612	1646	1619	+0.4%	-1.6%
SJMN91	MAP	0.190	0.199	0.217	+14.2%**	+9.0%*
51-100	Recall	1350	1427	1493	+10.6%**	+4.6%

We also compare our method with the translation-based semantic smoothing method (TSS). For each collection, we tune the translation coefficient λ to maximize the MAP of the TSS. The optimal λ is about 0.1 for all six collections, which is much smaller than the optimal value for the WSS ($\lambda = 0.4$). It is also a kind of indication that the word-word translation is much noisier than the concept-word translation. The comparison result is given in Table 5.

From the experimental results, we can see that the WSS greatly outperforms the TSLM and has considerable gains over the TSS, especially on the measure of MAP. Considering that the baseline model is already very strong, we think that the Wikipedia concept language model is very promising to improve IR performance.

4.4 Effect of Document Smoothing

To test the robustness of the Wikipedia concept language model, we also change the settings of the translation coefficient. The variance of MAP with the translation coefficient is shown in Fig. 2. In a wide range from 0 to 0.7, the Wikipedia concept language model always performs better than the baseline on all six collections. This shows the robustness of the model. For all six curves in Fig. 1, the best performance is achieved at the setting point of $\lambda = 0.4$. After that point, the performance is downward. A possible explanation is that the extracted Wikipedia concepts do not capture all points of the document, but the TSLM captures those missing points. For this reason, when the influence of the translation model is too high in the mixture model, the performance is downward and even worse than that of the baseline.

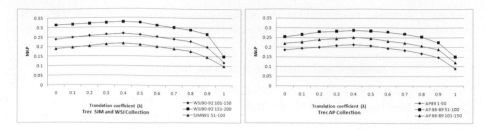

Fig. 2. The variance of MAP with λ, which controls the influence of the Wikipedia concept translation component in the mixture language model

5   Conclusions and Future Works

In this paper, we proposed a Wikipedia concept language model for ad hoc text retrieval. This new model decomposed a document into a set of weighted Wikipedia concepts and then mapped those unambiguous Wikipedia concepts into individual query terms. Because the Wikipedia concept itself contained unambiguous semantic meaning, the document model expansion based on Wikipedia concepts would be more accurate as compared to the document model expansion based on word or phrase mapping proposed in previous work such as [1], [2], [7], [9], [13], and thus improved the retrieval performance.

Further out, Wikipedia contains so much meaningful information, and our method only utilizes part of its resources. Other information in Wikipedia can be minded. For example, the link relation and the anchor text in Wikipedia is such a meaningful resource. Since anchor texts of links are also synonymies of the titles of linked articles, they can be used to do semantic mapping.

For future works, we will incorporate the link relation in Wikipedia into the Wikipedia concept language model. In addition, we will apply the Wikipedia concept language model to applications other than IR. Traditional text mining problems such as text clustering and text classification are also based on document models. Thus, it is natural to extend the application of the Wikipedia concept language model to those areas. In the future, we will further evaluate its effectiveness in those areas.

Acknowledgements

This work was supported by the Major Program of National Natural Science Foundation of China (No. 90920005), the National Natural Science Foundation of China (No. 60773167), and the Natural Science Foundation of Hubei Province (No. 2009 CDB145).

References

1. Berger, A., Lafferty, J.: Information Retrieval as Statistical Translation. In: Proc. 22nd Ann. Int'l ACM Conf. Research and Development in Information Retrieval (SIGIR 1999), pp. 222–229 (1999)

2. Cao, G., Nie, J.Y., Bai, J.: Integrating Word Relationships into Language Models. In: Proc. 28th Ann. Int'l ACM Conf. Research and Development in Information Retrieval (SIGIR 2005), pp. 298–305 (2005)
3. Dempster, A.P., Laird, N.M., Rubin, D.B.: Maximum Likelihood from Incomplete Data via the EM Algorithm. J. Royal Statistical Soc. 39, 1–38 (1977)
4. Gabrilovich, E., Markovitch, S.: Feature generation for text categorization using world knowledge. In: International Joint Conference on Artificial Intelligence, Edinburgh, Scotland (2005)
5. Gabrilovich, E., Markovitch, S.: Overcoming the brittleness bottleneck using wikipedia: enhancing text categorization with encyclopedic knowledge. In: National Conference on Artificial Intelligence (AAAI), Boston, Massachusetts (2006)
6. Gabrilovich, E., Markovitch, S.: Computing Semantic Relatedness using Wikipedia-based Explicit Semantic Analysis. In: Proceedings of the 20th International Joint Conference on Artificial Intelligence, pp. 6–12 (2007)
7. Jin, R., Hauptmann, A., Zhai, C.: Title Language Model for Information Retrieval. In: Proc. 25th Ann. Int'l ACM Conf. Research and Development in Information Retrieval (SIGIR 2002), pp. 42–48 (2002)
8. Lafferty, J., Zhai, C.: Document Language Models, Query Models, and Risk Minimization for Information Retrieval. In: Proc. 24th Ann. Int'l ACM Conf. Research and Development in Information Retrieval (SIGIR 2001), pp. 111–119 (2001)
9. Liu, X., Croft, W.B.: Cluster-Based Retrieval Using Language Models. In: Proc. 24th Ann. Int'l ACM Conf. Research and Development in Information Retrieval (SIGIR 2001), pp. 186–193 (2001)
10. Ponte, J.M., Croft, W.B.: A language modeling approach to information retrieval. In: Proceedings of the Twenty First ACM-SIGIR, Melbourne, Australia, pp. 275–281. ACM Press, New York (1998)
11. Wang, P., et al.: Improving Text Classification by Using Encyclopedia Knowledge. In: Seventh IEEE International Conference on Data Mining, pp. 332–341 (2007)
12. Wang, P., Domeniconi, C.: Building semantic kernels for text classification using Wikipedia. In: The 14th ACM SIGKDD International Conference on Knowledge Discovery & Data Mining, KDD 2008 (2008)
13. Zhou, X., Hu, X., et al.: Topic Signature Language Models for Ad Hoc Retrieval. IEEE Transactions on Knowledge and Data Engineering 19(9), 1276–1287 (2007)
14. Zhai, C., Lafferty, J.: A Study of Smoothing Methods for Language Models Applied to Ad Hoc Information Retrieval. In: Proc. 24th Ann. Int'l ACM Conf. Research and Development in Information Retrieval (SIGIR 2001), pp. 334–342 (2001)
15. Zhai, C., Lafferty, J.: Two-Stage Language Models for Information Retrieval. In: Proc. ACM Conf. Research and Development in Information Retrieval, SIGIR 2002 (2002)
16. Zhai, C., Lafferty, J.: Model-Based Feedback in the Language Modeling Approach to Information Retrieval. In: Proc. 10th Int'l Conf. Information and Knowledge Management (CIKM 2001), pp. 403–410 (2001)
17. Van Rijsbergen, C.J.: Information Retrieval, 2nd edn. Dept. of Computer Science, University of Glasgow (1979)

A Performance Prediction Approach to Enhance Collaborative Filtering Performance

Alejandro Bellogín and Pablo Castells

Universidad Autónoma de Madrid
Escuela Politécnica Superior
Francisco Tomás y Valiente 11, 28049 Madrid, Spain
{alejandro.bellogin,pablo.castells}@uam.es

Abstract. Performance prediction has gained increasing attention in the IR field since the half of the past decade and has become an established research topic in the field. The present work restates the problem in the area of Collaborative Filtering (CF), where it has barely been researched so far. We investigate the adaptation of clarity-based query performance predictors to predict neighbor performance in CF. A predictor is proposed and introduced in a kNN CF algorithm to produce a dynamic variant where neighbor ratings are weighted based on their predicted performance. The properties of the predictor are empirically studied by, first, checking the correlation of the predictor output with a proposed measure of neighbor performance. Then, the performance of the dynamic kNN variant is examined on different sparsity and neighborhood size conditions, where the variant consistently outperforms the baseline algorithm, with increasing difference on small neighborhoods.

Keywords: Recommender systems, collaborative filtering, neighbor selection, performance prediction, query clarity.

1 Introduction

Collaborative Filtering (CF) is a particularly successful form of personalized Information Retrieval, or personalized assistance over item choice problems in general [12,19]. CF has the interesting property that no item description is needed to recommend them, but only information about past interaction between users and items. Besides, it has the salient advantage that users benefit from other users' experience (opinions, votes, ratings, purchases, tastes, etc.), and not only their own, whereby opportunities for users' exposure to novel and unknown experiences with respect to previous instances are furthered, in contrast to other approaches that tend to reproduce the user's past, insofar as they examine the records of individual users in isolation.

CF is also based on the principle that the records of a user are not equally useful to all other users as input to produce recommendations [12]. A central aspect of CF algorithms is thus to determine which users form the best basis, and to what degree, to generate a recommendation for a particular user. Such users are usually referred to as *neighbors*, and their identification is commonly based on notions of similarity to the target user. The similarity of two users is generally assessed by examining to what

C. Gurrin et al. (Eds.): ECIR 2010, LNCS 5993, pp. 382–393, 2010.

degree they displayed similar behaviors (selection, rating, purchase, etc.) in their inter-action with items in the retrieval space. This basic approach can be complemented with alternative comparisons of virtually any user features the system may have access to, such as personal information, demographic data, or similar behaviors in external sys-tems. Thus, the more similar a neighbor is to the active user, the more his tastes are taken into account as good advice to make up recommendations. For instance, a com-mon CF approach consists of predicting the utility of an item for the target user by a weighted average of the ratings of all his neighbors, where the ratings are weighted by the similarity between each neighbor and the user. It is also common to set a maximum number of most similar users to restrict the set of neighbors to the k nearest, in order to avoid the noisy disruption of long tails of dissimilar users in the recommendation.

Similarity has indeed proved to be a key element for neighbor selection in order to provide accurate recommendations. Neighbor trustworthiness and expertise have also been researched as relevant complementary criteria to select the best possible collabo-rative advice [14, 19]. We believe however that further neighbor and data characteris-tics (individual or relative to the target user) can be exploited to enhance the selection and weighting of neighbors in recommendations. For instance, the size, heterogeneity, and other characteristics of the associated evidence (set of common known items, ratings, etc.), can be key to assess the significance of observations, the reliability of the evidence and the confidence of predictions, and the part of such elements in rec-ommendations could be adjusted accordingly. Observations on users with little ex-perience in common (where two or three coincidences of mismatches may lead to extreme similarity values) is far from being as significant as that on other users with a large subset of comparable history, and this difference should be accounted for in the CF algorithm. This type of issue is often mentioned and occasionally dealt with in the CF literature, but usually by hand-crafted solutions and manual tuning, rather than principled ways [4,12].

In this context, we research into notions of *neighbor goodness*, when seen as input for recommendation to a given user, where "goodness" should account for any aspect, besides similarity, that correlates with better results when the neighbor is introduced (or boosted) in computing a recommendation. Our proposed approach investigates the adaptation of performance prediction techniques developed in the IR field to assess neighbor goodness, where the latter is seen as an issue of *neighbor performance*. Spe-cifically, we propose a neighbor goodness predictor inspired on query clarity. We ana-lyze its correlation with an objective neighbor performance metric, and further measure the usefulness of the predictor by using it in a dynamic enhancement of a user-based k nearest neighbors (kNN) CF formulation, where neighbor ratings are weighted by their neighbor goodness. We show empiric evidence confirming that measurable improve-ments result from this approach.

The rest of the paper is organized as follows. Section 2 provides an overview of the state of the art in performance prediction in IR. In Section 3, the proposed approach is described, including the definition of the predictors and the formulation of rating prediction in CF as an aggregation operation with dynamic weights. Section 4 reports on the experimental work, where the proposed techniques are evaluated on a public dataset. Finally, Section 5 provides conclusions drawn from this work, along with potential lines for the continuation of the research.

2 Performance Prediction in Information Retrieval

Performance prediction in IR has been mostly addressed as a query performance issue, which refers to the performance of an IR system in response to a specific query. It also relates to the appropriateness of a query as an expression for a user information need. Dealing effectively with poorly-performing queries is a crucial issue in IR, and performance prediction provides tools that can be useful in many ways [22,23]. From the user perspective, it provides valuable feedback that can be used to direct a search, e.g. by rephrasing the query or providing relevance feedback. From the perspective of an IR system, performance prediction provides a means to address the problem of retrieval consistency: a retrieval system can invoke alternative retrieval strategies for different queries according to their expected performance (query expansion or different ranking functions based on the predicted difficulty). From the perspective of a system administrator, she can identify queries related to a specific subject that are difficult for the search engine, and e.g. expand the collection of documents to better answer insufficiently covered subjects. For distributed IR, performance estimations can be used to decide which search engine and/or database to use for each particular query, or how much weight to give it when its results are combined with those of other engines.

The prediction methods documented in the literature use a variety of available data as a basis for prediction, such as a query, its properties with respect to the retrieval space [7], the output of the retrieval system [5], or the output of other systems [3]. According to whether or not the retrieval results are used in the prediction, the methods can be classified into pre- and post-retrieval approaches [10]. The first type has the advantage that the prediction can be taken into account to improve the retrieval process itself. However, these predictors have the potential handicap, with regards to their accuracy, that the extra retrieval effectiveness cues available after the system response are not exploited [24]. In post-retrieval prediction, predictors make use of retrieved results [2,23,24]. Broadly speaking, techniques in this category provide better prediction accuracy. However, computational efficiency is usually a problem for many of these techniques, and furthermore, the predictions cannot be used to improve the retrieval strategies, unless some kind of iteration is applied, as the output from the retrieval system is needed to compute the predictions in the first place.

Pre-retrieval query performance has been studied mainly based on statistic methods, though linguistic approaches have also been researched [17]. Simple statistic approaches based on IDF, and variations thereof, have been proposed [11,13,18], showing moderate correlation with query performance though. He & Ounis propose the notion of query scope as a measure of the specificity of a query, which is quantified as the percentage of documents in the collection that contain at least one query term [11]. Query scope is effective in predicting the performance of short queries, though it seems very sensitive to query length [16].

More effective predictors have been defined on formal probabilistic grounds based on language models by the so-called *clarity score*, which captures the (lack of) ambiguity in a query with respect to the collection, or a specific result set [7,23,24] (the second case thus falling in the category of post-retrieval prediction). In this work, query ambiguity is meant to be "the degree to which the query retrieves documents in the given collection with similar word usage" [6]. Query clarity measures the degree of dissimilarity between the language associated with the query and the generic language of the collection as a whole. This is measured as the relative entropy, or Kullback-Leibler divergence, between the query and collection language models (with unigram distributions).

Analyzing the entropy of the language model induced by the query is indeed a natural approach since entropy measures how strongly a distribution specifies certain values, in this case, terms. In its original formulation [7], query clarity is defined as follows:

$$\text{clarity}(q) = \sum_{w \in V} \text{p}(w \mid q) \log_2 \frac{\text{p}(w \mid q)}{\text{p}_c(w)}$$

$$\text{p}(w \mid q) = \sum_{d \in R} \text{p}(w \mid d)\text{p}(d \mid q), \quad \text{p}(q \mid d) = \prod_{w_q \in q} \text{p}(w_q \mid d)$$

$$\text{p}(w \mid d) = \lambda \, \text{p}_{ml}(w \mid d) + (1 - \lambda) \, \text{p}_c(w)$$

(1)

with w being any term, q the query, d a document or its model, R the set of documents in the collection that contain at least one query term (it is also possible to take the whole collection here), $\text{p}_{ml}(w \mid d)$ the relative frequency of term w in document d, $\text{p}_c(w)$ the relative frequency of the term in the collection as a whole, λ a free parameter (set to 0.6 in [7]), and V the entire vocabulary.

It was observed that queries whose likely relevant documents are a mix of disparate topics receive a lower score than those with a topically-coherent result set. A strong correlation was also found between query clarity and the performance of the result set. Because of that, the clarity score method has been widely used for query performance prediction in the area. Some applications include query expansion (anticipating poorly performing queries as good candidates to be expanded), rank fusion, link extraction in topic detection and tracking [15], and document segmentation [8]. A prolific sequel of variants and enhancements on the notion of clarity followed the original works [8,11].

3   Neighbor Performance in Collaborative Filtering

Starting from the work on performance prediction in IR, our research addresses the enhancement of neighbor selection techniques in CF by introducing the notion of neighbor performance, as an additional factor (besides similarity) to automatically tune the neighbor's participation in the recommendations, according to the expected goodness of their advice.

Our approach investigates the adaptation of the query clarity technique from IR to CF, as a basis for finding suitable predictors. This involves finding a meaningful equivalence or translation of the retrieval spaces involved in ad-hoc IR (queries, words, documents) into the corresponding elements of a CF setting (users, items, ratings), in order to provide a specific formulation. Moreover, in order to validate any proposed predictor, we should consider a measurable definition of what neighbor performance means, in order to check the correlation between predicted outcomes and objective measurements. We further test the effectiveness of the defined predictors by introducing and testing a dynamic variant of memory-based, user-based CF, in which the weights of neighbors are dynamically adjusted based on their expected effectiveness.

3.1 Assessing Neighbor Performance

The purpose of predictors in the proposed approach is to assess how useful specific neighbors' ratings are as a basis for predicting ratings for the active user in the basic CF formula. A performance predictor for a neighbor needs thus to be contrasted to a measure of how "good" is the neighbor's contribution to the global community of users in the system. In contrast with query performance prediction, where a well-established array of metrics can be used to quantify query performance, there is not, to the best of our knowledge, an equivalent function for CF neighbors (let alone a standard one) in the literature. We therefore need to introduce some sound candidate metric.

The measure we propose, named *neighbor goodness* (NG, how "good a neighbor" a user is to her surroundings), is defined as the difference in performance of the recommender system when including vs. excluding the user (her ratings) from the dataset (the performance of an item could be analogously defined in item-based CF). For instance, based on the mean average error (MAE) standard metric, NG can be defined as follows:

$$NG(u) = \frac{1}{\left|\mathcal{R}_{\mathcal{U}-\{u\}}\right|} \sum_{v \in \mathcal{U}-\{u\}} CE_{\mathcal{U}-\{u\}}(v) - \frac{1}{\left|\mathcal{R}_{\mathcal{U}-\{u\}}\right|} \sum_{v \in \mathcal{U}-\{u\}} CE_{\mathcal{U}}(v)$$

$$= \frac{1}{\left|\mathcal{R}_{\mathcal{U}-\{u\}}\right|} \sum_{v \in \mathcal{U}-\{u\}} \left[CE_{\mathcal{U}-\{u\}}(v) - CE_{\mathcal{U}}(v) \right] \tag{2}$$

where $\mathcal{U}$ is the set of all users, $\mathcal{R}$ is the set of all user-item pairs in $\mathcal{U} \times \mathcal{I}$ with known ratings, $\mathcal{R}_{\mathcal{X}} = \{ (u,i) \in \mathcal{R} \mid u \in \mathcal{X} \}$ is the subset of $\mathcal{R}$ restricted to users in $\mathcal{X} \subset \mathcal{U}$, and $CE_{\mathcal{X}}(v)$ is the cumulative error of the recommender system on user v considering only the ratings of users in $\mathcal{X} \subset \mathcal{U}$, that is: $CE_{\mathcal{X}}(v) = \sum_{i \in \mathcal{I}, r(v,i) \neq \varnothing} \left| \tilde{r}_{\mathcal{X}}(v,i) - r(v,i) \right|$, $\tilde{r}_{\mathcal{X}}(v,i)$

denoting the rating predicted by the system when taking $\mathcal{X}$ as the CF user community.

Note that the first term $\frac{1}{\left|\mathcal{R}_{\mathcal{U}-\{u\}}\right|} \sum_{v \in \mathcal{U}-\{u\}} CE_{\mathcal{U}-\{u\}}(v)$ in equation (2) is just the MAE of the

system when leaving out user u. The second term $\frac{1}{\left|\mathcal{R}_{\mathcal{U}-\{u\}}\right|} \sum_{v \in \mathcal{U}-\{u\}} CE_{\mathcal{U}}(v)$ includes u in

the computation of the recommendations of which the errors are measured and summed, but excludes the error on u itself from the sum, since we mean to measure strictly the effect of u on its neighbors, and not the reverse.

This measure thus quantifies how much a user affects (contributes to or detracts from) the total amount of MAE of the system, since it is computed in the same way as MAE, but leaving out the user of interest –in the first term, it is completely omitted; in the second term, the user is only involved as a neighbor. In this way, we measure how a user contributes to the rest of users, or put informally, how better or worse is the world, in the sense of how well recommendations work, with and without the user.

3.2 Predicting Good Neighbors

Now, inspired by the clarity score defined for query performance [7], we consider its adaptation to predict neighbor performance in collaborative recommendation. In essence, the clarity score captures the lack of ambiguity (uncertainty) in a query, by computing the distance between the language models induced by the query and the collection. Cronen-Townsend et al showed that clarity is correlated with performance, because the less ambiguous a query, the more chances are that the system will return a good result in response [7]. Cronen-Townsend's experiments thus seem to confirm the underlying hypothesis that the system performance is largely influenced by the amount of uncertainty involved in the inputs it takes to build the retrieval result. That is, the uncertainty should correlate negatively with the performance level one may a priori expect.

CF systems rank and recommend items without an explicit user query. However, the system uses other inputs that may also determine the resulting performance. In analogy to the work on query clarity, we may hypothesize that the amount of uncertainty involved in a user neighbor may be a good predictor of his performance. In this case, the uncertainty can be understood as the ambiguity of the user's tastes, and it can be approximated as an adaptation of equation (1) to compute the clarity of users.

There are many possible ways to map the terms in equation (1) to elements of CF in meaningful ways, many of which we have studied before reaching the formulation proposed herein, which goes as follows. First, whereas the clarity measure follows a language modeling approach where three probability spaces are involved: queries, documents, and words, we map and fold this triadic approach into a dyadic one, involving only a set of users and a set of items. We have tested alternative triadic approaches, such as considering sets of features as the equivalent of document words, but they yield lower empiric performance, likely because the relation between a query and its constituent words, which is structural (i.e. a query is modeled as equivalent to the conjunction of its terms), does not map well to the relation between users and features (or even items and features), which is considerably looser in our experimental domain (based on MovieLens[1] and IMDb[2]).

In the dyadic approach, we have investigated two possible models, one in which the clarity of a user is measured against the set of all items, and one in which it is defined in terms of the set of all users. We shall follow here the latter option, which has shown a similar but slightly more consistent behavior than the former in our experiments:

$$\text{clarity}(u) = \sum_{v \in \mathcal{U}} p(v \,|\, u) \log_2 \frac{p(v \,|\, u)}{p_c(v)}$$

The conditional probability between users in the above formula can be rewritten in terms of conditional probabilities involving users and items:

$$p(v \,|\, u) = \sum_{i \in \mathcal{I}, r(u,i) \neq \varnothing} p(v \,|\, i) p(i \,|\, u)$$

---

[1] http://www.grouplens.org/node/73
[2] http://www.imdb.com

Now p(*v*|*i*) and p(*i*|*u*) can be computed by linearly smoothing probability estimates from observed evidence, as follows:

$$p(v \mid i) = \lambda_1 \, p_{ml}(v \mid i) + (1 - \lambda_1) p_c(v)$$
$$p(i \mid u) = \lambda_2 \, p_{ml}(i \mid u) + (1 - \lambda_2) p_c(i)$$

where we assume a uniform distribution for p_c, and we estimate p_{ml} based on rating data:

$$p_{ml}(v \mid i) = \frac{r(v,i)}{\sum_{u \in \mathcal{U}} r(u,i)}, \quad p_c(v) = \frac{1}{|\mathcal{U}|}$$

$$p_{ml}(i \mid u) = \frac{r(u,i)}{\sum_{j \in \mathcal{I}} r(u,j)}, \quad p_c(i) = \frac{1}{|\mathcal{I}|}$$

The same as query clarity captures the lack of ambiguity in a query, user clarity thus computed is expected to capture the lack of ambiguity in a user's tastes. Analogously, item clarities could be defined with respect to the space of users or the space of items in item-oriented CF, but we shall focus here only on the user-oriented approach. Having thus defined the notion of user clarity, the question is whether it can serve as a good neighbor performance predictor, and as such, whether its predictive power can be leveraged to dynamically weight the contribution of neighbors in CF in a way that improves the quality of recommendations. We address this in the next section.

3.3   Rating Prediction as a Dynamic Aggregation of Utilities

The same as performance prediction in IR has been used to optimize rank aggregation, in our proposed view each user's neighbor is seen as a retrieval subsystem (or criteria) whose output is to be combined to form the final system output (the recommendations) to the user.

A common utility-based formulation for rating prediction in memory-based CF, in a user-based, mean-centered variant [1], can be expressed as:

$$\tilde{r}(u,i) = \overline{r}(u) + C \sum_{v \in N[u]} \text{sim}(u,v) \cdot (r(v,i) - \overline{r}(v)), \qquad C = \frac{1}{\sum_{v \in N[u]} |\text{sim}(u,v)|} \qquad (3)$$

where N[*u*] is the set of neighbors of the active user, $\overline{r}(u)$ is the average of all ratings by user *u*, and *C* is a normalizing constant to keep the rating values within scale. Note that this particular formulation of memory-based CF is chosen here without loss of generality, as our approach can be developed in equivalent terms for alternative CF variants (not mean-centered, item-based, etc. [1]).

The term $\tilde{r}(u,i)$ in equation (3) can be seen as a retrieval function that aggregates the output of several utility subfunctions $r(v,i) - \overline{r}(v)$, each corresponding to a recommendation given by a neighbor of the target user. The combination of utility values is defined as a linear combination (translated by $\overline{r}(u)$) of the neighbor's ratings, weighted by their similarity sim(u,v) (scaled by C) to the target user. The computation of utility values in CF can thus be viewed as a case of rank aggregation in IR, and as such, a case for the enhancement of the aggregated result by predicting the performance of the recommendation outputs being combined. In fact, the similarity value can be seen as a prediction of how useful the neighbor's advice is expected to be for the active user, which has proved to be quite an effective approach. The question is whether other performance factors, beyond similarity can be considered in a way that further enhancements can be drawn.

We thus aim to investigate whether CF results can be further enhanced by introducing, in addition to a similarity function, further effectiveness predictors, such as the user clarity value defined in the previous section, into the weights of the linear combination of neighbor ratings. The idea can be expressed as rewriting equation (3) as:

$$\tilde{r}(u,i) = \overline{r}(u) + C \sum_{v \in N[u]} \gamma(v,u,i) \cdot \mathrm{sim}(u,v) \cdot \left(r(v,i) - \overline{r}(v)\right)$$

where $\gamma(v,u,i)$ is a predictor of the performance of neighbor v.

In the general case, γ can be sensitive to the specific target user u, the item i, and in general it could even take further inputs from the recommendation space and context. As a first step, we explore the simple case when the predictor only examines the data related to the neighbor user v, and in particular, we consider $\gamma(v,u,i) = \mathrm{clarity}(v)$. In the next section we show the experiments we have set up in order to observe the effect of the introduction of this predictor in the computation of collaborative recommendations.

4   Experimental Work

The experiments reported here have been carried out using the MovieLens dataset, and more specifically the so-called "100K" set. The main variable with respect to which the behavior of the proposed method is tested is the amount of sparsity, which we relate to the number of available ratings in the dataset based on which recommendations are computed. To this purpose, we split the dataset into different training / test cuts (10% to 90% in increments of 10%), with ten random splits per sparsity level. The neighborhood size is another parameter with respect to which the results are examined.

We first check the direct correlation between the user clarity predictor proposed in section 3.2 and the NG performance metric defined in 3.1, computed with a standard CF algorithm using the Taste library[3]. NG quantifies how a user affects the total amount of MAE, so that a well performing user should relate to high values of this measure (and vice-versa), reflecting to what degree the whole community gets better (or worse) results when the user is included as a potential neighbor. In the computation of clarity values, the λ_1 and λ_2 parameters were set to 0.6, as in [7].

The values shown in Table 1 show a positive direct correlation, meaning that the higher the value of the measure (well performing user), the higher the value of the predictor (clear user), thus confirming the predictive power of clarity for neighbor performance. An exception to this is when only 10% of ratings are used, where the correlation appears as negative. This lower end value results from data splits in which users have

---

[3] http://lucene.apache.org/mahout/taste.html

about ten ratings each on average in the training set, which seems insufficient to draw reliable predictions from. The correlation by the Spearman and Kendall functions yields similar results. While being indicative of a positive trend, and not far from previous results in query performance [5], observed correlation values still leave room for further elaboration and refinements of the proposed predictor and alternative ones, as well as the NG metric itself, in order to match the best findings in query performance [7].

Table 1. Pearson correlation values between user clarity and the NG performance metric at nine training/test split levels on the MovieLens rating dataset. The percentages indicate the ratio of rating data used for training in the CF algorithm.

% training	10%	20%	30%	40%	50%	60%	70%	80%	90%
correlation	-0.10	0.10	0.18	0.18	0.18	0.17	0.17	0.15	0.15

The second experiment consists of measuring final performance improvements when dynamic weights are introduced in a user-based CF. That is, the dynamic aggregation of neighbor ratings based on a prediction of their performance, when seen as individual recommenders (as defined in section 3.3), is tested against the basic CF algorithm without dynamic weights. Again, we have used the Taste implementation of user-based heuristic CF, both as a baseline, and as the algorithm into which the clarity-based enhancement is introduced.

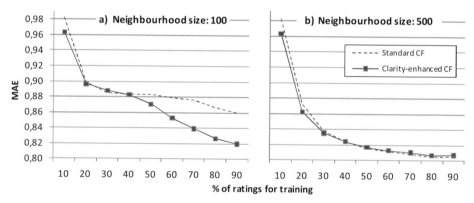

Fig. 1. Performance comparison of CF with clarity-based neighbor weighting, and standard CF, using neighborhoods of a) 100 users and b) 500 users

Figure 1 shows the results for the clarity predictor, when taking neighborhood sizes of 100 and 500 users respectively. Each graphic shows performance values (MAE) for the nine splits described above. Our method clearly improves the baseline (by up to 5% for 60-80% cuts) when smaller neighborhoods (100 users) are used, and gets almost equal performance with neighborhoods of size 500 users. This shows that our method works particularly well when limited neighborhoods are used, and the improvement fades down to the baseline as they are enlarged. This means that our method is more efficient than the static option with respect to this variable, i.e. that it is able to get better results out of more economic neighborhood sizes.

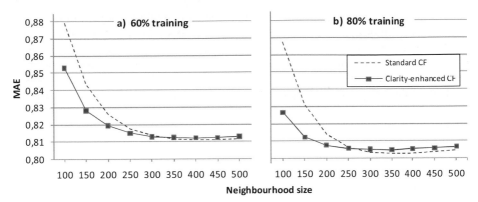

Fig. 2. Comparison of standard CF with dynamic, clarity-based neighbor weighting, and standard CF, using neighborhoods varying from 100 to 500 users, at a) 60% cut, b) 80% cut.

Enlarging neighborhoods comes at an important computational cost in a CF system. Computational cost being one of the well-known problems in the field [9], achieving equal (or improved) performance at a lower cost is a relevant result. Let us recall that the total number of users in this dataset is 943, which means that 100 users is about a 10% of the total user community. CF systems described in the literature commonly take neighborhood sizes of 5 to 500 users for this dataset, 50 to 200 being the most common range [19,21].

The trend in the evolution of performance with neighborhood size is clear in Figure 2, showing the effect of the clarity-based enhancement at different sizes, setting the sparsity cut at a) 60% and b) as a double check, 80% (which are standard ranges in the CF literature). It can be seen that the shape of the curves in both figures is very similar, evidencing the consistent superiority of clarity-enhanced CF with small to medium (i.e. usual) neighborhood sizes (e.g. over 5% improvement at size = 100 users with 80% training data).

5   Conclusions

Our work explores the use of performance prediction techniques to enhance the selection and weighting of neighbors in CF. The proposed approach consists of the adaptation of performance predictors originally defined for ad-hoc retrieval, into the CF domain, where users and items (and ratings), instead of documents and queries, make up the problem space. A predictor is proposed and used to introduce dynamic weights in the combination of neighbor ratings in the computation of collaborative recommendations, in an approach where the better the expected performance of a neighbor is, the higher weight is assigned to her ratings in the combination. The reported experimental results show performance improvements as a result of this dynamic weights adjustment approach, which supports the predictive power of clarity-based techniques in CF as a basis for this kind of adjustment. The results are particularly positive in small neighborhood situations.

Future work includes the exploration of alternative variants of the clarity-based predictor, as well as new predictors based on other techniques which have achieved good results in IR. We also aim to research neighbor selection methods based on external information sources, such as social network data. Our research so far has focused on the user-based kNN approach to CF, as it is particularly intuitive for the formulation of the researched problem, and lends itself well to exploit user qualities implicit in the data, or obtainable from external sources, linking to interesting problems in adjacent areas (e.g. social dynamics). We plan nonetheless to study the proposed approach under alternative baseline CF formulations, such as item-based kNN and factor models.

Beyond the current research presented here, recommender systems, and personalized IR at large, are particularly propitious areas for the introduction of performance prediction techniques, because of the naturally arising need for combination of multiple diverse evidence and strategies, and the uncertainty (and thus the variable accuracy) involved in the exploitation of implicit evidence of user interests. For instance hybrid recommender systems combine a content-based and a collaborative approach. Performance predictors could be researched to weight the influence of each component in the final recommendations (e.g. CF is sensitive to gray sheep or new item situations, while content-based filtering is not). Personalized ah-hoc retrieval is another interesting problem for this approach, where the weight of a query vs. implicit evidence from user history can be dynamically adjusted depending on the predicted effectiveness of each side. To the best of our knowledge, the introduction of performance predictors in these areas has been barely addressed, if at all, as a formal problem.

Acknowledgments. This work was supported by the Spanish Ministry of Science and Innovation (TIN2008-06566-C04-02) and the Ministry of Industry, Tourism and Commerce (CENIT-2007-1012).

References

1. Adomavicius, G., Tuzhilin, A.: Toward the next generation of recommender systems: A survey of the state-of-the-art and possible extensions. IEEE Transactions on Knowledge and Data Engineering 17(6), 734–749 (2005)
2. Amati, G., Carpineto, C., Romano, G.: Query difficulty, robustness, and selective application of query expansion. In: McDonald, S., Tait, J.I. (eds.) ECIR 2004. LNCS, vol. 2997, pp. 127–137. Springer, Heidelberg (2004)
3. Aslam, J.A., Pavlu, V.: Query hardness estimation using Jensen-Shannon divergence among multiple scoring functions. In: Amati, G., Carpineto, C., Romano, G. (eds.) ECiR 2007. LNCS, vol. 4425, pp. 198–209. Springer, Heidelberg (2007)
4. Baltrunas, L., Ricci, F.: Locally adaptive neighborhood selection for collaborative filtering recommendations. In: Nejdl, W., Kay, J., Pu, P., Herder, E. (eds.) AH 2008. LNCS, vol. 5149, pp. 22–31. Springer, Heidelberg (2008)
5. Carmel, D., Yom-Tov, E., Darlow, A., Pelleg, D.: What makes a query difficult? In: 29th annual international ACM SIGIR conference on Research and development in information retrieval (SIGIR 2006), pp. 390–397. ACM Press, New York (2006)
6. Cronen-Townsend, S., Zhou, Y., Croft, W.: Precision prediction based on ranked list coherence. Information Retrieval 9(6), 723–755 (2006)
7. Cronen-Townsend, S., Zhou, Y., Croft, B.W.: Predicting query performance. In: 25th annual international ACM SIGIR conference on Research and development in information retrieval (SIGIR 2002), pp. 299–306. ACM Press, New York (2002)
8. Diaz, F., Jones, R.: Using temporal profiles of queries for precision prediction. In: 27th annual international conference on Research and development in information retrieval (SIGIR 2004), pp. 18–24. ACM Press, New York (2004)

9. Goldberg, K., Roeder, T., Gupta, D., Perkins, C.: Eigentaste: A constant time collaborative filtering algorithm. Information Retrieval 4(2), 133–151 (2001)
10. Hauff, C., Azzopardi, L., Hiemstra, D.: The combination and evaluation of query performance prediction methods. In: Boughanem, M., et al. (eds.) ECIR 2009. LNCS, vol. 5478, pp. 301–312. Springer, Heidelberg (2009)
11. He, B., Ounis, I.: Inferring query performance using pre-retrieval predictors. In: Apostolico, A., Melucci, M. (eds.) SPIRE 2004. LNCS, vol. 3246, pp. 43–54. Springer, Heidelberg (2004)
12. Herlocker, J., Konstan, J.A., Riedl, J.: An empirical analysis of design choices in neighborhood-based collaborative filtering algorithms. Information Retrieval 5(4), 287–310 (2002)
13. Jones, K.S.: A statistical interpretation of term specificity and its application in retrieval. Journal of Documentation 28(1), 11–20 (1972)
14. Kwon, K., Cho, J., Park, Y.: Multidimensional credibility model for neighbor selection in collaborative recommendation. Expert Systems with Applications 36(3), 7114–7122 (2009)
15. Lavrenko, V., Allan, J., Deguzman, E., Laflamme, D., Pollard, V., Thomas, S.: Relevance models for topic detection and tracking. In: 2nd int. conference on Human Language Technology Research, pp. 115–121. Morgan Kaufmann Publishers, San Francisco (2002)
16. Macdonald, C., He, B., Ounis, I.: Predicting query performance in intranet search. In: ACM SIGIR Workshop on Predicting Query Difficulty – Methods and Applications (2005)
17. Mothe, J., Tanguy, L.: Linguistic features to predict query difficulty. In: ACM SIGIR Workshop on Predicting Query Difficulty – Methods and Applications, Salvador, Brazil (2005)
18. Plachouras, V., He, B., Ounis, I.: University of Glasgow at TREC2004: Experiments in Web, Robust and Terabyte tracks with Terrier. In: 13th Text Retrieval Conference (TREC 2004), Gaithesburg, Maryland (2004)
19. O'Donovan, J., Smyth, B.: Trust in recommender systems. In: 2005 International Conference on Intelligent User Interfaces (IUI), pp. 167–174. ACM Press, New York (2005)
20. Wang, J., de Vries, A.P., Reinders, M.J.T.: Unifying user-based and item-based collaborative filtering approaches by similarity fusion. In: 29th annual international ACM SIGIR conference on Research and development in information retrieval (SIGIR 2006), pp. 501–508. ACM Press, New York (2006)
21. Xue, G.R., Lin, C., Yang, Q., Xi, W., Zeng, H.J., Yu, Y., Chen, Z.: Scalable collaborative filtering using cluster-based smoothing. In: 28th annual international ACM SIGIR conference on Research and development in information retrieval (SIGIR 2005), pp. 114–121. ACM Press, New York (2005)
22. Yom-Tov, E., Fine, S., Carmel, D., Darlow, A.: Learning to estimate query difficulty: including applications to missing content detection and distributed information retrieval. In: 28th annual international ACM SIGIR conference on Research and development in information retrieval (SIGIR 2005), pp. 512–519. ACM Press, New York (2005)
23. Zhou, Y., Croft, B.W.: Ranking robustness: a novel framework to predict query performance. In: 15th ACM conference on Information and knowledge management (CIKM 2006), pp. 567–574. ACM Press, New York (2006)
24. Zhou, Y., Croft, B.W.: Query performance prediction in web search environments. In: 30th annual international ACM SIGIR conference on Research and development in information retrieval (SIGIR 2007), pp. 543–550. ACM Press, New York (2007)

Collaborative Filtering: The Aim of Recommender Systems and the Significance of User Ratings

Jennifer Redpath, David H. Glass, Sally McClean, and Luke Chen

School of Computing and Mathematics, University of Ulster, Newtownabbey,
Co. Antrim, BT37 0QB, UK
{redpath-j,dh.glass,si.mcclean,l.chen}@ulster.ac.uk

Abstract. This paper investigates the significance of numeric user ratings in recommender systems by considering their inclusion / exclusion in both the generation and evaluation of recommendations. When standard evaluation metrics are used, experimental results show that inclusion of numeric rating values in the recommendation process does not enhance the results. However, evaluating the accuracy of a recommender algorithm requires identifying the aim of the system. Evaluation metrics such as *precision* and *recall* evaluate how well a system performs at recommending items that have been previously rated by the user. By contrast, a new metric, known as *Approval Rate*, is intended to evaluate how well a system performs at recommending items that would be rated highly by the user. Experimental results demonstrate that these two aims are not synonymous and that for an algorithm to attempt both obscures the investigation. The results also show that appropriate use of numeric rating values in the process of calculating user similarity can enhance the performance when *Approval Rate* is used.

Keywords: Recommender Systems, Collaborative Filtering, Information Retrieval.

1 Introduction

Recommender systems [12] are typically employed to assist users with the discovery of items that may be of interest. A commonly-used approach for generating personalised recommendations is User-Based Collaborative Filtering (*CF*) [3], in which a database of user ratings profiles is searched for the closest matches to the active user, the user's Neighbourhood (*NN*). The top-*N* items in the neighbourhood are selected to form the recommendations. These recommendations are evaluated in various ways, with a popular approach being the *Leave-N-Out* protocol [2]. The items that are withheld from the recommender algorithm are referred to as the *hidden set*.

The central components of a recommender system are the dataset and the recommender algorithm. The dataset can be thought of as a set of items that are available for recommendation to a set of users. Information stored on the users is usually in the form of ratings collected, either implicitly or explicitly, although there could also be

C. Gurrin et al. (Eds.): ECIR 2010, LNCS 5993, pp. 394–406, 2010.
© Springer-Verlag Berlin Heidelberg 2010

other information, such as user demographics. If many of the hidden items occur in the recommendations, the system is considered to be successful.

Explicit user rating data usually takes the form of ranking an item on a nominal scale. These scales are subjective and systems that operate them use different granularities of ratings; for example 1-to-5 used by MovieLens [13] and 1-to-7 with Ringo [16]. Currently a considerable amount of significance is placed on ratings data. The recommendation procedure can incorporate or exclude ratings data, by using numeric values or treating them as unary ratings respectively, in three main areas of the process, namely the similarity measure used in neighbourhood generation, the recommendation selection method and the final evaluation technique. Throughout the remainder of this paper, when reference is made to "excluding ratings data", this signifies the use of unary ratings, as opposed to numeric values.

In terms of similarity measures, recommender systems typically use a distance or vector based measure such as Cosine, Pearson's Correlation or Jaccard's Coefficient, to calculate the relationship between two users. Some of these measures use the rating data to measure the divergence between user rating patterns, while others disregard the rating values and simply calculate the fraction of vector overlap.

A popular approach for generating recommendations is to select the Top-N items from a set of candidate items; these are items that are rated by the NN but not by the active user. The existing recommendation selection methods either use purely the ratings data, a simple incremental count or a combination of both.

Recommender Algorithms can be evaluated in terms of predictive, classification or rank accuracy [5]. This study initially focuses on the classification accuracy, whereby the algorithm is evaluated based on the number of items it correctly classifies as good. Two common techniques for this type of evaluation are; (1) consider any recommendation that corresponds to a hidden item or (2) only allow items with a rating above a predefined threshold [18] to contribute to the measurement of accuracy.

This paper investigates the significance of including or excluding numeric user ratings at each of these three stages in the recommendation process. Preliminary investigations suggested that numeric user ratings had little or no impact, which raised a question about the aim of recommendation. In order to address this question we have proposed an alternative evaluation technique which incorporates the *Approval Rate* metric, and undertaken detailed experiments using both the evaluation techniques described above and our new technique. Experimental results indicate that the significance of ratings depends on how the recommender algorithm is evaluated and hence on the goal of recommendation.

The remainder of this paper is organised as follows; Section Two presents and describes the three stages of the recommendation process; user similarity calculations, recommendation generation and evaluation techniques, including the new *Approval Rate* metric. This is followed by an outline of the experiments in Section Three. The results and analysis are detailed in Section Four, demonstrating clearly how algorithms can perform well in conjunction with one aim while achieving poorly with the other. Section Five presents a discussion of related work and rationale for using the *Approval Rate* alongside established evaluation techniques. Finally the conclusions and possible future work are described in Section Six.

2 Recommendation Process

The three areas in the recommendation process that ratings can be included or excluded from are the similarity measure, the recommendation selection method and the evaluation technique. Each of these is investigated and alternative approaches suggested, allowing for both a completely rating free procedure and one that considers them at all stages.

2.1 Similarity Measures

Three of the most commonly used similarity measures in the literature, Cosine Similarity, Pearson's Correlation and Jaccard's Coefficient, were implemented. These are defined in Table 1 along with a simple probability measure which was included to provide a baseline comparison. The items in the dataset can be formally denoted as a universe of n items, $I = \{i_j: 1 \leq j \leq n\}$. Each active user u has rated a subset of I, denoted I_u, and is represented by a n-dimensional rating vector, u, where $u_j = 0$ if $i_j \notin I_u$. Each of the similarity measures uses these rating vectors to decide if two users exhibit similar behaviours.

Table 1. Similarity Measures Definitions

Measures	Definition
Probability (*BASE_PROB*)	$$sim(u,v) = \frac{|Iu \cap Iv|}{|Iv|} \tag{1}$$
Cosine (*COS*)	$$sim(u,v) = \frac{u \cdot v}{\|u\| \, \|v\|} \tag{2}$$
Pearson (*PEAR*)	$$sim(u,v) = \frac{\sum_{v \in S} (r_{u,i} - \bar{r}_u)(r_{v,i} - \bar{r}_v)}{\sqrt{\sum_{v \in S} (r_{u,i} - \bar{r}_u)^2} \sqrt{\sum_{v \in S} (r_{v,i} - \bar{r}_v)^2}} \tag{3}$$ Where $r_{u,i}$ is the rating by user u for item i, $\bar{r}_u$ is the mean rating for user u and $S = I_u \cap I_v$.
Jaccard (*JACC*)	$$sim(u,v) = \frac{|Iu \cap Iv|}{|Iu \cup Iv|} \tag{4}$$

Cosine Similarity (*COS*) is widely used in recommendation algorithms (Equation 2); it is the measure of the cosine angle between two user rating vectors, in this case u and v. A value in the range of -1 to 1 can be generated, where -1 signifies exact opposites, 1 exactly the same and 0 indifference. The *COS* denominator is able to account for the number of movies each user has rated individually. This ensures that users with a high number of ratings do not swamp the neighbourhood selections. Unfortunately no normalisation techniques are employed, meaning that all user rating collections are treated equally, and as a rating scale of 1-to-5 is used, a negative similarity value never occurs. Furthermore, it is possible to get a similarity of 1 (meaning complete agreement) when two users have rated their movies with opposite values, Table 2. To account for this situation it is proposed that Cosine is used with an exchanged rating scale of -2-to-2 (*EXCH_COS*), allowing these users to be allocated a negative similarity value.

Pearson's Correlation (*PEAR*), by contrast, addresses the normalisation issue. It is defined in equation 3, where $r_{u,i}$ is the rating by user u for item i, $\bar{r}_u$ and $\bar{r}_v$ are the mean ratings of users u and v over their co-rated items, and $S = I_u \cap I_v$. A problem with *PEAR* is that the number of items rated by the individuals is ignored, meaning that users with a large number of movie ratings are more frequently selected for the *NN*. In addition published research identifies that if user ratings are uniform, as in table two, then similarity cannot be calculated as the denominator will be zero [1].

Jaccard's Coefficient (equation 4) ignores the user ratings and uses set relations to calculate the similarity. The magnitude of the intersection set, i.e. the number of co-rated items, is divided by the number of elements in the union set.

Table 2. Cosine Example Using rating Scale 1-to-5

User	Movie 1 Rating	Movie 2 Rating	Calculation
u	1	1	‖u‖ = 1.4142
v	5	5	‖v‖ = 7.0710
			u.v = 10
			sim(u,v) = 1

The probability metric (Equation 1) simply divides the intersection set by the size of *v*'s rating vector. This measure only takes the possible neighbours' rating vector sizes into account, whereas all the other measures consider both vectors.

Each of the measures discussed were used to explore the significance of user ratings and the two distinct aims of recommender systems. To investigate whether the rating values have a positive or negative effect on recommendation results, three measures that exclude ratings were implemented first of all. *JACC* and *BASE_PROB* do not consider rating values as standard and so were used directly. An unrated version of *COS* was also used by setting all ratings to 1 (*NORAT_COS*). This approach could not be used with *PEAR* because it would then be undefined.

The next stage was to consider how much the non co-rated movies contribute to the outcome. To address this issue, three measures that involve only co-rated (see Figure 1) movies were implemented. *PEAR* already applies to only co-rated cases so it was used. The standard *COS* measure considers all the movies rated individually by both users in addition to the co-rated, but it can also be applied to only the co-rated cases. This restriction was applied to *COS* and *EXCH_COS* and denoted by *COS_CO* and *EXCH_CO*.

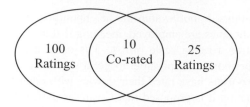

Fig. 1. Overlapping User Rating Vectors

Finally, three approaches were implemented which consider all the ratings without being restricted to the co-rated. *COS* and *EXCH_COS* were used along with a combined measure; *NORAT_COS* and *EXCH_CO* (*COMB1*). *NORAT_COS* will account for the vector overlap and the differences in rating vector sizes, while *EXCH_CO* will represent the variation in rating patterns for the co-rated items.

2.2 Recommendation Selection Methods

Two popular approaches were implemented for selecting recommendations; Most-Frequently-Rated (*MFR*) [15] and Most-Highly-Rated (*MHR*) [13]. Each selects the Top-*N* items for recommendation from the candidate set.

MFR is a simple incremental count of movie occurrences; movies are counted regardless of the rating attributed to them and all *NN* contribute equally. *MHR* calculates the sum of ratings for each movie and again all *NN* contribute equally.

During the experiments two selection techniques which consider *NN* similarity values were tested; Highest-Sum-of-Similarities *(HSS)* and Highest-Weighted-Rating *(HWR)*. With *HSS* ratings are ignored but *NN* similarity values are used, the sum of *NN* similarity for each rated movie is calculated. *HWR* considers both ratings and similarity values by calculating the sum of ratings multiplied by corresponding *NN* similarities. It was found that *HSS* resulted in no improvement over *MFR* and similarly *HWR* resulted in no improvement over *MHR*, which is probably due to the small amount of variation between *NN* similarity values. For this reason results are only presented for MFR and MHR.

2.3 Evaluation Techniques and Metrics

There is a wide variety of metrics used in the literature to assess the accuracy of a recommender algorithm. In fact, [7] reports that this abundance of metrics is what make it so difficult for researchers to evaluate recommender systems.

A user centric approach to recommendation is to provide a small number of good predictions, to prevent the user from having to search through a large set in order to find a handful of good ones. The recommended items at the top of the list are the items of most interest to the user.

Some researchers suggest that the only means for assessing a system is via on-line user studies [4]. These can record how the system recommendations are used by the individual, in other words did the user continue to use the system? Unfortunately these studies are expensive in terms of time and finding a diverse test group. It must also be noted that a test group of users can be more sympathetic than real users and this must be taken into account.

This investigation uses two widely accepted information retrieval metrics: *precision* and *recall*. *Precision* is the ratio of true positives *(TP)* to the number *N* of recommendations generated.

$$Precision = \frac{TP}{N} \tag{5}$$

Recall is the ratio of *TP* to the number of hidden items *H*.

$$Recall = \frac{TP}{H} \tag{6}$$

The interpretation of the evaluation method's result depends on whether the intention is to maximise the prediction of previously watched items, or to uncover the greatest number of movies the user has expressed a favoured opinion towards. In order to distinguish between these two different assessments, two evaluation methods are used: *Complete*, where all recommended items found in the hidden set are counted as true positives and *Positive*, where only items found with a rating above a predefined threshold are counted. In this paper the rating threshold is set to 3.

Recall is a good indicator of how well an algorithm performs at finding hidden items the user has previously rated. Unfortunately it cannot distinguish algorithms in terms of how good they are at recommending movies that a user is likely to rate highly. In other words, it is possible that an algorithm might be good at finding hidden items, but not so good at recommending movies a user would like and vice versa. Even when the *Positive* evaluation method is used rather than *Complete*, it is still possible that an algorithm will perform well primarily because it is good at finding hidden items rather than because it is good at recommending movies a user will rate highly. This provides the motivation for introducing, the *Approval Rate* metric (Equation 7), which is the proportion of *TP*s that are rated highly

$$Approval\,Rate = TP_p \,/\, TP_c \tag{7}$$

where *TPp* and *TPc* are the number of TPs found during *Positive* and *Complete* evaluation, respectively. Since this measure represents the proportion of highly rated movies found in the hidden set, it can be taken as an estimate of the probability that a recommendation not found in the hidden set would be rated highly by the user. It is clear that an algorithm could perform well in terms of this metric even if it is not good at finding hidden items since, even if it only finds a small proportion of hidden items, those it does find tend to be rated highly.

3 Experiments

The basic aim of this study was to discover what significance user rating values have on the three main stages of the recommendation process. To investigate this, tests were developed that either excluded the values from the rating vectors completely, included them only at certain points or incorporated them at every stage. Table 3 shows the available choices for each of the recommendation stages along with the alternatives for dealing with the rating values. The *Unrated Similarity* column in Table 3 represents experiments where ratings were excluded completely from the similarity calculations. Three similarity measures were considered in this case (*NORAT_COS*, *JACC* and *BASE_PROB*) and for each of them experiments were carried out for both of the recommendation selection methods and all three of the evaluation techniques. The remaining columns differ from *Unrated Similarity* only in terms of how the similarity is calculated. *The Co-Rated Similarity* represents experiments where only co-rated movies were included in the calculation of similarity, but in this case the ratings for these movies were included. The final column represents experiments where ratings were also included in the similarity calculation, but in this case all movies rated by at least one of the two users were taken into account.

Other factors known to affect the recommendation quality are neighbourhood size, recommendation set size, and data sparsity [18]. As these were not under review they were set to constant values and the same dataset was used. The Neighbourhood size, k, was set to 30, and the Recommendation Set size, N, to 20. There is a threshold of 5 applied to all co-rated only approaches and the *BASE_PROB*, meaning that any users with less than 5 co-rated movies receive a similarity value of 0.

The well known MovieLens dataset [11] was used. The hidden items were selected by randomly sampling 10 items from each user's profile. There were 6,040 visitors with 1,000,199 ratings on 3,706 movies. When using the *Complete* evaluation technique there were 60,400 possible TPs to find, whereas with the Positive approach only 37,575 movies remained. While using the Positive approach only movies rated above the positive threshold were included, this meant that 27 users were completely omitted as none of these had any highly rated movies in their hidden set.

Table 3. Choices Available for each Recommendation Stage

Recommendation Stage	Unrated Similarity	Co-Rated Similarity	Rated Similarity
Similarity Measure	NORAT_COS, JACC, BASE_PROB	EXCH_CO, COS_CO, PEAR	COS, COMB1, EXCH_COS
Recommendation Selection	MFR, MHR	MFR, MHR	MFR, MHR
Evaluation Technique	Complete, Positive, Approval Rate	Complete, Positive, Approval Rate	Complete, Positive, Approval Rate

4   Results and Analysis

When using the *Complete* evaluation method, *precision* is always half the *recall*, since the recommendation and hidden sets have fixed values of 20 and 10 respectively, but when using the *Positive* evaluation the number of hidden items changes per person. In order to provide a direct comparison, *recall* is used as the main focus in the results and analysis for these two evaluation methods. Figures 2 and 3 display the *recall* results using the *Complete* and *Positive* evaluation techniques respectively. The *Approval Rate* for all similarity approaches is presented in Figure 4. Significant differences between the mean *recall* values were tested for using paired sample T tests, alpha (α) was set to 0.05.

4.1 Complete Evaluation

The Unrated similarity measures *NORAT_COS* and *JACC* performed equivalently (*MFR* p value = 0.99; *MHR* p value = 0.26). The *BASE_PROB* is slightly lower, which may be because this approach penalises possible neighbours with large rating vectors and therefore does not capitalise on candidate items. There is little difference between the different recommendation selection methods. Including ratings (*MHR*) has no positive effect, and in fact excluding them yields a slight improvement (*MFR*).

When including only the Co-rated movies all approaches experience a decrease in the *recall* (9 – 10%) compared to both the Unrated (25 - 28%) and Rated (23 – 27%) experiments. *PEAR* performs marginally worse than the other two approaches, and *EXCH_CO* performs slightly better than *COS_CO*. In this section including the ratings during the recommendation selection method, using *MHR* results in no significant improvement over ignoring them (*MFR*) (*COS_CO* p value = 0.22; *EXCH_CO* p value = 0.45; *PEAR* p value = 0.24).

The Rated section presents the standard Cosine (*COS*) measure with a rating scale of 1-to-5, Cosine with an exchanged scale of -2-to-2 (*EXCH_COS*) and *COMB1* that combines *NORAT_COS* with *EXCH_CO*. As with the Unrated, all the approaches perform better than the Co-rated ones. Interestingly, *NORAT_COS* attains a slightly higher recall value than *COS* for both *MFR* and *MHR*, but the differences are not significant (*MFR* p value = 0.24; *MHR* p value = 0.34). In addition *COS* with *MFR* sees a slight improvement over *COS* with *MHR*. Both these findings strengthen the argument that, when trying to maximise on finding the greatest number of previously watched movies, ratings are of no significance.

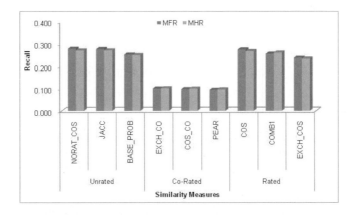

Fig. 2. *Complete* Evaluation for All Similarity Measures

4.2 Positive Evaluation

When using the *Positive* technique only movies found in the hidden set with a rating above a predefined threshold, in this case 3, are counted as *TP*s. All approaches perform better in the *Positive* evaluation, when compared to *Complete*.

Cosine using rating scale 1-to-5 (*COS_CO*), Cosine using rating scale -2-to-+2 (*EXCH_CO*) and Pearson's Correlation (*PEAR*) are the similarity measures used in the Co-rated section, where only the movies rated by both users are included in the similarity calculation. They all use a co-rated similarity threshold of 5, meaning that any two users with less than 5 co-rated movies receive a similarity of 0. Each of these methods experiences a *recall* increase of 3 - 4% on average (for example, *NO-RAT_COS* increases from 28 - 32%), over the same methods evaluated with *Complete*. Just as in the *Complete* case, including ratings (*MHR*) in the *COS_CO* (p value = 0.06) or *EXCH_CO* (p value = 0.12) similarity measures does not yield significantly better results than excluding them (*MFR*), however with *PEAR* (p value = 0.02) it does result in significantly different outcomes.

When using the *Positive* evaluation all methods using *MHR* attain a higher *recall* value than when using the *MFR* method, which is in contrast with the results for the *Complete* case. This suggests that if ratings are considered in the evaluation they should also be considered in the recommendation selection method.

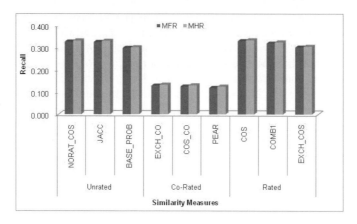

Fig. 3. *Positive* Evaluation for All Similarity Measures

4.3 Approval Rate Evaluation

The *Approval Rate* evaluation measures the percentage of highly rated movies found within the *TP*s when using the *Positive* compared to the *Complete* evaluation technique. Figure 4 shows that the Rated approaches do not perform as well in terms of the *Approval Rate* metric as the Co-rated approaches, although they perform better than the Unrated approaches.

A Cosine similarity measure is presented for each proposed method of dealing with ratings, *NORAT_COS*, *COS_CO* and *COS*, making it possible to track the influence ratings have over the percentage of highly rated movies found. *NORAT_COS* attains

73% with *MFR* and 75% with *MHR*, *COS_CO* achieves 82% and 83% for *MFR* and *MHR* respectively, while *COS* manages 73% with *MFR* and 76% with *MHR*.

Interestingly, *EXCH_COS*, a rated approach that employs an exchanged rating scale, results in a higher percentage of "good" recommendations than COS, on average 4.5% for both *MFR* and *MHR*.

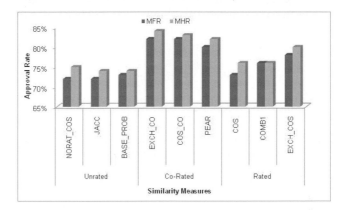

Fig. 4. *Approval Rate* Evaluation for All Similarity Measures

When the *Approval Rate* is measured, it shows that including the ratings in the similarity measure and the recommendation selection method has an impact on the proportion of *TPs* that the user has rated highly. In addition the results indicate that although using only the co-rated movies results in much poorer performance in terms of the overall number of TPs found, it results in a much better performance in terms of the proportion of movies rated at the higher end of the scale.

5   Discussion of Related Work

The experiments described above address two important issues in recommender systems: first, the significance of user ratings in the recommendation process and, second, the aim of recommendation systems as characterised by evaluation metrics. With regard to the former, there have been a plethora of investigations comparing content-based (without ratings) with ratings-based algorithms. To the best of the authors' knowledge, no systematic investigation into the inclusion versus the exclusion of user rating values in the context of user-based CF has been conducted. For this reason, the discussion of related work will focus on evaluation.

Ideally the evaluation of a recommender algorithm should be conducted by the end users, but that is not always feasible due to lack of time and resources. Comparative studies have been conducted [5, 6, 8, 2] to evaluate different collaborative recommender algorithms. These studies use a mixture of evaluation techniques and metrics making it hard to identify the best algorithm.

Recommendations also need to be evaluated in line with the purpose of the system. One technique does not suit all, as an investigation by McNee et al (2002) discovered.

Users react differently to recommendation lists depending on their current task. The *Complete* and *Positive* evaluation techniques are useful for evaluating systems that wish to build trust with new users, who are likely to trust a system which "does not recommend movies the user is sure to enjoy but probably already knows about" [5]. In comparison, the *Complete* and *Positive* evaluation techniques along with *Approval Rate* gives a better indication of how well a system provides established users with "good" recommendations. Swearingen and Sinha (2002) defined 'good' recommendations as those recommendations which interested the user. Good recommendations were further broken down into useful recommendations, recommendations which interested the user, but which the user had not experienced before and trust-generating recommendations, recommendations which the user has experienced before, but which they also liked.

The research community recognises that the problems with current evaluation methods and measures must be addressed. In fact, McNee et al (2006) state that the "recommendations that are most accurate according to the standard metrics are sometimes not the recommendations that are most useful to users." In response to this, new and modified metrics have started to appear in the literature [8; 19]. McLaughlin and Herlocker (2004) introduce a modified precision, whereby any recommendation generated which is not rated by the test user counts negatively towards the result. [19] introduced the Intra-List Similarity Metric to measure the diversity of recommendations within a list.

Mean Absolute Error (MAE) is a prediction evaluation metric used for measuring the average absolute deviation from each predicted rating to each known rating. McLaughlin and Herlocker (2004) argue that because *MAE* evaluates each predicted recommendation independently the results are biased towards algorithms that predict all items well, as opposed to predicting the top items well. The top items, usually 10 or 20, are the items of most interest to the user. Some researchers avoid using *MAE* as it does not measure the classification accuracy of an algorithm.

The *Approval Rate* evaluation metric proposed in this paper is intended to evaluate an important feature of recommendation systems. Ultimately the aim of a recommendation system is to recommend items that the user will like, i.e. will tend to rate highly. By contrast, standard metrics such as *recall*, *precision* and others related to them tend to measure how good an algorithm is at finding items that a user has already rated (or items that a user has already rated highly). The results presented here strongly suggest that this cannot be equated with the probability that a user will like an item that has been recommended. The *Approval Rate* metric attempts to measure this in terms of the proportion of hidden items found that were rated highly by the user, which can be considered as an estimation of the probability that a user will rate an item highly. *MAE* addresses this issue to some extent, but as noted above, *MAE* may be biased to algorithms that predict all items well rather than the top items. The *Approval Rate* approach does not suffer from this limitation.

6   Conclusions and Future Work

This study compared results obtained using nine similarity measures, two recommendation selection methods and three evaluation techniques, including the proposed

Approval Rate metric. The results have enabled the significance of user ratings in recommendation to be evaluated and alternative aims of a recommender algorithm to be clearly distinguished. In fact, the results show that the two issues are related because the significance of user ratings depends on the aim of recommendation. If the aim is to find hidden items that the user has already rated, as captured by the *precision* and *recall* metrics, then inclusion of ratings when calculating similarity values has no impact. By contrast, if the aim is to maximise the proportion of hidden items found that are highly rated, as captured by the new *Approval Rate* metric proposed here, then including ratings when calculating similarity is very important. Furthermore, the best results are obtained in the latter case when only co-rated items are taken into account when calculating similarity, whereas this restriction yields much worse results in the former case. It has also been argued that the aim captured by the new metric represents an important aspect of recommendation since it provides an estimate of the probability that a user would rate a recommended item highly.

Possible future work includes clustering users into groups and producing recommendations using different techniques on a group-by-group basis [14]. It is speculated that improvements could be made when including the ratings data if user groups were taken into consideration. Another direction concerns alleviating problems of data sparsity [18]. Work is currently underway into developing a hybrid similarity measure that uses the ratings for co-rated items and calculates a content divergence for all other movies rated. It is hypothesized that the content divergence will provide a more detailed measure of similarity for users in sparse areas of the dataset.

References

1. Ahn, H.J.: A New Similarity Measure for Collaborative Filtering to Alleviate the New User Cold-starting Problem. Inf. Sci. 178(1), 37–51 (2008)
2. Breese, J., Heckerman, D., Kadie, C.: Empirical Analysis of Predictive Algorithms for Collaborative Filtering. In: UAI 1998, pp. 43–52 (1998)
3. Goldberg, D., Nichols, D., Oki, B.M., Terry, D.: Using Collaborative Filtering to Weave an Information Tapestry. Commun. ACM 35(12), 61–70 (1992)
4. Hayes, C., Massa, P., Avesani, P., Cunningham, P.: An On-line Evaluation Framework for Recommender Systems. In: Workshop on Personalization and Recommendation in E-Commerce (2002)
5. Herlocker, J., Konstan, J., Terveen, L., Riedl, J.: Evaluating Collaborative Filtering Recommender Systems. ACM TOIS 22(1), 5–53 (2004)
6. Huang, Z., Zeng, D., Chen, H.: A Comparison of Collaborative-Filtering Recommendation Algorithms for E-commerce. IEEE IS 22(5), 68–78 (2007)
7. Konstan, J.A., Riedl, J.: Research Resources for Recommender Systems. In: CHI 1999 (1999)
8. McLaughlin, M.R., Herlocker, J.L.: A Collaborative Filtering Algorithm and Evaluation Metric that Accurately Model the User Experience. In: SIGIR 2004, pp. 329–336. ACM, New York (2004)
9. McNee, S.M., Riedl, J., Konstan, J.A.: Being Accurate is Not Enough: How Accuracy Metrics have hurt Recommender Systems. In: CHI 2006, pp. 1097–1101. ACM, New York (2006)

10. McNee, S.M., Albert, I., Cosley, D., Gopalkrishnan, P., Lam, S.K., Rashid, A.M., Konstan, J.A., Riedl, J.: On the Recommending of Citations for Research Papers. In: ACM CSCW 2002, pp. 116–125. ACM Press, New York (2002)
11. MovieLens Dataset (2006), http://www.grouplens.org/
12. Resnick, P., Varian, H.: Recommender Systems. CACM 40(3), 56–58 (1997)
13. Resnick, P., Iacovou, N., Sushak, M., Bergstrom, P., Riedl, J.: GroupLens: An Open Architecture for Collaborative Filtering of Netnews. In: ACM CSCW 1994, pp. 175–186. ACM Press, New York (1994)
14. Riedl, J., Jameson, A.: Advanced topics in recommendation. In: The 12th IUI 2007, pp. 11–11. ACM, New York (2007)
15. Sarwar, B., Karypis, G., Konstan, J., Riedl, J.: Analysis of Recommendation Algorithms for E-commerce. In: EC 2000, pp. 158–167. ACM, New York (2000)
16. Shardanand, U., Maes, P.: Social Information Filtering: Algorithms for Automating 'Word of Mouth'. In: SIGCHI 1995 Conference on Human Factors in Computing Systems, pp. 210–217. ACM Press, New York (1995)
17. Swearingen, K., Sinha, R.: Interaction Design for Recommender Systems. In: Designing Interactive Systems. ACM Press, New York (2002)
18. Symeonidis, P., Nanopoulos, A., Papadopoulos, A., Manolopoulos, Y.: Collaborative Filtering Process in a Whole New Light. In: IDEAS 2006, pp. 29–36. IEEE Computer Society, Washington (2006)
19. Ziegler, C., McNee, S.M., Konstan, J.A., Lausen, G.: Improving recommendation lists through topic diversification. In: WWW 2005, pp. 22–32. ACM, New York (2005)

Goal-Driven Collaborative Filtering – A Directional Error Based Approach

Tamas Jambor and Jun Wang

Department of Computer Science
University College London
Malet Place, London, WC1E 6BT, UK
t.jambor@ucl.ac.uk, wang.jun@acm.org

Abstract. Collaborative filtering is one of the most effective techniques for making personalized content recommendation. In the literature, a common experimental setup in the modeling phase is to minimize, either explicitly or implicitly, the (expected) error between the predicted ratings and the true user ratings, while in the evaluation phase, the resulting model is again assessed by that error. In this paper, we argue that defining an error function that is fixed across rating scales is however limited, and different applications may have different recommendation goals thus error functions. For example, in some cases, we might be more concerned about the highly predicted items than the ones with low ratings (precision minded), while in other cases, we want to make sure not to miss any highly rated items (recall minded). Additionally, some applications might require to produce a top-N recommendation list, where the rank-based performance measure becomes valid. To address this issue, we propose a flexible optimization framework that can adapt to individual recommendation goals. We introduce a *Directional Error Function* to capture the cost (risk) of each individual predictions, and it can be learned from the specified performance measures at hand. Our preliminary experiments on a real data set demonstrate that significant performance gains have been achieved.

1 Introduction

Collaborative filtering (CF) is concerned with predicting how likely a specific user will like certain information items (books, movies, music items, web pages, etc). As the term "collaborative" probably implies, the prediction has to rely on a collection of other (similar) users' preferences, which have been collaboratively collected. One of the popular applications of collaborative filtering is personalized content recommendation. A typical example is movie recommendation, where a user is explicitly asked to rate what he or she liked or disliked in the past. After rating a few movie items, the recommendation engine would be able to produce a prediction about the users ratings of unseen movie items by looking at other (similar) users past ratings for the movies items in question. In this case, users have to explicitly provide their ratings for movie items beforehand, e.g., give 1 star for the lowest rating (most hated) and 5 stars for the highest rating (most liked). As a major recommendation technique, collaborative filtering has been widely used in practice.

C. Gurrin et al. (Eds.): ECIR 2010, LNCS 5993, pp. 407–419, 2010.

The first Netflix competition [1] posed a challenge to develop systems that could beat the accuracy of Neflix in-house recommender by 10 percent. One of the importance of this challenge is that it specified an evaluation metric that is to be used to measure the efficiency of the system. Therefore forcing developers to think along the line of this measure. This would result in outcomes that have the same shortfalls as the measure [2]. This paper attempts to take another point of view of designing recommender systems. The aim is to introduce a design pattern that takes into account user preferences which would define the system itself. Different measures emphasize different qualities with respect to how closely they are correlated with certain objectives that the system would achieve. Therefore the measure itself gives a good indication of the qualities that the algorithm should possess. This approach offers a different solution. It enables the algorithm to be adjusted to user needs flexibly given that these needs are already defined and do not change during the session. It attempts to optimize the algorithm to these user needs instead of a measure which results in greater flexibility and better user experience. To achieve this, we first critically examine the issues of using squared errors as a cost function in collaborative filtering. Based on this discussion, we propose a goal-driven optimization framework where the users' or system's goal can be specified as a weight function. This weight function will be optimized by a genetic algorithm [3]. Experimental results on a real data set confirm our insights with improved performance.

The paper is organized as follows. We will discuss the related work in Section 2, present our theoretical development in Section 3, give our empirical investigation on recommendation in Section 4, and conclude in Section 5.

2   Related Work

The term, collaborative filtering, was first coined in [4] where the authors developed an automatic filtering system for electronic mail, called Tapestry. If we look at the collaborative filtering problem from a conceptual level, it is very much like Web retrieval in that it needs to calculate the correspondence (called relevance) between a user information need (in our case, a user preference or predefined preferable topics) and an information item (e.g., a movie or a book) [5]. In text retrieval, the correspondence is usually calculated by looking at content descriptions, e.g., how many and how frequent the query terms occur with a document. In contrast, when we make personalized recommendations, users unseen preferences can be predicted by aggregating the opinions and preferences of previous users.

Originally, the idea of collaborative filtering was derived from heuristics, assuming that users who have similar preferences in the past are likely to have similar preferences in the future, and the more similar they are, the more likely they would agree with each other in the future. The preference prediction is therefore calculated by weighted-averaging of the ratings from similar users.

In the memory-based approaches, all user ratings are indexed and stored into memory, forming a heuristic implementation of the "Word of Mouth" phenomenon. In the rating prediction phase, similar users or (and) items are sorted based on the memorized ratings. Relying on the ratings of these similar users or (and) items, a prediction of an item rating for a test user can be generated. Examples of memory-based collaborative filtering include user-based methods [6], item-based methods [7] and combined methods [8].

In the model-based approaches, training examples are used to generate an abstraction (model parameters) that is able to predict the ratings for items that a test user has not rated before. In this regard, many probabilistic models have been proposed. For example, to consider user correlation, [9] proposed a method called personality diagnosis (PD), treating each user as a separate cluster and assuming a Gaussian noise applied to all ratings. On the other hand, to model item correlation, [10] utilizes a Bayesian Network model, in which the conditional probabilities between items are maintained. Some researchers have tried mixture models, explicitly assuming some hidden variables embedded in the rating data. Examples include the cluster model [10] and the latent factor model [11]. These methods require some assumptions about the underlying data structures and the resulting 'compact' models solve the data sparsity problem to a certain extent. However, the need to tune an often significant number of parameters has prevented these methods from practical usage.

Alternatively, collaborative filtering can be considered as a matrix factorization problem and it has emerged as the clear favorite in the Netflix competition [12]. In general, the approach aims to characterize both items and users by vectors of factors inferred from item-rating patterns. The approximation is usually found such that it minimizes the sum of the squared distances between the known entries and their predictions. One possibility of doing so is by using a Singular Value Decomposition (SVD) [13]. The main reason of its success may be due to the fact that the objective function of the approach is equivalent to the performance measure (Root Mean Squared Error) that has been targeted in the competition. The drawback of using the RMSE performance measure is studied in [14]. However, it is also important to point out that many collaborative filtering models use the squared errors as the objective function, either implicitly or explicitly. Part of this paper is intended to increase the awareness and provide a future study about the issues of using RMSE as the objective function.

On the other hand, researchers also argued that understanding collaborative filtering as a rating prediction problem has some significant drawbacks. In some of the cases, a better view of the task is of generating a top-N list of items that the user is most likely to like [15,16]. Thus, this paper attempts to develop a flexible goal-driven optimization framework so that the algorithm can be tailored to meet individual system requirements.

3   A Goal-Driven Optimization Framework

3.1   Problem Definition and Analysis

One of the most used performance (error) measures for rating-based recommender systems is RMSE (Root Mean Squared Error), which measures the difference between ratings predicted by a recommendation algorithm and ratings observed from the users. It is defined as the square root of the mean squared error: $\sqrt{E\left((\hat{r}-r)^2\right)}$, where E denotes the expectation, r is the true rating and $\hat{r}$ is the predicted value. In [14], researchers have already systematically examined many performance measures for collaborative filtering. In this section, we critically examine the RMSE measure as an objective function.

The RMSE metric measures recommendation error across different rating scales, and the error criterion is uniform over all the items. RMSE squares the error before submitting it which puts more emphasis on large errors. Naturally large errors can occur at the end of the rating scales. To see this, suppose we have a recommendation algorithm which predicts the rating of an item randomly from rating 1 to 5. Fig. 1(a) shows that it is more likely to get higher error at both the ends of the rating scale if a random algorithm is used.

Thus, the question arises if we should adopt RMSE as the measure of customer satisfaction. It measures the error across the system even for items that are not that important for users to be correctly predicted. Therefore the system might not want to penalize predictions that are not important for the user. If the user is only interested to get relevant recommendations, RMSE as a measure is not sufficient. Even if the user is interested in items that he or she would dislike, it is arguable whether the middle range of the rating spectrum is interesting to the user at all. If we take rating three out of five as middle, that range cannot help to explain why an item was recommended, neither can it explain why the item was not recommended.

Another issue may also arise if one directly optimizes RMSE. This is due to the fact that the training samples are not uniformly distributed across rating scales. To demonstrate this, Fig. 1(b) shows RMSE improvement in percentage over random recommendation by using a common recommendation algorithm (in this case an SVD-based approach is used to optimize the metric [12]). Since users are likely to rate items that they liked, in most cases, they give them a rating of four. So the algorithm has more data to make a prediction at that range. This is the reason why we have higher improvements for rate four.

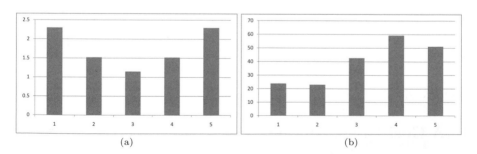

Fig. 1. (a) The expected value of RMSE per rating. (b) Percentage of RMSE improvement over random recommendation by a common collaborative filtering algorithm that optimizes RMSE.

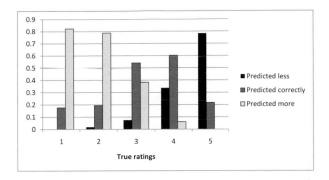

Fig. 2. Asymmetric Rating-prediction error offset

Predicted \ Rated	1 or 2	3	4 or 5
1 or 2			
3			
4 or 5			

Fig. 3. Directional Risk Preference of Recommendation Prediction

Improving accuracy on items that the user would like may be desirable from a user point of view, but if the prediction falls into the middle range the error does not matter as much as if the prediction falls into the lower range. It is similar with items that are rated low, reducing the error rate is more desirable as the error rate increases since the item gets a higher prediction. In addition to that, highly accurate predictions on uninteresting items (perhaps rated 3 out of 5) can drown out poor performance on highly/lowly ranked items. Therefore depending on the rating we need to pay attention to the direction of the offset between the rating and the prediction. Fig. 2 shows that the SVD algorithm tends to overpredict items in the middle range. Also, it is more likely to overpredict lower rated items than underpredict higher rated items.

Therefore a distinction can be made between items that are interesting to the user and items that are neutral. Within the interesting category we can differentiate between liked and disliked items. To decided which one is the most important to us, let us consider two different type of recommendations. Since the performance of a recommender system should be evaluated with respect to a specific task, it is useful to define the two main tasks that a typical recommender system fulfills. If the output of the recommender system is the first n items then RMSE is not an appropriate objective to optimize, since it is not important to measure the system performance on items that do not fall into the first n good

Table 1. The two dimensional weighting function, where p is the predicted value of the item and r is the ground truth

	$r = 1, 2$	$r = 3$	$r = 4, 5$
$p <= 2.5$	w_1	w_2	w_3
$2.5 < p <= 3.5$	w_4	w_5	w_6
$p > 3.5$	w_7	w_8	w_9

items. As long as the system correctly identified that these items do not fall into the n good items the accuracy is irrelevant. Users might be interested in exploring movies, looking through the database or checking particular movies. In this case everything matters, because users are interested in the justification on how movies are made. Clearly, in both scenarios we can differentiate between two separate kind of risks. First, the risk of recommending something that is not relevant to the users, second, the risk of not recommending something that is relevant to the users. These are two kind of errors, that should be separated when it comes to measuring the error rate. For example assume that the system predicts a movie four, and the user watches that movie, which he or she would have rated only three. This is clearly different from the case where the system rates a movie three, which would have been rated four is the user took the time and watched it. Since the error of the algorithm in the second scenario would never be found out, because the user would never watch a movie that is rated three, from a user point of view this error would be hidden. Therefore the system that makes errors like that would not considered better by the user than a system that makes errors illustrated in the first scenario.

Therefore it is important to introduce two concepts here. *Taste boundaries* and the *direction of taste*. Taste boundary could be defined as the interval that is between liked and disliked items. In a rating scale from one to five this boundary would be three. Direction would represent whether the predicted rating is towards the taste boundary or not, at one level, on another level it would represent whether the error is large enough to cross the taste boundary or not. In other words, whether the algorithm suggests that the user would like the item when it is not the case and vice versa. These boundaries are illustrated in the matrix shown in Fig. 3. It shows that we would like to minimize errors where the prediction is correct and as we go further from the correct prediction we take higher risks depending on the direction (the risk is illustrated by the size of the arrows). Fig. 3 can also be applied to a ranking problem since higher predicted items represent higher risk. For example in ranking an error should be penalized more if an item is ranked higher than if it happens the other way around.

3.2   Optimizing the Weighted Errors

Based on our discussion, we should penalize more for more risk given a specified recommendation goal. Also, risk is directional as shown in Fig. 3. Previous recommender systems considered the absolute value of the error, taking equally into account negative distance and positive distance from the ground truth. Here, we propose an optimization framework that would differentiate between negative and positive distance between the prediction and the ground truth rating, assigning a higher penalty for positive distance than negative distance. To achieve this, we assign a weight for each type of error. As shown in Fig. 3, the weights are two dimensional, depending on both the prediction and the ground truth. Mathematically, we optimize the following proposed objective function in order to obtain the parameters of a recommendation model:

$$\hat{\theta} = \underset{\theta}{\operatorname{argmin}} \sum_{u,i} w\big(f(\theta), r_{u,i}\big)\big(f(\theta) - r_{u,i}\big)^2 \qquad (1)$$

where $f(\theta)$ denotes the recommendation model parameterized by θ. u and i are the user and item index respectively. $w\big(f(\theta), r_{u,i}\big)$ denotes the cost (or risk) weighting function. We can solve the optimization problem by applying a Gradient Descent method [17], which requires to differentiate the objective function as follows:

$$\sum_{u,i} 2w(f(\theta) - r_{u,i})f'(\theta)d\theta + w'\big(f(\theta) - r_{u,i}\big)^2 f'(\theta)d\theta \qquad (2)$$

A discrete form of the weighting function is adopted in this paper (see in Table 1) – the risk preference of the system is thus captured by the nine weights. Because the weight w is constant for the three regions defined by f, the expression can be further approximated by:

$$\sum_{u,i} 2w(f(\theta) - r_{u,i})f'(\theta)d\theta \qquad (3)$$

To demonstrate the optimization framework, we adopt an incremental SVD (Singular Value Decomposition) factorization method [12], which is defined as follows:

$$\underset{q,p}{\operatorname{argmin}} \sum_{u,i} w(r_{u,i} - q_i^T p_u)^2 + \lambda(||q_i||^2 + ||p_u||^2) \qquad (4)$$

where $f = q_i^T p_u$, where q and p are the model parameters. This algorithm factors the matrix using only user and item pairs where $r_{u,i}$ is known. As mentioned earlier we introduced a weight w depending on the given criteria that is added to the equation. Therefore w is introduced, aiming to control the magnitude how conservative the system is to be in a given rating sector.

The system learns the model by fitting the previously observed rating. In order to avoid overfitting the second half of the equation regularizes the learning parameters and the constant λ is set to control the extent of regularization. Stochastic gradient descent is used to optimize the equation [12] introduced by [18].

The next question is how to obtain the optimal weighting w given a recommendation goal. Normally, a recommendation goal can be defined by a performance metric. For example, if the output is a ranked recommendation list, rank-based

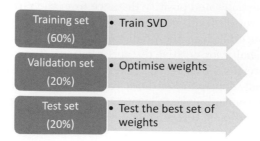

Fig. 4. Two-level optimization

metrics such as NDCG [19] might be suitable. We adopt a Genetic algorithm to obtain the optimal weights. Genetic algorithms are search algorithms that work via the process of natural selection. They begin with a sample set of potential solutions which then evolves toward a set of more optimal solutions. Within the sample set, solutions that are poor tend to die out while better solutions remain in the population, thus introducing more solutions into the set. The genetic algorithm does its best when there is a smooth slope of fitness over the problem space towards the optimum solution. This approach requires a two-level optimization illustrated in Fig. 4.

4   Evaluation

4.1   Experiment Setups

We empirically investigated the relationships between the taste boundaries and the CF performance, using the MovieLens dataset. This publicly available dataset consist of 100,000 ratings for 1682 movies by 943 users. We divided the dataset into three parts (Fig. 4) making sure that ratings from any given user are in all of the sets. Every user in the dataset rated at least 20 movies and the movies from each user distributed randomly when the dataset was divided. This is an important criterion since the performance measures that are discussed below consider users as a point of evaluation. The result is cross-validated using a five-fold cross-validation method and the outcomes are averaged.

The algorithm measures system effectiveness based on two assumptions. First, we consider recommendation as a ranking problem. Second, we define risk in nine different sectors (Fig. 3) which can be adjusted based on the desired outcome of the system. Even if the goal is to measure the effectiveness of the system across all users and items, from a user point of view there are items that are more important than others. If we consider recommendation as a ranking problem, it is sensible to optimize the algorithm using some of the measures from IR. Therefore two main concepts from IR should be defined in the domain of recommendations. Relevance shows whether an item is relevant to the query issued. However, the query is hidden in a recommender system, since it is defined by the user's preference which is usually not expressed explicitly. In this paper we make an assumption that users would watch a movie if it is rated four or five on a five point scale (relevant), but we acknowledge that this might be different for individual users. Therefore relevance is defined on a binary scale. Movies that are rated four or five are considered relevant, the rest of the movies are considered irrelevant. Retrieved items represent a list of items that are presented to the user. This concept might be important if the task of the system is to return the first-N relevant items. In order to reach the desired effect we evaluated the system using measures from IR. For a given user the algorithm ranks unseen movies such that the movies he or she likes most are suggested first. The following performance measures are used in this experiment.

The Mean reciprocal rank (MRR) [20] is the average of the reciprocal ranks of results for all users in the dataset. This measure only takes into account the first relevant item in the list. So the algorithm would achieve a high score if all the items that are relevant are predicted correctly.

Mean average precision (MAP) [20] obtains the precision score after each relevant document is retrieved. The mean of this score is calculated for all users to obtain the MAP score. The algorithm would achieve a higher score if it improves the precision in the retrieved list. So in this case all the documents that are retrieved count toward the score.

Normalized discounted cumulative gain (NDCG) [20] measures the gain based on the items position in the recommended list. This measure was introduced in [21]. It penalizes the system if it returns highly relevant documents lower in the ranking list but penalizes less if the lower end of the ranking list was retrieved incorrectly. NDCG is normalized by the perfect permutation of all the documents in the set. One of the problems if we apply it to recommendation is that the average number of ratings by user is relatively low. The aspect of picking the right k elements from a big dataset is lost here. Therefore all users are evaluated on a fixed number of items which is set. So most of the time all of the considered items are retrieved. Thus, there is no penalty on having the wrong elements within the retrieved documents, the only penalty can arise from the wrong order. In this experiment we used the formula defined in [22].

Since we used a small set it was also important to define the best solution to the problem and use a measure that is relative to the best solution. This is particularly important for MRR. As mentioned above in collaborative filtering the algorithm is tested in a relatively small set compared to sets used in IR. Therefore it is more likely that the algorithm is tested on users where all the items are non-relevant. So the algorithm does not have a chance to return relevant documents from a set where there is not any relevant documents. This would decrease the performance of the algorithm. Therefore, in this experiment the algorithm disregards users where there is no relevant documents in the test set. This is a reasonable assumption, because if the algorithm runs on the whole database it is very likely that there will be at least one item that is relevant to the user.

Table 2. Baseline SVD

	$r = 1, 2$	$r = 3$	$r = 4, 5$
$p <= 2.5$	0.05175	0.01935	0.0106
$2.5 < p <= 3.5$	0.0904	0.1461	0.1391
$p > 3.5$	0.02995	0.10125	0.4115

Table 3. SVD with weights where $w7 > w8 > w4$

	$r = 1, 2$	$r = 3$	$r = 4, 5$
$p <= 2.5$	0.0759	0.04075	0.0264
$2.5 < p <= 3.5$	0.0837	0.16765	0.23815
$p > 3.5$	0.0125	0.0583	0.29665

4.2 Results

Introducing weight for the error would penalize unwanted categories therefore the recommender prefers to have higher error rate in categories that are not penalized. For example if an item was rated five by the baseline recommender with the ground truth of one, the weighted recommender in Table 3 would more likely to rate it less than three instead, reducing the error, but increasing the error on items that are not that important (e.g. items where the ground truth is one).

The first experiment aimed to demonstrate that introducing weight in different sectors would reduce the number of items that fell into those sectors. We introduced weights in sectors where the algorithm would make a higher prediction than the ground truth ($w4, w7, w8$) and set the magnitude of the weight in the order of risk illustrated in Fig. 3. Table 3 shows that the probability that an item would fall into those sectors is reduced. However, this is a trade-off since it reduces the number of items in the sector (items that are rated five and predicted five). On the positive side, it increases the accuracy in the middle and lower range.

In the second experiment weights are set to one by default at points where the prediction should be the most accurate as defined by Fig. 3. In this case only weights that fall into the interval where prediction were higher than ground truth considered ($w4, w7, w8$). The rationale behind this choice is that the combination of this force (enabling the algorithm to modify only these weights) and the measure would result in an optimal solution for the user where higher rated items are considered more important and items that are overpredicted are penalized. Table 4 shows the result of the four performance measures that are used in our experiment. The first column indicates the score that is computed using the optimal weights and the second column is the baseline score (without weights). This shows that weights in fact improved the algorithm. The samples in the table are tested and found statistically significant. The reason why this method would provide a more robust recommendation from the user point of view is that it is reorganizing the ranking in a way that would take into account our initial criteria defined by the weights and re-rank it in a way that is ideal for these criteria. The advantage of the second approach is that it dynamically chooses the parameters

Table 4. Experimental results

	Measure(Test)	Baseline(Test)
MAP	0.450	0.447
MRR	0.899	0.889
NDCG@10	0.726	0.720
NDCG@5	0.574	0.570
NDCG@3	0.450	0.447

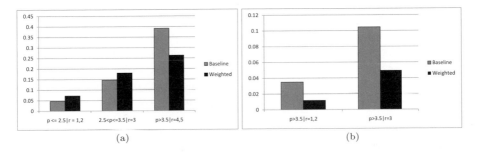

(a)   (b)

Fig. 5. (a) The probability of correct predictions within sectors described in Table 1. (b) The probability of predicting an item relevant when it is not.

for a given measure, however, as we will discuss below the measures do not cover all the possibilities given in our initial criterion. In contrast, the first approach can be tuned to reach a result that satisfies these criteria, but it cannot reach an optimal solution for all users.

Essentially this approach aims to minimize the error for the predefined sectors which inevitably results in the increase of error in other sectors. Fig. 5(a) shows the probability that true ratings are correctly predicted within our predefined taste boundary by the optimized versus the baseline approach using the weights obtained in the second experiment (Table 4). As expected the baseline approach predicts higher ratings better than our optimized approach, since the optimized approach does not penalize this type of error (high ratings predicted less), whereas we have some improvement in the lower range where we aimed to reduce the error. This approach takes the low risk approach therefore it hurts the performance at the higher range of the spectrum where it is less risky to predict something less, in exchange it reduces the error for item that are rated low. This means that it is less likely that users get items that are not relevant to them (Fig. 5(b)).

It is also important to investigate how the improvement of this evaluation metrics can be translated into improvement in user experience. Using MRR as a measure would reduce the probability that an irrelevant item would be presented to the user at the first position in the list. That implies that it would reduce the chance that lower rated items are rated higher for all items and it would also reduce the chance that higher rated items rated lower given that they are relevant items. The only place where it does not fit to our initial specification is that it does not differentiate between item and item within the irrelevant category, therefore there is not any difference in the score if an item rated one or and item rated three was ranked higher. Therefore parameter $w4$ does not add anything extra to this measure since it only penalizes low rated (one or two) items being predicted as uninteresting items. The same applies to NDCG, however it is a more subtle measure so it is able to differentiate between the order of the items in the ranked list. Therefore an NDCG score can tell us how well relevant items are ranked, which would be an optimal solution for the user.

5   Conclusion and Future Work

This paper presented a simple approach to optimize the outcome of collaborative filtering algorithm from the user point of view. This approach put an emphasis on the risk of making an incorrect recommendation. It considered recommendation in a more flexible way by taking into account predefined taste boundaries where users receive a list of items and are only interested in those items that are presented to the user. Another criterion is that items that are presented to the user are only interesting if they are within our predefined taste boundary. Therefore the algorithm aims to optimize its performance on those items. This approach can be fine-tuned further by considering how the items would be presented to the user. For example if a user would like to have just one recommendation, the algorithm is best optimized by MRR, or if the user would like to have more items recommended it would be better to optimized it by NDCG. The choice of parameters can be tailored to users need penalizing sectors that are more important to predict correctly to a specific user. For example calculating the mean of all the ratings for a particular user would suggest where the taste boundaries lie, so it can be determined for each individual user.

As it was discussed above all the measures only care about relevant items, but for our purposes it is also important to minimize error on disliked items (rated one or two). So we would like to measure how the algorithm performs on both sides of the rating scale. In both cases the middle range (items rated three) would be considered non-relevant. These two scores could be combined taking the high rated list more into account than the low rated one.

It is a widely discussed topic that accuracy alone is not a sufficient to measure whether a recommender system provides an effective and satisfying experience [2]. It is also important to note that a data is not homogeneous. In terms of prediction we can differentiate between easy and difficult items as well as easy and difficult users.

References

1. Bennett, J., Lanning, S.: The netix prize. In: Proceedings of KDD Cup and Workshop, vol. 2007 (2007)
2. Herlocker, J., Konstan, J., Terveen, L., Riedl, J.: Evaluating collaborative filtering recommender systems. ACM Transactions on Information Systems 22(1), 5–53 (2004)

3. Mitchell, M.: An introduction to genetic algorithms. The MIT press, Cambridge (1998)
4. Goldberg, D., Nichols, D., Oki, B.M., Terry, D.: Using collaborative filtering to weave an information tapestry. Commun. ACM 35(12), 61–70 (1992)
5. Manning, C.D., Raghavan, P., Schtze, H.: Introduction to Information Retrieval. Cambridge University Press, Cambridge (2008)
6. Herlocker, J.L., Konstan, J.A., Borchers, A., Riedl, J.: An algorithmic framework for performing collaborative filltering. In: SIGIR 1999 (1999)
7. Deshpande, M., Karypis, G.: Item-based top-N recommendation algorithms. ACM Trans. Inf. Syst. 22(1), 143–177 (2004)
8. Wang, J., de Vries, A.P., Reinders, M.J.T.: Unifying user-based and item-based collaborative filtering approaches by similarity fusion. In: SIGIR 2006: Proceedings of the 29th annual international ACM SIGIR conference on Research and development in information retrieval, pp. 501–508. ACM Press, New York (2006)
9. Pennock, D.M., Horvitz, E., Lawrence, S., Giles, C.L.: Collaborative filtering by personality diagnosis: A hybrid memory and model-based approach. In: UAI 2000 (2000)
10. Breese, J., Heckerman, D., Kadie, C.: Empirical analysis of predictive algorithms for collaborative filtering. In: Proceedings of the 14th Annual Conference on Uncertainty in Artificial Intelligence, UAI 1998 (1998)
11. Canny, J.: Collaborative filtering with privacy via factor analysis. In: SIGIR 2002 (2002)
12. Koren, Y., Bell, R., Volinsky, C.: Matrix factorization techniques for recommender systems. Computer 42(8), 30–37 (2009)
13. Weimer, M., Karatzoglou, A., Smola, A.: Adaptive collaborative filtering. In: RecSys 2008: Proceedings of the 2008 ACM conference on Recommender systems, pp. 275–282. ACM, New York (2008)
14. Herlocker, J.L., Konstan, J.A., Terveen, L.G., Riedl, J.T.: Evaluating collaborative filtering recommender systems. ACM Trans. Inf. Syst. 22(1), 5–53 (2004)
15. Wang, J., Roberston, S.E., de Vries, A.P., Reinders, M.J.T.: Probabilistic relevance models for collaborative filtering. Journal of Information Retrieval (2008)
16. Liu, N.N., Yang, Q.: Eigenrank: a ranking-oriented approach to collaborative filtering. In: SIGIR 2008: Proceedings of the 31st annual international ACM SIGIR conference on Research and development in information retrieval, pp. 83–90. ACM, New York (2008)
17. Duda, R.O., Hart, P.E., Stork, D.G.: Pattern Classification. Wiley Interscience, New York (2001)
18. Funk, S.: Netix update: Try this at home (2006), http://sifter.org/~simon/journal/20061211.html
19. Jrvelin, K., Keklinen, J.: Cumulated gain-based evaluation of ir techniques. ACM Trans. Inf. Syst. 20(4), 422–446 (2002)
20. van Rijsbergen, C.J.: Information Retrieval. Butterworths, London (1979)
21. Jarvelin, K., Kekalainen, J.: IR evaluation methods for retrieving highly relevant documents. In: Proceedings of the 23rd annual international ACM SIGIR conference on Research and development in information retrieval, pp. 41–48. ACM, New York (2000)
22. Vassilvitskii, S., Brill, E.: Using web-graph distance for relevance feedback in web search. In: Proceedings of the 29th annual international ACM SIGIR conference on Research and development in information retrieval, pp. 147–153. ACM, New York (2006)

Personalizing Web Search with Folksonomy-Based User and Document Profiles

David Vallet[1,2], Iván Cantador[1,2], and Joemon M. Jose[1]

[1] University of Glasgow, Glasgow, UK
{dvallet,cantador,jj}@dcs.gla.ac.uk
[2] Universidad Autónoma de Madrid, Madrid, Spain
{david.vallet,ivan.cantador}@uam.es

Abstract. Web search personalization aims to adapt search results to a user based on his tastes, interests and needs. The way in which such personal preferences are captured, modeled and exploited distinguishes the different personalization strategies. In this paper, we propose to represent a user profile in terms of social tags, manually provided by users in folksonomy systems to describe, categorize and organize items of interest, and investigate a number of novel techniques that exploit the users' social tags to re-rank results obtained with a Web search engine. An evaluation conducted with a dataset from Delicious social bookmarking system shows that our personalization techniques clearly outperform state of the art approaches.

1 Introduction

The huge and ever increasing volume and complexity of information available in the Web constitutes a difficult challenge for content retrieval technologies. In a traditional Web search system, such as Google[1] or Yahoo![2], a user expresses his information needs by providing a textual query consisting in a limited number of keywords. The search system takes as input this query, and attempts to retrieve the Web documents that best match its keywords. Queries are usually short – containing no more than 3 keywords on 85% of the times – and ambiguous [7], and often fail to reflect the user's needs. Nonetheless, although the information contained in these keywords rarely suffices for the exact determination of the user's wishes, this approach represents a simple way of interaction users are accustomed to. There is thus a need to investigate ways to enhance information retrieval, without altering the way the users specify their requests. It is in such scenario where personalized information retrieval techniques can help the users, by tailoring the search results based on both the users' short and long term preferences [4]. However, to achieve that goal, information about the users' tastes and interests has to be found in other sources.

With the advent of the Web 2.0, social tagging systems have been exponentially grown both in terms of users and contents. These systems allow the users to provide annotations (*tags*) to resources, expressing personal descriptions and opinions about

---

[1] Google, http://www.google.com/
[2] Yahoo! Search, http://search.yahoo.com/

C. Gurrin et al. (Eds.): ECIR 2010, LNCS 5993, pp. 420–431, 2010.
© Springer-Verlag Berlin Heidelberg 2010

the resources for organizational and sharing purposes. For instance, in Last.fm[3], the users annotate their favorite songs; in Flickr[4], the users store and annotate their own photo streams; and in Delicious[5], the users bookmark and annotate interesting Web pages. Apart from facilitating the organization and sharing of content, these 'social tagging' data, also known as *folksonomies*, can be considered as a fairly accurate source of user interests. Several studies have proven that a user profile can be effectively harvested from these systems [1, 11], and later exploited on different personalization services, such as tag recommendation [3], item recommendation [9], and personalized search [6, 9, 13], to name a few.

In this work, we present two novel personalization techniques that exploit a user profile defined within a social tagging system to re-rank the document lists retrieved by a traditional Web search engine. In particular, we investigate whether a folksonomy-based user profile defined in Delicious social bookmarking system can really enhance the results provided by Yahoo! Search engine. To evaluate such techniques, we propose an automatic mechanism that generates test datasets from social tagging corpora. The results obtained in our experiments show that our personalization techniques clearly outperform state of the art approaches.

The rest of the paper has the following structure. In Section 1, we describe works that are related to our research. In Section 2, we present the state of the art and own personalization approaches we evaluate and compare. In Section 3, we propose an evaluation framework and an experimental methodology for folksonomy-based Web search personalization techniques. We present the evaluation results in Section 4. Finally, in Section 5, we provide some conclusions and possible future work lines.

2   Related Work

Personalized retrieval models that exploit user profiles based on social tags have been investigated in previous works.

Shepitsen et al. [9] present a strategy that clusters the entire space of tags to obtain sets of (semantically) related tags. Representing coherent topic areas, the obtained clusters are used to provide personalized item recommendations. Rather than item recommendation, the techniques presented in this paper follow personalized retrieval models applicable to Web search, where lists of search results are re-ranked according to the user's preferences.

Hotho et al. propose the FolkRank algorithm [4], an adaptation of the PageRank algorithm to the folksonomy structure. FolkRank performs a weight-spreading ranking scheme on folksonomies. It transforms the hypergraph between the sets of users, tags and resources into an undirected, weighted, tripartite graph. On this graph, it applies a version of PageRank that takes into account the obtained edge weights. Among other applications, FolkRank provides a popularity measure of a document that seems to be better than PageRank, as it exploits the user generated folksonomy, rather than the Web links. Bao et al. [2] also investigate the use of popularity measures derived from the folksonomy structure, but focusing its application in a Web

---

[3] Last.fm - Personal online radio, http://www.last.fm/
[4] Flickr - Photo sharing, http://www.flickr.com/
[5] Delicious - Social bookmarking, http://delicious.com/

search system. They introduce two importance score values, SocialSimRank and SocialPageRank, which measure the relevance of a document to a query, and the popularity of a document, respectively. They conclude that these measures provide a better performance than traditional measures, such as term matching and PageRank. Similar to the studies of Hotho et al. and Bao et al., we exploit the folksonomy structure, but focus on offering a personalized search to the user, rather than improving the overall rank of documents.

Noll and Meinel [6] present a personalization model that exploits the user's and documents' related tags, improving a Web search system during their user evaluation. Xu et al. [13] also present a user-document similarity function that relates the user's and documents' tags, and enrich the user's profile representation following a tag expansion strategy, which is applied over a restricted corpus. Our personalization approaches follow the same personalization model as Xu et al.'s and Null and Meinel's, but utilize different techniques to calculate the user-document similarities. We shall evaluate and compare our proposals against the approaches presented by these authors.

3   Web Search Personalization Based on Folksonomies

A folksonomy $\mathcal{F}$ can be defined as a tuple $\mathcal{F} = \{\mathcal{T}, \mathcal{U}, \mathcal{D}, \mathcal{A}\}$, where $\mathcal{T} = \{t_1, \ldots, t_L\}$ is the set of tags that comprise the vocabulary expressed by the folksonomy, $\mathcal{U} = \{u_1, \ldots, u_M\}$ and $\mathcal{D} = \{d_1, \ldots, d_N\}$ are respectively the set of users and the set of documents[6] that annotate and are annotated with the tags of $\mathcal{T}$, and $\mathcal{A} = \{(u_m, t_l, d_n)\} \in \mathcal{U} \times \mathcal{T} \times \mathcal{D}$ is the set of assignments (annotations) of each tag t_l to a document d_n by a user u_m.

The profile of user u_m is then defined as a vector $\overrightarrow{u_m} = (u_{m,1}, \ldots, u_{m,L})$, where $u_{m,l} = |\{(u_m, t_l, d) \in \mathcal{A} | d \in \mathcal{D}\}|$ is the number of times the user has annotated documents with tag t_l. Similarly, the profile of document d_n is defined as a vector $\overrightarrow{d_n} = (d_{n,1}, \ldots, d_{n,L})$, where $d_{n,l} = |\{(u, t_l, d_n) \in \mathcal{A} | u \in \mathcal{U}\}|$ is the number of times the document has been annotated with tag t_l. In our Web search scenario, the set of documents $\mathcal{D}$ represents the resources present in the Web, and are identified by an URL. Users are identified by a user id.

In this work, we exploit folksonomy-based user and document profiles in order to personalize the results of a Web search system. A non-personalized Web search system S provides a ranked list of documents $S(q) \subseteq \mathcal{D}$ that satisfy a given query topic q. The ranking follows an ordering $\tau = [d_1 \geq d_2 \geq \cdots \geq d_k]$, in which $d_i \in \mathcal{D}$ and $\geq$ is the ordering relation implemented by the search system. Upon this ranked document list, we define a personalization approach S' that provides a ranked list of documents $S'(q, u) \subseteq \mathcal{D}$ by reordering the results $S(q)$ according to the preferences of user u. More formally, it provides an ordering $\tau' = [d_1 \geq d_2 \geq \cdots \geq d_k]$ such that the ordering relation is defined by $d_i \geq d_j \Leftrightarrow \text{sim}(u, d_i, q) \geq \text{sim}(u, d_j, q)$, where $\text{sim}(u, d, q)$ is a similarity function between user u and document d, taking into consideration the ranking of d in S(q).

---

[6] In more general definitions of *folksonomy*, annotated items, which are not necessarily textual, are usually called "resources". As we deal with the exploitation of folksonomies in a Web Search scenario, we instead use the term (Web) "documents" to reference resources.

The subsequent subsections present the different personalization techniques we propose and evaluate. The first two techniques are obtained from the state of the art, and are based on the Vector Space Model (VSM). The third technique is a personal adaptation of the VSM to social tagging profiles. The last technique is a novel personalization approach that follows a probabilistic model. More specifically, it is an adaptation of the Okapi BM25 ranking model.

For a better understanding, Table 1 gathers the definition of common elements appearing in the models of the above techniques. It is worth noting that whereas in the classic VSM the document collection is the only source for the calculation of term frequencies and inverse document frequencies, in a folksonomy-based framework, we can also consider how informative the tags (terms) are in the user set. Thus, the user-based tag frequency $tf_{u_m}(t_l)$ measures how relevant a tag t_l is to a user u_m, and the user-based tag inverse frequency $iuf(t_l)$ measures how common or popular a tag t_l is across all users $\mathcal{U}$. The presented approaches can also be differentiated by how these local and global importance values are exploited.

Table 1. Elements that are used by the folksonomy-based personalization models

Element	Definition
User-based tag frequency	$tf_{u_m}(t_l) = u_{m,l}$
Document-based tag frequency	$tf_{d_n}(t_l) = d_{n,l}$
User-based inverse tag frequency	$iuf(t_l) = \log\dfrac{M}{n_u(t_l)}, n_u(t_l) = \left|\{u_m \in \mathcal{U} \mid u_{m,l} > 0\}\right|$
Document-based inverse tag frequency	$idf(t_l) = \log\dfrac{N}{n_d(t_l)}, n_d(t_l) = \left|\{d_n \in \mathcal{D} \mid d_{n,l} > 0\}\right|$
User size	$\lvert u_m \rvert = \sum_{l=1}^{L} u_{m,l}$
Document size	$\lvert d_n \rvert = \sum_{l=1}^{L} d_{n,l}$

3.1 Cosine Similarity Based Personalization

The approach presented by Xu et al. [13] uses the classic cosine similarity measure to compute the similarity between user and document profiles. As weighting scheme, it uses $tf\text{-}idf$ [7]. Following our notation, their approach can be defined as follows:

$$\cos_{tf\text{-}idf}(u_m, d_n) = \frac{\sum_l \left(tf_{u_m}(t_l) \cdot iuf(t_l) \cdot tf_{d_n}(t_l) \cdot idf(t_l)\right)}{\sqrt{\sum_l (tf_{u_m}(t_l) \cdot iuf(t_l))^2} \cdot \sqrt{\sum_l (tf_{d_n}(t_l) \cdot idf(t_l))^2}},$$

where the numerator is the dot product of the $tf\text{-}iuf$ and $tf\text{-}idf$ vectors associated with the user and the document, respectively. The denominator is the user and document length normalization factors, calculated as the magnitude value of those vectors.

---

[7] Xu et al. do not specify if they use the user-based or the document-based inverse tag frequency weights, or both. We chose to use both, as it gave the best performance values.

Xu et al. use a weighting scheme based on the BM25 model, this variation will be henceforth denoted as $\cos_{bm25}(u_m, d_n)$. See Section 0 for more details on this approach.

3.2 Scalar Tag Frequency Based Personalization

The approach presented by Noll and Meinel [6] is similar to the cosine-based approach, but does not make use of the user and document length normalization factors, and only uses the user tag frequency values. The authors normalize all document tag frequencies to 1, since they want to give more importance to the user profile when computing the similarity measures. Following the notation given in Table 1, their similarity measure can be defined as follows:

$$tf(u_m, d_n) = \sum_{l:d_{n,l}>0} tf_{u_m}(t_l) \,.$$

3.3 Scalar *tf-if* Based Personalization

Next, we present our first proposed personalization approach. Similarly to Xu et al. [13], we use the *tf-idf* weighting scheme. We eliminate however the user and document length normalization factors. In the classic VSM, the finality of the length normalization factor is to penalize the score of documents that contain a high amount of information, and might have matched the query only by chance. In terms of a social tagging system, a high amount of related tags is correlated with the popularity of the documents among users. Hence, if we used a length normalization factor, we would penalize the score of popular documents. As several works point out, this popularity value is a good source of relevancy [2, 4]. Thus, it would not be advisable to penalize popular documents. Note that eliminating the user length normalization factor does not have any effect, as it is constant in all user-document similarity calculations.

The main difference between our approach and Noll and Meinel's [6] is that we incorporate both the user and document tag distribution global importance factors, i.e. *iuf* and *idf*, following the VSM principle that as more rare a tag is, the more important it is when describing either a user's interests or a document's content. We do not normalize the content of the documents, as we believe that the distribution of tags on a document may give insights on how important a tag is to describe its content. This personalization approach can thus be defined as following:

$$tf\text{-}if(u_m, d_n) = \sum_l \left(tf_{u_m}(t_l) \cdot iuf(t_l) \cdot tf_{d_n}(t_l) \cdot idf(t_l) \right) \,.$$

3.4 BM25 Based Personalization

The novel personalization approach presented in this section differs from the previously presented ones in that it follows a probabilistic model, rather than the classic VSM. We adapt the Okapi BM25 ranking model [8] to a personalization ranking of similarity between a user and a document. The BM25 model computes a ranking score function of a document given a query. We can then adapt this model in two different ways: 1) by assuming that the user profile takes part as a query indicating the user's interests, or 2) by assuming that the document takes part as a query, and is matched against all user profiles. The former option will be henceforth denoted as $bm25_{u_m}$ and the latter as $bm25_{d_n}$. We first define both score functions for a single tag t_l:

$$bm25_{u_m}(t_l) = iuf(t_l) \cdot \frac{u_{m,l} \cdot (k_1 + 1)}{u_{m,l} + k_1 \left(1 - b + b \cdot |u_m| \big/ avg(|u_m|)\right)},$$

$$bm25_{d_n}(t_l) = idf(t_l) \cdot \frac{d_{n,l} \cdot (k_1 + 1)}{d_{n,l} + k_1 \left(1 - b + b \cdot |d_n| \big/ avg(|d_n|)\right)},$$

where b and k_1 are set to the standard values of 0.75 and 2, respectively. Then, we define the two variations of this personalization approach:

$$bm25_{u_m}(d_n, u_m) = \Sigma_{(l \mid d_{n,l} > 0)} \, bm25_{u_m}(t_l),$$

$$bm25_{d_n}(d_n, u_m) = \Sigma_{(l \mid u_{m,l} > 0)} \, bm25_{d_n}(t_l) .$$

Xu et al. [12] compute a VSM based cosine similarity measure with a weighting scheme inspired by the BM25 retrieval model. Following the notation of this section, this measure can be defined as follows:

$$\cos_{bm25}(d_n, u_m) = \frac{\Sigma_l \left(bm25_{u_m}(t_l) \cdot bm25_{d_n}(t_l) \right)}{\sqrt{\Sigma_l (bm25_{u_m}(t_l)^2)} \cdot \sqrt{\Sigma_l (bm25_{d_n}(t_l)^2)}} \quad \text{8}$$

4 Evaluating Folksonomy-Based Personalization Approaches

Noll and Meinel [6] evaluated their personalization approach combined with a Web search engine. They adopted a user centered evaluation by creating a set of predefined queries, and by asking users to evaluate the results. More specifically, users were asked to evaluate which result list they preferred: either the Web search ranking or the personalized ranking. Xu et al. [13] used the social bookmarking information to create an automatic evaluation framework. The main advantage of their framework is that the experiments could be reproduced. However, they did not explore the performance of their personalization approaches when combined with a Web search engine. They combined their approach with a search system that was limited to the bookmarks pertinent to their test beds, ranging from 1K to 15K Web documents. The goal of our evaluation framework falls in the middle of these two approaches: 1) as Noll and Meinel, we are more interested in testing our personalization approach in a real Web search environment; and 2) as Xu et al., we adopt an automatic evaluation framework with a test bed of topics and relevance judgments extracted from the social bookmarking information. In this section, we describe our evaluation framework, highlighting the main differences between it and the previously presented.

---

[8] Xu et al. use a slightly modified version of the *idf* measure: $\log((M - n(t_l) + 0.5)/(n(t_l) + 0.5))$, using $n_u(t_l)$ and $n_u(t_l)$ on the $bm25_{u_m}(t_l)$ and $bm25_{d_n}(t_l)$, respectively. The reported results make use of this measure.

4.1 Topic and Relevance Judgment Generation

We split the tagging information of a given user into two parts. The first part forms the user profiling information, whereas the second is used for the automatic topic generation process. Hence, the subset of tag assignments used in the topic generation process is not included in the user profile, constitutes our test dataset, and thus is not part of our training dataset. This splitting process is applied to all users belonging to the initial test bed collection. Figure 1 outlines how the partition is made.

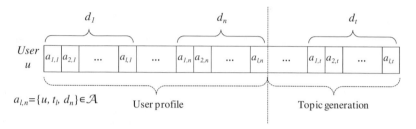

Fig. 1. Partitioning of user tag assignments into user profile and information intended for topic generation

As shown in the figure, the topic creation process attempts to create a new topic from each document $d \in [d_{n+1}, ..., d_t]$ belonging to the test split part. A topic is defined by extracting the top most popular tags related to a document d. We use the most popular tags as they are more objective to describe the document contents than those assigned by a single or few users. These tags are used to launch a Web search, and we collect the retrieved result list.

We then study how the different personalization approaches re-rank the returned result list. As document d was contained in the original user profile, we can assume that the document is relevant to the user. Thus, a good personalization approach would always rank the document in the top positions of the result list.

We use the Mean Reciprocal Rank (MRR) [12] metric to measure the performance of the personalization techniques. This measure assigns a value of performance for a topic of $1/r$, where r is the position of the relevant d in the final personalized result list. We also provide the $P@N$ (Precision at position N) metric, which has a value of 1 iff $r \leq N$. These values are averaged over all the generated topics.

The topic generation and evaluation can be summarized in the following methodology. For each document $d \in [d_{n+1}, ..., d_t]$: 1) we generate a topic description using the top k most popular tags associated to the document; 2) we execute the topic on a Web search system and return the top R documents as the topic's result list; 3) if document d is not found in the result list, we discard the topic for evaluation; 4) we apply the different personalization techniques to the result set; and 5) we compute MRR and P@N values.

In our experiments, we used a query size of $k = 3$ tags, and a result list size of $R = 300$ documents. Several studies point out an average user query size of 2-3 keywords in Web search [7]. We thus opted for a query size of 3 in order to emulate a user using a Web search system, and to evaluate if user profiles obtained from the social tagging actions of the users could be successfully exploited to improve a Web search system. We also investigated the generation of query topics with 2 keywords obtaining performance results similar to those obtained with topic sizes of 3 keywords. There is of course a chance that document d does not appear in the result list. In this case, the document is discarded for topic generation. With these settings, 24.2% of the topics were successfully generated, and the average position of document d on the result list was 62.2.

As mentioned before, Xu et al. also presented an automatic topic generation methodology based on the users' tagging data. However, there are some key differences between their evaluation framework and ours. First, they applied the personalization techniques to a custom search engine that only retrieves documents that belong to the same test bed. Our methodology, on the other hand, makes use of Web search system to return the topic document. In this way, we intend to have a more realistic set up. Second, they used each tag of the user profiles as a query topic, thus resulting on queries with a single keyword. This resulted on too broad queries, which are not suitable for a free Web search system. We rather choose to use more specific queries of three keywords, which are generated based on the social tagging information associated to a document that was originally in the user profile. Third, their approach assumed that a returned document was relevant to the user if it was tagged by him with the same tag that belonged to the topic query. Our ground truth is more restrictive, as we only consider as relevant the document that generated the topic query. By doing this we can ensure that the document is relevant to both the topic query, as the query keywords represent the people's view of the document's content, and to the user, as the document belongs to the user's profile. In summary, we consider that our approach is more suited to evaluate folksonomy-based personalization of a Web search system. Nonetheless, we do believe that both approaches may complement each other in order to give more insights on the performance of personalization strategies.

4.2 Experimental Setup

We created a test bed formed by 2,000 Delicious users. Delicious is a social bookmarking site for Web pages. As of the 26th of November of 2008, delicious had 5.3 million users[9], up from 1 million users registered on September of 2006[10]. With over 180 million unique URLs, Delicious can be considered a fairly accurate "people's view" of the Web. This vast amount of user information has been previously successfully exploited to improve Web search [2], to provide personal recommendations [4, 9], and to personalize search results [6, 13], among others.

Due to limitations of Delicious API, we only extracted the latest 100 bookmarks of each user, from which we use 90% of the bookmarks to create the user profile, and the remaining 10% to generate the evaluation topics as described in Section 4.1. The test bed contained 161,542 documents and 69,930 distinct tags. We did not apply any pre-processing steps to the user tags. Users used an average of 5.6 tags to describe each bookmark. As experimental Web search system, we use Yahoo!'s open Web search platform, Yahoo! Boss[11]. After the topic generation process, we ended up with

---

[9] http://blog.delicious.com/blog/2008/11/delicious-is-5.html

[10] http://blog.delicious.com/blog/2006/09/million.html

[11] http://developer.yahoo.com/search/boss/

6,109 evaluation topics. For each document in the topic result set, we downloaded the 100 most recent bookmarks. Those documents had an average of 24.3 distinct associated tags. On average, 20.13% of the documents of the result list had been bookmarked at least once by a user.

5   Experiment Results

We present the results of the proposed personalization techniques within the evaluation framework explained in Section 4. We first provide the performance of the approaches when applied in isolation to the search results returned by the Web search system. Then, we show their performance when taking into consideration the result ranking provided by the Web search system.

5.1   Results of Personalization Approaches

In this section, we analyze the performance of the personalization approaches when only the personalization scores are used to reorder the results returned by the Web search system, i.e., when the ranking given by the search system is not taken into account. Table 2 shows Mean Reciprocal Ranking (MRR) and Precision (at 5, at 10, at 20) values of the personalization approaches.

Table 2. Personalization approaches performance. Values with an asterisk indicate a statistically significant higher value than the tf approach (Wilcoxon test, $p < 0.05$). Values in bold indicate the highest values with statistical significance. The column $comb$ refers to the rank-based combination of $bm25_{u_m}$ and $tf\text{-}if$ approaches

Metric	$\cos_{tf\text{-}idf}$ [13]	$\cos_{bm25}$ [13]	$bm25_{d_n}$	tf [6]	$bm25_{u_m}$	$tf\text{-}if$	$comb$
MRR	0.0809	0.0912	0.2878	0.2845	0.3055*	0.3084*	**0.3241***
P@5	0.0915	0.1111	0.4502	0.4554	0.4601	0.4839*	**0.4924***
P@10	0.1838	0.2252	0.6290	0.6369	0.6363	0.6595*	**0.6702***
P@20	0.3812	0.4259	0.7816	0.7967	0.7900	0.8082*	**0.8093***

The cosine similarity approaches presented by Xu et al. [13], $\cos_{tf\text{-}idf}$ and $\cos_{bm25}$, have much lower performance values than the rest of the approaches, even though Xu et al. report for them a performance better than the tf approach, presented by Noll and Meinel [6]. A possible reason for this contradiction is the difference between Xu et al.'s and our evaluation setups. On one hand, the length normalization factor used in the cosine similarity function penalizes those documents with a high amount of assigned tags, i.e., those documents that are more popular, in favor of documents that have fewer related tags. This penalization factor may be self-defeating according to different studies [2, 4] which suggest that a popular document has a higher chance to be relevant to a user. We noticed that the documents returned by the Web search system were highly diverse in terms of popularity, and thus the discrimination of popular documents had a sensible negative impact. On the other hand, Xu et al. make use of a controlled document collection, no larger than 15K documents, which may not have these characteristics.

The $bm25_{u_m}$ approach obtains a performance statistically significant higher than the $bm25_{d_n}$ approach. This implies that, in a folksonomy model, it is better to assume that the user acts as a document in terms of the probabilistic model's relevance computation. Compared to the other personalization approaches, the $bm25_{u_m}$ approach has a better performance in terms of the MRR metric, outperforming the tf approach, which is the best found in the state of the art. However, it has a performance statistically significant lower than the $tf\text{-}if$ approach. The performance of $tf\text{-}if$ approach is higher than both the if and the $bm25$ approaches, with statistical significant differences on all the used metrics. These results highlight the importance of incorporating the global frequencies calculated for a given tag, i.e., the tag user inverse frequency iuf and the tag document inverse frequency idf.

Moreover, since the $bm25_{u_m}$ and the $tf\text{-}if$ approaches are based on different models, the probabilistic and the vector space models, respectively, we investigate the performance of a combination of both approaches. We use a simple, parameter free aggregation strategy, CombSUM with rank-based normalization [10], to merge their rankings. The obtained performance results are presented in the last column of Table 2, and are encouraging: this strategy is the highest performing approach, achieving a 13.91% improvement on MRR, and a 8.12% improvement in terms of P@5, with respect to the best performing state of the art approach, tf, indicating that both our approaches complement each other. We also computed Kendall's tau over the ranks produced by our combination and tf approaches, in order to check if these techniques were personalizing results differently. The average Kendall's tau over all topics was 0.185 (SD = 0.187, p < 0.05) which lead us to think that this was the case. Other combinations did not result in a performance improvement.

5.2 Results of Folksonomy-Based Personalized Web Search

We now investigate the performance of the personalization approaches when used in combination with a Web search system. In order to do this, we merge the result lists returned by the Web search system (denoted as $S(q)$ in Section 3) with the result lists produced by the personalization approaches, i.e., the results evaluated in the previous section.

Table 3. Personalization approaches performance when combined with the Web search engine results. Values with an asterisk indicate a statistically significant higher value than the Web search ranking (Wilcoxon test, p < 0.05). Values marked with a † also indicate a statistically significant higher value than the tf approach. Values in bold are the highest significant values.

Metric	baseline	$\cos_{tf\text{-}idf}$	$\cos_{bm25}$	$bm25_{d_n}$	tf	$bm25_{u_m}$	$tf\text{-}if$	comb
MRR	0.3292	0.1626	0.1810	0.3750^*	0.3905^*	0.3931^*	$0.4019^\dagger$	$\mathbf{0.4073^\dagger}$
P@5	0.4523	0.2354	0.2696	0.5435^*	0.5554^*	0.5593^*	$0.5652^\dagger$	$\mathbf{0.5705^\dagger}$
P@10	0.5793	0.3968	0.4325	0.6720^*	0.6859^*	0.6790^*	$0.6903^\dagger$	$\mathbf{0.6955^\dagger}$
P@20	0.7078	0.5945	0.6181	0.7903^*	0.7952^*	0.7983^*	0.7980^*	$\mathbf{0.8006^\dagger}$

As a baseline, we use the Web search system. In order to make a fairer comparison, we eliminate from its result lists those documents that were not bookmarked by any user. The final ranked lists are combinations of both the non-personalized and the personalized rank lists using CombSUM with rank based normalization as aggregation method [10]. Table 3 shows the performance values of the personalization approaches combined with the Web search. Values are correlated with those presented in Table 2.

The cosine similarity personalization approaches degrade the performance of the Web search, while the rest of approaches outperform the baseline. The two approaches proposed in this work, tf-if and $bm25_{u_m}$, perform better than both the baseline and tf. Again, the combination of tf and $bm25_{u_m}$ personalization approaches yield the best performance, both in terms of MRR and precision. This demonstrates the complementarily of both approaches, whose combination achieves 23.72% and 4.3% improvements with respect to the baseline and tf approaches, respectively.

6   Conclusions and Future Work

In this paper, we have presented two novel techniques that exploit user and document profiles defined in a social tagging system to personalize the result rankings of a Web search system. The first personalization approach is based on the vector space information retrieval model, and incorporates the concepts of tag inverse document frequency and tag inverse user frequency, which are global measures that rely on the tag distribution within the folksonomy-based user and document profiles. The second personalization approach is an adaptation of the BM25 probabilistic model to folksonomy systems based on the above user and document representations.

We have also proposed a novel evaluation framework and a topic generation methodology which allow the automatic evaluation of folksonomy-based Web search personalization approaches. The results obtained with the evaluations conducted over a dataset from Delicious social bookmarking system show that our techniques outperform the state of the art folksonomy-based personalization approaches. Furthermore, we demonstrate how our two personalization techniques can complement each other, achieving the best overall performance when combined by a well-known rang aggregation strategy. We claim that the key points of the achieved performances are 1) the use of the proposed global tag importance measures, 2) the removal of length normalization factors in personalization formulas, and 3) the adaptation of the probabilistic model.

The presented techniques can be applied to any Web search system, providing personalization capabilities to any user who has a profile in a social tagging service. Thus, with no extra effort, a user can personalize and enhance the results provided by a certain Web search engine. In our evaluations, we obtained a performance increase of 23.7% over Yahoo! Search, demonstrating the feasibility of this personalization paradigm.

The approaches evaluated in this paper exploit the folksonomy's user, tag and document distribution. However, there are also specific techniques which exploit the folksonomy structure in order to expand the folksonomy-based profiles. The main problem is that, to date, these techniques are not easily scalable to the Web, and have to be evaluated in small controlled collections [13]. Thus, we were unable to incorporate them into our Web search personalization framework. In the future, we will investigate a scalable expansion strategy that could allow its application to personalization approaches focused on Web search.

We will also study how our personalization techniques can be combined with folksonomy-based popularity measures presented in the state of the art [2, 4]. Although our techniques include some basic document popularity factors, they could be complemented by the above more complex measures.

Acknowledgements. This research was supported by the European Commission under contract FP6-027122-SALERO and by the Spanish Ministry of Science and Education (TIN2008-06566-C04-02).

References

1. Au-Yeung, C.M., Gibbins, N., Shadbolt, N.: A study of user profile generation from folksonomies. In: Proc. of the Social Web and Knowledge Management Workshop (2008)
2. Bao, S., Xue, G., Wu, X., Yu, Y., Fei, B., Su, Z.: Optimizing web search using social annotations. In: Proc. of WWW 2007, pp. 501–510. ACM Press, New York (2007)
3. Chirita, P.A., Costache, S., Nejdl, W., Handschuh, S.: P-tag: large scale automatic generation of personalized annotation tags for the web. In: Proc. of WWW 2007, pp. 845–854. ACM, New York (2007)
4. Hotho, A., Jäschke, R., Schmitz, C., Stumme, G.: Information retrieval in folksonomies: search and ranking. In: Sure, Y., Dominguc, J. (eds.) ESWC 2006. LNCS, vol. 4011, pp. 411–426. Springer, Heidelberg (2006)
5. Micarelli, A., Gasparetti, F., Sciarrone, F., Gauch, S.: Personalized search on the World Wide Web. In: Brusilovsky, P., Kobsa, A., Nejdl, W. (eds.) Adaptive Web 2007. LNCS, vol. 4321, pp. 195–230. Springer, Heidelberg (2007)
6. Noll, M.G., Meinel, C.: Web search personalization via social bookmarking and tagging. In: Aberer, K., Choi, K.-S., Noy, N., Allemang, D., Lee, K.-I., Nixon, L.J.B., Golbeck, J., Mika, P., Maynard, D., Mizoguchi, R., Schreiber, G., Cudré-Mauroux, P. (eds.) ISWC 2007. LNCS, vol. 4825, pp. 367–380. Springer, Heidelberg (2007)
7. Jansen, B.J., Spink, A., Bateman, J., Saracevic, T.: Real life information retrieval: a study of user queries on the web. SIGIR Forum 32(1), 5–17 (1998)
8. Robertson, S.E., Walker, S.: Some simple effective approximations to the 2-Poisson model for probabilistic weighted retrieval. In: Proc. of SIGIR 2004, pp. 345–354. Springer, Heidelberg (2004)
9. Shepitsen, A., Gemmell, J., Mobasher, B., Burke, R.: Personalized recommendation in social tagging systems using hierarchical clustering. In: Proc. of RecSys 2008, pp. 259–266. ACM Press, New York (2008)
10. Shaw, J.A., Fox, E.A.: Combination of multiple searches. In: Text REtrieval Conference, pp. 243–252 (1993)
11. Szomszor, M., Alani, H., Cantador, I., O'hara, K., Shadbolt, N.: Semantic modelling of user interests based on cross-folksonomy analysis. In: Sheth, A.P., Staab, S., Dean, M., Paolucci, M., Maynard, D., Finin, T., Thirunarayan, K. (eds.) ISWC 2008. LNCS, vol. 5318, pp. 632–648. Springer, Heidelberg (2008)
12. Voorhees, E.: The TREC-8 question answering track report. In: The 8th Text REtrieval Conference (TREC 8), pp. 77–82 (1999)
13. Xu, S., Bao, S., Fei, B., Su, Z., Yu, Y.: Exploring folksonomy for personalized search. In: Proc. of SIGIR 2008, pp. 155–162. ACM Press, New York (2008)

Tripartite Hidden Topic Models for Personalised Tag Suggestion

Morgan Harvey[1], Mark Baillie[1], Ian Ruthven[1], and Mark Carman[2]

[1] University of Strathclyde, CIS Department, Glasgow, UK
{morgan,mb,ir}@cis.strath.ac.uk
[2] University of Lugano, Faculty of Informatics, Lugano, Switzerland
mark.carman@usi.ch

Abstract. Social tagging systems provide methods for users to categorise resources using their own choice of keywords (or "tags") without being bound to a restrictive set of predefined terms. Such systems typically provide simple tag recommendations to increase the number of tags assigned to resources. In this paper we extend the latent Dirichlet allocation topic model to include user data and use the estimated probability distributions in order to provide personalised tag suggestions to users. We describe the resulting tripartite topic model in detail and show how it can be utilised to make personalised tag suggestions. Then, using data from a large-scale, real life tagging system, test our system against several baseline methods. Our experiments show a statistically significant increase in performance of our model over all key metrics, indicating that the model could be successfully used to provide further social tagging tools such as resource suggestion and collaborative filtering.

1 Introduction

Social tagging systems provide a new way for Internet users to organise and share their own digital content and content from other users. Users are able to annotate each resource with any number of free-form tags of their own choosing without having to adhere to an *a-priori* set of keywords. The result of which is a personalised categorisation system defined by its users that can assist in locating resources in the future. This freedom to categorise resources in any way a user chooses is seen as an important advantage for such systems, tags become more personally meaningful and the initial categorisation process is made easier.

Unfortunately this ease of use and freedom of word choice comes at a significant cost. If each user is free to choose whatever tags she wishes then it is unlikely that other users will choose exactly the same tags to describe the same resource or indeed to tag similar resources they have found. Many studies have shown that obtaining high consistency among different taggers is very difficult to achieve and can be affected by many factors including vocabulary use, personal understanding of the resource and language [14, 6]. These factors result in the categorisation scheme displaying a number of highly undesirable characteristics

C. Gurrin et al. (Eds.): ECIR 2010, LNCS 5993, pp. 432–443, 2010.
© Springer-Verlag Berlin Heidelberg 2010

such as polysemous and synonymous terms which make searching or browsing through the collection difficult and inaccurate.

This lack of a consistent and shared vocabulary also results in a large number of unique or "singleton" tags appearing in the folksonomy. Sigurbjörnsson and van Zwol investigated [10] the characteristics of a large sample of the Flickr database (which can be taken as a good reference point for most large-scale tagging systems) and found that the tag frequency closely follows a Zipfian distribution. This is where a small number of tags are used very frequently with tag use quickly tailing off leaving the so called "long tail" of infrequently used tags. Generally speaking, the tags at the extreme ends of the distribution are not particularly useful; the high-frequency tags are too generic and the singleton tags tend to be either compound phrases or misspellings and are likely to only be useful in very specific cases. The distribution of tags per resource was also found to follow a power law with a small number of resources being very thoroughly annotated and a large majority (64%) having only 1, 2 or 3 tags.

To assist the user when tagging new resources, most of these systems offer some form of tag recommendation to increase the chance that a given resource is tagged and also to increase the average number of tags assigned to each resource in the system. Despite their clear utility in improving social tagging system the literature on - particularly personalised - tag recommendation is still quite sparse. Existing approaches tend be based on a mixture of the most popular tags and tags which the user has used previously. Recently more sophisticated systems have been proposed, focussing on methods derived from collaborative filtering and simple co-occurence data or making use of information other than the tags provided by users (for example the HTML content of web pages) [9, 2].

In this paper we model the complete tripartite structure of a folksonomy by extending the latent Dirichlet allocation topic model and use this to provide personalised tag suggestions. Previous work by Wu et. al. [13] modelled broad folksonomic data using a simple Separable Mixture Model representation which reportedly worked well, however it makes the assumption that the probabilities of a user, a tag and a resource are all independent given a dimension d_α. It is also not an entirely generative model and does not make use of a Bayesian hierarchical structure when inferring parameters, meaning that it could easily suffer from problems of over-fitting. A similar model proposed by Plangprasopchok and Lerman [8] was used to recommend resources to tagging system users. Our model improves on these by taking a fully Bayesian view of the problem at hand, thus providing a more statistically principled and scalable solution.

We go on to propose how the model can be used to improve on the performance of existing tag suggestion algorithms and describe appropriate experiments to test our hypothesis. We evaluate the precision and success of our models based purely on actual data from a live system, rather than via a user study. We then present and describe in detail the results from the experiments and discuss various advantages of our complete tagging model approach over function-specific algorithms such as those previously proposed. Finally we conclude with discussion of the results and explore directions for future work based on this model.

2   Hidden Topic Models and Modelling Folksonomies

Topic models attempt to probabilistically uncover the underlying semantic structure of a collection of resources based on analysis of only the vocabulary words present in each resource, this latent structure is modelled over a number of topics which are assumed to be present in the collection.

2.1   Latent Dirichlet Allocation

Latent Dirichlet allocation (LDA) [1] is a generative topic model which has attracted a lot of interest from both the machine learning and language processing community. LDA represents documents as random mixtures over latent topics which are random mixtures over observed words in the vocabulary. The model possesses a number of advantageous attributes; it is fully generative meaning that it is easy to make inferences on new documents or terms and overcomes the overfitting problem present in models such as Probabilistic Latent Semantic Indexing (pLSI) [5]. Also since in LDA each document is a mixture over latent topics it is far more flexible than models that assume each document is only drawn from a single topic.

In LDA each individual word token w_n in the corpus $\mathbf{w}$ is assumed to have been derived from a single latent topic z_n, drawn from a distribution over topics for its parent document d_n. The probability of a word w in the vocabulary given a topic t is denoted by $\phi_{w|t} = P(w_n = w|z_n = t)$ and the probability of a topic given a document is denoted $\theta_{t|d} = P(z_n = t|d_n = d)$. Thus probability of a corpus $\mathbf{w}$ given (the matrices of) all term and topic probabilities Φ and Θ is:

$$P(\mathbf{w}|\Phi, \Theta) = \prod_{n=1}^{N} \sum_{t=1}^{T} \phi_{w_n|t}\theta_{t|d_n}$$

where N is the length (in words) of the corpus $\mathbf{w}$ and T is the number of latent topics. In order to make the model fully Bayesian, symmetric Dirichlet priors with hyperparameters α and β are placed over the distributions θ_d and ϕ_t.

Exact inference of the LDA model is intractable, however a number of methods of approximating the posterior distribution have been proposed including mean field variational inference [1] and Gibbs sampling [3]. Gibbs sampling is a Markov chain Monte Carlo method where a Markov chain is constructed that slowly converges to the target distribution of interest over a number of iterations. Each state of the Markov chain is (in this case) an assignment of a discrete topic (from 1 to T) to each z_n, i.e. to each observed word in the corpus. In Gibbs sampling the next state in the chain is reached by sampling all variables from their distribution when conditioned on the current values of all the other variables. Therefore for LDA each Gibbs sample is obtained by the following:

$$P(z_n = t|\mathbf{z}_{-n}, \mathbf{w}) \propto \frac{N_{-n,t}^{(w_n)} + \beta}{N_{-n,t}^{(\cdot)} + V\beta} \frac{N_{-n,t}^{(d_n)} + \alpha}{N_{-n}^{(d_n)} + T\alpha}$$

where $\mathbf{z}_{-n}$ denotes the assignment of topics to all word positions (except the current topic z_n). V is the vocabulary size. $N_{-n,t}^{(w_n)}$ is the number of times word w_n is assigned to topic t and $N_{-n,t}^{(\cdot)}$ is the total number of words assigned to topic t (both excluding z_n). $N_{-n,t}^{(d_n)}$ is the number of times topic t occurs in document d_n (excluding z_n) and $N_{-n}^{(d_n)}$ is the total number of words in document d_n (less 1). After the sampling algorithm has been run over each word position in the corpus an appropriate number of times (i.e. until the chain has converged to a stationary distribution) we sample from the distribution to obtain estimates for our parameters Φ and Θ via the following equations:

$$\phi_{w|t} = \frac{N_t^{(w)} + \beta}{N_t^{(\cdot)} + V\beta}$$

$$\theta_{t|d} = \frac{N_t^{(d)} + \alpha}{N^{(d)} + T\alpha}$$

The priors α and β essentially act as a pseudo count indicating a relation to smoothing in language models. This allows the model to fall back on the priors in the event of sparse data. We can now use our estimated parameters Φ and Θ to compute a variety of useful distributions such as which documents are similar to each other, which words are similar to each other and by sampling over new data we can easily incorporate new documents into our model without having to re-run the entire algorithm.

This model is not, however, particularly suited to tagging data as it is missing some potentially useful information: the identity of the user who made each annotation. We therefore present a new model, influenced by LDA, which will include this useful information therefore improving the accuracy of tag predictions and allowing for other useful tasks such as determining which users are similar and providing personalised search.

2.2   Tripartite Topic Model (TTM)

Annotations in social tagging systems typically consist of 3 parts: the resource being tagged, the user who tagged the resource and the tag itself. In [7] this is modelled as a tripartite graph with 3 disjoint sets of nodes: resources $\mathcal{D} = \{d_1, \ldots, d_D\}$, users $\mathcal{U} = \{u_1, \ldots, u_U\}$ and tags $\mathcal{W} = \{w_1, \ldots, w_V\}$ with the edges between these nodes representing the individual annotations. Each assignment of a tag to a resource by a user is denoted as the relation $\mathcal{Y}$ and is typically called a tag assignment. Therefore the complete folksonomy is a quadruple $\mathcal{F} := (\mathcal{U}, \mathcal{W}, \mathcal{D}, \mathcal{Y})$. The resources are typically identifiers linking each unique resource id to a single web resource such as an image - as on Flickr - or a bookmark - as on social bookmarking sites such as del.icio.us.

In [13] it is noted that tags are usually semantically related if they are used to describe the same resources many times. Correspondingly, resources are similar if they are annotated with the same tags and users share similar interests if their annotations share many related tags. These relationships can be mapped onto a conceptual space of T dimensions (or in the topic modelling case, topics), that represent categories of knowledge, where each entity's component on a given dimension measures how similar it is to that category. This provides a framework for discovery of meaningful relationships between entities. In order to include information about which user was responsible for each annotation (word position in the corpus) we change the $\theta_{t|d}$ distribution to be the probability of a topic t given a resource d and a user $u \in \mathcal{U}$, denoted $\theta_{t|d,u}$. The matrix Θ becomes a tensor and the probability of the corpus $\mathbf{w}$ is simply:

$$P(\mathbf{w}|\Phi,\Theta) = \prod_{n=1}^{N} \sum_{t=1}^{T} \phi_{w_n|t}\theta_{t|d_n,u_n}$$

where u_n is the user who submitted the tag at position n in the corpus $\mathbf{w}$ and d_n was the resource being tagged.

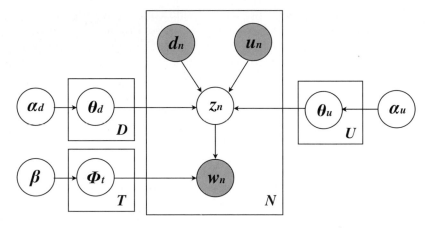

Fig. 1. Plate model graphical representation of Tripartite Topic Model. T is the number of latent topics; D the number of resources and N is the number of word tokens and U is the number of users.

The representation of users and resources over topics can be seen as a large, extremely sparse, 3D tensor $\in \mathbb{N}^{D \times \mathcal{U} \times T}$. Due to the size and sparsity of this tensor it would take an enormous amount of time to fully sample a conditional distribution and therefore we have two options. We can either sample from the distribution via a method such as Gibbs sampling or we can split the tensor Θ into two matrices Θ_d and Θ_u and make the Naive Bayes assumption that the distributions of documents and users are conditionally independent given the topic assignments $\mathbf{z}$. That is for each position in the corpus the probability of a topic given the resource the tag is assigned to is independent of the probability of the topic given the user who assigned the tag. Therefore the probability of a given topic assignment given θ_d and θ_u is:

$$p(z|\theta_d, \theta_u) = \frac{p(z)p(\theta_d, \theta_u|z)}{p(\theta_d, \theta_u)} = \frac{p(z)p(\theta_d|z)p(\theta_u|z)}{p(\theta_d, \theta_u)}$$

$$= \frac{p(z)[\frac{p(\theta_d)p(z|\theta_d)}{p(z)}][\frac{p(\theta_u)p(z|\theta_u)}{p(z)}]}{p(\theta_d, \theta_u)} \propto \frac{p(z|\theta_d)p(z|\theta_u)}{p(z)}$$

We place a Dirichlet distribution as a prior on θ_u, so now we have hyperparameters β, α_d and α_u on ϕ_t, θ_d and θ_u respectively. Figure 1 shows a Bayesian network plate diagram of the complete model.[1] Now that we have an appropriate representation for the probability of a topic given a user and the probability of a topic given a resource a Gibbs sample can be obtained as follows:

$$P(z_n = t | \mathbf{z}_{-n}, \mathbf{w}) \propto \frac{N^{(w_n)}_{-n,t} + \beta}{N^{(\cdot)}_{-n,t} + V\beta} \left(\frac{N^{(d_n)}_{-n,t} + \alpha_d}{N^{(d_n)}_{-n} + T\alpha_d} \frac{N^{(u_n)}_{-n,t} + \alpha_u}{N^{(u_n)}_{-n} + T\alpha_u} \right) \left(\frac{N_{-n,t}}{N_{-n}} \right)^{-1}$$

Where for each tag annotation for each resource u_n is the known user who annotated the resource with that tag. Estimates for the ϕ and θ_d matrices are the same as for LDA and the new θ_u matrix can be calculated from samples of the Gibbs sampling Markov chain in a similar manner. The resulting model of the complete folksonomy can then be used to uncover relationships between users, tags and resources and therefore make useful inferences about new data. In this case we use the model to offer users intelligent tag predictions based on their own prior tagging behaviour as well as the behaviour of the community.

2.3   Suggesting New Tags

Given some initial tags provided by the user for a given resource and the output from our topic model, we want to predict which tags the user will enter next and offer them as suggestions. To do this we can estimate a distribution over the latent topics for the pseudo-document q comprising the tags supplied by the user. This can be calculated as a point estimate:

$$P(t|q) = \frac{N^{(q)}_t + P(t)\alpha_d}{N^{(q)} + \alpha_d}$$

To calculate a value for the topic distribution $P(t|q)$, we need to estimate a value for $N^{(q)}_t$, the count for the topic t in the pseudo-document q. We can calculate the expected value for $N^{(q)}_t$ by summing over all terms in the query as follows:

$$E[N^{(q)}_t] = \sum_{w \in q} P(t|w)N^{(q)}_w$$

---

[1] We note that the conditional independence assumption between d_n and u_n given z_n implies that one of the arrows joining the three variables should be reversed. We have, however, left both arrows pointing to z_n in order to simplify the description of the generative process.

Where $P(t|w)$ can be calculated using the ϕ_t distribution from our model via Bayes' rule, and $N_w^{(q)}$ is the number of times tag w appears in the query q. Now that we have a distribution for the pseudo-document over the latent topics we can estimate the probability of observing a new term:

$$P(w|q) = \sum_{t}^{T} P(w|t)P(t|q)$$

This gives us the probability of a term in the corpus given the pseudo-document, so if we calculate this for $\forall w \in \mathcal{W}$ and then order these by probability in descending order we can choose the top n terms in this ranked list as tag suggestions. In our tripartite model we also want to include the user's personal preferences in these suggestions. Based on the matrix Θ_u from our model, the personalised distribution over terms can be calculated thus:

$$P(w|q, u) = \sum_{t}^{T} P(w|t) \frac{P(t|q)\theta_{t|u}}{P(t)}$$

Where u is the user who generated the tags for the pseudo-document. This final distribution over terms indicates each term's probability given the previously observed terms (the terms of the psuedo-document) and the topical interests of the user. These probabilities can then be used as a multiplier on traditional tag suggestion methods (such as the one outlined in the description of baseline method 3 below) and provides a smoothed, personalised weighting for each term.

3   Experimental Set-Up

In our experiment we compare our method and LDA with 3 "baseline" methods through empirical evaluation based on held-out data from a real-life data set obtained from a large online social tagging system. In this section we outline our experimental set-up in more detail, explain the various methods for tag suggestion and briefly describe the data set and the settings of parameters for our tripartite model.

3.1   Evaluation Method

In order to evaluate the accuracy of the tags suggested by our model we need some form of relevancy judgement, for example a list of all accurate and useful tags for each resource. One method for doing this which has been utilised previously is a user study where users are asked if they think that tags suggested for each resource are relevant or not. We have chosen not to use this method as we are interested in personalised results, therefore only the user(s) who originally tagged the resource can really say whether a tag is relevant or not. In this case a user study would likely provide an over-estimate of the quality of the results and therefore we choose to evaluate our system based only on the tags provided by the user on the live system. Given a set of l tags for a given resource we choose m tags as input for our suggestion algorithm and use the remaining l-m as the set of relevant suggestions. These resource are chosen from a set of held-out resources (i.e. resources that have not been used to train the model) and will give an estimate of the quality of the suggested tags which we believe will more accurately reflect the performance of a live system.

Since we are interested in the ability of our system to return a good ranked list of suggested items we use the following evaluation metrics:

P@k - "precision at rank k" the ratio of suggested tags that are relevant, averaged over test resources. We report P@k for k=1, k=5 and k=20.

S@k - "success at rank k" the ratio of times where there was at least 1 relevant tag in the first k returned. We report S@k for k=1, k=5 and k=20. S@1 and P@1 are the same and are therefore not reported separately.

MRR = "mean reciprocal rank" the multiplicative inverse of the rank of the first relevant suggested tag, averaged over test resources.

Note that we have chosen to evaluate the precision and success metrics for k values up to 20 as this is the number of tags usually suggested on social tagging web sites. k values of 1 and 5 are the most commonly reported in other literature and people tend to pay more attention to the first few results in a ranked list. When training the systems we hold out 20% of resources chosen at random and then "bin" these resources into two sets; one (hereafter referred to as set1) containing documents with between 4 and 8 annotations and the other (set2) with 9 or more annotations.

3.2   Baseline Systems

So that we compare the results of our algorithms to the algorithms already used in social tagging systems we run the above tests on 3 "baseline" methods, LDA as well as on our Tripartite Topic Model (TTM). The first 2 of these methods simulate the tags that would be suggested on sites such as Flickr and Delicious and the final baseline method represents a slightly more sophisticated algorithm that has been proposed in previous literature [10, 9].

TopSys the simplest set of suggestions; the top k tags in the system by frequency of use.

TopUser the most frequently used tags by the user who tagged the resource, if more than 1 user has tagged the resource the union of all users' tags is used.

CoTag tag co-occurence using asymmetric normalisation, as used in previous research to find like terms in a folksonomy [9]. Where $Sim(i,j) = \frac{|i \cap j|}{|i|}$.

3.3   Data Set

We conducted our tests on data provided by Bibsonomy[1] - a social bookmark and publication sharing system and a good example of a large, broad folksonomy

---

[1] Knowledge and Data Engineering Group, University of Kassel: Benchmark Folksonomy Data from BibSonomy, version of June 30th, 2007. http://bibsonomy.org/

and as such is ideal for our research aims. The Bibsonomy data set shares similar characteristics with other large folksonomic data sets noted in previous research, most notably the tag use frequency follows a power law, as does the number of annotations per resource.

To filter out noise and to provide useful data for our evaluation methods we discard any resources that have less than 4 annotations and remove any tags that are used to annotate less than 5 resources. This results in a data set of 36167 resources from 992 users with a total vocabulary of 5116 terms, 28143 (77.8%) of the resources fit into set1, the remaining 8024 (22.2%) fit into set2. To select test data we use stratified random sampling resulting in a total of 7235 (20%) held-out resources with 5630 (79.4%) from set1 and 1605 (20.6%) from set2. In order to ensure that the results returned are not simply due to the held-out resources chosen we perform all tests over 10 different folds. The unfiltered data set displays similar characteristics to those of other folksonomies analysed in related literature; the mean number of tags per resource is 3.27 (median 2), 68.6% of all resources have less than 3 tags. Our filtered data therefore represents only 31.4% of the total Bibsonomy data set, highlighting the wide applicability of a good tag suggestion system.

3.4   Choosing T and Determining Convergence

One very influential parameter that must be set in any latent topic model is the value of T; the number of latent topics in the model. While there has been some work published on algorithms which attempt to estimate this value automatically using Dirichlet Processes [12], it is generally acceptable to use empirical methods to determine the optimal value to use. In our analysis we run our tag suggestion algorithm and compare the precision and success values for each model, we are looking for vales of T where the delta improvement in the precision is small as this is where the optimal value of T lies. When run on the Bibsonomy data set we find a correlation between both metrics for the values over T, with both indicating that around 200 latent topics provides the most optimal fit for the data using our model. We therefore use this parameter setting for our experiments.

We use the Rao-Blackwellised Gibbs sampling method proposed by Styvers and Griffths [4, 3]. It is important when using methods such as Gibbs sampling to estimate a posterior distribution that the Markov chain is given enough time to "burn-in", i.e. when it begins to approach a stationary distribution [11]. To determine when our chain is beginning to approach a stationary distribution we can calculate the log likelihood of the model given the currently sampled estimate every n iterations. If the chain is converging correctly we should find that these values should initially decrease quite rapidly then as the chain approaches convergence the change (delta) in log likelihoods should become smaller until the deltas become negligible. For all of our topic model estimations we discarded the first 500 iterations of the chain and then averaged over samples for every 25 iterations of the chain thereafter until reaching 2000 iterations.

4   Experimental Results

In this section we present the results from the series of experiments described in the previous section. First we look at overall performance of the 5 tag suggestion methods for a "typical" scenario of a user providing 2 tags for our methods to base their suggestions on, we highlight the difference in performance over the two resource "sets" and comment on how this is likely to relate to real performance. We then look at how varying the number of user tags provided affects the quality of tags suggested by our tripartite model.

4.1   Tag Suggestion Performance

The results of the tag suggestion tests using the 5 different methods on resources from set1 (sparsely annotated resources) are presented in Table 1. The results for TopSys over all metrics are extremely poor (as expected), the results for TopUser are slightly better but still well below those returned by the other, more sophisticated methods. Statistically significant improvements of the TTM method over both CoTag and basic LDA are observed for all metrics. These results show that our tripartite model is able to fit the available training data better than the other methods and therefore provides more useful and accurate suggestions. The larger improvements in precision and MRR indicate that the TTM method is suggesting fewer incorrect tags and is returning more relevant tags at a higher rank than the other methods.

Table 1. Results for sparsely annotated resources. P@20 has been excluded from this table as it is not relevant for resources from set1. Last column shows the percentage improvement of TTM over CoTag, * indicates a statistically significant result, 2-sample t at 95% confidence.

	TopSys	TopUser	CoTag	LDA	TTM	% change
S@1	0.0490	0.2269	0.3449	0.3197	0.3736	+8.32*
S@5	0.1540	0.4495	0.5648	0.5494	0.6270	+11.01*
P@5	0.0353	0.1329	0.1786	0.1705	0.2029	+13.61*
S@20	0.3552	0.6853	0.7637	0.7583	0.8238	+7.87*
MRR	0.1023	0.2718	0.3608	0.3574	0.4056	+12.42*

The results from resources from set2 (densely annotated resources), presented in Table 2, show that while the TTM still outperforms other methods over all metrics, the improvements are smaller. In this case the difference in performance between CoTag and TTM is statistically significant for all metrics except for S@20. In keeping with the results from set1 the greatest improvements are in precision and MRR, however all improvements over LDA and CoTag are smaller with the success metric being fairly similar for all 3 methods. This is likely because the small number of resources where the systems are unsuccessful are annotated with terms that have either not been used together before or do not

Table 2. Results for densely annotated resources. Last column shows the percentage improvement of TTM over CoTag, * indicates a statistically significant result, 2-sample t at 95% confidence.

	TopSys	TopUser	CoTag	LDA	TTM	% change
S@1	0.1576	0.3499	0.6312	0.5879	0.6437	+1.98*
S@5	0.3829	0.5882	0.7811	0.7693	0.8132	+4.11*
P@5	0.1258	0.2436	0.4007	0.3796	0.4236	+5.71*
S@20	0.6593	0.8246	0.9376	0.9329	0.9516	+1.49
P@20	0.0749	0.1391	0.2022	0.1972	0.2181	+7.86*
MRR	0.2244	0.2788	0.3857	0.3890	0.4125	+6.95*

exist at all in the training set. In this case the scope for performance improvement over the CoTag method is very small.

4.2   Varying the Number of Input Tags

Selected results from CoTag and TTM are presented in Table 3 for varying numbers of input tags. They indicate that the performance of CoTag at 2 and 3 input tags is significantly better than with 1, however there is little difference between the performance with 2 or 3 tags. TTM performs well when only given a single input tag to infer suggestions from and its performance in terms of precision and MRR increases as the number of input tags increases. Success@k metrics are not significantly different over varying numbers of input tags.

Table 3. Results from densely annotated resources from fold 10 for varying number of input tags. Number of input tags in square brackets.

	CoTag[1]	CoTag[2]	CoTag[3]	TTM[1]	TTM[2]	TTM[3]
S@1	0.6058	0.6398	0.6186	0.6464	0.6594	0.6492
S@20	0.8986	0.9322	0.9388	0.9366	0.9522	0.9520
P@20	0.1936	0.2022	0.1948	0.2214	0.2245	0.2302
MRR	0.3494	0.4032	0.4061	0.3966	0.4172	0.4243

5   Conclusions and Future Work

In this paper we have proposed a new probabilistic latent topic model to deal with data from social tagging systems. The model allows us to estimate topic distributions over users and documents and term distributions over topics. The model is applied to data from the broad-folksonomy social tagging system Bibsonomy and is used to suggest new tags to users based on a small number of tags that they have entered as well as their past annotations. We have shown that this model suggests more relevant tags than current systems by comparing these to held-out tags from annotated resources. In terms of precision, the use of our model improves upon the suggestions provided by the CoTag method on sparsely annotated resources by between 7.87 and 13.6%, improves upon basic LDA by 11.4 to 19.1% and vastly outperforms the more common TopSys and TopUser methods. The results are particularly promising for sparsely annotated resources which are extremely common in tagging systems, indicating that the tripartite model's suggestions would work well in a live system.

TTM provides a complete model of the data collected from a folksonomy and therefore could easily be utilised in future work for other useful estimations and is not merely suited to tag suggestion. For example the model could be used to find similar user groups by clustering based on values from the user-topic matrix. The tag suggestion algorithm could be adapted to suggest new resources rather than tags and therefore provide a form of personalised collaborative filtering over resources in the tagging system. Since topic models do not require explicit co-occurence between terms (tags) in order for them to share semantic similarity, our model could be utilised to improve searching in folksonomic systems, which at the current time are heavily restricted by the vocabulary problem. Further research into these possibilities is left as future work.

References

[1] Blei, D., Ng, A., Jordan, M.: Latent dirichlet allocation. Journal of Machine Learning Research (3), 993–1022 (2003)
[2] Garg, N., Weber, I.: Personalized tag suggestion for flickr. In: WWW (2008)
[3] Griffiths, T., Steyvers, M.: Finding scientific topics. PNAS (2004)
[4] Heinrich, G.: Parameter estimation for text analysis. Technical report, Fraunhofer IGD (2008)
[5] Hofmann, T.: Unsupervised learning by probabilistic latent semantic analysis. Machine Learning 42(1/2), 177–196 (2001)
[6] Hooper, R.S.: Indexer consistency tests—origin, measurements, results and utilization. Technical report, IBM, Bethesda (1965)
[7] Hotho, A., Jaschke, R., Schmitz, C., Stumme, G.: Information retrieval in folksonomies: Search and ranking. In: Sure, Y., Domingue, J. (eds.) ESWC 2006. LNCS, vol. 4011, pp. 411–426. Springer, Heidelberg (2006)
[8] Plangprasopchok, A., Lerman, K.: Exploiting social annotation for automatic resource discovery. In: AAAI 2007 (2007)
[9] Schmitz, P.: Inducing ontology from flickr tags. In: WWW (2006)
[10] Sigurbjörnsson, B., van Zwol, R.: Flickr tag recommendation based on collective knowledge. In: WWW (2008)
[11] Smith, A.F.M., Roberts, G.O.: Bayesian computation via the gibbs sampler and related markov chain monte-carlo methods (with discussion). Journal of the Royal Statistical Society 55, 3–23 (1993)
[12] Teh, Y.W., Jordan, M.I., Beal, M.J., Blei, D.M.: Hierarchical dirichlet processes. JASA 101(476), 1566–1581 (2006)
[13] Wu, X., Zhang, L., Yu, Y.: Exploring social annotations of the semantic web. In: WWW (2006)
[14] Zunde, P., Dexter, M.E.: Indexing consistency and quality. American Documentation 20(3), 259–267 (1969)

Extracting Multilingual Topics
from Unaligned Comparable Corpora

Jagadeesh Jagarlamudi and Hal Daumé III

School of Computing, University of Utah
{jags,hal}@cs.utah.edu

Abstract. Topic models have been studied extensively in the context of monolingual corpora. Though there are some attempts to mine topical structure from cross-lingual corpora, they require clues about document alignments. In this paper we present a generative model called JointLDA which uses a bilingual dictionary to mine multilingual topics from an unaligned corpus. Experiments conducted on different data sets confirm our conjecture that jointly modeling the cross-lingual corpora offers several advantages compared to individual monolingual models. Since the JointLDA model merges related topics in different languages into a single multilingual topic: a) it can fit the data with relatively fewer topics. b) it has the ability to predict related words from a language different than that of the given document. In fact it has better predictive power compared to the bag-of-word based translation model leaving the possibility for JointLDA to be preferred over bag-of-word model for Cross-Lingual IR applications. We also found that the monolingual models learnt while optimizing the cross-lingual copora are more effective than the corresponding LDA models.

1 Introduction

With the increasing amount of text published in varied languages, comparable corpora - documents written in different languages but talking about same topics - are increasingly available. This situation raises the need for novel ways of organizing a multilingual corpus based on common topics/events, which could potentially be useful for many cross-lingual applications like Cross-Lingual Information Retrieval (CLIR) [1] and Cross-Lingual Text Classification [2]. Though there have been many attempts to mine the topical structure from a document corpus [3,4,5] most of these approaches operate in a monolingual scenario.

Topic models like LDA [6] use co-occurrence information to group similar words into a single topic. In case of cross-lingual corpus, two related words in different languages (like English and Spanish) will rarely co-occur in a monolingual document and hence these models fail to group such pair of words into a single topic. As an illustration, we picked a sample of the Europarl [7] English (176777 tokens) and Spanish (227487 tokens) parallel corpus and ran LDA[1] [8]

---

[1] We used collapsed Gibbs sampler for inference.

C. Gurrin et al. (Eds.): ECIR 2010, LNCS 5993, pp. 444–456, 2010.

Table 1. Few topics that were identified by LDA on Europarl parallel corpus. The language of most probable words (E for English and S for Spanish) in each topic is also indicated.

Topic 3 (E)	Topic 16 (S)	Topic 6 (S)	Topic 18 (E)	Topic 10 (S)	Topic 12 (E)
water	directiva	política	european	consejo	council
food	ambiente	europea	union	kosovo	mr
safety	agua	social	europe	europea	european
environmental	medio	desarrollo	states	unión	kosovo
community	enmiendas	unión	president	pregunta	union
environment	aguas	polticas	policy	señoría	question
fisheries	pesca	pases	mr	situación	peace
disaster	propuesta	mujeres	economic	ayuda	government
fishing	principio	trabajo	countries	usted	situation
states	costes	objetivos	political	sr	cyprus

with 20 topics. Not surprisingly we found ten out of the 20 topics have English words as high probable words and the rest of the topics have Spanish words as high probable words. Table 1 shows six of the 20 topics that were identified.

There is a striking similarity between the topics in different languages. For example, pairs of topics {10,12}, {3,16} and {6,18} are essentially same but realized in different languages. This leads to two primary concerns:

1. Because there are different possible realizations of a topic based on language, similar documents in different languages will have different document-topic probability distributions. This makes the task of finding similar documents across languages harder which is inherent in cross-lingual IR applications.
2. If we can generate a multilingual topic by combining two related monolingual topics then it may be possible to achieve same level of modeling capability with fewer topics.

This motivated us to explore techniques to identify multilingual topic-word distributions from an unaligned cross-lingual corpora. The main desirable property of any such approach is to identify topics that distribute their probability mass on related words from different languages. Thus two similar documents, irrespective of their language, will have similar topical distributions. In addressing this task we also explore some interesting questions that arise because of the availability of cross-lingual corpora. For example, [9] shows that bursty patterns can be effectively mined by using cross-lingual documents when compared to mining only from monolingual documents. We would like to see if a similar phenomenon happens in the topic models as well, i.e. "does the availability of related information in different language, i.e. in a completely different style, help in mining any better topical structure?" Another question, related to the ability to compress the data, is "does the additional, but related, data in different language require twice the number of topics to achieve the same level of accuracy (in terms of predictability on an unseen data)?"

There have been some attempts to mine topical structure from cross-lingual corpus, but those approaches assume either explicit or some indirect clues about document alignment. In one of the early approaches for CLIR [10], the authors form an artificial document by concatenating the aligned documents in different languages. A term by document matrix of these new documents is used to learn the lower dimensional representation using Latent Semantic Indexing. Documents across language are compared in this subspace. [9] propose a generative model to mine correlated bursty topic patterns from news articles of different languages. In their approach authors use time index to link documents in different languages. In CorrLDA [11] authors propose an asymmetric model to match words and pictures, even in this model both the image and its corresponding words are generated simultaneously. Recently [12] propose an extension of LDA to mine multilingual topics from Wikipedia articles by forcing aligned articles to share at least one topical distribution. All these approaches critically require alignments at the document level to mine the multilingual topic models and hence can't be applied to a comparable corpora.

In this paper we explore the use of bilingual dictionary to identify the common structure and hence our model *does not* require document alignments. We propose an extension of the LDA model, called JointLDA, which uses bilingual dictionary to generate documents in different languages.

2   Joint Model of Cross-Lingual Corpora

In this section, we describe the details of JointLDA model for cross-lingual corpora. First we propose a model assuming every word is found in the dictionary and then extend it to handle out-of-dictionary words. Neither of these models needs document alignments.

Similar to LDA model [6], a document is assumed to be a mixture over T topics where the mixture weights (θ_d) is drawn from a Dirichlet distribution with symmetric prior (α). But we introduce an additional layer of hidden variables, called *concepts*, in defining topic distributions. Each topic is now a mixture over these concepts rather than words. The topic distribution (ϕ_k) is also drawn from a Dirichlet distribution with a different symmetric prior (β). Finally, a concept can be realized in different ways depending on the choice of the document language (l_d). This additional layer of language independent abstraction over the words allows the model to capture common topics in different languages effectively. In this paper we use bilingual dictionary entries[2] as substitute for these concepts. To understand the process consider generating an English document, first choose a topic mixture say 70% of sports and 30% of entertainment. Now choose a topic for the first word say 'sports' and then choose a concept from the sports topic, let it be 'player:jugador'. Since we are generating an English document we will pick the word 'player' from this concept and discard the Spanish

---

[2] Bilingual dictionary entry (or simply dictionary entry) is used to refer to a pair of words from different language that are possible translations of each other.

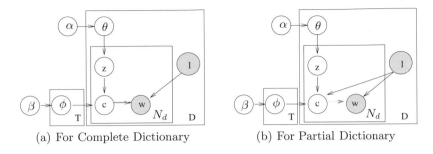

(a) For Complete Dictionary   (b) For Partial Dictionary

Fig. 1. The graphical representation of JointLDA model

word. If we were to generate Spanish document we would pick 'jugador'. This process repeats as many times as the number of words in the document.

Formally the model is described as follows (Fig. 1(a)):

1. For each topic k=1...T, choose $\phi_k \sim \text{Dir}(\beta)$.
2. For each document d, choose $\theta_d \sim \text{Dir}(\alpha)$ and language $l_d \sim \text{Binomial}(\frac{1}{2})$.
   - For each token $i = 1 \cdots N_d$:
     (a) Select a topic $z_i \sim \text{Multinomial}(\theta_d)$.
     (b) Select a concept (dictionary entry) $c_i \sim \text{Multinomial}(\phi_{z_i})$.
     (c) Select a word from $p(w_i|c_i, l_d)$.

Note that given a dictionary entry and language there is only one possibility for a word and hence $p(w_i|c_i, l_d) = 1$. Note that the model doesn't require translation probability for a pair of words[3].

2.1 Handling Out-of-Dictionary Words

Since the coverage of bilingual dictionary is limited, new words will always appear. The model as described above, does not describe the generation of such words. Neglecting these words will leave a major portion of the document unexplained, especially when the dictionary is small. As a result the model will not learn good topic distributions. In order to overcome this, we will handle out-of-dictionary words by adding some artificial dictionary entries to the dictionary. For each of the out-of-dictionary source[4] (target) word we create an artificial dictionary entry of the form w : _NA_ (_NA_ : w). The only difference between an artificial entry and an actual bilingual dictionary entry is that the former is restricted to generate a word in only one language while the latter can generate both source and target language words. Note that if there is any common word between the vocabulary of both these languages that is not found in the dictionary then we create two unrelated artificial entries. In the extreme case

---

[3] Hence techniques like [13] can be used when the dictionary is not available
[4] For clarity, one of the languages is referred as source and the other as target language.

where the dictionary has only artificial entries, the one-to-one relationship between artificial entries and words forces the topic distribution to a distribution over words. In this case JointLDA model reduces to LDA model.

Although artificial entries explain the generation of out-of-dictionary words they lead to deficient topic-word probability distributions. To understand this, consider $p(w|k, l; \theta, \phi)$

$$= \sum_{c \in C} p(w, c|k, l) = \sum_{c \in C} p(w|c, l)p(c|k) = \sum_{c \in C_b \cup C_s \cup C_t} p(w|c, l)p(c|k)$$

where C_b, C_s and C_t are dictionary entries that can generate both language words, only source language and only target language words respectively. Now with out loss of generality fix the language to be source. Then, for any dictionary entry $c \in C_t$ and $\forall w$, $p(w|c, l=\text{src}) = 0$ (because it can not generate a source language word) and hence

$$p(w|k, l=\text{src}) = \sum_{c \in C_b \cup C_s} p(w|c, l_s)p(c|k) \Rightarrow \sum_w p(w|k, l_s) = \sum_{c \in C_b \cup C_s} p(c|k) \leq 1$$

This is because of our assumption that choosing a dictionary entry is independent of the document language, which is a reasonable assumption in the absence of artificial entries. But in the presence of them, while generating a source (target) language word the model should not choose a dictionary entry that can generate only target (source) language word otherwise it fails to generate source (target) language word.

Here we propose a refined model called JointLDA model (Fig. 1(b)) which carefully chooses a dictionary entry based on (document) language.

1. For each topic $k = 1...T$, choose $\phi_k \sim \text{Dir}(\beta)$.
2. For each document d, choose $\theta_d \sim \text{Dir}(\alpha)$ and language $l_d \sim \text{Binomial}(\frac{1}{2})$.

   – For each token $i = 1 \cdots N_d$:

   (a) Select a topic $z_i \sim \text{Multinomial}(\theta_d)$.
   (b) Select a concept (dictionary entry) $c_i \sim \text{Multinomial}(\phi_{z_i}) \cdot \psi(c_i, l_d)$.
   (c) Select a word from $p(w_i|c_i, l_d)$.

Where the function $\psi(c_i, l_d)$ is 1 if the dictionary entry c_i can generate a word from language l_d and 0 otherwise. Note that the effect of language variable in sampling dictionary entry is only to constrain the model to choose a dictionary entry that can generate a given language word. Intuitively, once language variable is observed, this is same as renormalizing the probability mass across a subset of dictionary entries and sampling a dictionary entry from that set.

We use collapsed Gibbs Sampling [8] for estimating the parameters (θ, ϕ). In each iteration the topic and dictionary entry assignments for each token are sampled from the probability distribution given by:

$$p(z_i = k, c_i = j|\mathbf{w}, \mathbf{z}_{-i}, \mathbf{c}_{-i}, \mathbf{l}) \propto \frac{n_{-i,k}^{d_i} + \alpha}{n_{-i,(\cdot)}^{d_i} + T\alpha} \cdot \frac{n_{-i,k}^{j} + \beta}{n_{-i,k}^{(\cdot)} + C\beta} \cdot p(w_i|c = j, l_d)$$

Where $n^j_{-i,k}$ ($n^{(\cdot)}_{-i,k}$) denote the number of times the dictionary entry $c = j$ (any dictionary entry) is used along with topic k for sampling any word excluding the token w_i. Similarly, $n^{d_i}_{-i,k}$ ($n^{d_i}_{-i,(\cdot)}$) is the number of tokens in document d_i that are assigned to topic k (any topic) excluding the token w_i. Note that the above probability is non-zero only for dictionary entries that can generate the word w_i[5] and hence this is a very small subset compared to the total number of dictionary entries. As a result the running time complexity of the joint model is comparable to that of LDA model.

3   Experiments

We ran our model on cross-lingual corpora from two language pairs: English-Spanish (datasets with prefix ENES-) and English-German (prefix ENDE-). We collected two types of data sets for each language pair. The first one is a subset of articles from Europarl corpus (denoted by ENES-P and ENDE-P with 529707 and 386648 tokens respectively). The second one consists of a set of aligned Wikipedia articles in both the pairs of languages (ENES-W and ENDE-W with 282446 and 489840 tokens). Though the first data set is parallel, the Wikipedia articles are related only at the topic level and aligned articles differ in document lengths. The article alignments are used only to facilitate comparison with other models and are hidden to JointLDA model. The dictionaries required for JointLDA are also generated from Europarl corpus using GIZA++ [14]. For language pairs with similar script (like English and Spanish) the common script can be exploited to get initial dictionary [13]. But for generality of our results we ignore this in our experiments. In all our experiments the vocabularies of each language are disjoint, i.e. a common word in different languages is treated differently.

Table 2 shows four out of 20 topical dictionary entries (ranked according to $p(c|k)$ within each topic) that were identified by JointLDA on Wikipedia articles (ENES-W). Since a dictionary entry can generate either of the words depending on language variable, a multilingual topic (as shown in the table) is essentially merged version of two monolingual topics into a single topic. The dictionary entries within each topic are related and as a result a topic-word distribution will have related words from both the languages. The word "speer" in topic 1 occurred in the vocabulary of both the languages and the dictionary doesn't provide any evidence about them being translations. Yet JointLDA model grouped the artificial entries corresponding to these words into the same topic. Also notice that JointLDA is able to group related words in different languages (aramaic & arameo in topic 16 and comunión & communion in topic 17) into a single topic though they are not directly related by any dictionary entry.

---

[5] For this reason, both $\psi(c = j, l_d)$ and $p(w_i|c = j, l_d)$ terms can be omitted during sampling.

Table 2. Few topics that were identified by JointLDA on Wikipedia articles (ENES-W). Entries with _NA_ are artificial entries (Sec. 2.1).

Topic 1	Topic 16	Topic 17	Topic 13
_NA_:speer	arabic:árabe	church:iglesia	aol:aol
hitler:hitler	art:arte	anglican:anglicano	apple:apple
archery:archery_NA_	words:palabras	churches:iglesias	ii:ii
arc:arco	word:palabra	english:inglés	language:lenguaje
attack:ataque	form:forma	ad:ad	assembly:asamblea
speer:_NA_	language:lengua	prayer:oracipn	games:juegos
arrow:flecha	aramaic:_NA_	sick:enfermos	software:software
racing:carreras	arabic:árabes	_NA_:comunión	code:código
german:alemán	dialects:dialectos	communion:_NA_	amway:_NA_
hand:mano	forms:formas	roman:romano	atari:_NA_
target:objetivo	letter:letra	catholic:católica	amd:_NA_
allosaurus:_NA_	_NA_:arameo	regular:regulares	users:usuarios

3.1   Perplexity Evaluation

Perplexity is a standard way to evaluate the predictive power of a generative model on an unseen data. We compare our model with LDA and CorrLDA[11] models in terms of perplexity scores. In each data set 75% of document tokens are randomly chosen for training while the rest of the tokens are used for computing the perplexity. For all the models, Collapsed Gibbs Sampling [8] is used to estimate the parameters on the training data and the parameter estimates for testing are obtained from a single sample of Gibbs iteration. The article alignments in each of the data sets are available only for CorrLDA model and are hidden to JointLDA model.

For JointLDA, the perplexity is given by $\exp(-\frac{1}{N}\sum_{w_i} p(w_i|d_i, l_d))$ where $p(w|d, l_d) = \sum_k p(w|k, l_d)p(k|d)$ and $p(w|k, l_d)$ is the sum of $p(c|k, l_d)$ over all the dictionary entries that can generate the word w. While computing the perplexity values for the LDA, we have used the normal $p(w|d) = \sum_k p(w|k)p(k|d)$ (run labelled as LDA) as well as the probability of test word conditioned on its language: $p(w|l_d, d) = \sum_k p(w|k, l_d)p(k|d)$ where $p(w|k, l_d)$'s are obtained by renormalizing topic word probabilities specific to the given language (LDA_Cond run). The results are shown in Fig. 2, the set of figures in first column report perplexity scores on the Europarl data sets while the second column report the scores on the Wikipedia articles. In all the cases, LDA_Cond model results in a better perplexity scores than the normal LDA model which is intuitive as the uncertainty in the possible words decrease dramatically when language is known.

Figures 2(a), 2(b) show the effect of jointly modeling the cross-lingual corpus versus individual models (with 20 topics). We run JointLDA with different initializations of dictionary: a) for every source language word two target language words are selected at random and are added as translations ('JointLDA_2 Rand') b) with different levels of threshold on the conditional translation

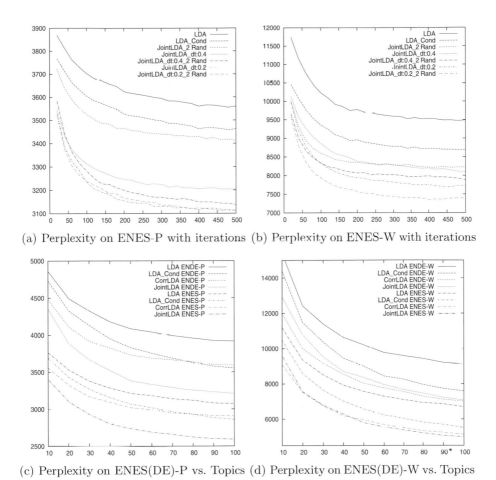

(a) Perplexity on ENES-P with iterations (b) Perplexity on ENES-W with iterations

(c) Perplexity on ENES(DE)-P vs. Topics (d) Perplexity on ENES(DE)-W vs. Topics

Fig. 2. Perplexity scores on both data sets, the first column being Europarl data set and the second column is the Wikipedia articles

probabilities[6] given by GIZA++ ('JointLDA_dt:0.4/0.2'- dictionary threshold of 0.4 and 0.2 respectively) c) combine both the dictionary translations and random translations ('JointLDA_dt:0.4/0.2_2 Rand'). The fact that 'JointLDA_2 Rand' run performed better than the 'LDA_Cond' model indicates that having bilingual information helps. From the rest of the curves (for example, 'JointLDA_2 Rand' vs. 'JointLDA_dt:0.4') it is very evident that the quality of translations does effect and aid the model in identifying better multilingual topics. But, note that there is an increase in performance when the translation probability threshold is decreased from 0.4 to 0.2. This is because of the increased number of bilingual

---

[6] Notice that JointLDA doesn't use translation probability and hence all translations with probability greater than the threshold are treated equally likely.

Table 3. Number of bilingual and total (including artificial) dictionary entries vs. size of of vocabulary

	Bilingual	Total	Vocab Size
ENES-P	16922	32731	38605
ENDE-P	14976	38585	40979
ENES-W	22400	53638	70843
ENDE-W	26515	88854	92086

dictionary entries as the threshold decreased. In general, we observed that as the number of dictionary entries increase, number of free parameters increase and hence model finds a better fit for the document corpus. But, the reader should not attribute the lower perplexity scores of JointLDA (compared to LDA_Cond) to this fact, because in all our data sets we found that the total number of free parameters per topic when the dictionary is loaded with translation threshold of 0.2 (third column of table 3) is less than that of LDA (the vocabulary size – last column of table 3). In rest of the experiments it is assumed that a threshold of 0.2 is used while loading the dictionary unless explicitly mentioned. With a closer look, we found that JointLDA efficiently uses dictionaries in predicting infrequent words and out-of-training words more accurately compared to other models. From figures 2(a), 2(b) it is clear that jointly modeling cross-lingual corpora is better than individually modeling. For brevity we don't include the graphs for English-German data set but they look similar.

Figures 2(c), 2(d) show the ability of the models to fit the data with respect to the number of topics required. When the data is parallel, JointLDA is able to achieve the same modeling capability with nearly half of the number topics as needed by the other models. This is completely justifiable because in any parallel data nearly half of the information is redundant and is simply expressed in different form. If a model can identify this redundancy it needs fewer topics. As the data set becomes comparable (less parallel) it needs more than half of topics, but significantly less than the number of topics required by LDA_Cond. Though CorrLDA performs competitively with JointLDA on Wikipedia data set, it estimates different topic-word distributions for each language and fails to identify the relatedness between topics of different language. It also uses the alignment information between training documents in different languages, which is not required for JointLDA.

One of the hoped advantages of modeling the cross-lingual corpus together is that by using the extra information written in another language, the model will learn better monolingual models. Here we compare the monolingual models learnt by the JointLDA while optimizing the cross-lingual corpus to the monolingual models that LDA learn only on the monolingual data. Fig. 3 shows the perplexity values on monolingual part of each test set (indicated by EN, ES and DE). When the data is parallel JointLDA efficiently uses the cross-lingual corpora to mine better monolingual models and when the data is not parallel (e.g. Wikipedia article) its monolingual models are not as effective.

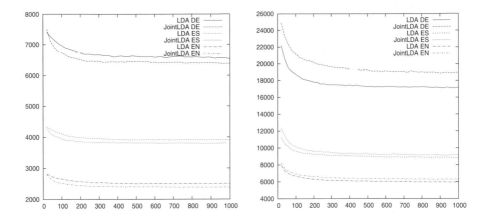

Fig. 3. Comparison of monolingual models learnt by JointLDA vs. the monolingual models of LDA on parallel (left figure) and comparable (right figure) corpora

Table 4. Test set perplexity given an aligned article in different language

	JointLDA	WordTrans
ENES-P	5732.503	3244.35
ENDE-P	4936.483	3771.34
ENES-W	7867.091	11930.3
ENDE-W	12750.12	18078.42

3.2   Perplexity of the Aligned Test Set

The traditional perplexity measures only the ability to predict a test word given a document of same language. Apart from this, a cross-lingual model should also be able to predict related words from different languages. In order to measure this aspect we compute a modified perplexity score using topic distribution of corresponding aligned document. We also report $\exp(-\frac{1}{N}\sum_{w_i} p(w_i|d_i^a, l_{d_i}))$ where d_i^a denote the aligned document (of d_i) in other language. For comparison, we use bag-of-word based translation model (referred as WordTrans) smoothed using appropriate unigram language model [15] which is proved to give good results in CLIR [1]. Under this model:

$$p(w_t|d_s) = (1 - \lambda) \sum_{w_s} p(w_t|w_s)p(w_s|d_s) + \lambda p(w_t|C_t)$$

where $p(w_t|C_t)$ is the unigram probability of the word in the target language corpus. Table 4 shows the perplexity scores of JointLDA (with 100 topics and 1000 iterations) in comparison with WordTrans model. The better performance of WordTrans model on first two data sets is due to the fact that the dictionary is also learnt from Europarl data set. Also note that WordTrans model uses the translation probabilities given by GIZA++, where as JointLDA model does not. But on the Wikipedia articles, JointLDA model achieves lower perplexity scores which indicate better predictability than a bag-of-word translation model. This leaves a possibility for JointLDA to be preferred over bag-of-word translation for applications like CLIR and Cross-lingual Text Categorization [2].

4   Discussion

As discussed in section 2, the JointLDA model is not limited to cross-lingual scenario. We claim that the model is applicable in a wide range of situations where some initial matching is available between the observations. For example, we can apply the JointLDA model to monolingual data by using synonyms (extracted from WordNet) as concepts. The generative story for the document corpus remains same and the probability of a word is given by:

$$p(w|d; \theta, \phi) = \sum_k p(k|d)p(w|k, d) = \sum_{k,c} p(k|d)p(c|k)p(w|c)$$

But, unlike cross-lingual situation, a synonym can generate both words so the parameters $p(w|c)$'s also need to be estimated during the inference process. When we tested this model on the English corpus of Wikipedia articles we found that JointLDA not only achieves lower perplexity scores (compared to LDA) on the whole test set but it also models infrequent words very well, which are typically excluded during the preprocessing stage of topic modeling algorithms.

Another line of approach to mine multilingual topics would be to use LDA to find monolingual topics in one language and use the dictionary to translate the topics into the other language. The disadvantage of this strategy is its inherent bias towards one language. It forces the topics in second language to be consistent with the identified topics in first language rather than letting them to evolve from the data. Comparison with WordTrans model in Sec. 3.2 confirms that, such a translation of topics would fail to predict unseen data when the data becomes less parallel.

Recently [16] has proposed MuTo model to extract multilingual topics from cross-lingual corpora. At any stage MuTo considers a matching between vocabularies of both languages and hence it doesn't allow any source word to pair up with multiple target language words. This underlies a strong assumption that a word is used in only one sense in the entire corpus. Where as JointLDA model deals with sense ambiguity by allowing a word to be paired with multiple target language words. Another major difference is that, in MuTo all unmatched words come from a single topic distribution. Which implies that when the dictionary size is small MuTo reduces to a simple unigram model while JointLDA reduces to the LDA model. Thus JointLDA can be seen as a generalization of the MuTo model.

5 Conclusion and Future Work

In this paper we have proposed generative model called JointLDA, which can extract multilingual topics from an unaligned cross-lingual corpora. Unlike other models, JointLDA model doesn't require document alignments among training documents for inference. It needs parallel data only to learn dictionaries and these dictionaries can be used again for a different document corpus. In order to facilitate comparison with other models and to compute the perplexity on the aligned test set we used aligned documents. The experiments conducted on different data sets showed that jointly modeling the cross-lingual corpus has several advantages compared to modeling the individual monolingual corpora.

It may appear that the model relies heavily on the availability of dictionary but the topics mined by JointLDA (Table 2) do contain translations that are not part of the initial dictionary. So we believe that it may be possible to start with a small but good quality translations and learn pairs of related words to be added to the dictionary at regular intervals. We leave this for future work.

References

1. Xu, J., Weischedel, R., Nguyen, C.: Evaluating a probabilistic model for cross-lingual information retrieval. In: SIGIR 2001, pp. 105–110. ACM, New York (2001)
2. Bel, N., Koster, C.H.A., Villegas, M.: Cross-lingual text categorization. In: Koch, T., Sølvberg, I.T. (eds.) ECDL 2003. LNCS, vol. 2769, pp. 126–139. Springer, Heidelberg (2003)
3. Blci, D.M., Lafferty, J.D.: A correlated topic model of science. Annals of Applied Statistics, 17 35 (August 2007)
4. Blei, D.M., Lafferty, J.: Topic models. Text Mining: Theory and Applications. Taylor and Francis, Abington (2009)
5. Steyvers, M., Griffiths, T.: Probabilistic topic models. Latent Semantic Analysis: A Road to Meaning (2005)
6. Blei, D.M., Ng, A.Y., Jordan, M.I.: Latent dirichlet allocation. Journal of Maching Learning Research 3, 993–1022 (2003)
7. Koehn, P.: Europarl: A parallel corpus for statistical machine translation. In: MT Summit (2005)
8. Griffiths, T.L., Steyvers, M.: Finding scientific topics. Proceedings of National Academy of Sciences USA 101(suppl. 1), 5228–5235 (2004)
9. Wang, X., Zhai, C., Hu, X., Sproat, R.: Mining correlated bursty topic patterns from coordinated text streams. In: KDD 2007: Proceedings of the 13th ACM SIGKDD, pp. 784–793. ACM, New York (2007)
10. Dumais, S.T., Landauer, T.K., Littman, M.L.: Automatic cross-linguistic information retrieval using latent semantic indexing. In: Working Notes of the Workshop on Cross-Linguistic Information Retrieval, SIGIR, Zurich, Switzerland, pp. 16–23. ACM, New York (1996)
11. Blei, D.M., Jordan, M.I.: Modeling annotated data. In: SIGIR 2003, pp. 127–134. ACM, New York (2003)
12. Ni, X., Sun, J.T., Hu, J., Chen, Z.: Mining multilingual topics from wikipedia. In: 18th International World Wide Web Conference, April 2009, pp. 1155–1155 (2009)

13. Koehn, P., Knight, K.: Learning a translation lexicon from monolingual corpora. In: Proceedings of the ACL 2002 workshop on Unsupervised lexical acquisition, Morristown, NJ, USA, pp. 9–16. Association for Computational Linguistics (2002)
14. Och, F.J., Ney, H.: A systematic comparison of various statistical alignment models. Computational Linguistics 29(1), 19–51 (2003)
15. Zhai, C., Lafferty, J.: A study of smoothing methods for language models applied to ad hoc information retrieval. In: SIGIR 2001, pp. 334–342. ACM Press, New York (2001)
16. Boyd-Graber, J., Blei, D.M.: Multilingual topic models for unaligned text. In: Uncertainty in Artificial Intelligence (2009)

Improving Retrievability of Patents in Prior-Art Search

Shariq Bashir and Andreas Rauber

Institute of Software Technology and Interactive Systems
Vienna University of Technology, Austria
{bashir,rauber}@ifs.tuwien.ac.at
http://www.ifs.tuwien.ac.at

Abstract. Prior-art search is an important task in patent retrieval. The success of this task relies upon the selection of relevant search queries. Typically terms for prior-art queries are extracted from the claim fields of query patents. However, due to the complex technical structure of patents, and presence of terms mismatch and vague terms, selecting relevant terms for queries is a difficult task. During evaluating the patents retrievability coverage of prior-art queries generated from query patents, a large bias toward a subset of the collection is experienced. A large number of patents either have a very low retrievability score or can not be discovered via any query. To increase the retrievability of patents, in this paper we expand prior-art queries generated from query patents using query expansion with pseudo relevance feedback. Missing terms from query patents are discovered from feedback patents, and better patents for relevance feedback are identified using a novel approach for checking their similarity with query patents. We specifically focus on how to automatically select better terms from query patents based on their proximity distribution with prior-art queries that are used as features for computing similarity. Our results show, that the coverage of prior-art queries can be increased significantly by incorporating relevant queries terms using query expansion.

1   Introduction

Patent retrieval falls into the recall-oriented application domain, where not missing a relevant patent is considered more important than retrieving only set of relevant patents at top rank results. This is particularly important in *prior-art search*, where missing one patent could result in a multimillion dollar lawsuit because of a patent infringement. The goal of searching a patent database for prior-art is to find all previously published patents on a given topic [9,10,11]. It is a common task for patent examiners and attorneys to decide, whether a new patent application is novel or contains technical conflicts with some already patented invention. Patent applications have complex structures and technical contents, which can create significant challenges for retrieval systems [7]. The vocabulary of patent applications is quite diverse, which leads to an extremely

C. Gurrin et al. (Eds.): ECIR 2010, LNCS 5993, pp. 457–470, 2010.
© Springer-Verlag Berlin Heidelberg 2010

large dictionary. Writers are suspected to intentionally use many vague terms and expressions in order to avoid narrowing the scope of the invention. Combinations of general terms often have a special meaning that also has to be captured. Patent applications further contain many acronyms and new terminology. Furthermore, in order to pass the patent examination, writers tend to develop their own terminologies, which can cause serious terms mismatch problems [6]. The combination of these factors make prior-art search significantly different with other search tasks, such as web search.

Current prior-art retrieval systems use keyword search, where patent users (e.g. patent examiners or attorneys) extract relevant keywords from query patents, particularly from the claim field, to formulate their queries [9,10,11]. Here, *query patents* are new applications, which are examined for novelty. The success of the search highly depends upon the quality of queries terms selected from query patents. However, due to the above mentioned problems, selecting relevant keywords can be a difficult task. Some documents are retrieved by many queries, whereas others may never show up within the top-c documents retrieved for any reasonable query up to a certain length.

Retrievability measurement [1] is used to analyze the bias of retrieval systems and their capability of potentially retrieving each document in the corpus. Bias of retrieval systems denotes the characteristic of a system to give preference to certain features of documents, when it ranks results of any given query. For example, *PageRank* favors *"popular"* documents by evaluating the number of in-links of web pages in addition to pure content features. Similarly, *TFIDF* and *OKAPI-BM25* favor large terms frequencies and thus longer documents over shorter ones. Given the fact that only a limited number of top-ranked documents can be evaluated for any given query, we may arrive at the situation, that some set of documents cannot be retrieved for any plausible query by e.g., using all possible queries up to a certain length. To measure retrievability, a large set of potential queries (e.g. all combinations of all important keywords up to a pre-specified query length) are passed to a retrieval system, and the number of documents that can and that cannot be retrieved is evaluated. The resulting figure provides an estimate on the amount of bias introduced by a certain retrieval system, indicating its suitability for recall-oriented applications.

With prior-art queries generated only from query patents, our experiments using retrievability measurement indicate a large bias toward a subset of patents in state of the art retrieval systems [2]. A large subset of patents either has very low retrievability scores or could not be accessible via any query. In order to increase the coverage of prior-art queries, in this paper we reformulate (expand) queries using Query Expansion (QE) with Pseudo Relevance Feedback (PRF) [3]. Prior-art queries extracted from query patents may not contain all terms. Therefore, missing terms can be extracted from PRF documents. For selecting better patents for PRF, in this paper we propose a novel approach where relevant patents for PRF are identified based on their similarity with query patents via specific terms. The success of this approach, highly depends upon the selection of those terms from query patents, that produce the best PRF candidates. For example, those terms which appear closely with terms of prior-art queries in the same *claim, paragraph, sentence* or *phrase* can identify better patents for PRF as compared to using all terms of a query patent. This term selection problem can be considered a term classification problem. In this paper, we try to separate positive terms from others - according to their potential impact on the retrieval effectiveness. Finally, in order to evaluate how far this novel prior-art retrieval approach can increase the retrievability of patents in collection, we compare it with state of the art retrieval systems including different QE approaches. Our experiments indicate that patent retrievability can be improved significantly using QE with more sophisticated PRF patents selection.

The remainder of the paper is organized as follows. Section 2 reviews related work in the field of prior-art patents retrieval. In Section 3, we introduce retrievability measurement for recall-oriented applications, which is used as a basis of our experiments. In Section 4, we explain the working of our prior-art retrieval approach, introducing the features used for classification (Section 4.1), learning accuracy (Section 4.2) and expanding queries using language modeling approach (Section 4.3). In Section 5, we evaluate the performance of our retrieval approach with other state of the art retrieval models using collection of patent documents under retrievability measurement.

2   Related Work

Osborn et al. [17] introduce a system that integrates a series of shallow natural language processing techniques into a vector-based document information retrieval system for searching relevant patents. Their methods are mainly based on the patterns of part-of-speech tags, where firstly phrases from the patents are extracted, and then these phrases are used as indexed features. Their methods use all contents of a patent for constructing the query vector, but ignore the structure information in the patent. Larkey [12] uses a probabilistic information retrieval system for searching and classifying US patents. In their approach, instead of searching in full patents, they select certain sections *(patent fields)* and portions of sections for reducing text and selecting dominant terms. Next, weights to different terms in these reduced patents are assigned based on their relative importance to different sections and term frequencies for efficient retrieval. Mase et al. [15] propose a two-stage retrieval strategy for patents search. In stage1, general text analysis and retrieval methods are applied to improve recall; while in stage2, top c patents retrieved from stage1 results are rearranged for improving precision by applying different text analysis and retrieval methods using the claim field.

Fujii [8] applies link analysis techniques to the citation structure for efficient patent retrieval. In their method, they first perform text based retrieval for obtaining *top c* patents, and then citation scores of these *top c* patents are computed based on PageRank and topic-sensitive citation-based methods. Finally, both the text-based and citation-based scores are combined for better ranking of these patents. Custis et al. [6], evaluate query expansion methods for legal domain applications. For this purpose, they systematically introduce query document terms mismatch into a corpus in a controlled manner and then measure the performance of retrieval systems as the degree of terms mismatch changes. A recent approach on patent retrieval considers a novel search scenario, in which users can pose full patents as a query instead of selecting relevant keywords from them in prior-art search [19]. They also explore the effect of different fields of patents as a search feature and further consider how these fields can be combined with learning techniques. A considerable improvement in relevance judgment results is reported using this approach.

3   Retrievability Measurement

"Retrievability" measures [1], how likely each and every document $d \in D$ can be retrieved within the top c ranked results for all queries in Q. More formally, retrievability $r(d)$ of $d \in D$ can be defined as follows.

$$r(d) = \sum_{q \in Q} f(k_{dq}, c) \tag{1}$$

Here, $f(k_{dq}, c)$ is a generalized utility/cost function, where k_{dq} is the rank of d in the result set of query $q \in Q$, c denotes the maximum rank that a user is willing to proceed down the ranked list. The function $f(k_{dq}, c)$ returns a value of 1 if $k_{dq} \leq c$, and 0 otherwise.

Retrievability inequality can be further analyzed using the **Lorenz Curve**. Documents are sorted according to their retrievability score in ascending order, plotting a cumulative score distribution. If the retrievability of documents is distributed equally, then the Lorenz Curve will be linear. The more skewed the curve, the greater the amount of inequality or bias within the retrieval system. The **Gini coefficient** G is used to summarize the amount of bias in the Lorenz Curve, and is computed as follows.

$$G = \frac{\sum_{i=1}^{n}(2 \cdot i - n - 1) \cdot r(d_i)}{(n-1)\sum_{j=1}^{n} r(d_j)} \tag{2}$$

where $n = |D|$ is the number of documents in the collection. If $G = 0$, then no bias is present because all documents are equally retrievable. If $G = 1$, then only one document is retrievable and all other documents have $r(d) = 0$. By comparing the Gini coefficients of different retrieval methods, we can analyze the retrievability bias imposed by the underlying retrieval system on the given document collection.

Retrievability of patents is analyzed on a large collection of queries. Clearly, it is impractical to calculate the absolute $r(d)$ scores because the set of all query terms Q would be extremely large and require a significant amount of computation time as each query would have to be issued against the index for a given retrieval system. In order to perform measurements in a practical way, a subset of all possible queries is commonly used that is sufficiently large and contains relatively probable queries [1]. In query generation, we try to reflect the way patent examiners use for generating prior-art queries from the claim fields of query patents [9,10,11]. We first extract all frequent terms from the patents that have a term frequency greater than a minimum threshold (≥ 3). For generating longer queries, single frequent terms are combined into *two*, *three* and *four* term combinations. For those patents that contain a large number of single frequent terms, the different term combinations become very large. To generate a similar number of queries for every patent, we put an upper bound of *90* queries generated from every patent.

4   Selecting Pseudo-relevance Feedback Documents

In QE with PRF, it is normally assumed that the set of top documents retrieved by user queries is relevant for relevance feedback, and that learning expansion terms from these feedback documents can increase the effectiveness of search [14]. However, from the queries extracted from query patents, our retrievability results show a large bias toward some subset of patents [2,3]. These high retrievable patents can skew the results, and due to this a large subset of patents either could become very low retrievable or could not be retrievable via any query. For selecting relevant patents for PRF, we consider a novel approach, where patents for PRF are identified based on their similarity with query patents over a subset of terms, rather than the overall document similarity. The success of this approach depends on two main factors. Firstly, appropriate terms need to be identified in the query patent via which to retrieve the best-matching documents for PRF that can help in improving retrievability during QE. As experiments below will show, these are terms that co-occur closely with the query terms stemming from the query patent. Secondly, we analyze which fields of a query patent (*title, abstract, description of patent, background summary*, and the *claim* field) should be considered for query expansion.

In a nutshell, the experimental set-up works as follows: Retrievability is analyzed for all patents in the corpus. 500 low-retrievable patents are identified to focus on improving their retrievability via PRF. For these we identify the 35 most similar documents to be used for PRF using the SMART similarity measure [16]. Using Language Modeling (LM) we then expand the queries based on the PRF documents and check whether retrievability of the original query patent has increased. The variable step is the identification of the PRF documents. Here, in each iteration we eliminate one term of the (low-retrievable) query patent, resulting in different documents being ranked high for PRF. Terms that, when removed, lead to a ranking of PRF documents that reduce retrievability by the resulting expanded queries, are obviously helpful in determining better documents for PRF. These terms serve as a training set to automatically learn which terms help in identifying better PRF documents. The characteristics of these positive terms are analyzed and described via a range of features that serve as a basis for automatically identifying them via a machine learning approach. Thus, during deployment, starting from a query patent only such terms are used to identify the best matching documents for PRF.

After applying QE using LM, we furthermore analyze, which sections of a patent these terms come from that are used for expanding queries that lead to higher retrievability. The individual steps are described in more detail below.

4.1   Query Patent Term Selection Features

Following the process outlined above, we obtain a set of terms that, when included in identifying PRF documents, lead to higher or lower retrievability. We now want to describe these terms via a set of features, focussing mostly on their positional relation to the original query terms. Our terms classification feature set thus consists of several measures that capture the proximity of terms in query patent with all terms in the original queries, based on [4,5,18,21]. These features measure the closeness or compactness of the selected QP term with terms in the query. The underlying intuition is that, the more compact the terms are, the more likely it is that they are topically related, and thus higher the possibility that the terms help improve PRF. The feature set consists of 6 features, which are further computed seperately on all 6 individual fields of patents *(Title, Abstract, Claim, Background Summary, Description,* and *wholepatent)* capturing the positional relationship between the terms in the initial query and in the documents, plus a number of other features capturing the importance of candidate terms for query expansion. The total dimensionality of the feature set is 42.

The following sample query patent $\widehat{QP}$ will be used to explain how each of the features explained below can be calculated for a query $\widehat{q}$ containing query terms a and b with query patent term m.

Term Positions =	1	2	3	4	5	6	7	8	9	10	11	12	13	14	15	16
Query Patent $(\widehat{QP})$ =	a	b	e	m	a	b	s	n	x	h	i	a	j	b	k	m
Query $(\widehat{q})$ =		a	b													

(f1): Average Minimum Distance to Single Query Term. This feature is defined as the average of the shortest distance between the occurrences of term t in QP with the terms of the query [21]. This feature rewards terms of a query patent that appear very close to query terms, e.g. in the same *phrase, sentence, paragraph* or *claim*.

Let $q = \{q_1, q_2, ..., q_m\}$ be the set of different query terms in a query q. $O_{q_i} = \{o_{i_1}, o_{i_2}, ..., o_{i_n}\}$ is the set of term occurrence positions of the query term q_i in QP p. $PD(q_i, t; p)$ denotes the distance between query term q_i and term t in p. Following [18], $f1$ is the distance between the closest occurring positions of term q_i and t, and can be measured through their occurring positions in p.

$$f_1(t) = \frac{\sum_{q_i \in q} PD(q_i; t|p)}{|q|} \tag{3}$$

$$PD(q_i; t|p) = min_{o_{i_k} \in O_{q_i}, o_{t_k} \in O_t} \left\{ abs(o_{i_k} - o_{t_k}) \right\} \tag{4}$$

where $|q|$ is the length of query q, and O_{q_i} and O_t are the sets of occurrences positions of terms q_i and t in p. In the example, the minimum value of $f1(m)$ using this feature with query terms a and b is $= ((5 - 4) + (6 - 4))/2 = 1.5$.

(f2): Pair-wise Terms Proximity Based on Minimum Distance. $f1$ captures the average minimum QP term distance with single terms of a query. However, a better feature could be with pairs of query terms [21]. $f2$ considers the minimum distance between the selected QP term and pairs of terms in the query. This feature is calculated as follows.

$$f_2(t) = min_{\hat{p}(q_i,q_j) \in q, t \neq q_i, t \neq q_j, q_i \neq q_j} \{PD2(q_i, q_j; t|p)\} \tag{5}$$

$$PD2(q_i, q_j; t|p) = min_{o_{i_k} \in O_{q_i}, o_{j_k} \in O_{q_j}, o_{t_k} \in O_t} \{PD(q_i; t|p) + PD(q_j; t|p)\} \tag{6}$$

$\hat{p}(q_i, q_j)$ enumerates all possible terms pairs in q. $PD2(q_i, q_j; t|p)$ denotes the pair-wise distance between terms pair q_i and q_j in the query and term t in p. Similar to $f2$, it is the distance between the closest occurring positions of terms q_i, q_j and t. In the example, the minimum value of $f2(m)$ using this feature is $= (5 - 4) + (6 - 4) = 3$.

(f3): Pair-wise Terms Proximity Based on Average Distance. Instead of relying only on the minimum distance, this feature calculates the average distance between pairs of query terms q_i, q_j and term t in p [5]. This feature promotes those terms of a query patent that consistently occur closer to query term pairs in localized areas, e.g. in the same *claim, paragraph, sentence* or *phrase*. Given the set $\hat{p}(q_i, q_j)$ of all possible query terms pairs, the value of this feature can be calculated as follows. In the example, since there is only one query terms pair, therefore the pair-wise average proximity distance of $f3(m)$ using this feature is $= ((5 - 4) + (6 - 4))/1 = 3$.

$$f_3(t) = \frac{\sum_{\hat{p}(q_i,q_j) \in q, t \neq q_i, t \neq q_j, q_i \neq q_j} \{PD2(q_i, q_j; t|p)\}}{|\hat{p}(q_i, q_j)|} \tag{7}$$

(f4): Query Terms Difference Average Position. This feature considers the difference between the average positions of individual query terms with terms t in p. This feature first calculates the average positions of individual terms of query with t using the position vectors, and then these average positions are used for calculating average proximity positions [5]. This feature captures where terms of query and t are occurring together. In the example, the value of $f4(m)$ using this feature is $= abs((1+5+12)/3 - (4 | 16)/2) | abs((2 | 6 | 14)/3 | (4 | 16)/2)/2 = abs(6 - 10) + abs(7.33 - 10)/2 = 3.34$.

$$f_4(t) = \frac{\sum_{q_i \in q} abs\{\sum_{o_{i_k} \in O_{q_i}} \frac{o_{i_k}}{|O_{q_i}|} - \sum_{o_{t_k} \in O_t} \frac{o_{t_k}}{|O_t|}\}}{|q|} \tag{8}$$

(f5): Co-Occurrence with Single Query Term. In learning better expansion terms using classification approach all terms are considered useful for expansion, that co-occur frequently with query terms [4]. f(5) captures this co-occurrence.

$$f_5(t) = log \frac{1}{|q|} \sum_{q_i \in q} \frac{c(q_i, t|p)}{tf(q_i|p)} \tag{9}$$

where $c(q_i, t|p)$ is the frequency of co-occurrences of query term q_i and the term t within text windows of p. $tf(q_i|p)$ denotes the term frequency of q_i in p. The window size is empirically set to 20 terms. In the example, if window size is set to 3 then the number of co-occurrence of term m with query term a is 1 and with query term b is 3. Using equation 9, the value of $f5(m)$ is $= log(1/2 * ((1/3) + (3/3))) = -0.41$.

(f6): Co-occurrence with Pairs Query Terms. The previous feature considers only the co-occurrence of term t with individual terms of query q. This feature captures a stronger co-occurrence relation of term t with pairs of terms of the query [4]. Given the set $\hat{p}(q_i, q_j)$ of all possible pairs of query terms, the value of this feature can be calculated as follows.

$$f_6(t) = log \frac{1}{|\hat{p}(q_i, q_j)|} \sum_{\hat{p}(q_i, q_j) \in q, t \neq q_i, t \neq q_j, q_i \neq q_j} \frac{c(q_i, q_j, t|p)}{tf(q_i|p) + tf(q_j|p)} \quad (10)$$

$c(q_i, q_j, t|p)$ denotes the frequency of co-occurrences of term t with terms pair q_i and q_j of query q, within text windows of p. The window size is empirically set to 20 terms. In the example, if window size is set to 3 then the number of co-occurrence of term m with query term pairs a and b is 1. Using equation 10 the value of $f6(m)$ is $= log(1/2 * (1/6)) = -2.48$.

Other Features. Some other features that we consider for term classification purpose and use in our experiments are: (a) Sum of query terms and t in p, (b) Product of term frequencies of individual query terms and t in p. If the value of this feature is high, the probability of closer occurrence of term t with query terms will be high. (c) $fullcover(q, t, p)$ is the length of the patent segment that covers all occurrences of query terms with term t, (d) idf of term t, (e) $tfidf$ value of term t, and (f) length of patent.

4.2   Terms Classification and PRF Patents Selection

We use neural networks with radial basis function (RBF) to train a model for identifying terms that help in returning documents for PRF that improve the overall retrievability of a given patent. PRF documents are selected calculating the similarity between the query patent (with different terms removed) and the patent corpus using the SMART similarity measure [16], using the top 35 patents for PRF with subsequent query expansion via LM as described below.

The calculation of the majority of features describing the terms is based upon the proximity distribution of query terms and terms in patents. Thus 3000 queries that can retrieve the 500 low-retrievable patents are randomly selected for training the model, and further 800 queries are used for testing the accuracy of learning model. Using the RBF classifier, we obtain a classification accuracy of 72%.

4.3   Expanding Queries

We use *Language Modeling* (LM) [13] to select the most dominant terms from PRF patents for expanding the queries. Under LM each term w is ranked according to the sum of divergences between its prevalence in each relevance feedback patent it occurs and the importance of the term in the whole collection.

$$score(w) = \sum_{d \in K} P(d)P(w|d)P(q|d) \qquad (11)$$

K is the set of PRF patents selected using the approach above. We assume that $P(d)$ is uniform over the set. After this estimation, the most $e = 35$ terms (words) from $P(w|K)$ are chosen for expanding the queries. The values $P(w|d)$ and $P(q|d)$ can be calculated following Equations 12 and 13.

$$P(q|d) = \prod_{i=1}^{m} P(q_i|d) \qquad (12)$$

where q_i is the ith query term, m is the number of terms in a query q, and d is a document model. Dirichlet smoothing [20] is used to estimate non-zero values for terms in the query which are not in a patent document. It is applied to the query likelihood language model as follows.

$$P(w|d) = \frac{|d|}{|d| + \lambda} P_{ML}(w|d) + \frac{\lambda}{|d| + \lambda} P_{ML}(w|D) \qquad (13)$$

$$P_{ML}(w|d) = \frac{freq(w,d)}{|d|}, P_{ML}(w|D) = \frac{freq(w,D)}{|D|} \qquad (14)$$

where $P_{ML}(w|d)$ is the maximum likelihood estimate of a term w in document d, D is the entire collection, and λ is the smoothing parameter [20]. $|d|$ and $|D|$ are the lengths of a patent document d and collection D, respectively, $freq(w,d)$ and $freq(w,D)$ denote the frequency of a term w in d and D, respectively.

5   Experiments

For experiments, we use a collection of freely available patents from the US patent and trademark office, downloaded from *(http://www.uspto.gov/)*. We collect all patents that are listed under *United State Patent Classification* (USPC) classes *422 (Chemical apparatus and process disinfecting, deodorizing, preserving, or sterilizing)*, and *423 (Chemistry of inorganic compounds)*. There are a total of $54,353$ patents in our collection, with an average patent size of $3,317.41$ terms (without stop words removing).

In query generation, we only consider the *claim field* of every patent, as this is the section that most professional patent searchers use as their basis for query formulation [9,11]. For retrieval we index the full text of all patents *(Title, Abstract, Background Summary, Claim, Description)*. This reflects the

Table 1. Queries sets properties used for Retrievability Measurement

Queries Set	Total Queries	Average Retriev-ability	Average Queries/Patent
2 Terms Queries	4, 308, 562	512.83	78.27
3 Terms Queries	2, 908, 972	373.49	53.42
4 Terms Queries	2, 876, 587	282.24	51.78

Table 2. Gini coefficient (G) values of different retrieval systems, for different *rank cut-off factors (c)*. As c increases, G steadily decreases indicating that lower bias is experienced when considering longer ranked lists.

Retr. Model/ Rank cut-off	Two Terms Queries					Three Terms Queries					Four Terms Queries				
	30	40	60	80	100	30	40	60	80	100	30	40	60	80	100
TFIDF	0.48	0.51	0.50	0.49	0.49	0.50	0.51	0.51	0.48	0.48	0.63	0.62	0.62	0.62	0.62
BM25	0.58	0.54	0.51	0.50	0.50	0.56	0.53	0.52	0.50	0.50	0.67	0.65	0.64	0.63	0.63
Exact Match	0.79	0.77	0.75	0.71	0.67	0.90	0.87	0.83	0.76	0.70	0.91	0.88	0.84	0.78	0.72
LM	0.53	0.53	0.53	0.52	0.51	0.62	0.62	0.63	0.61	0.60	0.71	0.71	0.72	0.70	0.68
QP-with-TS	0.39	0.39	0.38	0.38	0.37	0.38	0.37	0.37	0.36	0.36	0.54	0.53	0.53	0.52	0.51
QP-without-TS	0.50	0.57	0.55	0.55	0.54	0.58	0.55	0.55	0.55	0.54	0.66	0.65	0.63	0.63	0.62

default setting in a standard full text retrieval engine. Before indexing, we remove stop words and stem the words. For indexing and querying we use the *Apache LUCENE*[1] IR toolkit. Four state-of-the art retrieval systems along with our proposed prior-art retrieval approach are used for evaluating retrievability inequality and prior-art queries coverage. The retrieval systems that we evaluate are; **TFIDF**, OKAPI retrieval function **(BM25)**, **Exact Match** model, Language Modeling with term smoothing **(LM**[2]**)** [20], our PRF patents selection approach based on query patents similarity with terms selection **(QP-with-TS)**, and PRF patents selection based on query patents similarity using all terms of query patents **(QP-without-TS)**. In QP-with-TS we select terms from all fields of query patents. We create a set of queries from each query patent using two terms as well as three and four terms combinations (Section 3). Table 1 shows the properties of different query sets.

Figure 1 shows the retrievability inequality of different retrieval systems using Lorenz Curves with *rank cut-off* factor of 30. For other *rank cut-off* parameters configurations, we show the retrievability inequality of different retrieval approaches using *Gini coefficient values* in Table 2. We can see that as the *rank cut-off* factor increases, the Gini coefficient tends to decrease slowly on all different queries sets. This indicates that the retrievability inequality within the collection is mitigated by the willingness of the user to search deeper down into the ranking, as expected. If users are willing to examine only the top documents, then they will face a greater degree of retrieval bias.

---

[1] http://lucene.apache.org/java/docs/

[2] In LM, top 40 patents are used for PRF, and top 35 terms from PRF patents are used for expansion.

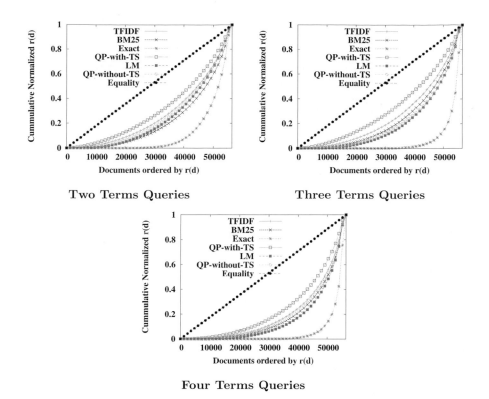

Fig. 1. Lorenz Curves visualizing the retrievability inequality of different retrieval systems, with *rank cut-off factor (c)* = 30. *Equality* refers to an optimal system which has no bias.

On almost all rank cut-off factors with different lengths of query sets, the Lorenz curve and Gini coefficient values of our prior-art retrieval approach *(QP-with-TS)* are less skewed and have lower Gini coefficient values than other retrieval approaches. This indicates that our prior-art retrieval approach makes individual patents more easily retrievable. The *Exact Match* method, which is widely used in professional patent retrieval systems, consistently shows the worst performance. In *LM* approach, considering top documents in queries relevant for PRF also does not perform too well with respect to providing potential access to all patents. The performance results of *QP-without-TS* are worse than *QP-with-TS*. This happens because in patent retrieval domain a large diversity exists in patent lengths. Therefore, when calculating PRF patents similarity using all terms of query patent *(whole patent)*, longer patents have more chance that they can increase their similarity values with query patents. This results in a higher bias in retrievability scores. The retrievability results of *TFIDF* are better than other retrieval approaches.

The Gini coefficient values of Table 3 summarize our analysis which field of query patent is better for terms extraction in *QP-with-TS* approach. Given the

Table 3. Effect of different fields of patents on Gini coefficient for query patent terms selection, with different *rank cut-off factors (c)*

Patent Field/	Two Terms Queries					Three Terms Queries					Four Terms Queries				
Rank cut-off	30	40	60	80	100	30	40	60	80	100	30	40	60	80	100
Whole Patent	**0.39**	**0.39**	**0.38**	**0.38**	**0.37**	**0.38**	**0.37**	**0.37**	**0.36**	**0.36**	**0.54**	**0.53**	**0.53**	**0.52**	**0.51**
Description	0.44	0.44	0.44	0.43	0.43	0.40	0.40	0.39	0.39	0.39	0.56	0.55	0.55	0.54	0.53
Claim	0.51	0.51	0.50	0.50	0.49	0.46	0.46	0.45	0.44	0.44	0.60	0.59	0.59	0.58	0.58
Abstract	0.47	0.47	0.46	0.46	0.46	0.44	0.44	0.44	0.44	0.43	0.58	0.58	0.57	0.57	0.57
Summary	0.47	0.47	0.46	0.46	0.44	0.43	0.43	0.43	0.42	0.42	0.59	0.58	0.57	0.56	0.55

results shown in Table 3, it is clear that query patent terms extracted from the description field can better decrease the retrievability inequality as compared to other fields. However, the performance results of selecting terms from the whole patent are much better than individual fields' results. From the results, it is interesting to notice, that the performance results of summary and abstract fields are much better than claim field results, which is mostly used in prior-art retrieval. This is because writers in claim fields, for protecting the invention, may tend to use language that extends the scope of patents, which may create serious term mismatch problems. Other fields like description, abstract and background summary are mainly written for the technical use, where authors briefly try to describe their description of invention relative to the field.

6   Conclusions

This paper evaluates the coverage of prior-art queries extracted from query patents using retrievability measurement. In experiments, retrievability shows large bias toward subset of patents using state of the art retrieval systems. Due to bias, a large number of patents either have very lower retrievability scores or could not be retrievable via any query. The main reason behind low retrievability is the presence of large terms mismatch in patents. For increasing the retrievability of patents, we consider query expansion approach with pseudo relevance feedback (PRF). In this way, missing terms of query patents are retrieved from related PRF patents. For better identifications of PRF patents, a novel approach is presented where patents for PRF are identified based on their similarity with query patents via selected terms. We identify relevant terms from query patents based on their proximity distribution with prior-art queries. Using this approach, an increase in the retrievability of individual patents is obtained, which indicates that this prior-art retrieval approach provides better opportunity for retrieving individual patents in search space.

Acknowledgments

We would like to thank anonymous reviewers for providing helpful comments on improving the quality of paper. We would also like to acknowledge Higher Education Commission of Pakistan (HEC) and Austria Exchange Service (OAD) for providing scholarship and support to Shariq Bashir for conducting his PhD.

References

1. Azzopardi, L., Vinay, V.: Retrievability: an evaluation measure for higher order information access tasks. In: Proc. of CIKM 2008, Napa Valley, California, USA, October 26-30, pp. 561–570 (2008)
2. Bashir, S., Rauber, A.: Analyzing Document Retrievability in Patent Retrieval Settings. In: Bhowmick, S.S., Küng, J., Wagner, R. (eds.) DEXA 2009. LNCS, vol. 5690, pp. 753–760. Springer, Heidelberg (2009)
3. Bashir, S., Rauber, A.: Improving retrievability of patents with cluster-based pseudo-relevance feedback documents selection. In: Proc. of CIKM 2009, Hong Kong, China, November 2-6, pp. 1863–1866 (2009)
4. Cao, G., Nie, J.-Y., Gao, J., Robertson, S.: Selecting good expansion terms for pseudo-relevance feedback. In: Proc. of SIGIR 2008, Singapore, pp. 243–250 (2008)
5. Cummins, R., O'Riordan, C.: Learning in a pairwise term-term proximity framework for information retrieval. In: Proc. of SIGIR 2009, Boston, MA, USA, pp. 251–258 (2009)
6. Custis, T., Al-Kofahi, K.: A new approach for evaluating query expansion: query-document term mismatch. In: Proc. of SIGIR 2007, Amsterdam, The Netherlands, July 23-27, pp. 575–582 (2007)
7. Fall, C.J., Torcsvari, A., Benzineb, K., Karetka, G.: Automated categorization in the international patent classification. ACM SIGIR Forum 37(1), 10–25 (Spring 2003)
8. Fujii, A.: Enhancing patent retrieval by citation analysis. In: Proc. of SIGIR 2007, Amsterdam, The Netherlands, pp. 793–794 (2007)
9. Itoh, H., Mano, H., Ogawa, Y.: Term distillation in patent retrieval. In: ACL 2003: Proceedings of the ACL-2003 workshop on Patent corpus processing, Sapporo, Japan, pp. 41–45 (2003)
10. Konishi, K.: Query terms extraction from patent document for invalidity search. In: Proc. of NTCIR 2005: NTCIR-5 Workshop Meeting, Tokyo, Japan (2005)
11. Konishi, K., Kitauchi, A., Takaki, T.: Invalidity patent search system at NTT data. In: Proc. of NTCIR-4 Workshop Meeting, Tokyo, Japan (2004)
12. Larkey, L.S.: A Patent Search and Classification System. In: Proc. of 4th ACM Conference on Digital Libraries, Berkeley, CA, USA, pp. 179–187 (1999)
13. Lavrenko, V., Croft, W.B.: Relevance based language models. In: Proc. of SIGIR 2001, New Orleans, Louisiana, USA, pp. 120–127 (2001)
14. Lee, K.S., Croft, W.B., Allan, J.: A cluster based resampling method for pseudo-relevance feedback. In: Proc. of SIGIR 2008, Singapore, pp. 235–242 (2008)
15. Mase, H., Matsubayashi, T., Ogawa, Y., Iwayama, M., Oshio, T.: Proposal of two-stage patent retrieval method considering the claim structure. ACM Transactions on Asian Language Information Processing 4(2), 190–206 (2005)
16. Murata, M., Kanamaru, T., Shirado, T., Isahara, H.: Using the k-nearest neighbor method and SMART weighting in the patent document categorization subtask at NTCIR-6. In: Proc. NTCIR-6 Workshop Meeting, Tokyo, Japan (2007)
17. Osborn, M., Strzalkowski, T., Marinescu, M.: Evaluating Document Retrieval in Patent Database: A Preliminary Report. In: Proc. of CIKM 1997, Las Vegas, Nevada, USA, pp. 216–221 (1997)
18. Tao, T., Zhai, C.: An exploration of proximity measures in information retrieval. In: Proc. of SIGIR 2007, Amsterdam, The Netherlands, pp. 295–302 (2007)

19. Xue, X., Croft, W.B.: Transforming patents into prior-art queries. In: Proc. of SIGIR 2009, Boston, MA, USA, pp. 808–809 (2009)
20. Zhai, C., Lafferty, J.: A study of smoothing methods for language models applied to information retrieval. ACM Trans. Inf. Syst. 22(2), 179–214 (2004)
21. Zhao, J., Yun, Y.: A proximity language model for information retrieval. In: Proc. of SIGIR 2009, Boston, MA, USA, pp. 291–298 (2009)

Mining OOV Translations from Mixed-Language Web Pages for Cross Language Information Retrieval

Lei Shi

Yahoo Software R&D (Beijing),
22FL C Building SP Tower, Tsinghua Science Park
Bejing, China 100086
`lshi@yahoo-inc.com`

Abstract. Translating Out-Of-Vocabulary (OOV) terms is crucial for Cross Language Information Retrieval (CLIR). In this paper, we propose a method that automatically acquires a large quantity of OOV translations from the web. Different from previous approaches that rely on a finite set of hand-crafted extraction rules, our method adaptively learns translation extraction patterns based on the observation that translation pairs on the same page tend to appear following similar layout patterns. The learned patterns are leveraged in a discriminative translation extraction model that treats translation extraction from a mixed language bilingual web page as a sequence labeling task in order to exploit useful relations among translation pairs on the page. Experiments demonstrate that our proposed method out-performs earlier work with marked improvement on OOV translation mining quality.

Keywords: Cross Language Information Retrieval, OOV Mining.

1 Introduction

One of the major reasons that CLIR does not perform as well as monolingual IR is the presence of OOV terms in the queries, which cannot be translated with a regular dictionary. An analysis [3] of a query log in a real world Chinese search engine reveals that 82.9% of the top 19,124 high frequency query terms were not included in the LDC Chinese-English dictionary. Due to the fact that the average length of web queries is short, even a single occurrence of an OOV term in the query may severely deteriorate the relevance of the retrieved documents by CLIR systems.

To deal with the OOV issue, manually compiling a wide-coverage and up-to-date bilingual lexicon requires substantial human effort. For this reason, web mining of term translations has drawn intensive attention. Recent research on web mining of term translations has primarily focused on mixed language bilingual web pages where terms and their translations co-occur in the same page written in a baseline language (fig 1) due to its enormous availability on the web. A major approach to leveraging such resources [19] identifies term translations using one or a fixed set of handcrafted layout patterns of translation pairs on the bilingual web page, e.g. a term followed by its translation within parenthesis, 蜘蛛侠(Spider-man). Since web pages are created

C. Gurrin et al. (Eds.): ECIR 2010, LNCS 5993, pp. 471–482, 2010.

by different people, such method suffers from low precision and coverage due to the variation of writing styles and layout arrangement of web documents. Some patterns valid in a set of bilingual web pages may extract noises in others. And a few hand-crafted rules can never be complete to cover all bilingual pages due to the diversity and sheer volume of the web.

In this paper, we propose a method that can adaptively learn a page's specific extraction patterns to facilitate term translation mining. It is based on the observation that when presenting foreign terms and their translations, web authors exhibit a very strong tendency to use the same pattern in the page (fig. 1). We built a statistical model to adaptively learn extraction patterns. And a discriminative translation extraction model is proposed to leverage the learned pattern together with other features for translation identification. Experiments show that our method can greatly improve mining precision and coverage over previous work.

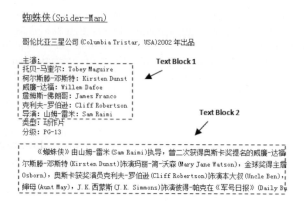

Fig. 1. Translation pairs in one page

In the rest of the paper, we first review related work in Section 2. Section 3 describes the adaptive pattern learning. Section 4 deals with the discriminative translation extraction model. Experiments are presented in Section 5. And finally, we conclude the paper in Section 6.

2   Related Work

Several systems [11,16] have been built to learn word translations from parallel text. Though of high quality, large scale OOV translation mining from parallel text is hindered by the insufficient availability and limited domain coverage of parallel corpus. [5] learned bilingual knowledge from comparable corpus. Compared with parallel text, comparable corpus is easier to obtain. But since comparable texts are not strict mutual translations, learning OOV translations from comparable corpus suffers from low accuracy. [9] proposed to mine term translations from anchor texts pointing to the same web pages, but except some popular terms e.g. celebrity names, popular organization names etc. most terms cannot be identified by this approach. The above methods all rely primarily on translation knowledge that resides between different web pages. But such resources are still too sparse on the web to build a wide coverage bilingual lexicon.

Besides the resources mentioned above, much attention has been paid to acquire term translations from bilingual web pages, where terms and translations co-occur on the same page. Many systems [3,7,12] extract term translations from search snippets of the bilingual pages. They typically require the term to be provided by the user and the translation is extracted from the returned snippets by leveraging co-occurrence frequency of the translation with the term. These methods works quite well for mining high frequency OOV translations but often fail for low frequency ones which compose the majority of OOVs. As patterns are generally more reliable than frequency counts, pattern based mining schemes have been discussed in [18,19]. [19] used two categories of patterns, but these hand-crafted rules are neither accurate nor complete due to the variation of page layout and writing styles. To replace manual pattern definition, [17] proposed to automatically gather surface patterns from the snippets by searching known translation pairs on the web. But it is questionable to apply all the collected patterns on snippets of different bilingual pages in that different pages may have different patterns. Therefore, a method that can adaptively learn and use patterns specific to individual pages to collectively extract term translations is called for.

3 Adaptive Pattern Learning

To adaptively learn the extraction patterns, we first generate a set of candidate patterns based on the layout format between the foreign terms and their surrounding context and then we employ a statistical model to estimate the probability of each candidate pattern based on several features of the bilingual page. In the following subsections, we first introduce candidate pattern generation and then the statistical model.

3.1 Candidate Pattern Extraction

A candidate pattern is generated based on a candidate translation pair in the following manner: the English term and its Chinese candidate translation are first converted into the tokens <eng> and <chi>. Tokens between <eng> and <chi> are kept as their original form. For the token prior and next to the pair, if it is a delimiter then the original form is saved, while Chinese and English characters are converted to <b_chi> and <b_eng> respectively. Html tags can be used in the pattern as well, with only tag labels saved. For instance, <a href="abcd.com"> is saved as <a>. Below shows some examples of how candidate patterns are converted from candidate translation pairs. The underlined is the Chinese candidate translation for the English term in the examples. The reason why tokens between the pair are saved as its original form instead of a general token of <b_chi> or <b_eng> is because normally term and their translation are close in distance and the tokens between them offers very indicative clues about whether the pair are true translations. Replacing the original tokens into <b_chi> or <b_eng> would make it too general to be distinctive.

周润发"yun-fat chow"   ➔   <chi>"<eng>"

他曾是香港电视广播有限公司 (英文是TVB)的签约艺人 ➔ <b_chi><chi>(英文是<eng>)

To determine whether a pair of English term and Chinese candidate translation match a pattern, the English term and the candidate translation are matched with <eng> and <chi> in the pattern first, and then the tokens between, prior to and next to the pair are matched with their corresponding parts in the pattern. The token <b_chi> and <b_eng> can be matched with any Chinese and English words. If all tokens of the pair can be matched with the pattern, then we consider it a fit or match with the pattern, or otherwise a rejection.

3.2 Statistical Pattern Learning Model

The statistical model is used to choose the best ones from the candidate patterns generated in the last step. We employ the Maximum Entropy (ME) model [1] to estimate the probability of each candidate pattern based on several features derived from the candidate pattern and the bilingual web page. The candidate pattern with the highest probability is considered as translation extraction pattern for the given page. Below lists all the features we use.

- Number of bilingual term pairs that match the pattern. Usually the more matches, the more likely it is a good pattern
- The ratio of matching bilingual pairs that are translations according to an existing dictionary
- The ratio of matching pairs not included in an existing dictionary. This feature indicates poor quality of the candidate pattern.
- Average translation probability of matching bilingual pairs. Term translation probability is computed with IBM model 1 [2]. We used the GIZA++ tool to train word translation probabilities on five million Chinese-English parallel sentences mined from the web [15].
- Average transliteration score of matching bilingual pairs. Transliteration score measures pronunciation similarities since many named entities are translated based on pronunciation instead of meaning. Refer to section 4.3 for details of calculating the transliteration score.
- Average length similarity of matching bilingual pairs. If the candidate pattern is the one that could extract real translations, the extract term pairs' length should be proportional. Refer to section 4.3 for details
- Web co-occurrence counts of the fitting candidate translation pairs

The ME model is trained with iterative scaling on 200 training bilingual web pages where the correct patterns are manually annotated among the set of automatically generated candidate patterns. The Maximum Entropy model is formulated as

$$P(c \mid d) = \frac{1}{Z(d)} \exp(\sum_i \lambda_i f_i(d, c))$$

where d is the page and p is a candidate pattern. $f(d,c)$ is a feature function with feature weight λ. $Z(d)$ is the normalization factor over all candidates.

4   Discriminative Translation Extraction Model

A straightforward approach to using the pattern learned from the bilingual page is to apply it on the page as a regular expression rule and the matches are regarded as translations. However, though translation pairs on the same page tend to follow similar pattern, there are still many variations in many pages where it is harsh to assume everything not matching the pattern is not translation and everything that matches the pattern is indeed translations. Hence, such rigid rules, though simple, are not flexible enough to accommodate layout variations in the bilingual page. Instead, we employ a discriminative translation extraction model that, in addition to the patterns, incorporates many relevant features for more accurate translation identification.

In the discriminative translation extraction framework, given a bilingual web page such as figure 1, we can define the translation extraction task as for each English term in the page to choose its translation from among a set of Chinese candidates generated from its surrounding context window. So the whole process is split into two steps. The first step generates a set of translation candidates for each foreign term in the page from their surrounding context windows in the same way as in the pattern learning phase. In the second step, the discriminative model is applied to the candidates in order to choose the most likely ones with the highest probability.

$$\bar{t} = \arg\max_c P(c \mid T_e)$$

where $c \in \{T_c^0, T_c^1, ..., T_c^n, N/A\}$. N/A means none of the candidates is translation.

4.1   Modeling Translation Extraction as Sequence Labeling

Choosing the translation from the candidates can be regarded as a classification problem and it can be addressed by a Maximum Entropy classifier. So extracting all translation pairs from a bilingual page can be regarded as multiple independent classifications, as the classifier is applied to each foreign term on this page. However, such a view ignores several observed relations among translation selection of different English terms that could help us for better translation extraction. One relation is that a Chinese term in the bilingual page tends not to serve as the translation for more than one English term in the same page. For example in fig. 1, if our model already chooses "玛丽-简-沃森" as the translation for "Mary Jane Watson", then it is unlikely to be the translation for any other English terms in this page any more. Another useful relation is that the relative spatial order of the translations tends to be the same as the order of English terms. For example, if an English term E_a occurs after E_b, then E_a's translation tends to appear after E_b's translation as well. So instead of using ME to estimate the probabilities of candidates for each foreign term in the page individually, we model translation extraction as a sequence labeling task to estimate a single probability for the whole sequence of candidate selections, which is able to incorporate relations of candidate selections for different foreign terms in the page.

We use Conditional Random Field (CRF) for this. CRF has demonstrated many advantages over other sequence models such as the Hidden Markov Model and the Maximum Entropy Markov Model [8]. In the CRF model, both the features described in section 4.3 (which we call local features) and the above described transition features are used.

4.2 Translation Mining with Hidden Conditional Random Fields

As described in previous sections, the pattern with the highest probability is used as the pattern feature in the CRF translation extraction model. There are two major drawbacks of this. As the pattern is a crucial feature for translation identification, a wrong choice of the best pattern in the pattern learning step would have big impact on translation extraction accuracy. Another issue is that choosing only the best single pattern cannot handle pages with multiple patterns (such as figure 1) which are quite common on the web.

To deal with these issues, we extend the regular CRF by introducing a hidden variable h to represent the patterns used for translation pairs on the page since the patterns are not directly observed. Such an extended CRF with hidden variables is called the Hidden Conditional Random Field (HCRF) [14] and it has been successful in many difficult tasks such as object recognition [14] and phone classification [6] etc. By allowing a model with hidden variables, no a-priori pattern classification is needed. But rather the hidden variable h's value can choose from the finite set of extracted candidate patterns so that all candidate patterns have the potential of influencing term translation extraction. So the HCRF model is able to accommodate multiple valid patterns on a single page and avoid taking prudent classification over patterns. In our HCRF model, the conditional probability of sequence labels over input data takes the form:

$$P_\lambda(y \mid x) = \sum_h P_\lambda(y, h \mid x) = \frac{1}{Z_x} \exp(\sum_h \sum_{t=1}^{T} \sum_k \lambda_k f_k(y_{t-1}, y_t, h, x, t))$$

where the probability of a sequence labeling is the summation of the probability of labeling translations with all the candidate patterns h. Arbitrary feature function can be defined on the observed input x, the label and transition y_t and y_{t-1}, as well as the hidden variable h. We can inherit all the regular CRF feature function described in section 4.3 except the pattern feature. Instead, the pattern feature in HCRF is defined as taking the pattern probability calculated by the ME based pattern learning model (in section 3.2) if the pair fit the pattern as the value of h, or 0 if it does not fit h. Following previous work on CRFs, we use the following objective function in training the model parameter λ :

$$L(\lambda) = \sum_{i=1}^{n} \log P(y_i \mid x_i, \lambda) - \frac{1}{2\sigma^2} \|\lambda\|^2$$

where n is the total number of training sequences. The first term in the above equation is the log-likelihood of the data; the second term is the log of a Gaussian prior with variance σ^2. We use Quasi-Newton optimization technique for the optimal parameter values $\lambda^* = \arg\max_\lambda L(\lambda)$. Our annotated data to train the HCRF model consists of 500 bilingual web pages where the sequences of translation pairs are manually labeled. Since the pattern h is a hidden variable, it not necessary to explicitly label patterns in the training data. Efficient inference is dynamic programming similar to CRF with an additional summation over features involving hidden variables.

4.3 Translation Extraction Model Features

In our discriminative model, the observed data are the text on the web pages, and we want to identify translations (or *N/A*) for the set of English terms from their translation candidate sets. Though the adaptively identified pattern is a very important feature, our discriminative extraction model can incorporate other relevant feature to further improve extraction accuracy. Below we describe the set of features used in our CRF model for term translation extraction.

Pattern Feature: The pattern feature is one of the most important features to identify translation pairs in our method. It is a binary feature with 1 indicates that the layout of pair matches the pattern picked by the previous pattern identification step, otherwise the feature value is 0.

Dictionary Feature: Two terms are likely to be translations if many of their composing words are translations according to an existing dictionary. So we divide the number of word pairs found in the dictionary with the total number of words of both terms as the score of the feature.

Transliteration Score: Transliteration score is useful for personal names, location, and organizations etc that are transliterated based on pronunciation rather than meaning. The transliteration probability of a pair S^e and S^f is defined as

$$\Pr(S^e \mid S^f) = P_u(S^e \mid A, S^f) = P_u \sum_{A} \prod_{c^e \in S^e}^{c^f \in S^f} P(c^e \mid c^f)$$

where A is the alignment of their sound letters c^e and c^f. $P(c^e \mid c^f)$ is the transformation probability of the aligned sound letters and is estimated via the Expectation-Maximization algorithm [4] on 592607 proper name transliteration pairs.

Length Similarity: The length of a term is often proportional to that of its translation. We consider the normalized length difference $\delta(x, y)$ of the terms x and y to be normal distribution N(0,1)

$$\delta = \frac{y - x \times c}{\sqrt{(x+1)\sigma^2}}$$

where c is a constant indicating the average length ratio between target and source terms. σ^2 is the variance of length difference.

Distance Feature: Since translations usually occur close to the original term, we count the number of words between the term and the translation candidate as the distance feature.

Web Co-occurrence: Similar to the search snippet based scheme, we search the term in documents written in the target language on the Web. The returned snippets are saved. We count the number of occurrence of the candidate translation in the snippets. The more count, the more likely it is the translation of the term.

SCPCD: SCPCD is proposed in [3] to identify lexical boundaries of terms. It combines Symmetric Conditional Probability (SCP) and Context Dependency (CD) as their product. SCP is defined as:

$$SCP(w_1,...,w_n) = \frac{freq(w_1,...,w_n)^2}{\frac{1}{n-1}\sum_{i=1}^{n-1} freq(w_1,...,w_i) freq(w_{i+1},...,w_n)}$$

where $w_1,...,w_n$ is the word n-gram and $freq(w_1,...,w_n)$ is the frequency of the n-gram on the returned snippets. SCP measures whether the n-gram should be regarded as a term. Context dependency (CD) measures whether the n-gram could be merged with its context to form an independent term. It is defined as

$$SCP(w_1,...,w_n) = \frac{LC(w_1,...,w_n)RC(w_1,...,w_n)}{freq(w_1,...,w_n)^2}$$

where $LC(w_1,...,w_n)$ or $RC(w_1,...,w_n)$ is the number of unique left (or right) adjacent words.

Context Similarity: It is based on the idea that the term and its translation should have similar context in their own languages. The context vector is constructed by collecting the context words weighted by their *tf-idf* scores in monolingual search snippets. The similarity between a query term s and the translation candidate t is estimated with the cosine measure of their context vectors:

$$S_{cv}(s,t) = \cos ine(cv_s, cv_t)$$

First proposed in [5] to mine term translations from comparable text, context similarity is complementary to the web co-occurrence feature when co-occurrence count of the pair is low.

Transition Features: The transition features are used specifically for the CRF model aiming to capture the non-overlapping and spatial order relations between translation pairs as described in Section 4.3. The spatial order feature is set to 1 if the translation of the current English term appears before that of the previous English term, otherwise 0. So in testing time, if the feature value is 1, the translation candidate will receive a penalty by the model since the model rarely sees reversed spatial order of the translation layouts with feature to be 1 in the training samples. Similarly, the overlapping feature is set to 1 if the translation candidate of the current English term has overlapping characters with that of the previous English term which rarely happens in training data as well. The only exception is the case of *N/A*. If an English term chooses *N/A*, the selected translation of the previous English term is then used for the next English term to compute its transition feature. This is recursive in case the previous English term also chooses *N/A*. We implemented it in our CRF decoder that saves the label of previous terms for all terms choosing *N/A*.

5 Experiments and Evaluation

In the following subsections, we first report the results of our OOV translation mining on Chinese bilingual web pages and compare their quality with that by other methods. Second, we did experiments on the contribution of different features in the translation extraction model. Finally, OOV translations mined with different methods are tested in translating real queries in a CLIR system, where translation precision and recall as well as the cross language retrieval relevance are benchmarked.

5.1 Translation Mining Accuracy

We used two data sets to evaluate OOV translation accuracy. The first data set consists of 525,332 web pages we crawled from the web, of which 104,521 are bilingual pages with term translations. The second data set are 2000 bilingual web pages randomly selected from the first set and human annotators are hired to label all the 92813 term translations in these pages so that we can more accurately evaluate translation mining recall. We implemented several different OOV mining methods to mine term translations from both data sets and their performance is shown in table 1.

In table 1, "Fixed Patterns" denote the method that applies a fixed set of hand-crafted regular expression rules following [12,17] to extract translations from bilingual pages. The ones beginning with "Adaptive" are those employ adaptively learned patterns from each page and "RegEx", "ME", "CRF" and "HCRF" represents different means to leverage the patterns, with "RegEx" as regular expression, "ME" as Maximum Entropy, "CRF" as Conditional Random Field and "HCRF" as Hidden Conditional Random Fields. By comparing "Fixed Pattern" with "Adaptive(RegEx)", we can find that using adaptively pattern learning can greatly improve both mining precision and coverage over previous methods with a fixed set of rules. Mining accuracy is further improved by considering other features such as translation probability etc. with the discriminative translation extraction model. CRF outperform ME in that it is able to model relations between different term translation pairs on the page and HCRF offers the best performance (especially recall) since it is able to accommodate multiple patterns in a single bilingual page.

Table 1. Mining Results by Different Methods

Method	Large		Small	
	Precision	# of mined translations	Precision	Recall
Fixed Pattern	75.3%	1,724,383	75.1%	43.9%
Adaptive(RegEx)	83.5%	3,352,875	82.5%	73.0%
Adaptive(ME)	87.1%	3,921,578	87.3%	86.2%
Adaptive(CRF)	89.9%	4,132,922	90.2%	88.6%
Adaptive(HCRF)	91.5%	4,534,560	91.8%	93.5%

5.2 Evaluation of Different Features for the Translation Extraction Model

Our discriminative translation extraction model leverages several features including the learned pattern feature to facilitate OOV translation mining. We did several experiments to evaluate the contribution of each feature in the model. We trained many Maximum Entropy models by omitting one feature at a time and keep the remaining. Mining precision and recall are evaluated on the 2000 manually annotated bilingual web pages.

Table 2. Models Trained after Omitting Different Features

Feature	Precision	Recall
All	87.3%	86.2%
Pattern	69.4%	62.1%
Translation	83.2%	75.1%
Length	85.3%	83.4%
Transliteration	84.1%	80.2%
Dictionary	84.0%	78.1%
Distance	86.5%	85.1%

In table 2, the first lines show the performance of the model with all the features, and the remaining lines represent the models that omit a certain feature. Comparison shows that the pattern feature has the biggest impact on translation mining performance. And we see performance drop when we omit any one of the features which indicates that incorporating all these different features in the discriminative model helps improve OOV mining performance than just the pattern feature alone.

5.3 Cross Language Information Retrieval

We tested the effectiveness of the mined bilingual lexicon for translating queries in CLIR. Rather than evaluating its translation quality on unknown query terms directly, we instead measure the retrieval relevance of the translated queries given that the higher query translation quality usually results in better relevance of the retrieved documents. Since we only focus on translating OOV terms, after translating the terms covered by the regular dictionary, we compare the CLIR performance before and after OOV terms are translated with the bilingual lexicons mined by different methods from the large bilingual page set.

The test collection used in this evaluation task is the TREC-5 and TREC-6 Chinese collection. There are 54 topics in both English and Chinese which are mutual translations. In order to imitate typical web queries, we take the "title" section in each English topic as our queries and their average length is 5 words. These queries are translated into Chinese and Lemur [20] is employed as the IR engine to retrieve TREC Chinese documents. Since the manual relevance judgments of documents for each topic are available, we are able to measure the precision and recall for the returned retrieval results of the test queries. In the first run (Mono), we perform a mono-lingual retrieval using the Chinese titles of the TREC topics as queries. In the second run (Dict), the LDC dictionary is employed to translate the English queries. All the English queries consist of 445 words, and there are 27 OOV terms. For these OOV terms, say "Dalai Lama(达赖喇嘛)",

"Pinatubo(皮那图博)", they are translated by the bilingual lexicons mined by different approaches in subsequent runs.

Table 3 presents a comparison of recall precision values of our English-Chinese CLIR experiments. The results show that query translation with the regular dictionary (DICT) without translating OOVs has the lowest retrieval relevance with a mean average precision of 0.1347 which amounts to 64.4% of the monolingual IR. All other methods that translate OOV terms show improvement in CLIR performance over the dictionary based method. The performance of the best method "Adaptive (HCRF)" is significantly higher than that of the "Dict", even though OOV terms do not take a significant portion of all the query terms in this task. It clearly demonstrates that OOVs bears much information in the queries and correctly translating these terms with our proposed mining method is very effective in improving CLIR performance. Adaptive HCRF method produces the best result with 17.83% mean average precision which is 85.2% of the monolingual result. The improvement is achieved by correctly translating 24 of the 27 OOV terms in 20 queries

Table 3. English-Chinese CLIR results

Recall	Dict	Fixed Patterns	Adaptive (RegEx)	Adaptive (ME)	Adaptive (CRF)	Adaptive (HCRF)	Mono
0.0	0.6271	0.6890	0.7030	0.7101	0.7230	0.7321	0.7654
0.1	0.4284	0.5210	0.5420	0.5498	0.5623	0.5711	0.5873
0.2	0.3349	0.4233	0.4421	0.4587	0.4780	0.4897	0.4942
0.3	0.2659	0.3420	0.3429	0.3432	0.3877	0.4132	0.4058
0.4	0.2179	0.2748	0.2966	0.3103	0.3250	0.3510	0.3494
0.5	0.1801	0.2303	0.2501	0.2683	0.2802	0.2981	0.3046
0.6	0.1440	0.1678	0.1987	0.2210	0.2277	0.2451	0.2620
0.7	0.1112	0.1301	0.1625	0.1719	0.1847	0.1982	0.2180
0.8	0.0765	0.0921	0.1181	0.1301	0.1365	0.1443	0.1563
0.9	0.0274	0.0298	0.0367	0.0392	0.0427	0.0481	0.0662
1.0	0.0055	0.0059	0.0061	0.0065	0.0068	0.0071	0.0109
Avg.P	0.1347	0.1498	0.1599	0.1649	0.1731	0.1783	0.2093
% Mono	64.4%	71.6%	76.1%	78.8%	82.7%	85.2%	100%

6 Conclusions

In this paper, we propose a novel method for acquiring OOV translations from bilingual web pages. This method is able to adaptively learn translation extraction patterns from individual bilingual web pages and exploit the learned patterns in a probabilistic framework for term translation extraction. Introducing hidden variables in the conditional random fields allows the model to accommodate multiple patterns in a single page and avoid prudent classification of patterns. Experiments show marked improvement on OOV translation mining coverage over earlier methods while maintaining high accuracy.

References

1. Berger, A., Pietra, S.A.D., Pietra, V.J.D.: A Maximum Entropy Approach to Natural Language Processing. Computational Linguistics 22(1), 39–71 (1996)
2. Brown, P.F., Della Pietra, V.J., Della Pietra, S.A., Mercer, R.L.: The Mathematics of Statistical Machine Translation: Parameter Estimation. Computational Linguistics (1993)

3. Cheng, P.-J., Teng, J.-W., Chen, R.-C., Wang, J.-H., Lu, W.-H., Chien, L.-F.: Translating unknown queries with web corpora for cross-language information retrieval. In: Proceedings of ACM-SIGIR 2004 (2004)
4. Dempster, A.P., Laird, N.M., Rubin, D.B.: Maximum Likelihood from Incomplete Data via the EM algorithm. Journal of the Royal Statistical Society
5. Fung, P., Yee, L.Y.: An IR Approach for Translating New Words from Nonparallel, Comparable Texts. In: Proceedings of COLING-ACL, pp. 414–420 (1998)
6. Gunawardana, A., Mahajan, M., Acero, A., Platt, J.C.: Hidden Conditional Random Fields for Phone Classification. In: International Conference on Speech Communication and Technology
7. Huang, F., Zhang, Y., Vogel, S.: Mining Key Phrase Translations from Web Corpora. In: Proceedings of EMNLP 2005 (2005)
8. Lafferty, J., McCallum, A., Pereira, F.: Conditional Random Fields: Probabilistic Models for Segmenting and Labeling Sequence Data. In: Proceedings of ICML 2001 (2001)
9. Lu, W.-H., Chien, L.-F., Lee, H.-J.: Anchor Text Mining for Translation of Web Queries: A Transitive Translation Approach. ACM Transactions on Information Systems (2004)
10. Mahajan, M., Gunawardana, A., Acero, A.: Training Algorithms for Hidden Conditional Random Fields. In: Proceedings of IEEE International Conference on Acoustic, Speech and Signal Processing, ICASSP (2006)
11. McEwan, C.J.A., Ounis, I., Ruthven, I.: Building Bilingual Dictionaries from Parallel Web Documents. In: Crestani, F., Girolami, M., van Rijsbergen, C.J.K. (eds.) ECIR 2002. LNCS, vol. 2291, p. 303. Springer, Heidelberg (2002)
12. Nagata, M., Saito, T., Suzuki, K.: Using the Web as a Bilingual Dictionary. In: Proceedings of Workshop on Data-driven Methods in Machine Translation, pp. 95–102 (2001)
13. Quattoni, A., Wang, S., Morency, L.P., Collins, M., Darrell, T.: Hidden Conditional Random Fields. IEEE Transactions on Pattern Analysis and Machine Intelligence
14. Shi, L., Niu, C., Zhou, M., Gao, J.F.: A DOM Tree Alignment Model for Mining Parallel Data from the Web. In: Proceedings of ACL 2006 (2006)
15. Turcato, D.: Automatically Creating Bilingual Lexicons for Machine Translation from Bilingual Text. In: Proceedings of Coling/ACL 1998 (1998)
16. Wu, J.C., Lin, T., Chang, J.S.: Learning Source-Target Surface Patterns for Web-based Terminology Translation. ACL Interactive Poster and Demonstration Sessions (2005)
17. Zhang, Y., Vines, P.: Detection and Translation of OOV Terms Prior to Query Time. In: Proceedings of ACM-SIGIR 2004 (2004)
18. Zhang, Y., Vines, P.: Using the Web for Automated Translation Extraction in Cross-Language Information Retrieval. In: Proceedings of ACM-SIGIR 2004 (2004)
19. http://www.lemurproject.org

On Foreign Name Search

Jason Soo[1] and Ophir Frieder[2]

[1] Information Retrieval Laboratory
Illinois Institute of Technology
[2] Department of Computer Science
Georgetown University
soo@ir.iit.edu, ophir@cs.georgetown.edu

Abstract. We address foreign name search in a highly diverse user community. User sophistication ranges from highly experienced archivists to apprehensive users who shy away from technology; apprehensive users dominate system use. Thus, all system interfaces must assume minimal dependency on the user.

Our foreign names search approach, called SEGMENTS, is language independent; thus, there is no need to determine the language of origin from the diverse candidate set of thirteen languages. We compare SEGMENTS against traditional n-gram and Soundex based solutions. Actual and synthetic queries are used to search a names data set resident in the United States Holocaust Memorial Museum. We also search a subset of the 1990 United States Census Bureau Surnames data set to evaluate the performance of SEGMENTS on a predominately language specific (English) collection. Our results demonstrate statistically significant performance gains over both traditional approaches. The described approach supports search efforts at the United States Holocaust Memorial Museum.

1 Introduction

Name identification significantly impacts accuracy in the general search case; however, in historical document search, their identification is paramount. Complicating name search is the variance of accepted spellings for the same sounding name, for example Laurence and Lawrence. To circumvent spelling issues, phonetic search techniques are often used [17]. Common phonetic techniques are based on Soundex; JewishGen [7] uses the Daitch-Mokotoff (D-M) Soundex variant [10], a de facto standard by Jewish genealogical organizations.

A difficulty with using phonetic search stems from the reliance on the user to formulate an approximate sound, and hence spelling, of the foreign name. For example, to an English speaker, *"Roz'ishts' ávarati"* or *"Rozhyshche"* is likely to be difficult to pronounce. Furthermore, in searching name indices from historical documents, particularly for personal-data related applications such as JewishGen genealogy [7] or Yizkor Books [3], often the user knows that the name of interest has an "esto" in it (from a name like: Nové Mesto nad Váhom) but is uncertain about the remainder of the name or even the language that the name

C. Gurrin et al. (Eds.): ECIR 2010, LNCS 5993, pp. 483–494, 2010.

is in. This occurs fairly often since a variety of communities existed in each location. For example, for the German speaking community in Czechoslovakia during the 1930's, "Nové Mesto nad Váhom" was called "Neustadt an der Waag", and "Bratislava" was called "Pressburg" by German speakers and "Pozsony" by Hungarian speakers. In the hope that some fragment of the name will match, which obviously is not always the case, n-gram based solutions [8] are deployed.

Earlier efforts [2,4,6] have demonstrated that efficient simple rules can outperform many traditional approaches. Our language-independent name search approach, called SEGMENTS, follows this trend. That is, we search a collection of foreign names by segmenting the input string according to a set of simple rules. The search results obtained using the individual segments are merged, and a confidence for the merged list is derived. If the confidence is insufficient, namely below a predefined threshold, we invoke an n-gram search.

Extracting from a database of names derived from various documents and texts resident at and/or accessed by the United States Holocaust Memorial Museum, we favorably compare the search accuracy of SEGMENTS against traditional n-gram and D-M Soundex based solutions. Actual user queries as well as synthetic queries generated using single and multiple character addition, deletion, replacement, and inversion are used in our evaluation. We also show favorable results using a subset of the 1990 United States Census Bureau collection of surnames [16]. A subset rather than the entire collection was chosen so as to mirror the size of the United States Holocaust Memorial Museum data set used. SEGMENTS is used in support of search efforts for the United States Holocaust Memorial Museum.

2   Yizkor Book Metadata Search: A SEGMENTS Application

The Yizkor Book Metadata Search project, an effort led by the Archives Section of the United States Holocaust Memorial Museum, aims to create an online metadata global directory of Yizkor Books. Briefly, Yizkor Books memorialize life before, during, and after the Holocaust describing everyday events including births, marriages, and deaths. Many texts were written by survivors or their relatives or friends as a tribute to those who perished. Most texts are written using multiple languages. Thirteen languages are used: Czech, Dutch, English, French, German, Hebrew, Hungarian, Lithuanian, Polish, Romanian, Serbo-Croatian, Spanish, and Yiddish. Given the diversity of languages, only a language independent approach is viable.

Yizkor books are scattered globally, but currently, a global directory of these books is unavailable. Some Yizkor Book repositories go as far as to provide download-ready, scanned copies of the books, for example those residing within the New York City Public Library [11]; however, locating some of these digital repositories, particularly the lesser ones, is accomplished mainly by word of mouth.

The preliminary architecture of Yizkor Books metadata search system was initially described in [15]. Here, in Figure 1, we illustrate the currently deployed

architecture. As shown, metadata are generated internally by USHMM staff or
fellows, generated externally by a diversity of users including historians, geneal-
ogists, librarians, etc. or are downloaded directly from Yizkor Book repositories
and sent via the Internet to the metadata search engine. Once collected, they
are organized, temporarily housed in a verification repository (not illustrated),
and eventually stored in the Yizkor Books metadata repository. To guarantee
correctness, a temporary verification repository is used as an intermediary. That
is, all metadata are inspected and verified for accuracy by authorized personnel
prior to insertion into the metadata repository. Thus, there are always "humans
in the loop".

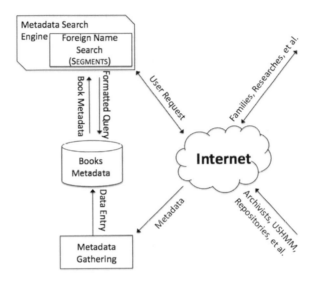

Fig. 1. Yizkor Overview

User queries from a diverse audience are issued and sent likewise via the Internet to the metadata search engine. SEGMENTS is the foreign name search
component within the search engine. The queries are translated to the appropriate internal format, issued against the repository, and corresponding metadata
are returned. The candidate results are then routed to the requesting party.

3   Algorithm

Our SEGMENTS approach operates as shown in Figure 2. Initially the user generated query is issued against the name index derived from the collection. No
attempt is made to identify the language. If an exact match is found, then the
matching name or names and their corresponding information are returned with
a confidence of 1. Otherwise, multiple substrings are derived applying simple

4   Soundex

To illustrate how SEGMENTS avoids problems faced by Soundex and D-M Soundex, a brief overview of these algorithms is necessary.

4.1   Soundex Overview

Soundex masks like-sounding characters by replacing them with integer representations, where said integers map to a set of characters. For example, in Soundex the integer 5 represents either "m" or "n". Furthermore, Soundex does not encode the first letter of the given query. Consider the word "Slovakia", which Soundex encodes as "S412" [10]. Details of the Soundex algorithm are omitted since only basic knowledge is required to understand the pitfalls.

4.2   Soundex Pitfalls

SEGMENTS addresses multiple known Soundex pitfalls. A subset of these known problems [12], which SEGMENTS resolves are:

1. **Dependence on initial letter.** If the first letter of the user's query is incorrect, Soundex will never find the correct result [12]. SEGMENTS however has 2 rules which will find the correct match.

2. **Noise intolerance.** [5,9,13] find that 80%-95% of misspellings within large documents are 1) one character insertions 2) one character deletions 3) one character replacements or 4) adjacent character swapping. Soundex, as demonstrated by [12], is unable to reliably resolve such noise. SEGMENTS however has demonstrated its tolerance for noise as shown in Tables 3 and 4.

3. **Poor precision.**
   One of the strengths of Soundex is the encoding of words as integers representing character groups. This representation however leads to ambiguity and ultimately degrades precision. For example, Soundex encodes the misspelled string "disapont" as "D215". A query would then be run for "D215" which would return: *disband, disbands, disbanded, disbanding, disbandment, disbandments, dispense, dispenses, dispensed, dispensing, dispenser, dispensers, dispensary, dispensaries, dispensable, dispensation, dispensations, deceiving, deceivingly, despondent, despondency, despondently, disobeying, disappoint, disappoints, disappointed, disappointing, disappointedly, disappointingly, disappointment, disappointments, disavowing* [9]. Should a query be correctly spelled, Soundex will still return several matches for the same reason. SEGMENTS does not suffer from the same ambiguity due to the voting process of the substring matching algorithms used. Since the above words all have the same encoding, ranking is usually done by frequency [14].

4.3   D-M Soundex

D-M Soundex, the Eastern European derivative, adjusts the elements of the character sets for language localization. It also improves upon the Soundex algorithm in the following select ways (localization changes/improvements are omitted)[10]:

1. Encoding of the initial letter. Consider the previous example "Slovakia", which Soundex encodes as "S412". In D-M Soundex "Slovakia" becomes "487500". Notice the extended length of the D-M Soundex encoding, that is the second improvement.
2. The first six (rather than four) significant codes are created. For example, Peters and Peterson have an identical encoding in Soundex ("P362"), but different in D-M Soundex ("739400", "739460").

These changes partially improve some of the downfalls of Soundex (4.2). For example, the first pitfall noted, *Dependence on initial letter*, is clearly solved by the first improvement. The second improvement aids the third pitfall, *Poor precision*, but adds to the time and space complexity. The *noise intolerance* pitfall however is not addressed, and is the root cause of the majority of misspellings. In fact, generally if any word contains more than four (Soundex) or six (D-M Soundex) consonants, all characters thereafter are ignored [1]. Therefore, neither Soundex or D-M Soundex are viable solutions for users, regardless of their knowledge of a language.

5   Evaluation

To evaluate our proposed approach, we randomly selected a subset of roughly 1,000 names from the Jewish Census residing at the United States Holocaust Memorial Museum. Names averaged 8 characters in length, with a median length of 8, a max length of 23, and a standard deviation of 2.8 characters. Using a set of 250 actual queries, we favorably compared the performance of SEGMENTS against the popular D-M Soundex approach and a traditional n-gram solution.

Two metrics were used, namely the percentage of names correctly identified and the average rank of those names found. A name was defined as found if it ranked in the top 60 entries (first three screens with 20 names listed per screen). Although we evaluated multiple n-values in the n-gram approach, we present results for only n=3, as it consistently supported the highest percentage found. Average rank rather than MRR is presented as it better illustrates the difference. Undetected entries are ignored in terms of the average rank computation. The statistical t-test was used to verify significance.

In Table 2, we present our findings using the collected actual 250 queries. Three measures (percentage found, average rank, and common average rank) for each of three approaches (D-M Soundex, 3-Grams, and SEGMENTS), when appropriate, are shown. The percentage of correctly identified names is presented in the percentage found column. Since the percentage of correctly identified names is higher when using SEGMENTS rather than 3-grams, a common column is presented to provide direct comparison of SEGMENTS average rank when only

Table 2. U.S. Holocaust Memorial Museum Live Query Performance Evaluation

	Percentage Found	Average Rank	Average Rank (Common)
D-M Soundex	27.56	1.03	N/A
3-Grams	62.17	12.00	N/A
SEGMENTS	78.19	8.26	7.39

considering names also correctly identified by 3-grams. The percentage of correctly identified names is presented in the percentage found column. Both SEGMENTS and 3-Grams statistically significantly (p<0.01) outperform D-M Soundex in terms of the percentage of correctly identified names. SEGMENTS likewise statistically significantly (p<0.01) outperforms 3-Grams in terms of the percentage of correctly identified names.

In the second column, the average rank (position) of those items found is presented. As shown, the average rank of SEGMENTS is superior to that of 3-Grams. The D-M Soundex average rank is nearly perfect; that is, names correctly identified are almost always positioned first in the rankings. However, this statistic is clearly misleading since roughly only a quarter of the names are correctly identified. What is true, however, is that whenever D-M Soundex recognizes the name, it perfectly identifies it, and this behavior is one possible explanation for the popularity of the D-M Soundex approach. Regardless of its popularity, the poor accuracy provided by D-M Soundex should prohibit its adoption.

All names identified by the 3-Gram approach are likewise identified by SEGMENTS. Given the difference in the percentage detected, clearly SEGMENTS detects additional otherwise unidentified names. To demonstrate how these additional names affect the average rank, we define the common percentage found metric. The common percentage found indicates the rank of those names found by both 3-Grams and SEGMENTS. As shown, the common percentage found is roughly one rank higher than the average rank. This demonstrates that the additionally identified names are, as expected, typically harder to match and increase the average rank.

The above analysis uses 250 actual queries and best represents typical use. However, to systematically evaluate our approach, we repeated the above evaluation, but this time, with an organized set of synthetically generated queries. That is, we randomly added, removed, replaced, and inverted characters in random locations, an approach commonly done to evaluate potential input errors [9]. Deletions were limited so that terms remained at least 4 characters. Three runs were made for each configuration; the averages are reported in Table 3.

In Table 3, the rows represent the various experiments conducted, namely insertion, deletion, replacement, and inversion of 1 to 4 characters. The position of the character(s) in the string is randomly generated using a uniform distribution. In the cases of multiple character inversions, randomly chosen pairs of characters are exchanged sequentially. In the single character inversion case, a

Table 3. U.S. Holocaust Memorial Museum Synthetic Query Performance Evaluation

	D-M Soundex (%)	N-Gram (%)	N-Gram (rank)	SEGMENTS (%)	SEGMENTS (rank)
INSERT					
1 char	41.44	94.94	2.55	100	1.71, 1.71
2 char	19.50	91.72	3.45	99.32	2.61, 2.43
3 char	10.67	87.82	4.11	97.52	3.18, 3.02
4 char	6.93	83.85	5.00	95.23	3.87, 3.79
DELETE					
1 char	42.12	93.33	3.45	99.97	2.51, 2.53
2 char	20.41	84.87	4.81	97.96	4.72, 3.95
3 char	11.12	74.68	5.77	92.71	6.42, 4.84
4 char	9.82	70.31	5.95	86.51	7.12, 5.14
REPLACE					
1 char	31.52	92.33	3.27	100	2.15, 2.01
2 char	16.35	80.95	4.49	93.90	4.19, 3.31
3 char	9.29	69.28	5.20	85.61	5.60, 3.87
4 char	5.87	57.81	5.98	75.18	6.85, 4.84
INVERT					
Adj. char	58.01	84.89	4.88	98.00	3.77, 3.00
2 char	17.31	54.59	6.78	71.61	7.38, 5.39
3 char	9.18	42.89	7.40	57.59	8.55, 6.22
4 char	7.09	34.64	8.49	46.76	9.29, 7.26

single adjacent pair of characters is selected. This special case was chosen so as to match the described errors in [9].

As shown, once again, SEGMENTS sustains a statistically significant ($p<0.01$) performance improvement over both D-M Soundex and the 3-gram solutions in terms of the percentage of names correctly identified. Likewise, once again, SEGMENTS correctly identifies all names detected by the 3-gram approach as well as some additional names. Hence, in the SEGMENTS (RANK) column, there are two entries: the first entry represents the average rank for all names identified; the second entry is the common average rank. The SEGMENTS approach always sustains a better average ranking when considering only those entries correctly identified by both approaches. In most cases, SEGMENTS also continues to sustain a better average ranking overall including those entries not found by the n-gram approach. In the few cases that SEGMENTS does not support a higher overall average ranking, the difference is relatively minimal. This occurs when the difference in percentage detection is significant. For all tests conducted, for all names identified by both the 3-Gram and SEGMENTS approaches, SEGMENTS was statistically significantly ($p<0.01$) superior.

Similar evaluation was performed using a subset of the 1990 United States Census Bureau Surnames data set [16]. Based on provided statistics, we tested the 1,000 most frequent surnames. A subset was chosen so as to mirror the size of the United States Holocaust Memorial Museum data set used. Names have a mean and median of 6 characters in length, a max of 11, and a standard

Table 4. 1990 U.S. Census Bureau Synthetic Query Performance Evaluation

	Soundex (%)	D-M Soundex (%)	N-Gram (%)	N-Gram (rank)	SEGMENTS (%)	SEGMENTS (rank)
INSERT						
1 char	27.89	28.23	95.69	1.65	100	1.71, 1.65
2 char	8.28	7.76	87.76	2.36	98.02	2.90, 2.36
3 char	2.67	2.29	79.11	3.14	94.73	3.10, 3.13
4 char	1.23	0.48	70.36	3.68	88.79	3.31, 3.68
DELETE						
1 char	45.28	41.87	95.90	2.08	100	2.05, 2.08
2 char	18.49	16.02	80.72	3.47	98.00	5.03, 3.47
3 char	7.11	5.32	62.79	4.14	90.18	7.40, 4.14
4 char	2.48	1.03	48.14	3.96	77.47	8.93, 3.96
REPLACE						
1 char	23.87	23.84	88.60	2.25	99.97	2.46, 2.25
2 char	8.75	8.54	64.75	3.26	84.79	6.02, 3.26
3 char	3.95	3.50	45.72	4.09	67.19	8.07, 4.08
4 char	1.88	1.95	29.23	4.86	51.23	10.75, 4.86
INVERT						
Adj. char	58.34	49.22	72.43	3.28	93.86	5.90, 3.28
2 char	21.65	18.52	31.39	4.16	48.00	10.45, 4.16
3 char	14.25	11.41	27.02	4.48	41.34	11.13, 4.48
4 char	11.96	10.37	21.02	4.74	32.67	11.44, 4.73

deviation of 2.4 characters. We, once again, synthetically altered all 1,000 names to generate queries, as previously described. We justify replacing user query logs with machine altered queries on the grounds that given a user who is proficient in a particular language, their queries have a higher probability to contain typos rather than true syntactical errors. As such, random manipulation of query terms results in near real-world examples. Furthermore, such input error testing is commonly done [9]. The results are shown in Table 4. As seen, the relative performance of these algorithms are similar to those obtained using the United States Holocaust Memorial Museum data set. Note Soundex and D-M Soundex have similar performance; hence, all remarks pertaining to D-M Soundex apply to Soundex.

Thus, for both actual queries and for systematically generated synthetic queries, SEGMENTS supports a statistically significant (p<0.01) performance improvement over both D-M Soundex and a traditional 3-gram solution.

6   Conclusion

To support foreign name identification in an environment in which users recall only distorted portions of desired names, we developed a language-independent, fusion-based, segment-oriented, n-gram supported, search system called SEGMENTS. Initially, SEGMENTS searches a name index for an exact match. If a name or names are found, they are returned to the user with a confidence of 1. Otherwise, a set of candidate substrings are generated using a set of simple parsing rules. These generated substrings are searched as candidate queries against the name index, and all partially matching names are returned. A confidence for each partial match is computed, and a global confidence for all derived potential result names is likewise computed. The global confidence is compared against a pre-established threshold, and if this threshold is met, a ranked list of derived name candidates, along with the individual confidence of each candidate, is returned to the user. Name candidates are ranked according to their confidence. If, however, the global confidence fails to exceed the pre-established threshold, a traditional n-gram solution is run to derive an additional set of potential result candidates. A global confidence is similarly computed for these candidates. The global confidences for both approaches are compared, and the results corresponding to the higher of the two confidences are returned to the user.

We evaluated our approach using a Jewish Census data set resident at the United States Holocaust Museum and using the 1990 Surnames Census data set from the United States Census Bureau. For our Jewish Census evaluation, to determine expected "real-world" performance, we collected user queries and used them as our initial query test set. As user queries, however, do not necessarily systematically evaluate the approaches under consideration, we likewise created a synthetic query mix derived based on the prior art. That is, actual queries were used to access realistic typical behavior; synthetic queries were used to systematically "stress test" the search system. Our results demonstrate the significantly higher accuracy of our approach as compared to both the D-M Soundex approach presently used initial the JewishGen genealogy search and a traditional n gram approach using a variety of n values.

Our approach is in current use to enhance search functionally for the United States Holocaust Memorial Museum Yizkor Books effort.

References

1. Beider, A., Morse, S.: Beider-Morse Phonetic Matching: An Alternative to Soundex with Fewer False Hits. Avotaynu: the International Review of Jewish Genealogy (Summer 2008)
2. Aljlayl, M., Frieder, O.: On Arabic Search: Improving the Retrieval Effectiveness via a Light Stemming Approach. In: ACM Eleventh Conference on Information and Knowledge Management (CIKM), Washington, DC (November 2002)
3. Amir, M.: From Memorials to Invaluable Historical Documentation: Using Yizkor Books as Resource for Studying A Vanished World. In: Annual Convention of the Association of Jewish Libraries, La Jolla, California (June 2001)
4. Aqeel, S., Beitzel, S., Jensen, E., Grossman, D., Frieder, O.: On the Development of Name Search Techniques for Arabic. Journal of the American Society of Information Science and Technology 57(6) (April 2006)
5. Damerau, F.: A technique for computer detection and correction of spelling errors. Communications of the ACM 7(3), 171–176 (1964)
6. Guy, F., Oard, D.: The TREC-2001 Cross-Language Information Retrieval Track: Searching Arabic using English, French or Arabic Queries. NIST TREC, Gaithersburg, Maryland (November 2001)

7. JewishGen, September 1 (2009), http://jewishgen.org
8. Manning, C., Raghavan, P., Schutze, H.: Introduction to Information Retrieval. Cambridge University Press, Cambridge (2008)
9. Mitton, R.: Spellchecking by Computers. Journal of the Simplified Spelling Society, J20 (1996)
10. Mokotoff, G.: Soundexing and Genealogy (2007), http://www.avotaynu.com/soundex.html (September 1, 2009)
11. New York Public Library Yizkor Books, September 1 (2009), http://www.nypl.org/research/chss/jws/yizkorbookonline.cfm
12. Patman, F., Shaefer, L.: Is Soundex Good Enough for You? On the Hidden Risks of Soundex-Based Name Searching. Language Analysis Systems, Inc., Herndon (2003)
13. Pollock, J., Zamora, A.: Automatic spelling correction in scientific and scholarly text. Communications of the ACM 27(4) (April 1984)
14. Snae, C., Bruckner, M.: Novel Phonetic Name Matching Algorithm with a Statistical Ontology for Analysing Names Given in Accordance with Thai Astrology. Issues in Informing Science and Information Technology (2009)
15. Soo, J., Cathey, R., Frieder, O., Amir, M., Frieder, G.: Yizkor Books: A Voice for the Silent Past. In: ACM Seventeenth Conference on Information and Knowledge Management (CIKM), Napa Valley, California (October 2008)
16. United States Census Bureau 1990 Surnames, September 1 (2009), http://www.census.gov/genealogy/names/dist.all.last
17. Zobel, J., Dart, P.: Phonetic String Matching: Lessons from Information Retrieval. In: ACM Nineteenth Conference on Research and Development in Information Retrieval (SIGIR), Zurich, Switzerland (August 1996)

Promoting Ranking Diversity for Biomedical Information Retrieval Using Wikipedia

Xiaoshi Yin[1,2], Xiangji Huang[2], and Zhoujun Li[1]

[1] College of Computer Science and Technology,
Beihang University Beijing, China
[2] School of Information Technology,
York University Toronto, Canada
xiaoshiyin@cse.buaa.edu.cn, jhuang@yorku.ca, lizj@buaa.edu.cn

Abstract. In this paper, we propose a cost-based re-ranking method to promote ranking diversity for biomedical information retrieval. The proposed method concerns with finding passages that cover many different aspects of a query topic. First, aspects covered by retrieved passages are detected and explicitly presented by Wikipedia concepts. Then, an aspect filter based on a two-stage model is introduced. It ranks the detected aspects in decreasing order of the probability that an aspect is generated by the query. Finally, retrieved passages are re-ranked using the proposed cost-based re-ranking method which ranks a passage according to the number of new aspects covered by the passage and the query-relevance of aspects covered by the passage. A series of experiments conducted on the TREC 2006 and 2007 Genomics collections demonstrate the effectiveness of the proposed method in promoting ranking diversity for biomedical information retrieval.

Keywords: Ranking diversity, re-ranking, biomedical IR.

1 Introduction

Given a query, information retrieval (IR) system should return a ranked list of retrieved documents (or passages) to users. Retrieved documents are ranked in the order of their probabilities of relevance to the query. Traditional retrieval models assume that the relevance of a document is independent of the relevance of other documents. This assumption may result in high redundancy and low diversity in a ranked list, since documents that are similar in content tend to appear over and over again. However, in reality, a user may want to see the top ranked documents concerning different aspects of his/her information need instead of reading relevant documents that only deliver redundant information. A better information retrieval system thus should return ranked lists that take relevance as well as ranking diversity into account.

Current genomic research is characterized by immense volume of data, accompanied by a tremendous increase in the number of genomic and biomedical related publications. This wealth of information has led to an increasing

C. Gurrin et al. (Eds.): ECIR 2010, LNCS 5993, pp. 495–507, 2010.
© Springer-Verlag Berlin Heidelberg 2010

amount of interest and need for applying information retrieval techniques to access the scientific literature in genomics and related biomedical disciplines. In many cases, the desired information of a question (query) asked by biologists is a list of a certain type of entities covering different aspects that are related to the question [1], such as genes, proteins, diseases, mutations, etc. Hence it is important for a biomedical IR system to be able to provide comprehensive and diverse answers to fulfill biologists' information needs. In the most recent TREC Genomics tracks, the "aspect retrieval" was investigated. Its purpose is to study how a biomedical retrieval system can support a user gather information about the different aspects of a topic[1]. Biomedical IR systems are required to return relevant information at the passage level, while relevance judges not only rated the passages, but also grouped them by aspect. Aspects of a retrieved passage could be a list of named entities or MeSH terms, representing answers that cover different portions of a full answer to the query. Aspect Mean Average Precision (Aspect MAP) was defined in the Genomics tracks to capture similarities and differences among retrieved passages. It indicates how comprehensive the question is answered. Relevant passages that do not contribute any new aspects to the aspects retrieved by higher ranked passages are removed from the ranking, and will not be used to accumulate Aspect MAP [5]. Therefore, Aspect MAP is a measurement for the redundancy and diversity in the IR ranked list.

In order to promote diversity and reduce redundancy in the ranked list for biomedical information retrieval, we propose a cost-based re-ranking method. First, aspects that are covered by each retrieved passage are detected using Wikipedia. Retrieved passages are presented by a list of Wikipedia concepts[2]. Second, since for a given query, relevant aspects and non-relevant aspects are both detected, a two-stage model is employed for aspect filtering. The aspect filter ranks the detected aspects in decreasing order of the probability that an aspect is generated by the query. Third, in order to take relevance as well as diversity into account for passage re-ranking, a cost-based re-ranking method is proposed. The ranking score of a retrieved passage is decided by two factors: the number of new aspects delivered by the passage and the query-relevance of aspects delivered by the passage.

Experiments conducted on two TREC Genomics collections and two very different IR baseline runs demonstrate the effectiveness of our approach. The evaluation results show that our approach can achieve 20% improvement over the highest Aspect MAP reported in the TREC 2007 Genomics track, and 14% improvement over the highest passage level MAP reported in the TREC 2007 Genomics track [1]. When experiments are conducted on the TREC 2006 Genomics collection, our approach can also make 9% performance improvement over the baseline run in terms of Aspect MAP and 6.6% improvement over the baseline run in terms of Passage MAP.

---

[1] The "aspect retrieval" is also defined in the TREC Interactive track with a similar objective [2][3][4].

[2] The title of a Wikipedia article is regarded as a Wikipedia concept.

The rest of this paper is organized as follows. Section 2 gives a brief survey of prior work. The methods for detecting aspects using Wikipedia, filtering aspects and re-ranking retrieved passages are presented in Section 3, Section 4 and Section 5 respectively. Experiments and result analyses are provided in Section 6. Finally, conclusions and future work are given in Section 7.

2   Related Work

Our work is inspired by several recent papers that concerned with promoting diversity and novelty in the information retrieval ranked list. Carbonell *et al.* introduced the maximal marginal relevance (MMR) method, which attempts to maximize relevance while minimizing similarity to higher ranked documents [6]. In order to measure the redundancy between documents, Zhang *et al.* presented four redundancy measures, which are "set difference", "geometric distance", "distributional similarity" and "a mixture model" [7]. They modeled relevance and redundancy separately. Since they focused on redundant document filtering, experiments in their study were only conducted on a set of relevant documents. However, in reality, non-relevant documents are always returned by IR systems along with relevant documents. Redundancy and relevance should be both considered. Zhai *et al.* validated a subtopic retrieval method based on a risk minimization framework [8]. Their subtopic retrieval method combines the mixture model novelty measure with the query likelihood relevance ranking. Different from their method, our approach directly classifies aspects into "new relevant aspects", "redundant relevant aspects", "new non-relevant aspects" and "redundant non-relevant aspects". The combination of diversity and relevance is conducted at the aspect level.

In biomedical information retrieval, the Genomics aspect retrieval was firstly proposed in the 2006 Genomics track and further investigated in the 2007 Genomics track. Many research groups joined these annual campaigns to evaluate their systems and methodologies. However, to the best of our knowledge, there is not too much previous work conducted on Genomics aspect retrieval for promoting diversity in the ranked list. University of Wisconsin re-ranked the passages using a clustering-based approach named GRASSHOPPER to promote ranking diversity [9]. GRASSHOPPER is an alternative to MMR and variants with a principled mathematical model and strong empirical performance on artificial data set [10]. Unfortunately, for the Genomics aspect retrieval, this re-ranking method hurt their system's performance and decreased the Aspect MAP of the original results [9]. Later in the TREC 2007 Genomics track, most teams tried to obtain the aspect level performance through their passage level results, instead of working on the aspect level retrieval directly [1,11,12]. Another recent study concerning on the Genomics aspect retrieval was conducted by Huang *et al.* [13]. Their experimental results demonstrated that the hidden property based re-ranking method can achieve promising and stable performance improvements. Our approach, in contrast, presents aspects covered by retrieved passages explicitly. Genomics aspects delivered by a specific passage thus become readable to users.

3   Aspect Detection Using Wikipedia

3.1   Why Wikipedia?

Wikipedia is a free online encyclopedia edited collaboratively by large numbers of volunteers. The exponential growth and the reliability of Wikipedia make it a potentially valuable knowledge resource. How to utilize Wikipedia as an external knowledge base to facilitate information retrieval became a hot research topic over the last few years [14,15,16,17,18].

However, as far as we are aware, there is no work done on investigating how to use Wikipedia for improving biomedical literature retrieval. The main reason for this is that some domain-specific thesauri are available for biomedical retrieval (e.g. UMLS, MeSH and the Gene Ontology). Nonetheless, these domain-specific thesauri only provide synonyms, hypernyms, hyponyms of a specific term without any other context. Therefore, it is hard to tell which lexical variants of a specific term should be used for retrieving users' information need. Previous studies of using domain-specific thesauri to promote biomedical retrieval accuracy usually assigned lexical variants manually to achieve retrieval performance improvements [19][20][21]. The use of lexical variants of medical terms in automatic indexing is questionable in some cases [22] and the retrieval results of using domain-specific knowledge base are somewhat conflicting [11,23,24].

Wikipedia on the other hand not only provides concepts (entities) and lexical variants of a specific term, but also provides abundant contexts. With the help of enriched entity pages, it is possible to identify which concepts and lexical variants are related under a specific context. As Wikipedia articles are constantly being updated and new entries are created everyday [14], we can expect that Wikipedia covers the great majority of medical terms. Another reason for using Wikipedia instead of domain-specific thesauri is that it contains plenty of linkage information among semantic related entities. Each link in Wikipedia is associated with an anchor text, which can be regarded as a descriptor of its target article. Anchor texts provide alternative names, morphological variations and related phrases for the target articles. Anchors also encode polysemy, because the same anchor may link to different articles depending on the context in which it is found [25].

3.2   Aspect Detection

In order to evaluate how comprehensive a biomedical related question is answered by a information retrieval system, TREC Genomics track proposed the aspect level retrieval. The aim of the aspect retrieval task is to promote diversity and novelty of retrieval ranking. Passages that are relevant as well as covers new aspects of a complete topic answer should be ranked higher than passages that are relevant to the query but only deliver redundant aspects. Therefore, intuitively, detecting aspects that are covered by each retrieved passage is the first step for promoting diversity and novelty of retrieval results.

Wikipedia, as introduced above, can be used for aspect detection. There are three steps involved: (1) identifying candidate phrases in the given retrieved passage, (2) mapping them to Wikipedia articles, (3) selecting the most salient concepts. The outcome is a set of concepts representing the aspects mentioned in the input passages [17,25]. The Wikify service provided by the Wikipedia Miner[3] is used to automatically detect aspects covered by a retrieved passage. An example is shown in Table 1. Terms that can be linked to their corresponding Wikipedia concepts are displayed in bold font.

Table 1. An example for aspect detection using Wikipedia

query: What serum [PROTEINS] change expression in association with high disease activity in lupus?
retrieved passage: The association aCL anti- 223 2 GPI **lupus nephritis** strengthened strong association seen aCL positivity conjunction positivity **anti-dsDNA** anti-C1q **antibodies** examined presence levels **serum** act useful markers severity **renal disease** lupus nephritis patients likely positive **autoantibodies** non-nephritis **SLE** PAPS patients Further prospective studies monitoring levels **autoantibodies** useful determining disease activity predicting development nephritis **SLE**
detected aspects: Lupus nephritis; Systemic lupus erythematosus; Antibody; Kidney; Autoantibody;

4 Aspect Filtering

Previous work in reducing redundancy and promoting diversity of IR ranked lists usually assumed that retrieved documents (or passages) are relevant to the query or previous ranked documents are relevant to the query [7,9,6]. However, this assumption may not hold in practice because non-relevant documents could also be retrieved. Therefore, some detected aspects of retrieved passages may be irrelevant to the specific query and should not be used for re-ranking.

Detected aspects are firstly divided into two categories: query background aspects and passage aspects. Aspects appear repeatedly among retrieved passages (aspects with high passage frequencies) are regarded as query background aspects. Detected aspects except query background aspects are regarded as passage aspects. Since query background aspects can be viewed as query-specific stopwords, they should not be used for re-ranking either. Passage aspects on the other hand determine whether a passage is relevant as well as novel. The aspect filter introduced here aims to sort the passage aspects according to the probability that an aspect is generated by the query. A two-stage model that combines a relevance model and a co-occurrence model [26][27] is used for ranking detected passage aspects. More formally, the two-stage model is defined as:

$$P(a_i|Q) = \sum_{psge} P(a_i, psge|Q) = \sum_{psge} P(psge|Q)P(a_i|psge, Q) \tag{1}$$

where $P(psge|Q)$ is the relevance model presenting whether a retrieved passage $psge$ is relevant to the query Q; $P(a_i|psge, Q)$ is the co-occurrence model presenting whether an aspect a_i is associated with the query.

The relevance model can be estimated using the baseline ranking scores of retrieved passages. To estimate the co-occurrence model, a linear interpolation of $P(a_i|psge, Q)$ and the query background information are used as:

---

[3] http://wikipedia-miner.sourceforge.net

$$P(a_i|psge, Q) = (1 - \mu)\frac{freq(a_i, psge)}{\sum_{a' \in A} freq(a', psge)} + \mu\frac{1}{df(a_i)}\sum_{psge' \in P}\frac{freq(a_i, psge')}{\sum_{a' \in A} freq(a', psge')} \qquad (2)$$

where A denotes the set of detected passage aspects for the query; P denotes the set of retrieved passages for the query; $freq(x, y)$ denotes the frequency of x in y and $df(a_i)$ denotes the passage frequency of an aspect a_i. We use a Dirichlet prior for the smoothing parameter μ.

$$\mu = \frac{\kappa}{\sum_{a' \in A; psge' \in P} freq(a', psge') + \kappa} \qquad (3)$$

where κ is the average aspect frequency of all detected aspects in the retrieved passages.

The outcome of the aspect filter is a ranked list of detected passage aspects and each passage aspect a_i is associated with its ranking score $P(a_i|Q)$.

5   A Re-ranking Method for Promoting Diversity

Retrieved passages are presented by passage aspects[4] explicitly, and each passage aspect is associated with its ranking score calculated by Equation 1. We now consider how to utilize passage aspects for re-ranking so as to promote ranking diversity and novelty.

We consider that a user would prefer a ranked list of passages such that the first ranked passage should be relevant to the query and could also cover as many aspects as possible. Therefore, the first ranked passage is:

$$passage_1 = \arg\max_{psge_i}\{\sum_{a_j \in A_{p_i}} P(a_j|Q)\} \qquad (4)$$

where $psge_i$ is a retrieved passage and A_{p_i} is the set of aspects that are detected from $psge_i$.

For other retrieved passages, the passage ranking should depend on which passages the user has already seen. Suppose that we have ranked top $i - 1$ passages, and now we need to decide which passage should be ranked at the ith position in the ranking list. The passage which can deliver the most new and query-relevant aspects (aspects that are relevant to the query and are not covered by previous ranked passages) should be considered as the passage at the ith position in the ranking list. In order to take both passage novelty and relevance into account, we rank passages using the following cost function:

$$c(psge_i|psge_1, ..., psge_{i-1}) = \lambda_1 P(NEW_R|psge_i) + \lambda_2 P(NEW_{NR}|psge_i)$$
$$+ \lambda_3 P(RDD_R|psge_i) + \lambda_4 P(RDD_{NR}|psge_i) \qquad (5)$$

---

[4] Since only passage aspects are used for re-ranking, we use "passage aspect" and "aspect" interchangeably in the rest of this paper.

where $P(NEW_R|psge_i)$ is the probability that a passage delivers relevant as well as new aspects; $P(NEW_{NR}|psge_i)$ is the probability that a passage delivers new but non-relevant aspects; $P(RDD_R|psge_i)$ is the probability that a passage delivers relevant but redundant aspects; $P(RDD_{NR}|psge_i)$ is the probability that a passage delivers non-relevant and redundant aspects; and $\lambda_i (i = 1, 2, 3, 4)$ are cost constants.

We consider that there is no cost associated to retrieving relevant as well as non-redundant passages, we thus set $\lambda_1 = 0$. Since a passage that covers new non-relevant aspects and a passage that only covers old non-relevant aspects are both not of interest to users, we set $\lambda_2 = \lambda_4$. Because λ_3 indicates the cost of users seeing a query-relevant but redundant passage, we assume that $\lambda_3 > 0$. We denote λ_2 as λ_{NR} and denote λ_3 as λ_R, then Equation 5 can be written as:

$$c(psge_i|psge_1, ..., psge_{i-1}) = (\lambda_R - \lambda_{NR})P(RDD_R|psge_i) + \lambda_{NR}(1 - P(NEW_R|psge_i)) \quad (6)$$

Therefore, the ranking of a retrieved passage is decided by:

$$c(psge_i|psge_1, ..., psge_{i-1}) \stackrel{rank}{=} (\frac{\lambda_R}{\lambda_{NR}} - 1)P(RDD_R|psge_i) - P(NEW_R|psge_i) \quad (7)$$

The cost score is effected by the ratio $\frac{\lambda_R}{\lambda_{NR}}$[5]. As the cost of a user seeing a non-relevant passage should be larger than the cost of a user seeing a redundant relevant passage, we consider that $\frac{\lambda_R}{\lambda_{NR}} < 1$. $P(RDD_R|psge_i)$ and $P(NEW_R|psge_i)$ in Equation 7 can be calculated as:

$$P(RDD_R|psge_i) = \sum_{a_j \in A_{p_i}, a_j \in A_0} P(a_j|Q) \quad (8)$$

$$P(NEW_R|psge_i) = \sum_{a_j \in A_{p_i}, a_j \notin A_0} P(a_j|Q) \quad (9)$$

where A_0 is the set of aspects that have been seen in the top $i - 1$ passages. $P(RDD_R|psge_i)$ reflects the commonness between $psge_i$ and previously ranked passages, while $P(NEW_R|psge_i)$ reflects the new information delivered by $psge_i$. Therefore, when $\frac{\lambda_R}{\lambda_{NR}} < 1$, the passage that contains more common query-relevant aspects as well as more unseen query-relevant aspects has lower cost score and should be ranked higher.

6   Experiments

6.1   Test Collections and Evaluation Measures

In order to evaluate the proposed method for promoting diversity in ranking for biomedical information retrieval, we use the TREC 2006 and 2007 Genomics

---

[5] We refer $\frac{\lambda_R}{\lambda_{NR}}$ as "cost ratio" in the rest of this paper.

track collections as the test corpus. These two TREC collections are full-text biomedical corpus consisting of 162,259 documents from 49 genomics-related journals indexed by MEDLINE. Documents in the collections are in HTML format, and can be identified by their PMIDs [5][1]. 28 official topics from the TREC 2006 Genomics track and 36 official topics from the TREC 2007 Genomics track are used as queries. Topics are in the form of questions asking for lists of specific entities that cover different portions of full answers to the topics [5][1].

There were three levels of retrieval performance that were measured in the TREC 2006 and 2007 Genomics tracks: passage retrieval, aspect retrieval and document retrieval. Each was measured by some variants of mean average precision (MAP). Passage MAP, Passage2 MAP[6], Aspect MAP and Document MAP were four evaluation measures corresponding to the three levels of retrieval performance. The definitions of these MAPs can be found in [5] and [1]. In this paper, we mainly focus on Aspect MAP, Passage2 MAP (for the 2007's topics) and Passage MAP (for the 2006's topics), since our objective is to promote diversity in the ranked list of retrieved passages. Moreover, aspect retrieval and passage retrieval were also the major tasks in these two Genomics tracks.

6.2   Information Retrieval Baseline Runs

For the 2007's topics, two IR baseline runs are used. The first baseline run is NLMinter developed by the U.S. National Library of Medicine [11]. This baseline run achieved the best performance in the TREC 2007 Genomics track in terms of Aspect MAP, Passage2 MAP and Document MAP [1][11]. NLMinter merged retrieval results from five IR systems (Essie, Indri, Terrier, Theme and EasyIR) and further integrated an interactively created filter [11]. The second baseline run is an Okapi run, which is solely based on the probabilistic weighting model BM25 [28]. The performance of the Okapi run is above average among all results reported in the TREC 2007 Genomics track [1]. For the 2006's topics, we only test our method based on the Okapi run as the retrieval ranked list of NLMinter is not available. The Okapi run's performance on the 2006's topic is also above average among all results reported in the TREC 2006 Genomics track [5]. The performances of baseline runs are shown in Table 3 and Table 4. The best and mean results reported in the 2006 and 2007 Genomics tracks are shown in Table 2.

Table 2. The best and mean results in the Genomics tracks

	2007 Genomics track				2006 Genoics track		
	Aspect	Passage	Passage2	Document	Aspect	Passage	Document
Best MAP	0.2631	0.0976	0.1148	0.3286	0.4411	0.1486	0.5439
Mean MAP	0.1326	0.0560	0.0398	0.1862	0.1643	0.0392	0.2887

---

[6] Passage2 MPA was defined in the TREC 2007 Genomics track, which is an alternative measure to the Passage MAP defined in the TREC 2006 Genomics track.

Table 3. Re-ranking performances on the TREC 2007 Genomics collection

MAP	Aspect		Passage		Passage2		Document	
NLMinter	0.2631		0.0968		0.1148		0.3286	
aspect-cost, filter	0.3166	(+20.3%)	0.1052	(+8.7%)	0.1312	(+14.3%)	0.3502	(+6.6%)
aspect-cost, no filter	0.2299	(-12.6%)	0.0878	(-9.3%)	0.1080	(-5.9%)	0.2950	(-10.2%)
MAP	Aspect		Passage		Passage2		Document	
Okapi	0.1428		0.0633		0.0641		0.2025	
aspect-cost, filter	0.1642	(+15%)	0.0651	(+2.8%)	0.0679	(+5.9%)	0.2116	(4.5%)
aspect-cost, no filter	0.1394	(-2.4%)	0.0571	(-9.8%)	0.0569	(-11.2%)	0.2001	(-1.2%)

Table 4. Re-ranking performances on the TREC 2006 Genomics collection

MAP	Aspect		Passage		Document	
Okapi	0.2176		0.0362		0.3476	
aspect-cost, filter	0.2374	(+9.1%)	0.0386	(+6.6%)	0.3549	(+2.1%)
aspect-cost, no filter	0.2081	(-4.4%)	0.0334	(-7.7%)	0.3389	(-2.5%)

6.3 Re-ranking Performances

Re-ranking results of using the proposed method are shown in Table 3 and Table 4. The values in the parentheses are the relative rates of improvement over the original results. We can see that our approach (with the aspect filter) makes consistent improvements over both baseline runs. Improvements can be achieved on all levels of MAP measures. For the efficiency reason, we only re-ranked the top 100 passages. However, significant improvements over all baseline runs in terms of Aspect MAP still can be made (Wilcoxon test at significance level 0.05). Re-ranking performances are also effected by the cost ratio $\frac{\lambda_R}{\lambda_{NR}}$. In Tables 3 and 4, we only show the re-ranking results with $\frac{\lambda_R}{\lambda_{NR}} = 0.2$. The impact of the cost ratio will be discussed in the next subsection.

6.4 Result Analyses

Effect of the aspect filter. As shown in Table 3 and Table 4, re-ranking without the aspect filter hurts the baseline runs' performances. This is because detected aspects are not ranked by the filter and are all treated as relevant aspects. Experimental results demonstrate the effectiveness of the aspect filter. The use of the aspect filter can make a positive impact to the passage re-ranking.

Impact of the cost ratio. The cost ratio $\frac{\lambda_R}{\lambda_{NR}}$ indicates the relative cost of seeing a relevant passage with redundant aspects compared with seeing a passage only with non-relevant aspects. This tradeoff between retrieving passages with new aspects and avoiding the retrieval of non-relevant passages has an impact on the retrieval performance. As the aspect level retrieval and the passage level retrieval were two major tasks in the TREC 2006 and 2007 Genomics tracks, system performances at these two levels with different cost ratios are shown in

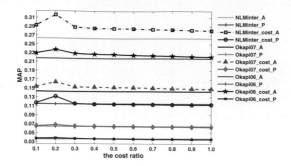

Fig. 1. Impact of the cost ratio. ("baseline_A" and "baseline_P" denote baseline run's aspect level and passage level retrieval performance respectively. "baseline_cost_A" and "baseline_cost_P" denote system's retrieval performance after applying the cost-based re-ranking method.)

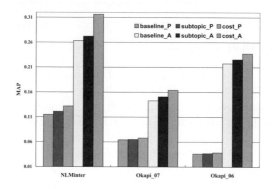

Fig. 2. Comparison with the subtopic retrieval method. ("baseline_A" and "baseline_P" denote baseline's retrieval performance at the aspect level and the passage level respectively. "subtopic_A", "cost_A", "subtopic_P" and "cost_P" denote system's retrieval performance at these two levels with the use of the subtopic method and the cost-based re-ranking method respectively.)

Figure 1. We can see that when $\frac{\lambda_R}{\lambda_{NR}} < 0.3$, performance improvements can be achieved at both the passage level and the aspect level retrieval. However, when $\frac{\lambda_R}{\lambda_{NR}} \geq 0.4$, improvements on the aspect level retrieval tend to decrease as the cost ratio increases, and systems' retrieval performances at the passage level are not as good as the baseline runs. Overall, in all experiments we conducted, the proposed re-ranking method performs the best when $\frac{\lambda_R}{\lambda_{NR}} = 0.2$.

Comparison with the subtopic retrieval method. Mixture model is another novelty measure. It assumes that the new document is generated by a two-component mixture model, in which one component is the old reference topic model and the other is a background language model [8]. Previous studies demonstrated the effectiveness of the mixture model. It outperforms several commonly used novelty measures, such as "set difference", "geometric distance", "distributional similarity", etc. [7][8]. The subtopic retrieval method based on a risk minimization framework proposed by Zhai *et al.* in [8] combined relevance scores from a retrieval baseline with novelty scores from the mixture model at document (passage) level. This subtopic retrieval method has been shown to be effective in promoting novelty and diversity in the ranked list [8]. In order to further evaluate our proposed method, we compare our method with Zhai's subtopic retrieval method.

The comparison results shown in Figure 2 illustrate that our proposed re-ranking method is more effective in promoting diversity for biomedical information retrieval. In our method, aspects covered by retrieved passages are presented by corresponding Wikipedia concepts. Only detected aspects (concepts) are used for computing novelty and diversity. On the other hand, Zhai's subtopic retrieval method employs the mixture model as the novelty measure, which uses the content (terms) of retrieved passages to compute novelty and diversity. Because of the frequent use of (possibly non-standardized) acronyms, the presence of homonyms and synonyms in biomedical literatures, using Wikipedia (with enriched contexts provided by entity pages) to detect aspects and presenting them with Wikipedia concepts could result in better biomedical IR performances.

7   Conclusions and Future Work

In this paper, we propose a cost-based re-ranking method to promote diversity in the ranked list for biomedical information retrieval. Our contribution is three-fold. First, to the best of our knowledge, this is the first study of using Wikipedia to facilitate biomedical IR. The major advantage of using Wikipedia instead of using domain-specific thesauri is that Wikipedia not only provides concepts (entities) and lexical variants of a specific term, but also provides abundant contexts. With the help of enriched entity pages in Wikipedia, it is possible to identify which concepts and lexical variants are related under a specific context. Second, an aspect filter based on the two-stage model is introduced. It sorts passage aspects according to their query relevance, which is defined as the probability that the aspect is generated by the query. Experimental results show the effectiveness of the aspect filter. Third, a cost-based re-ranking method is proposed, which takes both passage novelty and relevance into account. The ranking score of a retrieved passage is decided by the number of new aspects delivered by the passage and the query-relevance of aspects delivered by the passage. Experiments conducted on two TREC Genomics collections demonstrate that the proposed re-ranking method is effective in promoting ranking diversity for biomedical IR.

For our future work, we plan to improve the efficiency of the proposed re-ranking method. Instead of detecting aspects covered by retrieved passages, we could conduct the aspect detection as a preprocessing step. We also plan to further evaluate the cost-based re-ranking method on other test collections. For example, we could use the ClueWeb09 collection to investigate whether the proposed method is still effective for promoting ranking diversity in Web search.

Acknowledgment

This research is supported in part by the research grant from the Natural Science & Engineering Research Council of Canada, the Early Research Award/Premiers Research Excellence Award and the award from the China Scholarship Council.

References

1. Hersh, W., Cohen, A., Ruslen, L., Roberts, P.: TREC 2007 Genomics track overview. In: Proc. of TREC-16 (2007)
2. Over, P.: TREC-6 Interactive track report. In: Proc. of TREC-6 (1998)
3. Over, P.: TREC-7 Interactive track report. In: Proc. of TREC-7 (1999)
4. Hersh, W., Over, P.: TREC-8 Interactive track report. In: Proc. of TREC-8 (2000)
5. Hersh, W., Cohen, A., Roberts, P., Rekapalli, H.: TREC 2006 Genomics track overview. In: Proc. of TREC-15 (2006)
6. Carbonell, J., Goldstein, J.: The use of MMR, diversity-based reranking for re-ordering documents and producing summaries. In: Proc. of the 21st ACM SIGIR (1998)
7. Zhang, Y., Callan, J., Minka, T.: Novelty and redundancy detection in adaptive filtering. In: Proc. of the 25th ACM SIGIR (2002)
8. Zhai, C., Cohen, W.W., Lafferty, J.: Beyond independent relevance: methods and evaluation metrics for subtopic retrieval. In: Proc. of the 26th ACM SIGIR (2003)
9. Goldberg, A.B., Andrzejewski, D., Gael, J.V., Settles, B., Zhu, X., Craven, M.: Ranking biomedical passages for relevance and diversity: University of Wisconsin, Madison at TREC Genomics 2006. In: Proc. of TREC-15 (2006)
10. Zhu, X., Goldberg, A., Gael, J.V., Andrzejewski, D.: Improving diversity in ranking using absorbing random walks. In: Human Language Technologies 2007: The Conference of the North American Chapter of the Association for Computational Linguistics; Proc. of the Main Conference (2007)
11. Demner-Fushman, D., Humphrey, S.M., Ide, N.C., Loane, R.F., Mork, J.G., Ruch, P., Ruiz, M.E., Smith, L.H., Wilbur, W.J., Aronsona, A.R.: Combining resources to find answers to biomedical questions. In: Proc. of TREC-16 (2007)
12. Zhou, W., Yu, C.: TREC Genomics track at UIC. In: Proc. of TREC-16 (2007)
13. Huang, X., Hu, Q.: A bayesian learning approach to promoting diversity in ranking for biomedical information retrieval. In: Proc. of the 32nd ACM SIGIR (2009)
14. Yu, Y., Jones, G.J., Wang, B.: Query dependent pseudo-relevance feedback based on Wikipedia. In: Proc. of the 32nd ACM SIGIR (2009)
15. Ye, Z., Huang, X., Lin, H.: A graph-based approach to mining multilingual word associations from Wikipedia. In: Proc. of the 32nd ACM SIGIR (2009)
16. Medelyan, O., Witten, I., Milne, D.: Topic indexing with Wikipedia. In: Proc. of AAAI Workshop on Wikipedia and Artificial Intelligence (2008)
17. Milne, D.N., Witten, I.H., Nichols, D.M.: A knowledge-based search engine powered by Wikipedia. In: Proc. of the 16th ACM CIKM (2007)
18. Gabrilovich, E., Markovitch, S.: Computing semantic relatedness using Wikipedia-based explicit semantic analysis. In: Proc. of the 20th IJCAI (2007)
19. Hersh, W., Buckley, C., Leone, T., Hickam, D.: OHSUMED: An interactive retrieval evaluation and new large test collection for research. In: Proc. of the 17th ACM SIGIR (1994)

20. Srinivasan, P.: Optimal document-indexing vocabulary for MEDLINE. Information Processing and Management 32(5), 503–514 (1996)
21. Savoy, J.: Bibliographic database access using free-text and controlled vocabulary: An evaluation. Information Processing and Management 41(4), 873–890 (2005)
22. Cimino, J.J.: Vocabulary and health care information technology: State of the art. Journal of the American Society for Information Science 46(10), 725–800 (1995)
23. Stokes, N., Li, Y., Cavedon, L., Huang, E., Rong, J., Zobel, J.: Entity-based relevance feedback for genomic list answer retrieval. In: Proc. of TREC-16 (2007)
24. Huang, X., Zhong, M., Si, L.: York University at TREC 2005: Genomics track. In: Proc. of TREC-14 (2005)
25. Huang, A., Milne, D., Frank, E., Witten, I.H.: Clustering documents with active learning using Wikipedia. In: Proc. of the 8th IEEE ICDM (2008)
26. Cao, Y., Liu, J., Bao, S., Li, H.: Research on expert search at Enterprise track of TREC 2005. In: Proc. of TREC-14 (2005)
27. Zhu, J., Huang, X., Song, D., Ruger, S.: Integrating multiple document features in language models for expert finding. Knowledge and Information Systems (2009)
28. Beaulieu, M., Gatford, M., Huang, X., Robertson, S., Walker, S., William, P.: Okapi at TREC-5. In: Proc. of TREC-5 (1997)

Temporal Shingling for Version Identification in Web Archives

Ralf Schenkel

Saarland University, Saarbrücken, Germany
schenkel@mmci.uni-saarland.de

Abstract. Building and preserving archives of the evolving Web has been an important problem in research. Given the huge volume of content that is added or updated daily, identifying the right versions of pages to store in the archive is an important building block of any large-scale archival system. This paper presents temporal shingling, an extension of the well-established shingling technique for measuring how similar two snapshots of a page are. This novel method considers the lifespan of shingles to differentiate between important updates that should be archived and transient changes that may be ignored. Extensive experiments demonstrate the tradeoff between archive size and version coverage, and show that the novel method yields better archive coverage at smaller sizes than existing techniques.

1   Introduction

Building and preserving archives of the evolving Web has been an important problem in research. Given the huge volume of content that is added or updated daily, identifying the right versions of pages to store in the archive is an important building block of any large-scale archival system. This is especially true in a context where not a carefully designed crawler (such as variants of [6,15]) chooses the pages to add to the archive, but unsynchronized clients push snapshots of pages to the archive. This technique has been increasingly used in the recent past, for example by Iterasi[1] and freearchive.org[2], but also by research prototypes such as the distributed archive EverLast [1]; it has the potential for a much higher coverage of versions than any centralized crawler can have, allowing for between-update times of seconds or minutes. Such a high frequency is necessary to extensively cover sites with a high update rate such as news portals. As an example, Figure 1 shows the lifetimes of consecutive versions of the heise.de news portal, where a new version is started by adding or removing a news item from the page (Section 5.1 shows how we extracted this information); here, most versions are valid for a few minutes only.

A key issue in such systems is selecting the 'right' snapshots to store in the archive, namely those that contain 'important' updates or, to put it in slightly

---

[1] http://www.iterasi.net
[2] http://www.freearchive.org

C. Gurrin et al. (Eds.): ECIR 2010, LNCS 5993, pp. 508–519, 2010.

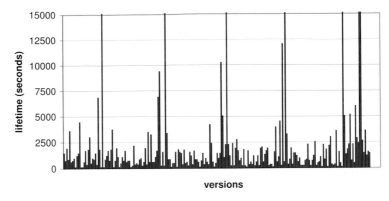

Fig. 1. Lifetimes of the sampled versions from the heise.de portal over one week

more formal terms, diverge enough in content from their predecessor in the series of snapshots. Effectively extracting this sequence of versions (which we call *timeline* in this paper) is the focus of this paper. While there have been many proposals for measuring divergence (or, in fact, similarity) of documents, for example[2,3,8,9,11,11,16], we will see later that they do not always identify good versions, and especially frequently identify too many versions.

The techniques presented in this paper improve on these existing techniques by considering how a page changes over time. We observe that for many important classes of pages (such as news portals or blogs), content that is 'important' for a page typically stays relatively long on a page, whereas content unrelated to the actual page (such as advertisements or headlines of other recent articles) is often transient and quickly replaced. Our novel method, coined temporal shingling, extends the well-established shingling [4] and SpotSigs [16] techniques by considering the lifespan of shingles to differentiate between important updates that should be archived and transient changes that may be ignored.

This paper makes the following important and novel contributions: (1) It formalizes the problem of timeline extraction for a Web page in an archive, (2) it introduces a temporal extension of the well-known shingling method for effectively extracting such a timeline, (3) it shows how a benchmark for timeline extraction can be defined, and (4) it evaluates the effectiveness of the proposed method with this benchmark, comparing it to the two established methods shingling [4] and SpotSigs [16].

2   Related Work

2.1   Similarity Measures for Documents

Identifying similar documents in a large collection has been extensively studied in the literature. An important subset of that work uses variants of shingling, i.e., n-grams of the documents' contents, to determine the similarity of documents,

'important' parts of a page only. Starting at an antecedent, a spot signature includes the following c words at a distance of d each. Similarity and divergence of two pages can be defined similarly to shingling, replacing shingle sets by spot sets.

Both methods come with additional optimizations to improve the efficiency of duplicate checking, for example by combining different shingles into super shingles, which often incur a slight decrease in precision. This paper cosniders only the basic methods as performance is not as critical in our setting (where we consider timelines of single pages only) as in the original setting of these papers (where Web-scale duplicate elimination was the target). However, we will evaluate the effect of considering only a (randomly chosen) subset of all shingles in Section 6.3. For both methods, it is possible to use different similarity measures such as cosine similarity or the Dice coefficient.

4.2   Temporal Extension

Shingling and SpotSigs have been designed to find different pages with highly similar content. Our problem, timeline extraction for a single page, is different in the sense that it considers a sequence of snapshots of the same page over time. We can therefore consider how the set of shingles of a page developed over time. Our basic assumption is the following: Content that is 'important' for a page (such as a pointer to a new news item) typically stays relatively long on a page, whereas content unrelated to the actual page (such as advertisements or headlines of other recent articles) is often transient and quickly replaced. We therefore propose to consider how long a shingle or a spot signature stays on a page (the *lifetime* or *age* of a shingle or a spot signature), and to filter out any of them that do not survive long enough. We now formally define this, restricting the discussion to shingles; extending it to spot signatures is straight forward.

Given a series of snapshots $S(p)$ of page p, we compute for each snapshot p_j the corresponding shingle set $SH(p_j)$. Given a timestamp $t_{p,i}$ from the set of timestamps $TS(p)$ of $S(p)$, we now define for a shingle $s \in SH(p_i)$ its lifetime $l(s, t_{p,i})$. The *left boundary* $lb(s, t_{p,i})$ is the largest $t_{p,j} \in TS(p)$ such that $j \leq i$ and $s \in S(p_j)$, but $s \notin S(p_{j-1})$ or $-\infty$ if that does not exist, i.e., the timestamp prior or equal to $t_{p,i}$ where s first occurs in the shingle set of a snapshot of p. Analogously, we define the *right boundary* $rb(s, t_{p,i})$ as the timestamp after $t_{p,i}$ when s does not occur in the shingle set of a snapshot of p for the first time, or ∞ if this does not exist. Note that the definition makes sure that shingles existing at the beginning or the end of the snapshot series are not penalized. The lifetime $l(s)$ of a shingle is then defined as

$$l(s, t_{p,i}) = rb(s, t_{p,i}) - lb(s, t_{p,i})$$

Note that this definition includes the timestamp because a shingle may enter and leave the shingle set more than once in a series of snapshots. We now can define the lifetime-restricted shingle set of a snapshot p_j as

$$SH_\theta(p_j) = \{s \in SH(p_j) | l(s, t_{p,j}) > \theta\}$$

that includes only shingles with a lifetime greater than θ. Using this, we can define a lifetime-aware version of the similarity of two snapshots p_i and p_j ($i \leq j$) as

$$s_\theta(p_i, p_j) = \frac{|SH_\theta(p_i) \cap SH_\theta(p_j)|}{|SH_\theta(p_i) \cup SH_\theta(p_j)|}$$

and an analogous lifetime-aware divergence $d_\theta(p_i, p_j)$.

5   Defining a Benchmark

The problem of accurately identifying the version timelines of Web pages has not yet been widely present in the literature, and hence there is no standard benchmark for it. However, its evaluation is closely related to the evaluation of policies for page refresh policies for Web crawlers, so we adopt methodology used there for the evaluation of our method. More specifically, our experimental setup follows the setup by Cho and Garcia-Molina in [6] which use frequent recrawls of certain Web sites and an evaluation metric that measures cache freshness, with certain adaptions to the problem twist considered here.

5.1   Dataset

The most difficult challenge when building a dataset for the experiments is to determine the set of "important" versions of a Web page that should be represented in the archive. For general Web pages, this would require humans which determine, given a set of snapshots of a page, when the page has changed "enough" to be considered as new version. Such a 'gold standard' has been used, for example, to assess the quality of algorithms for duplicate detection [16]. However, in our case, we think that such a tedious work would be difficult and error-prone, because it is much more difficult to determine 'how' different two snapshots are than to rate them as highly similar.

Our experiments are therefore performed on the index pages of 30 English and German news portals, including BBC, CNN, NYTimes, Spiegel, and Stern. Choosing this special kind of pages has been motivated by the following observations: (1) they are updated with fresh content at a very high frequence,

Table 1. Distribution of the number of versions per page

version count range	number of pages
0–49	1
50–99	4
100–199	8
200–299	6
300–399	7
400–499	2
500–599	2

usually every few minutes; (2) they also contain 'noise' like advertisments and user comments; (3) they come with RSS feeds that identify new content that has been posted to those sites. For each such page p, we collected a snapshot of p and the corresponding RSS feed every two minutes for a period of about a week, resulting in 5,000 snapshots each. Following the notation introduced earlier, we label the set of snapshots of page p as $S(p)$ and the corresponding timestamps of the snapshots $TS(p)$.

To build the groundtruth $GT^+(p)$ for a page p, i.e., the correct timeline of p, we extracted the set of posted items, including the corresponding uri and posting date, from the snapshots of the RSS feed for p. In a post-processing step, we had to modify the found uris for some pages in order to convert them to the format used on the page itself. While this sometime required only syntactic changes like removing suffixes, it sometimes included requesting the uri from the server, which answered with a redirection uri in the right format. We then removed all uris that did not occur in at least one snapshot of p, in order to remove any links to subportals, for example. The denote the remaining set of links as $L(p)$.

Table 2. Distribution of the lifetime of versions

lifetime range (seconds)	number of versions
121–240	1118
241–480	1970
481–960	1950
961–1920	1324
1921–3840	509
3841–7680	187
7681–15360	102
15361–30720	70
30721–61440	17
61441–122880	1

Our initial assumptions was that the posting timestamps from the RSS feed should accurately reflect the timestamps when the page was updated, too, because any update visible in the feed should become visible on the page at the same time. However, while this was true for most of the updates, for some we either found major delays until the link was found in a snapshot of the page (in the order of several thousand seconds), or the link actually appeared on the page significantly before it appeared on the feed. At the same time, not all feeds come with timestamps of the updates. We therefore use an approximation to derive the timeline of a page from its RSS feed: We assume that a new version starts whenever one of the links from $L(p)$ appears in a snapshot of the page for the first time. This defines the set of timestamps that forms the groundtruth $GT^+(p)$ for a page p.

We identified 234.8 versions per page on average; Table 1 details the distribution of the number of versions per page. The average lifetime of a version was 2499.2 seconds; Table 2 details the distribution of the lifetimes.

So far, we considered only timestamps when information was added to a site, as this is usually an event that should be reflected as quickly as possible in an archive. However, for some applications, it may also be interesting to know when information was *removed* from a site. Analogously to the previous approach, we can define the set $GT^-(p)$ of timestamps when at least one of the links from $L(p)$ *dissapears* from p. Finally, we define $GT(p) := GT^+(p) \cup GT^-(p)$ as the timestamps where p was significantly changed.

5.2   Evaluation Metric

We need to evaluate two dimensions of performance for our methods for time-line identification: (1) the quality of the timelines, i.e., how well the computed timelines reflect the real timelines, and (2) the space overhead, i.e., how much larger the computed timelines are compared to the real timelines.

Quality. We evaluate the *quality* of a timeline $T(p)$ using a variant of the *freshness* measure introduced in [6] to measure how accuratly a search engine captures the changing versions of a set of Web pages in a time interval. This can be immediately applied in our setting to measure the *coverage* of the generated timeline, i.e., how accurately the timeline reflects important versions of a Web page in a time interval. To do so, we first need to define a binary measure $F(p, t)$ that captures if the timeline reports the correct version of page p at time t:

$$F(p,t) = \begin{cases} 1 \text{ if } p \text{ is up-to-date in } T(p) \text{ at time } t \\ 0 \; otherwise \end{cases}$$

Here, "up-to-date at time t" means the following: Let $t_g \in GT(p)$ be the greatest timestamp from the groundtruth for p for which $t_g \leq t$, i.e., the timestamp of the last major change of p before or at time t. Then p is up-to-date in $T(p)$ at time t if there is a $t' \in T(p)$ with $t_g \leq t' \leq t$, i.e., the last major change (at time t_g) is reflected in the timeline at time t because it contains a snapshot of p at time t' that includes the update at t_g. (Note that this definition of up-to-dateness assumes that content added at time $t_g \in GT(p)$ does not disappear before the next timestamp in the groundtruth—which is reasonable given that news portals typically keep their articles for some time before removing them from the index page.)

With this definition, we can now compute–again following [6]–the coverage for page p in a time interval $T = [b, e]$ as

$$coverage(p, T) = \frac{1}{|T|} \int_b^e F(p, t)$$

and the coverage for a set $P = \{p_1, \ldots, p_m\}$ of pages as

$$coverage(P, T) = \frac{1}{|P|} \sum_{p \in P} coverage(p, T)$$

Space Overhead. In our archiving scenario, each entry of a timeline corresponds to a snapshot of a page kept in the archive. To keep the archive as small as possible to keep storage costs low, we want the timeline to be as small as possible. To measure this, we compute how much larger the timeline is compared to the corresponding groundtruth, i.e., $o(T(p)) = \frac{T(p)}{GT(p)}$. For a set of pages, we compute the micro average to take into account that the timelines have different sizes:

$$overhead(T(p_1) \ldots T(p_m)) = \frac{\sum T(p_i)}{\sum GT(p_i)}.$$

6 Experiments

6.1 Standard Parameters

We compare Temporal Shingling with standard shingling [4] and SpotSigs [16]. Both techniques were applied on the textual content of pages after stripping all tags and comments. Shingling used $k = 4$, SpotSigs used one of the English antecedents lists from the original paper (it, a, there, was, said, the, is) for English pages and a similar list of German antecedents (der, die, das, ein, war, waren, sie, es, er, ist, sagte, da) for German pages, with a setting of $c = 3$ and $d = 2$. The lifetime threshold θ for temporal shingling was set to 1200 seconds. We evaluated all methods for 50 divergence thresholds between 0 and 1.

Figure 3 shows the results of this experiment. This scatter plot depicts the coverage vs. the overhead required by the different methods with different thresholds. It is evident that both temporal versions (filled circles and triangles) clearly outperform their non-temporal counterparts (non-filled circles and triangles). For example, for a coverage of 0.875, temporal shingling requires an overhead of approximately 2, whereas standard shingling has almost twice as much overhead. The chart additionally shows that standard SpotSigs (unfilled triangles) is slightly more space-efficient than standard Shingling (unfilled circles), especially for medium coverage. On the other hand, it cannot identify all versions, no matter how low we set the divergence threshold. This is probably due to the nature of some news portals in our collection that add only headlines without the antecedents needed by SpotSigs.

6.2 Influence of Temporal Threshold

This experiment considers varying lifetime thresholds θ while leaving the other parameters unchanged. Figure 4 shows the results for temporal shingling with different values for θ, again compared to standard shingling. It is evident that $\theta = 1200s$ is the best choice. Lower thresholds lead to increased overhead because the divergence function is more and more dominated by transient updates. Higher thresholds, on the other hand, lead to decreased coverage: When only shingles with a lifetime of more than 86400s (one full day) are considered (the red plusses in the chart), no more than 55% of all versions can be identified because many

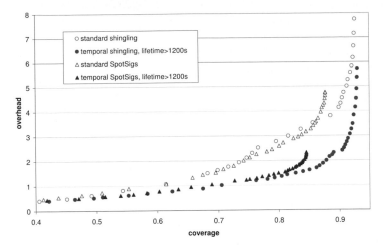

Fig. 3. Results with standard parameters

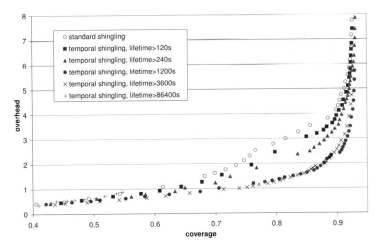

Fig. 4. Results with different lifetime thresholds for standard shingling

news items come and go within a day and are therefore never identified with such a high threshold.

Figure 5 shows the result of the same experiment for temporal SpotSigs, where the results are mainly comparable. Note that the effect of reduced coverage with high thresholds is even more visible here.

6.3 Partial Shingling

This experiment evaluates the effectiveness of temporal shingling when only a (randomly chosen) subset of all shingles are considered, which can save both

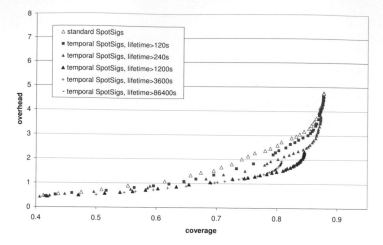

Fig. 5. Results with different lifetime thresholds for SpotSigs

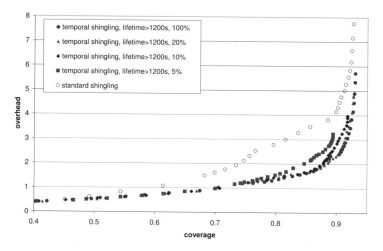

Fig. 6. Results with partial shingles and $\theta = 1200s$

intermediate memory and execution time. Figure 6 shows the result of this experiment, where we considered temporal shingling restricted to 20%, 10%, and 5% of all shingles (which were selected by restricting the hash code of a shingle). It is evident that coverage is slightly reduced with the partial shingle sets, which can be explained by the fact that some small but important updates of news portals introduce only a few shingles to a page, all of which may be dropped for small percentages. On the other hand, the coverage is still almost 90% with just 5% of shingles and an overhead of slightly more than three, whereas standard shingling with the full shingle sets needs an overhead of about four for the same coverage. Additional experiments with partial temporal SpotSigs (not shown here for space reasons) demonstrated similar effects.

7 Conclusions and Future Work

Identifying timelines of Web pages in a large archive is an important problem. This paper has developed a temporal extension for shingling-based methods that significantly outperforms existing methods, namely shingling and SpotSigs, in terms of coverage and overhead. Additionally, the paper has shown how an unsupervised benchmark for this problem can be defined.

While the work presented in this paper has already demonstrated promising results, there are still a number of possible extensions and improvements. First, the presented solution is purely offline, i.e., it requires a complete series of snapshots as input; our future work will develop an online version of the algorithm. Second, parameter tuning could be done on a per-page base instead of globally, for example by determining a good lifetime threshold for each page. Last, extended notions of timelines could be considered, for example finding the most important n versions of a page.

References

1. Anand, A., et al.: EverLast: a distributed architecture for preserving the web. In: JCDL, pp. 331–340 (2009)
2. Brin, S., Davis, J., Garcia-Molina, H.: Copy detection mechanisms for digital documents. In: SIGMOD Conference, pp. 398–409 (1995)
3. Broder, A.Z.: Identifying and filtering near-duplicate documents. In: Giancarlo, R., Sankoff, D. (eds.) CPM 2000. LNCS, vol. 1848, pp. 1–10. Springer, Heidelberg (2000)
4. Broder, A.Z., Glassman, S.C., Manasse, M.S., Zweig, G.: Syntactic clustering of the web. Computer Networks 29(8-13), 1157–1166 (1997)
5. Charikar, M.: Similarity estimation techniques from rounding algorithms. In: STOC, pp. 380–388 (2002)
6. Cho, J., Garcia-Molina, H.: Effective page refresh policies for web crawlers. ACM Trans. Database Syst. 28(4), 390–426 (2003)
7. Cho, J., Garcia-Molina, H.: Estimating frequency of change. ACM Trans. Internet Techn. 3(3), 256–290 (2003)
8. Chowdhury, A., et al.: Collection statistics for fast duplicate document detection. ACM Trans. Inf. Syst. 20(2), 171–191 (2002)
9. Conrad, J.G., et al.: Online duplicate document detection: signature reliability in a dynamic retrieval environment. In: CIKM, pp. 443–452 (2003)
10. Henzinger, M.R.: Finding near-duplicate web pages: a large-scale evaluation of algorithms. In: SIGIR, pp. 284–291 (2006)
11. Hoad, T.C., Zobel, J.: Methods for identifying versioned and plagiarized documents. JASIST 54(3), 203–215 (2003)
12. Kolcz, A., Chowdhury, A., Alspector, J.: Improved robustness of signature-based near-replica detection via lexicon randomization. In: KDD, pp. 605–610 (2004)
13. Manber, U.: Finding similar files in a large file system. In: USENIX Winter, pp. 1–10 (1994)
14. Manku, G.S., Jain, A., Sarma, A.D.: Detecting near-duplicates for web crawling. In: WWW, pp. 141–150 (2007)
15. Olston, C., Pandey, S.: Recrawl scheduling based on information longevity. In: WWW, pp. 437–446 (2008)
16. Theobald, M., Siddharth, J., Paepcke, A.: SpotSigs: robust and efficient near duplicate detection in large web collections. In: SIGIR, pp. 563–570 (2008)

Biometric Response as a Source of Query Independent Scoring in Lifelog Retrieval

Liadh Kelly and Gareth J.F. Jones

Centre for Digital Video Processing,
Dublin City University, Dublin 9, Ireland
{lkelly,gjones}@computing.dcu.ie
http://www.cdvp.dcu.ie/iCLIPS

Abstract. Personal lifelog archives contain digital records captured from an individual's daily life, e.g. emails, web pages downloaded and SMSs sent or received. While capturing this information is becoming increasingly easy, subsequently locating relevant items in response to user queries from within these archives is a significant challenge. This paper presents a novel query independent static biometric scoring approach for re-ranking result lists retrieved from a lifelog using a BM25 model for content and content + context data. For this study we explored the utility of galvanic skin response (GSR) and skin temperature (ST) associated with past experience of items as a measure of potential future significance of items. Results obtained indicate that our static scoring techniques are useful in re-ranking retrieved result lists.

Keywords: Lifelog retrieval, affective response, query independent weights.

1 Introduction

Advances in digital technologies mean a wealth of personal information is now becoming available in digital format. This information can be gathered together and stored in a personal lifelog (PL) [3]. Lifelog archives can contain everything from items read, written, or downloaded; to footage from life experiences, e.g. photographs taken, music heard, details of places visited, details of people met, etc, along with details of location and social context. Finding important relevant items from within these archives in response to user queries poses significant challenges. Any additional information which can assist in identifying important items is thus potentially very important. Such information could be used in the re-ranking of information retrieval (IR) result sets. One potential source of useful information is the user's biometric response associated with previous experience with an item. In this study we explore two biometric responses associated with items, namely galvanic skin response (GSR) and skin temperature (ST).

Previous work has shown an individual's biometric response to be related to their overall arousal levels [11]. Significant or important events tend to raise an individual's arousal level, causing a measurable biometric response [12]. Events

C. Gurrin et al. (Eds.): ECIR 2010, LNCS 5993, pp. 520–531, 2010.
© Springer-Verlag Berlin Heidelberg 2010

that can be recalled clearly in the future are often those which were important or emotional in our lives [7]. It has been demonstrated that the strength of the declarative or explicit memory for such emotionally charged events has a biological basis within the brain. Specifically involving interaction between the amygdala and the hippocampal memory system [6]. Variations in arousal level elicit physiological responses such as changes in heart rate or increased sweat production. Thus one way of observing an arousal response is by measuring the skin conductance response (SCR) (also referred to as GSR). The GSR reflects a change in the electrical conductivity of the skin as a result of variation in the activity of the sweat glands. It can be measured even if this change is only subtle and transient, and the individual concerned is not obviously sweating [7]. Arousal response can also be observed through ST. With increased arousal levels, sympathetic nervous activity increases, resulting in a decrease of blood flow in peripheral vessels. This blood flow decrease causes a decrease in ST [15]. Current technologies enable the capture of a number of biometric measures on a continuous basis. For example using a device such as the BodyMedia SenseWear Pro II armband [4] which can continuously record the wearer's GSR and ST.

We propose that lifelog items which are important to an individual at the time they were experienced may be useful to the individual again in the future, and further that such incidents are associated with emotional responses that can be detected by measuring an individual's biometric response when accessing these items. Thus recording GSR and ST as part of a lifelog may enable us to identify important items which would be most important in the context of this paper, to a given future information searching task. In particular we hypothesize that adding a query independent boost (static score) to important items in lifelog IR result lists, where important items are detected based on recorded GSR or ST levels associated with past accesses to the items, may improve retrieval performance.

In this paper we report our findings to date which may guide future research in this area. We describe our study to investigate the utility of biometric response in re-ranking traditional information retrieval result lists. We find positive results for a technique for adding static biometric scores to the results of content or content+context result lists obtained using a BM25 retrieval model.

The next section discusses related work and highlights the contributions of this paper. Section 3 describes the test-set gathered for this study. Section 4 presents our experimental set-up and results are discussed in Section 5. We conclude the paper with a discussion of findings and directions for future work.

2   Towards Static Biometric Scores

While observed biometric response has been used to detect tasks or items in different test sets which are of current relevance or importance to the individual, for example movie scene selection [16] and elicitation of topical relevance in multimedia systems [2], to our knowledge previous research has not investigated the exploitation of observed biometric response as an implicit indicator of future item importance. This we believe is an important previously unexploited opportunity to gain passive feedback from subjects for improving the retrieval performance of future searches in both lifelogging and other domains. In a preliminary study we found correlation between lifelog items coincident with maximum observed GSR and with minimum observed ST at the time of item creation/access and current importance of the items [10] . This finding motivates our current study to investigate the utility of adding a static biometric relevance score generated using a BM25 model for content or content+context retrieval.

There are many examples of the use of various types of static scores to boost user query driven scores in different domains. Examples here include the well known PageRank which uses the webs link structure to create static scores for web pages [13], using web page features such as document length and anchor text as static scores [14], and using links created between computer files to infer static file importance scores [17].

Various approaches can be used for integrating static scores with query dependent scores. We explore using a linear combination of the query dependent score and static biometric score. We also investigate various approaches for transforming the biometric response into a static score. In particular raw biometric scores and various nonlinear transformations of the biometric readings are explored. A particularly promising technique is presented in [5] where a sigmoid functional form is used to transform PageRank, link indegree, ClickDistance and URL length into static scores. This technique forms part of our investigation.

3   Test-Set

In order to explore our hypothesis, a suitable test-set must be available. As part of our ongoing work on PLs we are gathering long term multimedia lifelog collections, stored locally on individuals PCs, from a small group of subjects [8]. For the current investigation we augmented these lifelogs for 3 postgraduate students within our research group (1 male, 2 females; from Asian and Caucasian ethnic groups), for a 1 month period, with capture of their GSR and ST data.

GSR, ST and energy expenditure were collected using a BodyMedia armband [4] worn on the upper arm. Based on results from initial calibration experiments, GSR data was capture once per second and ST data once every ten seconds[1].

A problem in analysis of biometric data for the purposes of this experiment is to identify variation in biometric data which are likely to be associated with meaningful variations in arousal levels, as opposed to physical activity. Energy expenditure (sampled once per minute) correlates well with physical activity levels. Thus measured energy expenditure can be used to differentiate between high GSR and low ST biometric data levels, resulting from physical activity and those arising from events experienced from the environment.

In addition to the biometric data, our 1 month experimental lifelogs contained data of computer activity and SMSs sent and received. See [8] for full details on data capture. For this experiment we used the content of SMSs and computer

---

[1] Due to an error in settings ST was sampled once per minute for Subject 3.

items (e.g. word document) created/accessed annotated with context types[2]: word in file name; extension type; month; day of week; weekday or weekend; is beginning of week; is mid-week; is end week; is morning; is afternoon; is evening; is night. Lucene [1], an open source search engine, was used to index items and their associated context data into different fields (e.g. day of week field, etc).

4   Experiment Procedure

In this section we describe the setup of our study to examine the utility of GSR and ST biometric data at the time of previous computer item or SMS access, in re-ranking the output of a user query driven IR result list. We begin this section by describing our test case and result set generation approach, and follow with details of our investigation and static scoring approaches.

4.1   Test Case Generation

If PLs are to be recorded and accessed over an extended period it is important that users are able to reliably retrieve content recorded in the distant past. A user is likely to remember a significant amount of content and context data soon after an event occurred, however with time memory fades and it is anticipated that less will be remembered a substantial delay after the event occurred [9]. Query generation in the PL domain is challenging. We wished to mimic the 'real' re-finding requirements of individuals, and details they are likely to recall about required items as closely as possible. In generating the test cases for this experiment the following approach was used to generate 50 queries per subject:

- After 8 months lifelog collection build up (5 months after the one month biometric data capture period) subjects listed lifelog retrieval tasks they might want to perform in the future. Typical test cases generated in this manner were: 'show me documents I created associated with conference X'.
- Subjects then entered their list of task descriptions along with keywords and remembered context, e.g. extension type, into a provided form.

4.2   Result Set Generation

Pooled result lists were created by entering content (keywords) only, context only, content+extension type, and content+context query types into two good standard retrieval systems, namely the vector space model (VSM) and BM25, to retrieve as many relevant items from subjects' collections as possible. The BM25 k and b parameters tuned to 1.5 and 1 respectively using the full set of each user's queries. The Lucene implementation of the VSM and an in-house developed implementation of BM25 for Lucene were used to process these queries. Queries combining content and context are straightforward concatenations of the content data score with the individual context types scores. The results from each of the 8 IR techniques were pooled and presented to subjects for relevance judgment (i.e. 0 = irrelevent; 1 = relevant). These judged sets were used for determining the utility of our techniques.

---

[2] In experiment section context data refers to all these context types.

4.3   Investigation

We investigated if GSR and ST at the time of item experience could be used to re-rank the output of IR in response to a user query. Queries used for this investigation were those contained in the subset of the 50 test cases generated for each subject which contained items occurring during the biometric capture month. Subject 1 had 22 such tasks, Subject 2 had 8 and Subject 3 had 36.

While VSM was found to enrich the pooled result lists generated in Section 4.2, comparison showed BM25 to perform better in retrieval. Hence our in-house developed version of the BM25 system for Lucene was used to obtain queried content and queried content+context retrieval scores in this experiment. For content+context querying, the relevance scores obtained for the items content and each of the item's context types were summed. The weight (w) assigned to each field and BM25 k and b parameters were tuned using the full set of the 3 subjects biometric month test cases. Only the top 1000 results were taken in each case for efficiency, without a serious degradation in performance. Static biometric scores were added to the content and content+context scores (techniques used to obtain static biometric scores are described in Section 4.4). In each case the rank of the relevant items in the result set was noted. For content only retrieval 4116 items were retrieved for Subject 1 (relevant: 90, rel_ret: 40), 84 items were retrieved for Subject 2 (relevant: 16, rel_ret: 0), and 16768 items were retrieved for Subject 3 (relevant: 556, rel_ret: 480). For content+context retrieval 11912 items were retrieved for subject 1 (relevant: 90, rel_ret: 90), 3385 items were retrieved for subject 2 (relevant: 16, rel_ret: 16), and 28132 items were retrieved for subject 3 (relevant: 557, rel_ret: 530).

4.4   Static Relevance Scores

Each retrieved item for content only retrieval was annotated with the maximum observed GSR and associated energy expenditure (engGSR) and with the minimum observed ST and associated energy expenditure (engST), across all accesses to the item. Items with no associated biometric readings, due to biometric recording devices being removed for data downloading purposes, subjects' need for mental break from wearing of devices, etc, were assigned default biometric values. The default value used was the average of the GSR, engGSR, ST and engST readings associated with retrieved items.

Increases in physical activity (detected through increases in energy expenditure) cause GSR levels to increase and ST levels to decrease. To discern changes in GSR caused by changes in arousal level as opposed to changes in physical activity, we also tagged items with GSR divided by engGSR. As stated in Section 1 the lower the ST level the greater the arousal level, hence the inverse of ST and the inverse of ST divided by engST levels (to account for changes in physical activity) associated with retrieved items were also tagged to items. The GSR, inverse ST, GSR divided by engGSR, and inverse ST divided by engST values associated with retrieved items were normalised using min-max normalisation.

To allow for investigation of the approach, mentioned in Section 2, which calculates static relevance scores for features where lower values indicate greater importance, we also normalised the ST values and ST values multiplied by energy expenditure using min-max normalisation and tagged these values to items.

The same process was also applied to tag items retrieved from content+context retrieval with GSR and ST levels. The following approaches for calculating static relevance scores using the normalised biometric data tags were investigated:

$$STbase = w \cdot \frac{\frac{1}{ST}}{engST} \tag{1}$$

$$logST = w \cdot log(\frac{1}{ST}) \tag{2}$$

$$logSTdivEng = w \cdot log(\frac{\frac{1}{ST}}{engST}) \tag{3}$$

$$sigmST = w \cdot \frac{s^a}{k^a + s^a}, \; where \; s = \frac{1}{ST} \tag{4}$$

$$sigmSTdivEng = w \cdot \frac{s^a}{k^a + s^a}, \; where \; s = \frac{\frac{1}{ST}}{engST} \tag{5}$$

$$sigmIncST = w \cdot \frac{k^a}{k^a + s^a}, \; where \; s = ST \tag{6}$$

$$sigmIncSTmultEng = w \cdot \frac{k^a}{k^a + s^a}, \; where \; s = ST \times engST \tag{7}$$

$$GSRbase = w \cdot \frac{GSR}{engGSR} \tag{8}$$

$$logGSR = w \cdot log(GSR) \tag{9}$$

$$logGSRdivEng = w \cdot log(\frac{GSR}{engGSR}) \tag{10}$$

$$sigmGSR = w \cdot \frac{s^a}{k^a + s^a}, \; where \; s = GSR \tag{11}$$

$$sigmGSRdivEng = w \cdot \frac{s^a}{k^a + s^a}, \; where \; s = \frac{GSR}{engGSR} \tag{12}$$

Equations 1 and 8 are our baseline static scoring approaches, used to examine the effect of the raw ST and GSR values divided by energy expenditure on re-ranking result lists. The remaining equations investigate the use of non-linear transformations of the biometric score. Equations 2, 3, 9, 10 examine the effect of using logs of ST and GSR. The performance of our biometric scores using the transformation approach presented in [5] is examined with Equations 4, 5, 11, 12. This approach is used to generate static relevance scores for features where higher values indicate greater importance. An approach for calculating static relevance scores for features where lower values indicate greater importance is also provided in [5]. This techniques performance using our ST data is investigated with Equations 6 and 7. The effect of accounting for energy expenditure is investigated in Equations 1, 3, 5, 7, 8, 10, 12. Following parameter tuning using the full set of the 3 subjects' biometric month test cases, the static score's weight of importance (w) and parameters k and a were set for each equation.

The static scoring techniques presented in this section are added to content and content+context relevance scores generated using BM25 model, described in Section 4.3. The next section discusses results obtained using these approaches.

5   Experiment Results and Analysis

Average precision (AveP), P@5 and P@10 were investigated. P@5 and P@10 show how effective our techniques were at moving relevant items towards the top of the result lists. Table 1 shows the percentage improvement over the content only baseline for content+static_score retrieval averaged over Subjects 1 and 3 (relevant items were not retrieved for Subject 2 using content only retrieval). Percentage improvement for content+context+static_score retrieval over the content+context baseline, averaged over all 3 subjects, are also presented in Table 1. Table 2 presents the individual breakdown of results for each subject. Overall results suggest that adding either a GSR or ST static score to content or content+context IR scores is useful for re-ranking PL text-based collections. In particular, the use of ST as a static score yields the greatest overall improvement in performance. In this section we analyse the results obtained and suggest a general function for calculating query independent biometric scores for re-ranking BM25 model generated content and content+context PL result lists.

5.1   Overall Static Score Performance

Considering both content and content+context retrieval the addition of a static score using *sigmSTdivEng* resulted in the greatest percentage improvement from the content and content+context baselines. 0%, 5% and 28% improvement in average precision, P@5 and P@10 respectively were observed for content only retrieval using this technique. While content+context retrieval yielded 4%, 34% and 6% improvement for average precision, P@5 and P@10 respectively. The lower percentage in improvement for content+static_ST_score may be explained by the lack of retrieved items for content only IR for Subject 2 who benefited the most from the addition of a static ST score (see Table 2).

The superior performance of *sigmSTdivEng* to *STbase* and to the approaches which calculate logs of the ST scores is consistent with the findings noted in [5] where this transformation was used with the greatest success in calculating weights for static features of web pages. When using this transformation to calculate query independent ST scores we took the inverse of ST, since lower ST levels indicate greater importance. However [5] also presents a transformation for calculating static scores for situations where lower static values indicate greater importance. This technique did not perform as well overall on our test-set, see results *sigmIncST* and *sigmIncSTmultEng* in Table 1.

While adding GSR static scores also improved retrieval performance, albeit not to the same extent as the addition of ST static scores, no clear best approach

Table 1. Average percentage improvement, rounded to nearest whole number, by adding a static score (staticS) to the content (C) and content+context (CC) baselines

Static Technique	C+staticS			CC+staticS		
	AveP	P@5	P@10	AveP	P@5	P@10
STbase	-2%	-3%	-4%	3%	35%	-1%
logST	0%	1%	1%	1%	3%	6%
logSTdivEng	0%	1%	0%	4%	34%	7%
sigmST	0%	6%	34%	0%	3%	6%
sigmSTdivEng	0%	5%	28%	4%	34%	6%
sigmIncST	1%	2%	2%	1%	1%	8%
sigmIncSTmultEng	0%	-1%	-2%	-1%	0%	6%
GSRbase	0%	-1%	0%	-2%	32%	6%
logGSR	1%	3%	2%	-2%	35%	7%
logGSRdivEng	0%	1%	2%	-1%	34%	8%
sigmGSR	1%	3%	2%	-2%	35%	8%
sigmGSRdivEng	0%	1%	2%	-1%	34%	8%

for adding GSR static scores was detected. Greatest improvement observed for content only retrieval by the addition of a static GSR score was 1%, 3% and 2% for AveP, P@5 and P@10 respectively using *logGSR* or *sigmGSR*, as shown in Table 1. In all cases average precision was decreased by the addition of a static GSR score, while P@5 increased by 32-35% and P@10 increased by 6-8% for content+context retrieval. Additionally unlike ST static scores, the performance of GSR static scores were not greatly altered overall by factoring in energy expenditure (see Table 1).

5.2   Performance across Individual Subjects

Exploring the individual results of each subject (see Table 2), we find that results for Subject 2 were greatly improved by the addition of a static score to the base content+context score. This improvement was observed equally for *logST-divEng* and *sigmSTdivEng*, and to a lesser extent for *GSRbase*, *logGSRdivEng* and *sigmGSRdivEng*, with a 100% improvement in precision @5 in all cases.

Subject 3 benefited the least from the introduction of a static ST score, with 1% increase in P@5 being observed. Similar to Subject 2, greatest improvement was noted using the *logSTdivEng* and *sigmSTdivEng* transformations to calculate static ST scores. However, no variation in performance was observed between the five static GSR scoring techniques for this subject.

Subject 1 benefited more than Subject 3 from the addition of static scores to the base content and content+context retrieval scores. For Subject 1, in contrast to Subjects 2 and 3, greater performance was observed when we did not divide by energy expenditure while calculating ST static scores (see results for *logST* and *sigmST* in Table 2). Greatest improvement in performance was observed by the addition of ST static scores to the base content only score using *sigmST* for this subject (12% improvement for P@5 and 69% improvement for P@10). Of

Table 2. Subjects' percentage improvement, rounded to nearest whole number, for Average precision (AveP), P@5 and P@10 by adding a static score to the content and content+context baselines

Static Technique	Subject 1			Subject 2			Subject 3		
	AveP	P@5	P@10	AveP	P@5	P@10	AveP	P@5	P@10
Content + static score									
STbase	0%	-4%	-3%	-	-	-	-4%	-1%	-6%
logST	2%	0%	0%	-	-	-	0%	1%	-1%
logSTdivEng	0%	0%	0%	-	-	-	0%	1%	0%
sigmST	0%	12%	69%	-	-	-	0%	1%	-1%
sigmSTdivEng	-1%	9%	56%	-	-	-	0%	1%	0%
sigmIncST	1%	4%	3%	-	-	-	0%	0%	0%
sigmIncSTmultEng	1%	-4%	-3%	-	-	-	0%	1%	0%
GSRbase	-1%	-4%	0%	-	-	-	0%	1%	0%
logGSR	2%	4%	3%	-	-	-	0%	1%	0%
logGSRdivEng	1%	0%	3%	-	-	-	0%	1%	0%
sigmGSR	2%	4%	3%	-	-	-	0%	1%	0%
sigmGSRdivEng	1%	0%	3%	-	-	-	0%	1%	0%
Content + Context + static score									
STbase	1%	4%	-3%	9%	100%	0%	-1%	1%	-1%
logST	2%	8%	0%	1%	0%	20%	0%	1%	-1%
logSTdivEng	0%	0%	0%	11%	100%	20%	0%	1%	0%
sigmST	0%	8%	0%	1%	0%	20%	0%	1%	-1%
sigmSTdivEng	0%	0%	-3%	11%	100%	20%	0%	1%	0%
sigmIncST	1%	4%	3%	1%	0%	20%	0%	0%	0%
sigmIncSTmultEng	1%	0%	-3%	-4%	0%	20%	0%	1%	0%
GSRbase	-2%	-4%	-3%	-4%	100%	20%	0%	1%	0%
logGSR	1%	4%	0%	-7%	100%	20%	0%	1%	0%
logGSRdivEng	0%	0%	3%	-4%	100%	20%	0%	1%	0%
sigmGSR	1%	4%	3%	-7%	100%	20%	0%	1%	0%
sigmGSRdivEng	0%	0%	3%	-4%	100%	20%	0%	1%	0%

the static GSR scoring approaches *sigmGSR* proved most useful for this subject. On biometric data analysis we found that Subject 1 had unexplained periods of particularly high energy expenditure relative to the other subjects which caused energy expenditure to be less useful for this subject.

On analysing the minor improvement in results observed for Subject 3 we found this subject to have a higher average precision for content only retrieval (=0.4715) and for content+context retrieval (=0.4593), than Subject 1 (content only = 0.3317, content+context = 0.3421) and Subject 2 (content+context = 0.1005). The higher default AveP values for this subject may partially explain the minor improvements introduction of static scores made for this subject. Conversely Subject 2 with the lowest default average precision benefited the most from the introduction of static scores. Subject 3's ST BM25+static_score values might have been further affected by the fact that ST was only sampled once per minute for this subject (compared to once every 10 seconds for the other subjects), as discussed in Section 3. Finally the high percentage of content and content+context query results which were assigned the default ST and GSR scores might also have impacted on Subject 3's results. 58% of content+context and 58% of content only retrieved items for Subject 3 did not have ST and GSR values associated with them, and hence were assigned the default ST and GSR values. This compares with 35% of content+context retrieved results and 29% of content only retrieved items for Subject 1 and with 41% of content+context retrieved items for Subject 2.

5.3   Biometric Static Scoring Function

Overall static ST scores provided greatest improvement. *logSTdivEng* and *sigm-STdivEng* performed best for Subjects 2 and 3 where the engST range tagged to retrieved lifelog items was quite narrow. *sigmST* or *logST* worked better for Subject 1 whose data contained unusually high engST readings. Analyses of engST readings which were captured by the biometric device and tagged to items retrieved for content+context retrieval revealed the following for Subject 1 (rounded to 2 decimal places): average = 2.19; median = 1.45; max = 12.10; min = 1.16. For Subject 2: average = 1.13; median = 1.07; max = 3.77; min = 1.03. For Subject 3: average = 1.34; median = 1.23; max = 4.88; min = 1.09.

Median and min values for Subject 1, 2 and 3 are in line. However, Subject 1 has a much larger max engST reading, of 12.10, than Subjects 2 and 3. Subject 1's median and max values indicate that they had infrequent unexplained periods of unusually high energy expenditure, which would somewhat degrade the static ST biometric scoring performance observed when dividing ST by engST. Of greater consequence though perhaps is the fact that their average engST is also higher as a result of the unusually high energy expenditure readings. This higher average engST was the default engST value assigned to items in the result list with no recorded biometric data. In all probability this negatively affected the results in Table 2 for Subject 1 where engST was factored. This analysis, leads us to the following approach for calculating the static ST score to add to the base BM25 retrieval score:

if (median engST * 5) $\leq$ max engST : use *sigmST*,
else : use *sigmSTdivEng*.

5.4   Concluding Remarks

While this study was performed on a limited number of subjects, it provides preliminary support for the use of biometric static scores, in particular ST, to boost relevant retrieved items in lifelogs and supports investment of further research in this space. We are interested in examining the scalability of our approach using larger numbers of subjects and in determining if the results presented in this section can be improved using alternate approaches. In particular, given the findings from our analysis of energy expenditure values, future work will explore the use of alternate approaches for calculating the default values assigned to items missing biometric response, for example using median values instead of averages. The affect on performance when GSR and ST readings are combined will also be looked at. As well as the use of alternate biometric readings for calculating static scores, for example using heat flux and heart rate.

6   Conclusions

In this paper we set out to investigate the role of biometric response in lifelog item retrieval. We presented a novel approach for calculating static relevance scores to boost results in a query driven IR result list using an individual's biometric response at the original time of item access. Results obtained support the use of this approach. Greatest improvement in performance was found by the addition of a static skin temperature (ST) score. From these results a general function for calculating query independent ST scores was derived.

While these results are promising, it is acknowledged that this study was conducted on a limited number of subjects over a relatively short period of time. Further experiments with larger numbers of subjects are required to establish the scalability of the technique presented in this paper. However, due to the large psychological burden placed on subjects wearing the biometric devices for extended periods of time, and the difficulty in gaining participants willing to partake in experiments which log their personal data, this initial study formed a good means to establish if further research in this domain is warranted. Given the results presented in this paper we believe it is worth investing in further research in this space using larger collections of subjects.

Technological developments are enabling individuals to store increasing amounts of digital data pertaining to their lives. As these personal archives grow ever larger, reliable ways to help individuals locate required items from these lifelogs becomes increasingly important. The results of these experiments indicate that static biometric scores, in particular ST, serve as a useful tool for aiding extraction of important items from long-term lifelogs. Additionally, beyond the lifelogging domain, we envisage several possible applications of the technique presented in this paper both in the archive searching and recommendation spaces. Indeed in a future where biometric recording is prevalent, the same patterns of biometric response may be observed across individuals for the same items in shared archives (e.g. digital libaries, photo archives, retail websites), which might allow such items to be given query independent boosts for all users of the archive. Current research exploring development of less cumbersome biometric recording devices, for example research at MIT Media Lab, provides indication that reliable unobtrusive biometric devices embedded in individuals clothes or braclets for example will be widely available for use by such tools.

Acknowledgments. This work is funded by a grant under the Science Foundation Ireland Research Frontiers Programme 2006 Grant No: 06/RFP/CMS023.

References

1. Apache Lucene, `http://lucene.apache.org/java/docs/`
2. Arapakis, I., Konstas, I., Jose, J.: Using Facial Expressions and Peripheral Physiological Signals as Implicit Indicators of Topical Relevance. In: Proceedings of the ACM International Conference on Multimedia, pp. 461–470 (2009)
3. Bell, G.: Challenges in Using Lifetime Personal Information Stores based on MyLifeBits. Alpbach Forum (2004)
4. BodyMedia, `http://www.bodymedia.com`
5. Craswell, N., Robertson, S., Zaragoza, H., Taylor, M.: Relevance Weighting for Query Independent Evidence. In: Proceedings of the Annual ACM SIGIR Conference on Research and Development in Information Retrieval, pp. 416–423 (2005)
6. Ferry, B., Roozendaal, B., McGaugh, J.: Basolateral Amygdala Noradrenergic Influences on Memory Storage Are Mediated by an Interaction between beta- and alpha1-Adrenoceptors. Journal of Neuroscience 19(12), 5119–5123 (1999)
7. Gazzaniga, M.S., Ivry, R.B., Mangun, G.R.: Cognitive Neuroscience, 2nd edn. Norton (2002)
8. Kelly, L., Byrne, D., Jones, G.J.F.: The role of places and spaces in lifelog retrieval. In: PIM 2009 - Proceedings of Personal Information Management, Workshop at ASIST 2009 (2009)
9. Kelly, L., Chen, Y., Fuller, M., Jones, G.J.F.: A study of remembered context for information access from personal digital archives. In: 2nd International Symposium on Information Interaction in Context (IIiX), pp. 44–50 (2008)
10. Kelly, L., Jones, G.J.F.: Examining the utility of affective response in search of personal lifelogs. In: 5th Workshop on Emotion in HCI, British HCI Conference 2009 (2009)
11. Lang, P.J.: The emotion probe: Studies of motivation and attention. American Psychologist 50(5), 372–385 (1995)
12. McGaugh, J.: Strong memories are made of this Memory and Emotion: The Making of Lasting Memories. Columbia University Press (2003)
13. Page, L., Brin, S., Motwani, R., Winograd, T.: The PageRank Citation Ranking: Bringing Order to the Web. Technical report (January 1998)
14. Richardson, M., Prakash, A., Brill, E.: Beyond PageRank: Machine Learning for Static Ranking. In: Proceedings of the International World Wide Web Conference, pp. 707–715 (2006)
15. Sakamoto, R., Nozawa, A., Tanaka, H., Mizuno, T., Ide, H.: Evaluation of the driver's temporary arousal level by facial skin thermogram-effect of surrounding temperature and wind on the thermogram. IEEJ Trans. EIS 126(7), 804–809 (2006)
16. Soleymani, M., Chanel, G., Kierkels, J.J., Pun, T.: Affective ranking of movie scenes using physiological signals and content analysis. In: Proceedings of the 2nd ACM workshop on Multimedia Semantics, pp. 32–39 (2008)
17. Soules, C.A.N.: Using context to assist in personal file retrieval. PhD thesis, School of Computer Science, Carnegie Mellon University, Pittsburgh, PA, USA (2006)

Enabling Interactive Query Expansion through Eliciting the Potential Effect of Expansion Terms

Nuzhah Gooda Sahib[1], Anastasios Tombros[1], and Ian Ruthven[2]

[1] Department of Computer Science, School of Electronic Engineering and Computer Science,
Queen Mary University of London, London E1 4NS, UK
{nuzhah,tassos}@dcs.qmul.ac.uk
[2] Department of Computer and Information Sciences, University of Strathclyde,
Glasgow G1 1XQ, UK
ian.ruthven@cis.strath.ac.uk

Abstract. Despite its potential to improve search effectiveness, previous research has shown that the uptake of interactive query expansion (IQE) is limited. In this paper, we investigate one method of increasing the uptake of IQE by displaying summary overviews that allow searchers to view the impact of their expansion decisions in real time, engage more with suggested terms, and support them in making good expansion decisions. Results from our user studies show that searchers use system-generated suggested terms more frequently if they know the impact of doing so on their results. We also present evidence that the usefulness of our proposed IQE approach is highest when searchers attempt unfamiliar or difficult information seeking tasks. Overall, our work presents strong evidence that searchers are more likely to engage with suggested terms if they are supported by the search interface.

1 Introduction

One of the main difficulties that users of IR systems face is how to choose appropriate search terms to express their information need in the form of a query. One of the ways to assist users is query expansion, where an IR system tries to improve user queries by adding related terms to the original query. Automatic query expansion (AQE) is where the retrieval system automatically adds terms to the query, whereas with interactive query expansion (IQE) the user indicates which terms should be added to their query. Ruthven [14] argues that AQE is often favourable as the system has access to more statistical information on the relative utility of expansion terms, while Koenemann and Belkin [11] provide evidence that IQE is often preferred by users as it offers them more control.

Despite the searchers' preference for interactive query expansion, a number of studies, for example, [1][12][14], have shown that the uptake of IQE is limited, despite it having good potential. The situation is such mainly because of the following two reasons.

Firstly, searchers are unsure as to why expansion terms have been suggested to them and how they relate to their initial query. Despite numerous assumptions in relations to searchers' understanding of semantics between their search area and

C. Gurrin et al. (Eds.): ECIR 2010, LNCS 5993, pp. 532–543, 2010.
© Springer-Verlag Berlin Heidelberg 2010

suggested expansion terms [12], research [4][10] has shown that searchers can fail to manifest this understanding, especially when they do not have sufficient background knowledge in their area of search. Secondly, searchers are also unsure as to what effect the selection of some of the expansion terms will have on their search results [17]. Searchers are encouraged to use expansion words to reformulate their queries but they are still uncertain as to how this reformulation of their initial query will help them to retrieve better results for their search task.

In this paper, we address the second issue by providing searchers with additional information to allow them to make an assessment of the impact of adding a term to their query. Our proposed approach presents the additional information in the form of query biased summaries. Summaries are constructed and displayed in real time depending on searcher's actions, whether they add or remove expansion terms. The summary, therefore, is an overview of the documents that will be retrieved in response to the query that the searcher is currently creating through IQE.

In the remainder of this paper, we first discuss related work in section 2, and we describe our proposed approach in section 3. Then in sections 4 and 5 we present the design and results from the user study and discuss our findings. We conclude in section 6.

2   Related Work

Query expansion is a technique where a shortlist of terms are added to a query in order to enhance the effectiveness of searches [8]. Query expansion exists in two types, namely, automatic query expansion (AQE) and interactive query expansion (IQE). With AQE, additional terms are added to the initial query by the system with no user involvement, whereas with IQE, searchers are presented with a list of suggested terms and they are left to make their choices as to which term to add to their initial query.

There have been a number of studies conducted to compare AQE and IQE. For example, Koenemann and Belkin [11] showed how IQE can outperform AQE for certain tasks and Beaulieu [3] demonstrated that AQE produced more effective results in an operational environment. However, as mentioned by Ruthven in [14], the discrepancy in findings results from the difference in methodology, user experiments, search tasks and interface design. Ruthven [14] demonstrated how a comparison between query expansion and no query expansion resulted in 50% of improved queries when AQE was used. However, when he studied AQE and IQE, he found that it was difficult for users to make good IQE decisions in order to outperform the AQE decisions of a system. Therefore, it seems like the potential benefits of IQE are not being exploited.

It is difficult for searchers to realise the effectiveness of IQE without training and experience in terms of selection and search strategies. In reality, IQE is based on the hypothesis that it is reasonable to assume that searchers will be able to distinguish between good and bad terms, if they are given a list of terms [12]. This underlying assumption was however proved wrong in studies where experiments involving real searchers to use IQE failed to show any significant improvement on AQE [2][6]. Yet, there is evidence that users prefer to use IQE even if it does not improve performance because they feel more in control [11].

White and Marchionini [17] investigated the effectiveness of providing suggested terms as the user enters their query (real time query expansion, RTQE). Although they reported an increase in the uptake of IQE with RTQE, they highlighted that the technique could also lead searchers in the wrong direction, especially when the search area is unfamiliar. As searchers were neither able to predict the effect of adding an expansion term nor preview the documents that would be generated, they could add erroneous terms to their initial query.

Therefore, it appears that IQE could improve search effectiveness, but users have difficulties in exploiting its potential. Simple term listing interfaces, that is, those that present terms in isolation, are not adequate to provide support in making good expansion decisions. Kelly et al. [10] argue that searchers are more likely to use system-generated suggestions if they are presented as query suggestions rather than single term suggestions because searchers look for additional information that allows them to fully understand the relationships between the terms and their information needs.

Interfaces should aim at supporting the identification of relationship between relevant material and suggested expansion terms [14]. By showing a clear relationship between relevant information and suggested terms, we address what Beaulieu and Jones [7] referred to as *'functional visibility'*. They suggest that not only should searchers be aware of the features that the system provides, but they should also know what effect these features could have on their tasks.

3   Summarising the Effect of Interactive Query Expansion

In this section we first describe the search interface for our proposed IQE approach (section 3.1), and then the summarisation and query expansion methods (section 3.2).

3.1   Interface

One of the reasons for the lack of uptake in IQE is poor interface design that does not support the user in making good expansion decisions. With current IQE systems, searchers only receive feedback on the effect of any system-suggested terms they manually add to their queries, only after issuing a new query. For this reason, we propose a search interface (Figure 1) that allows searchers to view the impact of their expansion decisions in real time.

In our proposed search interface, searchers can add expansion term from the list of 10 suggested terms (label 3 in Figure 1) to their initial query by clicking on the term(s), and can also remove a selected term by clicking on it. Selected terms are appended at the end of the initial query in the query box and searchers can also manually modify their queries. Every time an expansion term is selected or deselected, the summary box (label 4 in Figure 1) is updated to display a summary of the document set that *would* be retrieved if the searcher had searched for their initial query plus the selected terms.

This is intended to give the searcher a sense of what information will be retrieved as the result of making IQE decisions. By demonstrating the effect of adding an expansion term, or a set of expansion terms, on the information that will be retrieved, the interface provides immediate feedback to the searcher with the intention of helping him make better expansion decisions.

The interface also includes a query box (label 1), and a list of retrieved document titles and snippets (label 2). The list of retrieved documents and the expansion terms are updated for each new executed query (i.e. only when the searcher presses the 'search' button on the interface). Retrieval is performed by using a standard *tf-idf* scoring function in Lucene[1]. Retrieved documents are displayed in groups of 10 and searchers can view further retrieved document by clicking on the page numbers at the bottom of the interface. The full text of a document can be viewed in a separate window by clicking on the document title.

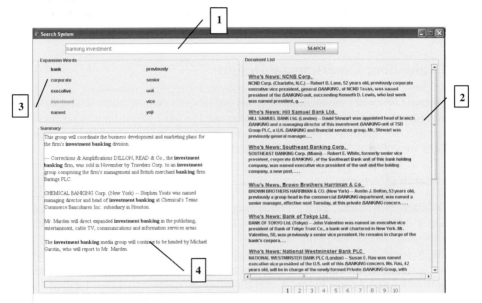

Fig. 1. Screenshot of search interface

3.2 Summarisation Method and Expansion Term Generation

We compute and display summaries in real time depending on the searcher's actions, whether suggested expansion terms are added or removed. Every time an expansion term is selected or deselected (label 3 in Figure 1), the user's current query is modified accordingly (label 1) (i.e. by respectively appending or removing the term). This query is used to retrieve a set of documents. Note, however, that these documents are not displayed on the interface; they are only used 'internally' for the summarisation process. Retrieval is again done by using a standard *tf-idf* scoring function in Lucene.

We then extract the sentences from the top 25 retrieved documents, index them, match them against the current user query using *tf-idf* scoring in Lucene, and use the top 5 retrieved sentences to construct the summary; our summaries are therefore query-biased.

Through the summary (label 4), we aim to show searchers the impact of their choice of expansion terms so that they can make informed expansion decisions. The

---

[1] Apache Lucene: http://lucene.apache.org/

summary, therefore, can be seen as an overview of the documents that will be retrieved in response to the query that the searcher is currently creating through IQE. For this purpose, we display top-ranking sentences [15][18] instead of the complete set of retrieved documents so that searchers have a quick overview of the impact of their choice of expansion terms on the results and do not have to engage in browsing the result set to determine whether they have made the right expansion decisions.

In this study, we use pseudo relevance feedback to create the list of suggested terms. From the set of documents retrieved for the initial query, we assume the top 10 documents as relevant and use them to generate potential expansion terms. We used the *wpq* weight [13] to weight and rank expansion terms and only the 10 highest ranked terms were displayed on the interface.

4   Evaluation: Experimental Design

This section describes the methodology for our laboratory-based user study. In section 4.1 we outline the different search interfaces used in our evaluation and we include in section 4.2 the description of the tasks and collections. We give details about the participants in section 4.3 and describe, in section 4.4, our data collection methods.

4.1   Search Interfaces

During each experiment, each searcher used three different search interfaces and performed a different search task on each. All interfaces included variations of features from the interface shown in the Figure 1.

The *Non-QE System* (S_1) was our baseline, as the interface resembles a standard search system that the majority of searchers would be familiar with. It contains only components 1 and 2 from Figure 1, and therefore no query expansion facility. Participants submitted their own queries and the system would present documents in order of retrieval score.

The *IQE System* (S_2) provided a query expansion feature where subjects could choose the terms to reformulate their query from a list generated by the system. Additionally to S_1, this system displays a list of 10 terms (label 3 in Figure 1) that searchers can use to reformulate their queries, which can then be resubmitted to the system. This search interface was chosen in order to investigate the uptake of IQE in the absence of query biased summaries. We used S_2 also to find out whether searchers continue to manually reformulate their queries despite terms being suggested to them by the system.

Our final system, the *SummaryIQE System* (S_3), included IQE and a summary component (label 4 in Figure 1) as described in section 3. Our motivation for S_3 is to study whether query biased summaries are effective at helping searchers to engage more with the suggested expansion terms.

4.2   Collection of Documents and Tasks

The experiments were carried out on news articles from the Wall Street Journal (WSJ) (1990 – 92) and Associated Press (AP) (1988) collections, taken from the TREC-5 collection [16]. Table 1 summarises the collection statistics.

Table 1. Collection statistics

	WSJ	AP
Number of documents	74,520	79,919
Median number of terms/document	301	438
Mean number of terms/document	508.4	468.7

The tasks performed by searchers were derived from TREC-5 topics [16], and required subjects to gather information about the topic area in a maximum of 15 minutes. The tasks were classified in three categories (Easy, Medium, Hard) depending on their hardness as calculated in TREC-5 [16]. System and tasks were allocated using a Greco-Latin square design so that the order of task and system is not repeated in one block of participants.

We chose to use the TREC collections and tasks because this set up allowed us to vary the hardness of the tasks that searchers were asked to complete. We were particularly interested in hard tasks and how searchers faired in them because one of the assumptions behind IQE is that searchers will prefer more system support through IQE when search tasks are more difficult [5]. We also use searchers' perception of task difficulty and familiarity in our analysis (section 5.2.3).

4.3 Subjects

A total of 18 subjects took part in our experiments and the average age was 23.4 years. All subjects were university students and 14 had a computer science background. Subjects had on average 8.8 years of online search experience. Half of the searchers had used suggested terms provided by some online search engines and all subjects used online search systems on a daily basis.

4.4 Data Collection

All subjects completed a total of 8 questionnaires, each at a different stage in the experiment. There was a pre-experiment questionnaire, which gathered demographic information. Before each task, subjects were briefed about the interface they were about to use and were asked to fill a questionnaire about their familiarity with the topic and the perceived difficulty of the task. A questionnaire after each task included questions about the subjects' search experience with the interface.

For every task, participants were asked to find as many useful documents as they could in a maximum of 15 minutes and to write down the documents' title on their task sheet. After the three tasks, subjects completed a final questionnaire, which required them to rate the three search interfaces and the tasks. There were also open ended questions for general comments. Finally, there was an informal interview with subjects.

To gather information about the subjects' interaction with each search interface, system logging was used. The logs recorded the queries being searched and the expansion terms the searcher added or removed when S_2 and S_3 were being used. The unique identifier for a document would also be recorded in the logs if the searchers viewed the full text of that document.

5   Evaluation: Results and Discussion

In this section we present the findings from our user experiments. In section 5.1 we study whether summaries are perceived by searchers as useful in helping them make sense of term suggestions and we investigate, in section 5.2, whether any increased use of IQE leads to better performance. We focus particularly on the support provided to searchers for hard and unfamiliar tasks. We provide both quantitative and qualitative analysis of our findings. Statistical testing was carried out at $p < 0.05$ with a Wilcoxon signed-ranks test, and we use M and SD to denote the mean and standard deviation respectively.

5.1   Usefulness of Summaries for Predictive IQE

Firstly, we report on the searchers' perceptions of the interfaces reporting data from the post experiment questionnaire and the informal interviews about subjects' overall impression of the systems. When asked which system they preferred overall, 61% of participants preferred S_3 compared to 22% for S_2 and 11% for S_1, while 6% of all subjects did not have a preference. When asked which system was more supportive for completing search tasks, 50% of the participants felt most supported by S_3 (S_1 =11%, S_2 =28% and 11% had no preference) and 33% of searchers thought S_3 was easiest to use (S_1 =22%, S_2 =22% and 22% had no preference).

One of the core aims of our research was to make IQE decisions easier to make by increasing the information available to searchers when selecting expansion terms. For S_2 and S_3, the two IQE systems, subjects were asked whether the suggested terms were useful in completing the search task, whether they understood how to use the terms and whether they found them intuitive. A 5-point Likert scale from 1 (Strongly disagree) to 5 (Strongly agree) was used for the participant responses. Table 2 shows a summary of their response for each system.

Table 2. Subjects' views about suggested terms. Bold indicates higher response

	IQE system (S_2)		SummaryIQE System (S_3)	
	M	SD	M	SD
Suggested terms were useful in completing tasks	3.2	1.0	**3.8**	0.8
Terms were intuitive	3.3	0.7	**3.9**	0.8
Understood how to use suggested terms	3.7	0.8	**4.2**	0.7

Participants' responses were significantly higher on their ability to understand how to use the suggested expansion terms ($p=0.046$) on S_3 compared with S_2 and significantly higher on the intuitiveness of the suggested terms with terms on S_3 being seen as significantly more intuitive than those on S_2 ($p=0.013$). Our experimental systems used pseudo-relevance feedback to create the list of expansion terms, so it is unlikely that the suggested terms were actually better or more intuitive terms; rather the summaries made the terms appear more intuitive because participants could see the context in which the terms appeared and could compare this to their own information need. If the term appeared in a context similar to what the participants were looking for, they could use the terms to reformulate their query. This was reflected in the qualitative feedback from the interviews, where the participants reported summaries as being useful in describing the context of expansion terms and making it easier to specify the query.

On S_2 a number of participants commented on not knowing how to use the expansion terms. They claimed that, although a simple listing of the terms provided them with alternative search terms, they were not supported in making use of the suggested terms. They could not, for example, see the benefits or drawbacks of adding more than one term at a time. If they added several terms to their query, performed a search and did not find the results useful, they could not know which of the terms they added led to this kind of results.

During the informal interview, 10 out of 18 searchers commented on the usefulness of the summaries in providing them with an idea of how their results will be impacted depending on different combinations of expansion terms. Subjects in general felt that if they could see how expansion terms appear in documents, then they could make a more informed decision about the usefulness of suggested terms in their respective search tasks. They compared this feature to S_2, where they had to add the expansion terms to their queries and search for the new query to find out how their results are affected. Some searchers reported on finding this process frustrating.

5.2   Do Summaries Increase the Uptake of IQE and the Effectiveness of QE?

Given that summaries were seen as useful in section 5.1 we now turn to how the summaries were used by the experimental participants and the effect of query expansion on their search performance.

5.2.1   Query Reformulation

We studied the subjects' interaction with each of the search systems through the logs and determined the number of times a query was reformulated (i.e. the number of times a query was issued to the system other than the first query). There were no significant differences in the number of times participants reformulated their query on the three systems, although participants tend to reformulate their queries a higher number of times using S_2, on average 7.44 reformulations per task compared to only 5.89 reformulations on S_1 and S_3. This echoes the findings mentioned in 5.1, where participants reported having to engage in more trial-and-error behaviour when using S_2. On S_3 the participants could explore the possible consequences of query expansion without running a new query: the number of times searchers added expansion terms on S_3 was significantly higher than on S_2 (p=0.026) and searchers experimented with suggested terms more, by adding and removing terms, on S_3 than on S_2 before using them to reformulate their queries (p= 0.009).

The content of queries was also different. On S_3, 46.23% of query reformulations include one or more of the suggested expansion terms; on S_2 only 30.59% of the reformulations include expansion terms. This shows that users of S_2 are reformulating their queries more often manually than with suggested terms (p= 0.046), by adding terms they pick up from reading the full text of documents, from the task description or from their own knowledge of the topic. With S_3, subjects are using a mixed-mode approach, where they combine system-offered expansion terms with their own expansion terms.

5.2.2 Number of Relevant Documents Found by Searchers

The number of relevant documents for each subject was computed from the list of documents that searchers noted on each of their task sheets. Subjects found more relevant documents on S_2 (M=5.78, SD= 2.05) and S_3 (M=5.61, SD=2.33) compared to S_1 (M=4.67, SD=1.46), however there was no statistical difference between the systems. The value of the expansion terms is therefore somewhat mixed. Participants had additional support from the IQE systems and this led to finding more relevant documents, but not to a significant degree over the baseline system. However, 77.8% of users felt that they were more supported in their tasks by the two IQE systems.

The participants spent more time interacting with S_3: 10 out of 18 subjects took the maximum time allocated when using the system while only 4 and 5 users spent 15 minutes on their tasks with S_1 and S_2 respectively. Participants also felt that the time available on S_3 was the least sufficient among the three (S_3=3.9, S_1=4.1, S_2=4.3).

The fact that users had to spend longer time on their selection of terms by reading the summary might explain the need for extra time. Also, the fact that searchers felt that the task time was not sufficient implies that they did not have enough time to view as many documents as they would have liked. Despite finding slightly less relevant documents with S_3, searchers were most satisfied with the information they found with S_3 (S_1=3.44, S_2=3.78, S_3=3.83) and significantly more satisfied with the information they found on S_3 compared to S_1 (p=0.037).

5.2.3 Task Difficulty and IQE

A commonly reported finding in IQE studies is that IQE is more useful in complex or difficult tasks [5]. In the post experiment questionnaire, searchers were asked to rate the tasks they had performed in terms of their difficulty (*1 – Easiest to 3 – Most Difficult*). We identified the hardest task for each participant and studied their performance on the system which they used for that task. Table 3 shows the number of documents that searchers found relevant for their hardest task and the number of query reformulations on each system.

Table 3. Average number of relevant documents and query reformulations for hard tasks on each system

	No-QE System	*IQE System*	*SummaryIQE System*
Number of relevant documents	3.2	4.7	6.0
Number of query reformulations	11.3	5.8	7.4

There was no statistical difference between the means in Table 3. However, for their hardest tasks, searchers used suggested terms more for query reformulations with the *SummaryIQE System* (S_3) than they did with the *IQE System* (S_2). With S_2,

42.86% of query reformulation included one or more suggested expansion terms compared to 48.65% with S_3.

We also used the pre-task questionnaire to identify tasks perceived as unfamiliar by searchers. The question about task familiarity was posed using a 5-point Likert scale [Unfamiliar (1) to Familiar (5)]. We considered 1 and 2 to represent unfamiliarity with the topic area and identified searchers' unfamiliar tasks and the systems they performed these tasks on. We present the results in Table 4.

Table 4. Average number of relevant documents and query reformulations for unfamiliar tasks

	No-QE System	*IQE System*	*SummaryIQE System*
Number of relevant documents	4.6	6.5	5.2
Number of query reformulations	6.6	10.0^2	5.5

Therefore, searchers tend to reformulate their queries most on the *IQE System* for unfamiliar tasks, which is in agreement with [9]. However, no significant statistical differences were found between the three systems. We also studied the percentage of query reformulations for unfamiliar tasks that included suggested terms and found that only 10% of query reformulations included suggested terms with the *IQE System* compared to 45.45% with the *SummaryIQE System*.

In both cases, we observe a higher uptake of IQE with S_3. However, the difference is more acute for unfamiliar tasks. These findings are in line with the data gathered during informal interviews at the end of user experiments despite no significant statistical difference. In fact, a number of searchers mentioned that they were more likely to use IQE when the tasks were hard or unfamiliar. It appears that the summaries in our system led to an increase in the use of suggested terms to reformulate queries. We attribute the higher percentage for the use of IQE for the unfamiliar tasks to the fact that in an unknown environment, searchers are more likely to try different terms that have been suggested to them by the system because they do not have prior knowledge of the search area. On the other hand, searchers might think a task is difficult because they are unable to find documents relevant to the search topic and not because they are unfamiliar with the topic itself.

6 Conclusions

We present in this paper a study that tried to address the problem of limited uptake of IQE. Our proposed approach provides users with query biased summaries that are updated in real time depending on a searcher's choice of expansion terms. Summaries aim at providing searchers with additional information on how their choice of expansion terms affects the results retrieved by the system.

Our study shows strong evidence that summaries are effective in supporting searchers to complete their information seeking tasks. We reported an increase in the

---

[2] This result for *IQE System* is affected by an outlier of 29 query reformulations for one user.

uptake of IQE when searchers were provided with additional information through the query biased summaries. We studied whether summaries supported searchers for the tasks they perceived as most difficult and unfamiliar, and despite no significant differences in searchers' performance in those tasks, our results showed that 45% of query reformulations for unfamiliar tasks include suggested expansion terms when summaries were provided compared to 10% in the absence of summaries. In addition, there were significant statistical differences between the *IQE System* and the *SummaryIQE System* when it comes to how intuitive users found suggested terms. This is due to the fact that users feel more supported when they can see the context in which suggested terms appear in documents that would be retrieved if they submitted their initial query with the term. Our search interface was preferred by 61% of searchers because of the additional support. We report therefore, that summaries appear to be effective in helping searchers to complete their tasks. The additional support provided by the query biased summaries was also important for difficult and unfamiliar tasks.

If the reasons for the limited uptake of IQE are overcome, its potential effectiveness could provide users of IR systems with enhanced support in formulating their queries. We view this study as a step towards this direction.

References

1. Anick, P.: Using Terminological Feedback for Web Search Refinement – A Log-based Study. In: Proceedings of the 26th ACM SIGIR Conference, pp. 88–95. ACM Press, New York (2003)
2. Araya, J.E.: Interactive query formulation and feedback experiments in information retrieval, PhD thesis, Cornell University, Ithaca, New York (1990)
3. Beaulieu, M.: Experiments with interfaces to support query expansion. Journal of Documentation 53, 8–19 (1997)
4. Blocks, D., Binding, C., Cunliffe, D., Tudhope, D.: Qualitative evaluation of thesaurus-based retrieval. In: Agosti, M., Thanos, C. (eds.) ECDL 2002. LNCS, vol. 2458, pp. 346–361. Springer, Heidelberg (2002)
5. Fowkes, H., Beaulieu, M.: Interactive searching behaviour: Okapi experiment for TREC-8. In: Proceedings of the IRSG 2000 Colloquium on IR Research (2000)
6. Hancock-Beaulieu, M., Fieldhouse, M., Do, T.: An evaluation of interactive query expansion in an online library catalogue with a graphical user interface. Journal of Documentation 51(3), 225–243 (1995)
7. Beaulieu, M., Jones, S.: Interactive searching and interface issues in the Okapi best match probabilistic retrieval system. Interacting with Computers 10(3), 237–248 (1998)
8. Harman, D.: Towards Interactive Query Expansion. In: Proceedings of the 11thACM SIGIR Conference, pp. 321–331. ACM Press, New York (1988)
9. Kelly, D., Cool, C.: The effects of topic familiarity on information search behaviour. In: Proceedings of the 2nd ACM/IEEE-CS Joint Conference on Digital Libraries, pp. 74–75. ACM Press, New York (2002)
10. Kelly, D., Gyllstrom, K., Bailey, E.W.: A Comparison of Query and Term Suggestion Features for Interactive Searching. In: Proceedings of the 32nd ACM SIGIR Conference, pp. 371–378. ACM Press, New York (2009)

11. Koenemann, J., Belkin, N.J.: A case for interaction: a study of interactive information retrieval behaviour and effectiveness. In: Proceedings of the Human Factors in Computing Systems Conference (CHI 1996), pp. 205–212 (1996)

12. Magennis, M., van Rijsbergen, C.J.: The Potential and Actual Effectiveness of Interactive Query Expansion. In: Proceedings of the 20th ACM SIGIR Conference, pp. 324–332. ACM Press, New York (1997)

13. Robertson, S.E.: On term selection for query expansion. Journal of Documentation 46(4), 359–364 (1990)

14. Ruthven, I.: Re-examining the Potential Effectiveness of Interactive Query Expansion. In: Proceedings of the 26th ACM SIGIR Conference, pp. 213–220. ACM Press, New York (2003)

15. Tombros, A., Sanderson, M.: The Advantages of Query Biased Summaries in Information Retrieval. In: Proceedings of the 21st ACM SIGIR Conference, pp. 2–10. ACM Press, New York (1998)

16. Voorhees, E.M., Harman, D.: Overview of the Fifth Text REtrieval Conference (TREC-5). In: Proceedings of Text Retrieval Conference, TREC-5 (1995)

17. White, R.W., Marchionini, G.: Examining the effectiveness of real time query expansion. Information Processing and Management 43(3), 685–704 (2006)

18. White, R.W., Ruthven, I., Joemon, J.M.: Finding Top Relevant Documents using Top Ranking Sentences: An Evaluation of Two Alternative Scheme. In: Proceedings of the 25th ACM SIGIR Conference, pp. 57–64. ACM Press, New York (2002)

Evaluation of an Adaptive Search Suggestion System

Sascha Kriewel and Norbert Fuhr

University of Duisburg–Essen

Abstract. This paper describes an adaptive search suggestion system based on case–based reasoning techniques, and details an evaluation of its usefulness in helping users employ better search strategies. A user experiment with 24 participants was conducted using a between–subjects design. One group received search suggestions for the first two out of three tasks, while the other didn't. Results indicate a correlation between search success, expressed as number of relevant documents saved, and use of suggestions. In addition, users who received suggestions used significantly more of the advanced tools and options of the search system — even after suggestions were switched off during a later task.

1 Introduction

While nowadays information search technology is pervasive, and it is used by the larger public instead of just search professionals or search intermediaries, the effective use of these information retrieval (IR) technologies by end–users remains a challenge [10,26]. Many modern information retrieval systems provide functionalities beyond those of low–level search actions, but rarely do they support users in finding the right tactic for the right situation [8], i.e. help searchers choose the most appropriate or useful action in a specific situation during a long search session to implement an overall search strategy.

Novices often try to express several concepts at once, and either over– or under–specify their requests [23]. If they reformulate their query, they mostly use parallel movements or add terms to their original query [24]. They have difficulties finding a set of search terms that captures their information need and brings a useful result set [11]. Rarely do they employ anything resembling sophisticated search strategies [9], and often they even use counter–productive moves. Search experts on the other hand have access to effective information finding strategies, that enable them to interact with the system to achieve good retrieval results, depending on the history of their interaction so far, and the result currently displayed [11]. Furthermore, users rarely know how and when to use advanced search features or tools to the best effect [21].

This is especially problematic when users work on complex search problems, such as researching viewpoints to support an argumentation. Users engage in long interactive search sessions and don't expect to find all their answers as a result of a single query. And while there has been extensive work in supporting

C. Gurrin et al. (Eds.): ECIR 2010, LNCS 5993, pp. 544–555, 2010.

users in executing specific moves, tactics or stratagems (to use the terminology introduced by Bates [4]), such sessions are rarely supported by common systems.

This is where automatic search assistance can be of help. In this paper a system for situational search suggestions is described, and an evaluation of its usefulness is presented. Three main questions were examined: do users who receive suggestions search more successfully over the course of the complete session, do they use more advanced options, and can they learn from these suggestions.

2   Providing Strategic Help

Search strategies can be seen as plans for completing search tasks, encompassing many tactics and stratagems used in the process of an information search [4]. To use such a strategy, users need to select the appropriate stratagems or tactics supported by the system. However, searching is an opportunistic process and search goals can shift throughout the task [27]. It is often more fruitful for a searcher to follow promising opportunities that arise from previous results, instead of sticking to a straight path towards the perfect result set. Järvelin [17] showed that search sessions using individually ineffective queries may actually be more effective than long single queries, depending on how costs and benefits of interactive IR sessions are considered in the evaluation of search success.

Therefore, users need to recognize these strategic opportunities during a session and use the strategic options available to maneuver out of dead–ends. Unfortunately, users rarely exploit the advanced capabilities and features of modern search systems, even if these would improve their search. They might not be aware of their existence, might not understand them, or don't know in which situation they could be effectively employed. Search systems providing strategic assistance can improve search effectiveness by suggesting the use of these advanced features or options automatically [15].

Drabenstott [10] studied the use of domain expert strategies by nondomain experts, and concluded that there is a "need for system features that scaffold nondomain experts from their usual strategies to the strategies characteristic of domain experts." Wildemuth [26] pointed out that current IR systems offer little support to help searchers formulate or reformulate effective search strategies.

In [8] Brajnik et al. describe a strategic help system based on collaborative coaching, which tries to assist users by providing them with suggestions and hints during so–called critical or enhanceable situations. The system uses a hand-crafted knowledge base of 94 production rules to provide suggestions based on the tactics and stratagems proposed by Bates [3,4]. The strategic help module was integrated into FIRE, a user-interface to a Boolean IR system. Only six people participated in the user evaluation, but the results showed promise for the usefulness of strategic suggestions.

Bhavnani et al. [7] introduce the idea of Strategy Hubs, which provide domain specific search procedures for web search. While domain specific strategies may not be transferable to a general search system, they also noted the importance of procedural search knowledge in addition to system and domain knowledge, and the need to make such knowledge explicit within the search system.

Jansen and McNeese evaluated the effectiveness of automated search assistance in a within–subject study with 40 participants (although only 30 of them actually used the search assistance provided), and found that automated assistance can improve the performance of the searcher depending on how performance is measured. However, it was also noted, that about half of the participants actually did not perform better, and that automated assistance should to be targeted and possibly personalized to achieve the best results [16].

3   The Suggestion Tool

We developed a suggestion tool using case–based reasoning techniques for the DAFFODIL system to support users with useful strategic search advice. DAFFODIL [12] is a digital library search system offering a rich set of tools that experienced users can employ in their search. However, previous user experiments showed that inexperienced users have problems utilizing these tools [19].

For the experiment described in this paper, a stripped down version of DAFFODIL was used (shown in fig. 1). It contains a search tool for querying digital libraries using a fielded search form (the available fields are "title", "author", "year", and "free–text"). The search supports Boolean queries and phrases. Results from different sources are combined and ranked. A search history shows all previous queries of the user, and allows for re–use and editing of queries. Relevant documents (as well as queries, terms, or authors) can be stored in a clipboard. A detail viewer shows a short summary of a result document, including abstract and a full–text link where available. Details can be shown either for documents in the result list or for documents stored in the clipboard. The document details are interactive, and links for authors, keywords, or terms can be used to run new queries, modify the current one, or call other tools. In addition, a number of support tools were provided:

- extraction of popular terms and authors from the result list
- extraction of terms and authors from relevant documents in the clipboard
- display of related terms for the current query
- a thesaurus with synonyms, super– or subordinate concepts, and definitions
- an author network showing co–author relationships for a selected author
- a classification browser for the search domain

The suggestion system itself has been described previously [20], and consists of three main components. The *observing agent* triggers after every search to collect information about the current query, the returned results, and the search history of the user. From these, a case is constructed that is used by the *reasoning agent* to match against previously stored cases. Those suggestions that were found to be appropriate solutions for similar cases are collected and ranked.

The *suggestion tool* adapts these weighted suggestions for the current situation and offers them to the user. Users can either ignore the offer, or call up a ranked list of suggestions with a graphical bar for each list entry indicating the estimated likelihood of the suggestion being useful in the current situation. The suggestions come with a short explanation of their purpose and usage scenario, and most allow for direct execution by simple double–click.

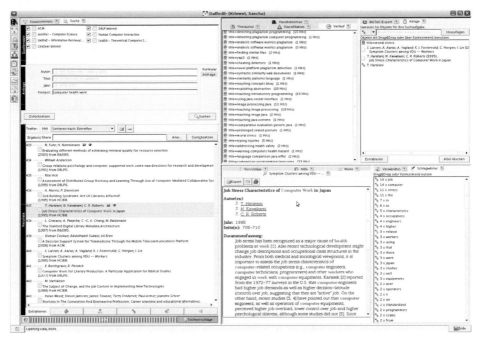

Fig. 1. The DAFFODIL desktop: search tool (left), search history (top center), clipboard (top right), detail view (bottom center), and Related Terms (bottom right)

3.1   Suggestions

The suggestions provided by the suggestion tool were compiled from the information science literature, e.g. from [3,4,14,13], and supplemented by suggestions specific to problems discovered during user experiments with the DAFFODIL system [19,25]. 22 suggestions have been included, which can be grouped as

- terminological suggestions (use of spelling variants, related terms, synonyms, subordinate or superordinate concepts)
- suggestions regarding use of operators and search fields (use of disjunction instead of conjunction, use of phrase operators, use of title or free–text field, restricting by years or authors)
- strategic suggestions (creating conceptual facets, pearl growing using a document previously saved as relevant, avoiding overspecification, author search)
- suggestions for advanced tools of the DAFFODIL system (computation of co–author network, browsing the classification, filters to extract result list, extracting terms from relevant documents or result sets).

3.2   Finding Suggestions

Techniques from case–based reasoning are used to find and rank the most appropriate suggestions for the user's search situation. Belkin and colleagues [5]

previously proposed a case–based reasoning approach for designing an interactive IR system. They applied this in the MERIT system to provide a script–based
interaction with the system, based on cases that are associated with specific regions of the ISS space defined in [6].

In the scenario described in this paper cases are previous search situations of
other users, and their solutions are specific suggestions that were found useful by
those users. Users can judge suggestions as useful or not useful for their current
situation, thereby adding to the case base. The cases are composed of a number
of aspects which are used for comparing cases and to compute similarities. These
aspects all fall within one of three categories: numeric values (e.g., number of
results), sets of terms (e.g., the query terms), or vectors of term weights (e.g.,
popular terms extracted from a result set with number of occurrences per term).

After every search action of the user, a new situation is compiled and the
reasoning component retrieves the most similar cases for each available solution
from the database, with a cut–off value of 50% — i.e. no cases with a similarity
score of less than 0.5 compared to the current situation are used.

The similarity $sim_T(a, b)$ of two situations a and b is computed as the *weighted
mean* (using weights w_k) of the individual similarity scores sim_k of the various
aspects of a situation.

For selecting and ranking the suggestions that will be presented to the user as
search advice, both positive and negative cases are used. Positive cases are cases
similar to the current situation for which a user has rated a specific suggestion as
useful. Correspondingly, negative cases are those where a user has rated a specific
suggestion as not useful. The use of negative cases is common for applications
of case–based reasoning in medicine, where positive and negative indicators for
treatments or diagnoses. Ontañón and Plaza [22] propose a justification based
method for deriving a confidence score. A similar method has been used here.

$$p_v := s_v * \frac{\sum_{b_i \in F^+(v)} sim_T(a, b_i)}{\sum_{b_i \in F^+(v)} sim_T(a, b_i) + \sum_{b_i \in F^-(v)} sim_T(a, b_i)} \tag{1}$$

Be s_v the similarity of the most similar case for a suggestion v, $F^+(v)$ the set
of similarities of all positive cases for the suggestion v and $F^-(v)$ the set of
similarities for all negative cases for v. The total weight for this suggestion is
then computed as shown in eqn. 1. The suggestions are ranked by decreasing
weights p_v.

3.3   Pilot Evaluation

A small scale pilot evaluation was conducted in 2007, during which 12 users
worked with the suggestion tool on a simulated, complex work task. The evaluation pointed out a number of possible improvements, but overall user reception
of the suggestions was positive. The users found the automated, non–intrusive
advice to be helpful, and implemented suggested tactics with success to further
their search task [20]. Several problems that were found during the evaluation
were fixed and the list of available suggestions has been extended. The system
has been in use since then and user ratings have expanded the case base.

4   Experiment

A user experiment was conducted to evaluate if an adaptive suggestion tool using case-based reasoning techniques can help users search better and learn how to use the advanced capabilities of a complex search system.

4.1   Research Questions

Three major research questions drove the experiment: (1) Do the suggestions lead to a more successful search? (2) Do the suggestions help users to use advanced search techniques and more of the advanced tools of the system? (3) Do users utilise these techniques and tools on their own, after having encountered them during their use of the suggestion tool?

For the first question, the number of relevant documents saved by the users was considered. For the second and third question, the search logs were analysed and the use of different DAFFODIL tools and search fields was counted.

4.2   Participants

For the study, 24 volunteers were recruited from among students of computer science, communication science, and related subjects at the University of Duisburg–Essen. The students, 20 male and 4 female, received extra course credit for their participation. The age of the participants ranged from 22 to 48, with an average age of 27.25 years (standard deviation 5.41, mode 24). Among other prestudy questions, the participants were asked to report their previous search experience. On a five-point Likert scale, two rated themselves as inexperienced (2), six as moderately experienced (3), 15 as experienced (4), and one as expert (5).

Since studies [1] have shown that self-evaluation of search skill can be a poor indicator for search performance (at least for some tasks), the participants were also asked to describe their use of search systems in years of experience and frequency of use. The average experience in years was 4.75 with a standard deviation of 2.45. Not all participants provided a useful estimation of their frequency of search engine use, but 6 reported less then daily use. Of the daily users that provided a more exact estimate, the frequency ranged between 1-35 distinct uses of search engines per day (average 11 times per day, standard deviation 11.84).

When asked about search systems with which they are familiar, 23 of the participants named Google, 4 named Yahoo, and 2 MSN Search or Bing. Three of the students reported that they used desktop search systems like Google Desktop or Beagle for Linux. Three students used digital library search systems.

4.3   Setup

For the study two systems were used, both based on the DAFFODIL system described in section 3. The systems were identical, except for the inclusion of the suggestion tool in one of them. Both contained a search tool configured to search six digital libraries from the computer science domain.

Table 3. User actions taken during tasks (suggestions for tasks 1 and 2)

Action	avg. (stdv.) with sugg.	avg. (stdv.) without sugg.	Welch's t test t	df	p
Basic actions / Suggestions					
Run query	46.50 (9.76)	51.33 (19.89)	-0.82	17.01	0.423
View details	79.17 (45.47)	51.67 (28.50)	1.78	18.49	0.092
Save document	38.17 (14.86)	23.25 (14.18)	2.52	21.95	*0.019*
Request suggestions	8.33 (3.11)	— (—)	—	—	—
Execute suggestion	9.50 (4.01)	— (—)	—	—	—
Advanced actions	10.33 (6.53)	3.75 (3.89)	3.02	18.00	*0.007*
Extract terms	4.91 (4.60)	0.50 (1.73)	2.99	12.58	*0.011*
Use thesaurus	1.17 (1.75)	0.92 (1.56)	0.37	21.73	0.716
Change query from other tool	4.67 (2.87)	2.33 (2.31)	2.19	21.03	*0.039*
all adv. actions (only task 1)	3.42 (2.39)	1.16 (1.34)	2.84	17.26	*0.011*
all adv. actions (only task 2)	4.25 (1.24)	1.33 (1.44)	3.24	16.52	*0.005*
all adv. actions (only task 3)	2.67 (1.78)	1.25 (1.29)	2.48	21.52	*0.021*

Table 4. Average number of relevant documents saved

Task	avg. (stdv.) with sugg.	avg. (stdv.) without sugg.	Welch's t test t	df	p
Health	8.25 (5.64)	3.00 (2.95)	2.86	16.61	*0.011*
Plagiarism	14.92 (5.30)	8.25 (7.62)	2.49	19.63	*0.022*
Java	2.42 (1.78)	1.42 (1.62)	1.44	21.81	0.165
Total	25.58 (9.61)	12.67 (8.95)	3.41	21.89	*0.002*

It seems that, although the search suggestions did help users to search more successfully, the assistance doesn't help for all tasks.

Research question 2. The second question concerned the actions that users perform during their search. The hypothesis was that users in the assisted group would employ advanced tools and search options of DAFFODIL significantly more often than users in the unassisted group.

Evaluation of the search logs showed no statistically significant differences between both groups with regards to moves to change the query by adding or removing terms or boolean operators, correcting the spelling of a term, re–using a previous query, replacing one or more terms, or replacing the entire query (see table 2). However, the differences for the later two moves were only barely not significant: users in the *unassisted* group replaced query terms on average 16.92 times per search versus 11.08 times ($p = 0.055$) and replaced the whole query 1.31 per search versus 0.83 times ($p = 0.062$).

On the other hand, assisted users restricted their queries with the year field, switched between free–text and title fields, or used the author field significantly more often then users in the other group (7.08 vs. 0.92, $p = 0.001$). In fact, even though users in both groups received the same written introduction to the fielded search interface, and all search fields were visible at all times, users in the unassisted group generally ignored all other fields but the one chosen for their first query (either "title" or "free–text") and used this for all searches in all tasks.

The assisted group used less query moves in total, but instead employed more advanced actions using the other tools of the system. No significant differences could be observed in the use of the thesaurus for looking up more general or specific terms, or to get word definitions or synonyms. However, users in the assisted group made significantly more use of the extraction capabilities of DAFFODIL, to get new terms or common authors from result sets or relevant documents (4.91 vs. 0.5, $p = 0.01$). They also used more terms from other tools (like the displays of related terms, extracted terms, the thesaurus, or the classification) directly in their query (4.67 vs. 2.33, $p = 0.039$).

Research question 3. The last question examined was, if users of the suggestion tool would independently use advanced search actions after having seen them suggested for previous tasks. While this can really only be answered with a long–term experiment, a first test was done as part of this user experiment. For the assisted group, the suggestion tool was switched off for the last experiments, so that no suggestions were displayed to the users.

While there was a clear drop in advanced actions used by the group between tasks 2 and 3 (see table 3), they still used significantly more advanced actions than the unassisted group (2.67 vs. 1.25, $p = 0.02$). Similarly, the usage of additional search fields was higher among the users who had previously received search suggestions (1.75 vs. 0.25, $p = 0.004$).

This could be interpreted as there being a slight learning effect from receiving situationally appropriate suggestions, that led those users to employ them independently. Of course, the three tasks were performed in immediate succession, so it remains unclear if they would retain this knowledge for future search tasks. On the other hand, a more pronounced effect might have been observed during a long term study, where searchers use the suggestion tool over a much longer period of time than just during two 20 minute search tasks.

5   Summary and Conclusion

In this paper a system for providing adaptive search suggestions based on the situations of previous users was presented. The system uses case–based reasoning techniques and although it was implemented within the framework of the DAF-FODIL search system for digital libraries, the basic principle is not dependent on the search system used, and has also been adapted for Google web search (implemented as an add–on for the open source browser Firefox) [2].

A user experiment was conducted to evaluate the suitability of the suggestion system for helping nonexpert users to search more successfully and to utilise more advanced tools and search tactics.

The results of the experiment were positive. Users with assistance were able to search more successfully, as measured by the number of relevant documents saved — at least for some tasks. It seems that not all tasks benefit from the suggestions in their current form. If this results from a lack of suitable cases, is due to the specific search task, or points to a more general problem of the approach needs further examination. For those tasks where users were helped by the suggestions, the differences were very significant ($p = 0.011$ and $p = 0.022$).

Furthermore, users who received suggestions used more of DAFFODIL's tools and a greater variety of search fields in their tasks. This hold true even for the task which they performed unassisted. It would be interesting to examine how these differences in tool use change during a longer experiment covering multiple search sessions over several weeks. Additional analyses on the search and tool use logs could be done, in particular to identify search tactics and strategies from the query logs (as done e.g. in [9]). It is an open question if the use of search suggestions leads users to more sophisticated search strategies.

References

1. Aula, A., Nordhausen, K.: Modeling successful performance in web searching. Journal of the American Society for Information Science and Technology 57(12), 1678–1693 (2006)
2. Awasum, M.: Suggestions for Google websearch using a firefox add–on. bachelor thesis, University of Duisburg-Essen (2008)
3. Bates, M.J.: Information search tactics. Journal of the American Society for Information Science 30(4), 205–214 (1979)
4. Bates, M.J.: Where should the person stop and the information search interface start? Information Processing and Management 26(5), 575–591 (1990)
5. Belkin, N.J., Cool, C., Stein, A., Thiel, U.: Cases, scripts, and information-seeking strategies: On the design of interactive information retrieval systems. Expert Systems with Applications 9(3), 379–395 (1995)
6. Belkin, N.J., Marchetti, P.G., Cool, C.: BRAQUE: Design of an interface to support user interaction in information retrieval. Information Processing and Management 29(3), 325–344 (1993)
7. Bhavnani, S.K., Christopher, B.K., Johnson, T.M., Little, R.J., Peck, F.A., Schwartz, J.L., Strecher, V.J.: Strategy hubs: next-generation domain portals with search procedures. In: Proceedings of the conference on Human factors in computing systems, pp. 393–400. ACM Press, New York (2003)
8. Brajnik, G., Mizzaro, S., Tasso, C., Venuti, F.: Strategic help in user interfaces for information retrieval. Journal of the American Society for Information Science and Technology 53(5), 343–358 (2002)
9. Carstens, C., Rittberger, M., Wissel, V.: How users search in the german education index - tactics and strategies. In: Proceedings of the workshop Information Retrieval at the LWA 2009 (2009)
10. Drabenstott, K.M.: Do nondomain experts enlist the strategies of domain experts. Journal of the American Society for Information Science and Technology 54(9), 836–854 (2003)
11. Fields, B., Keith, S., Blandford, A.: Designing for expert information finding strategies. In: Fincher, S., Markopoulos, P., Moore, D., Ruddle, R.A. (eds.) BCS HCI, pp. 89–102. Springer, Heidelberg (2004)
12. Fuhr, N., Klas, C.-P., Schaefer, A., Mutschke, P.: Daffodil: An integrated desktop for supporting high-level search activities in federated digital libraries. In: Agosti, M., Thanos, C. (eds.) ECDL 2002. LNCS, vol. 2458, pp. 597–612. Springer, Heidelberg (2002)

13. Harter, S.P.: Online information retrieval: concepts, principles, and techniques. Academic Press Professional, Inc., San Diego (1986)
14. Harter, S.P., Peters, A.R.: Heuristics for online information retrieval: a typology and preliminary listing. Online Review 9(5), 407–424 (1985)
15. Jansen, B.J.: Seeking and implementing automated assistance during the search process. Information Processing and Management 41(4), 909–928 (2005)
16. Jansen, B.J., McNeese, M.D.: Evaluating the effectiveness of and patterns of interactions with automated searching assistance. Journal of the American Society for Information Science and Technology 56(14), 1480–1503 (2005)
17. Järvelin, K.: Explaining user performance in information retrieval: Challenges to ir evaluation. In: Azzopardi, L., Kazai, G., Robertson, S., Rüger, S., Shokouhi, M., Song, D., Yilmaz, E. (eds.) ICTIR 2009. LNCS, vol. 5766, pp. 289–296. Springer, Heidelberg (2009)
18. Klas, C.-P., Albrechtsen, H., Fuhr, N., Hansen, P., Kapidakis, S., Kovács, L., Kriewel, S., Micsik, A., Papatheodorou, C., Tsakonas, G., Jacob, E.: A logging scheme for comparative digital library evaluation. In: Gonzalo, J., Thanos, C., Verdejo, M.F., Carrasco, R.C. (eds.) ECDL 2006. LNCS, vol. 4172, pp. 267–278. Springer, Heidelberg (2006)
19. Klas, C.-P., Fuhr, N., Schaefer, A.: Evaluating strategic support for information access in the DAFFODIL system. In: Heery, R., Lyon, L. (eds.) ECDL 2004. LNCS, vol. 3232, pp. 476 487. Springer, Heidelberg (2004)
20. Kriewel, S., Fuhr, N.: Adaptive search suggestions for digital libraries. In: Goh, D.H.-L., Cao, T.H., Sølvberg, I.T., Rasmussen, E. (eds.) ICADL 2007. LNCS, vol. 4822, pp. 220–229. Springer, Heidelberg (2007)
21. Markey, K.: Twenty-five years of end-user searching, part 1: Research findings. Journal of the American Society for Information Science and Technology 58(8), 1071–1081 (2007)
22. Ontañón, S., Plaza, E.: Justification-based multiagent learning. In: Mishra, N., Fawcett, T. (eds.) The Twentieth International Conference on Machine Learning (ICML 2003), pp. 576–583. AAAI Press, Menlo Park (2003)
23. Pollock, A., Hockley, A.: What's wrong with internet searching. D-Lib Magazine (March 1997)
24. Rieh, S.Y., Xie, H.(I.): Patterns and sequences of multiple query reformulations in web searching: a preliminary study. In: Proceedings of the 64th Annual Meeting of the American Society for Information Science and Technology, vol. 38, pp. 246–255 (2001)
25. Schaefer, A., Jordan, M., Klas, C.-P., Fuhr, N.: Active support for query formulation in virtual digital libraries: A case study with DAFFODIL. In: Rauber, A., Christodoulakis, S., Tjoa, A.M. (eds.) ECDL 2005. LNCS, vol. 3652, pp. 414–425. Springer, Heidelberg (2005)
26. Wildemuth, B.M.: The effects of domain knowledge on search tactic formulation. Journal of the American Society for Information Science and Technology 55(3), 246–258 (2004)
27. Xie, H.I.: Shifts of interactive intentions and information-seeking strategies in interactive information retrieval. Journal of the American Society for Information Science 51(9), 841–857 (2000)

How Different Are Language Models and Word Clouds?

Rianne Kaptein[1], Djoerd Hiemstra[2], and Jaap Kamps[1,3]

[1] Archives and Information Studies, University of Amsterdam, The Netherlands
[2] Database Group, University of Twente, Enschede, The Netherlands
[3] ISLA, Informatics Institute, University of Amsterdam, The Netherlands

Abstract. Word clouds are a summarised representation of a document's text, similar to tag clouds which summarise the tags assigned to documents. Word clouds are similar to language models in the sense that they represent a document by its word distribution. In this paper we investigate the differences between word cloud and language modelling approaches, and specifically whether effective language modelling techniques also improve word clouds. We evaluate the quality of the language model using a system evaluation test bed, and evaluate the quality of the resulting word cloud with a user study. Our experiments show that different language modelling techniques can be applied to improve a standard word cloud that uses a TF weighting scheme in combination with stopword removal. Including bigrams in the word clouds and a parsimonious term weighting scheme are the most effective in both the system evaluation and the user study.

1 Introduction

This paper investigates the connections between tag or word clouds popularised by Flickr and other social web sites, and the language models as used in IR. Fifty years ago Maron and Kuhns [14] suggested a probabilistic approach to index and search a mechanised library system. Back then, documents were indexed by a human cataloguer who would read a document and then assign one or several indexing terms from a controlled vocabulary. Problems with this approach were the ever increasing amount of documentary data and the semantic noise in the data. The correspondence between a document and its index terms is not exact, because the meaning of terms are a function of their setting. The meaning of a term in isolation is often quite different when it appears in an environment (sentence, paragraph, etc.) of other words. Also, word meanings are individual and can vary from person to person. Because of these problems, Maron and Kuhns [14] proposed to, instead of having a human indexer decide on a yes-no basis whether or not a given term applies for a particular document, assign weights to index terms to more accurately characterise the content of a document. Since then the information retrieval community has developed many models to automatically search and rank documents. In this paper we focus on the language modelling approach. We choose this approach because it is conceptually simple and it is based on the assumption that users have some sense of the frequency of words and which words distinguish documents from others in the collection [16].

The new generation of the Internet, the social Web, allows users to do more than just retrieve information and engages users to be active. Users can now add tags to categorise

C. Gurrin et al. (Eds.): ECIR 2010, LNCS 5993, pp. 556–568, 2010.

Fig. 1. Word cloud from 10 results for the topic "diamond smuggling"

web resources and retrieve your own previously categorised information. By sharing these tags among all users large amounts of resources can be tagged and categorised. These generated user tags can be visualised in a tag cloud where the importance of a term is represented by font size or colour. Terms in a tag cloud usually link to a collection of documents that are associated with that tag. To generate tag clouds the tripartite network of users, documents and tags [10] can be exploited. Of course, the majority of documents on the web is not tagged by users. An alternative to clouds based on user-assigned tags, is to generate tags automatically by using statistical techniques. Clouds generated by automatically analysing the document contents are referred to as 'word clouds'. Word clouds have for example been generated for the inaugural speeches of American presidents [e.g., 13]. Word clouds can be used in the same way as tag clouds, but are especially useful to get a first impression of long documents, such as books, or parliamentary proceedings. Also word clouds can be used to summarize a collection of documents, such as clustered or aggregated search results. Figure 1 shows a word cloud summarising top 10 retrieved documents.

This paper investigates the connections between tag or word clouds and the language models of IR. Our main research question is: do words extracted by language modelling techniques correspond to the words that users like to see in word clouds? We discuss related work on tag clouds and language modelling in Section 2 with the goal of determining which specific techniques have been explored in both approaches. We decide to focus on four different features of word clouds, i.e. pseudo-relevance vs. relevance information, stemming, including bigrams, and term weighting schemes. Each of them is investigated in a separate section (Sections 4 through 7). In Section 3 we describe our experimental set-up. We use an IR test collection to evaluate the effectiveness of the technique for language models, and we conduct a user study establishing user preferences over the resulting word clouds as a means to convey the content of a set of search results. Finally, in Section 8 we draw our conclusions.

2   Related Work

In this section, we will discuss related work on tag/word clouds and language modelling, with the aim of determining a number of techniques applicable for both types of approaches. The first appearance of a tag cloud is attributed to Douglas Coupland's novel Microserfs [4]. In this novel the main character Daniel writes a program to take terms out of his journal entries and create snapshots of keywords, which are called 'subconscious files.' The first widespread use of tag clouds was on the photo-sharing site Flickr. Other sites that contributed to the popularisation of tag clouds were Del.ici.ous and Technorati. Nowadays tag clouds are often considered as one of the typical design

elements of the social Web. Evaluation of tag clouds appears scarcely in scientific literature, in the blogosphere however there is a lot discussion on the usefulness of tag clouds
[2]. Part of the evaluation of tag clouds are the effects of visual features such as font
size, font weight, colour and word placement [1, 6, 18]. Font size and font weight are
considered the most important visual properties. Font sizes are commonly set to have a
linear relationship to the log of the frequency of occurrence of a tag. Colour draws the
attention of users, but the meaning of colours needs to be carefully considered. The position of the words is important, words in the top of the tag cloud attract more attention.
An alphabetical order of the words helps users to find information quicker. Rivadeneira
et al. [18] identify four tasks tag clouds can support. In our experiments we will evaluate
our word clouds on the basis of these tasks:

- Search: locating a specific term that represents a desired concept.
- Browsing: casually explore the cloud with no specific target in mind.
- Impression Formation or Gisting: use the cloud to get a general idea on the under-
  lying data.
- Recognition / Matching: recognise which of several sets of information the tag
  cloud is likely to represent.

Some recent information retrieval papers discuss the use of tag or word clouds for various applications. PubCloud uses clouds for the summarisation of results from queries
over the PubMed database of biomedical literature [9]. A stopword list is used to remove
common words, and a Porter Stemmer is applied [17]. Colours are used to represent re-
cency, and font size represents frequency. Mousing over a tag displays a list of words
that share the same prefix and a hyperlink links to the set of PubMed abstracts containing the tag. In Dredze et al. [5] summary keywords are extracted from emails. Common
stopwords and e-mail specific stopwords such as 'cc,' 'to' and 'http' are removed. Latent semantic analysis and latent Dirichlet allocation outperform a baseline of TF-IDF
on an automated foldering and a recipient prediction task.

On the Internet tools like Wordle [22] and ManyEyes [13] create visually pleasing
word clouds from any document. To create word clouds these tools remove stopwords
and use term frequencies to determine font sizes. Information retrieval systems mainly
remove stopwords to reduce index space and speed up processing. Since the discrimination value of stop words is low, removing these terms will not have a large effect on
retrieval performance. Modern IR systems and web search engines exploit the statistics
of language and do not use stopword lists, or very small stopword lists (7-12 terms)
[12]. For word clouds however it is essential to have a good stopword list. Both Wordle and ManyEyes also have an option to include multi-word phrases. Popular social
tagging sites like Flickr and Technorati allow multi-word tags. Most first-generation
tagging systems did not allow multi-word tags, but users find this a valuable feature.

Term frequencies are most commonly used to create tag clouds. For information retrieval term frequencies are also a commonly used method of term weighting, but in
addition some alternative weighting schemes have been developed. It was recognised
early that more weight should be given to query terms matching documents that are
rare within a collection, and therefore the inverse document frequency (IDF) was introduced [20]. The idf factor varies inversely with the number of documents n in which
a term occurs in a collection of N documents. Since then many variants with different

normalisation steps have been developed to improve retrieval results. Several relevance feedback approaches attempt to filter out background noise from feedback documents. Zhai and Lafferty [23] apply an Expectation-Maximization model to concentrate on words that are common in the feedback documents but are not very common in in the complete collection. This same idea is used to create parsimonious models of documents in [8].

This section aimed to determine a number of techniques applicable for both language modelling and word cloud generation. The innovative features of tag clouds lie in the presentation and the willingness of users to assign tags to resources. Considering other technical features of tag clouds, we have not found features in tag clouds that have not been explored in the language modelling approach to information retrieval. From the techniques in the literature we will investigate the four features we think are the most interesting for creating word clouds, i.e., using relevance or pseudo-relevance information, stemming, including bigrams and term weighting schemes. In Sections 4 to 7 each of these features will be discussed and evaluated using the set-up that is discussed in the next section.

3   Evaluation of Word Clouds

In this section, we will detail our experimental set-up. Since there is no standard evaluation method for word clouds, we created our own experimental test bed. Our experiments comprise of two parts, a system evaluation and a user study. For both experiments we use query topics from the 2008 TREC Relevance Feedback track.

3.1   System Evaluation

We test our approaches using the 31 topics that have been evaluated using Pool10 evaluation, which is an approximation of the normal TREC evaluation strategy, and allows for ranking of systems by any of the standard evaluation measures [3]. We execute two experiments that correspond to tasks tag clouds can support, as described in the previous section. In the first experiment we evaluate the tasks 'Impression Formation' and 'Recognition' by using the words of the clouds for query expansion. Our assumption is that the quality of the query expansion equates the quality of the used model. The weights that are used to determine font size, are now used to represent the weight of query expansion terms. Prominent words carry more weight, but less prominent items can still contribute to the performance of the complete cloud, which is also the case in the two tasks. Our query expansion approach is similar to the implementation of pseudo-relevance feedback in Indri [21]. We keep the original query, and add the expansion terms with their normalised probabilities. We use the standard evaluation measures MAP and P10 to measure performance.

In our second experiment we evaluate the 'Search' task. In this task you want to locate a specific term that represents a desired concept. In our experiment the desired concept is the topic, and all terms that represent this topic are therefore relevant. We consider a word representative of the topic if adding the word to the original query leads to an improvement of the retrieval results. We take the feedback sets and 31 queries that we also used in the previous experiment. We let each model generate a word cloud consisting of 25 terms. For each topic we generate 25 queries where in each query a word from the word cloud is added to the original query. No weights are assigned to the expansion terms. For each query we measure the difference in performance caused by adding the expansion term to the original query. Our evaluation measure is the percentage of 'relevant' words in the word cloud, i.e. the percentage of words where adding them to the query leads to an improvement in retrieval results. Additionally, we also calculate the percentage of 'acceptable' words that can be added to the query without a large decrease (more than 25%) in retrieval results.

3.2   User Study

In addition to the system-based approach for evaluation, we evaluate the word clouds from a user's point of view. In this user study we are focusing on the question which words should appear in a word cloud. We set the size of the word cloud to 25 terms. We do not want to investigate the optimal size for word clouds, this size suffices to show users the differences between the different types of word clouds. The only visual feature we are considering is font size, other features, such as lay-out, colours etc. are not considered. We present a word cloud as a list of words in alphabetical order. The test persons first read a TREC topic consisting of the query title (keywords that are used for search), query description (one line clarification of the query title) and narrative (one paragraph that explains which documents are relevant). For each topic users rank four groups of word clouds. In each group we experiment with a different feature:

- Group 1: Pseudo relevance and relevance information
- Group 2: Stemming
- Group 3: Including bigrams
- Group 4: Term weighting scheme

Test persons may add comments to each group to explain why they choose a certain ranking. Each test person gets 10 topics. In total 25 topics are evaluated, each topic is evaluated by at least three test persons and one topic is evaluated by all test persons. 13 test persons participated in the study. The test persons were recruited at the university in different departments, 4 females and 9 males with ages ranging from 26 to 44.

3.3   Baseline

Each group of clouds includes the standard word cloud which acts as a baseline to which the other clouds are compared. Since stopwords have high frequencies, they are likely to occupy most places in the word cloud. We therefore remove an extensive stopword list consisting of 571 common English words. Only single words (unigrams) are included in the standard cloud. Stemming is applied and words are conflated as described later in Section 5. The standard word cloud uses a TF weighting scheme which equals term frequency counting. The probability of a word occurring in a document is its term frequency divided by the total number of words in the document. For all models we have a restriction that a word has to occur at least twice to be considered. To create a word cloud all terms in the document are sorted by their probabilities and a fixed number of

Fig. 2. Word cloud from 100 relevant results

Table 1. Effectiveness of feedback based on pseudo-relevance vs. relevance information. Evaluated after removing the used 100 relevant documents from runs and qrels.

Approach	MAP	P10	% Rel. words	% Acc. words
Pseudo	0.0985	0.1613	35	73
Rel. docs	**0.1161** ⁻	**0.2419** ⁻	**50**	**85**

the 25 top ranked terms are kept. Since this results in a varying probability mass depending on document lengths and word frequencies, we normalise the probabilities in order to determine the font size. The standard cloud uses pseudo-relevant documents to generate the word cloud. The top 10 documents retrieved by a language model run are concatenated and treated as one long document. Throughout this paper we will use the topic 766 'diamond smuggling' to show examples. In the earlier Figure 1 the standard TF word cloud of this topic was shown.

4   Clouds from Pseudo Relevant and Relevant Results

In this section, we look at the impact of using relevant or pseudo-relevant information to generate language models and tag clouds. In the first group a TF cloud made from 10 pseudo-relevant documents is compared to a cloud of 100 relevant documents. By making this comparison we want to get some insights on the question if there is a mismatch between words which improve retrieval performance, and the words that users would like to see in a word cloud. Our standard word cloud uses pseudo-relevant results because these are always available. The cloud in Fig. 2 uses 100 pages judged as relevant to generate the word cloud.

Results. When we look at the query expansion scores, shown in Table 1[1], a query based on the 100 relevant documents is on average better than the query based on 10 pseudo-relevant documents, and also there are more relevant and acceptable words in the clouds based on the 100 relevant documents. The test persons in our user study however clearly prefer the clouds based on 10 pseudo-relevant documents: 66 times the pseudo-relevant document cloud is preferred, 36 times the relevant documents cloud is preferred, and in 27 cases there is no preference (significant at 95% using a two-tailed sign-test).

There seem to be three groups of words that often contribute positively to retrieval results, but are not appreciated by test persons. First, there are numbers, usually low

---

[1] In all tables significance of increase/decrease over baseline according to t-test, one-tailed is shown: no significant difference($^-$), significance levels 0.05($^\circ$), 0.01($^\bullet$), and 0.001($^\bullet$).

numbers from 0 to 5, which occur frequently in relevant documents. Without context these numbers do not provide any information to the user. Numbers that represent years can sometimes be useful. The second group are general and frequently occurring words which do not seem specific to the query topic. e.g. for the query 'hubble telescope repairs' adding the word 'year' ,'up' or 'back' results in improved retrieval results. The third group consists of words that test persons don't know. These can be for example abbreviations or technical terms. In this user study the test persons did not create the queries themselves, therefore the percentage of unknown words is probably higher than in a normal setting. In addition for most of the test persons English is not their first language. In some cases also the opposite effect takes place, test persons assume words they don't know (well) are relevant, while in fact the words are not relevant. Words appreciated by test persons and also contributing to retrieval performance are the query title words and keywords from the description and the narrative. The query description and narrative are in a real retrieval setting usually not available. Most of the informative words are either a synonym of a query word, or closely related to a query word.

These findings agree with the findings of a previous study, where users had to select good query expansion terms [19]. Also here reasons of misclassification of expansion term utility are: users often ignore terms suggested for purely statistical reasons, and users cannot always identify semantic relationships.

5   Non-stemmed and Conflated Stemmed Clouds

In this section, we look at the impact of stemming to generate conflated language models and tag clouds. To stem, we use the most common English stemming algorithm, the Porter stemmer [17]. To visualize terms in a word cloud however, Porter word stems are not a good option. There are stemmers or lemmatizers that do not affect the readability of words, the simple S-removal stemmer for example conflates plural and singular word forms by removing the suffix -s according to a small number of rules [7]. The Porter stemmer is more aggressive, reducing for example 'immigrant' to 'immigr,' and 'political' to 'polit'. A requirement for the word clouds is to visualize correct English words, and not stems of words which are not clear to the user. Using word stems reduces the number of different terms in a document, because different words are reduced to the same stem. Since these words are very closely correlated, it is useful to aggregate them during the generation of terms for the word clouds. The question remains however which words should be visualised in the word cloud. In our experiments we consider non-stemmed word clouds and conflated word clouds where word stems are replaced by

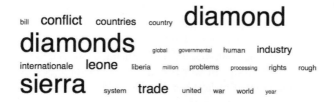

Fig. 3. Word cloud of 10 results using plain (non-stemmed) words

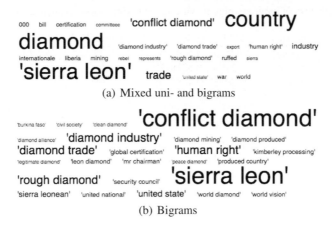

(a) Mixed uni- and bigrams

(b) Bigrams

Fig. 4. Word cloud of 10 results with unigrams and bigrams

the most frequently occurring word in the collection that can be reduced to that word stem. The standard word cloud is conflated, in Figure 3 a non-stemmed word cloud is displayed. The non-stemmed cloud contains both 'diamond' and 'diamonds,' while the corresponding conflated cloud (see Fig. 1) only contains 'diamond'. The conflated cloud does bring up a small conflation issue. The non-stemmed cloud contains the word 'leone' (from Sierra Leone), but in the conflated cloud this is undesirably conflated to 'leon'. We opted for the collection-wise most frequent expansion since it is easy to process, but with hindsight choosing the most frequent word in the specific document(s) would have been preferred.

Results. The effect of stemming is only evaluated in the user study. We did not do a system evaluation, because we do not have a non-stemmed index available. Looking at pairwise preferences, we see that it often makes only a small difference to the word clouds to conflate words with the same stem: 38 times the conflated cloud is preferred, 20 times the non-stemmed cloud is preferred, and 71 times there is no preference (significant at 95% on a two-tailed sign-test). Often the difference is so small that it is not noticed by test persons. A disadvantage of the conflated cloud is that sometimes words are conflated, but then expanded to an illogical word. For example for the query 'imported fire arms' in the word cloud 'imported' is changed into 'importante'. A disadvantage of the non-stemmed cloud is that users do not like to see two words that are obviously reduced to the same stem, like 'ant' and 'ants'. These kind of words also appear next to each other, because of the alphabetical order of the words.

6   Bigrams

In this section, we look at the impact of adding bigrams to generate more informative language models and tag clouds. For users, bigrams are often easier to interpret than single words, because a little more context is provided. We have created two models that incorporate bigrams, a mixed model that contains a mix of unigrams and bigrams,

Table 2. Effectivenss of unigram, bigram, and mixed tokenizations evaluated over the full qrels

Approach	MAP	P10	% Rel. words	% Acc. words
Unigrams	0.2575	0.5097	**35**	**73**
Mixed	**0.2706** ⁻	**0.5226** ⁻	31	71
Bigrams	0.2016°	0.4387 ⁻	25	71

Table 3. Pairwise preferences of test person over unigram, bigram, and mixed tokenizations

Model 1	Model 2	# Preferences			Sign test 95%
		Model 1	Model 2	Tied	
bigram	mixed	49	54	26	–
mixed	unigram	71	33	25	0.95
bigram	unigram	62	46	21	–

and a bigram model that consists solely of bigrams. To incorporate bigrams, we use the TF model with some adjustments. In the bigram model each term now consists of two words instead of one word. Bigrams containing one or two stopwords are excluded. The most frequently occurring bigram will receive the highest probability. In the mixed model, a term can either consist of one or two words. Both unigrams and bigrams contribute to the total term count. Again all terms containing one or two stopwords are excluded from the model. The probability of occurrence of a term, either bigram or unigram, is its frequency count, divided by the total term count. We want to avoid however that unigrams which occur usually as part of a bigram, receive too much probability. Therefore, we subtract from each unigram that occurs as part of a bigram, the probability of the most frequently occurring bigram that contains the unigram. Since the probabilities of the unigrams and the bigrams are estimated using the same approach, the resulting probabilities are comparable. So, we can create word clouds and query expansions that are a mix of unigrams and bigrams. To include a bigram as a query expansion term we make use of the proximity operator available in Indri [15]. The terms in the bigram must appear ordered, with no terms between them. For the user study we placed bigrams between quotes to make them more visible as can be seen in Figure 4, bigrams can also be differentiated by using different colours.

Results. In Table 2 the system evaluation results are shown. For query expansion, the model that uses a mix of unigrams and bigrams performs best with a MAP of 0.2706. Using only bigrams leads to a significant decrease in retrieval results compared to using only unigrams. Looking at the percentages of relevant and acceptable words, the unigram model produces the most relevant words. The mixed model performs almost as good as the unigram model.

In the user study, the clouds with mixed unigrams and bigrams and the clouds with only bigrams are selected most often as the best cloud as can be seen in Table 3. There is no significant difference in preference between mixed unigrams and bigrams, and only bigrams. Users do indeed like to see bigrams, but for some queries the cloud with only bigrams contains too many meaningless bigrams such as 'http www'. An advantage of the mixed cloud is that the number of bigrams in the cloud is flexible. When bigrams occur often in a document, also many will be included in the word cloud.

7 Term Weighting

In this section, we look at the impact of term weighting methods to generate language models and tag clouds. Besides the standard TF weighting we investigate two other variants of language models to weigh terms, the TFIDF model and the parsimonious model. In the TFIDF algorithm, the text frequency (TF) is now multiplied by the inverse document frequency (IDF). Words with an inverse document frequency of less than 10 are excluded from the model. In Figure 5(a) the example word cloud of the TFIDF model is shown. The last variant of our term weighting scheme is a parsimonious model [8]. The parsimonious language model concentrates the probability mass on fewer words than a standard language model. Only terms that occur relatively more frequent in the document as in the whole collection will be assigned a non-zero probability. The model automatically removes both common stopwords and corpus specific stopwords, and words that are mentioned occasionally in the document. The parsimonious model estimates the probability $P(t|R)$ using *Expectation-Maximization*:

$$\text{E-step: } e_t = tf(t, R) \cdot \frac{(1 - \lambda)P(t|R)}{(1 - \lambda)P(t|R) + \lambda P(t|C)} \tag{1}$$

$$\text{M-step: } P(t|R) = \frac{e_t}{\sum_t e_t}, \text{ i.e. normalize the model} \tag{2}$$

λ determines the weight of the background model $P(t|C)$. There is no fixed number of terms that are kept, but in the M-step terms that receive a probability below a threshold of 0.001 are removed from the model. In the next iteration the probabilities of the

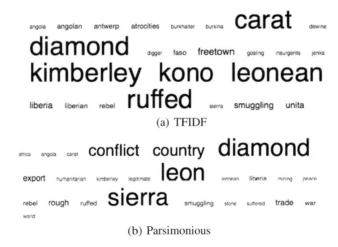

(a) TFIDF

(b) Parsimonious

Fig. 5. Word cloud of 10 results with TFIDF and parsimonious term weighting

Table 4. Effectiveness of term weighting approaches evaluated over the full qrels

Approach	MAP	P10	% Rel. words	% Acc. words
TF	0.2575	0.5097	**35**	**73**
TFIDF	0.1265•	0.3839°	22	67
Pars.	**0.2759°**	**0.5323¯**	31	68

[16] Ponte, J., Croft, W.: A language modeling approach to information retrieval. In: SIGIR 1998, pp. 275–281 (1998)

[17] Porter, M.: An algorithm for suffix stripping. Program 14(3), 130–137 (1980)

[18] Rivadeneira, A.W., Gruen, D.M., Muller, M.J., Millen, D.R.: Getting our head in the clouds: toward evaluation studies of tagclouds. In: Proceedings CHI 2007, pp. 995–998. ACM, New York (2007)

[19] Ruthven, I.: Re-examining the potential effectiveness of interactive query expansion. In: SIGIR 2003, pp. 213–220. ACM, New York (2003)

[20] Spärck Jones, K.: A statistical interpretation of term specificity and its application in retrieval. Journal of Documentation 28, 11–21 (1972)

[21] Strohman, T., Metzler, D., Turtle, H., Croft, W.B.: Indri: a language-model based search engine for complex queries. In: Proceedings of the International Conference on Intelligent Analysis (2005)

[22] Wordle (2009), `http://wordle.net`

[23] Zhai, C., Lafferty, J.: Model-based feedback in the language modeling approach to information retrieval. In: Proceedings CIKM 2001, pp. 403–410. ACM, New York (2001)

Colouring the Dimensions of Relevance*

Ulises Cerviño Beresi[1], Yunhyong Kim[1], Mark Baillie[2],
Ian Ruthven[2], and Dawei Song[1]

[1] The Robert Gordon University, School of Computing
[2] The Strathclyde University, Department of Computer and Information Sciences

Abstract. In this article we introduce a visualisation technique for analysing relevance and interaction data. It allows the researcher to quickly detect emerging patterns in both interactions and relevance criteria usage. The concept of "relevance criteria profile", which provides a global view of user behaviour in judging the relevance of the retrieved information, is developed. We discuss by example, using data from a live search user study, how these tools support the data analysis.

1   Introduction

In this paper we examine the multi-dimensionality of relevance judgment processes (cf. Borlund's proposed evaluation method[3]): we use Barry and Schamber's relevance criteria classes, Section 2, to encode verbal data gathered from users in a search task, and, define relevance criteria profiles and a session visualisation method (Section 3) which we use to analyse how relevance criteria are used to judge document relevance. We conclude the paper with some final remarks and recommendations for future work (Section 4).

2   Relevance Criteria in (I)IR

Researchers suggest that *"a finite range of [relevance] criteria exists and that these criteria are applied consistently across types of information users, [...]"*[2], defined as an overlap of taxonomies identified within two studies [1,4]. We adopt the taxonomy and extend it by re-introducing, from [1], three forms of information novelty, users's background knowledge and their ability to understand the information. As a sample, we only list the relevance criteria related to the examples discussed in this paper:

- Depth/Scope/Specificity: related to the range of focus and detail, e.g. how specific it is to the user's needs.
- Currency: whether the information is current or up to date.
- Tangibility: related to tangibility of content, and the inclusion of hard data/facts.

---

* This research is funded in part by EPSRC Autoadapt project, grant no: EP/F035705/1.

C. Gurrin et al. (Eds.): ECIR 2010, LNCS 5993, pp. 569–572, 2010.

- Affectiveness: related to affective or emotional response to the information aroused in the user.
- Ability to Understand: user's judgement regarding his/her ability to understand the information presented.
- Document novelty: the extent to which the document itself is new to the user.

3   Relevance Criteria Profiles and Session Visualisation

Relevance Criteria Profiles. The user data were quantified to produce a user *relevance criteria profile* (RCP). RCPs are defined as a set of counts where each count corresponds to one of the relevance criteria. Counts are defined as the number of utterances made by a user or a group of users that are classified as the corresponding criterion. To *normalise* RCPs, we divide the count of the ith relevance criterion by the total number of user utterances classified as one of the relevance criteria in the encoding schema. That is, the normalised value of relevance criterion i is defined as $rc_i' = rc_i / \sum_{j=0}^{N} rc_j$, where rc_i is the count from the basic RCP, for relevance criterion i, and N is the total number of relevance criteria. Normalising makes RCPs comparable.

Session Visualisation. The visualisation proposed in this section is intended to provide a bird's eye view of the relevance judgement process within each search session. The session is visualised as a sequence of *relevance criteria piles*. Each pile (sample in Figure 3) can be viewed as a summarisation, according to relevance criteria, of the utterances taking place between interactions. The pile consists of blocks, each block representing a relevance criterion (each criterion represented by a unique colour). Each block is annotated with the polarity (negativity or positivity) of the utterances, where a negative instance is indicated by a minus sign next to the corresponding block. For example, "this document is too old" might be a negative instance of Currency, while "this document is up-to-date" might be a positive instance. The order of the blocks from the bottom of the pile to the top correspond to the order in which each relevance criteria is first mentioned (subsequent utterances classified as the same criterion and polarity as a previous block are not repeated). The colours are assigned to criteria according to the sequence of colours recommended by Ware's study of effective colour coding [5]. The most frequent relevance criterion is assigned the first colour in the sequence, the second most frequent criterion the second and so on.

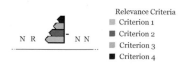

Relevance Criteria
◾ Criterion 1
■ Criterion 2
◾ Criterion 3
■ Criterion 4

Fig. 1. Example of a relevance criteria pile with four criteria

By using sequential ordered piles we can analyse whether a user's relevance judgement process exhibits dependencies between relevance criteria. Delimiting processes by interactions will lead to piles not necessarily being aligned with final relevance judgements. However, a finer granularity in interaction encodings in the visualisation may improve alignment.

Relevance Criteria Profiles and Visualisation in Action. In this section, we discuss RCPs and session visualisations from real data collected between January to August of 2008. A total number of 21 research scientists, affiliated to one of three school in the Robert Gordon University (the School of Computing, the Information Management Group and the School of Pharmacy) participated. The participants searched outside their research field for literature related to their own area of research, and verbalised their thoughts throughout the session. This was recorded and processed to produce the RCPs.

The aggregated RCP for the group of all users shows that *tangibility* and *depth/scope/specificity* are the most frequently mentioned criteria (595 and 406 times respectively). This tendency exhibited as an aggregated group is not preserved, however, when we consider user behaviour according to research background. The differences resulting from research background on the usage of relevance criteria is observable in Figure 2, where we present the normalised RCPs across groups of users clustered according to their school affiliation. In the figure we can observe that members form the School of Computing show a preference for hard, tangible data while this preference is not as evident in the case of members of the Information Management Group. Members from the School of Pharmacy prefer properties such as length and depth over tangibility.

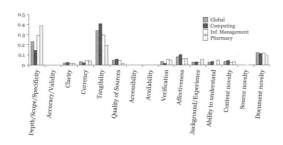

Fig. 2. A comparison of normalised school RCPs and the global RCP

An example of a session visualised is presented in figure 3. The first pile shows that the session begins with a mention of *document novelty*, after which the user navigates away. The second pile ends with a negative mention of *ability to understand* (top most block), again leading the user to navigate away from the document. In the next pile, the user again encounters a known document (black box denoting *document novelty*). The judgement process continues, however, suggesting that the importance of *document novelty* can be overridden, for example, by an affective response to the document. The visualisation also

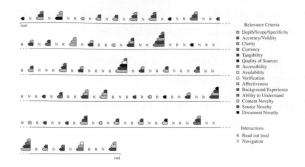

Fig. 3. Example search session visualised: participant 2

highlights the fact that negative relevance judgements are likely to be preceded by a pattern of the form *negative relevance criteria → navigation*. Out of the 14 sequences of the form *negative relevance criteria → navigation*, 12 are indeed implicit negative relevance judgements.

4   Concluding Remarks

In this article, we presented relevance criteria profiles (RCP) and a visualisation technique for relevance judgement processes within a search session. We demonstrated how we might use these to analyse data, and how comparing RCPs can help uncover similarities and differences in relevance criteria usage across users and groups. We also illustrated the potential of visualising relevance judgement processes throughout a session might highlight patterns that lead to final relevance judgements. The next step would be to carry out a fuller comparison of users' relevance criteria, and session visualisations across different groups of users. User interactions might also be further classified to align visualisations with relevance judgements. A comprehensive analysis would likely require aligning relevance criteria, judgement processes, final judgements, and click-through data.

References

1. Barry, C.L.: User-defined relevance criteria: an exploratory study. Journal of the American Society for Information Science 45(3), 149–159 (1994)
2. Barry, C.L., Schamber, L.: Users criteria for relevance evaluation: A cross-situational comparison. Information Processing and Management 34(2-3), 219–236 (1998)
3. Borlund, P.: The IIR evaluation model: a framework for evaluation of interactive information retrieval systems. Information Research 8(3), 8–3 (2003)
4. Schamber, L.: Users'criteria for evaluation in a multimedia environment. In: Proceedings of the 54 Annual Meeting of the American Society for Information Science, vol. 28, pp. 126–133 (1991)
5. Ware, C.: Color sequences for univariate maps: Theory, experiments and principles. IEEE Computer Graphics and Applications 8(5), 41–49 (1988)

On Improving Pseudo-Relevance Feedback Using Pseudo-Irrelevant Documents

Karthik Raman[1], Raghavendra Udupa[2],
Pushpak Bhattacharya[1], and Abhijit Bhole[2]

[1] Indian Institute of Technology Bombay, Mumbai
{karthikr,pb}@cse.iitb.ac.in
[2] Microsoft Research India, Bangalore
{raghavu,v-abhibh}@microsoft.com

Abstract. Pseudo-Relevance Feedback (PRF) assumes that the top-ranking n documents of the initial retrieval are relevant and extracts expansion terms from them. In this work, we introduce the notion of pseudo-irrelevant documents, i.e. high-scoring documents outside of top n that are highly unlikely to be relevant. We show how pseudo-irrelevant documents can be used to extract better expansion terms from the top-ranking n documents: good expansion terms are those which discriminate the top-ranking n documents from the pseudo-irrelevant documents. Our approach gives substantial improvements in retrieval performance over Model-based Feedback on several test collections.

Keywords: Information Retrieval, Pseudo-Relevance Feedback, Query Expansion, Pseudo-Irrelevance, Linear Classifier.

1 Introduction

Pseudo-Relevance Feedback (PRF) is a well-studied query expansion technique which assumes that the top ranking $n(> 0)$ documents of the initial retrieval are relevant and extracts expansion terms from them [1]. While several algorithms have been proposed for extracting expansion terms from the top ranking n documents of the initial retrieval, none of them leverage the empirical fact that many of the high scoring documents are actually irrelevant [2], [3]. In this paper, we make use of such documents to improve PRF substantially.

In order to make use of high-scoring irrelevant documents in PRF, we need to solve the following two problems:

1. **IDENTIFY:** Identifying irrelevant documents in the pool of high-scoring documents of the initial retrieval.
2. **EXTRACT:** Extracting good expansion terms from the pseudo-relevant documents with the help of irrelevant documents.

Identifying irrelevant documents among the top-ranking n documents of the initial retrieval can automatically improve retrieval performance. For instance, if we could remove the irrelevant documents from the top 10 results, we would be

C. Gurrin et al. (Eds.): ECIR 2010, LNCS 5993, pp. 573–576, 2010.
© Springer-Verlag Berlin Heidelberg 2010

automatically improving precision at 10 substantially in most cases. Further, we could use only the relevant among the top-ranking n documents for feedback and improve retrieval performance. Unfortunately, identifying irrelevant documents among the top-ranking n documents is not easy. However, given the set of top-ranking n documents, it is possible to identify high-scoring documents outside of the top n that are highly dissimilar to the top-ranking n documents. Most but not all of these documents are irrelevant in most cases in practice. We call such documents as *Pseudo-Irrelevant* documents. We propose a novel algorithm for identifying pseudo-irrelevant documents from the initial retrieval (Section 2).

Once the pseudo-irrelevant documents have been identified, extracting good expansion terms from the top-ranking n documents boils down to the problem of identifying terms that discriminate the top-ranking n documents from the pseudo-irrelevant documents. To see this, note that good expansion terms should a) increase the scores of the top-ranking n documents and documents similar to them and b) not increase the scores of high-scoring irrelevant documents. By selecting terms that discriminate the top-ranking n documents from the pseudo-irrelevant documents, we achieve both of these objectives. We propose a novel algorithm for extracting discriminative terms (Section 3).

2   Identifying Pseudo-Irrelevant Documents

Let F_R denote the set of top-ranking n documents of the initial retrieval, F_I denote the set of pseudo-irrelevant documents, X denote the the set of high-scoring documents outside of top n and Y denote the set of documents that are similar to any document in F_R. We note that pseudo-irrelevant documents are by definition a) high-scoring documents outside of top n and b) highly dissimilar to any of the top-ranking n documents. Therefore, an intuitive approach to find pseudo-irrelevant documents is to first intersect X with Y and then remove the intersection from the former: $F_I = X - (X \cap Y)$.

The above approach works only if we can extract the set Y from F_R. But we note that it is easy to form the set Y. For each document D in F_R, we only need to find documents in the collection that are similar to it. We form a query Q_D out of D by taking terms that have a collection frequency ≥ 5 and $Idf \geq \log 10$ and retrieve the top-ranking 10 documents for Q_D. These documents are deemed similar to D.

3   Extracting Discriminative Expansion Terms

As mentioned in Section 1, the expansion terms we are interested in are those which discriminate F_R from F_I. Such terms can be found via a classification problem in which each document in F_R is a $+ve$ instance and each document in F_I is a $-ve$ instance and the goal of classification is to learn a discriminant function w that correctly classifies the training instances. Feature vector for each instance is formed as follows: each term that appears in the document forms a feature provided it is not a stop-word and its collection frequency ≥ 5 and $Idf \geq \log 10^1$. The value of a feature is the $tf * idf$ score of the associated term.

We learn a linear discriminant function w from the labeled instances by training a Logistic Regression classifier [4]. The linear discriminant function associates a weight w_i to the term t_i, $i = 1, \ldots, N$. The linear discriminant function classifies a vector x as $+ve$ if $w^T x > 0$ and as $-ve$ if $w^T x \leq 0$. Ideally, $w^T x > 0$ for all documents in F_R and $w^T x \leq 0$ for all documents in F_I. Thus, terms $\{t_i : w_i > 0\}$ can be viewed as relevant expansion terms as their presence in a document contributes to the document being classified as $+ve$. Similarly, terms $\{t_i : w_i < 0\}$ can be viewed as non-relevant expansion terms as their presence in a document contributes to the document being classified as $-ve$. We pick the largest weighted $k > 0$ terms as the set of relevant expansion terms.

4   Empirical Investigation

4.1   Experimental Setup

We employed a KL-divergence based retrieval system with two stage Dirichlet smoothing for the initial retrieval [5]. We used model-based feedback (Mixture Model) as a representative PRF technique [3]. We formed the expanded query by interpolating the feedback model with the original query model with the interpolation factor being 0.5. For extracting the expansion terms, we used the top 10 documents fetched by the initial retrieval. We removed stop-words from topics as well as documents and stemmed the remaining words using the Porter stemmer. We trained the discriminant function using LibLinear with the default parameter settings [4].

We used the following test collections in our experiments:

1. CLEF:
   (a) LATimes 94, Topics 1 - 140 (CLEF 2000-2002).
   (b) LATimes 94 + Glasgow Herald 95, Topics 141-200 (CLEF 2003), 251-350 (CLEF 2005-2006).
2. TREC:
   (a) Associated Press 88-89, Topics 51 - 200 (TREC Adhoc Tasks 1, 2, 3).
   (b) Wall Street Journal, Topics 51 - 200 (TREC Adhoc Tasks 1, 2, 3).

4.2   Results

Table 1 compares the performance of the initial retrieval (LM), Mixture Model (MF), and our approach (PI). The performance measures are Mean Average Precision (MAP) and Precision at 5 (P@5). We see that our approach gives improvements in both MAP and P@5 for all the collections over the initial retrieval. Further, improvements in MAP over MF is substantial on the CLEF collections. In the case of TREC collections, the improvement in P@5 over MF is substantial although the MAP is the same. These preliminary results suggest that pseudo-irrelevant documents can improve retrieval performance of PRF.

---

[1] This throws away most of the noisy terms from the document which would otherwise interfere in the learning of the discriminant function.

Table 1. Retrieval Performance Comparison

Collection	LM		MF		PI	
	MAP	P@5	MAP	P@5	MAP	P@5
CLEF 00-02	0.43	0.49	0.44	0.50	**0.47**	0.50
CLEF 03-06	0.38	0.42	0.41	0.43	**0.43**	**0.47**
AP	0.28	0.47	0.33	0.50	0.33	**0.52**
WSJ	0.27	0.48	0.30	0.52	0.30	**0.53**

4.3   Analysis

While pseudo-irrelevant documents have a positive effect on the performance of expanded retrieval for all the collections we experimented with, the effect is rather varied. CLEF collections seem to benefit from our approach with respect to MAP whereas the TREC collections seem to benefit with respect to P@5. We investigated the causes for this varied effect and found out that TREC collections had more relevant documents per topic on an average than the CLEF collections (25 as against 100). As a consequence, the percentage of true irrelevant documents in the set of pseudo-irrelevant documents is much higher for CLEF topics (90%) than for TREC topics (75%). As a result, our extraction algorithm was losing some valuable expansion terms for some TREC topics as these terms were present in some of the pseudo-irrelevant documents. We are currently investigating ways of improving the percentage of true irrelevant documents amongst pseudo-irrelevant documents.

5   Future Work

In a related experiment, we observed that the mean pseudo-irrelevant distribution is close to the distribution of true irrelevant documents in the feedback document set. This opens up the possibility of leveraging pseudo-irrelevant documents in identifying irrelevant documents among the feedback documents. Such an approach is very likely to benefit PRF beyond what is reported in this work.

References

1. Efthimiadis, E.N.: Query expansion. Annual Review of Information Science and Technology, vol. 31, pp. 121–187 (1996)
2. Lavrenko, V., Croft, W.B.: Relevance based language models. In: Proceedings of SIGIR 2001, pp. 120–127 (2001)
3. Zhai, C., Lafferty, J.: Model-based feedback in the language modeling approach to information retrieval. In: Proceedings of CIKM 2001, pp. 403–410 (2001)
4. Fan, R.E., Chang, K.W., Hsieh, C.J., Wang, X.R., Lin, C.J.: Liblinear: A library for large linear classification. Journal of Machine Learning Research (August 2008)
5. Zhai, C.: Statistical language models for information retrieval a critical review. Found. Trends Inf. Retr. (3)

Laplacian Co-hashing of Terms and Documents

Dell Zhang[1], Jun Wang[2], Deng Cai[3], and Jinsong Lu[1]

[1] School of Business, Economics and Informatics
Birkbeck, University of London
Malet Street, London WC1E 7HX, UK
dell.z@ieee.org, jingsong.lu@gmail.com
[2] Department of Computer Science
University College London
Gower Street, London WC1E 6BT, UK
jun.wang@cs.ucl.ac.uk
[3] State Key Lab of CAD&CG, College of Computer Science
Zhejiang University
100 Zijinggang Road, 310058, China
dengcai@gmail.com

Abstract. A promising way to accelerate similarity search is *semantic hashing* which designs compact binary codes for a large number of documents so that semantically similar documents are mapped to similar codes within a short Hamming distance. In this paper, we introduce the novel problem of *co-hashing* where both documents and terms are hashed simultaneously according to their semantic similarities. Furthermore, we propose a novel algorithm Laplacian Co-Hashing (LCH) to solve this problem which directly optimises the Hamming distance.

1 Introduction

The technique of *similarity search* (aka *nearest neighbour search*) is at the core of Information Retrieval (IR) [1]. A promising way to accelerate similarity search is *semantic hashing* [2] which designs compact binary codes for a large number of documents so that semantically similar documents are mapped to similar codes (within a short Hamming distance). Existing methods for semantic hashing include stacked Restricted Boltzmann Machine (RBM) [2] and Spectral Hashing (SpH) [3] etc. In this paper, we introduce the novel problem of *co-hashing* where both documents and terms are hashed simultaneously according to their semantic similarities. This problem is different from all existing work that is about hashing one type of items only, i.e., either documents or terms. The co-hashing technique can not only provide a high hashing quality by exploiting the duality of document hashing and term hashing, but also facilitate applications that require documents and terms to be matched in the same semantic space.

Given a document collection that contains m terms and n documents, it can be represented by a $m \times n$ term-document matrix X whose rows correspond to terms and columns to documents. A non-zero entry in this matrix, say X_{ij}, indicates the presence of term i in document j, and its value reflects the strength

C. Gurrin et al. (Eds.): ECIR 2010, LNCS 5993, pp. 577–580, 2010.

of their association (e.g., using the TF×IDF weighting scheme [1]). Suppose that the desired length of code is l bits. We use $\mathbf{a}_i \in \{-1,+1\}^l$ and $\mathbf{b}_j \in \{-1,+1\}^l$ to represent the binary codes for term i and document j respectively, where the p-th element of $\mathbf{a}_i$ or $\mathbf{b}_j$ is $+1$ if the p-th bit of code is on or -1 otherwise. Let $A = [\mathbf{a}_1, \ldots, \mathbf{a}_m]^T$ and $B = [\mathbf{b}_1, \ldots, \mathbf{b}_n]^T$. So the problem of co-hashing can be formally defined as finding the compact binary code matrices A and B that best preserve the semantic similarity structure of a given term-document matrix X.

2   Approach

The well-known technique Latent Semantic Indexing (LSI) [1] maps both documents and terms to l-dimensional real-valued vectors in the same semantic space through truncated Singular Value Decomposition (SVD): $X \approx U_l \Sigma_l V_l^T$, where the row vectors of U_l and V_l provide the l-dimensional representation for terms and documents respectively. Therefore one simple method for co-hashing is to perform LSI first, and then binarize U_l and V_l via thresholding to get A and B. This method is named binarized-LSI in [2]. However, it can be shown that the binary codes $\mathbf{a}_1, \ldots, \mathbf{a}_m$ and $\mathbf{b}_1, \ldots, \mathbf{b}_n$ obtained in this way optimises the objective function $\sum_i \sum_j \left(\mathbf{a}_i^T \Sigma_l \mathbf{b}_j - X_{ij} \right)^2$, which is not directly related to our goal of retaining the semantic similarity structure in the Hamming space.

We propose a novel algorithm **Laplacian Co-Hashing (LCH)** which directly optimises the Hamming distance. Let's consider the document collection as an undirected *bipartite* graph [4]: the first m vertices represent the terms while the remaining n vertices represent the documents; an edge (i,j) exists if term i occurs in document j and it is weighted by X_{ij}; there are no edges between terms or between documents. The adjacency matrix of this graph can be written as $W = \begin{bmatrix} 0 & X \\ X^T & 0 \end{bmatrix}$, and its degree matrix is given by $D = \begin{bmatrix} D_1 & 0 \\ 0 & D_2 \end{bmatrix}$ where D_1 and D_2 are diagonal matrices such that $D_1(i,i) = \sum_j X_{ij}$ and $D_2(j,j) = \sum_i X_{ij}$. The Hamming distance between $\mathbf{a}_i$ and $\mathbf{b}_j$, is given by the number of bits that are different between them, which can be calculated as $\frac{1}{4}\|\mathbf{a}_i - \mathbf{b}_j\|^2$. To meet the similarity-preserving criterion, we seek to minimise the weighted average Hamming distance (as in [3]) $\frac{1}{4}\sum_i \sum_j W_{ij}\|\mathbf{a}_i - \mathbf{b}_j\|^2$ because it incurs a heavy penalty if a pair of strongly associated term and document are mapped far apart in the Hamming space. The above objective function can be rewritten in matrix form as $\frac{1}{4}\mathrm{Tr}(Z^T L Z)$, where $Z = \begin{bmatrix} A \\ B \end{bmatrix}$ and $L = D - W = \begin{bmatrix} D_1 & -X \\ -X^T & D_2 \end{bmatrix}$ is the *graph Laplacian* [4]. We found the above objective function actually proportional to that of a well-known manifold learning algorithm, Laplacian Eigenmap (LapEig) [5], except that LapEig does not have the constraints $\mathbf{a}_i \in \{-1,+1\}^l$ and $\mathbf{b}_i \in \{-1,+1\}^l$. So, if we relax the discreteness condition but just keep the similarity-preserving requirement, we can get the optimal l-dimensional real-valued vectors $\tilde{\mathbf{a}}_i \in \mathbb{R}^l$ and $\tilde{\mathbf{b}}_j \in \mathbb{R}^l$ by solving the LapEig problem:

$$\underset{\tilde{Z}}{\arg\min} \, \mathrm{Tr}(\tilde{Z}^T L \tilde{Z}) \quad \text{subject to} \quad \tilde{Z}^T D \tilde{Z} = I \text{ and } \tilde{Z}^T D 1 = 0$$

where $\text{Tr}(\tilde{Z}^T L \tilde{Z})$ gives the real relaxation of the weighted average Hamming distance, and the two constraints prevent collapse onto a subspace of dimension less than l. The solution of this optimisation problem is given by $\tilde{Z} = [\mathbf{z}_1, \ldots, \mathbf{z}_l]$ whose columns are the l eigenvectors corresponding to the smallest eigenvalues (except 0) of the generalized eigenvalue problem $L\mathbf{z} = \lambda D\mathbf{z}$. Taking advantage of the special structure of L and D, we can rewrite the above equation as

$$\begin{bmatrix} D_1 & -X \\ -X^T & D_2 \end{bmatrix} \begin{bmatrix} \mathbf{s} \\ \mathbf{t} \end{bmatrix} = \lambda \begin{bmatrix} D_1 & 0 \\ 0 & D_2 \end{bmatrix} \begin{bmatrix} \mathbf{s} \\ \mathbf{t} \end{bmatrix}.$$

Assuming that D_1 and D_2 are non-singular, we get

$$D_1^{\frac{1}{2}}\mathbf{s} - D_1^{-\frac{1}{2}}X\mathbf{t} = \lambda D_1^{\frac{1}{2}}\mathbf{s} \quad \text{and} \quad -D_2^{-\frac{1}{2}}X^T\mathbf{s} + D_2^{\frac{1}{2}}\mathbf{t} = \lambda D_2^{\frac{1}{2}}\mathbf{t}.$$

Introducing $\mathbf{u} = D_1^{\frac{1}{2}}\mathbf{s}$ and $\mathbf{v} = D_2^{\frac{1}{2}}\mathbf{t}$, we get

$$D_1^{-\frac{1}{2}}X D_2^{-\frac{1}{2}}\mathbf{v} = (1-\lambda)\mathbf{u} \quad \text{and} \quad D_2^{-\frac{1}{2}}X^T D_1^{-\frac{1}{2}}\mathbf{u} = (1-\lambda)\mathbf{v}.$$

Letting $\check{X} = D_1^{-\frac{1}{2}}X D_2^{-\frac{1}{2}}$, we finally have

$$\check{X}\check{X}^T\mathbf{u} = (1-\lambda)^2\mathbf{u} \quad \text{and} \quad \check{X}^T\check{X}\mathbf{v} = (1-\lambda)^2\mathbf{v},$$

which means that $\mathbf{u}$ and $\mathbf{v}$ are the left and right singular vectors of $\check{X}$ respectively, while $1-\lambda$ is the corresponding singular value. Let $\check{\Sigma}_l = diag(\sigma_1, \ldots, \sigma_l)$ denote the diagonal matrix that contains the largest singular values (except 1) of $\check{X}$; and let $\check{U}_l = [\mathbf{u}_1, \ldots, \mathbf{u}_l]$ and $\check{V}_l = [\mathbf{v}_1, \ldots, \mathbf{v}_l]$ denote respectively the matrices that consist of the corresponding l left and right singular vectors of $\check{X}$. Thus we have $\tilde{A} = D_1^{-\frac{1}{2}}\check{U}_l$ and $\tilde{B} = D_2^{-\frac{1}{2}}\check{V}_l$ as the real approximation to A and B. Moreover, given a previously unseen document $\mathbf{q}$, the real-valued l-dimensional vector $\tilde{\mathbf{q}}$ that approximates its binary code can be computed using the "fold-in" formula: $\tilde{\mathbf{q}} = \check{\Sigma}_l^{-1}\check{U}_l^T D_1^{-\frac{1}{2}}\mathbf{q}/(\sum_i q_i)$. The same method can be used for previously unseen terms as well. To convert the obtained l-dimensional real-valued term vectors or document vectors into binary codes, we threshold the p-th element of each term vector or document vector at the *median* of $D_1^{-\frac{1}{2}}\mathbf{u}_p$ or $D_2^{-\frac{1}{2}}\mathbf{v}_p$. In this way, the p-th bit will be on for half of the corpus and off for the other half. Furthermore, as the eigenvectors given by LapEig are orthogonal to each other, different bits in the generated binary codes will be uncorrelated. Therefore this thresholding method gives each distinct binary code roughly equal probability of occurring in the corpus, thus achieves the best utilization efficiency of the hash table.

3 Experiments

We have conducted experiments on two datasets, 20NG[1] and TDT2[2]. The 20NG corpus consists of 18846 documents that are evenly distributed across 20 semantic classes. The original 'bydate' split leads to 11314 (60%) documents for

---

[1] http://people.csail.mit.edu/jrennie/20Newsgroups/
[2] http://www.nist.gov/speech/tests/tdt/tdt98/index.htm

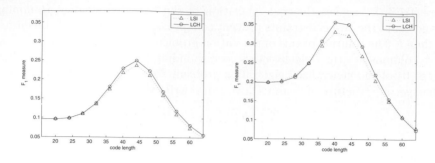

Fig. 1. The results of binarized-LSI and LCH on 20NG (left) and TDT2 (right)

training and 7532 (40%) documents for testing. The TDT2 corpus consists of 11201 documents that are classified into 96 semantic classes. In our experiments, those documents appearing in more than one class were removed, and only the largest 30 classes were kept, thus leaving us with 9394 documents in total. We randomly selected 5597 (60%) documents for training and 3797 (40%) documents for testing. Each dataset is pre-processed by stop-word removal, Porter stemming, selection of 2000 terms with the highest document frequencies (as in [2]), and TFxIDF weighting [1]. For each dataset, we use each document in the test set as a query to retrieve documents in the training set, and then compute the F_1 measure that is *micro*-averaged over all test queries [1]. To decide whether a retrieved document is relevant to the query document, we simply check whether they have the same class label (as in [2]). Figure 1 compares LCH with binarized-LSI [2] in terms of their F_1 measures with Hamming distance ≤ 16, using different length of codes. On both datasets our proposed algorithm LCH outperforms the baseline algorithm binarized-LSI consistently. We think the superior performance of LCH is attributed to its ability of directly optimising the Hamming distance.

References

1. Manning, C.D., Raghavan, P., Schütze, H.: Introduction to Information Retrieval. Cambridge University Press, Cambridge (2008)
2. Salakhutdinov, R., Hinton, G.: Semantic hashing. International Journal of Approximate Reasoning 50(7), 969–978 (2009)
3. Weiss, Y., Torralba, A.B., Fergus, R.: Spectral hashing. In: Proceedings of NIPS 2008, Vancouver, Canada, pp. 1753–1760 (2008)
4. Dhillon, I.S.: Co-clustering documents and words using bipartite spectral graph partitioning. In: Proceedings of KDD 2001, San Francisco, CA, USA, pp. 269–274 (2001)
5. Belkin, M., Niyogi, P.: Laplacian eigenmaps for dimensionality reduction and data representation. Neural Computation 15(6), 1373–1396 (2003)

Query Difficulty Prediction for Contextual Image Retrieval

Xing Xing[1], Yi Zhang[1], and Mei Han[2]

[1] School of Engineering, UC Santa Cruz, Santa Cruz, CA 95064
[2] Google Inc., Mountain View, CA 94043

Abstract. This paper explores how to predict query difficulty for contextual image retrieval. We reformulate the problem as the task of predicting how difficult to represent a query as images. We propose to use machine learning algorithms to learn the query difficulty prediction models based on the characteristics of the query words as well as the query context. More specifically, we focus on noun word/phrase queries and propose four features based on several assumptions. We created an evaluation data set by hand and compare several machine learning algorithms on the prediction task. Our preliminary experimental results show the effectiveness of our proposed features and the stable performance using different classification models.

Keywords: Query difficulty, Contextual image retrieval.

1 Introduction

Given a word/phrase in a document as a query, a contextual image retrieval system tries to return images that match the word/phrase in the given context. Psychological studies for decades [2] have justified the effectiveness of pictorial illustration on improving people's understanding and learning from texts. A contextual image retrieval system annotates the word/phrase in a document with appropriate images and can help readers learn new concepts. For example, a non-English speaker can easily understand the meaning of *panda* if she has seen a picture of it when reading an article that contains the word *panda*.

Although this idea sounds intriguing, image search engines also return useless or even misleading images, either because the image retrieval algorithm is not perfect, or because the query is inherently hard to be represented by images. For example, *honesty* is very difficult to be explained by an image. Some people may suggest a picture of little Washington with a chopped-down cherry tree, while others may disagree with it for the reason that it may be only a legend instead of fact or users may have no idea about this story. If the word in the context cannot be represented by an image, it is better not to annotate it with images, thus avoiding confusing the users with poor image retrieval results. To decide when to use a contextual image retrieval system to provide image annotations, we explore the task of predicting the inherent difficulty of describing the query as images in this paper. As a starting point to study this problem, we focus on

C. Gurrin et al. (Eds.): ECIR 2010, LNCS 5993, pp. 581–585, 2010.

noun word/phrase queries and exploit the query context from the text to make the prediction. Each query is represented as a vector, where each dimension corresponds to a feature proposed based on our intuition about image query difficulty. Machine learning algorithms are used to train the query difficulty classification models using cross validation. The trained models are compared on a evaluation data set.

2   Representing Contextual Query as a Vector of Features

To make it possible to build a contextual image difficulty classifier, we first represent each query as a vector, where each dimension corresponds to one feature. Good features are critical for this application. To find features for all the noun queries, we explore linguistic features and heuristics. In our preliminary research, we start with the following four features, and each is based on one heuristic assumption.

Concreteness. To capture the concreteness of a word, we use whether this word is physical or abstract given its context as a feature. The assumption is that a physical query usually corresponds to some physical existence and hence is easier to be illustrated with images than an abstract query. Since the same word can be either concrete or abstract given different context, we use word sense disambiguation (WSD) and WordNet [5] to compute the query concreteness. Semcor [6] is used as our training corpus to train the WSD models. First, we extract the context features for the training queries, including part-of-speech of neighboring words, single words in the surrounding context and local collocations [4]. Next, we use the maximum entropy modeling approach [3] to train the WSD models. Then, we extract the same feature for the noun queries in the document and obtain the sense-level disambiguation results using the trained WSD models. Finally, the sense level hypernymy in WordNet is used to get the query concreteness. If the disambiguated sense of the query in the context is traced back to a physical entity, its concreteness is 1; otherwise its concreteness is 0.

However, not all queries can be disambiguated this way since WSD training data semantically aligned with WordNet are still limited concerning the coverage of the whole vocabulary. For a word without WSD training data but still available through WordNet, the ratio of its physical senses to all the senses in WordNet is used as the concreteness measure. For a word absent from Wordnet, we look up its first dictionary entry in Dictionary.com and use the average concreteness value of all the nouns in the definitions and explanations as its concreteness measure. For example, *endotherm* is not available in WordNet, while its first dictionary result from Dictionary.com is *a warm-blooded animal*. In this definition, *animal* is the only noun and its concreteness is computed to be 1, therefore the concreteness for *endotherm* is considered to be 1. If the word is neither covered by Wordnet nor Dictionary.com, its concreteness is set to be 0.

Commonness. To capture the commonness of a word, we use the word usage frequency on the web as a feature. More specifically, we use the Google unigram frequency count publicly available in [1] to approximate the commonness. The assumption is that the most frequently used nouns are usually simple words and might be easier to be illustrated by an image. Although this assumption is not always true, there might be a correlation between query commonness and query difficulty.

Also, we use this dataset to validate the correctness of the query. Since the token counts are generated from approximately 1 trillion word tokens of publicly accessible Web pages, we assume that all the nouns are included. If the query is not present in this dataset, we consider it to be an invalid word and assume it cannot be represented by an image.

Ambiguity. To capture the ambiguity of a word, we use the number of noun senses for the word in WordNet as a feature. The assumption is that ambiguous words are more difficult to describe pictorially than unambiguous words. For words not covered by WordNet, the first dictionary result from Dictionary.com is used and the ambiguity is approximated by the number of entries in the result. If the word is absent from Dictionary.com, its ambiguity is set to 1.

Figurativeness. To capture the figurativeness of a word, we send the word to Google image search and Google web search and then use the statistics of the retrieval results as a feature. The ratio of the total number of images retrieved from Google image search to the total number of webpages retrieved from Google web search is computed as the figurativeness measure. The idea is that current commercial image search engines mainly rely on text features to do image retrieval and therefore the percentage of the webpages containing the images for the word may indicate the difficulty to obtain an image to represent the word.

3   Experiments

We randomly select several paragraphs from four literature books and one science book: *The Wind in the Willows, The Story of My Life, A Connecticut Yankee in King Arthur's Court, Don Quixote* and *CPO Focus on Life Science*. Two evaluators manually label the 675 noun words in the texts into two classes: class 0 refers to difficult queries which are unable to be represented by images in the context and class 1 refers to the opposite. The evaluation for each query is based on the top 100 images crawled from Google image search. If the evaluator finds any images that can clearly explain the word in the context, it is considered as an easy query; otherwise it is a difficulty query. We use this approach because our contextual image retrieval system uses search engine to collect candidate images for all the noun words/phrases, and then applies our contextual image retrieval algorithm to select the most appropriate images for the query.

Evaluator 1 labels 357 easy queries and 318 difficult queries, while evaluator 2 labels 408 easy queries and 267 difficult queries. They agree on 81.5% (550) of the

Table 1. Experimental Results

Classifier	Class	Precision			Recall		
		All	Concrete	C&F	All	Concrete	C&F
J48	0	0.717	0.706	0.52	0.739	0.761	0.278
	1	0.808	0.818	0.611	0.791	0.772	0.816
NBTree	0	0.664	0.655	0.503	0.8	0.809	0.37
	1	0.832	0.835	0.619	0.709	0.694	0.738
Bagging	0	0.679	0.684	0.519	0.743	0.791	0.413
	1	0.802	0.831	0.632	0.747	0.738	0.725
Average	0	0.687	0.682	0.514	0.761	0.787	0.354
	1	0.814	0.828	0.621	0.749	0.735	0.76

queries, among which 230 are difficult queries, such as *absence, benefit* and *chatter*, and 320 are easy queries, such as *destruction, earth* and *face*. The Cohen's kappa coefficient between the evaluators is 0.6229. We use the mutually agreed 550 queries as the ground truth in our experiments. We first represent each query as a vector of 4 features described before and then compare three machine learning algorithms on the task of predicting difficulty level for these queries, which are C4.5 decision tree (J48), naive Bayes tree (NBTree) and bagging predictors (Bagging). 10-fold cross-validation is used in our experiments, and precision and recall are used to evaluate the performance of these algorithms.

As shown in Table 1, the prediction performance doesn't vary much using different machine learning algorithms. We also compare the performance with various feature combinations: using all proposed features (All), using concreteness (Concrete), and using commonness and figurativeness (C&F). The result is also shown in Table 1. It is expected that concreteness is the most important feature to predict query difficulty, however it is unexpected that the other features are almost useless given concreteness. Further investigation shows that ambiguity is irrelevant and the other two features are useful only if concreteness is absent. The irrelevance of the ambiguity may be caused by several reasons. First, our objective is to predict the possibility to represent the queries as images, while the assumption that ambiguity is related to this possibility is wrong. Second, the measure of ambiguity may be poor, since a word with many senses or dictionary entries may not be ambiguous given the context. Commonness and figurativeness are inherently inferior to concreteness, because contextual information is not captured by these two features. Although common words may be easier to be illustrated with images, they are also likely to have more senses. Without word sense disambiguation, features derived from the whole web, as commonness and figurativeness, are different from commonness and figurativeness of the word in the context.

4 Discussions

Predicting query difficulty is an important problem for a contextual image retrieval system, since the queries are not explicitly provided by users, and thus automatically suggesting queries and returning poor retrieved images bother the users a lot. This paper is a first step towards solving this problem and the results are promising using the proposed features and machine learning algorithms. We find that features/classifiers based on the context of the query are better than those without contextual information. Although we study the query difficulty prediction for contextual image search, we expect the proposed context based features arc useful for query difficulty prediction of other contextual search engines.

In our preliminary research, all the features are based on the text. How to incorporate visual features to supplement existing textual features still needs to be investigated, especially when certain queries only have one or two qualified images among all the candidate images. An assumption is that the image results of easy queries may be more homogeneous. Similarly, the context information may be critical when this assumption is used. For example, the query *bearing* means *the manner of Don Quixote* instead of *a machine part* in our experimental data, thus it is labeled as a difficult query in the context. However, all the images retrieved for *bearing* are images of *a machine part*, thus the image features developed based on this assumption may not be useful under this circumstance. Another direction for improvement may exploit the query relationships. For example, some abstract words related to *human beings*, such as *admire*, *worship* and *courage*, can be explained by a picture of body language, facial expression or human behavior, while other abstract words related to *measurement*, such as *weight*, *mile* and *hour*, are still difficulty queries. This may indicate that abstract queries, when closely related to some concrete entities, may be able to use some related pictures of these entities to represent.

References

1. Brants, T., Franz, A.: Web 1t 5-gram version 1 (2006), http://www.ldc.upenn.edu/
2. Carney, R.N., Levin, J.R.: Pictorial illustrations still improve students' learning from text. Educational Psychology Review 14(1), 5–26 (2002)
3. Le, Z.: Maximum entropy modeling toolkit for c++, http://homepages.inf.ed.ac.uk/s0450736/maxent_toolkit.html
4. Lee, Y.K., Ng, H.T.: An empirical evaluation of knowledge sources and learning algorithms for word sense disambiguation. In: EMNLP 2002, NJ, USA (2002)
5. Miller, G.A.: Wordnet: a lexical database for english. Commun. ACM 38(11), 39–41 (1995)
6. Miller, G.A., Chodorow, M., Landes, S., Leacock, C., Thomas, R.G.: Using a semantic concordance for sense identification. In: Proc. of Workshop on Human Language Technology (1994)

Estimating Translation Probabilities from the Web for Structured Queries on CLIR

Xabier Saralegi and Maddalen Lopez de Lacalle

Elhuyar Foundation, R & D,
20170 Usurbil, Spain
{xabiers,maddalen}@elhuyar.com

Abstract. We present two methods for estimating replacement probabilities without using parallel corpora. The first method proposed exploits the possible translation probabilities latent in Machine Readable Dictionaries (MRD). The second method is more robust, and exploits context similarity-based techniques in order to estimate word translation probabilities using the Internet as a bilingual comparable corpus. The experiments show a statistically significant improvement over non weighted structured queries in terms of MAP by using the replacement probabilities obtained with the proposed methods. The context similarity-based method is the one that yields the most significant improvement.

Keywords: Cross-lingual Information Retrieval, Structured Query Translation, Web as Corpus.

1 Introduction

Several techniques have been proposed for dealing with translation ambiguity for the query translation task on CLIR, such as structured query-based translation (also known as Pirkola's method) [1], word co-occurrence statistics [2] and statistical translation models [3]. Structured queries are adequate for less resourced languages, rare pairs of languages or certain domains where parallel corpora are scarce or even non-existent. The idea behind this method is to treat all the translation candidates of a source word as a single word (*syn* operator) when calculating TF and DF statistics. This produces an implicit translation selection during retrieval time. There are many works dealing with structured queries, and some variants are proposed. [4] for example, proposes that weights or replacement probabilities be included in the translation candidates (*wsyn* operator). One drawback with this approach is that it needs parallel corpora in order to estimate the replacement probabilities.

Following this line of work, we propose a simple method based on the implicit translation probabilities of a dictionary, and also a more robust one which uses translation knowledge mined from the web. We have analyzed different ways of accessing web data: **Web As Corpus** tools, **News** search engines, and **Blog** search engines. Our aim is to examine how the characteristics of each access strategy influence the representation of the constructed contexts, and also, how far these strategies are adequate

C. Gurrin et al. (Eds.): ECIR 2010, LNCS 5993, pp. 586–589, 2010.
© Springer-Verlag Berlin Heidelberg 2010

for estimating translation probabilities by means of the cross-lingual context similarity paradigm. All experiments have been carried out taking Spanish as source language and English as target.

2   Obtaining Translation Probabilities from a Dictionary

The first method proposed for estimating translation probabilities relies on the hypothesis that, in a bilingual MRD (D), the position (pos) of the translation (w) among all the corresponding translation candidates f for a source word (v) is inversely proportional to its translation probability ($p(w|v)$). If we assume that it is an exponential decay relation, we can model the translation probability through this formula:

$$p(w|v) = 1/ \sum_{(v,f) \in D} \left(\frac{1}{pos(D,v,f)} \right) \cdot pos(D,v,w)$$

(1)

The principal problems of these assumptions are, firstly, that translations are not ordered in all MRD (partially or at all) by frequency of use, and secondly, that the proposed relation above does not fit all translation equivalents. So, we propose a method that is useful for ordering the translations of an MRD as well as for estimating more accurate translation probabilities, as presented in the following section.

3   Translation Probabilities by Context Similarity

The idea is to obtain translation probabilities by using the web as a bilingual comparable corpus. This strategy is based on estimating the translation probability of the translation candidates taken from the MRD in accordance with the context similarity of the translation pairs [5]. The hypothesis is that the more similar the contexts are, the more probable the translation will be. The computation of the context similarity requires a large amount of data (contexts of words), which has to be representative and from comparable sources. The Internet is a good source of large amounts of texts, and that is why, we propose that different search-engines be analyzed to obtain these contexts. These search engines have different features, such as domain, coverage and ranking, which affect both the degree of comparability and the representativeness of the contexts, as follows:

WebCorp: This Web Concordancer is based on main search APIs. Therefore, navigational queries and popular ones are promoted. These criteria can reduce the representativeness of the contexts retrieved. Since we take a maximum number of snippets for each query, the selected contexts depend on the ranking algorithm. It guarantees good recall, but perhaps poor precision. Thus, the comparability degree between contexts in different languages can be affected negatively.

Google News Archive: The content is only journalistic. It seems appropriate if we want to deal with journalism documents but not with other registers or more specialized domains. In short, it offers good precision, enough recall and a good degree of comparability.

Google Blog search: The language used is more popular, and although the register is similar to that of journalism, the domain is more extensive. This could offer good recall but not very comparable contexts.

The method to estimate the translation probabilities between a source word (v) and its translations $f\left(\left(v,f\right)\in D\right)$ starts by downloading, separately, the snippets of both words as returned by the search engines mentioned above. Then, we set up context vectors for the source $\vec{v}$ and the translation word $\vec{w}$ by taking keywordness (using log-likelihood scores) of the content words (nouns, adjectives, verbs and adverbs selected by using *Treetagger*) belonging to all their snippets. The next step is to translate the Spanish context vector $\vec{v}$ into English $tr\left(v\right)$. This is done by taking the first translation from a Spanish-English MRD (D) (34,167 entries). Cross-lingual context similarity is calculated according to cosine measure which is transformed into translations probabilities:

$$p\left(w|v\right)=\frac{\cos\left(\overrightarrow{tr\left(v\right)},\vec{w}\right)}{\displaystyle\sum_{\left(v,f\right)\in D}\left(\cos\left(\overrightarrow{tr\left(v\right)},\vec{f}\right)\right)} \tag{2}$$

We analyze the differences between the translation rankings obtained with the different search engines and those in the original dictionary. We computed Pearson's correlation for the translation rankings obtained for the polysemous content words in all 300-350 Spanish CLEF topics. The correlation scores (cf. Table 1) show that the different characteristics of each search engine produce translation rankings which are quite different from those in the dictionary (**Dic.**) and also from each other.

Table 1. Mean of Pearson's correlation coefficients for translation rankings compared to each other

	WebCorp	News	Blog
Dic.	0.42	0.31	0.40
WebCorp		0.44	0.54
News			0.49

4   Evaluation and Conclusions

We evaluated 50 queries (title+description) taken from 300-350 CLEF topics against collections from CLEF 2001 composed by LA Times 94 and Glasgow Herald 95 news. Previously, nouns, adjectives, verbs and adverbs were selected manually both in Spanish and English topics. *Indri* was used as the retrieval model and the queries were translated using several methods: taking the first translation of the MRD (**First**); taking all the translations and grouping them by the *syn* operator (**All or Pirkola**); and weighting the translations by using the *wsyn* operator and the methods described in sections 2 (**Dic.**) and 3 (**Webcorp**, **News** and **Blog**). The results are shown in Table 2.

In the first column we show the MAP results obtained with each method, with the English monolingual results first. In the second column we show the percentage of the

Table 2. MAP for 300-350 topics

Method	MAP	% Monolingual	% Improv. over All
Monolingual (en)	0.3651		
First	0.2462	67.43	
All	0.2892	79.21	
Dic.	0.2951	80.83	2.04
WebCorp	0.2943	80.55	1.76
News	0.2993	82.63	3.49
Blog	0.2960	81.07	2.35

cross lingual MAP with respect to the monolingual result. We can see that using all translations with their replacement probability estimated according to the dictionary order produces better results than using only the first translation or using all translations, with a significant improvement (according to the Paired Randomization Test with α=0.05) over the **All** method. So, exploiting the translation knowledge latent in the position of the translations improves the MAP when provided by the dictionary. Otherwise, the web-based estimation techniques also improve significantly over the **First** and **All** strategies (α=0.05). However, there is no significant improvement over the **Dic.** method. It seems that context similarity calculated from **Blog** or **News** sources is more suited to estimating translation probabilities since they significantly outperform **WebCorp** in terms of MAP. Therefore, comparability between sources of both languages, domain precision and informational snippets seem to be important factors in order to obtain useful context for context-similarity, although deeper analyses must be carried out to determine the importance of each more precisely. Finally, we conclude that translation knowledge obtained from the Internet, offers an adequate means, and by means of cross-lingual context similarity, it is useful for estimating replacement probabilities. Moreover, it could be an alternative when parallel corpora or MRDs with translations sorted according frequency of use are not available.

References

1. Pirkola, A.: The Effects of Query Structure and Dictionary Setups in Dictionary-Based Cross-Language Information Retrieval. In: SIGIR 1998, pp. 55–63 (1998)
2. Ballesteros, L., Croft, W.B.: Resolving Ambiguity for Cross-Language Retrieval. In: SIGIR 1998, pp. 64–71 (1998)
3. Hiemstra, D., De Jong, F.: Statistical Language Models and Information Retrieval: natural language processing really meets retrieval. University of Twente (2001)
4. Darwish, K., Oard, D.W.: Probabilistic structured Query Methods. In: SIGIR 2003 (2003)
5. Fung, P., Yuen Yee, L.: An IR Approach for Translating New Words from Nonparallel, Comparable Texts. In: COLING-ACL (1998)

Using Weighted Tagging to Facilitate Enterprise Search

Shengwen Yang, Jianming Jin, and Yuhong Xiong

Hewlett-Packard Labs China
{shengwen.yang,jian-ming.jin,yuhong.xiong}@hp.com

Abstract. Motivated by the success of social tagging in web communities, this paper proposes a novel document tagging method more suitable for the enterprise environment, named weighted tagging. The method allows users to tag a document with weighted tags which are then used as an additional source for the query matching and relevance scoring to improve the search results. The method enables a user-driven search result ranking by adapting the relevance score of a search result through weighted tags based on user feedbacks. A prototype intranet search system has been built to demonstrate the viability of the method.

Keywords: Weighted Tagging, Enterprise Search, User-Driven Ranking.

1 Introduction

Social tagging is a very popular feature of web 2.0 which has been widely used in web communities in recent years. Among of them, social bookmarking services such as Delicious [1] and Diigo [2] have attracted attention of millions of Internet users. Tags accumulated through these social bookmarking services can be used to guide the web navigation and improve the web search. Many research works (e.g. [3][4][5]) have been reported in this area.

Motivated by the success of social tagging in the Internet, the similar idea has also been applied to the intranets of enterprises and some application systems have been reported in literatures (e.g. [6][7]). However, the existing enterprise tagging systems usually adopt the tagging method used by the web communities and they do not consider the intrinsic differences between the Internet environment and the enterprise environment.

First, the Internet is an open environment with a very large user base, while the enterprise web is a close environment with a far smaller user base. This means that a web page in the Internet might be tagged by a very large number of Internet users and a tag may be used by hundreds or thousands of users to tag a web page, but a document within an enterprise might be tagged only by a few users. Thus there is very abundant tag count information in the Internet environment which can be seen as an indicator of importance of a tag to a web page. However this information is missing in the enterprise environment and we need a way to obtain the importance of a tag. Second, the sources of Internet users are very diverse, while the enterprise users are always limited to employees. This means that the tagging behavior of Internet users is uncontrollable and there is risk of malicious tagging, but the tagging behavior of

C. Gurrin et al. (Eds.): ECIR 2010, LNCS 5993, pp. 590–593, 2010.

enterprise users is more manageable. Thus we can allow enterprise users to provide additional information on the importance of tags.

Based on the above analysis, this paper proposes a novel document tagging method more suitable for the enterprise environment, named *weighted tagging*, and aims to facilitate the enterprise search by using weighted tags collected through the method as an additional source for query matching and relevance scoring.

The rest of this paper is organized as follows. We describe the weighted tagging in Section 2, present the user-driven ranking and weight calculation in Section 3, and conclude this paper in Section 4.

2   Weighted Tagging for Enterprise Environment

Social tagging is a popular Web 2.0 technique which allows the Internet users to store, organize, search, and manage bookmarks of web pages with the help of tags. A tag is a word or phrase used by a user to describe a web page. By introducing a weight attribute for a tag, we extend the definition of tags and propose the concept of *weighted tags*. A weighted tag is a pair of tag and weight, denoted as $<t, w>$, where t is the tag text, a word or phrase describing a document, and w is the tag weight, a real value indicating the degree of relevance between the tag text and the target document. To distinguish with the general tagging method, the tagging method using weighted tags is called *weighted tagging* in this paper.

Weighted tagging brings new opportunities for facilitating the enterprise search because weighted tags associated with documents provide an additional information source for query matching and relevance scoring. On the one hand, the tag text can be seen as a brief description of the document. As users may choose any word or phrase to tag a document, the tag text may be comprised of words or phrases that don't exist in the document content. Thus, it increases the opportunity for a document to match with queries which are semantically relevant to but non-overlapping literally with the document. On the other hand, as the tag weight indicates the degree of semantic relevance between the tag text and the target document, it can be used as a factor to score the relevance between a query and a document.

There are two means by which users can tag a document. The first is a browser plug-ins, which is commonly used by most of social bookmarking services. This case occurs when a user views a web page. The user can do online tagging very conveniently without leaving the current page. The second is a web widget, which is embedded in a search result page. This case occurs when a user submits a query and browses the search results. With the weighted tagging, a search result page will not only show the title, snippet, and URL of a hit page, it will also show the tags associated with the page in the form of a tag cloud [8]. Figure 1 shows an example search result.

How to assign a proper weight for a tag is a key issue in the weighted tagging. In this paper, we weight a tag according to its purpose. If the tag is informational, i.e. providing keyword or category information, it will be assigned a default weight value of 1. These tags can be used to navigate the documents or filter the search results. If the tag is promotional, i.e. providing additional relevance scoring information for promoting or demoting a document in search results, it will be assigned a weight value calculated by considering the user feedback on the desired rank of a document under a query. Actually this leads to a way of user-driven ranking, as described next.

Fig. 1. An example search result with the tag information. The underlined tags are informational (categorical tags) and the following two blue tags are promotional. The rotation button nearby a promotional tag allows users to provide feedbacks on the rank of the page (e.g. top 5 or top 10). Users can also promote a page with more than one tag.

3   User-Driven Ranking and Weight Calculation

How to utilize the power of user participation to improve the search is an interesting problem. Google SearchWiki[9] allows a user to customize search by re-ranking, deleting, adding, and commenting on search results. We can also achieve a kind of user-driven ranking with weighted tagging. If a user submits a query and finds that a relevant page is ranked lower or an irrelevant document is ranked higher in the search results, he or she can tag the document with one or more query words and then promote or demote the document by providing a feedback on the desired rank position. The system will calculate a proper weight for each tag based on the feedback. These weighted tags will affect the ranking of coming searches.

The weight calculation based on feedbacks can be derived from the formula scoring the relevance between a document and a query. With the weighted tagging, a document can be seen as comprising of at least two fields, document content and weighted tags. Thus we can compute a relevance score between a document d and a query q based on each filed and then combine them to get a final score as follows (where $0 \leq \alpha \leq 1$ is a mixture factor):

$$score_{final}(q,d) = \alpha \cdot score_{content}(q,d) + (1-\alpha) \cdot score_{tag}(q,d) \qquad (1)$$

The $score_{content}$ and $score_{tag}$ in the above formula can be computed with any text similarity measures, e.g. cosine similarity [10]. The trick here is to treat the tag weight as the term frequency when computing the similarity based on weighted tags.

According to the formula (1), the weight of a tag can be calculated based on the user feedback by solving the following inequation

$$score_{content}(q,d) + score_{tag}(q,d) \geq score_{content}(q,d_k) + score_{tag}(q,d_k) \qquad (2)$$

Where k is the user feedback on the desired rank of d, and d_k is the document currently positioned at k given the query q. For a one-word query, the formula (2) can give a unique solution to the weight. If a multi-word query, it can also give a unique solution by assuming that these multiple tags are equally weighted, or we can allow users to provide further information on the relative importance among these tags.

4   Conclusions and Future Work

This paper proposes a novel document tagging method, named weighted tagging, which is more suitable for use in the enterprise environment. The method not only allows users to tag a document with informational words or phrases (e.g. keywords or categories), which can be used for document navigation or search result filtering, but also allows users to re-rank search results by providing feedbacks on ranks of documents in search results, which enables a user-driven search result ranking.

A prototype intranet search system has been built to demonstrate the viability of the method. The prototype system is built on Apache Lucene [11] and indexes more than 3 millions web pages crawled from a corporate intranet. For each web page, a few informational tags are extracted with some automatic keyword extraction methods. The system allows a user to remove existing tags, add new tags, and promote or demote the rank of a search result.

Currently the system is for experimental purpose. There are several problems to be solved to make it fully functional, such as the real-time update of index supporting concurrent operations, conflict resolution of user intentions, prevention of possible malicious operations, etc. Besides, we plan to systematically evaluate the performance of the method in future as more data are collected through the trial use.

References

[1] Delicious, http://www.delicious.com
[2] Diigo, http://www.diigo.com
[3] Xu, S.L., Bao, S.H., Cao, Y.B., Yu, Y.: Using social annotations to improve language model for information retrieval. In: CIKM 2007 (2007)
[4] Bao, S.H., Xue, G.R., Wu, X.Y., Yu, Y., Fei, B., Su, Z.: Optimizing web search using social annotations. In: WWW 2007 (2007)
[5] Heymann, P., Koutrika, G., Garcia-Molina, H.: Can Social Bookmarking Improve Web Search? In: WSDM 2008 (2008)
[6] Millen, D.R., Feinberg, J., Kerr, B.: Dogear: Social Bookmarking in the Enterprise. In: CHI 2006 (2006)
[7] Dmitriev, P.A., Eiron, N., Fontoura, M., Shekita, E.: Using annotations in enterprise search. In: WWW 2006 (2006)
[8] Tag Cloud, http://en.wikipedia.org/wiki/Tag_cloud
[9] Google SearchWiki, http://googleblog.blogspot.com/2008/11/searchwiki-make-search-your-own.html
[10] Cosine Similarity, http://en.wikipedia.org/wiki/Cosine_similarity
[11] Apache Lucene, http://lucene.apache.org/

An Empirical Study of Query Specificity

Avi Arampatzis[1] and Jaap Kamps[2]

[1] Electrical and Computer Engineering, Democritus University of Thrace, Greece
[2] Media Studies, University of Amsterdam, The Netherlands

Abstract. We analyse the statistical behavior of query-associated quantities in query-logs, namely, the sum and mean of IDF of query terms, otherwise known as *query specificity* and *query mean specificity*. We narrow down the possibilities for modeling their distributions to gamma, log-normal, or log-logistic, depending on query length and on whether the sum or the mean is considered. The results have applications in query performance prediction and artificial query generation.

1 Introduction and Definitions

Inverse document frequency (IDF) is a widely used and robust term weighting function capturing *term specificity* [1]. Analogously, *query specificity* (QS) or query IDF can be seen as a measure of the discriminative power of a query over a collection of documents. A query's IDF is a log estimate of the inverse probability that a random document from a collection of N documents would contain all query terms, assuming that terms occur independently. The mean IDF of query terms, which we call *query mean specificity* (QMS), is a good pre-retrieval predictor for query performance, better than QS [2]. For a query with k terms $1, \ldots k$, QS and QMS are defined as

$$\mathrm{QS}_k = \log \left(\prod_{i=1}^{k} \frac{N}{\mathrm{df}_i} \right) = \sum_{i=1}^{k} \log \frac{N}{\mathrm{df}_i} , \qquad \mathrm{QMS}_k = \mathrm{QS}_k / k ,$$

where df_i is the document frequency (DF), i.e. the number of collection documents in which the term i occurs.

We analyse statistical properties of QS and QMS, for all queries in a search engine's query-log and per query length, with an empirical brute-force approach. The proposed models provide insight on engine performance for given query sets. The models can also be combined with query-length models, e.g. [3], for generating artificial queries. Artificial queries have applications in areas such as score normalization for distributed retrieval or fusion [4], pseudo test collection construction [5], and efficiency testing.

2 Distributions of QS and QMS

The distribution of any of QS, QS_k, QMS, QMS_k, is a combined result of a query set and a document collection, i.e. the source of DFs, the query set is submitted to. We use two query sets: the AOL log consisting of 21M queries from AOL search (March–May 2006), and the MSN log consisting 15M queries from the MSN search engine

C. Gurrin et al. (Eds.): ECIR 2010, LNCS 5993, pp. 594–597, 2010.
© Springer-Verlag Berlin Heidelberg 2010

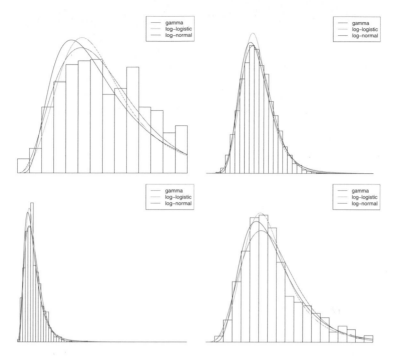

Fig. 1. Empirical distributions of QS and QMS for AOL queries using DFs from NYT, and the fitted gamma, log-logistic, and log-normal for query length $k = 1$ (top left), $k = 5$ (top right), all-lengths QS (bottom left), and all-lengths QMS (bottom right)

(May 2006). Since we have no actual term statistics, we rely on two other sources of DF: Web term document frequencies from the Berkeley Digital Library Project (32M terms from 50M Web pages); and New York Times document frequencies based on the 1998–2000 NY Times articles in AQUAINT (506,433 terms from 314,452 articles).

Per query length k, QS_k is a linear transformation of QMS_k, thus we only need to analyse one of the two and we opt for the latter. We analyse the distributions of QS and QMS irrespective of query length, separately. The quantity QMS_k, as defined in the previous section, has a discrete distribution with a support over $\binom{N+k-1}{k}$ real numbers in $[0, \log N]$. For large N, however, it can be approximated by a continuous distribution with support in $[0, \log N]$, especially for large k where the cardinality of the support set is higher. Thus, as $N \to +\infty$, we are looking for a suitable known continuous distribution supported on the semi-infinite interval $[0, +\infty)$.

By examining histograms of empirical data, we see that the distributions are unimodal with positive skew. Three distributions which seem capable of matching well the shape of the empirical data are: gamma, log-logistic, and log-normal. They all have support in $[0, +\infty)$. Figure 1 shows the fits for $k = 1$, $k = 5$, and QS and QMS over all lengths. We eliminated the following possibilities which gave consistently worse fits: Weibull, inverse gamma, chi-square, and inverse chi-square. We also tried the beta distribution with a bounded support in $[0, 1]$ for normalized QMS and QMS_k, but it

Table 1. χ^2 goodness-of-fit test (upper one-sided at .05 significance) for observed QS and QMS against 3 fitted theoretical distributions: gamma (G), log-logistic (LL), log-normal (LN). The results are presented across all combinations of query set, collection, and query length k. First, large sets of observed data, per length or for all lengths, are uniformly down-sampled to 1,000 points. Then, each set is binned into bins of $0.3\,\sigma$ width, where σ is the standard deviation of the observed data. Bins with expected frequencies < 5 are combined; this may result to slightly different number of bins for the same dataset across candidate distributions. A plus in a cell means that the null hypothesis that the data follow the candidate cannot be rejected, while for a minus it is rejected. The leading numbers are ranks of the quality of fits according to the comparison of their χ^2 with the observed data. This is a loose (although indicative) comparison due to the possibly slightly different degrees of freedom of their χ^2 distributions, a result of the bin-combining.

		AOL/Web			AOL/NYT			MSN/Web			MSN/NYT		
	k	G	LL	LN	G	LL	LN	G	LL	LN	G	LL	LN
	1	1-	3-	2-	1-	2-	3-	1-	2-	3-	1-	2-	3-
	2	2+	3-	1+	1+	2-	3-	3-	2-	1+	1+	2-	3-
	3	2+	3+	1+	1+	3-	2-	1+	2+	3-	1+	2-	3-
QS_k or QMS_k	5	3-	2+	1+	1+	3-	2+	2+	1+	3+	1+	3+	2+
	7	3-	2-	1+	1+	3-	2+	2+	3+	1+	1+	3-	2+
	10	3-	1-	2-	3-	2+	1+	3-	2-	1-	3-	2+	1+
	15	3-	1-	2-	3-	1-	2-	3-	1-	2-	3-	1+	2-
	QS	1-	2-	3-	1-	3-	2-	1-	2-	3-	1-	2-	3-
	QMS	3-	2-	1-	3-	2-	1-	3-	2-	1-	2-	1-	3-

was consistently worse as well. The inverse Gaussian gave very similar shapes to the log-normal, but we eliminated it due to consistently better fits of the latter.

The goodness-of-fit results are summarized in Table 1. For $k = 1$, the data are messy and difficult to model. This may be due to their discrete nature that comes more into effect for small k, or due to unusual terms like full URLs. However, the gamma seems more flexible than the alternatives. The good fits come at lengths 2 to 7, where the gamma and the log-normal provide better approximations than the log-logistic. At larger k, the log-normal and log-logistic provide better fits than the gamma. Since short queries are more frequent, we are inclined to suggest modeling QS_k and QMS_k with a gamma. The gamma shape of the short lengths and the fact that short queries dominate the aggregate, influence strongly QS, where the gamma is the best fit throughout, but not QMS, where the log-normal fits best and the log-logistic is not bad either.

Since we have not arrived to a single model distribution, we analyse statistics of the datasets, shown in Table 2, rather than a specific distribution's parameters. Using the median as central tendency is more suitable than the mean, since the data are skewed. Given that QMS is correlated with query performance, the fact that the median and standard deviation are declining with increasing k suggests that performance may be declining with query length. But this may not be the case, since past research has found that such correlations may be weakening with increasing k [6]. According to the median QMS of the aggregates, AOL and MSN queries would perform better on Web than on NYT. This a multiplicative result of having larger normalized QMS on the Web than on the NYT (as expected for Web query sets), and N being larger for Web than for NYT. The expected result that larger collections improve performance is apparent. Comparing

Table 2. Median and standard deviation of observed *normalized* QS and QMS, across all combinations of query set, collection, and query length k. The median and std. dev. of QS_k, which are not shown, are k times those of QMS_k. In order to enable comparisons across collections of different size, we scale the data by dividing them by $\log N$ per collection. N equals 49,602,191 for Web and 314,452 for NYT. This procedure normalizes QMS, QMS_k in $[0, 1]$, and QS, QS_k in $[0, k_{max}]$, where k_{max} is the maximum observed query length.

	k	AOL/Web median	AOL/Web std.dev.	AOL/NYT median	AOL/NYT std.dev.	MSN/Web median	MSN/Web std.dev.	MSN/NYT median	MSN/NYT std.dev.
	1	0.557	0.234	0.476	0.235	0.501	0.241	0.448	0.241
	2	0.402	0.129	0.394	0.151	0.376	0.122	0.375	0.155
QMS_k	3	0.360	0.102	0.346	0.116	0.343	0.099	0.344	0.121
	5	0.316	0.082	0.291	0.096	0.300	0.067	0.279	0.092
	7	0.283	0.065	0.258	0.086	0.269	0.058	0.244	0.083
	10	0.253	0.068	0.215	0.072	0.244	0.055	0.205	0.074
	15	0.237	0.060	0.189	0.070	0.224	0.047	0.177	0.061
QS		0.901	0.506	0.837	0.511	0.788	0.464	0.807	0.452
QMS		0.395	0.184	0.365	0.180	0.384	0.186	0.362	0.186

the two sets of queries with each other, the QMS indicates a similar performance. This is also expected; we do not see why one query-set would be better than the other.

3   Conclusions

We empirically investigated the distributions of query specificity and mean specificity for query-logs. We have not arrived to a single model, but narrowed down the possibilities considerably. Per query length, both specificity and mean specificity are well approximated with a gamma distribution for short to medium queries, and with a log-normal or log-logistic distribution for long queries. Irrespective of query length, specificity can be approximated with a gamma, and mean specificity by either a log-normal or log-logistic. For all practical purposes, these distributions provide good approximations of all queries in a query-log or per length. We have interpreted the results from a query performance perspective, which may suggest ways to improve performance by a directed expansion of collection coverage or support in query formulation. Further, the proposed models can be applied for artificial query generation.

References

1. Spärck Jones, K.: A statistical interpretation of term specificity and its application in retrieval. Journal of Documentation 28, 11–21 (1972)
2. Cronen-Townsend, S., Zhou, Y., Croft, W.B.: Predicting query performance. In: SIGIR, pp. 299–306. ACM, New York (2002)
3. Arampatzis, A., Kamps, J.: A study of query length. In: SIGIR, pp. 811–812. ACM, New York (2008)
4. Arampatzis, A., Kamps, J.: A signal-to-noise approach to score normalization. In: CIKM, pp. 797–806. ACM, New York (2009)
5. Tague, J., Nelson, M., Wu, H.: Problems in the simulation of bibliographic retrieval systems. In: SIGIR, pp. 236–255 (1980)
6. He, B., Ounis, I.: Query performance prediction. Inf. Syst. 31(7), 585–594 (2006)

Semantically Enhanced Term Frequency

Christof Müller and Iryna Gurevych

Ubiquitous Knowledge Processing Lab, Computer Science Department,
Technische Universität Darmstadt, Germany
http://www.ukp.tu-darmstadt.de

Abstract. In this paper, we complement the term frequency, which is used in many bag-of-words based information retrieval models, with information about the semantic relatedness of query and document terms. Our experiments show that when employed in the standard probabilistic retrieval model BM25, the additional semantic information significantly outperforms the standard term frequency, and also improves the effectiveness when additional query expansion is applied. We further analyze the impact of different lexical semantic resources on the IR effectiveness.

Keywords: Information Retrieval, Semantic Relatedness.

1   Introduction and Approach

The majority of information retrieval (IR) models is based on the bag-of-words paradigm. The performance of these models, however, is limited among other things due to the polysemy and synonymy of terms. The importance of the semantic relations or associations between terms has therefore long been recognized. Several approaches have been proposed to improve IR effectiveness employing methods like query expansion, document expansion [1], or topic models [2]. In this paper, we complement the term frequency (tf), which is widely used in IR models, with information about the semantic relatedness (SR) of query and document terms. Our hypothesis is that this additional knowledge will enable IR models to estimate the document relevance more accurately, as thereby also information about the meaning of non-query terms in the documents is taken into account. In our experiments, we evaluate this approach using a standard probabilistic model, i.e. BM25 [3]. For computing SR, we use *Explicit Semantic Analysis* (ESA), which was introduced by Gabrilovich and Markovitch [4].

The BM25 model estimates the relevance of a document d and a query q as

$$r(d,q) = \sum_{t_q \epsilon q} \frac{(k_1 + 1)\, tf_{t_q,d}}{k_1 \left((1 - b) + b \frac{l}{l_{avg}} \right) + tf_{t_q,d}} \cdot \frac{(k_3 + 1)\, tf_{t_q,q}}{k_3 + tf_{t_q,q}} \cdot \log \frac{(N - df_{t_q} + 0.5)}{df_{t_q} + 0.5}$$

where t_q is a term in q, $tf_{t_q,q}$ or $tf_{t_q,d}$ is its tf in q or d, l is the document length, l_{avg} is the average document length in the collection, N is the collection size, and df_{t_q} is the number of documents containing t_q. k_1, k_3, and b are parameters. In order to enhance tf, we substitute $tf_{t_q,d}$ in the above equation with $tf_{t_q,d} + s \cdot \sum_{t_d \epsilon d, t_d \neq t_q} tf_{t_d,d} \cdot sr\,(t_q,t_d)$ where $sr\,(t_q,t_d)$ is an estimation of the

C. Gurrin et al. (Eds.): ECIR 2010, LNCS 5993, pp. 598–601, 2010.

SR of query term t_q and document term t_d, and s is a parameter that controls the impact of SR on the final tf value. For $sr(t_q, t_d)$, we use the score computed by ESA after applying a predefined threshold to take into account only strong SR values. We additionally experiment with binary values, setting $sr(t_q, t_d)$ to either 0 or 1 depending on whether the ESA score is below or above the threshold.

ESA requires a lexical semantic resource (LSR) for which Gabrilovich and Markovitch originally employed Wikipedia (WP). Terms are represented as vectors of their tf.idf values in WP articles which are taken as textual representations of concepts. Texts are represented as the centroid vectors of the terms' concept vectors. The SR of a pair of terms or texts is then computed by using the cosine similarity measure. ESA has shown very good effectiveness for assessing the SR of terms and texts. However, when applied to IR, retrieval effectiveness was improved only when ESA scores were linearly combined with the relevance scores of bag-of-words based IR models [5,6]. Egozi et al. also found that additionally employed centroid vectors of small passages of the documents were crucial for the IR effectiveness. They argue that otherwise "ESA tries to "average" several topics and the result is less coherent". In our approach, the ESA scores are directly integrated into the bag-of-words based model and no passage-based index is built, as the ESA scores are computed for pairs of query and document terms.

2 Experiments and Discussion

Besides using the German WP as LSR, we follow Zesch et al. [7] and employ Wiktionary (WKT) and GermaNet[1] (GN), as they have shown good performance on estimating the SR of terms. We use the German GIRT corpus (consisting of titles, abstracts, and meta data of social science texts) and a collection of German newspaper articles (NEWS) as test collections and LSRs.[2] With the goal of reducing noise and computational costs, we set small values for concepts to zero if they are below an empirically set pruning threshold after the concept vector was normalized by its length. We tested several thresholds and found that 0.015 performs best for all LSRs. Table 1 shows the document collections used as LSR, the number of contained terms and documents (concepts), and the average number of concepts per term with non-zero values before and after pruning.

Besides standard preprocessing steps like tokenization and stop word removal, we perform stemming and compound splitting. We employ the IR framework Terrier[3] and select the parameters b and k_1 of BM25, as well as the parameters of SR by using simulated annealing on a training set.[4] We perform two sets of experiments where we enhance tf with SR (i) only for documents that contain at least one query term, and (ii) for all documents. For the optimized parameter configurations, we additionally apply query expansion (QE) with the

---

[1] http://www.sfs.uni-tuebingen.de/GermaNet
[2] Both collections were used at CLEF. See http://clef-campaign.org for details.
[3] http://ir.dcs.gla.ac.uk/terrier
[4] Training set: 75 topics of CLEF'03–05 for GIRT; 100 topics of CLEF'01–02 for NEWS test set: 75 topics of CLEF'06–08 for GIRT; 60 topics of CLEF'03 for NEWS.

Table 1. Document collections used as LSR for ESA

LSR	#terms	#concepts	avg. #concepts per term	
			unpruned	*pruned*
GIRT	348,308	151,319	34.03	19.32
NEWS	874,637	294,339	41.17	21.94
WP	4,185,730	530,886	31.38	12.63
WKT	195,705	113,341	6.11	6.05
GN	44,879	42,014	5.86	5.85

term weighting model Bo1 which uses the Bose-Einstein statistics and is one of the most effective weighting models based on the *Divergence From Randomness* framework [8]. The two parameters for QE, i.e. the number of terms to expand a query with and the number of top-ranked documents from which these terms are extracted, are also optimized using simulated annealing. We use SR however only for the initial step of retrieving documents from which to extract expansion terms and not for the final retrieval with the expanded query. Otherwise, computing SR for the large number of (possibly erroneous) expansion terms (up to 100) causes a strong topic drift and retrieval effectiveness decreases.

Table 2 shows the mean average precision (MAP) for each LSR on the two test collections with and without QE using the topics of the test set. Except for WKT on the GIRT collection, the enhancement of tf with SR increases MAP for all LSRs on both test collections. Especially for the NEWS test collection, the improvements are statistically significant. We found that using the original ESA score consistently performs better than substituting it with a binary value. Without employing QE, the test collection itself is the best performing LSR. SR shows similar improvements for BM25 as when employing QE, but only when the test collection is used as LSR. Other resources either have a lower coverage of query and document terms or contain too general term relations which can cause a topic drift for some of the queries. Anderka and Stein [9] found that ESA also performs well with other document collections than WP, which do not necessarily fulfill the requirement that each document describes exactly one concept as was originally the idea behind ESA. Our experimental results suggest a similar conclusion, as the NEWS and GIRT collections perform similar or even better than WP. We also linearly combined the relevance scores of the BM25 model and the original ESA model of Gabrilovich and Markovitch [4], but found that our approach consistently results in a higher MAP for all configurations.

Employing SR improves QE for all configurations. The quality of expansion terms is increased as SR ranks relevant documents higher during the initial retrieval step, and the top-ranked documents contain a larger number of terms that are strongly related to the query terms. Of all LSRs, WKT and GN contain the lowest number of terms and term relations, as they do not consist of long texts, but rather short lexicographic entries. In our experiments, they show the lowest MAP, except for GN on the GIRT collection, where QE is surprisingly improved by GN the most.

The consideration of documents that do not contain any of the query terms when computing SR, improves the retrieval effectiveness in most cases, although not dramatically. This comes, however, with an increase in computational costs. The efficiency of our approach can in general be increased by precomputing or caching the SR values.

Table 2. MAP and difference to BM25 without SR in percent. Statistically significant improvements (paired t-test, $\alpha = 0.05$) are marked with *. Highest MAP is in bold.

LSR	GIRT MAP	% diff	with QE MAP	% diff	NEWS MAP	% diff	with QE MAP	% diff
—	0.3609	—	0.4076	—	0.3487	—	0.4156	—
computing SR for documents that contain at least one query term								
GIRT	0.3986	+10.45*	0.4110	+0.83	0.3933	+12.79*	0.4421	+6.38*
NEWS	0.3693	+2.33	0.4128	+1.28	0.4116	+18.04*	0.4435	+6.71*
WP	0.3742	+3.69	0.4126	+1.23	0.3881	+11.30*	0.4458	+7.27
WKT	0.3575	−0.94	0.4076	0.00	0.3814	+9.38*	0.4308	+3.66
GN	0.3612	+0.08	0.4092	+0.39	0.3712	+6.45*	0.4326	+4.09
computing SR for all documents								
GIRT	**0.4148**	**+14.93***	0.4135	+1.45	0.3871	+11.01*	0.4210	+1.30
NEWS	0.3754	+4.02*	0.4145	+1.69	**0.4166**	**+19.47***	**0.4494**	**+8.13***
WP	0.3850	+6.68*	0.4118	+1.03	0.3901	+11.87*	0.4472	+7.60*
WKT	0.3548	−1.69	0.4073	−0.07	0.3817	+9.46*	0.4351	+4.69
GN	0.3621	+0.33	**0.4158**	**+2.01**	0.3726	+6.85*	0.4391	+5.65

The results of our experiments need to be further analyzed and substantiated on other test collections. However, they are very promising, and as tf is widely used, this approach allows a simple while effective integration of SR into existing IR models. Preliminary results on employing the enhanced tf in the PL2 model suggest similar performance improvements as were yielded in the BM25 model.

Acknowledgements. This work was supported by the Volkswagen Foundation (grant I/82806) and the German Research Foundation (grant GU 798/1-3). We thank Aljoscha Burchardt, György Szarvas, and the anonymous reviewers for their helpful comments.

References

1. Tao, T., Wang, X., Mei, Q., Zhai, C.: Language Model Information Retrieval with Document Expansion. In: Proc. of HLT-NAACL 2006 (2006)
2. Yi, X., Allan, J.: A Comparative Study of Utilizing Topic Models for Information Retrieval. In: Proc. of ECIR 2009 (2009)
3. Sparck Jones, K., Walker, S., Robertson, S.E.: A probabilistic model of information retrieval: development and comparative experiments. Information Processing and Management 36(6) (2000)
4. Gabrilovich, E., Markovitch, S.: Computing Semantic Relatedness using Wikipedia-based Explicit Semantic Analysis. In: Proc. of IJCAI 2007 (2007)
5. Egozi, O., Gabrilovich, E., Markovitch, S.: Concept-Based Feature Generation and Selection for Information Retrieval. In: Proc. of AAAI 2008 (2008)
6. Müller, C., Gurevych, I.: Using Wikipedia and Wiktionary in Domain-Specific Information Retrieval. In: Peters, C., Deselaers, T., Ferro, N., Gonzalo, J., Jones, G.J.F., Kurimo, M., Mandl, T., Peñas, A., Petras, V. (eds.) Evaluating Systems for Multilingual and Multimodal Information Access. LNCS, vol. 5706, pp. 219–226. Springer, Heidelberg (2009)
7. Zesch, T., Müller, C., Gurevych, I.: Using Wiktionary for Computing Semantic Relatedness. In: Proc. of AAAI 2008 (2008)
8. Amati, G.: Probability Models for Information Retrieval based on Divergence from Randomness. PhD thesis, Dept. of Computing Science, Univ. of Glasgow (2003)
9. Anderka, M., Stein, B.: The ESA Retrieval Model Revisited. In: Proc. of SIGIR 2009 (2009)

Crowdsourcing Assessments for XML Ranked Retrieval

Omar Alonso[1], Ralf Schenkel[1,2], and Martin Theobald[1]

[1] Max-Planck Institute für Informatik, Saarbrücken, Germany
[2] Saarland University, Saarbrücken, Germany

Abstract. Crowdsourcing has gained a lot of attention as a viable approach for conducting IR evaluations. This paper shows through a series of experiments on INEX data that crowdsourcing can be a good alternative for relevance assessment in the context of XML retrieval.

1   Introduction

IR benchmarks with large document collections, numerous topics and pools of manually assessed relevance judgments are an integral building block for qualitative comparisons of retrieval approaches, and different evaluation communities such as TREC, CLEF and INEX have emerged. While TREC relies on own retired employees from NIST for the assessment phase, more recent community-based efforts like CLEF and INEX [3] distribute the task of relevance assessments among their benchmark participants, i.e., actual scientists in areas like databases, information retrieval, and related fields. Due to the high load of assessment work per participant, with more than 100 new topics per track and year, each topic typically is assessed only by one or at most two different assessors in INEX. Moreover, assessing a topic is a fairly time-consuming and tiring process, and our own experience indicates that actual relevance judgments often degrade to a rather cursory level after several hours of work—even when done by the participants themselves. This situation calls for the use of crowdsourcing platforms, which become increasingly popular for a variety of human intelligence tasks. Crowdsourcing has been employed successfully for a wide range of scientific applications, including user studies and producing relevance assessments. Examples of such research are the studies on rater-expert comparison for TREC [1] and NLP tasks such as affect recognition and word sense disambiguation[2]. This paper shows that crowdsourcing can be a good alternative for relevance assessment also in the context of XML retrieval.

2   Experimental Design

We used the crowdsourcing platform Amazon Mechanical Turk (AMT, http://www.mturk.com/) for our evaluation study. AMT is a common marketplace for work that requires human intelligence with a huge community of workers. By

C. Gurrin et al. (Eds.): ECIR 2010, LNCS 5993, pp. 602–606, 2010.

creating a so-called HIT (Human Intelligence Task), one can upload a task to the website and state how much one will pay a worker to complete the work. Our research question can be framed as: *"Is it possible to perform relevance assessments for XML retrieval with AMT?"*. The key question is whether assessments by workers of diverse and unknown skill and cultural background are comparable to those performed by the INEX assessors—at least over the typical set of common-sense retrieval topics posed over the INEX Wikipedia collection [5].

The capabilities of the AMT dashboard are significantly limited compared to the assessment tools used by INEX since 2005. At INEX, a pool of candidate documents is formed for each topic, based on the results submitted by the participants. An assessor then looks at each document in a pool and highlights relevant passages. The relevance of an element is derived as the fraction of its content that has been highlighted. In contrast, the AMT interface offers only simple HTML forms, allowing for the use of check boxes and text fields. Instead of the superior passage-based assessment [3], we thus need to take one step back to element-based assessment, where elements of different granularities are shown as a whole and one at a time to the assessors. We selected ten topics from the 2007 Ad-hoc track [4], including "Beijing Olympics", "Hip Hop Beats", and "Salad Recipes", aiming for a good coverage of common-sense retrieval tasks.

Each turker (i.e., AMT user) was given some basic instructions, a description of the topic taken from the `description` field in the original INEX topic [4], and one result element to assess. They were then asked to provide a binary relevance judgment (relevant/not relevant) for this element with respect to the topic via a check box. Each topic/element pair was judged by 5 turkers, who had been initially required to take a qualification test as a mechanism to control the quality of their work. We also asked users to optionally justify their answers in an additional text field. We paid 1 (US) cent for such a task.

After some initial experiments for testing and calibration purposes (and to get turkers interested in our setting), we extracted, for each topic, 40 relevant element results of type `article`, `body`, `sec` (section) and `p` (paragraph) from the INEX qrels. We additionally injected, for each topic, 10 non-relevant elements drawn from the qrels of a randomly chosen other topic as noise. Elements were then presented in random order to the turkers, i.e., overlapping elements from the same document were intentionally not presented consecutively. Requiring five assessors per element and thus resulting in 2,500 single votes, this task took approximately one week on AMT.

3   Analysis and Results

3.1   Comparison against INEX Assessors

For the first set of experiments, we derived, for each result, an assessment from the turkers' answers. We defined a result to be relevant (R) if more turkers said "relevant" than "non-relevant", and non-relevant (NR) otherwise. We then measured the agreement of the INEX assessments and the derived assessments, separated for relevant and non-relevant elements (as the Jaccard coefficient of

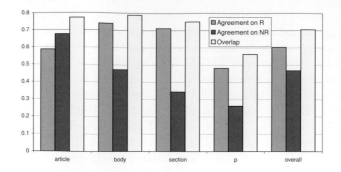

Fig. 1. Agreement of INEX and turker-based assessments, grouped by tag

the two sets, similarly to [3]), as well as the overlap (i.e., the fraction of elements where both INEX assessors and turkers gave the same relevance assessment).

Figure 1 shows the overall results of this step and, additionally, results grouped by tag. Interestingly, the agreement on relevant articles (0.58) is almost identical to that reported for inter-assessor agreement at INEX [3], which indicates that turkers' performance is comparable to that of an average INEX assessor. In addition, the general trend is that agreement and overlap decreases when elements get smaller (from articles over sections to paragraphs). We explain this by the fact that we don't show the paragraph in context, but isolated, so its relevance may not be evident to the assessor without the context.

3.2   Agreement by Relevance Grade

The second experiment tries to see in more detail how good turkers and INEX match, especially considering that INEX provides information about the fraction of an element's content that was considered relevant by the assessor ("rsize"). We grouped elements into four buckets, based on their rsize:

- *non-relevant* elements have $rsize = 0$
- *partly relevant* elements have $rsize \leq 10\%$
- *mostly relevant* elements have $10\% < rsize < 100\%$
- *fully relevant* elements are completely relevant

For each element, we computed the difference (delta) of "relevant" votes and "non-relevant" votes by turkers. Figure 2 shows how the delta values distributed over the four buckets of elements, where the line in each bar denotes the split of negative (where turkers mostly voted for non-relevance) and positive deltas.

We see that there is a wide agreement for partly and mostly relevant elements, but turkers frequently disagree from the INEX assessors for non-relevant and the fully relevant elements. A further analysis that additionally groups by tag (not shown here) shows that the turkers assign a wrong "R" mostly for articles and bodies, whereas they assign a wrong "NR" mostly for sections and paragraphs. While the explanation for latter is the same as above, the background of mis-assessed articles is different. We manually checked the content of these articles,

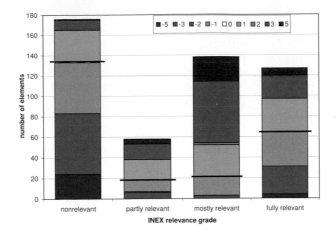

Fig. 2. Distribution of vote deltas for elements with different INEX relevance grade

and it turned out that most of them may be considered relevant if one considers simply the keywords of the query (but does not analyze the text very deeply). For example, the "hip hop beat" query cares only about explicit descriptions of beats, not about hip hop in general; this can easily be done wrongly (and has been done so several times).

3.3   Worker Feedback

As mentioned earlier, our experiments included a feedback input box where turkers can justify answers. This was particularly useful for detecting text that was not long enough to get assessed properly. Examples of such feedback were: "Short and irrelevant", "Not enough info given to assess", "Single sentence".

4   Conclusions and Future Work

Our findings are that a crowdsourcing-based evaluation approach for XML retrieval can produce relevance assessments which are comparable to those from INEX assessors, usually being specialists. Our future work will thus include building up a broader set of evaluated topics, which will then also allow for a more detailed evaluation of the correlation of *system ranks* based on the different assessments produced by INEX and AMT.

References

1. Alonso, O., Mizzaro, S.: Can we get rid of TREC assessors? Using Mechanical Turk for relevance assessment. In: SIGIR IR Evaluation Workshop (2009)
2. Snow, R., et al.: Cheap and fast–but is it good?: evaluating non-expert annotations for natural language tasks. In: EMNLP (2008)

3. Piwowarski, B., Trotman, A., Lalmas, M.: Sound and complete relevance assessment for XML retrieval. ACM Trans. Inf. Syst. 27(1), 1–37 (2008)
4. Fuhr, N., Kamps, J., Lalmas, M., Malik, S., Trotman, A.: Overview of the INEX 2007 Ad Hoc Track. In: Fuhr, N., Kamps, J., Lalmas, M., Trotman, A. (eds.) INEX 2007. LNCS, vol. 4862, pp. 1–23. Springer, Heidelberg (2008)
5. Denoyer, L., Gallinari, P.: The Wikipedia XML corpus. SIGIR Forum 40(1), 64–69 (2006)

Evaluating Server Selection for Federated Search

Paul Thomas[1] and Milad Shokouhi[2]

[1] CSIRO
paul.thomas@csiro.au
[2] Microsoft Research
milads@microsoft.com

Abstract. Previous evaluations of server selection methods for federated search have either used metrics which are unconnected with user
satisfaction, or have not been able to account for confounding factors
due to other search components.

We propose a new framework for evaluating federated search server
selection techniques. In our model, we isolate the effect of other confounding factors such as server summaries and result merging. Our results suggest that state-of-the-art server selection techniques are generally effective but result merging methods can be significantly improved.
Furthermore, we show that the performance differences among server
selection techniques can be obscured by ineffective merging.

1 Introduction

Federated search attempts to provide a single interface to several independent
search engines. A *broker* translates a user's query, forwards it to one or more
search engines, and presents any results in a single display [2].

Brokers commonly include a process of *server selection* to determine which
search engines should receive each query. By using the most appropriate set of
search engines, a broker can reduce direct costs, network traffic, and processing
requirements; there is also some evidence that it can increase effectiveness even
over that of a single, non-federated, system [1]. A large number of server selection
methods have been proposed in the literature, and with thorough evaluations
it should be possible to answer two questions: first, what methods should be
considered for a new broker? Second, is the selection method in an existing
broker performing well?

Selection evaluation unfortunately remains problematic. There are at least
three desiderata for evaluation: (1) the measure should somehow reflect user
experience; (2) it should be possible to identify how well server selection is doing,
without confounding factors of other broker tasks; and (3) it should be possible
to measure effectiveness across a range of external factors.

The first of these criteria has been satisfied by using a metric such as precision
at ten documents (P@10), measured on a broker's final list of results [e.g. 4].
Although this does reflect user-visible quality, it does not allow us to find out
how effective the broker's selection (or any other) component is—as well as
selection, P@10 results are influenced by query translation, merging results, and

C. Gurrin et al. (Eds.): ECIR 2010, LNCS 5993, pp. 607–610, 2010.

other steps. Since the effect of each component cannot be separated it is also impossible to understand which parts of a broker should receive attention.

Other evaluations have satisfied the second criterion by using a metric which considers only the effectiveness of selection—commonly $\mathcal{R}_k$, which is an analogue of recall measured just after selection but before any further processing by search engines or the broker [3]. However, there is no evidence to suggest that $\mathcal{R}_k$ correlates with any user-visible effectiveness and some authors have seen $\mathcal{R}_k$ decline as P@10 is stable [8]. Further, since $\mathcal{R}_k$ does not depend on the quality of retrieval at each search engine, it is not possible to understand how selection effectiveness relates to search engine effectiveness—violating our third criterion.

We propose a novel method which isolates the effects of broker and search engine components, and which satisfies the criteria above.

2   Evaluating Server Selection

To satisfy the first of our two desirable characteristics, we propose evaluating systems with respect to P@10.

To reconcile the remaining characteristics, we propose two steps. First, we attempt to remove the confounding effect of other broker tasks. This isolates any performance impact due to the selection process itself; with confounding effects removed, any change in performance is due to selection alone. Second, we consider a range of possible cases for those aspects outside the broker's control.

There are several subsystems in a broker: creating summaries of a server's holdings (probably by taking samples of documents); selection; query translation; parsing result sets; and merging results into a final list. Outside the broker's control, but having considerable impact, is the retrieval at each selected engine.

Removing the Effect of Server Summaries. Errors or inaccuracies in server summaries will inevitably impact server selection [7]. To remove the effect of server summaries, we assume that selection should only consider those servers where an accurate summary would match the query; for the methods in our experiments, this is equivalent to using full term statistics for the "summaries".

Removing the Effect of Query Translation and Result Set Parsing. We simply assume these steps are possible without any loss of fidelity. (This is a common assumption in the federated search literature, although not often acknowledged.)

Removing the Effect of Merging. After servers are selected, and have retrieved results, there is a further possible confounding effect from the process of merging the returned results. Most merging algorithms are "stable"—that is, if there is no overlap between returned result lists then they will respect each server's ordering. Since servers may be ineffective and return irrelevant documents before relevant ones, we control for the confounding effect by assuming "perfect" merging, which instead puts all relevant documents at the head of the merged list.

Best Possible Performance. We evaluate selection using each method's *best possible performance* (BPP). BPP at a given point in a broker's working is the performance measure it could achieve if every subsequent component works without error, given a particular configuration of those things outside the broker's control (Figure 1(a)). Observe that initially, BPP is fixed at the best possible performance. After building summaries, BPP is still maximum (since, at least in theory, the rest of the broker can still get the right documents), but any later drop reflects the performance of successive steps. The drop in BPP at each step isolates the effectiveness of that component (our second desideratum); so the drop at point (a) is the loss in effectiveness due to server selection, which a broker can control, and that at point (b) is that due to retrieval, which it cannot.

Here, we are interested in selection alone so we remove the confounding effect of other components, as above; in effect, they make no difference to BPP. We use P@10 as the performance measure and vary the query, the testbed, and the effectiveness of each search engine (this satisfies our second desideratum).

3   Experiments

In these initial experiments, we use the "uniform" testbed of Xu and Croft, which features 100 "collections" formed from TREC ad-hoc data [9], and TREC topics 51–100. Each server is assumed to return up to ten documents, and up to ten servers are selected for each query. We consider the CRCS(e) [5] and ReDDE [6] selection algorithms, as well as SUSHI [8] selecting for precision ("SUSHI") and recall ("SUSHI.R").

What is the Impact of Selection? Our first experiments consider the BPP of a broker using each of the three algorithms, assuming perfect retrieval—all relevant documents are returned before any non-relevant—as well as perfect performance of other broker components. Since all topics have at least ten relevant documents, any drop in P@10 from 1.0 can be attributed to inaccuracy in server selection.

The left-hand boxes in Figure 1(b) illustrate best possible P@10 in this scenario. ReDDE, CRCS, and SUSHI.R are all capable of near-perfect selection; SUSHI, which selects many fewer servers, performs much more poorly.

Previous experiments have shown no difference between these methods on P@10, but have not accounted for errors introduced by other processing. The difference between the BPP reported here and the P@10 figures reported elsewhere [5; 6; 8]—around 0.4 ± 0.05—is due to ineffective merging. As well as obscuring the high performance of these selection algorithms, relatively poor merging has also obscured the differences between SUSHI and other methods.

How Robust are the Techniques? The right-hand side (grey) boxes in Figure 1(b) illustrate BPP after selection and retrieval when selected servers use a standard language modelling retrieval engine instead of perfect retrieval. Comparing the two sets of data allows a comparison of broker effectiveness, for each method, as one aspect of the broker's environment changes.

Here it is clear that retrieval effectiveness does impact BPP for each of the four algorithms, but (again with the exception of SUSHI) P@10 scores are still fairly high. This is expected—ad-hoc retrieval should not be so poor as to make a large impact. The BPP here is still well above P@10 figures for whole systems, however, so it still seems that poor merging is responsible for much of the poor effectiveness observed in earlier work.

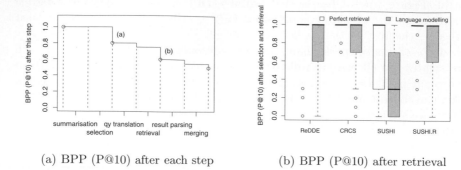

(a) BPP (P@10) after each step   (b) BPP (P@10) after retrieval

Fig. 1. Using Best Possible Performance (BPP) to investigate search effectiveness

4   Conclusions

By considering a broker's best possible performance, it is possible to evaluate server selection with a user-visible metric; in isolation from other components; and with regard to a number of factors outside the broker's control. Early experiments suggest that modern selection algorithms perform very well, and that the performance seen in complete brokers is likely due to poor merging in particular.

References

[1] Abbaci, F., Savoy, J., Beigbeder, M.: A methodology for collection selection in heterogeneous contexts. In: Proc. IEEE Conference on Information Technology (2002)
[2] Callan, J.: Distributed information retrieval. In: Bruce Croft, W. (ed.) Advances in information retrieval. The information retrieval series, vol. 7. Springer, Heidelberg (2000)
[3] Gravano, L., García-Molina, H.: Generalizing GlOSS to vector-space databases and broker hierarchies. In: Proc. VLDB (1995)
[4] Powell, A.L., French, J.C., Callan, J., Connell, M., Viles, C.L.: The impact of database selection on distributed searching. In: Proc. ACM SIGIR (2000)
[5] Shokouhi, M.: Central-rank-based collection selection in uncooperative distributed information retrieval. In: Amati, G., Carpineto, C., Romano, G. (eds.) ECiR 2007. LNCS, vol. 4425, pp. 160–172. Springer, Heidelberg (2007)
[6] Si, L., Callan, J.: Relevant document distribution estimation method for resource selection. In: Proc. ACM SIGIR (2003)
[7] Thomas, P., Hawking, D.: Server selection methods in personal metasearch: A comparative empirical study. Information Retrieval 12, 581–604 (2009)
[8] Thomas, P., Shokouhi, M.: SUSHI: Scoring scaled samples for server selection. In: Proc. ACM SIGIR (2009)
[9] Xu, J., Bruce Croft, W.: Cluster-based language models for distributed retrieval. In: Proc. ACM SIGIR (1999)

A Comparison of Language Identification Approaches on Short, Query-Style Texts

Thomas Gottron[1] and Nedim Lipka[2]

[1] Institut für Informatik, Johannes Gutenberg-Universität Mainz,
55099 Mainz, Germany
gottron@uni-mainz.de

[2] Faculty of Media, Media Systems, Bauhaus University Weimar,
99421 Weimar, Germany
nedim.lipka@uni-weimar.de

Abstract. In a multi-language Information Retrieval setting, the knowledge about the language of a user query is important for further processing. Hence, we compare the performance of some typical approaches for language detection on very short, query-style texts. The results show that already for single words an accuracy of more than 80% can be achieved, for slightly longer texts we even observed accuracy values close to 100%.

1 Introduction

The difficulty of a Cross Language Information Retrieval (CLIR) system is to find relevant documents across language boundaries. This induces the need for a CLIR system to be capable of doing translations between documents and queries. If the system has to handle more than one language for queries or documents, it additionally needs to be able to detect the language of a text. This is necessary to correctly translate the query or document into the language of the respectively other.

To our knowledge, the focus in research on language detection is usually on analysing full documents, i.e. on reasonably long and well formulated texts. Queries, instead, are rarely formulated as full sentences and are usually very short (typically 2-4 words for web search). Nevertheless, recent systems [1] participating at CLEF detected the language of queries by applying tools intended for long texts. This leads us to the question: how well do these approaches work on texts in the style of queries?

Automatic language detection on written texts, also known as language identification and sometimes as language recognition, is a categorization task. The most distinguished related works are based on statistical learning algorithms and lexical text representations, particularly n-grams, cf. [2,3,4,5,6]. Dictionary-based approaches, concerning words as lexical representation, are discussed in [7,2]. Non-lexical representations, used in the field of language identification, are for example phoneme transcriptions [8] or the rate of compression [9].

The paper in hand studies the potential and reliability of some commonly used language detection approaches on very short, query-style texts. Lacking a large corpus of annotated multi-language queries, we based our experiments on news headlines of the Reuters CV1 and CV2 collection and single words extracted from bilingual dictionaries.

C. Gurrin et al. (Eds.): ECIR 2010, LNCS 5993, pp. 611–614, 2010.

2   *n*-Gram-Based Language Detection

As short, query-style texts provide too little data for approaches based on words or full sentences, we focus on methods based on character *n*-grams, for short *n*-grams. An *n*-gram consists of *n* sequential characters; usually its relative occurence in a text is determined.

One such language detection method, that is used quite often is the one of Cavnar and Trankle [3]. Following the observation that each languages has some characteristic *n*-grams which appear frequently, the idea is to compare the frequency-ranks of *n*-grams in a previously unseen text with those of reference texts for different languages. The text is then attributed to the language with the most similar frequency-rank according to an out-of-place measure. As this measure is problematic for the few entries in the frequency-rank list of short texts, we "normalised" the ranks in our implementation to values between 0 for the most frequent and 1 for the least frequent *n*-gram.

As language detection is a classification task, Naive Bayes is a classical approach to the problem. A Naive Bayes classifier uses conditional probabilities of observing features in a text to deduce a probability of a text to be written in a given language. In our case the *n*-grams serve as features.

Vojtek and Bielikova [4] use Markov processes to determine the language of a text. Here, the idea is to detect the language via the probabilities to observe certain character sequences. The probabilities depend on a limited number k of previously seen characters, which form the states of the Markov process. The conditional probabilities can be estimated via frequencies of k-grams and $k + 1$-grams from a reference text [2].

The last approach we look at in this context is based on the vector space of all possible *n*-grams. A text can be represented in this space as the vector of the frequencies of its *n*-grams. Its language can be determined by looking at the cosine-similarity of its vector representation with the vectors of reference texts in different languages.

3   Evaluation

All the algorithms we discussed in the previous section need to be trained on reference documents. We used the English documents from the Reuters collections CV1 [10] and the language annotated Danish, German, Spanish, French, Italian, Dutch, Norwegian, Portuguese and Swedish documents from CV2. Table 1 shows the detailed distribution of the individual languages among these 1,102,410 documents.

The texts of the news articles were used for training the language classifiers. In order to see the influence of the length of *n*-grams we varied the value of *n* between 1 and 5 characters. The relatively short and noisy news headlines were retained for classification. They are on average 45.1 characters and 7.2 words long, thus, longer than an average query on the web. However, the titles frequently contain named entities (*"Berlusconi* TV faces legal cliffhanger") or numerical values ("Dollar General Q2 *$0.24* vs *$0.20*"). These entities and a lack of stopwords render the headlines a quite suitable set of short, query-alike texts for language detection. For the evaluation of single word texts, we obtained terms from small, bilingual dictionaries from English to French, German, Spanish, Italian and Portuguese. We extracted the words, which were

Table 1. Distribution of languages in the Reuters corpus and among dictionary terms

Corpus	da	de	en	es	fr	it	nl	no	pt	sv
Reuters	11.184	116.209	806.788	18.655	85.393	28.406	1.794	9.409	8.841	15.731
Dictionaries	–	3.463	12.391	3.260	1.153	2.432	–	–	501	–

unambiguous from a language point of few (i.e. existed in only one language). This gave us a total of 20.048 words of on average 8.1 characters. Again, see table 1 for details about the indivdual languages.

The algorithms were implemented from scratch and trained on the Reuters articles. For the frequency-rank approach, we additionally used a readily trained implementation of the original algorithm, which we included in the evaluation process as LC4J[1]. We used each of the algorithms to detect the languages of the previously unused Reuters headlines and the words obtained from dictionaries.

Table 2 shows the accuracies[2] for detecting the language of the Reuters headlines and the dictionary entries across all algorithms and all settings for n. But, the values of LC4J need to be treated carefully: in many cases the algorithm could not detect any language at all. This might be, because the language models provided with the implementation are too sparse for short texts. The values given here are solely based on those cases where language detection was successful. When taking into account the unclassified documents, the accuracy drops drastically to 39.24% for the headlines and to 30.33% for the dictionary words.

Table 2. Accuracy of language classifiers (in %)

Data	Method	1-grams	2-grams	3-grams	4-grams	5-grams
Headlines	Naive Bayes	87.90	95.01	98.52	99.40	**99.44**
	Multinomial	65.42	90.08	97.63	99.17	99.22
	Markov	10.28	85.87	73.13	4.50	0.00
	Frequency-rank	6.07	14.90	59.93	25.91	3.47
	Vectorspace	54.68	47.21	61.04	69.67	75.37
	LC4J (where successful)	–	–	67.72	–	–
Dictionaries	Naive Bayes	52.26	64.40	73.49	79.13	**81.61**
	Multinomial	35.65	57.04	68.27	75.74	77.88
	Markov	19.95	57.34	55.14	21.52	2.95
	Frequency-rank	12.32	24.04	42.82	23.25	6.70
	Vectorspace	29.99	33.98	44.28	52.73	59.23
	LC4J (where successful)	–	–	49.93	–	–

The poor performance of the Markov process and our own frequency-rank implementation for higher values of n can be explained with data sparseness, too. The accuracy of Markov drops probably due to a higher number of n-grams not seen during training and an unequal language distribution in the training data. The frequency-rank

---

[1] http://olivo.net/software/lc4j/

[2] Confusion matrices with more detailed results are available in an online Appendix at http://www.informatik.uni-mainz.de/forschung/ir/ecir2010.php

approach instead suffers from the sparseness of n-grams in the query-like documents, resulting in skewed rankings. Even with the normalised ranking, the performance drops for larger values of n.

The best performing approach for short texts is the Naive Bayes classifier (and its Multinomial variation without the class distribution normalisation). For larger values of n they perform remarkably good and achieve an accuracy close to 100% on the headlines. This observation holds also when looking at individual languages. On a language level, the accuracy varies between 99.71% for Italian and 96.52% for Norwegian. The misclassifications of Norwegian headlines were mostly assigned to Danish. In general, the Scandinavian languages tend to be confused more than other languages. A similar observation was made for dictionary terms of Latin-based languages. Here the most mistakes occured between Spanish, Portuguese and Italian.

4   Conclusions

We looked into language detection for short, query-style texts. Comparing different approaches based on n-grams, it turned out, that Naive Bayes classifiers perform best on very short texts and even on single words. Errors tend to occur within language families, i.e. among Scandinavian or Latin languages.

Future work will comprise a closer look at an adaptation of the frequency-rank approach for short texts, a hierarchical approach to better distinguish between texts from the same language family and the evaluation on real-world multilingual user queries.

References

1. Oakes, M., Xu, Y.: A search engine based on query logs, and search log analysis at the university of Sunderland. In: CLEF 2009: Proceedings of the 10th Cross Language Evaluation Forum (2009)
2. Dunning, T.: Statistical identification of language. Technical Report MCCS-94-273, Computing Research Laboratory, New Mexico State University (1994)
3. Cavnar, W.B., Trenkle, J.M.: N-gram-based text categorization. In: SDAIR 1994, Proceedings of the 3rd Annual Symposium on Document Analysis and Information Retrieval (1994)
4. Vojtek, P., Bieliková, M.: Comparing natural language identification methods based on Markov processes. In: Computer Treatment of Slavic and East European Languages, 4th Int. Seminar, pp. 271–282 (2007)
5. Suen, C.Y.: N-gram statistics for natural language understanding and text processing. IEEE Transactions on Pattern Analysis and Machine Ingelligence PAMI-1(2), 164–172 (1979)
6. Sibun, P., Reynar, J.C.: Language identification: Examining the issues (1996)
7. Řehůřek, R., Kolkus, M.: Language identification on the web: Extending the dictionary method. In: Computational Linguistics and Intelligent Text Processing. LNCS, vol. 5449, pp. 357–368. Springer, Heidelberg (2009)
8. Berkling, K., Arai, T., Barnard, E.: Analysis of phoneme-based features for language identification. In: Proc. ICASSP, pp. 289–292 (1994)
9. Teahan, W.J.: Text classification and segmentation using minimum cross-entropy. In: RIAO 2000, vol. 2, pp. 943–961 (2000)
10. Lewis, D.D., Yang, Y., Rose, T., Li, F.: RCV1: A new benchmark collection for text categorization research. Journal of Machine Learning Research 5, 361–397 (2004)

Filtering Documents with Subspaces

Benjamin Piwowarski, Ingo Frommholz, Yashar Moshfeghi,
Mounia Lalmas, and Keith van Rijsbergen

University of Glasgow, Department of Computing Science,
Glasgow G12 8QQ, UK

Abstract. We propose an approach to build a subspace representation
for documents. This more powerful representation is a first step towards
the development of a quantum-based model for Information Retrieval
(IR). To validate our methodology, we apply it to the adaptive document
filtering task.

1 Introduction

We explore an alternative representation of documents where each document is
not represented as a vector but as a subspace. This novel way of representing
documents is more powerful than the standard one-dimensional (vector) representation. Subspaces are a core component of the generalisation of the probabilistic framework brought by quantum physics [1], which enables to combine
both geometry and probabilities.

Sophisticated document representations have already been explored. Melucci
proposed to use subspaces to describe the locus of relevant documents for capturing context in IR; however documents are still represented as vectors [2].
Zuccon et al. [3] showed that the cluster hypothesis still holds when representing
documents as subspaces. In our work, we propose a different approach to build
such subspaces, where we suppose that a document can be represented as a set
of information needs (IN), each being represented as a vector. We also show how
to build a user *profile* from relevance feedback that can be used to compute the
probability of a document to be relevant.

Knowing how to represent documents is the first step towards a working IR
system, and here we focus on how to build such a representation and leave out
(among others) the problem of the query (or topic) representation. This makes
information filtering a suitable task to investigate our proposed subspaces since
it does not necessitate to represent a profile from a set of keywords like in an
ad-hoc task. We evaluate our approach on the adaptive document filtering task
of TREC-11 [4].

2 Document Filtering with Subspaces

In the adaptive filtering task [4], for each topic, three relevant documents from
the training set are given to build a profile representation. Then, documents

C. Gurrin et al. (Eds.): ECIR 2010, LNCS 5993, pp. 615–618, 2010.

are filtered one by one in a specified order, and each time the system decides whether to retrieve the incoming document or not. Only when the document is retrieved by the system, its associated relevance assessment can be used to update the profile representation before the system evaluates the relevance of the next incoming document. This process simulates a user interactive relevance feedback, since the user can only judge a document if it is retrieved.

2.1   Building the Document Subspace

Our main hypothesis is that a document can be represented as the subspace S_d spanned by a set of vectors, where each vector corresponds to an IN covered by the document. In practice, we assume that we can decompose a document into text excerpts that are associated with one or more INs. For a document d, we denote $\mathcal{U}_d$ the set of such vectors.

There are various possibilities to define the excerpts and how to map an excerpt to a vector, ranging from extracting sentences, paragraphs to using the full document as the single excerpt. As a first approximation, we chose to use sentences as excerpts (simple heuristics were applied to detect sentence boundaries[1]), and to transform them into vectors in the standard term space after stop word elimination and stemming. The weighting scheme used to construct vectors was either tf or tf-idf (see Section 3).

To compute the subspace S_d from the set of vectors of $\mathcal{U}_d$ (which are then spanning the subspace), an eigenvalue decomposition is used. The eigenvectors associated with the set of non-null eigenvalues of the matrix $\sum_{u \in \mathcal{U}_d} u u^\top$ define a basis of the subspace spanned by the vectors from $\mathcal{U}_d$. As the vectors from $\mathcal{U}_d$ are extracted from a corpus, we are not interested in all the eigenvectors but only in those that are associated with high eigenvalues λ_i, since low eigenvalues might be associated with noise. We used a simple strategy to select the rank of the eigenvalue decomposition, where we only select the eigenvectors with eigenvalues superior to the mean of the eigenvalues.

2.2   Profile Updating and Matching

The representation of the filtering profile is closely related to the above described document representation. We rely on the quantum probability framework to compute the probability of a document matching this profile.

The profile is updated whenever a document is retrieved. At each step, we can construct two sets Ψ^+ and Ψ^- that correspond to the set of all the INs of the retrieved documents that are relevant (resp. non relevant). From the set Ψ^- we build a *negative* subspace N (as described in the previous section) and assume that vectors lying in this subspace correspond to non-relevant INs. This process is the underlying motivation of using a subspace for the negative sub-profile. We denote $N^\perp$ the subspace orthogonal to this negative subspace.

---

[1] We use `http://www.andy-roberts.net/software/jTokeniser/index.html`

To determine if a document d, represented as a projector D on the subspace S_d (constructed as described in section 2.1), is retrieved or rejected with respect to the profile, we first project each (unit) vector $\psi_i \in \Psi^+$ of the positive profile onto the subspace $N^\perp$, in order to remove its non-relevant part. The result is a vector ψ'_i . We then suppose that a relevant document should "contain" as much as possible of these vectors ψ'_i. It is possible to give a probabilistic definition of this containment, by letting the probability that the document contains the IN ψ'_i be $\Pr(D|\psi'_i) = {\psi'_i}^\top D \psi'_i$ which has a value between 0 and 1, since D is a projector and ψ'_i has a norm less than 1.

As we have no preference about which of the vectors ψ'_i should be contained, we assume that each of the vectors is picked with a uniform probability, so that the probability of the document being relevant is given by $\Pr(D) \propto \sum_i \Pr(D|\psi'_i) = \mathrm{tr}(\rho D)$ where ρ equals $\sum_i \psi'_i {\psi'_i}^\top$ and tr is the trace operator. We can compute the actual probability by dividing $\mathrm{tr}(\rho D)$ by $\mathrm{tr}(\rho)$, which is a normalisation constant. If the value $\Pr(D)$ is over a given threshold, we retrieve the document; otherwise, we reject it. For simplicity, we only use a fixed threshold in the experiments, whereas a better approach would be to use a threshold that depends on the topic and the current state of the profile.

3   Experiments

We experimented with the adaptive filtering task of TREC-11 [4] and followed the task guidelines. Note that we ignored documents for which there was no relevance judgement. One important issue is to set a threshold so that a document whose score (as determined by the profile) is above the threshold is retrieved. As we wanted to focus on showing how the subspace approach performs compared to a baseline, we used a fixed threshold. We tried several values for this threshold, and selected the best performing runs. Comparing to approaches reported in [4], we have the unfair advantage of reporting the best performing settings but at the same time are penalised by the fact that our threshold is constant.

Table 1. Mean F-0.5 measure for the TREC-11 adaptive filtering task, for the Subspace and the Rocchio-based approach. The corresponding threshold values are (a) 0.05 and (b) 0.10.

Subspace	Neg	$\overline{\text{Neg}}$	Rocchio
TF	0.44^a	0.30^b	0.35^a
TF-IDF	0.41^a	0.31^b	0.44^b

We report results using one of the official metrics, the mean of F-0.5 metric (harmonic mean biased towards precision) which is less sensitive to the threshold policy. As a simple baseline, we report results obtained using the Rocchio-based approach for user profiling [5], and use a constant threshold (for a fair comparison with our approach) and a cosine similarity measure between a profile and a document (since it allows to experiment with the tf and the tf-idf weighting schemes). For the subspace approach, we experimented with the following parameters: (1) Using a negative subspace as described above (Neg) or not ($\overline{\text{Neg}}$, where we do not project ψ_i onto $N^\perp$) (2) using a tf-idf or tf weighting scheme to construct the ψ_i vector. Note that as for [5], idf values were estimated using an external collection (in our case Wikipedia) and updated with statistics from filtered documents. Eventually, for all models, we did an exhaustive search using a 0.05 step for the threshold. Values reported in Table 1 should be regarded as the maximum achievable with a fixed threshold; due to the small scale of the experiment, we do not report here statistical significance.

Our best runs are able to compete with those reported in [4], although it should be noted that we selected our best performing run (but also the baseline) *a posteriori*. We can then outline two facts from the results. First, using negative subspace was beneficial both for tf and tf-idf schemes: Using orthogonality to define non relevance is thus meaningful. Second, our subspace approach is competitive with a Rocchio-based baseline without relying on idf values.

4   Conclusion

There is the view that using subspaces instead of one dimension space is essential for sophisticated IR tasks like e.g. interactive IR [6]. In this paper, we showed through document filtering experiments that both the subspace representation of documents and the way we construct it lead to positive results. To exploit this representation, we also showed how to construct a user *profile* as a weighted set of vectors (and not a single vector as in e.g. Rocchio). This profile was constructed from documents, and our future work is to show how to construct and update this profile through sophisticated user interaction (query formulation, clicks, etc.), thus further exploiting the proposed subspace document representation.

Acknowledgements. This research was supported by an Engineering and Physical Sciences Research Council grant (Grant Number EP/F015984/2). M. Lalmas is currently funded by Microsoft Research/Royal Academy of Engineering.

References

1. van Rijsbergen, C.J.: The Geometry of Information Retrieval. Cambridge University Press, New York (2004)
2. Melucci, M.: A basis for information retrieval in context. ACM TOIS 26(3) (2008)
3. Zuccon, G., Azzopardi, L., van Rijsbergen, C.J.: Semantic spaces: Measuring the distance between different subspaces. In: Third QI Symposium (2009)
4. Robertson, S., Soboroff, I.: The TREC 2002 filtering track report. In: NIST (ed.) TREC-11 (2001)
5. Zhang, Y., Callan, J.: Yfilter at TREC-9, pp. 135–140. NIST special publication (2001)
6. Piwowarski, B., Lalmas, M.: A Quantum-based Model for Interactive Information Retrieval (extended version). ArXiv e-prints (September 2009) (0906.4026)

User's Latent Interest-Based Collaborative Filtering

Biyun Hu, Zhoujun Li, and Jun Wang

School of Computer Science and Engineering, Beihang University
XueYuan Road No.37, HaiDian District, Beijing, China
hubiyun@cse.buaa.edu.cn, lizj@buaa.edu.cn, junwang8151@163.com

Abstract. Memory-based collaborative filtering is one of the most popular methods used in recommendation systems. It predicts a user's preference based on his or her similarity to other users. Traditionally, the Pearson correlation coefficient is often used to compute the similarity between users. In this paper we develop novel memory-based approach that incorporates user's latent interest. The interest level of a user is first estimated from his/her ratings for items through a latent trait model, and then used for computing the similarity between users. Experimental results show that the proposed method outperforms the traditional memory-based one.

Keywords: Latent Interest, Latent Trait Models, Memory-based Collaborative Filtering, Sparsity.

1 Introduction

Collaborative Filtering (CF) is a popular technique used to help recommendation system users find out the most valuable information based on their past preferences. These preferences can be explicitly obtained by recording the ratings that users have awarded on items, such as albums, movies, and etc. CF algorithms can be mainly divided into two categories: model-based and memory-based. Model-based approaches first learn a model from history dataset. The model is then used for recommending. On the other hand, Memory-based algorithms look into the similarity between users or items, and then use these relationships to make recommendations [1] [2]. However, the sparsity in the user-item matrix often prohibits one from obtaining the similarity, resulting in poor prediction.

To overcome the drawbacks, the paper proposes a user's latent Interest-based Collaborative Filtering (LICF) method. The method follows the computation procedure of Memory-based Collaborative Filtering (MCF) to predict unknown ratings, but the similarity between users is computed based on their latent interest. A user's latent interest is measured through a psychometric model.

The rest of the paper is organized as follows: Section 2 provides a brief review of the traditional MCF. The proposed LICF is presented in Section 3. Preliminary experimental results are reported in Section 4. Finally, Section 5 gives conclusions and future works.

C. Gurrin et al. (Eds.): ECIR 2010, LNCS 5993, pp. 619–622, 2010.

2 Memory-Based Collaborative Filtering

Memory-based collaborative filtering consists of three major steps, namely user similarity weighting, neighbor selection, and prediction computation.

The similarity weighting step requires all users in the dataset to be weighted with respect to their similarity with the active user (the user whom the recommendations are for). For two particular users, the Pearson correlation coefficient is often used to compute the similarity between them [1] [2].

The neighbor selection step requires that a number of neighbors of the active user be selected. These selected users have the highest similarity weights.

Predictions are computed using the weighted combination of neighbors' ratings:

$$p_{a,i} = \overline{r_a} + \frac{\sum_{u=1}^{k} (r_{u,i} - \overline{r_u}) \times sim_{a,u}}{\sum_{u=1}^{k} |sim_{a,u}|}, \tag{1}$$

where $p_{a,i}$ is the predicted rating for the unrated item i; $\overline{r_a}$ the average rating of active user a; $r_{u,i}$ the rating given to item i by user u; $sim_{a,u}$ the similarity between users a and u; and k the number of neighbors.

As shown above, MCF often uses the Pearson correlation coefficient to compute the similarity between users. However, in many commercial recommender systems, even active users may have rated well under 1% of the items [2]. The number of items that two users have rated in common may be even less. As a result the accuracy of recommendations may be poor.

3 User's Latent Interest-Based Collaborative Filtering

In this section, we propose a novel user's latent interest-based collaborative filtering method. It is a variant of the traditional memory-based method, but uses latent interest instead of rating to compute the similarity between users.

3.1 User's Latent Interest

In psychometrics, latent trait models are a series of mathematical models applied to data from questionnaires or tests for measuring latent traits, such as attitudes, interests, and etc [3]. For ease of understanding, first consider the dichotomous model (each item has only two possible responses, e.g. like/dislike, usually coded as rating 1/0). The basic idea is the following: the more disagreeable one item is, the more probable is only person with high interest gives like response to it; the more interested one person is, the more probable is that his/her responses are like for a lot of items. Let Pni1 be the probability of an observation in rating 1, and Pni0 be the probability of an observation in rating 0. The dichotomous model can be formulated as (2),

$$\log (Pni1/Pni0) = Bn - Di. \tag{2}$$

The parameters Bn (n=1,…, N) are called person parameters, and they are the quantitative measures of person interest; the parameter Di (i=1,…, I) are called item parameters, and they are the quantitative measures of item agreeability. Total scores of persons and items are used to obtain maximum likelihood estimates of persons and items parameters [5]. The model we used (formulated as (3)) is an extension of the dichotomous model (2).

$$\log (Pnij/Pnij\text{-}1) = Bn - Di - Fj. \tag{3}$$

In formula (3), Fj is the step difficulty or threshold between ratings j and j-1, where the ratings are numbered, 0, J [4]. The more Bn is larger than the sum of Di and Fj, the more probable person n will give a rating j. WINSTEPS Rasch Software [6] is used to estimate the parameters in Formula (3) in our paper.

3.2 User Similarity Computation

Utilizing user's latent interest, we compute the user similarity as follows.

$$SIMa,u = 1 - |Ba - Bu| / (MAXinterest - MINinterest), \tag{4}$$

where SIMa,u is the similarity based on user's latent interest, Ba is the interest of active user a, MAXinterest is the maximum value among all the users' interest values, and MINinterest is the minimum value.

3.3 Recommendation

The prediction is made by combining neighbors' ratings, so that the prediction function (5) is the same as (1), except that the similarity between users is based on user's latent interest (ref. Formula (4)).

$$P_{a,i} = \overline{r_a} + \frac{\sum_{u=1}^{k} (r_{u,i} - \overline{r_u}) \times SIMa,u}{\sum_{u=1}^{k} |SIMa,u|}, \tag{5}$$

4 Experiments

We use the MovieLens dataset [2]. The dataset contains 1,000,209 ratings of approximately 1,682 movies made by 943 users. Ratings are discrete values from 1 to 5. 80% of the ratings were random selected into a training set and the remaining into a test set.

The results for the LICF and MCF methods are reported in Fig.1. We report the Mean Absolute Error (MAE) metric [2]. MAE corresponds to the average absolute deviation of predictions to the actual ratings in the test set. A smaller MAE value indicates a better performance.

The results in Fig.1 show that the proposed approach gets a better result than the traditional memory-based CF. The LICF outperforms MCF by 12.3% at most when the number of neighbors is set to be 15. The improvement of recommendation accuracy in Fig.1 also shows that the LICF can find out more accurate neighbors.

LICF is sparsity free, in that it uses one's interest instead of ratings to compute the similarity between users. Every user's interest can be estimated from his/her ratings,

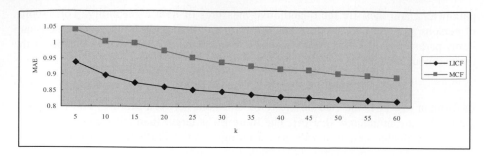

Fig. 1. Experimental results with different neighborhood size

thus even two users have no items rated in common can be compared utilizing their interest values, so that more neighbors can be found.

5 Conclusion

The Memory-based CF (MCF) is the state-of-the-art approach in recommendation systems, but it suffers from the data sparsity problem in computing the similarity between users. We proposed a novel memory-based CF that incorporates user's latent interest. The user interest is first estimated from a user's ratings for items by a psychometric model. We then compute the similarity between users based on their interest values. Experiments show that the proposed LICF outperforms the traditional MCF by finding out more neighbors. Other similarity computation methods based on user's latent interest need to be further investigated.

References

1. Resnick, P., Iacovou, N., Suchak, M., Bergstrom, P., Riedl, J.: Grouplens: An open architecture for collaborative filtering of netnews. In: Proc. ACM Computer Supported Cooperative Work, pp. 175–186 (1994)
2. Sarwar, B.M., Karypis, G., Konstan, J.A., Riedl, J.: Item-based collaborative filtering recommender algorithms. In: Proceedings of the WWW10, Hong Kong, China, pp. 285–295 (2001)
3. Item response theory,
   http://en.wikipedia.org/wiki/Item_response_theory#IRT_Models
4. Linacre, J.M.: Modeling Rating Scales. Annual Meeting of the American Educational Research Association, Boston (1990)
5. Rasch model estimation,
   http://en.wikipedia.org/wiki/Rasch_model_estimation
6. WINSTEPS Rasch Software, http://www.winsteps.com/winsteps.htm

Evaluating the Potential of Explicit Phrases for Retrieval Quality

Andreas Broschart[1,2], Klaus Berberich[2], and Ralf Schenkel[1,2]

[1] Saarland University, Saarbrücken, Germany
[2] Max-Planck-Institut für Informatik, Saarbrücken, Germany

Abstract. This paper evaluates the potential impact of explicit phrases on retrieval quality through a case study with the TREC Terabyte benchmark. It compares the performance of user- and system-identified phrases with a standard score and a proximity-aware score, and shows that an optimal choice of phrases, including term permutations, can significantly improve query performance.

1 Introduction

Phrases, i.e., query terms that should occur consecutively in a result document, are a widely used means to improve result quality in text retrieval [2,3,4,5,7], and a number of methods has been proposed to automatically identify useful phrases, for example [5,11]. However, there are studies indicating that phrases are not universally useful for improving results, but that the right choice of phrases is important. For example, Metzler et al.[6] reported that phrase detection did not work for their experiments in the TREC Terabyte track, and Mitra et al. [8] reported similar findings for experiments on news corpora.

This paper experimentally analyses the potential of phrase queries for improving result quality through a case study on the TREC Terabyte benchmark. We study the performance improvement through user-identified and dictionary-based phrases over a term-only baseline and determine the best improvement that any phrase-based method can achieve, possibly including term permutations.

2 Experimental Setup

We did a large-scale study on the effectiveness of phrases for text retrieval with the TREC GOV2 collection, a crawl of approximately 25 million documents from the .gov domain on the Web, and the 150 topics from the TREC Terabyte AdHoc tracks 2004-2006 (topics 701–850). All documents were parsed with stopword removal and stemming enabled. We compared different retrieval methods:

- A standard BM25F scoring model [9] as an established baseline for content-based retrieval, with both conjunctive (i.e., all terms must occur in a document) and disjunctive query evaluation.

C. Gurrin et al. (Eds.): ECIR 2010, LNCS 5993, pp. 623–626, 2010.

– Phrases as additional post-filters on the results of the conjunctive BM25F, i.e., results that did not contain at least one instance of the stemmed phrase were removed. As the TREC topics don't contain explicit phrases, we considered the following ways to find phrases in the queries:

- We performed a small user study where five users were independently asked to highlight any phrases in the titles of the TREC queries.
- As an example for a dictionary-based method for phrase detection, we matched the titles with the titles of Wikipedia articles (after stemming both), following an approach similar to the Wikipedia-based phrase recognition in [11].
- To evaluate the full potential of phrases, we exhaustively evaluated the effectiveness of all possible phrases for each topic and chose the best-performing phrase(s) for each topic.
- To evaluate the influence of term order, we additionally considered all possible phrases for all permutations of terms and chose the best-performing phrases, potentially after permutation of terms, for each topic.

– A state-of-the-art proximity score [1] as an extension of BM25F, including the modifications from [10]; this score outperformed other proximity-aware methods on TREC Terabyte according to [10].

We additionally report the best reported results from the corresponding TREC Terabyte tracks, limited to title-only runs.

3   Results

Our small user study showed that users frequently disagree on phrases in a query: On average, two users highlighted the same phrase only in 47% of the queries, with individual agreements between 38% and 64%. For each topic with more than one term, at least one user identified a phrase; for 43 topics, each user identified a phrase (but possibly different phrases). The same user rarely highlighted more than one phrase in a topic. Overall, our users identified 227 different phrases in the 150 topics.

Our experimental evaluation of query effectiveness focuses on early precision. We aim at validating if the earlier result by [8] (on news documents) that phrases do not significantly improve early precision is still valid when considering the Web. Table 1 shows precision@10 for using the phrases identified by the different users (as strict post-filters on the conjunctive BM25F run). Surprisingly, it seems to be very difficult for users to actually identify useful phrases, there hardly is any improvement. In that sense, the findings from [8] seem to be still valid today.

In the light of these results, our second experiment aims at exploring if phrase queries have any potential at all for improving query effectiveness, i.e., how much can result quality be improved when the 'optimal' phrases are identified. Tables 2 and 3 show the precision@10 for our experiment with the different settings introduced in the previous section, separately for each TREC year.

It is evident from the tables that an optimal choice of phrases can significantly improve over the BM25F baseline, with peak improvements between 12% and

Table 1. Precision@10 for user-identified phrases

	BM25F (conjunctive)	user 1	user 2	user 3	user 4	user 5
701-750	0.536	0.512	0.534	0.504	0.546	0.536
751-800	0.634	0.576	0.484	0.548	0.592	0.602
801-850	0.528	0.518	0.500	0.514	0.546	0.526
average	0.566	0.535	0.506	0.522	0.561	0.554

Table 2. Precision@10 for different configurations and query loads, part1

	BM25F (disjunctive)	BM25F (conjunctive)	best user phrases	Wikipedia phrases
701-750	0.548	0.536	0.546	0.566
751-800	0.630	0.634	0.592	0.564
801-850	0.538	0.528	0.546	0.526
average	0.572	0.566	0.561	0.552

14% when term order remains unchanged, and even 17% to 21% when term permutations are considered[1]. Topics where phrases were most useful include "pol pot" (843), "pet therapy" (793) and "bagpipe band" (794) (which were usually identified by users as well). On the other hand, frequently annotated phrases such as "doomsday cults" (745) and "domestic adoption laws" (754) cause a drastic drop in performance. Interesting examples for improvements when permuting terms are "hybrid alternative fuel cars" (777) where the best phrase is actually "hybrid fuel" (with a P@10 of 0.8, compared to 0.5 for the best in-order phrase and 0.2 for term-only evaluation), and "reintroduction of gray wolves" (797) with p@10 of 1.0 with the phrase "wolves reintroduction", compared to 0.6 otherwise.

Table 3. Precision@10 for different configurations and query loads, part2

	proximity score	best phrases	best phrases incl. permutations	best title-only TREC run
701-750	0.574	0.616	0.668	0.588
751-800	0.660	0.704	0.740	0.658
801-850	0.578	0.606	0.654	0.654
average	0.604	0.642	0.687	0.633

The best possible results are way above the best reported results for 2004 and 2005 and get close to the best result from 2006 (which was achieved, among other things, by the use of blind feedback)[2]. Wikipedia-based phrase recognition, a simple automated approach to phrase recognition, does not lead to significant improvements. Interestingly, the proximity-aware score yields significant

---

[1] Both significant according to a t-test with a p-value≤0.01.
[2] No significance tests possible as we don't have per-topic results for these runs.

improvements over the baseline[3]; as it automatically considers "soft phrases", there is no need to explicitly identify phrases here.

4   Discussion and Lessons Learned

The experimental analysis done in this paper yields the following results:

- We validated the common intuition that phrase queries can boost performance of existing retrieval models. However, choosing good phrases for this purpose is nontrivial and often too difficult for users, as the result of our user study shows.
- Existing methods for automatically identifying phrases can help to improve query performance, but they have their limits (like the methods based on Wikipedia titles evaluated here). While we expect that more complex methods such as the advanced algorithm introduced in [11] will get close to the upper bound, they need to include term permutations to exploit the full potential of phrases. The common intuition that term order in queries bears semantics does not seem to match reality in all cases.
- Proximity-aware scoring models where the user does not have to explicitly identify phrases can significantly improve performance over a non-proximity-aware scoring model.

References

1. Büttcher, S., Clarke, C.L.A., Lushman, B.: Term proximity scoring for ad-hoc retrieval on very large text collections. In: SIGIR, pp. 621–622 (2006)
2. Clarke, C.L.A., Cormack, G.V., Tudhope, E.A.: Relevance ranking for one to three term queries. In: RIAO, pp. 388–401 (1997)
3. Croft, W.B., Turtle, H.R., Lewis, D.D.: The use of phrases and structured queries in information retrieval. In: SIGIR, pp. 32–45 (1991)
4. Fagan, J.L.: Automatic phrase indexing for document retrieval: An examination of syntactic and non-syntactic methods. In: SIGIR, pp. 91–101 (1987)
5. Liu, S., Liu, F., Yu, C.T., Meng, W.: An effective approach to document retrieval via utilizing wordnet and recognizing phrases. In: SIGIR, pp. 266–272 (2004)
6. Metzler, D., Strohman, T., Croft, W.B.: Indri trec notebook 2006: Lessons learned from three terabyte tracks. In: TREC (2006)
7. Mishne, G., de Rijke, M.: Boosting web retrieval through query operations. In: Losada, D.E., Fernández-Luna, J.M. (eds.) ECIR 2005. LNCS, vol. 3408, pp. 502–516. Springer, Heidelberg (2005)
8. Mitra, M., Buckley, C., Singhal, A., Cardie, C.: An analysis of statistical and syntactic phrases. In: RIAO, pp. 200–217 (1997)
9. Robertson, S.E., Zaragoza, H., Taylor, M.J.: Simple BM25 extension to multiple weighted fields. In: CIKM, pp. 42–49 (2004)
10. Schenkel, R., Broschart, A., Hwang, S.-W., Theobald, M., Weikum, G.: Efficient text proximity search. In: Ziviani, N., Baeza-Yates, R. (eds.) SPIRE 2007. LNCS, vol. 4726, pp. 287–299. Springer, Heidelberg (2007)
11. Zhang, W., et al.: Recognition and classification of noun phrases in queries for effective retrieval. In: CIKM, pp. 711–720 (2007)

---

[3] p-value≤ 0.1 for TREC 2005 and ≤ 0.01 for the other two.

Developing a Test Collection for the Evaluation of Integrated Search

Marianne Lykke, Birger Larsen, Haakon Lund, and Peter Ingwersen

Royal School of Library and Information Science,
Department of Information Interaction and Information Architecture. Birketinget 6,
DK-2300 Copenhagen S, Denmark
{mln,blar,hl,pi}@db.dk

Abstract. The poster discusses the characteristics needed in an information retrieval (IR) test collection to facilitate the evaluation of *integrated search*, i.e. search across a range of different sources but with one search box and one ranked result list, and describes and analyses a new test collection constructed for this purpose. The test collection consists of approx. 18,000 monographic records, 160,000 papers and journal articles in PDF and 275,000 abstracts with a varied set of metadata and vocabularies from the physics domain, 65 topics based on real work tasks and corresponding graded relevance assessments. The test collection may be used for systems- as well as user-oriented evaluation.

Keywords: Test collection design; Task-based IR; Integrated search.

1 Introduction

As digital libraries offer access to increasingly large and diverse information sources there is recently a move from federated search, where a range of different sources are searched and the results presented for each source, to *integrated search* which instead presents the retrieved items in one, ranked list integrating results from different sources. Integrated search in this meaning is similar to universal search as found in some current web search engines mixing images, video and web results [1].

A main challenge in integrated search is that documents from different sources may be of different types, described on various levels of metadata and vocabularies. If for instance the domain is scientific publications, some documents may be available in full text, with and without metadata description, and some only as metadata records with or without abstracts. As all documents may be potentially relevant, treating all types in the same way in indexing and retrieval may overemphasise some types over others, e.g., resulting in the full text documents being more easily retrieved and higher ranked than documents only being described by metadata.

Evaluating different approaches to integrated search is currently difficult as no test collections exist with sufficiently different document types and comprehensive relevance assessments for each type. An appropriately designed test collection would be valuable and allow, e.g., the design of integrated search algorithms that better identify and rank relevant documents across the different types.

C. Gurrin et al. (Eds.): ECIR 2010, LNCS 5993, pp. 627–630, 2010.

In this poster we describe and analyse a new test collection that we have constructed for this purpose. We describe the development process, and analyse the resulting characteristics of the collection.

2   The Integrated Search Test Collection

IR systems evaluation is addressed from two quite different perspectives: the system-driven and the user-oriented perspectives [2]. Our approach is to develop a test collection that support both evaluation perspectives, by using a semi-laboratory/semi-real-life approach, using users' genuine information needs, and non-binary relevance judgements. Our aim has been to facilitate integration of the two perspectives at the study-design level [2], and to seek realism as well as experimental control.

A test collection for experiments with integrated search requires the following as a minimum: a corpus with several different document types, several levels of descriptions, appropriate search tasks, and relevance assessments with adequate amount of relevant documents for each type. In addition, it would be desirable to have documents without copyright restrictions (for acquiring a larger corpus), graded relevance assessments and tasks from users with real tasks/needs (for greater realism).

2.1   Document Collection

The scientific domain of physics comprises a realistic case with longstanding traditions for self-archiving of research publications in open access repositories and information sharing between scholarly and professional environments [3]. One of the largest repositories is arXiv.org, containing more than 500,000 papers covering the main areas of physics. We extracted two subsets from arXiv.org:

- 160,000+ full text papers in PDF including separate metadata
  (courtesy of Tim Brody, www.citebase.org)
- 274,000+ metadata records including abstracts for most documents
  (harvested using OAI PMH from www.arxiv.org)

The two subsets are very different in nature (see Table 1), the full text documents being much longer (4422 words on average) than the metadata records (272 words on average). In addition, we have added 18,000+ bibliographic book records classified as physics from the Danish national database covering all research and higher education libraries in Denmark (average length 189 words; no abstracts).

2.2   Search Tasks and Relevance Assessments

We extracted 65 natural search tasks from 23 lecturers, PhDs and experienced MSc students from three different university departments of physics. On average each test person provided three search task descriptions, which were captured in online forms via computers located in their own university environment. Prior to describing their tasks they were briefed about the project objectives and the structure and purpose of the form. After filling the forms they answered an online questionnaire concerning their personal data, domain and retrieval knowledge and experiences.

The search task description form had five questions, in line with [4]: a) *What are you looking for?* b) *Why are you looking for this?* c) *What is your background knowledge of this topic?* d) *What should an ideal answer contain to solve your problem or task?* e) *Which central search terms would you use to express your situation and information need?* Questions b) – c) correspond to questions asked in [4], with b) being about the underlying work task situation or context, and c) about the current knowledge state. Question a) asks about the formulation of the current information need, and d) correspond to the 'Narrative' section in a common TREC topic whilst e) asks for perceived adequate search terms.

The PDFs, abstract-only and library records were downloaded and indexed using the Fedora Generic Search Service with Lucene as search engine (see www.fedora-commons.org). A pool of up to 200 documents per task was retrieved for relevance assessments, separately for each document type and proportional to the corpus distribution where possible. In this way the (longer and more retrievable) PDFs would not be over represented. However, as shown in Table 2, in many tasks there were not enough abstract records to be retrieved, resulting in a higher proportion of PDFs. The searches were carried out manually by the research team as exhaustively as possible, based on the suggested search terms and other tokens in the original task descriptions.

Two months after task creation access to a web-based relevance assessment system was opened, allowing 1) access to the pool of documents to be assessed (sorted randomly within each document type), presented in overview form and with the possibility of opening full text PDFs where applicable; 2) assigning relevance scores according to the following 4-point scale: highly, fairly, marginally and non-relevant [5]. Documents could be re-assessed if the test person chose to. A post-assessment questionnaire on satisfaction with the assessment procedure and search results was filled for each task. Table 2 shows the relevance distributions over document types.

3   Discussion and Conclusion

The integrated search test collection has the following central features, Table 1. While there is no great variation in arXiv.org record size (abstracts) the library record size varies more owing to the presence/absence of table-of-contents data.

Table 1. Central features of the integrated search test collection

Features	Number	Mean no. of words per item
PDF items, arXiv.org	160,168	4422*
Abstracts, arXiv.org	274,749	272*
Library Records	18,222	189*
References (Citations)	3,748,555	(23.4 on avg. for 160,168 PDF items only)
Work Task situations	65	104.4*
a) Information need	65	17.7* (13 tasks with 5-10 terms)
b) Work task context	65	35.7*
c) Knowledge state	65	22.2*
d) Ideal information	65	19.3*
e) Search terms	65	9.4* (20 tasks with 3-6 terms)

* minus stop words (using the 318 word list at ir.dcs.gla.ac.uk/resources/linguistic_utils/).

Table 2. Relevance assessment statistics at document level. (*PDF*) are arXiv.org full text documents with metadata, abstract *and* PDF; (*ABS*) are arXiv.org records with metadata and abstract only. (*Meta*) are *all* arXiv.org records, with metadata and abstract, but omitting the PDF field. (*+gr*) implies positive relevance grades for the corresponding number of tasks.

Search – Collection types	*Tasks*	High (H)	Fair (F)	Marg. (M)	Non-rel.	Total	Books.	PDF	Abs. (Meta)
Total collect. 65 tasks		337	666	1875	8188	11066	992	5933	4141
Mean (65 tasks)		5.2	10.2	28.8	126	170.2	15.2	91.3	63.7
% (65 tasks)		3.0%	6.0%	16.9%	74.0%	-	9.0%	53.6%	37.4%
PDF, 44 tasks (2-3 +gr)		112	284	947	3213	4556	(← 29.5% H/F/M rel.)		
ABS, 48 tasks (2-3 +gr)		170	249	602	2336	3556	(← 30.4% H/F/M rel.)		
Books, 45 tasks (2-3 +gr)		53	130	241	568	992	(← 42.7% H/F/M rel.)		
Meta, 51 tasks (2-3 +gr)		284	532	1570	6130	8516	(← 28.0% H/F/M rel.)		

With reference to Table 1 we observe in line with [4] that the formulation of the information need (a) is short compared to underlying task description (b) and knowledge state (c), but longer than the suggested search terms (e). This phenomenon occurs regardless of the complexity or comprehensiveness of the described search task and knowledge state. Later analyses may reveal degrees of specificity, facets and vocabularies between different sections in the task descriptions, task categories and relevance assessments. Across the three document types 19 of the 65 tasks did not receive assessments covering all three positive degrees of relevance in all the types. By isolating tasks with at least one assessment of 2-3 different positive relevance grades we observe Table 2 that the collection allows for test using four separate sub-collections.

The strength of the collection is the amalgamation of realism and control, that the majority of the 65 real-life tasks have a fair number of relevant documents both across documents types and relevance grades (25.9% are relevant to some degree, with 6% and 3% being fairly and highly relevant respectively). The collection can therefore be used for integrated IR tests as intended. Secondly, search simulations of different aspects of information situations and work task contexts can be tested, e.g. to be pooled in a variety of combinations as evidence of the searcher situation.

References

[1] Sullivan, D.: Google Universal Search: 2008 Edition (2008),
   http://searchengineland.com/google-universal-search-2008-
   edition-13256
[2] Järvelin, K.: An analysis of two approaches in information retrieval: from frameworks to study designs. JASIST 58(7), 971–986 (2007)
[3] Gómez, N.D.: Physicists' information behaviour: a qualitative study of users. In: 70th IFLA Council and General Conference IFLA, Buenos Aires, August 22-27 (2004)
[4] Kelly, D., Fu, X.: Eliciting better information need descriptions from users of information search systems. Information Processing & Management 43(1), 30–46 (2007)
[5] Sormunen, E.: Liberal relevance criteria of TREC – Counting on negligible documents? In: Proceedings of SIGIR 2002, pp. 320–330. ACM Press, New York (2002)

Retrieving Customary Web Language to Assist Writers

Benno Stein, Martin Potthast, and Martin Trenkmann

Bauhaus-Universität Weimar, Germany
<first name>.<last name>@uni-weimar.de

Abstract. This paper introduces NETSPEAK, a Web service which assists writers in finding adequate expressions. To provide statistically relevant suggestions, the service indexes more than 1.8 billion n-grams, $n \leq 5$, along with their occurrence frequencies on the Web. If in doubt about a wording, a user can specify a query that has wildcards inserted at those positions where she feels uncertain.

Queries define patterns for which a ranked list of matching n-grams along with usage examples are retrieved. The ranking reflects the occurrence frequencies of the n-grams and informs about both absolute and relative usage. Given this choice of customary wordings, one can easily select the most appropriate. Especially second-language speakers can learn about style conventions and language usage.

To guarantee response times within milliseconds we have developed an index that considers occurrence probabilities, allowing for a biased sampling during retrieval. Our analysis shows that the extreme speedup obtained with this strategy (factor 68) comes without significant loss in retrieval quality.

1 Introduction

Writers who are in doubt about a certain expression often ask themselves: *"What wording would others use in a similar context?"* This question can be answered statistically when given a huge corpus of written text from which matching examples can be retrieved. In this paper we introduce NETSPEAK, which indexes a large portion of the Web, presumably the most comprehensive text corpus today.[1]

Related Work. Computer-aided writing has a long history dating back to the very beginning of personal computing, and so has research on this topic. This is why we can give only a brief overview. The main topics in writing assistance include spell checking, grammar checking, choosing words, style checking, discourse organization, and text structuring. Note that spell checkers and, to a certain extent, grammar checkers are currently the only technologies that reached a level of maturity to be shipped large-scale.

Search engines comparable to ours are for instance WEBCORP, WEBASCORPUS, PHRASESINENGLISH, and LSE.[2] All of them target exclusively researchers of linguistics. By contrast, our search engine targets the average writer, whose information needs and prior knowledge differs from those of a linguist. Moreover, NETSPEAK outperforms existing tools in terms of both retrieval speed and the extent of the indexed language resources. In [5] the authors propose an index data structure that supports linguistic queries; a comparison with our approach is still missing.

---

[1] NETSPEAK is available at http://www.netspeak.cc

[2] See http://www.webcorp.org.uk, http://webascorpus.org, http://phrasesinenglish.org, and [8].

C. Gurrin et al. (Eds.): ECIR 2010, LNCS 5993, pp. 631–635, 2010.
© Springer-Verlag Berlin Heidelberg 2010

Corpora of n-grams are frequently used in natural language processing and information retrieval for training purposes [7], e.g., for natural language generation, language modeling, and automatic translation. In particular, there is research on automatic translation within a language in order to correct writing errors [4]. We want to point out that our research is not directed at a complete writing automation since we expect a semi-automatic, interactive writing aid to be more promising in the foreseeable future.

2   NETSPEAK Building Blocks

The three main building blocks of NETSPEAK are (*i*) an index of frequent n-grams on the Web, (*ii*) a query language to formulate n-gram patterns, and (*iii*) a probabilistic top-k retrieval strategy which finds n-grams that match a given query and which allows to trade recall for time. The results are returned in a Web interface or as XML document.

Web Language Index.   To provide relevant suggestions, a wide cross-section of written text on the Web is required which is why we resort to the Google n-gram corpus [3]. This corpus is currently the largest of its kind; it has been compiled from approximately 1 trillion words extracted from the English portion of the Web, and for each n-gram in the corpus its occurrence frequency is given. Columns 2 and 3 of Table 1 give a detailed overview of the corpus. We applied two post-processing steps to the corpus at our site: case reduction and vocabulary filtering. For the latter, a white list vocabulary V was compiled and only these n-grams whose words appear in V were retained. V consists of the words found in the Wiktionary and various other dictionaries, as well as of these words from the 1-gram portion of the Google corpus whose occurrence frequency is above 11 000. See Table 1, Columns 4 and 5, for the size reductions after each post-processing step with respect to the original corpus.

In NETSPEAK the n-gram corpus is implemented as an inverted index, μ, which maps each word $w \in V$ onto a postlist π_w. For this purpose we employ a minimal perfect hash function based on the CHD algorithm [2]. π_w is a list of tuples $\langle d\hat{\,}, f(d) \rangle$, where $d\hat{\,}$ refers to an n-gram d on the hard disk that contains w, and where $f(d)$ is the occurrence frequency of d reported in the n-gram corpus. A tuple also stores information about w's position as well as other information omitted here for simplicity.

Query Language.   The query language of NETSPEAK is defined by the grammar shown in Table 2. A query is a sequence of literal words and wildcard operators, where the literal words must occur in the expression sought after, while the wildcard operators allow to specify uncertainties. Currently four operators are supported: the question mark,

Table 1. The Google n-grams before and after post-processing

Corpus Subset	Original Corpus		Case Reduction	Vocabulary Filtering
	# n-grams	Size		
1-gram	13 588 391	177.0 MB	81.34 %	3.75 %
2-gram	314 843 401	5.0 GB	75.12 %	43.26 %
3-gram	977 069 902	19.0 GB	83.24 %	48.65 %
4-gram	1 313 818 354	30.5 GB	90.27 %	49.54 %
5-gram	1 176 470 663	32.1 GB	94.13 %	47.16 %
Σ	3 354 253 200	77.9 GB	88.37 %	54.20 %

Table 2. EBNF grammar of the query language

Production Rule		
query	=	{ word | wildcard }$_1^5$
word	=	([" ' "] (letter { alpha })) | " , "
letter	=	"a" | ... | "z" | "A" | ... | "Z"
alpha	=	letter | "0" | ... | "9"
wildcard	=	"?" | "*" | synonyms | multiset
synonyms	=	"~" word
multiset	=	"{" word { word } "}"

which matches exactly one word, the asterisk, which matches any sequence of words, the tilde sign in front of a word, which matches any of the word's synonyms, and the multiset operator, which matches any ordering of the enumerated words. Of course other sensible operators are conceivable, which is part of our work in progress: constraints on particular parts of speech, person names, places, dates, and times.

Probabilistic Retrieval Strategy. Given the n-gram index μ and a query q, the task is to retrieve all n-grams D_q from μ that match q according to the semantics defined above. This is achieved within two steps: (*i*) computation of the intersection postlist $\pi_q = \bigcap_{w \in q} \pi_w$, and (*ii*) filtering of π_q with a pattern matcher that is compiled at runtime from the regular expression defined by q. Reaching perfect precision and recall is no algorithmic challenge unless retrieval time is considered. Note in this respect that the length of a postlist often amounts up to millions of entries, which is for instance the case for stop words. If a query contains only stop words, the retrieval time for D_q may take tens of seconds up to a minute, depending on the size of the indexed corpus. From a user perspective this is clearly unacceptable. In cases where a query also contains a rare word w', it is often more effective to apply the pattern matcher directly to $\pi_{w'}$, which is possible since $\pi_q \subseteq \pi_w$ holds for all $w \in q$. But altogether this and similar strategies don't solve the problem: the frequency distribution of the words used in queries will resemble that of written text, simply because of the NETSPEAK use case. Note that Web search engines typically get queries with (comparatively infrequent) topic words.

To allow for an adjustable retrieval time at the cost of recall we have devised a probabilistic retrieval strategy, which incorporates rank-awareness within the postlists. Our strategy hence is a special kind of a top-k query processing technique [1, 6]. The strategy requires an offline pre-processing of μ, so that (*i*) each postlist is sorted in order of decreasing occurrence frequencies, and (*ii*) each postlist is enriched by quantile entries κ, which divide the word-specific frequency distribution into portions of equal magnitude. Based on a pre-processed μ, the retrieval algorithm described above is adapted to analyze postlists only up to a predefined quantile. As a consequence, the portion of a postlist whose frequencies belong to the long tail of the distribution is pruned from the search. Note that the retrieval precision remains unaffected by this.

An important property of our search strategy is what we call *rank monotonicity*: given a pre-processed index μ and a query q, the search strategy will always retrieve n-grams in decreasing order of relevance, independently of κ. This follows directly from the postlist sorting and the intersection operation. An n-gram that is relevant for a query q is not considered if it is beyond the κ-quantile in some π_w, $w \in q$. The probability for this depends, among other things, on the co-occurrence probability between q's words. This fact opens up new possibilities for further research in order to raise the recall, e.g., by adjusting κ in a query-specific manner.

3   Evaluation

To evaluate the retrieval quality of our query processing strategy, we report here on an experiment in which the average recall is measured for a set of queries Q, $|Q| = 55\,702$, with respect to different pruning quantiles. The queries originate from the query logs

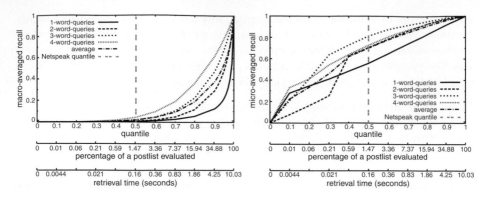

Fig. 1. Macro-averaged recall (left) and micro-averaged recall (right) over quantiles. The additional axes indicate how much of a postlist is evaluated and the required processing time.

of NETSPEAK; the service is in public use since 2008. We distinguish between macro-averaged recall and micro-averaged recall:

$$rec_{\mathrm{macro}}(\mu, q) = \frac{|D_q \cap D_q^*|}{|D_q^*|} \qquad rec_{\mathrm{micro}}(\mu, q) = \frac{\sum_{\langle d\hat{,} f(d)\rangle \in (\pi_q \cap \pi_q^*)} f(d)}{\sum_{\langle d\hat{,} f(d)\rangle \in \pi_q^*} f(d)}$$

As described above, D_q and π_q are the results retrieved from μ for query q under a top-k strategy, while D_q^* and π_q^* are the results if the postlists of μ are evaluated completely. While rec_{macro} considers only the result list lengths, rec_{micro} allots more weight to n-grams with high occurrence frequencies, since they are more relevant to the user. Figure 1 shows the obtained results for different query sizes.

Discussion and Conclusion. The macro-averaged recall differs significantly from the micro-averaged recall, which indicates that most of the relevant n-grams are retrieved with our strategy. The current NETSPEAK quantile of $\kappa = 0.5$ marks the best trade-off between recall and retrieval time. At quantile 0.5 only 1.47% of a postlist is evaluated on average, which translates into a retrieval speedup of factor 68. The average retrieval time at this quantile seems to leave much room in terms of user patience to evaluate more of a postlist, however, it does not include the time to generate and ship the result page. Short queries are more difficult to answer because the size of the expected result set is much larger on average than that of a long query. From an evaluation standpoint the micro-averaged view appears to be more expressive. Altogether, our retrieval strategy makes NETSPEAK a fast and reliable writing assistant.

Bibliography

[1] Bast, H., Majumdar, D., Schenkel, R., Theobald, M., Weikum, G.: IO-Top-k: Index-access Optimized Top-k Query Processing. In: Proc. of VLDB 2006 (2006)
[2] Belazzougui, D., Botelho, F.C., Dietzfelbinger, M.: Hash, Displace, and Compress. In: Fiat, A., Sanders, P. (eds.) ESA 2009. LNCS, vol. 5757, pp. 682–693. Springer, Heidelberg (2009)

[3] Brants, T., Franz, A.: Web 1T 5-gram Version 1. Linguistic Data Consortium (2006)
[4] Brockett, C., Dolan, W.B., Gamon, M.: Correcting ESL Errors Using Phrasal SMT Techniques. In: Proc. of ACL 2006 (2006)
[5] Cafarella, M.J., Etzioni, O.: A Search Engine for Natural Language Applications. In: Proc. of WWW 2005 (2005)
[6] Ilyas, I.F., Beskales, G., Soliman, M.A.: A Survey of Top-k Query Processing Techniques in Relational Database Systems. ACM Comput. Surv. 40(4), 1–58 (2008)
[7] Manning, C.D., Schütze, H.: Foundations of Statistical Natural Language Processing. MIT, Cambridge (1999)
[8] Resnik, P., Elkiss, A.: The Linguist's Search Engine: An Overview. In: Proc. of ACL 2005 (2005)

Enriching Peer-to-Peer File Descriptors Using Association Rules on Query Logs

Nazli Goharian[1], Ophir Frieder[1], Wai Gen Yee[2], and Jay Mundrawala[2]

[1] Department of Computer Science, Georgetown University
[2] Information Retrieval Laboratory, Illinois Institute of Technology
{nazli,ophir}@cs.georgetown.edu, {waigen,mundra}@ir.iit.edu

Abstract. We describe a P2P association rule mining descriptor enrichment approach that statistically significantly increases accuracy by greater than 15% over the non-enriched baseline. Unlike the state-of-the-art enrichment approach however, the proposed solution does not introduce additional network load.

1 Introduction

Peer-to-peer file sharing is a major Internet application, consuming an estimated 65% of the 2008 United States Internet traffic [1]. Such a major application requires accurate and efficient query processing capabilities.

In P2P file sharing, peers are both clients issuing search queries and servers responding to them. Files, typically audio or video in nature, are replicated across the network. Each copy of a file, namely a replica, is described by a user-provided textual descriptor of limited length. Files are searched for by comparing a query to a file descriptor. Unfortunately, file descriptors often are sparsely defined and result in few query matches. Thus, search accuracy is poor. Frustrated users ultimately issue additional queries, typically to no avail, and introduce unnecessary network traffic.

Poor search accuracy is due in part to the conjunctive query processing paradigm used in P2P file sharing. A relevant replica only matches a query if all query terms are in its descriptor. Given sparsely specified file descriptors and relatively long queries, query to file descriptor mismatch is likely.

To address the sparse descriptor problem, we propose to enrich each peer's replica descriptors with correlated terms. We do so using association rule mining. Descriptor enrichment increases the ability of a query to match relevant replicas, increasing overall retrieval accuracy. Our retrieval accuracy is comparable to that delivered using the state of the art enrichment method [2]. However, our technique is superior to [2] in that it can be performed by individual peers without coordination of others. The previous technique requires peers to share metadata, making them more prone to failures based on malicious nodes or connectivity patterns. Furthermore, they incur additional network traffic, which may reduce system performance. Our technique avoids these hazards.

2 Descriptor Enrichment

Each node maintains a log of queries. These queries are logged as the node routes queries, a process in which every node is assumed to participate. The node uses the

C. Gurrin et al. (Eds.): ECIR 2010, LNCS 5993, pp. 636–639, 2010.

contents of these logs to derive term correlations. Correlated terms are added to the terms already in the descriptors of the local replicas. These terms are added until a maximum size is reached. Introducing additional terms requires removing the least frequently used term.

We use the *Apriori Association Rule* algorithm to derive correlated terms (although other techniques are possible [3]). Terms that co-occur with a minimum *support* and *confidence* are identified. Support for a term set $\{t_i, t_j\}$ is defined as the ratio of queries in the query log that contain the term set. The confidence of the rule $t_i \rightarrow t_j$ indicates the ratio of the queries that contain t_i also containing t_j. Support and confidence are formally defined below, where σ is the counting operator and N is the total number of queries in a peer's query log.

$$ Support\ (t_i \rightarrow t_j) = \frac{\sigma\,(t_i \cup t_j)}{N} \quad Confidence\ (t_i \rightarrow t_j) = \frac{\sigma\,(t_i \cup t_j)}{\sigma\,(t_i)} $$

Using these discovered term correlations, each peer updates its replicas' descriptors to include the related terms.

3 Experimentation

To evaluate our proposed approach, we developed an experimental platform and assigned parameters matching those in the literature [2]. A set of 1,080 unique Web TREC 2GB (WT2G) documents are grouped into 37 interest categories, derived from the Web domains. Three to five categories are assigned to each 1,000 peers based on a Zipf distribution. The documents for a category are similarly distributed using a Zipf distribution. To each peer, 10 to 30 replicas are assigned at initialization based on the category assignments. The descriptor of each replica is initialized with 3 to 10 terms from the original document according to the term distribution within the document. Table 1 summarizes the experimental data and framework.

Queries are generated by each peer based on its interests. The terms used in each query are based on the term distribution of the corresponding Web TREC documents. The query lengths are based on distributions shown in prior work [2]: 28%, 30%, 18%, 14%, 4%, 3%, 2%, 1% of queries have 1, 2, 3, 4, 5, 6, 7, 8 terms, respectively.

Table 1. Statistics

Peers	1,000
Categories	37
Documents	1,080
Queries	10,000
Descriptor size (terms)	20
Initial descriptors size	3-10
Categories per peer	3-5
Files per peer at initialization	10-30
Trials per experiment	10

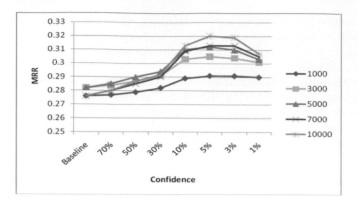

Fig. 1. MRR versus confidence and query log size

We apply the Apriori algorithm on each peer's query log to *discover* the correlations among terms. The correlations that meet the confidence thresholds are applied to the local replica descriptors to enrich them accordingly by adding the correlated terms to the existing file descriptor terms. We also varied the support threshold for our rules. A support of 0.3% yielded the best results. We omitted support results due to space constraints. Each peer creates a query log from the queries routed through it. Query logs range in size from 1,000 to 10,000 queries.

Accuracy was measured using Mean Reciprocal Rank (MRR):

$$MRR = \frac{1}{N_q} \sum_{i=1}^{N_q} \frac{1}{rank_i}$$

where N_q is the number of queries issued, and $rank_i$ is the rank of the desired result in query i's result set. If the desired result is not in the result set, then $1/rank_i = 0$.

In each trial, we recorded the MRR of 10,000 queries issued from random peers with and without enriched descriptors. We ran 10 trials and report the average MRR. We compute significance of the results using a paired T-test with a resulting statistical significance of greater than 99%.

4 Results

As shown in Figure 1, for all query log sizes, as confidence decreases, MRR increases up to a point (5% confidence) and eventually declines. The increase in MRR stems from the larger descriptors, and hence, higher probability of a match. On the other hand, too low of a confidence threshold results in the *over*-enrichment of the descriptor. Hence, too many descriptors match the query, including those of irrelevant results, which lowers MRR scores.

If confidence is set to 5%, the increase in MRR ranges from 5.5% (for a 1,000-query query log) to 16% (for a 10,000-query query log) over the baseline. The larger query log allows greater opportunities for deriving accurate term correlations.

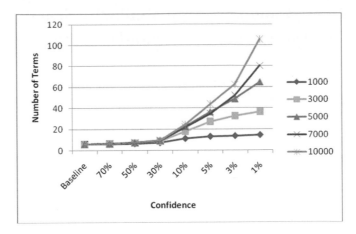

Fig. 2. Average descriptor size versus confidence and query log size

A cost measure for descriptor enrichment is descriptor size. With low confidence and large query logs, cost can be significant: descriptor size may increase from 6 terms to over 100 as shown in Figure 2. The steepest increases in descriptor size occur with confidence values below 5%. Fortunately, with 5% confidence, we yield the best MRR for all query log sizes and the descriptor size is a manageable 42 terms. Therefore, we recommend a confidence of 5%.

5   Conclusion

P2P file sharing search accuracy is limited due in part to conjunctive query processing strategy and the relative shortness of the file descriptors. By enriching file descriptors with correlated terms discovered by applying association rules on query logs, we improve accuracy by up to 15%. The benefit of this strategy over previous work is that it does not require additional network traffic to discover the correlations.

The described approach improves MRR performance by greater than 15% when a query log of 10,000 queries and a confidence of 5% are used for rule derivation. The cost at this point is manageable as well: 42 descriptor terms instead of 6. We claim that this is a small price to pay for query accuracy.

References

1. ipoque, ipoque Internet Study (2008),
   http://www.ipoque.com/resources/internet_studies
2. Jia, D., et al.: Distributed, Automatic File Description Tuning in P2P File-Sharing Systems. Peer-to-Peer Networking and Applications 1(2) (September 2008)
3. Han, J., Kamber, M.: Data Mining: Concepts and Techniques, 2nd edn. Morgan Kaufmann, San Francisco (2006)

Cross-Language High Similarity Search:
Why No Sub-linear Time Bound Can Be Expected

Maik Anderka, Benno Stein, and Martin Potthast

Bauhaus University Weimar, Faculty of Media, 99421 Weimar, Germany
<first name>.<last name>@uni-weimar.de

Abstract. This paper contributes to an important variant of cross-language information retrieval, called cross-language high similarity search. Given a collection D of documents and a query q in a language different from the language
of D, the task is to retrieve highly similar documents with respect to q. Use cases
for this task include cross-language plagiarism detection and translation search.

The current line of research in cross-language high similarity search resorts to
the comparison of q and the documents in D in a multilingual concept space—
which, however, requires a linear scan of D. Monolingual high similarity search
can be tackled in sub-linear time, either by fingerprinting or by "brute force n-
gram indexing", as it is done by Web search engines. We argue that neither fingerprinting nor brute force n-gram indexing can be applied to tackle cross-language
high similarity search, and that a linear scan is inevitable. Our findings are based
on theoretical and empirical insights.

1   High Similarity Search

In the literature the task of high similarity search is also referred to as near-duplicate
detection or nearest neighbor search. High similarity search techniques are applied in
many applications such as for duplicate detection on the Web, text classification and
clustering, plagiarism detection, or storage maintenance.

Without loss of generality we consider a document d represented under a bag of
words model, as an m-dimensional term vector $\mathbf{d}$. The similarity between a query document q and a document d is quantified with a measure $\varphi(\mathbf{q}, \mathbf{d}) \in [0; 1]$, with 0 and 1
indicating no and maximum similarity respectively. φ may be the cosine similarity.

Definition 1 (High Similarity Search). *Given a query document q and a (very large)
collection D of documents, the task of high similarity search is to retrieve a subset
$D_q \subset D$, containing the most similar documents with respect to q:*

$$d \in D_q \;\Rightarrow\; \varphi(\mathbf{q}, \mathbf{d}) \geq 1 - \epsilon \tag{1}$$

D_q is called ϵ-neighborhood of q: a document $d \in D_q$ is called near-duplicate of q.

Since $\mathbf{q}$ and $\mathbf{d}$ are term-based representations, a document d is considered as near-
duplicate of the document q if d and q share a very large part of their vocabulary. This
syntactic definition of near-duplicate cannot be applied between two languages L and
L', and cross-language near-duplicates need to be defined in a semantic manner. Consider for example a document d in language L that is a translation of a document q'

C. Gurrin et al. (Eds.): ECIR 2010, LNCS 5993, pp. 640–644, 2010.

in language L'. Then similarity can be measured by a multilingual retrieval model that maps q' and d into a common, multilingual concept space. A few multilingual retrieval models exist, for a comparative overview see [6]. One of the most promising multi-lingual retrieval models is Cross-language Explicit Semantic Analysis, CL-ESA [1,6], which exploits a document-aligned comparable corpus such as Wikipedia in order to represent documents written in different languages in a common concept space. The cross-language similarity between q' and d is computed as cosine similarity $\varphi_{cos}(\mathbf{q}'_{clesa}, \mathbf{d}_{clesa})$ of the CL-ESA representations of q' and d.

Definition 2 (Cross-language High Similarity Search). *Given a query document q' in language L' and a (very large) collection D of documents in language L, the task of cross-language high similarity search is to retrieve a subset $D_{q'} \subset D$, containing the most similar documents with respect to q':*

$$d \in D_{q'} \;\Rightarrow\; \varphi_{cos}(\mathbf{q}'_{clesa}, \mathbf{d}_{clesa}) \geq 1 - \epsilon \tag{2}$$

$D_{q'}$ *is called ϵ-neighborhood of q'; a document $d \in D_{q'}$ is called near-duplicate of q'.*

To determine the value of ϵ we empirically analyzed the similarity values of near-duplicates in both settings: monolingual high similarity search and cross-language high similarity search. The left plot in Figure 1 shows the distribution of similarities between randomly selected English Wikipedia articles and their revisions— which serve as near-duplicates—computed as defined in (1). The right plot in Figure 1 shows the distribution of cross-language similarities (*i*) between randomly selected aligned English and German Wikipedia articles, i.e., the articles describe the same concept in its respective language, and (*ii*) between randomly selected aligned English and German documents from the JRC-Acquis corpus, which contains professional translations. In both cases the aligned documents are considered as cross-language near-duplicates, the respective cross-language similarities are computed as defined in (2). The analysis shows that the absolute similarity values of near-duplicates heavily differ in monolingual high similarity search and cross-language high similarity search. In the former a reasonable ϵ to detect the near-duplicates has to be very small (~ 0.15), whereas, in the latter a reasonable ϵ to detect cross-language near-duplicates has to be much higher (~ 0.5). One explanation for the relatively small similarity values of the cross-language near-duplicates is that the CL-ESA model is not able to operationalize the concept of "semantic similarity" entirely; and, it is still questionable if this is possible at all. However, even if the absolute similarity values of cross-language near-duplicates are relatively small, cross-language high similarity search is still possible since the average cross-language similarities between randomly selected documents—which are not aligned—of Wikipedia as well as the JRC-Acquis is about 0.1.

2   Linear Scan

A naive approach to high similarity search is a linear scan of the entire collection, i.e., calculating $\varphi(\mathbf{q}, \mathbf{d})$ or $\varphi_{cos}(\mathbf{q}'_{clesa}, \mathbf{d}_{clesa})$ for all $d \in D$. The retrieval time is $O(|D|)$, which is unfeasible for practical applications when D is very large, e.g. the World Wide Web. However, there are several approaches that try to speed up the pairwise similarity

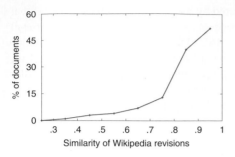 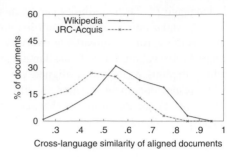

Fig. 1. The left plot shows the distribution of similarities between English Wikipedia articles and their revisions, the right plot shows the distribution of cross-language similarities between aligned English and German documents from Wikipedia and the JRC-Acquis corpus

calculation in practice, e.g., by distributing the similarity computation based on MapReduce [4], or by using a specialized inverted index in combination with several heuristics to reduce the number of required multiplications [2]. In low-dimensional applications ($m < 10$) similarity search can be accelerated by means of space- or data-partitioning methods, like, grid-files, kd-trees, or R-trees. However, if the dimensionality is larger than 10—which is usual in practical applications, where the documents are represented as high dimensional feature vectors—these methods are outperformed by a simple linear scan [7].

3   Fingerprinting

Hash-based search or fingerprinting does not depend on the dimensionality of the feature vectors and allows for monolingual high similarity search in sub-linear retrieval time. Fingerprinting approaches simplify a continuous similarity relation to the binary concept "similar or not similar". A multi-valued similarity hash-function h_φ is used to map a feature vector $\mathbf{d}$ onto a small set of hash codes $F_d := h_\varphi(\mathbf{d})$, called fingerprint of d. Two documents q and d are considered as similar if their fingerprints share some hash code: $F_q \cap F_d \neq \emptyset \Rightarrow \varphi(\mathbf{q}, \mathbf{d}) \geq 1 - \epsilon$, with $0 < \epsilon \ll 1$. The mapping between all hash codes $C := \bigcup_{d \in D} F_d$ and documents with the same hash code can be organized as a hash table $\mathcal{T} : C \rightarrow \mathcal{P}(D)$. Based on $\mathcal{T}$ the set D_q can be constructed in $O(|D_q|)$ runtime as $D_q = \bigcup_{k \in F_q} \mathcal{T}(k)$. In most practical applications $O(|D_q|)$ is bound by a small constant since $|D_q| \ll |D|$; the cost of a hash table lookup is assessed with $O(1)$. Many fingerprinting approaches are described in the literature, which mainly differ in the design of h_φ. For a comparative overview see [5].

A similarity hash-function h_φ produces with a high probability a hash collision for two feature vectors $\mathbf{q}$, $\mathbf{d}$, iff $\varphi(\mathbf{q}, \mathbf{d}) \geq 1 - \epsilon$, with $0 < \epsilon \ll 1$. This is illustrated by an empirical analysis of different fingerprinting approaches in a monolingual high similarity search scenario, see Figure 2. All approaches achieve reasonable precision and recall at high similarities (~ 0.9). As shown in Section 1) the similarity values

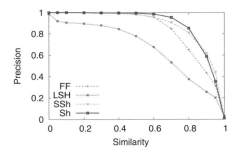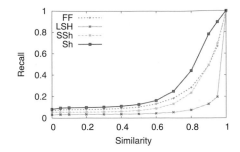

Fig. 2. Precision and recall over similarity for fuzzy-fingerprinting, FF, locality-sensitive hashing, LSH, supershingling, SSh, and shingling, Sh [5]

of cross-language near-duplicates are on average 0.5 (see Figure 1 where the recall of hash-based search drops dramatically. Hence, fingerprinting is not applicable to tackle cross-language high similarity search.

4 Brute Force Indexing

Web search engines solve the task of high similarity search very efficiently by indexing the collection D based on n-grams. Their "brute force n-gram indexing" strategy can be interpreted as a special case of fingerprinting if a query is considered as single n-gram with a reasonable large n, i.e. $n \in [5; 15]$. An example is the phrasal search functionality of a Web search engine. E.g. a Google query that is set into quotation marks is treated as a single n-gram. Consider a string-based hash function h, like MD5 or Rabin's hash function, that maps an n-gram onto a single hash code. The fingerprint F_d of a document d can be defined as $F_d = \bigcup_{c \in N_d} h(c)$, where N_d denotes the set of all n-grams of d. As described above, the mapping between hash codes and documents can be organized as a hash table $\mathcal{T}$. Since the query q is assumed to be a single n-gram, i.e. $|N_q| = 1$, the set D_q can be constructed as $D_q = \mathcal{T}(h(q))$. The runtime corresponds to a single hash table lookup, which is assessed with $O(1)$.

However, brute force n-gram indexing is not applicable to cross-language high similarity search since the hash codes $h(q')$ and $h(d)$ of a query q' in language L' and some document $d \in D$ in language L are not comparable.

5 Conclusion

For cross-language high similarity search no sub-linear time bound can be expected. We argued in this paper why—in contrast to monolingual high similarity search—neither fingerprinting nor brute force n-gram indexing can be used to model cross-language similarities that are close to 1. In our current research we use the LSH framework of Modwani to derive theoretical performance bounds for cross-language fingerprinting.

References

1. Anderka, M., Stein, B.: The ESA Retrieval Model Revisited. In: Proc. of SIGIR 2009 (2009)
2. Bayardo, R.J., Ma, Y., Srikant, R.: Scaling Up All Pairs Similarity Search. In: Proc. of WWW 2007 (2007)
3. Datar, M., Immorlica, N., Indyk, P., Mirrokni, V.S.: Locality-Sensitive Hashing Scheme Based on p-Stable Distributions. In: Proc. of SCG 2004 (2004)
4. Lin, J.: Brute Force and Indexed Approaches to Pairwise Document Similarity Comparisons with MapReduce. In: Proc. of SIGIR 2009 (2009)
5. Potthast, M., Stein, B.: New Issues in Near-duplicate Detection. In: Data Analysis, Machine Learning and Applications (2008)
6. Potthast, M., Stein, B., Anderka, M.: A Wikipedia-Based Multilingual Retrieval Model. In: Macdonald, C., Ounis, I., Plachouras, V., Ruthven, I., White, R.W. (eds.) ECIR 2008. LNCS, vol. 4956, pp. 522–530. Springer, Heidelberg (2008)
7. Weber, R., Schek, H.-J., Blott, S.: A Quantitative Analysis and Performance Study for Similarity-Search Methods in High-Dimensional Spaces. In: Proc. of VLDB 1998 (1998)

Exploiting Result Consistency to Select Query Expansions for Spoken Content Retrieval

Stevan Rudinac, Martha Larson, and Alan Hanjalic

Multimedia Information Retrieval Lab, Delft University of Technology
Mekelweg 4, 2628 CD Delft, The Netherlands
{s.rudinac,m.a.larson,a.hanjalic}@tudelft.nl

Abstract. We propose a technique that predicts both if and how expansion should be applied to individual queries. The prediction is made on the basis of the topical consistency of the top results of the initial results lists returned by the unexpanded query and several query expansion alternatives. We use the coherence score, known to capture the tightness of topical clustering structure, and also propose two simplified coherence indicators. We test our technique in a spoken content retrieval task, with the intention of helping to control the effects of speech recognition errors. Experiments use 46 semantic-theme-based queries defined by VideoCLEF 2009 over the TRECVid 2007 and 2008 video data sets. Our indicators make the best choice roughly 50% of the time. However, since they predict the right query expansion in critical cases, overall MAP improves. The approach is computationally lightweight and requires no training data.

Keywords: Speech-transcript-based video retrieval, query expansion, query performance prediction, list coherence, source selection.

1 Introduction

Query prediction makes possible effective use of query expansion [1,2]. We introduce a technique that decides whether a query should be expanded, and if so, which of several alternative expansion methods should be used. Our technique uses lightweight indicators that capture the topical consistency of the top documents in the results list.

Our use of query expansion is aimed at compensating for vocabulary mismatch between query and document in a spoken content retrieval task. Speech recognition transcripts are invaluable for retrieving spoken content, but domains dominated by unscripted or spontaneously produced speech continue to pose challenges [3,4]. In such domains, vocabulary mismatch is a substantial issue due to, first, elevated error rates and, second, more varied, less-focused vocabularies used in informal speech.

We address the mismatch issue by using query expansion, and deploy query-by-query prediction to control the risk of expansion introducing noise. Specifically, we assume that each query expansion produces a results list with a somewhat different topical consistency. We conjecture that the relative tightness of the topical clustering structure of top-ranked documents is a useful indicator of the noise-level and thereby the utility of a given results list. We use top-ranked documents in order to ensure that topical consistency is measured over a result-set with a maximal possible proportion of relevant items.

C. Gurrin et al. (Eds.): ECIR 2010, LNCS 5993, pp. 645–648, 2010.

Our proposal shares affinity with work on query prediction, such as [2,5], but also with work combining multiple results lists to improve spoken content retrieval [6]. The novelty of our technique lies in the use of lightweight indicators of document set topical structure that are based on pair-wise similarity among the top-N retrieved documents for the task of query prediction. We aim at robustness to parameter settings, which has been identified as an issue with query prediction approaches such as Clarity [2]. We investigate the coherence score introduced by [7] to capture the tightness of topical clustering structure as well as our own simplified version.

2   Algorithm

First, we experiment with a *coherence indicator* that selects the results list with the highest coherence among the top-N ranked documents, as expressed by the coherence score:

$$Co(TopN) = \frac{\sum_{i \neq j \in \{1,...,N\}} \delta(d_i, d_j)}{\frac{1}{2} N(N-1)} \tag{1}$$

Here, δ is a function defined on a pair of documents (d_i, d_j), which is equal to 1 if their similarity is higher than a pre-defined threshold θ and 0 otherwise. The threshold is set by measuring the pair-wise similarity value of particularly close pairs in the overall document collection. We chose the conventionally-used cosine similarity as our similarity metric, but other choices are also possible. Documents are represented as vectors of tf-idf weights. Particularly close pairs are pairs that are closer than an unusually large percent (in our case 90%) of other pairs in the collection.

Second, we experiment with a *max-AIS* and *mean-AIS*, two indicators that select the query expansion producing a results list in which top-N documents are characterized by high average item similarities (AIS) with their fellows.

$$AIS_{d_i} = \frac{\sum_{d_j \in TopN : d_j \neq d_i} sim(d_j, d_i)}{N - 1} \tag{2}$$

AIS is a variant of the average item distance (AID) used in [8] to identify representative documents in a results list. Max-AIS takes the highest AIS of any document in the top-N. Mean-AIS averages the AIS of the top-N documents. Again with the AIS-based indicators, we use cosine similarity. Exploratory experiments with an alternate similarity metric (vocabulary overlap) yielded similar results.

The rationale behind these indicators is that they distinguish between different lists by measuring the relative difference in the topical consistency of top-document sets without making reference to the background collection. They are appropriate for comparing results lists generated using different search strategies, which will be sufficiently similar in composition, due to their query bias.

3   Experimental Setup and Results

We perform experiments on the TRECVid 2007 and 2008 data sets, using as queries the 46 semantic theme labels accompanying this data by the VideoCLEF 2009 benchmark evaluation. Note that this task is video level retrieval and not shot level retrieval, so the document sets are quite small, but adequate for our purpose of dem-onstrating the viability of the method. We index the Dutch-language speech tran-scripts using the Lemur toolkit. The speech recognition word error rate is approxi-mately 60% [4]. Preprocessing involves stemming and stopword removal. We apply a stopword list that extends the Snowball list with additional words selected using a frequency-based heuristic. Indexation of the English translations of the transcripts is carried out in a similar manner.

For each query, we apply the indicators to four results lists, generated by the base-line (unexpanded) query and three query expansion methods. The queries are ex-panded via 1. Pseudo-relevance feedback (PRF), 2. WordNet[1] and 3. Google Sets[2]. PRF is performed using the Dutch-language index. For the WordNet expansion, we translate the query into English, expanding each with all of its synonyms, and perform retrieval using the English-language index. For the Google Sets expansion, we limit the number of expansion items (words or multiword phrases) to a maximum of 15. We choose the expansion method for which the indicator returns a maximum value, breaking ties with the baseline in favor of the baseline and otherwise randomly.

For our experiments, we first tested TRECVid 2007, performing exploratory ex-periments for tasks like setting θ for the coherence score and defining stopwords. Then, we tested our methods on TRECVid 2008 in order to determine how the results would extend without adaptation to unseen data. Results for individual query expan-sion methods in Mean Average Precision (MAP) are given in Table 1.

Table 1. MAP of the baseline and the query expansion methods used

	Baseline	PRF	WordNet	GoogleSets
TRECVid 2007	0.326	0.332	0.260	0.120
TRECVid 2008	0.245	0.265	0.268	0.142

Results for the three indicators that we used for selection of the query expansion method are reported in Table 2. Statistically significant improvement over the base-line is indicated with '^' (Wilcoxon Signed Rank test, p=0.05).

Table 2. MAP resulting from applying query expansion selection (for a range of Top-N values)

	TRECVid 2007					TRECVid 2008			
	N=5	N=10	N=15	N=20		N=5	N=10	N=15	N=20
Co	**0.358**	0.334	0.333	0.329	Co	0.300^	0.300^	0.316^	0.304^
Max-AIS	0.332	**0.352**^	0.344	0.335	Max-AIS	**0.317**^	0.328^	0.329^	0.328^
Mean-AIS	0.326	0.350^	**0.347**	**0.336**	Mean-AIS	0.306	**0.330**^	**0.330**^	**0.329**^

The results demonstrate that the ability of our indicators to select the query expan-sion methods is sufficient to achieve a modest, yet consistent overall improvement of MAP. The strongest performer is the simplified coherence indicator Mean-AIS.

---

[1] http://wordnet.princeton.edu/
[2] http://labs.google.com/sets

4 Discussion and Conclusion

We have proposed a lightweight approach to selection of results lists generated by query expansions that is computationally simple and requires no training data. Experiments on a small corpus of video data confirmed its potential to improve spoken content retrieval, which suggests that it makes a contribution to our goal of transcript error compensation. Our approach is capable of delivering improvement and the indicators based on simplified coherence (AIS) are competitive with coherence. The indicators are relatively robust across parameter settings and can be transferred to a new data set without adaptation or adjustment.

An oracle indicator that always chooses the correct strategy would have achieved a MAP of 0.392 on TRECVid 2007 and of 0.382 on TRECVid 2008. Although our indicators fall far short of this theoretical optimum, the results in Table 2 confirm the viability of our technique for query expansion selection. A spot check of query-by-query behavior reveals that our indicators choose the correct expansion method ca. 50% of the time. However, they do seem to be able to predict the right query expansion in critical cases, i.e., where expansion improves query AP substantially. Further, we have found evidence that our indicators make use of all query expansion methods. The GoogleSets method has a low overall MAP, but there are queries for which it does help and our indicators appear capable of predicting these cases.

Future work will investigate the possibilities for computing the proposed indicators on non-textual results lists and for working towards multi-modal prediction of query expansion performance in order to further improve the overall MAP when retrieving information from multi-modal data sets.

Acknowledgments. The research leading to these results has received funding from the European Commission's 7th Framework Programme (FP7) under grant agreement n° 216444 (EU PetaMedia Network of Excellence).

References

1. Yom-Tov, E., Fine, S., Carmel, D., Darlow, A.: Learning to estimate query difficulty. In: SIGIR 2005, pp. 512–519 (2005)
2. Hauff, C., Murdock, V., Yates, R.B.: Improved query difficulty prediction for the web. In: CIKM 2008, pp. 439–448 (2008)
3. Byrne, W., et al.: Automatic recognition of spontaneous speech for access to multilingual oral history archives. IEEE Trans. SAP 12(4), 420–435 (2004)
4. Huijbregts, M., Ordelman, R., de Jong, F.: Annotation of heterogeneous multimedia content using automatic speech recognition. In: Falcidieno, B., Spagnuolo, M., Avrithis, Y., Kompatsiaris, I., Buitelaar, P. (eds.) SAMT 2007. LNCS, vol. 4816, pp. 78–90. Springer, Heidelberg (2007)
5. Cronen-Townsend, S., Zhou, Y., Croft, W.B.: Predicting query performance. In: SIGIR 2002, pp. 299–306 (2002)
6. Olsson, J.S., Oard, D.W.: Combining Speech Retrieval Results with Generalized Additive Models. In: ACL 2008: HLT, pp. 461–469 (2008)
7. He, J., Weerkamp, W., Larson, M., de Rijke, M.: An effective coherence measure to determine topical consistency in user-generated content. International Journal on Document Analysis and Recognition 12(3), 185–203 (2009)
8. Rudinac, S., Larson, M., Hanjalic, A.: Exploiting visual reranking to improve pseudo-relevance feedback for spoken-content-based video retrieval. In: WIAMIS 2009, pp. 17–20 (2009)

Statistics of Online User-Generated Short Documents

Giacomo Inches, Mark J. Carman, and Fabio Crestani

Faculty of Informatics, University of Lugano, Lugano, Switzerland
{giacomo.inches, mark.carman, fabio.crestani}@usi.ch

Abstract. User-generated short documents assume an important role in online communication due to the established utilization of social networks and real-time text messaging on the Internet. In this paper we compare the statistics of different online user-generated datasets and traditional TREC collections, investigating their similarities and differences. Our results support the applicability of traditional techniques also to user-generated short documents albeit with proper preprocessing.

1   Introduction and Motivations

User-generated short documents are those produced online by visitors of blog or social networking websites, as well as by users of chat or instant messaging programs. The increasing popularity of these online services (Twitter, Facebook, IRC, MySpace) makes such generated content of great interest. From a commercial point of view, the analysis of these documents can highlight useful trends to focus online advertisement while, from a policing perspective, it may allow us to detect misbehavior or harassment.

While short documents and user-generated documents have been treated in recent works with different purposes (author identification, language analysis, gender prediction, documents clustering or law enforcement) to the best of our knowledge this is the first work which aims at understanding the differences between these and more traditional online collections, like the TREC "Ad-hoc" datasets.

In fact, it is important to asses the nature of similarity between user-generated short documents and more consolidated collections, to be able to infer the applicability of standard measures (of distance or similarity: BM25, Kullback-Leibler divergence, cosine with TFIDF weighting, etc.) and techniques (Probabilistic, Language or Topic Models) or to develop new ones which fit better the new datasets. For this purpose, we present the first results of our ongoing work, where we studied selected properties of 4 user-generated short document datasets and 3 more traditional ones, taken from the TREC Ad-hoc collections.

C. Gurrin et al. (Eds.): ECIR 2010, LNCS 5993, pp. 649–652, 2010.

2   Datasets

As representative of user-generated short documents, we used the training dataset[1] presented at the Workshop for Content Analysis in Web 2.0 [2]. This consists of 5 distinct collections of documents crawled from 5 different online sources: *Ciao* (a movie rating service), *Kongregate* (Internet Relay Chat of online gamers), *Twitter* (short messages), *Myspace* (forum discussions) and *Slashdot* (comments on news-posts). Since we were more interested in messages exchange between users, we did not consider the *Ciao* collection and left it to future study.

We compared these datasets with a subset of the standard TREC Ad-hoc collections[2], choosing 3 of the most representative ones: *Associated Press* (AP, all years), *Financial Times Limited* (FT, all years), *Wall Street Journal* (WSJ, all years). These collections contain news article (*AP* and *WSJ*: general news, *FT*: markets and finance) published in the corresponding newspapers.

We discuss in next section the properties of these collections, which are pre-sented in Table 1. Since the statistics of AP, WSJ and FT are similar to one anothers, we report only the values for the *WSJ* dataset.

Table 1. Statistics of collections (all values before stopwords removal unless indicated)

Collection	Collection size (# doc)	Avg. Doc. length (# word)	Vocabulary	% words			Slope $|\alpha|$
				stopwords	out-of-dictionary	singleton	
Kongregate	144,161	4.449	35,208	44.90	58.94	56.65	1.69
Twitter	977,569	13.989	364,367	44.99	68.37	66.95	1.54
Myspace	144,161	38.077	187,050	50.67	69.61	53.30	1.92
Slashdot	141,283	98.912	123,359	54.00	57.31	44.82	2.17
WSJ	173,252	452.005	226,469	41.45	67.57	34.33	2.70

3   Comparative Analysis of Datasets

We started our study by checking the average length of the documents present in each collection. We found that these user-generated documents are 5 to 100 times shorter then the ones in the traditional TREC collections (Table 1) and we performed a double analysis. First we indexed the documents without removing any stopwords, then we used a standard stopwords list to clean them.

We expected less terms to be discarded as stopwords, since we assume short documents (in particular the ones used to "chat" as *Twitter* or *Kongregate*) to be written "quicker and dirtier", with no care for orthography and using a lot of abbreviations. We found a proof of this when looking at the percentage of terms which occurred only once in the collection ("singleton terms"): short

---

[1] Dataset and details available at `http://caw2.barcelonamedia.org/`
[2] Datasets and details available at `http://trec.nist.gov/data/test_coll.html`

documents contain more singleton terms, which we can consider as spelling mistakes or mistyped words. This is more evident when we look at out-of-dictionary terms. These words are not contained in a standard dictionary and are identified as misspelled by a spell checker. Although the percentage of out-of-dictionary terms is similar across all datasets, we notice that for short documents collections this value is closer to the number of singleton words (from 2% to 16%), while for traditional TREC collections the distance is further (33%). This fact may indicate that in the short documents collections the presence of more singleton words could be considered as an indicator of a greater number of mistyped words. This is not the case of the traditional TREC collections, where the presence of singleton words is less evident and can be explained by the usage of particular terms such as geographical locations, foreign words or first names which are orthographically correct but not present in the spell checker used.

After this initial analysis, we took inspiration from the work of Serrano et al. [3] to investigate in more details the words distribution for each collection. We concentrate our study on two measures, the slope of the Zipf-Mandelbrot distribution and the vocabulary growth, also known as Heaps' law.

The Zipf-Mandelbrot law can be written as follows [1]:

$$\log f(w) = \log C - \alpha \ \log (r(w) - b) \qquad (1)$$

where $f(w)$ denotes the frequency of a word w in the collection and $r(w)$ is the ranking of the word (in terms of its frequency), while C and b are collection specific parameters. As can be seen in Fig. 1 (left), in a log-log scale and for large values of $r(w)$, the relationship between frequency and rank of a word can be approximated with a descending straight line of slope $-\alpha$. Values for the slope α are given in Table 1 and have been calculated with χ^2 metric [4]. As expected a linear graph is observed also for short documents. Moreover we noticed a dependence between the length of the documents and the slope: the collections containing longer documents tend to have a larger negative slope, which may mean that the words in them are repeated more frequently, while the collections containing shorter documents are less repetitive (as stated previously).

Fig. 1 (right) shows the vocabulary growth with respect to the size of the whole collection. The vocabulary of user-generated short documents grows much faster in comparison with that of longer documents. This means that the conversation between users (in *Kongregate* and *Twitter*) tends to vary greatly with the usage of ever more terms. This may be in part explained by the high percentage of singleton and out-of-dictionary mistyped words or abbreviation that are continuously introduced during the dialog. We also noticed a relationship between the decreasing value of the slopes of the Zipf's law and the growth of the vocabulary: *Twitter* has the minimum slope but the maximum vocabulary growth, to the contrary *WSJ* has the maximum slope and the minimum vocabulary growth. This could, again, be explained by the high frequency of mistyped terms in the vocabulary of user-generated short documents in comparison to the standard TREC documents.

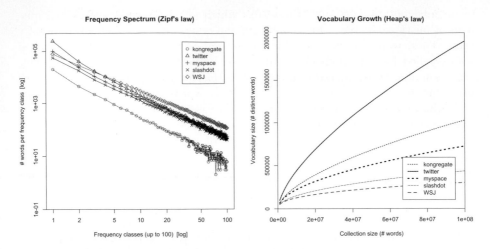

Fig. 1. Zipf's law (left) and Vocabulary Growth (right). We display only graphs after stopword removal (similar to others before).

4   Conclusions and Future Work

In this work we compared user-generated short documents and standard online datasets over an initial set of properties. We were able to identify the "messy" properties of user-generated short documents, which need therefore to be pre-processed before being treated with standard techniques. These seem to be easily applicable given the Zipf nature of the short documents. In the future we would like to compare user-generated short documents with a dictionary of common online abbreviation as well as mistyped words and to enlarge the number of collections analyzed (by adding *Ciao*, Blog or Tripadvisor collections). We would also like to study other measures (such as term distribution similarity and burstiness [3]) and to investigate further the differences between *discussion-style* and *chat-style* text content [5] that we noticed but did not discuss in here.

References

1. Manning, C.D., Schütze, H.: Foundations of Statistical Natural Language Processing. MIT Press, Cambridge (1999)
2. Codina, J., Kaltenbrunner, A., Grivolla, J., Banchs, R.E., Baeza-Yates, R.: Content analysis in web 2.0. In: 18th International World Wide Web Conference (2009)
3. Serrano, M., Flammini, A., Menczer, F.: Modeling statistical properties of written text. PLoS ONE 4(4), e5372 (2009)
4. Evert, S., Baroni, M.: zipfR: Statistical models for word frequency distributions. R package version 0.6-5 (2008)
5. Yin, D., Xue, Z., Hong, L., Davison, B.D., Kontostathis, A., Edwards, L.: Detection of harassment on web 2.0. In: CAW 2.0 2009: Proceedings of the 1st Content Analysis in Web 2.0 Workshop, Madrid, Spain (2009)

Mining Neighbors' Topicality to Better Control Authority Flow

Na Dai, Brian D. Davison, and Yaoshuang Wang

Department of Computer Science & Engineering, Lehigh University, USA
{nad207,davison,yaw206}@cse.lehigh.edu

Abstract. Web pages are often recognized by others through contexts. These contexts determine how linked pages influence and interact with each other. When differentiating such interactions, the authority of web pages can be better estimated by controlling the authority flows among pages. In this work, we determine the authority distribution by examining the topicality relationship between associated pages. In addition, we find it is not enough to quantify the influence of authority propagation from only one type of neighbor, such as parent pages in PageRank algorithm, since web pages, like people, are influenced by diverse types of neighbors within the same network. We propose a probabilistic method to model authority flows from different sources of neighbor pages. In this way, we distinguish page authority interaction by incorporating the topical context and the relationship between associated pages. Experiments on the 2003 and 2004 TREC Web Tracks demonstrate that this approach outperforms other competitive topical ranking models and produces a more than 10% improvement over PageRank on the quality of top 10 search results. When increasing the types of incorporated neighbor sources, the performance shows stable improvements.

1 Introduction

People inherit reputations from other members in their social network. For example, we say that a professor is an expert in artificial intelligence (AI), it is probable that her advisees get some prestige in the field of AI from her. However, if one of her advisees is an expert in system and computer network rather than AI, the chance that he inherits authority from his advisor is much less than other advisees who are well-known for AI, since the authorities of advisor and advisees within the same field are more likely to influence each other. In addition, the reputation of a researcher can come from other sources besides the advisor. For instance, collaborating with other famous researchers can improve one's reputation; advising a good student who becomes famous afterwards can also improve one's reputation.

Thus, one may think that one's reputation is aggregated from a comprehensive environment, and the more similar people are the greater the influence on one's reputation. These intuitions are also applicable to the interconnected pages on the web, especially benefiting the estimation of web page authority. However, traditional link analysis algorithms seldom consider these two issues at the same time. Algorithms such as PageRank and HITS assume that the authority flow among pages is equally distributed following limited directions. The authority propagation is independent of the contexts

C. Gurrin et al. (Eds.): ECIR 2010, LNCS 5993, pp. 653–657, 2010.
© Springer-Verlag Berlin Heidelberg 2010

correlating to the pages within limited relationships. Although topical ranking models [2,5] differentiate the page authority within diverse topical domains/communities, none of them directly compared the topicality between two linked pages in authority propagation.

Nie and Davison [4]'s Heterogeneous Topical Rank incorporates such topical context comparison in fine-grained local authority flows within pages. However, the direction of authority flows is only from source to target pages. Other approaches using relevance propagation use neighboring nodes' relevance score to influence the current node for given queries or perform content diffusion among pages to achieve better representation. Qin et al. [7] proposed a generic relevance propagation framework and did a thorough comparison study in this field. Shakery and Zhai [9] proposed a probabilistic framework which propagates page relevance scores for a given query through page inlink and outlinks. However, the relevance propagation sometimes has to proceed online, potentially making the computation complexity an issue. Their follow-up work [10] solved this problem by propagating terms though links and reducing the number of neighbors involved in the process.

In this work, we aim at better estimation of page authority based on the two assumptions proposed above. We 1) propose a model to propagate page authority based on the topicality concealed in the contexts among pages within diverse neighboring relationships; and 2) verify the superiority of our proposed model empirically by comparing it with many well-known algorithms.

2   Model

We extend traditional link analysis by allowing authority to flow among all possible neighboring pages, such that closely connected pages can directly influence the authorities of each other. The neighbor relationship is inferred from link structure in this work. Authorities from parent pages, children pages and sibling pages directly contribute to the current page. We also consider all pages on the web as a special type of neighboring pages of each individual since one is usually influenced by one's background. While the definition of neighboring pages isn't limited to the relationship inferred from link structure, we in this work focus on these four types of neighboring relationships described above. From another perspective, one page contributes its authority to all its neighbor pages on the web.

Once we determine where and how to distribute authority, the next step is to mine contexts between pages and use them to control authority flows. In traditional ranking models [2,5], a parent equally splits its authority to all its children. We hypothesize that the authority flow distribution should prefer more similar neighbors, hence we incorporate a preference factor α to control the authority that flows to a specific type of neighbor. The preference factor α is determined based on the relative similarity between the current page to this type of neighbor. In particular, we simplify the representation of page topicality by using a taxonomy from Open Directory Project (ODP) [6] (we select its 12 top-level categories), and use a the Rainbow Naive Bayes text classifier [3], trained on 19,000 pages from each category. Thus each page can be represented by a 12-dimensional topical distribution, demonstrating the probability that the page belongs to each topic. We then use cosine similarity to represent how much the current page and its one neighbor are similar, which is given by: $rel(p_1, p_2) = \frac{\sum_i T_i(p_1)T_i(p_2)}{|T(p_1)||T(p_2)|}$, where p_1 is p_2's neighbor, and $T(p_i)$ is the topical distribution of p_i. Since we distinguish the roles played by different types of neighbors, we use the average cosine similarity of the topicality between the current page and all neighbors within the same type to reflect how a page is similar to its specific type of neighbor. We use the centroid of topical distributions of all pages as the background distribution, and calculate the similarity between each page to this background as its similarity to the background neighbors. Specifically, we define the preference factor α with respect to each type as: $\alpha_i = \frac{avg\_rel(p_c, p_i)}{\sum_i avg\_rel(p_c, p_i)}$, where $i \in \{\text{Parent}(p_c), \text{Children}(p_c), \text{Sibling}(p_c), \text{Background}(p_c)\}$ and $avg\_rel(p_c, p_i)$ is the average similarity between the current page and the specified type of neighbor.

Once we determine the fraction of authority flowing to a specific type of neighbor, we next split the authority among all pages within the same type. To avoid an over-bias toward topical affinity, we distribute it uniformly within the neighbor type. When combining with the type-based preference factor, this step actually smoothes the authority propagation from the last step. We formalize the authority propagation model in Equation 1.

$$A^{(t+1)}(p_c) = \sum_i \alpha_i \left(\sum_{j : j \in i} \frac{A^{(t)}(p_j)}{O_i(p_j)} \right) \tag{1}$$

where $A(p_j)$ is the authority score and $O_i(p_j)$ is the number of links from the ith type of neighbor to page p_j. This iterative process converges and results in a static probability distribution over all pages on the web. We then order pages by this score, and linearly combine the ranks with those from query-specific IR scores based on page content.

3   Experimental Results

Our goal is to use the proposed model to better estimate page authority and improve web search quality. Our dataset is the TREC (http://trec.nist.gov/) GOV collection, which is a 2002 crawl in .gov domain, containing around 1.25 million web pages. We used the topic distillation tasks in TREC web tracks in 2003 (50 queries) and 2004 (75 queries) to evaluate our approach. We compare our approach with Okapi BM25 [8], PageRank (PR), GlobalHITS (GHITS), Topical-Sensitive PageRank (TSPR) [2], and Topical PageRank (TPR) [5]. All these link-based ranking models combine with Okapi BM25 linearly by ranks. The parameters used in Okapi BM25 are set to be the same as Cai et al. [1]. We show the performance comparison at the best rank combination of query-specific scores and authority scores. We use precision@10, Rprec, MAP, and NDCG@10 to measure ranking quality.

Table 1 shows the performance comparison on TREC 2003 and 2004. On TREC 2003, our APR models outperform all other approaches on all metrics. The authority propagation from children to parents contribute more than that among siblings. But when combining all neighboring resources, the performance has significant improvement over the ones by propagating among only subsets of neighbor pages. Single-tailed

Table 1. AuthorityPropagationRank and baselines performance

	TREC 2003				TREC 2004			
	P@10	MAP	R-prec	NDCG@10	P@10	MAP	R-prec	NDCG@10
BM25	0.1200	0.1485	0.1398	0.1994	0.1907	0.1364	0.1769	0.2330
PR	0.1380	0.1538	0.1621	0.2197	0.2267	0.1523	0.1844	0.2790
GHITS	0.1360	0.1453	0.1574	0.2031	0.2147	0.1483	0.1758	0.2669
TSPR	0.1420	0.1718	0.1779	0.2414	0.1907	0.1429	0.1769	0.2390
TPR	0.1440	0.1691	0.1864	0.2301	0.2173	0.1532	0.1832	0.2714
APR(P+C+B)	0.1480	**0.1761**	0.1864	0.2281	0.2267	0.1661	0.1895	0.2874
APR(P+S+B)	0.1420	0.1518	0.1808	0.2127	0.2253	0.1557	0.1885	0.2806
APR(P+S+C+B)	**0.1540**	0.1752	**0.1922**	**0.2342**	**0.2333**	**0.1664**	**0.1941**	**0.2957**

student t-tests over P@10 and NDCG@10 show that the APR models significantly out-perform PageRank and BM25 (p-value<0.01) at 95% confidence level.

On TREC 2004, our APR models show stable improvements over baselines. Even under the inconsistent performance on PageRank, the APR models show significant improvement over BM25 (p-value<0.0001), PageRank (p-value<0.03), TPR (p-value< 0.025), and PR (p-value<0.05) on P@10 and NDCG@10 at 95% confidence level.

4   Conclusion

We proposed an authority propagation model which propagates authorities among diverse types of neighboring pages based on topical affinity. Experimental results demonstrate its superiority on ranking performance over several representative link analysis algorithms. As future work, we expect to (1) generalize the concepts of neighboring nodes; (2) test the influence from the granularity of the taxonomy on ranking performance; and (3) investigate the sensitivity of the ways of interpreting the topicality context between pages with respect to ranking performance.

References

1. Cai, D., He, X., Wen, J.-R., Ma, W.-Y.: Block-level link analysis. In: Proc. 27th Annual Int'l ACM SIGIR Conf. on Research and Dev. in Information Retrieval (July 2004)
2. Haveliwala, T.H.: Topic-sensitive PageRank. In: Proc. of the 11th Int'l World Wide Web Conf., pp. 517–526. ACM Press, New York (2002)
3. McCallum, A.K.: Bow: A toolkit for statistical language modeling, text retrieval, classification and clustering (1996), http://www.cs.cmu.edu/~mccallum/bow
4. Nie, L., Davison, B.D.: Separate and inequal: Preserving heterogeneity in topical authority flows. In: Proc. 31st Annual Int'l ACM SIGIR Conf. on Research and Dev. in Information Retrieval, July 2008, pp. 443–450 (2008)
5. Nie, L., Davison, B.D., Qi, X.: Topical link analysis for web search. In: Proc. 29th Annual Int'l ACM SIGIR Conf. on Research & Dev. in Info. Retrieval, August 2006, pp. 91–98 (2006)
6. The dmoz Open Directory Project, ODP (2009), http://www.dmoz.org/

7. Qin, T., Liu, T.-Y., Zhang, X.-D., Chen, Z., Ma, W.-Y.: A study of relevance propagation for web search. In: Proc. 28th Annual Int'l ACM SIGIR Conf. on Research and Dev. in Information Retrieval, pp. 408–415 (2005)
8. Robertson, S.E.: Overview of the OKAPI projects. Journal of Documentation 53, 3–7 (1997)
9. Shakery, A., Zhai, C.: A probabilistic relevance propagation model for hypertext retrieval. In: Proc. of the 15th ACM Int'l Conf. on Information and Knowledge Management (CIKM), pp. 550–558 (2006)
10. Shakery, A., Zhai, C.: Smoothing document language models with probabilistic term count propagation. Inf. Retr. 11(2), 139–164 (2008)

Finding Wormholes with Flickr Geotags

Maarten Clements[1], Pavel Serdyukov[1],
Arjen P. de Vries[1,2], and Marcel J.T. Reinders[1]

[1] Delft University of Technology, Delft, The Netherlands
[2] CWI, Amsterdam, The Netherlands
m.clements@tudelft.nl

Abstract. We propose a kernel convolution method to predict similar
locations (wormholes) based on human travel behaviour. A scaling parameter can be used to define a set of relevant users to the target location
and we show how the geotags of these users can effectively be aggregated
to predict a ranking of similar locations. We evaluate results on world
and city level using several independent test collections.

1 Introduction

We define *wormholes* as similar, but not necessarily spatially close locations
on the planet. We hypothesize that users have a specific travel preference and
therefore visit locations that are to some extend similar. Furthermore, making
a photo at a visited location is an indication that the user likes that location.
Based on these hypotheses, the aggregated travel data of many users should be
able to reveal which locations are most similar to a given query location. In
photo sharing websites like Flickr[1], users can indicate the geographical location
of their pictures by placing them on a world map. We propose a method, similar
to neighborhood based collaborative filtering, to combine the users' *geotags* in a
prediction for similar locations.

The exploitation of geotags has shown to be effective for various tasks: global
event detection [1], mapping of popular tags and photos to geographical locations
[2,3], finding prominent landmarks and representative photos [4]. Furthermore,
several methods have been proposed to predict the geotags of a photo, based on
its textual tags [5], visual information [4] and individual user travel patterns [6].

As far as we know we propose a first attempt to predict similar locations
based on geotags alone. Textual tagging will always require manual effort and
cameras with GPS functionality will become mainstream in the coming years.
Therefore we believe that future data collections will contain more geotag data
than manual annotations, and data analysis will rely more strongly on geotags.
The contributions of this work can be summarized as follows:

1. We propose a weighing scheme to estimate the relevancy of a user to a given
   location at various scales.
2. We compare several methods to aggregate user information in a way that
   accurately predicts similar locations.

---

[1] http://www.flickr.com

C. Gurrin et al. (Eds.): ECIR 2010, LNCS 5993, pp. 658–661, 2010.

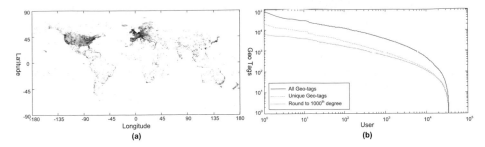

Fig. 1. a) A 2D histogram of the data clearly shows the most developed areas in the world. **b)** The distribution of the number of geotags per user. The large gaps between all geotags, unique geotags and unique geotags when rounded to a 1000th degree (max. 111 meter) shows that the data is strongly clustered.

3. We propose an evaluation technique for similar location prediction, based on an independent test set.

Using the public API of Flickr we have collected the top-100 most popular localities (cities, parks, etc.) for each day in 2008. The aggregated data contains 8,643 places including the number of photos geotagged in 2008 by all Flickr users. To retrieve the geotagged data, we now repeatedly follow the following procedure:

1. Select a location l from the full distribution, with the probability relative to the global popularity in 2008.
2. Get a photo i_l from this location.
3. Get all the photos from the user who made i_l.

Following this strategy, we have collected the geotags of 36,264 users. Together these users have uploaded 52,425,279 photos of which 22,710,496 have been geotagged, see Figure 1. This set contains over 20% of all public geotags available in Flickr in October 2009[2].

2   Wormhole Detection

From a given target location L we want to find the most similar locations around the world. For each user u, a weight $W_{L,u}$ is computed based on the distance of the nearest geotagged photo of the user to the target location, weighted by a normal distribution:

$$W_{L,u} = \exp\left(-\frac{\min_i(d(L, G_{u,i}))^2}{\sigma^2}\right), \tag{1}$$

where standard deviation σ is used as a scaling parameter and $d(L, G_{u,i})$ computes the euclidean distance between the ith geotag of a user $G_{u,i}$ and L. The

---

[2] According to: http://www.flickr.com/map

scale parameter σ can be compared to the number of users selected in neighbourhood based collaborative filtering algorithms, as it determines which users are similar enough to contribute to the prediction.

The wormholes from L are now derived by creating a 2000x4000 histogram H_L of all users' geotags, using $W_{L,u}$ as weight per user. For each geotag in the collection we add the weight of the corresponding user to the respective bin. The choice of grid size results in cells that are at most 10x10km (around the equator).

Using a grid is prone to errors when locations close to one of the cell boundaries are considered. To transfer the weight of a single grid cell to the neighbour cells, we perform a kernel convolution of the histogram with a Gaussian kernel (with the same σ as used in Equation 1). The difference between the resulting profile and a distribution based on all users ($W_{L,u} = 1$ for all u and convolution with same σ) gives a score that indicates the relevance of each position on earth with respect to the target location L.

We evaluate this method for increasing values of σ and 3 different aggregation methods:

1. N1: Add all the user's geotags with $W_{L,u}$ to the histogram.
2. N2: Normalize per user by dividing the weight by the number of photos of each user: $W_{L,u}/|I_u|$.
3. N3: Limit the contribution of each user to only one photo per histogram bin.

3   Results for Some Mountains, a Beach and a Cemetery

To evaluate our wormhole prediction method, we collect 156 test mountains from Wikipedia[3]. We use the 5 highest mountains per continent (excluding Antarctica and Oceania) as starting locations. We now evaluate the result by taking the top ranked grid cell and count it as positive if one of the test mountains is found within a radius of 3 cells around that cell. For mountain prediction this can be motivated because many people do not visit the actual summit, but hike around the slope, which can extend for several tens of kilometers. Figure 2a shows the mean average precision over the top-50 predicted peaks based on the five summits for increasing σ. The optimal performance is reached with normalization per grid cell and a kernel σ of 20km.

Next we evaluate the prediction of beaches by finding the wormholes from *Platja de Lloret de Mar*, just north of Barcelona. We collect a list of 7216 beaches from Geonames.org to evaluate the predicted locations. The results on beach prediction (Figure 2b) correspond to the results found on mountain prediction, N3 gives the optimal normalization and the optimal kernel width is around 20km.

Finally, we show that this method can be applied at multiple scales by predicting the wormholes from the famous cemetery *Père Lachaise* in Paris. Figure 2c shows the top-10 predicted locations. The highest ranked location not located at L is found at *Cimetière du Montparnasse*, another big cemetery in Paris, demonstrating that this method can find similar locations also at city scale.[4]

---

[3] http://en.wikipedia.org/wiki/List_of_peaks_by_prominence
[4] Additional results can be found at http://dmirlab.tudelft.nl/users/maarten-clements

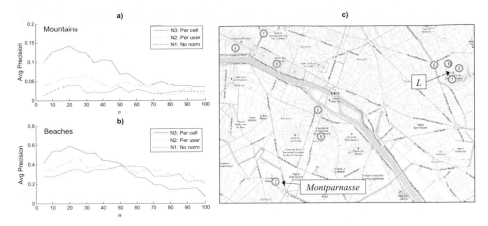

Fig. 2. a) Prediction of mountains based on the 5 major summits (σ in km). **b)** Prediction of beaches based on Platja de Lloret de Mar (σ in km). **c)** Wormholes in Paris from Père Lachaise ($\sigma = 60m$, Method: N3).

We have shown that geotags can effectively be used to predict similar locations with high precision. To limit the influence of individuals on the prediction only one geotag per grid cell should be considered per user. The kernel convolution method allows for detection of similar places at different scales, and can therefore be used for recommendations at global or city level.

References

1. Rattenbury, T., Good, N., Naaman, M.: Towards automatic extraction of event and place semantics from flickr tags. In: SIRIR 2007: Proceedings of the 30th annual international ACM SIGIR conference on Research and development in information retrieval, pp. 103–110. ACM, New York (2007)
2. Ahern, S., Naaman, M., Nair, R., Yang, J.H.: World explorer: visualizing aggregate data from unstructured text in geo-referenced collections. In: JCDL 2007: Proceedings of the 2007 conference on Digital libraries, pp. 1–10. ACM, New York (2007)
3. Kennedy, L., Naaman, M., Ahern, S., Nair, R., Rattenbury, T.: How flickr helps us make sense of the world: context and content in community-contributed media collections. In: MULTIMEDIA 2007: Proceedings of the 15th international conference on Multimedia, pp. 631–640. ACM, New York (2007)
4. Crandall, D., Backstrom, L., Huttenlocher, D., Kleinberg, J.: Mapping the world's photos. In: WWW 2009: Proceeding of the 18th international conference on World Wide Web, pp. 761–770 (2009)
5. Serdyukov, P., Murdock, V., van Zwol, R.: Placing flickr photos on a map. In: SIGIR 2009: Proceedings of the 32nd international ACM SIGIR conference on Research and development in information retrieval, pp. 484–491. ACM, New York (2009)
6. Kalogerakis, E., Vesselova, O., Hays, J., Efros, A.A., Hertzmann, A.: Image sequence geolocation with human travel priors. In: ICCV 2009: Proceedings of the IEEE International Conference on Computer Vision (2009)

Enhancing N-Gram-Based Summary Evaluation Using Information Content and a Taxonomy

Mijail Kabadjov[1], Josef Steinberger[1], Ralf Steinberger[1],
Massimo Poesio[2,3], and Bruno Pouliquen[1]

[1] Joint Research Centre, European Commission, Via E. Fermi 2749, Ispra (VA), Italy
{firstname.lastname}@jrc.ec.europa.eu
[2] University of Essex, Wivenhoe Park, Colchester CO4 3SQ, United Kingdom
poesio@essex.ac.uk
[3] Universitá di Trento, Povo, TN 38100, Italy

Abstract. In this paper we propose a novel information-theoretic metric for automatic summary evaluation when model summaries are available as in the setting of the AESOP task of the Update Summarization track of the Text Analysis Conference (TAC). The metric is based on the concept of information content operationalized by using a taxonomy. Hereby, we present and discuss the results obtained at TAC 2009.

1 Introduction

For years people have wanted to have means to be able to automatically evaluate the quality of system produced summaries in a robust and reliable manner. This need has been addressed recently by the Automatically Evaluating Summaries Of Peers (AESOP) task performed at NIST as part of the TAC 2009 Summarization track.

Up until now, work on summary evaluation has been largely dominated by n-gram co-occurrence approaches able to capture words and phrases overlap between system and reference summaries. An example of such approaches are ROUGE (Lin and Hovy, 2003) or from machine translation, BLEU (Papineni et al., 2001). Although these approaches are very robust and efficient, their main drawback is that if two summaries were produced using non- or almost non-overlapping vocabulary, yet conveying the same information, the similarity score such summaries would be assigned by purely n-gram based metrics would be too low and, hence, unrepresentative of the actual information they share.

In this paper we address the problem of evaluating system-produced summaries with respect to human summaries by going beyond exploiting co-occurrence at the string level and aiming at capturing more complex phenomena such as hypernymy and synonymy by using information content and a taxonomy. We contend that by combining n-gram co-occurrence with information content one obtains a better and more reliable evaluation metric than a metric based purely on n-gram co-occurrence.

The remainder of the paper is organized as follows: in section §2 we provide brief background on the subject; in section §3 we put forward our

C. Gurrin et al. (Eds.): ECIR 2010, LNCS 5993, pp. 662–666, 2010.
© Springer-Verlag Berlin Heidelberg 2010

information-content-based metric; in section §4 we present and discuss the evaluation results and finally conclude the paper with pointers to future work.

2   Background

TAC stands for Text Analysis Conference and is conducted by the US National Institute of Standards and Technology (NIST). In 2009 it consisted of three tracks: Knowledge Base Population, Recognizing Textual Entailment and Summarization. We participated in the Summarization track, which consisted of two main tasks: Update Summarization and Automatically Evaluating Summaries of Peers (AESOP).

The standard automatic evaluation metrics used at TAC are based on n-gram co-occurrence and are part of ROUGE (Recall-Oriented Understudy for Gisting Evaluation) (Lin and Hovy, 2003), whereas the standard human evaluation metrics are based on the pyramid method put forward by (Nenkova and Passonneau, 2004).

3   Information Content for Summary Evaluation

We propose to measure the amount of content shared between a pair of texts (e.g., summaries) on the basis of the average semantic similarity between the set of concepts within the first (model) text and the set of concepts within the second text. More formally,

$$avg\_sim(C_m, C_s) = \sum_{c_m \in C_m, c_s \in C_s} \frac{\max_{c_m, c_s} [sim(c_m, c_s)]}{|C_m|} \quad (1)$$

where C_m is the set of concepts contained in the model summary, C_s is the set of concepts within the system summary, $|C_m|$ denotes the size of C_m[1] and $sim(c_m, c_s)$ is Resnik's semantic similarity measure using a taxonomy (see (Resnik, 1995) for more details).[2]

The above information-content-based metric can be easily combined with surface level features such as unigram and bigram recall using a weighted linear combination as follows:

$$score = \alpha \cdot unigrams(M, Sys) + \beta \cdot bigrams(M, Sys) + \gamma \cdot avg\_sim(C_m, C_s) \quad (2)$$

where $unigrams(M, Sys)$ and $bigrams(M, Sys)$ represent the recall of unigrams and bigrams, respectively, of the system summary (Sys) with respect to the model summary (M). In order to ensure an overall score within $[0, 1]$ we set

---

[1] By dividing by the number of concepts contained in the model summary, we are aiming at a recall-like metric (as opposed to precision-like, i.e., by dividing by $|C_s|$).

[2] In our study we used the MeSH taxonomy (Medical Subject Headings), one of the main reasons behind our choice was simply that we had access to off-the-shelf tools for processing texts and identifying and grounding terms to the MeSH taxonomy.

$\alpha + \beta + \gamma = 1$. We estimated the optimal values for these weights on the training data[3] and the best combination obtained was $\alpha = 0.4$, $\beta = 0.4$, $\gamma = 0.2$.

Intuitively, such a hybrid evaluation metric captures on one hand higher level lexical relationships, such as hypernymy and synonymy, whilst on the other hand still retaining the capacity to capture direct lexical co-occurrence.

4   Evaluation

Herewith, we discuss the evaluation results obtained at the TAC 2009 AESOP task for the two metrics described in the previous section.[4]

Table 1 shows Pearson's correlations[5] of our metrics with the Pyramid and Overall Responsiveness scores for initial and update summaries including the model summaries (i.e., all peers). In addition, we give the corresponding scores for the two AESOP baselines, which are purely n-gram-based metrics.

Table 1. Pearson's correlations including models

Run No.	Initial Summaries		Update Summaries	
	Pyramid	Resp.	Pyramid	Resp.
run3, IC	0.771	0.758	*0.701*	0.686
run27, IC+n-grams	0.826	0.74	0.799	0.686
run1, ROUGE-SU4	0.734	0.617	0.726	0.564
run2, BE	0.586	0.456	0.629	0.447

Our metrics surpassed the two baselines in all cases except for IC's correlation with the pyramid score for the update summaries (see number in italics). *IC+n-grams* correlated better with the pyramid score, whereas *IC* with the overall responsiveness for both types of summaries. Overall, the metrics were ranked either 13th or 14th out of the 37 runs.[6]

In order to evaluate metric discriminative power, contingency tables with five distinctive cells were included.

In this analysis we propose to cast the discriminative power contingency tables in terms of precision, recall and balanced F1 measure (see (Steinberger et al., 2009) for more details). The $P/R/F1$ scores are shown in table 2.

On the basis of the $F1$ measure alone, neither of our metrics surpassed the baselines (see numbers in italics) and were ranked within the 3rd quartile of the 37 runs.

An interesting thing to note, though, is that our IC metric in all cases consistently yielded a precision of 1.0, though on the expense of recall. This suggests

---

[3] We used the TAC 2008 data for training.

[4] See http://www.nist.gov/tac/2009/Summarization/aesop.09.guidelines.html for details on the evaluation of the AESOP task.

[5] All with $p < 0.01$.

[6] See (Steinberger et al., 2009) for more details omitted here due to space constraints.

Table 2. Discriminative power including models

Run No.	Overall Responsiveness					
	Initial Summaries			Update Summaries		
	P	R	$F1$	P	R	$F1$
run3, IC	1.0	0.064	*0.112*	1.0	0.005	0.009
run27, IC+n-grams	0.963	0.473	*0.634*	0.94	0.28	0.432
run1, ROUGE-SU4	0.966	0.516	0.673	0.958	0.419	0.583
run2, BE	0.924	0.22	0.356	0.925	0.223	0.36

a precise evaluation metric most likely suffering from data sparseness resulting from either terms not found in the MeSH taxonomy (a widely pointed out drawback of lexical databases such as MeSH and WordNet), or terms not seen in the training corpus (possibly less severe, since we used a basic back-off scheme by assigning default weights to such terms, though proper fine-tuning might further alleviate this problem).

In the case when no summary models are included[7] the situation is similar; on $F1$ score, neither of our metrics surpassed the baselines.

Our best metric on overall responsiveness was *IC+n-grams* and was ranked 6th and 7th out of 37 at initial and update summaries, respectively.

5   Conclusion

In this work we presented an information-theoretic metric for automatic summary evaluation of system summaries versus human model summaries. The metric aims at capturing more complex phenomena such as hypernymy and synonymy and thus at complementing the currently dominant summary evaluation approaches based on n-gram co-occurrence. We presented and discussed the results obtained at TAC 2009 and we argued that by combining n-gram co-occurrence with information content one obtains a better and more reliable evaluation metric than a metric based on n-gram co-occurrence.

In future work we intend to explore more in depth the problem of automatic evaluation of the quality of system-produced summaries, and in particular the possibility of bypassing the need for a model summary by making the most of the TAC datasets and the baseline systems available.

Bibliography

Lin, C.-Y., Hovy, E.: Automatic evaluation of summaries using n-gram co-occurrence statistics. In: Proceedings of HLT-NAACL, Edmonton, Canada (2003)

Nenkova, A., Passonneau, R.: Evaluating content selection in summarization: The pyramid method. In: Proceedings of the Meeting of the North American Chapter of the Association for Computational Linguistics, NAACL (2004)

---

[7] Table omitted due to space constraints (see (Steinberger et al., 2009)).

Papineni, K., Roukos, S., Ward, T., Zhu, W.-J.: BLEU: a method for automatic evaluation of machine translation. Technical Report RC22176 (W0109-022), IBM Corporation (2001)

Resnik, P.: Using information content to evaluate semantic similarity in a taxonomy. In: Proceedings of the National Conference on Artificial Intelligence, AAAI (November 1995)

Steinberger, J., Kabadjov, M., Pouliquen, B., Steinberger, R., Poesio, M.: WB-JRC-UT's participation in TAC 2009: Update summarization and AESOP tasks. In: National Institute of Standards and Technology (eds.) Proceedings of the Text Analysis Conference, Gaithersburg, MD (November 2009)

NEAT: News Exploration Along Time

Omar Alonso[*], Klaus Berberich, Srikanta Bedathur, and Gerhard Weikum

Max-Planck Institute für Informatik, Saarbrücken, Germany

Abstract. There are a number of efforts towards building applications
that leverage temporal information in documents. The demonstration
of our NEAT (News Exploration Along Time) prototype system that we
propose here, is an attempt towards building an intuitive and exploratory
interface for search results over large news archives using timelines. The
demonstration uses the New York Times Annotated Corpus[1] as an illustrative example of such a news archive.

The NEAT system consists of two parts: the back-end server extracts
and stores in an index all the temporal information from documents,
and performs important phrase discovery from sentences that have time-
sensitive information. The front-end user interface, *anchors* the results
of a keyword search along the timeline where the user can explore and
browse results at different points in time. To aid in this exploration, the
interesting phrases discovered from the result documents are displayed
on the timeline to provide an overview.

Another key feature of NEAT, which distinguishes it from other
timeline-based approaches, is the adoption of semantic temporal annotations to anchor results on the timeline. An appropriate choice of
personally-identifiable temporal annotations can enable users to more effectively contextualize results. For example, Barack Obama was elected
in 2008 and Germany hosted the FIFA World Cup in 2006. We gathered temporal annotations at large-scale by *crowdsourcing* it over Amazon Mechanical Turk (AMT[2]). Each HIT (Human Intelligence Task) on
AMT consists of a request to expand a temporal expression (such as a
year, a time-interval, or decade, etc.) with an entity (e.g., a person, coun-
try, organization etc.). Based on the agreement level among workers, we
derive key entities for constructing a semantic temporal annotation layer
on top the timeline. The outcome is a manually annotated timeline that
can be very useful to anchor search results. Examples of annotations produced by crowdsourcing are (`1969: Woodstock, Moon landing`), (`1970:
Nixon`), and (`2003-2009: Iraq war`) to name a few with different time
granularities.

The demonstration consists of an exploratory search interface where
we show how queries can produce different timelines and how one can
use temporal information to discover interesting facts.

Keywords: Temporal information retrieval, information needs, Web
archives, user interfaces, exploratory search, crowdsourcing.

---

[*] Current affiliation: Microsoft Corp.
[1] http://corpus.nytimes.com
[2] http://www.mturk.com/

C. Gurrin et al. (Eds.): ECIR 2010, LNCS 5993, p. 667, 2010.

Opinion Summarization of Web Comments

Martin Potthast and Steffen Becker

Bauhaus-Universität Weimar, Germany
`<first name>.<last name>@uni-weimar.de`

Introduction. All kinds of Web sites invite visitors to provide feedback on comment boards. Typically, submitted comments are published immediately on the same page, so that new visitors can get an idea of the opinions of previous visitors. Popular multimedia items, such as videos and images, frequently get up to thousands of comments, which is too much to be read in reasonable time. I.e., visitors read, if at all, only the newest comments and hence get an incomplete and possibly misleading picture of the overall opinion. To address this issue we introduce OPINIONCLOUD, a technology to summarize and visualize opinions that are expressed in the form of Web comments.[1]

Related Work. Most of the related work pertains to opinion mining in product and movie reviews, where the summarization of reviews has been studied quite intensively [1, 2, 3, 4, 10]. Given a set of reviews on a particular product, the task is to synthesize a summary that contrasts certain product properties a reviewer considers to be positive or negative. In all papers that are referenced here, the generated summaries are lists of ranked sentences extracted from the reviews. Within our approach we focus on words, since extracting sentences is pointless for Web comments: unlike product reviews, Web comments cannot be expected to have a sensible structure or a sufficient writing quality to extract sentences. The difference between reviews and comments becomes apparent if one compares the reviews on products sold at Amazon with the comments on videos published at YouTube. We consider reviews as a special kind of comments, which nonetheless deserve a special treatment. Note further that Web comments in general have been studied far less frequently than reviews [5, 6, 9].

Summarization and Visualization. The summarization of a set of comments D divides into an offline step and an online step. Suppose that two dictionaries V^+ and V^- are given, comprising human-annotated terms that are commonly used to express positive or negative opinions [7]. In the offline step we use the well-known sentiment analysis approach described in [8] to extend V^+ and V^- to the application domain. The extension is necessary in order to learn terms that are not covered by the dictionaries. The semantic orientation, SO, of an unknown word w is measured by the degree of its association with known words from V^+ and V^-:

$$SO(w) = \sum_{w^+ \in V^+} \mathrm{assoc}(w, w^+) - \sum_{w^- \in V^-} \mathrm{assoc}(w, w^-),$$

where $\mathrm{assoc}(w, w')$ maps two words to a real number that indicates their association strength. If $SO(w)$ is greater than a threshold ε (less than $-\varepsilon$) w is added to V^+ (V^-); otherwise w is considered as neutral. As association measure the point-wise mutual information statistic is applied:

---

[1] OPINIONCLOUD is available at http://www.webis.de/research/projects/opinioncloud.

C. Gurrin et al. (Eds.): ECIR 2010, LNCS 5993, pp. 668–669, 2010.

$$\mathrm{PMI}(w, w') = \log_2 \frac{p(w \wedge w')}{p(w) \cdot p(w')},$$

where $p(w \wedge w')$ is the probability of observing w together with w', and $p(w)$ is the a-priori probability of w. In the online step, when a set of comments D is observed, a summary is visualized in the form of a tag cloud which contrasts the positive, neutral, and negative terms found using the sentiment dictionaries. Terms which do not appear in the dictionaries are considered as neutral by default. As is customary for tag clouds, the font size of a term grows proportionally with its frequency in the comments. Moreover, the percentages of positive and negative terms from all non-neutral terms is computed.

Implementation. The OPINIONCLOUD is implemented as a browser add-on which, whenever the user views a YouTube video or a Flickr image, downloads the recent comments and summarizes them on-the-fly. The summaries are injected into the Web page. The figures below show examples: the left summary contrasts positive and negative terms on a YouTube video, and the right summary shows the positive, neutral, and negative terms on a Flickr image. For a quick overview it suffices to look at the percentages on top of each cloud which, in this case, indicate that the opinions about the YouTube video are divided with a tendency of dislike, while the Flickr image is clearly appreciated. If a user is interested to know more about what visitors felt when viewing the item, the tag cloud provides the words organized according to their occurrence frequency. By clicking on a word the list of comments containing it is retrieved.

Bibliography

[1] Beineke, P., Hastie, T., Manning, C., Vaithyanathan, S.: An Exploration of Sentiment Summarization. In: Proc. of AAAI 2003 (2003)

[2] Lerman, K., Blair-Goldensohn, S., McDonald, R.: Sentiment Summarization: Evaluating and Learning User Preferences. In: Proc. of EACL 2009 (2009)

[3] Liu, B., Hu, M., Cheng, J.: Opinion Observer: Analyzing and Comparing Opinions on the Web. In: Proc. of WWW 2005 (2005)

[4] Lu, Y., Zhai, C., Sundaresan, N.: Rated Aspect Summarization of Short Comments. In: Proc. of WWW 2009 (2009)

[5] Mishne, G., Glance, N.: Leave a Reply: An Analysis of Weblog Comments. In: Proc. of WWE 2006 (2006)

[6] Potthast, M.: Measuring the Descriptiveness of Web Comments. In: Proc. of SIGIR 2009 (2009)

[7] Stone, P.J.: The General Inquirer: A Computer Approach to Content Analysis. MIT, Cambridge (1966)

[8] Turney, P.D., Littman, M.L.: Measuring Praise and Criticism: Inference of Semantic Orientation from Association. ACM Trans. Inf. Syst. 21(4), 315–346 (2003)

[9] Yee, W.G., Yates, A., Liu, S., Frieder, O.: Are Web User Comments Useful for Search? In: Proc. of LSDS-IR 2009 (2009)

[10] Zhuang, L., Jing, F., Zhu, X.Y.: Movie Review Mining and Summarization. In: Proc. of CIKM 2006 (2006)

EUROGENE: Multilingual Retrieval and Machine Translation Applied to Human Genetics

Petr Knoth[1], Trevor Collins[1], Elsa Sklavounou[2], and Zdenek Zdrahal[1]

[1] Knowlege Media Institute, The Open University, United Kingdom
{P.Knoth,T.D.Collins,Z.Zdrahal}@open.ac.uk
[2] Systran, Paris, France
sklavounou@systran.fr

The objective of Eurogene is to collect a critical mass of educational content in the field of human genetics in nine European languages and to build a platform that will support the retrieval, sharing and navigation over the learning content. The Eurogene platform is already operational and is being used by the genetics community. In this paper, a part of the Eurogene platform related to the retrieval and machine translation of domain specific content is described. Our contribution lies in an approach for domain-specific adaption of cross-language information retrieval (CLIR) and machine translation (MT). The CLIR system is based on a multilingual domain ontology which is also used as a synchronization component between CLIR and MT. The MT system is adapted to the target domain using the terminology represented in the ontology and using statistical training performed on a collection of parallel texts. In the statistical training phase, new translations of a term can be discovered and used for ontology updating. The paper is organized as follows. First, we describe the motivation for our approach and the multilingual domain ontology. Later, the CLIR and MT components and their domain adaption and synchronization are discussed.

Many of the important players in the information retrieval field (including Google and Yahoo!) offer CLIR, some of them also in combination with MT. While the performance of these systems may be sufficient for general queries, they are often less accurate for domain-specific queries. For monolingual domain IR systems where terminology is often part of a query, such as PubMed, large thesauri/ontologies carrying information about the relatedness of the domain terms have been successfully adopted to improve the performance of the system. There are also efforts to enable cross-lingual medical retrieval, however the combination of the domain-specific retrieval and machine translation is rarely available.

In Eurogene, the initial genetic ontology was developed by merging six monolingual ontologies, translating them into nine European languages (English is used as an interlingua) by domain experts and inferring an upper-level ontology using Unified Medical Language System (UMLS). The upper-level ontology helps to organize concepts from a relatively flat structure into a concept hierarchy, which is represented in the Simple Knowledge Organization System (SKOS) format. The ontology is used to allow domain CLIR and to provide validated terminological translations to the MT system in the form of rules extracted from the ontology.

C. Gurrin et al. (Eds.): ECIR 2010, LNCS 5993, pp. 670–671, 2010.
© Springer-Verlag Berlin Heidelberg 2010

The Eurogene system aggregates educational content in the form of documents (pdf, ppt, word etc.), images, videos and external resources. Textual resources are automatically annotated in terms of the ontology. Language dependent stemmers are applied in the annotation process. Multimedia resources are annotated manually in terms of the same ontology. This allows multimodal retrieval and navigation over the content.

The CLIR system is powered by Lucene extended with a dedicated query parser which allows the submission of complex queries. Queries can be expressed in any of the available languages, and the results can be filtered to a subset of the available languages. Textual resources can be machine translated using the domain adapted MT system. The retrieval system allows a user to visualize the concept hierarchy and to interactively control query expansion for synonyms and/or narrower terms, thus utilizing the benefits of ontology-based retrieval. The CLIR system also supports the discovery of related materials across languages. This is enabled by the fact that all resources are due to the ontology represented using language-independent vectors.

The MT system used is a hybrid MT system which combines rule-based and statistical-based MT. The MT system is adapted to the domain using rules automatically extracted from the ontology and using statistical training from parallel domain corpora. Parallel domain corpora is typically obtained when a presentation, book or other educational content is submitted to the system in more than one language. To build a statistical model using our system, approximately 20 thousands of parallel sentences are needed. We assume that this amount is typically not available when building a system, but may be gradually gathered. When this amount of parallel text becomes available, the statistical training may discover new terminological pairs. Pairs with high confidence can be used to provide a feedback mechanism for the ontology which can be then updated.

The synchronization of CLIR and MT systems works in the following way. The systems communicate using SOAP messages that allow the sending of extracted translation rules from CLIR, which is handling the ontology, to MT, and the sending of newly proposed translations from MT to CLIR. When newly proposed translations are received by CLIR, the ontology is updated. Domain experts then perform terminology validation which is supported by the system and results in sending new translation rules to the MT rule-base. The ontology can be also updated manually at any time which results in sending new translation rules to MT. This synchronization provides a mechanism for continuous semi-automatic adaption of both CLIR and MT systems.

The approach combines the advantages of ontology-based retrieval with hybrid machine translation. The approach is suitable for multilingual domain-specific collections where it should provide sufficient performance and should speed-up the adaption process using statistical approaches at the same time.

Acknowledgement

This work has been partially supported by the Eurogene (The First Pan-European Learning Service in the Field of Genetics, Contract no. ECP-2006-EDU-410018).

NETSPEAK—Assisting Writers in Choosing Words

Martin Potthast, Martin Trenkmann, and Benno Stein

Bauhaus-Universität Weimar, Germany
<first name>.<last name>@uni-weimar.de

NETSPEAK is a Web service which helps writers in finding alternative expressions for what they want to say.[1] It provides a large index of writing samples in the form of n-grams, $n \leq 5$, along with an efficient means to retrieve them by the use of wildcard queries. When in doubt about a phrasing, a user can get additional evidence by retrieving samples that match a given context. The figure below shows the results for a query where a user is interested in the two most frequently written words between "looks" and "me". The first two columns give an idea about the customariness of each result, and the user can select the one most appropriate for her sentence.

To provide a rich choice of writing samples we index the Google n-gram corpus which was compiled from a large portion of the English Web and which consists of more than 3 billion n-grams along with their occurrence frequencies [2]. We have developed a space-optimal inverted index based on minimal perfect hashing. The hash function maps the vocabulary V of the corpus to the storage positions of postlists. A hash function is perfect if it does not produce hash collisions for the key set V, and it is minimal if the number of storage positions required does not exceed $|V|$. The hash function is constructed with the CHD algorithm which produces a space overhead of $2.07 \times |V|$ bits [1]. Moreover, the index provides a top-k retrieval strategy to find the n-grams matching a query; details can be found in [3]. The table below shows selected performance data of our index. NETSPEAK is currently deployed on a cluster of 15 computers. In a load test the service was measured to process about 10 000 queries per second.

Netspeak		looks ? ? me	Search
Frequency		**Phrase**	
56 925	32.0 %	looks good to me	⊟
		Looks good to me right ◄ 2/8 ► now and I'm surprised by ...	
19 103	10.7 %	looks fine to me	⊞
12 647	7.1 %	looks ok to me	⊞
11 794	6.6 %	looks like to me	⊞
10 300	5.8 %	looks up at me	⊞
10 047	5.6 %	looks good on me	⊞
9 540	5.4 %	looks great to me	⊞
177 944	100.0 %		0.919 seconds

NETSPEAK Benchmarks	
Index size (compressed)	66.2 GB (31.8 GB)
Indexing time (single PC)	7:31 hours
Avg. retrieval time	0.164 seconds
Longest postlist (",")	178 691 474 entries
Avg. postlist length	5334 entries

Index Scalability	
Index size	# postlists × postlist size
# postlists (key set size)	max. unsigned 32 bit integer
maximal postlist size	max. allowed file size

Bibliography

[1] Belazzougui, D., Botelho, F.C., Dietzfelbinger, M.: Hash, Displace, and Compress. In: Fiat, A., Sanders, P. (eds.) ESA 2009. LNCS, vol. 5757, pp. 682–693. Springer, Heidelberg (2009)

[2] Brants, T., Franz, A.: Web 1T 5-gram Version 1. Linguistic Data Consortium (2006)

[3] Stein, B., Potthast, M., Trenkmann, M.: Retrieving Customary Web Language to Assist Writers. In: Proc. of ECIR 2010 (2010)

---

[1] NETSPEAK is accessible at http://www.netspeak.cc

C. Gurrin et al. (Eds.): ECIR 2010, LNCS 5993, p. 672, 2010.

A Data Analysis and Modelling Framework for the Evaluation of Interactive Information Retrieval

Ralf Bierig*, Michael Cole, Jacek Gwizdka, and Nicholas J. Belkin

School of Communication and Information, Rutgers University, USA

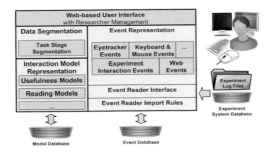

Fig. 1. System components of the data analysis and modelling framework

Over the last two decades, Interactive Information Retrieval (IIR) has established a new direction within the long tradition of IR that introduces the user at its center and poses new challenges for system evaluation. IR systems can improve performance by utilizing information about the entire interactive process of search. This approach has so far only been initially explored [1,2] with much potential for the future. This demonstration describes an extensible data analysis and modelling framework that enables researchers to integrate, explore and analyze interactive experiment data obtained from task-based IIR experiments and build and test models of interactive user behavior. Figure 1 shows the framework components: The *Event Representation* integrates experiment data through the *Event Reader Interface* through a configurable set of *Event Reader Import Rules* into a unified and extensible event data structure. An extensible list of event types ensures that researchers can adapt and extend the framework to process data from a variety of IIR experiments on a single platform. *Data Segmentation* divides experiment data into semantic units guided by research hypotheses. A segmentation can for example differentiate interaction data based on users' current stage in the search task. The *Interaction Model Representation* processes (segmented) event sequences to test specific research hypotheses (e.g. identifying users' perceived usefulness of content or determining reading behavior). The *Web-based User Interface* extends the system to an online service where researchers can generate, inspect and share event representations and create data segmentations and interaction models. The framework is currently applied for the analysis of three IIR experiments in our research project. It will become an open source project that allows for wider public access within the research community.

* This work is supported, in part, by the Institute of Museum and Library Services (IMLS grant LG-06-07-0105-07).

C. Gurrin et al. (Eds.): ECIR 2010, LNCS 5993, pp. 673–674, 2010.

References

1. Belkin, N., Keynote: Some(what) Grand Challenges for Information Retrieval. In: 30th European Conference on Information Retrieval Research. Glasgow, Scotland, 2008.
2. Belkin, N., Bierig, R., Buscher, G., Van Elst, L., Gwizdka, J., Jose, J., Teevan, J.,: Workshop on Understanding the User. In: 32nd Annual ACM SIGIR Conference, Boston, USA, 2009.

Requirements

The Demonstration requires an internet connection and a large screen (19" or higher) to show the user interface in one-to-one discussions. A projector would be beneficial for larger audiences. A notebook is provided by the demonstrator.

Author Index

Agarwal, Shivani   332
Alfonseca, Enrique   62
Alonso, Omar   13, 602, 667
Altingovde, Ismail Sengor   306
Aly, Robin   241
Anderka, Maik   640
Arampatzis, Avi   594
Araujo, Lourdes   26
Athanasakos, Konstantinos   217
Atilgan, Duygu   306
Azzopardi, Leif   153, 204, 357

Baillie, Mark   432, 569
Balog, Krisztian   319
Bashir, Shariq   457
Becker, Steffen   668
Bedathur, Srikanta   13, 667
Belkin, Nicholas J.   673
Bellogín, Alejandro   382
Bennett, Paul N.   140
Berberich, Klaus   13, 623, 667
Bhattacharya, Pushpak   573
Bhole, Abhijit   573
Bierig, Ralf   673
Boudin, Florian   50
Bron, Marc   319
Broschart, Andreas   623

Cai, Deng   577
Cantador, Iván   420
Carman, Mark J.   432, 649
Castells, Pablo   382
Cerviño Beresi, Ulises   569
Chang, Hau-Wen   229
Chen, Long   370
Chen, Luke   394
Clements, Maarten   658
Cole, Michael   673
Collins, Michael   332
Collins-Thompson, Kevyn   140
Collins, Trevor   670
Cox, Ingemar   265
Crestani, Fabio   649

Dai, Na   127, 653
Daumé III, Hal   444

Davison, Brian D.   127, 653
de Jong, Franciska   153, 204
de Rijke, Maarten   191, 319
de Vries, Arjen P.   658
Doherty, Aiden   241

Efthimiadis, Efthimis N.   100
Eriksson, Gunnar   38

Fang, Hui   344
Ferrarotti, Flavio   281
Frieder, Ophir   483, 636
Frommholz, Ingo   615
Fuhr, Norbert   293, 544
Fu, Ruoxun   265

Glass, David H.   394
Goharian, Nazli   636
Gooda Sahib, Nuzhah   532
Gottron, Thomas   611
Gurevych, Iryna   598
Gurrin, Cathal   1
Gwizdka, Jacek   673

Hall, Keith   62
Hanjalic, Alan   645
Han, Jiawei   166
Han, Mei   581
Hansen, Lars Kai   265
Harvey, Morgan   432
Hauff, Claudia   153, 204
He, Tingting   370
He, Yulan   1
Hiemstra, Djoerd   153, 204, 241, 556
Huang, Xiangji   495
Hu, Biyun   619

Inches, Giacomo   649
Ingwersen, Peter   627

Jagarlamudi, Jagadeesh   444
Jambor, Tamas   407
Jin, Jianming   590
Jones, Gareth J.F.   520
Jose, Joemon M.   217, 420

Kabadjov, Mijail   662
Kamps, Jaap   556, 594
Kaptein, Rianne   556
Karlgren, Jussi   38
Kazai, Gabriella   1
Kelly, Liadh   520
Khapra, Mitesh M.   75
Kim, Hung-sik   229
Kim, Hyun Duk   166
Kim, Yunhyong   569
Knoth, Petr   670
Kriewel, Sascha   544
Kruschwitz, Udo   1

Lalmas, Mounia   615
Lapata, Mirella   12
Larsen, Birger   627
Larson, Martha   645
Lee, Dongwon   229
Lee, Jeongkyu   229
Lipka, Nedim   611
Little, Suzanne   1
Li, Zhoujun   495, 619
Lopez de Lacalle, Maddalen   586
Lu, Jinsong   577
Lund, Haakon   627
Luo, Jing   370
Lykke, Marianne   627

Macdonald, Craig   87, 114
Marín, Mauricio   281
Martin-Brualla, Ricardo   62
Martinez-Romo, Juan   26
McClean, Sally   394
Mendoza, Marcelo   281
Morin, Emmanuel   253
Moshfeghi, Yashar   615
Müller, Christof   598
Mundrawala, Jay   636

Nie, Jian-Yun   50

Ounis, Iadh   87, 114

Paşca, Marius   62
Papapetrou, Odysseas   293
Peña Saldarriaga, Sebastián   253
Peng, Jie   87, 114
Piwowarski, Benjamin   615

Poblete, Barbara   281
Poesio, Massimo   662
Potthast, Martin   631, 640, 668, 672
Pouliquen, Bruno   662

Raman, Karthik   573
Rauber, Andreas   457
Redpath, Jennifer   394
Reinders, Marcel J.T.   658
Robledo-Arnuncio, Enrique   62
Roelleke, Thomas   1
Rudinac, Stevan   645
Rüger, Stefan   1
Ruthven, Ian   432, 532, 569

Sahlgren, Magnus   38
Sanderson, Mark   179
Santos, Rodrygo L.T.   87
Saralegi, Xabier   586
Schenkel, Ralf   508, 602, 623
Schütze, Hinrich   11
Serdyukov, Pavel   658
Shi, Lei   471
Shi, Lixin   50
Shokouhi, Milad   607
Siberski, Wolf   293
Sklavounou, Elsa   670
Smeaton, Alan   241
Smyth, Barry   10
Song, Dawei   569
Soo, Jason   483
Stamou, Sofia   100
Stathopoulos, Vassilios   217
Stein, Benno   631, 640, 672
Steinberger, Josef   662
Steinberger, Ralf   662

Täckström, Oscar   38
Tang, Jiayu   179
Theobald, Martin   602
Thomas, Paul   607
Tombros, Anastasios   532
Trenkmann, Martin   631, 672
Tsagkias, Manos   191
Tu, Xinhui   370

Udupa, Raghavendra   75, 573
Ulusoy, Özgür   306

Vallet, David   420
van Rijsbergen, Keith   1, 615
Viard-Gaudin, Christian   253

Wang, Jun   407, 577, 619
Wang, Yaoshuang   653
Weerkamp, Wouter   191
Weikum, Gerhard   13, 667

Xing, Xing   581
Xiong, Yuhong   590

Yang, Shengwen   590
Yee, Wai Gen   636
Yin, Xiaoshi   495

Zdrahal, Zdenek   670
Zhai, ChengXiang   166
Zhang, Dell   577
Zhang, Maoyuan   370
Zhang, Yi   581
Zheng, Wei   344
Zhu, Jianhan   265
Zuccon, Guido   357

Printing: Mercedes-Druck, Berlin
Binding: Stein + Lehmann, Berlin